Physical Chemistry
of Foods

FOOD SCIENCE AND TECHNOLOGY

A Series of Monographs, Textbooks, and Reference Books

Additional Volumes in Preparation

Physical Chemistry
of Foods

Pieter Walstra
Wageningen University
Wageningen, The Netherlands

MARCEL DEKKER, INC.　　　　　　　NEW YORK · BASEL

ISBN: 0-8247-9355-2

This book is printed on acid-free paper.

Headquarters
Marcel Dekker, Inc.
270 Madison Avenue, New York, NY 10016
tel: 212-696-9000; fax: 212-685-4540

Eastern Hemisphere Distribution
Marcel Dekker AG
Hutgasse 4, Postfach 812, CH-4001 Basel, Switzerland
tel: 41-61-260-6300; fax: 41-61-260-6333

World Wide Web
http://www.dekker.com

The publisher offers discounts on this book when ordered in bulk quantities. For more information, write to Special Sales/Professional Marketing at the headquarters address above.

Current printing (last digit):
10 9 8 7 6 5 4 3 2 1

PRINTED IN THE UNITED STATES OF AMERICA

Foreword

Knowledge of physical chemistry is of great importance to anyone who is interested in understanding the properties of food, improving its quality and storage stability, and controlling its behavior during handling. Yet, curricula in food science often do not contain a course in food physical chemistry, especially at the undergraduate level, and failure to acquire skills in this important area can hinder a food scientist's success in scientific endeavors. Some believe that an introductory course in physical chemistry offered by a department of chemistry will fill this void, but I disagree. If possible, an introductory course in physical chemistry should be a prerequisite for a course in food physical chemistry—the former providing a sound background in the principles of physical chemistry and the latter focusing on application of the principles most relevant to food.

Failure of many food science departments to offer a course on food physical chemistry is attributable mainly to the lack of an appropriate textbook. Whereas instructors in food science can select from several good textbooks on food microbiology, food engineering, food chemistry, and other more specialized topics, choices in food physical chemistry are severely limited. The publication of Pieter Walstra's excellent textbook on food physical chemistry is therefore an event of major importance to the field of food science. This book will stimulate universities that do not offer this subject to do so, and will improve the quality of instruction in universities that do. It will also be of great value to food researchers.

Professor Walstra is eminently qualified to write a book on food physical chemistry because of his in-depth knowledge of the subject and his understanding of, expertise in, and dedication to food science education. This book provides comprehensive coverage of food physical chemistry at a depth suitable for students in food science, and will serve as an excellent reference source for food researchers. I congratulate Professor Walstra for this fine accomplishment.

Owen Fennema
Professor
Department of Food Science
University of Wisconsin–Madison
Madison, Wisconsin

Preface

The scientific basis needed to understand and predict food properties and changes occurring in foods during processing, storage and use has been enormously expanded over the past half-century. This has caused a revolutionary change in the teaching of food science and technology, especially by application of the disciplines of organic chemistry, biochemistry, and microbiology. With the possible exception of rheology, application of the more physical disciplines has lagged behind. Many years' experience in food research and teaching has convinced me—although I am not a physical chemist—of the importance of physical chemistry and related theories for food science and technology. Moreover, great progress has been made during the past two decades in the study of physicochemical phenomena in foods; yet these aspects often remain greatly underexposed.

The main reason for this deficiency is, in my opinion, that the teaching of physical chemistry for food science majors is often inadequate. In most universities, students have one introductory course in basic physical chemistry; this is unsatisfactory for the following reasons:

1. Many of the subjects are of little or no importance for foods or are treated in too much theoretical detail (e.g., quantum mechanics, statistical thermodynamics, much of spectroscopy).

2. Many subjects of great importance to foods are not included. In order to keep the theory simple, the treatment is often restricted to gases and crystals and dilute, homogeneous, ideal solutions, whereas most foods are concentrated, inhomogeneous, highly nonideal systems. In particular, coverage of colloid and surface science is insufficient.

3. In most cases, the course is taken at the wrong time, that is, before the student is able to see the relevance of the various subjects for foods.

The primary aim of this book is to help in remedying this situation. It can be used as the basis of a course for food science majors at a not-too-early stage in the curriculum (some universities may prefer a graduate course). Hence it is written as a textbook. A second aim is its use as a reference book, since basic aspects of physical chemistry are not always taken into account in food research or process development. Therefore, I have tried to cover all subjects of importance that are not treated in most courses in food chemistry or food processing/engineering.

The selection of topics—including theories, phenomena, systems, and examples—is naturally colored by my experience and opinions. This means that the treatment is to some extent biased. In my view this is unavoidable, but I would like to hear from readers who feel that a topic has been omitted or overemphasized. Remarks about errors and unclear explanations are also welcome.

ACKNOWLEDGMENTS

The idea for this book was born when I was asked to give a course on the physical chemistry of foods in the Department of Food Science at the University of Guelph, Ontario, Canada, in 1993. This course was based largely on courses in food physics given at Wageningen University, developed in cooperation with my colleagues Dr. Ton van Vliet and Dr. Albert Prins. I also want to mention that Dr. Owen Fennema of the University of Wisconsin has greatly encouraged me in writing this text.

Several colleagues have made valuable suggestions, which have greatly improved the quality of this book. I am especially indebted to Dr. Eric Dickinson, Professor of Food Colloids at the University of Leeds, England, with whom I discussed plans and who has read and commented on virtually all my drafts. His contributions have been invaluable. Furthermore, several colleagues have scrutinized one or more draft chapters: Dr. C. van den Berg, Dr. B. H. Bijsterbosch, Dr. M. A. J. S. van Boekel, Dr. O. R. Fennema, Dr. G. J. Fleer, Dr. G. Frens, Dr. H. D. Goff, Dr. H. H. J. de Jongh,

Dr. J. Lucassen, Dr. E. H. Lucassen-Reynders, Dr. J. Lyklema, Dr. E. R. Morris, Dr. W. Norde, Dr. J. H. J. van Opheusden, Dr. L. van der Plas, Dr. M. J. W. Povey, Dr. J. Verhagen, and Dr. T. van Vliet. I express my gratitude to all these friends and colleagues.

Pieter Walstra

Contents

1

Introduction

1.1 PHYSICAL CHEMISTRY IN FOOD SCIENCE AND TECHNOLOGY

Food science and technology are concerned with a wide variety of problems and questions, and some will be exemplified below. For instance, food scientists want to understand and predict changes occurring in a food during processing, storage, and handling, since such changes affect food quality. Examples are

> The rates of chemical reactions in a food can depend on many variables, notably on temperature and water content. However, the relations between reaction rates and the magnitude of these variables vary widely. Moreover, the composition of the mixture of reaction products may change significantly with temperature. How is this explained and how can this knowledge be exploited?
>
> How is it possible that of two nonsterilized intermediate-moisture foods of about the same type, of the same water activity, and at the same temperature, one shows bacterial spoilage and the other does not?
>
> Two plastic fats are stored at room temperature. The firmness of the one increases, that of the other decreases during storage. How is this possible?

Bread tends to stale—i.e., obtain a harder and shorter texture—during storage at room temperature. Keeping the bread in a refrigerator enhances staling rate, but storage in a freezer greatly reduces staling. How is this explained?

The physical stability of a certain oil-in-water emulsion is observed to depend greatly on temperature. At 40°C it remains stable, after cooling to 25°C also, but after cooling to 10°C and then warming to 25°C small clumps are formed; stirring greatly enhances clump formation. What are the mechanisms involved and how is the dependence on temperature history explained?

Another emulsion shows undesirable creaming. To reduce creaming rate a small amount of a thickener, i.e., a polysaccharide, is added. However, it increases the creaming rate. How?

Food technologists have to design and improve processes to make foods having specific qualities in an efficient way. Examples of problems are

Many foods can spoil by enzyme action, and the enzymes involved should thus be inactivated, which is generally achieved by heat denaturation. For several enzymes the dependence of the extent of inactivation on heating time and temperature is simple, but for others it is intricate. Understanding of the effects involved is needed to optimize processing: there must be sufficient inactivation of the enzymes without causing undesirable heat damage.

It is often needed to make liquid foods with specific rheological properties, such as a given viscosity or yield stress, for instance to ensure physical stability or a desirable eating quality. This can be achieved in several ways, by adding polysaccharides, or proteins, or small particles. Moreover, processing can greatly affect the result. A detailed understanding of the mechanisms involved and of the influence of process variables is needed to optimize formulation and processing.

Similar remarks can be made about the manufacture of dispersions of given properties, such as particle size and stability. This greatly depends on the type of dispersion (suspension, emulsion, or foam) and on the specific properties desired.

How can denaturation and loss of solubility of proteins during industrial isolation be prevented? This is of great importance for the retention of the protein's functional properties and for the economy of the process.

How can one manufacture or modify a powdered food, e.g., spray-dried milk or dry soup, in such a manner that it is readily dispersable in cold water?

How does one make an oil-in-water emulsion that is stable during storage but that can be whipped into a topping? The first question then is: what happens during a whipping process that results in a suitable topping? Several product and process variables affect the result.

All of these examples have in common that knowledge of physical chemistry is needed to understand what happens and to solve the problem.

Physical chemistry provides quantitative relations for a great number of phenomena encountered in chemistry, based on well-defined and measurable properties. Its theories are for the most part of a physical nature and comprise little true chemistry, since electron transfer is generally not involved. Experience has shown that physicochemical aspects are also of great importance in foods and food processing. This does not mean that all of the phenomena involved are of a physical nature: it is seen from the examples given that food chemistry, engineering, and even microbiology can be involved as well. Numerous other examples are given in this book.

The problems encountered in food science and technology are generally quite complex, and this also holds for physicochemical problems.

In the first place, nearly all foods have a very wide and *complex composition*; a chemist might call them dirty systems. Anyway, they are far removed from the much purer and dilute systems discussed in elementary textbooks. This means that the food is not in thermodynamic equilibrium and tends to change in composition. Moreover, several changes may occur simultaneously, often influencing each other. Application of physicochemical theory may also be difficult, since many food systems do not comply with the basic assumptions underlying the theory needed.

In the second place, most foods are *inhomogeneous* systems. Consequently, various components can be in different compartments, greatly enhancing complexity. This means that the system is even farther removed from thermodynamic equilibrium than are most homogeneous systems. Moreover, several new phenomena come into play, especially involving colloidal interactions and surface forces. These occur on a larger than molecular scale. Fortunately, the study of mesoscopic physics—which involves phenomena occurring on a scale that is larger than that of molecules but (far) smaller than can be seen with the naked eye—has made great progress in recent times.

In the third place, a student of the physical chemistry of foods has to become acquainted with theories derived from a *range of disciplines*, as a look at the table of contents will show. Moreover, *knowledge of the system* studied is essential: although basic theory should have universal validity, the particulars of the system determine the boundary conditions for application of a theory and thereby the final result.

All of this might lead to the opinion that many of the problems encountered in food science and technology are so intricate that application of sound physical chemistry would hardly be possible and that quantitative prediction of results would often be impossible. Nevertheless, making use of the basic science involved can be quite fruitful, as has been shown for a wide variety of problems. Reasons for this are

> Understanding of basic principles may in itself be useful. A fortunate characteristic of human nature is the desire to explain phenomena observed and to create a framework that appears to fit the observations. However, if such theorizing is not based on sound principles it will often lead to wrong conclusions, which readily lead to further problems when proceeding on the conceived ideas with research or process development. Basic knowledge is a great help in (a) identifying and explaining mechanisms involved in a process and (b) establishing (semi-)quantitative relations.
>
> Even semiquantitative answers, such as giving the order of magnitude, can be very helpful. Mere qualitative reasoning can be quite misleading. For instance, a certain reaction proceeds much faster at a higher temperature and it is assumed that this is because the viscosity is lower at a higher temperature. This may be true, but only if (a) the reaction rate is diffusion controlled, and (b) the relative increase of rate is about equal to the relative decrease in viscosity. When the rate increases by a factor of 50 and the viscosity decreases by a factor of 2, the assumption is clearly wrong.
>
> Foods are intricate systems and also have to meet a great number of widely different specifications. This means that process and product development will always involve trial and error. However, basic understanding and semiquantitative relations may greatly reduce the number of trials that will lead to error.
>
> The possibilities for establishing quantitative relations are rapidly increasing. This is due to further development of theory and especially to the greatly increased power of computer systems used for mathematical modeling of various kinds. In other words, several processes occurring in such complex systems as foods—or in model systems that contain all the essential elements—can now be modeled or simulated.

Altogether, in the author's opinion, application of physical chemistry and mesoscopic physics in the realm of food science and technology is often needed—besides food chemistry, food process engineering, and food microbiology—to solve problems and to predict changes that will occur during manufacture, storage, and use of foods.

1.2 ABOUT THIS BOOK

1.2.1 What Is Treated

The book is aimed at *providing understanding*, hence it primarily gives principles and theory. Moreover, facts and practical aspects are included, because knowledge of the system considered is needed to apply theory usefully, and also because the text would otherwise be as dry as dust. Basic theory is given insofar as it is relevant in food science and technology. This implies that several physicochemical theories are left out or are only summarily discussed. It also means that many aspects will be treated that are not covered in standard texts on physical chemistry, which generally restrict the discussion to relatively simple systems. Since most foods are complicated systems and show nonideal behavior, treatment of the ensuing complexities cannot be avoided if the aim is to understand the phenomena and processes involved.

As mentioned, molecular and mesoscopic approaches will be needed. The first part of the book mainly considers *molecules*. We start with some basic thermodynamics, interaction forces, and chemical kinetics (Chapters 2–4). The next chapter is also concerned with kinetic aspects: it covers various transport phenomena (which means that a few mesoscopic aspects are involved) and includes some basic fluid rheology. Chapters 6 and 7 treat macromolecules: Chapter 6 gives general aspects of polymers and discusses food polysaccharides in particular, with a largish section on starch; Chapter 7 separately discusses proteins, highly intricate food polymers with several specific properties. Chapter 8 treats the interactions between water and food components and the consequences for food properties and processes.

Then *mesoscopic* aspects are treated. Chapter 9 gives a general introduction on disperse or particulate systems. It concerns properties that originate from the division of a material over different compartments, and from the presence of a large phase surface. Two chapters give basic theory. Chapter 10 is on surface phenomena, where the forces involved primarily act in the direction of the surface. Chapter 12 treats colloidal interactions, which primarily act in a direction perpendicular to the surface. Two chapters are concerned with application of these basic aspects in disperse systems: Chapter 11 with emulsion and foam formation, Chapter 13 with the various instabilities encountered in the various dispersions: foams, emulsions, and suspensions.

Next we come to *phase transitions*. Chapter 14 mentions the various phase transitions that may occur, such as crystallization, gas bubble formation, or separation of a polymer solution in two layers; it then treats the nucleation phenomena that often initiate phase transitions. Chapter 15 discusses crystallization, a complicated phase transition of great importance in foods. It includes sections on crystallization of water, sugars, and triacylglycerols. Chapter 16 introduces glass transitions and the various changes that can occur upon freezing of aqueous systems.

Finally, Chapter 17 is about *soft solids*, a term that applies to the majority of foods. It gives an introduction into solids rheology and fracture mechanics, but otherwise it makes use of many of the theories treated in earlier chapters to explain properties of the various types of soft solids encountered in foods.

1.2.2 What Is Not Treated

Some aspects are not covered. This includes analytical and other experimental techniques. A discussion of these is to be found in specialized books. Basic principles of some methods will be given, since this can help the reader in understanding what the results do represent. Possible pitfalls in the interpretation of results are occasionally pointed out.

Aspects that are generally treated in texts on food chemistry are for the most part left out; an example is the mechanism and kinetics of enzyme-catalyzed reactions. Some subjects are not fully treated, such as rheological and other mechanical properties, since this would take very much space, and several books on the subject exist.

Basic theory is treated where needed, but it does not go very deep: giving too much may cause more confusion than enlightenment. We will generally not go to atomic scales, which implies that quantum mechanics and electron orbitals are left out. We also will not go into statistical thermodynamics. Even classical chemical thermodynamics is restricted to a minimum. Theories that involve mathematical modeling or simulation, such as Brownian dynamics, are not discussed either. Equations will be derived only if it helps to understand the theory, and if the derivation is relatively simple.

1.2.3 For Whom It Is Intended

The book is written as a text, with clear and full explanations; illustration of trends rather than giving precise research results; not too many details, although details cannot always be left out; numerous cross-references in the text; no full account of literature sources, but a discussion of selected references at the end of each chapter. Worked out examples and questions are also given.

The questions not only serve to let the reader test whether he or she can make use of what has been treated but also serve as further illustrations. To that end, most questions are followed by worked out answers. By the nature of food science and technology, the questions often involve a number of different aspects, and the reader may not be familiar with all of them. Hence do not worry when you cannot immediately find a full answer, so long as you can understand the reasoning given.

The readers are assumed to be familiar with elementary mathematics (up to simple calculus) and with the basics of chemistry, and to have attended (introductory) courses in food chemistry and food engineering (or food processing).

The book tries to treat all physicochemical aspects of importance for foods and food processing. On the one hand, this means that it gives more than most teachers will want to treat in a course, so that a selection should be made. On the other hand, it makes the book also suitable as a work of reference. Some additional factual information is given in the appendix.

1.2.4 Equations

As mentioned, physical chemistry is a *quantitative* science, which implies that equations will frequently be given. It may be useful to point out that equations can be of various types. Some equations *define* a property, like "pressure equals force over area." Such an equation is by definition exact. Generally, the sign for "is defined as" ($\equiv$) is used rather than "is equal to" ($=$).

Most equations are meant to be *predictive*. According to their validity we can distinguish those that are assumed to be

Generally valid. For instance "force $=$ mass $\times$ acceleration" (although even this one breaks down in quantum mechanics).

Of restricted validity. The restriction is sometimes added to the equation, by indications like "for $x > 1$" or "if $z \to 0$." Another variant is that a "constant" in the equation has restricted validity.

Approximate. Then the $\approx$ sign is used.

A scaling relation, which means that only a proportionality can be given. This is done by using the $\propto$ sign (some use the $\sim$ sign), or by putting an "unknown constant" after the $=$ sign.

Finally, we have equations that describe an experimentally established *correlation*; such relations are generally not meant to provide much understanding. Ideally, the equation should include a measure of uncertainty, often a standard deviation, such as

$$y = 6.1 - 1.7x^2 \pm 0.5, \text{ for } 0 < x < 1.5$$

Some mathematical symbols are given in Appendix A.3.

You may want to *derive a predictive equation* yourself, as is often desirable in research, or even during study. One should always perform some checks on the correctness of the equation. An important check is whether it is dimensionally correct: Is the dimension on either side of the equal sign identical.

Note It is often easier to use the SI units for quantities (m, kg, s, etc.) rather than their true dimensions (L, M, T, etc.).

Furthermore, check whether the sign is correct. Calculate some results for cases where you know or can guess what the outcome should be. For instance, consider what the result is if a certain variable equals zero.

1.2.5 Some Practical Points

Nearly all rules of basic physicochemical theory are generally accepted, but this does not hold for the application of the theory to food systems and processes. The author has refrained from discussing such disagreements and has merely given what he feels is the best explanation. In a few cases, differences of opinion have been mentioned.

Definition of terms: in the subject index, a page number printed in bold indicates where a term is defined.

The text contains some Notes. These are interesting aspects or facts that are not part of the main treatment.

Throughout, SI units will be used, unless stated otherwise; see Appendix A.5. The SI rules for notation are also followed; see Appendix A.4. A short table of conversion factors is given in Appendix A.6.

Fundamental constants are discussed in the text and are tabulated in Appendix A.11.

After the number of some equations an asterisk (*) is added. This means that it is worthwhile to memorize this equation.

Each chapter (except this one) ends with a Recapitulation. It may serve to help the reader in rehearsing the main points made in the chapter. It can also be read in advance to see what topics are being discussed.

BIBLIOGRAPHY

Several textbooks of physical chemistry are available. A well-known, comprehensive, and authoritative one, although by no means is it easy, is

P. W. Atkins. Physical Chemistry. 6th ed. Oxford Univ. Press, Oxford, 1998.

A shorter textbook by the same author is

P. W. Atkins. The Elements of Physical Chemistry. 3rd ed. Oxford Univ. Press, Oxford, 2001.

Another example, also of high quality, is

R. Chang. Physical Chemistry for the Chemical and Biological Sciences. 3rd ed. University Science Books, Sausalito, CA, 2000.

It should be realized that all of these books give thorough introductions into several of the basic aspects discussed in the present text, as well as fundamentals of various, especially spectroscopic, techniques, but that the major part is of little importance for foods; heterogeneous systems are hardly discussed.

A very useful and comprehensive book on a wide range of experimental techniques is

E. Dickinson, ed. New Physico-Chemical Techniques for the Characterization of Complex Food Systems. Blackie, London, 1995.

A brief and clear explanation of various types of mathematical modeling, including Monte Carlo simulation, Brownian dynamics, and molecular dynamics, is given in Chapter 4, "Computer Simulation," of

E. Dickinson, D. J. McClements. Advances in Food Colloids. Blackie, London, 1995.

2

Aspects of Thermodynamics

Chemical thermodynamics can provide the food scientist with important quantitative knowledge. It treats—despite its name—equilibrium situations. Three components of *thermodynamic equilibrium* can be distinguished: (a) mechanical, implying that there are no unbalanced forces; (b) thermal, i.e., no temperature gradients; and (c) chemical, implying that no chemical reactions and no net transport of components occur. Thermodynamics may tell us whether there is equilibrium and, if not, in what direction the change will be, but nothing about the rate at which any reaction or other change may occur.

Thermodynamic theory does not involve molecular explanations and is thus model independent. Nevertheless, and even because of this, it is very useful because it is rigorous. In other words, it is always correct, provided, of course, that it is applied in a correct manner. Physical chemists mostly combine thermodynamic concepts with molecular theories and have thus developed powerful tools for studying matter.

Some of the theory will be briefly recalled in this chapter, and applications to foods will be illustrated. It is by no means an attempt to treat the rudiments of thermodynamics: the reader is advised to study a general textbook.

2.1 CONCEPTS

Physical chemists call the part of the universe that they want to consider the **system** and the remainder the surroundings. The system may be a collection of water molecules, an emulsion droplet, a beaker containing a solution, a loaf of bread, a yeast cell, etc. An *open* system can exchange mass and energy with its surroundings, a *closed* system can exchange no mass, and an *isolated* system neither mass nor energy. If the system is large enough, it has measurable properties, which are conveniently separated in two classes. *Intensive* parameters are independent of the amount of matter and thus include temperature, pressure, refractive index, mass density, dielectric constant, heat conductivity, pH, and other compositional properties, viscosity and so on. *Extensive* parameters depend on (and often are proportional to) the amount of matter and thus include mass, volume, energy, electric charge, heat capacity, etc.

Most systems that food scientists consider are heterogeneous, and this is further discussed in Chapter 9. As mentioned, thermodynamics is model independent, but it is necessary to consider the existence of more than one phase. A **phase** is defined as a (part of a) system that is (a) uniform throughout and (b) bounded by a closed surface, at which surface at least some of the intensive parameters change abruptly. For instance, density, refractive index, and viscosity change, whereas composition does not, as between water and ice; in many cases compositional parameters change as well, as between water and oil. In general, temperature does not alter abruptly, and pressure may or may not. Since most of the changes mentioned occur over a distance of several molecular layers, the criterion of abruptness implies that a very small region of material can never constitute a phase: the change has to occur over a distance that is small compared to the size (in every direction) of the region considered. That is why elements like a soap micelle or a layer of protein adsorbed onto a surface cannot be considered to constitute a phase. Another criterion is that the boundary or interface between the two phases contain energy, and that enlargement of the interfacial area thus costs energy, the amount of which can in many cases be measured (Section 10.1).

Thermodynamics is primarily concerned with energy and entropy. **Energy**, also called internal energy (U), comprises *heat* and *work*; it is measured in joules (J). Work may be mechanical, electrical, chemical, interfacial, etc. It may be recalled that work generally equals force times distance (in $N \cdot m = J$) and that force equals mass times acceleration (in $kg \cdot m \cdot s^{-2} = N$). According to the first law of thermodynamics, the quantity of energy, i.e., heat + work + potential energy, is always preserved.

Entropy (S) is a measure of disorder; it is given by

$$S = k_B \ln \Omega \tag{2.1}$$

where k_B is the Boltzmann constant $(1.38 \cdot 10^{-23} \text{J} \cdot \text{K}^{-1})$ and Ω is the number of ways in which the system can be arranged, also called the *number of degrees of freedom*. If the system consists of perfect spheres of equal size, this only relates to the positions that the spheres can attain in the volume available (translational entropy). This is illustrated in a simplified way in Figure 2.1 for a two-dimensional case, where spherical particles or molecules can be arranged in various ways over the area available. If the interparticle energy $= 0$, which means in this case that there is no mutual attraction or repulsion between the particles, the entropy is at maximum: the particles can attain any position available and are thus randomly distributed (and they will do so because of their thermal or Brownian motion). If there is net attraction (U is negative), they tend to be arranged in clusters, and S is much lower. If there is repulsion (U is positive), the particles tend to become evenly distributed and also in this case entropy is relatively small. If we have more realistic particles or molecules, there are more contributions to entropy. Anisometric particles may attain various orientations (orientational entropy) and most, especially large, molecules can attain various conformations (conformational entropy). If two or more kinds of molecules are present, each kind must be distributed at random over the volume to attain maximum entropy (mixing entropy).

The paramount thermodynamic property is the **free energy**. Two kinds are distinguished. The *Helmholtz (free) energy* is given by

$$A \equiv U - TS \tag{2.2}$$

Note that entropy is thus expressed in $\text{J} \cdot \text{K}^{-1}$; T is the absolute temperature (in K). At constant volume, every system will always change until it has obtained the lowest Helmholtz energy possible. This may thus be due to lowering of U or increase of S. Since we mostly have to do with constant pressure rather than constant volume, it is more convenient to use the *Gibbs (free) energy*. To that end we must introduce the **enthalpy** (H), defined as

$$H \equiv U - pV \tag{2.3}$$

where p is pressure (in Pa) and V volume (in m^3). For condensed (i.e., solid or liquid) phases at ambient conditions, any change in pV mostly is very small compared to the change in U. At constant pressure, every

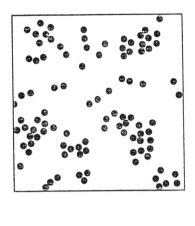

$U < 0$
More or less ordered
Entropy fairly small

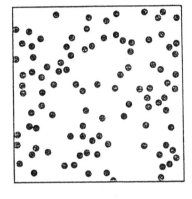

$U = 0$
Unordered
Arrangement determined
by entropy

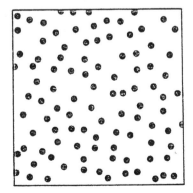

$U > 0$
More or less ordered
Entropy fairly small

FIGURE 2.1   Energy (U) and entropy. Depicted for a given number of molecules or particles in a given two-dimensional space.

system will change until it has obtained the lowest **Gibbs energy**:

$$G \equiv H - TS \tag{2.4}$$

Unless mentioned otherwise, we will always mean the Gibbs energy when speaking about free energy.

The free energy is thus the property determining what will happen. If we add some sugar to water, it will dissolve and the sugar molecules will distribute themselves evenly throughout the liquid, because that gives the lowest free energy. In this case the increase in entropy has a greater effect than the increase in enthalpy (in crystalline sugar, the molecules attract each other and the enthalpy is thus lower than in solution). If we have pure oil droplets in water, they will rise to the surface (lower potential energy) and then coalesce into one layer (lower interfacial area and thus lower surface free energy). If we bring water to a temperature of $-20°C$, it will crystallize (lower enthalpy, which in this case more than compensates for the decrease in entropy). If we have a solution of ethanol in water with air above it, the ethanol will divide itself in such a way over the phases that its partial free energy (or chemical potential: Section 2.2.1) is the same in both; the same applies for the water. All these processes occur spontaneously, and they will never reverse if the external conditions (temperature, pressure, volume available) are left unaltered.

All this applies, however, only to *macroscopic amounts of matter.* Thermodynamics is valid only for large numbers of molecules. If small numbers are considered, say less than a few times 100, exceptions to the rule stated above may occur; even at $10°C$, a few water molecules may temporarily become oriented as in an ice crystal, just by chance, but macroscopically ice will never form at that temperature.

Another remark to be made is that the absolute values of enthalpy and entropy are generally unknown. (Only a perfect crystal of one component at zero absolute temperature has zero entropy.) Quantitative results therefore mostly refer to some standard state (usually $0°C$ and 1 bar), where these parameters are taken to be zero. One always considers the change in thermodynamic properties, and that is quite sufficient. At constant pressure and temperature, the basic equation thus is

$$\Delta G = \Delta H - T\,\Delta S \tag{2.5}^*$$

The change may be from one state to another, say water plus crystalline sugar to a sugar solution, etc. If the change considered is reversible, we have at equilibrium $\Delta G = 0$ and thus $\Delta H = T_e \Delta S$. For example, at 273.15 K ($0°C$) and 1 bar there is equilibrium between (pure) water and ice. ΔH is here the enthalpy of fusion, which can readily be measured by calorimetry and which equals $6020 \, \text{J} \cdot \text{mol}^{-1}$. Consequently, the change in entropy of water molecules going from the solid to the liquid state equals $\Delta H / T_e = 22 \, \text{J} \cdot \text{mol}^{-1} \cdot \text{K}^{-1}$. This signifies that the molecules gain entropy on melting, in agreement with the nature of entropy: Eq. (2.1).

It is often tacitly assumed that ΔH and ΔS are independent of conditions like temperature, and this is indeed often true, as long as the temperature range is small (but not for water: see Section 3.2). Note that the effect of a change in entropy will be larger at a higher temperature. If the system does no work, a change in enthalpy can be measured as a change in heat, i.e., by calorimetry. Changes in entropy mostly cannot be measured directly.

According to thermodynamic theory, any system will spontaneously change until it has attained the state of lowest free energy. In an isolated system (no exchange of energy with the environment) this means that the entropy will increase until it has attained the highest possible value. Consequently, a system is *stable* if it is in a state of lowest free energy. In any other state the system would thus be *unstable*. This does not mean that we observe every unstable system to change. First, the system may be *metastable*. This means that it is in a local state of minimum free energy; at least one other state of still lower free energy does exist, but the system cannot reach it, because it then has to pass through a state of higher free energy. An example is the potential energy of a ball laying in a rut on top of a hill: the ball has to go over the rim of the rut—i.e., go to a state of higher energy—before it can roll down the hill, reaching its state of lowest potential energy. Second, the change may be too slow to be observable. In principle, the rate of change is proportional to the decrease in free energy involved and inversely proportional to a *resistance* to change. Envisage, for example, a stone lying on a sloping surface. Gravity will try to move it down the slope, but the frictional resistance between stone and surface may be too large to allow perceptible movement. Resistance to change can be due to a variety of causes: for chemical reactions a high activation free energy (Section 4.3.3), for evening of concentration in a liquid system a high viscosity (Section 5.1.2), etc. For most causes, the resistance tends to be smaller at a higher temperature, although there are some exceptions.

2.2 SOLUTIONS

Almost all foods are or contain solutions, and solution properties thus are paramount. In this section we will briefly discuss some properties of simple solutions of nonelectrolytes.

2.2.1 The Chemical Potential

In a homogeneous mixture each component i has a chemical potential μ_i, defined as the partial molar free energy of that component (i.e., the change in Gibbs energy per mole of component n_1 added, for addition of an infinitesimally small amount). It is given by

$$\mu_i \equiv \left(\frac{\partial G}{\partial n_i}\right)_{T,p,n_j} \equiv \mu_i^{\circ} + RT \ln a_i \tag{2.6}$$

where μ_i° is the standard chemical potential of the pure substance i and the subscript j refers to all other components. R is the universal gas constant, given by

$$R = k_B N_{AV} = 8.314 \text{ J} \cdot \text{K}^{-1} \cdot \text{mol}^{-1} \tag{2.7}^*$$

where Avogadro's number N_{AV} = the number of molecules in a mole $(6.02 \cdot 10^{23})$. It should be noted that Eq. (2.6) only applies at standard pressure.

Note If the pressure is raised, this increases the chemical potential, in first approximation by an amount pv_i, where v_i is the molar volume of the component.

a_i is the **activity** of i; it is sometimes called the thermodynamic or the effective concentration. The activity is directly related to concentration: for zero concentration $a = 0$; for the pure substance $a = 1$. For solutions called *ideal*, the activity equals concentration, if the latter is expressed as mole fraction (x). Ideality is, however, not often observed, except for a mixture of very similar compounds. Figure 2.2a gives an example for the mixture ethanol (2) and water (1), and it is seen that the deviation from ideality is large. It is also seen that for small mole fractions of ethanol its activity is proportional to its mole fraction (or to ethanol concentration expressed in another way). One then speaks of an *ideally dilute* system. This is further illustrated in Figure 2.2b, where a hypothetical example is given of the chemical potential of a solute (2) as a function of its mole fraction, for two components (1 and 2) that can be mixed in all proportions. For small x_2, the chemical potential is proportional to $\ln x_2$ and the slope is given by RT, all in agreement with Eq. (2.6), but the line does not extrapolate to μ° at $\ln x_2 = 0$ (i.e., $x_2 = 1$). For small x_2, the solution is thus ideally dilute, and we now have an *apparent standard chemical potential* $\mu^{\ominus}$ (pronounced mu plimsoll) of the solute in this particular solvent ($\mu^{\ominus}$ may also depend on pressure). For an ideally dilute solution the chemical potential thus is given by

$$\mu_2 = \mu_2^{\ominus} + RT \ln x_2 \qquad (2.6a)$$

Note in Figure 2.2a that at high concentration of ethanol the activity of the solvent (water) is proportional to its mole fraction, hence to $1 - x_2$; here the solution is ideally dilute for the solvent.

For those cases where x does not equal a, one arbitrarily introduces an *activity coefficient* γ, defined by

$$a_i \equiv x_i \times \gamma_i \qquad (2.8)^*$$

To be sure, it is implicitly assumed that the standard chemical potential is the apparent one ($\mu^{\ominus}$ rather than μ°) in the present solvent. Even in dilute solutions, γ often markedly deviates from unity.

The chemical potential determines the *reactivity* of a component, i.e., the composition of a mixture at chemical equilibrium and the driving force for a reaction, though not its rate. Transfer of a component from one phase or position to another one will always proceed in the direction of the lowest chemical potential, whether the transfer is by diffusion, evaporation, crystallization, dissolution, or some other process. If temperature and

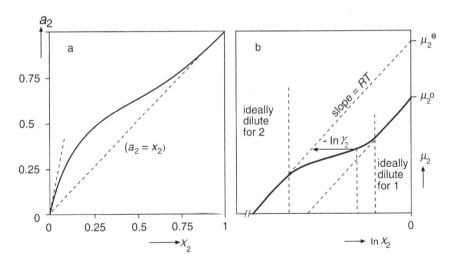

FIGURE 2.2 Thermodynamic aspects of mixtures. (a) Activity (a_2) of ethanol as a function of its mole fraction in an aqueous mixture (x_2). (b) Example of the chemical potential (μ_2) of a substance as a function of its mole fraction (x_2) in a mixture of substances 1 (solvent) and 2 (solute).

pressure are constant, it is convenient to use the activity. The activity rather than the concentration enters in relations on the solubility, the distribution of a component over various phases, the adsorption of a component onto a surface, and so on. If for some reason—say, the addition of another component—the activity coefficient becomes smaller without the concentration altering, the reactivity of the solute thus has become smaller and its solubility increased.

> *Note* Equations like (2.6) and (2.8) can also be put in a form where molarity, molality, or some other concentrative unit is used rather than mole fraction. This means that $\mu^{\ominus}$ has another value, but—more important—it also affects the value of the (apparent) activity coefficient. For a very dilute solution, the differences tend to be negligible, but in other cases, the concentrative unit to which the activity coefficient relates should be stated. Naturally, the various kinds of concentration can be recalculated into each other; see Appendix A.7.

2.2.2 Solubility and Partitioning

For a **mixture** of components that behaves ideally, it can be derived that there is no change in enthalpy when the components are mixed, i.e., no heat is released nor consumed. The decrease in free energy due to mixing then is purely due to an increase in entropy. Such a situation may occur for two components of very similar properties, for instance for a mixture of closely related triglycerides. However, if one of the components is a solid at the temperature of mixing, it has to melt, and this means an increase in enthalpy, equal to the enthalpy of fusion ΔH_f (the enthalpy of mixing is still assumed to be zero). This implies that there is a limited *solubility* (x_s), given by the Hildebrand equation,

$$\ln x_s = \frac{\Delta H_f}{R} \left(\frac{1}{T_f} - \frac{1}{T} \right) \tag{2.9}$$

where the solubility is expressed as a mole fraction and where the subscript f refers to fusion. Most solutions are far from ideal, and especially at high concentrations the activity coefficient may differ greatly from unity (often being larger). Even the introduction of activity rather than mole fraction in Eq. (2.9) is insufficient, since the change in enthalpy will generally include some enthalpy of mixing, which may be large. Nevertheless, a relation like Eq. (2.9) often holds, viz. a linear relation between log solubility and $1/T$. Examples are given in Chapter 15.

It may further be noted that the volume of a mixture of two components is generally not equal to the sum of the volumes of each. For many aqueous mixtures, the volume is decreased; one then speaks of *contraction*. For example, when mixing 10 ml (15.8 g) of ethanol with 80 ml of water, the mixture has a volume of 98.3 ml, which implies a contraction by 1.7%.

Partitioning. A substance may have limited solubility in two mutually immiscible solvents, for instance water and oil. This often happens in foods, for example with many flavoring and bactericidal substances. It then is important to know the concentration (or rather activity) in each phase. For low concentration, the partitioning or *distribution law of Nernst* usually holds:

$$\frac{c_\alpha}{c_\beta} = \text{constant} \qquad (2.10)^*$$

where c is concentration and α and β refer to the two phases.

This law is readily derived from thermodynamics, assuming both solutions to be ideally dilute. At equilibrium we must have for the solute (2) that $\mu_{2,\alpha} = \mu_{2,\beta}$. Although the standard chemical potential of the pure solute μ° is, of course, the same, the apparent standard chemical potential $\mu^\ominus$ (see Fig. 2.2) will generally be different. We thus have

$$\mu_{2,\alpha}^\ominus + RT \ln a_{2,\alpha} = \mu_{2,\beta}^\ominus + RT \ln a_{2,\beta}$$

from which follows

$$\frac{a_{2,\alpha}}{a_{2,\beta}} = \exp\left(\frac{\mu_{2,\beta}^\ominus - \mu_{2,\alpha}^\ominus}{RT}\right) \qquad (2.11)$$

which is constant at constant temperature. Since for dilute solutions the (apparent) activities mostly are proportional to the concentrations, Eq. (2.11) comes down to Nernst's law. Note that the partition ratio (c_α/c_β) will decrease with increasing temperature if the ratio is larger than unity and vice versa.

2.2.3 Determination of Activity

When preparing a solution, one usually knows the concentration of the solute. Most analytical methods also yield concentrations rather than activities. Often, the solute is allowed to react in some way, and although the reaction rate will be determined by the activity rather than the concentration, these reactions are generally chosen such that the reacting solute will be completely consumed, and a concentration results. Generally, spectroscopic methods give concentrations as well. Equilibrium methods, on the other hand, yield activities. A good example is measurement of an electric potential by means of an ion-selective electrode, as in pH measurement. Also a partition equilibrium between two phases yields activity.

This provides an easy way of determining the activity of a substance if it is volatile. It will then have the same activity in the gas phase as in the solution, and at ambient conditions a gas generally shows ideal behavior. The latter is true as long as the so-called ideal gas law,

$$p V = n R T \tag{2.12}^*$$

holds, where n is the number of moles in the system. The prime example is determination of the water activity of a solution. Because of the ideality in the gas phase, i.e., $a_1 = x_1$, the a_1 in the solution, mostly designated a_w, is equal to the relative humidity of the air with which the solution is in equilibrium, which can readily be measured. If the solute is also volatile, it is often possible to determine its activity in the gas phase, hence in the solution.

For a solution of one (nonvolatile) solute in water, whose water activity is known over a concentration range, the activity of the solute can be derived from the *Gibbs–Duhem relation*, which can for this case be written as

$$x_1 \cdot d \ln a_1 + x_2 \cdot d \ln a_2 = 0 \tag{2.13}$$

By (numerical or graphical) integration, a_2 can now be derived. Figure 2.3 gives as an example the activities of sucrose solutions. It is seen that the activities greatly deviate from the mole fractions at higher concentration. For example, at $x_2 = 0.1$, the activity coefficient of water $\approx 0.85/0.90 = 0.94$, that of sucrose $\approx 0.26/0.1 = 2.6$. For mixtures of more than two components, the activities cannot be derived in this way.

2.2.4 Colligative Properties

The lowering of the chemical potential of a solvent by the presence of a solute causes changes in a number of physical properties: vapor pressure, boiling point, freezing point, osmotic pressure, etc. In an ideally dilute solution the magnitudes of these changes all are proportional to the mole fraction of solute; they follow from the same cause and are thus called *colligative solution properties*. In Section 2.3, electrolyte solutions will be discussed, but it is convenient to recall here that solutes that largely

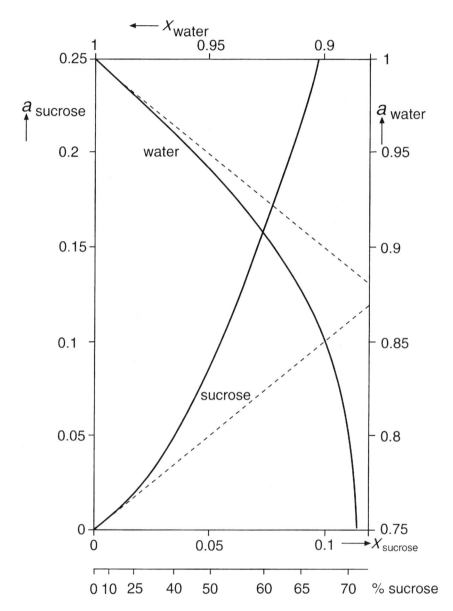

FIGURE 2.3 Activities (*a*) of sucrose and water in binary mixtures as a function of the mole fraction (*x*). The broken lines give the hypothetical activities for ideally dilute systems.

dissociate into two species—commonly ions—have an effective molarity that is about twice the nominal one, etc.; this effective concentration is called the *osmolarity*. We will here only consider water as the solvent and assume the solution to be ideally dilute. Subscript 1 refers to the solvent (water), 2 to the solute.

The lowering of vapor pressure at any temperature then follows from *Raoult's law*,

$$p_1 = x_1 p_1^o = (1 - x_2)p_1^o \qquad (2.14)$$

where p is vapor pressure and p_1^o is that of pure water.

The change in **boiling point** at standard pressure (1 bar) is given by

$$\Delta T_b = -\frac{T_{b,1}^2}{\Delta H_{v,1}} R \ln x_1 \approx -28 \ln x_1 \approx 28 x_2 \approx 0.51 \, m_2 \text{ (K)} \qquad (2.15)$$

where $T_{b,1}$ is the boiling point of the pure solvent, ΔH_v is the enthalpy of vaporization $(40.6 \, \text{kJ} \cdot \text{mol}^{-1}$ for water at $100°C)$, and m is the solute concentration in moles per liter. The approximations successively made, when going from the first to the last righthand term in the equation, all apply at infinite dilution. The boiling point elevation is often given as $K_b \cdot m_2$, where for water $K_b = 0.51 \, \text{K} \cdot \text{L} \cdot \text{mol}^{-1}$. It should be noted that its magnitude significantly depends on ambient pressure.

The change in **freezing point** is similarly given by

$$\Delta T_f = \frac{T_{f,1}^2}{\Delta H_{f,1}} R \ln x_1 \approx 103 \ln x_1 \approx -103 x_2 \approx -1.86 m_2 \text{ (K)} \qquad (2.16)$$

where ΔH_f is the enthalpy of fusion, $6020 \, \text{J} \cdot \text{mol}^{-1}$ for water. Note that the freezing point depression is considerably greater than the boiling point elevation, because the molar enthalpy of fusion is far smaller than the enthalpy of vaporization.

The **osmotic pressure** (Π) of a solution can be interpreted as the pressure that has to be applied to the solution to increase the chemical potential of the solvent to the value of the pure solvent at standard pressure. Π is thus higher for a higher solute concentration. If local differences in concentration exist, solvent (i.e., liquid) will move to the regions where Π is highest, to even out concentration gradients; this means that osmotic pressure is in fact a negative pressure.

The osmotic pressure becomes manifest and can thus be measured in a situation as depicted in Figure 2.4, where solvent and solution are separated by a *semipermeable membrane* that lets the solvent pass but not the solute(s). Solvent now moves to the solution compartment until the osmotic pressure is compensated by the difference in height of, hence in gravitational pressure in, both compartments. Incidentally, this implies that by application of a pressure to a solution that is higher than its osmotic pressure, solvent can be removed from the solution, thereby increasing the concentration of the solute; this is called reversed osmosis.

The osmotic pressure of an ideally dilute aqueous solution is given by

$$\Pi = -55,510RT \ln x_1 \approx 55,510RTx_2 \approx 10^3 m_2\,RT \qquad (2.17)$$

where the factor 55,510 represents the number of moles of water in a m^3. Since $m_2 = n/V$, where $n = $ the number of moles in the volume V, Eq. (2.17)

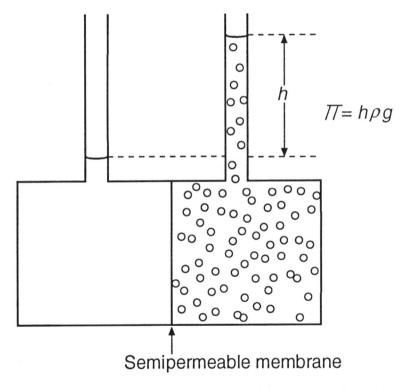

Semipermeable membrane

FIGURE 2.4 Measurement of osmotic pressure (Π). Solvent and solution are separated by a membrane that lets the solvent pass but not the solute. The small circles denote the solute molecules (or particles). $h = $ height (m), $\rho = $ mass density (10^3 kg $\cdot$ m^{-3} for water) and $g = $ acceleration due to gravity (9.81 m $\cdot$ s^{-2}).

can also be written as $\Pi V \approx nRT$; the equation is thus of the same type as Eq. (2.12).

The **validity** of the quantitative relations given is illustrated in Table 2.1 for the freezing point depression by some solutes, Eq. (2.16). It is seen that the agreement is reasonable, especially if the equations are used in their most rigid form. Nevertheless, deviations remain, which implies that at the concentration considered the solutions are not ideally dilute anymore. Comparison with Figure 2.3 shows that, for sucrose, deviations become large for concentrations over 20%. Actually, Eqs. (2.14–2.17) are all based on the assumption that $x_1 = a_1$, i.e., = the water activity a_w. By inserting a_w—which can often be measured—for x_1, a much better agreement will be obtained. As derived in Eq. (8.3), $- \ln a_w$ is a property colligative with the others mentioned.

The colligative properties are of importance by themselves, but they can also be used to determine the molar mass of a solute, since they all depend on the molar concentration and since the mass concentration generally is known. To this end, the determination of the freezing point often is most convenient. Because of nonideality, determinations should be made at several concentrations and the results extrapolated to zero. For determination of the molar mass of macromolecules, osmotic pressure measurement is to be preferred, since membranes exist that are not permeable for macromolecules, while they are for small-molecule solutes, and even small quantities of the latter have a relatively large effect on the colligative properties. Actually, a difference in osmotic pressure is thus determined, the difference being due to the macromolecules only.

TABLE 2.1 Estimates and Determined Values (ΔT_f, in K) of Freezing Point Depression of Some 20% (w/w) Aqueous Solutions

Solute	M	m	$1.86\,m$	$10^3\,x_2$	$-10^3 \ln x_1$	ΔT_f
Ethanol	46.1	4.20	7.8	9.2	9.6	10.9
Glycerol	92	2.27	4.22	4.80	4.92	5.46
d-Glucose	180	1.20	2.23	2.51	2.54	2.64
Sucrose	342	0.63	1.18	1.34	1.35	1.46
NaCl*	58.4	7.85	14.6	15.4	16.7	16.5

M = molar mass (Da), m = molarity, x_1 = mole fraction of water, x_2 = mole fraction of solute.
* Assuming effective molarity to be twice the nominal value.

2.2.5 Deviations from Ideality

In foods, we often have situations in which concentrations are markedly different from activities, or in other words, the activity coefficients may be far from unity. This may have important consequences for partition equilibria, for reaction equilibria, and often also for reaction rates. Below some important causes for deviations from ideality are listed.

1. Not all species are reactive. This is theoretically fairly trivial, but the practical implications may be considerable. An example is the presence of a reducing sugar, for instance D-glucose, in various forms. Here we have

$$\alpha\text{-glucose} \rightleftharpoons \text{open chain form} \rightleftharpoons \beta\text{-glucose}$$

and only in the open-chain form, which may be less than 1% of the glucose present, can the sugar participate in Maillard reactions or other reactions involving the aldehyde group. Another example is an organic acid, here denoted by HAc, which dissociates according to

$$\text{HAc} \rightleftharpoons \text{H}^+ + \text{Ac}^-$$

The dissociated form Ac^- can react with cations, whereas the undissociated form HAc may be active as an antimicrobial agent. The activities of each depend not only on the overall concentration but also on the dissociation constant (which depends on temperature), the pH, the presence and concentration of various cations, etc. It may be argued in these cases not that the activity coefficient is (much) smaller than unity but that we should take the concentration of the species involved in the reaction only. The result is, of course, the same, and we may speak of an apparent activity coefficient.

2. High concentration. At high concentrations of a solute, its activity coefficient nearly always deviates from unity. This may be for two reasons. First, the *solvent quality* affects the activity coefficient, and the effect increases with increasing solute concentration. Solvent quality depends on the interaction energy between solute and solvent molecules; this is further discussed in Section 3.2. A poor solvent tends to increase and a good solvent to decrease the activity coefficient of the solute (and thereby, for instance, to increase its solubility).

Second, *volume exclusion* occurs, which always causes an increase in the activity of a solute if the solute molecules are larger than the solvent molecules. At high concentration the amount of solvent available to the solute is effectively less than the nominal amount, which means that the solute concentration is effectively higher. This is easiest envisaged for spherical molecules of radius r; such a molecule takes up a volume equal to

$(4/3)\pi r^3$, but another molecule cannot come closer to it than a distance $2r$ (taking the position of the molecules to be at their centers), which implies that a volume of $(4/3)\pi(2r)^3$ is excluded for a second molecule. If the volume fraction occupied by the molecules is φ, this would imply that a volume of 4φ is not available as a solvent (not 8φ, because we then would count the excluded volume twice). The "effective" concentration of the solute would be increased by a factor $1/(1 - 4\varphi)$, if r_2 (solute molecules) $>> r_1$ (solvent molecules). Mostly r_1 and r_2 differ less; the numerical factor would then be < 4, and it is zero if $r_1 = r_2$. This reasoning is an oversimplification, because (a) only for small φ does the excluded volume indeed equal 4φ, whereas for larger φ it becomes relatively less; (b) for nonspherical molecules the excluded volume is less well defined, though generally higher; and (c) interactions between solute and solvent molecules—i.e., the solvent quality mentioned above—may modify the result. Nevertheless volume exclusion is a very real and important source for

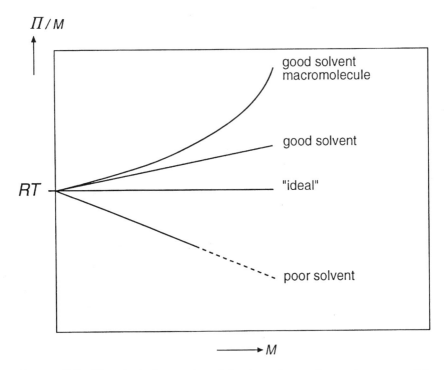

FIGURE 2.5 Hypothetical examples of the dependence of osmotic pressure (Π) divided by concentration against molar concentration (m).

deviations from ideality at high concentration, especially if the solute molecules are large.

The effect of concentration is often expressed in a **virial expansion**. For the osmotic pressure it reads

$$\Pi = RT(m + B m^2 + C m^3 + \cdots) \tag{2.18}$$

where m is in $mol \cdot m^{-3}$. B is called the second virial coefficient, C the third virial coefficient, and so on. In first approximation, B is due to both solvent quality and volume exclusion effects and it can either be negative or, more likely, positive; C is only caused by volume exclusion and would be zero if solvent and solute molecules have equal size. Figure 2.5 illustrates some trends. It is seen that for a not too high concentration the second virial coefficient may suffice (i.e., Π/m is linear with m), but this is rarely true for macromolecular solutes.

The same relation can be used for the chemical potential of the solvent (more precisely for $\mu_1^\circ - \mu_1$, which equals zero for $m = 0$) and for all

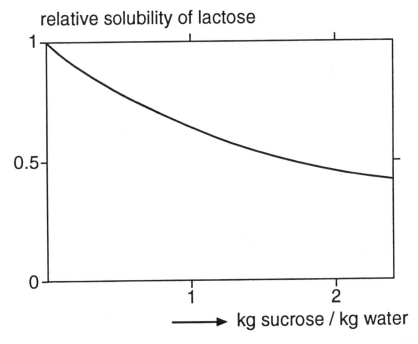

FIGURE 2.6 Effect of sucrose on the relative solubility of lactose in water at 50°C. After results by T. A. Nickerson and E. E. Moore. *J. Food Sci.* 37 (1972), 60.

TABLE 2.2 Effect of Sucrose on the Solubility of Lactose

(1) Sucrose concentration, kg/kg water	(2) Lactose solubility, kg/kg water	(3) Mole fraction of sugar	(4) Activity coefficient	(2) × (4)
0	0.45	0.023	1.26	0.57
0.40	0.38	0.042	1.55	0.59
0.66	0.34	0.050	1.73	0.59
1.00	0.28	0.063	1.98	0.55
1.50	0.21	0.083	2.28	0.48
2.32	0.19	0.167	?	?

colligative properties. From the results depicted in Figure 2.3, it can be derived that for sucrose solutions a second virial coefficient suffices up to a mole fraction of about 0.015 (about 25% w/w sucrose).

If we have more than one solute, the situation becomes far more complicated. For two solutes (subscripts 2 and 3) we have in principle

$$\Pi = RT(m_2 + m_3 + B_2 m_2^2 + B_3 m_3^2 + B_{23} m_2 m_3 + \cdots) \tag{2.19}$$

We now must also take into account a second virial coefficient for the interaction between both solutes B_{23}. If both solutes are very similar in chemical nature, B_{23} may be negligible. As an example we take the effect of sucrose on lactose. Figure 2.6 shows that high concentrations of sucrose considerably lower the solubility of lactose. This is to be expected, since (a) at high sucrose concentrations the activity coefficient of sucrose is greatly increased (see Figure 2.3); (b) because of the similarity between both sugars we may expect that also the activity coefficient for lactose is greatly increased; and (c) since the solubility is to be expressed as an activity, and activity equals concentration times activity coefficient, a higher activity coefficient means a lower solubility in terms of concentration. In Table 2.2 some calculations are given. Column (2) gives the solubility of lactose; in (3) the mole fraction of lactose + sucrose is given; in (4) a presumed activity coefficient is given, based on the assumption that all sugar behaves as if it were sucrose (thus derived from Figure 2.3); and finally the product of (2) and (4) should be the solubility of lactose in terms of activity and thus be constant, if the assumption is correct. It is seen that up to about 1 kg sucrose per kg water, this is nearly correct (the variation is probably within experimental error), but for higher concentration it is not so. Naturally, for solutes that differ considerably in chemical structure, much larger deviations may be expected. For very concentrated systems, as in partly dried foods, very large deviations between concentration and activity often occur.

3. Adsorption. Several kinds of molecules can adsorb onto various surfaces or bind onto macromolecules, thereby lowering their activity coefficients. Binding should not be interpreted here as forming a covalent or an ionic bond, for in such a case the concentration of the substance is indeed decreased. It is well known that for many flavor components the threshold concentration for sensory perception is far higher in a particular food than in water. This means that the activity coefficient is smaller in the food than in water, and a decrease by a factor of 10^3 is no exception. It may be recalled that many flavor components are fairly hydrophobic molecules, which readily adsorb onto proteins. Because of this, the so-called head space analysis for flavor components makes good sense, since the concentrations of the various components in the gas phase (which is, in principle, in equilibrium with the food) are indeed expected to be proportional to the activities in the food.

A surface active component naturally adsorbs onto many surfaces, say of oil droplets or solid particles or macromolecules, by which its activity is decreased (see Section 10.2). Another example is the binding of cations, especially of heavy metals, to proteins; when concentrating by ultrafiltration a protein solution that also contains some Cu, almost all of the Cu is concentrated with the protein, although Cu ions would be perfectly able to pass the ultrafiltration membrane.

In the simplest case, the amount adsorbed is proportional to the concentration (or rather activity) of the species in the solution, implying that the activity is a constant fraction of the concentration. If there is a limited number of binding or adsorption sites, as is often the case, we often have a Langmuir type adsorption isotherm, given as

$$\Gamma = \frac{\Gamma_\infty m}{C + m} \tag{2.20}$$

where Γ is the surface concentration of adsorbed material (in $mol \cdot m^{-2}$), Γ_∞ its value if all adsorption sites are occupied, and the constant C denotes the concentration at which $\Gamma = 0.5\Gamma_\infty$. Note that m ($mol \cdot m^{-3}$) refers to the concentration in solution; the total concentration equals $m + \Gamma A$, where A is the specific surface area. The activity coefficient then would be given by $m/(m + \Gamma A)$, which increases with total concentration, as illustrated in Figure 2.7b.

4. Self association. Amphiphilic molecules, i.e., molecules that consist of a hydrophilic (or polar) and a hydrophobic (or apolar) part, often tend to associate in an aqueous environment. Good examples are soaps and other small-molecule surfactants, which form micelles, i.e., roughly spherical aggregates in which the hydrophobic tails are in the interior and the hydrophilic heads to the outside; see Figure 2.8a. Micelles are often formed at a well-defined critical micellization concentration (CMC), above which almost all additional molecules are incorporated into micelles. This implies that properties that depend on the activity of the solute, like osmotic pressure and surface tension, hardly alter above the CMC, since the activity hardly increases anymore; see Figure 2.8b. The way in which the activity coefficient alters is illustrated in Figure 2.7c.

Incidentally, the presence of micelles may greatly enhance the apparent solubility of apolar components in an aqueous phase, since they tend to be incorporated into the interior of the micelles up to a certain level. The activity of such an apolar component remains of course very low, but it

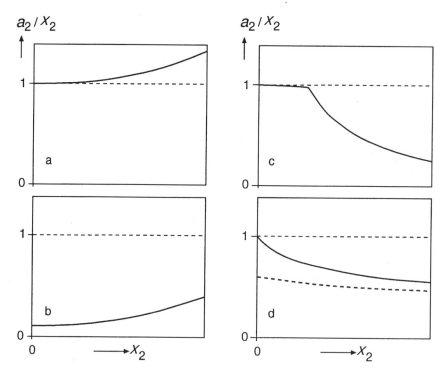

FIGURE 2.7 Hypothetical examples of nonideality of a solute (2) in aqueous solutions. The activity coefficient is given as a_2/x_2 versus the mole fraction x_2. (a) "High concentration." (b) Adsorption or binding. (c) Self-association, especially micellization. (d) Electric shielding; the broken line is for a case where other salts are present at constant concentration.

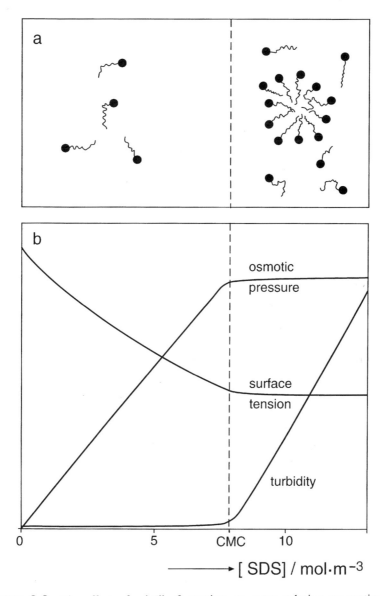

FIGURE 2.8 The effect of micelle formation on some solution properties. (a) Schematic picture of micelle formation. (b) Osmotic pressure, surface tension, and turbidity of solutions of sodium dodecyl sulfate (SDS) as a function of concentration (approximate). CMC = critical micellization concentration.

may still be available as a reactant, although it will react only sluggishly. But if no micelles were present, a very apolar "solute" would not be there and thus would not react at all.

Other kinds of self-association occur. Several proteins associate to form fairly large aggregates, thereby markedly decreasing their activity. Fatty acids present in an oil phase always dimerize through hydrogen bonds:

thereby greatly decreasing their activity and increasing their solubility (see Question 2, below).

5. Electric shielding. The activity of ions is diminished by electric shielding. This is discussed in Section 2.3. The total ionic strength rather than the concentration of the ions studied determines the activity coefficient; see Figure 2.7.d.

Question 1

Consider an aqueous solution of 5 millimolar (pure) sodium dodecyl sulfate (SDS). How large will the freezing point depression be? And the water activity? What will happen if the solution is cooled down to $-0.1°C$?

Answer

SDS will fully dissociate at low concentration. A solution of 5 mmol $\cdot$ L^{-1} will thus contain 0.01 mol ions per liter. Inserting $m_2 = 0.01$ in Eq. (2.16) leads to $-\Delta T_f = -0.0186$ K. Putting water activity $a_w = x_1$ then yields $a_w = 0.99982$. At 0.1 K below the freezing point of water, we thus have significant freeze concentration. According to Eq. (2.16), m_2 would then equal 0.054 molar, corresponding to 27 mmolar SDS, assuming it to be fully dissociated. However, Figure 2.8 shows that the critical micellization concentration (CMC) equals about 8 mmolar SDS. Hence micelles will be formed, and m_2 would hardly increase over the CMC. This then would mean that by far the largest part of the water must freeze at $-0.1°C$. The little bit of concentrated solution left will be strongly nonideal, and prediction of its properties would be quite difficult.

Question 2

An oil-in-water emulsion, with an oil volume fraction (φ) of 0.2, contains a certain fatty acid (HAc); total concentration is 0.01 mole per liter emulsion. The following data are provided: the association constant for dimerization of the acid in the oil phase $K_A = 10^4\ \text{L}\cdot\text{mol}^{-1}$; the partitioning constant between oil and water $K_N = 1$; the dissociation constant of the acid in water $K_D = 1.25\cdot 10^{-5}\ \text{mol}\cdot\text{L}^{-1}$; the pH is 4.90. What is the proportion of the acid in the aqueous phase and what is its (apparent) activity coefficient in the oil?

Answer

If dimerization in the oil and dissociation in the water phase would not be taken in to account, a partitioning constant of unity would imply the concentrations to be equal in both phases, and the quantity in the oil phase would be φ times the overall concentration, i.e., 2 mmol per liter emulsion. However, the situation is as follows:

$$\text{OIL} \qquad \text{HAc} \rightleftharpoons \tfrac{1}{2}\,(\text{HAc})_2 \qquad\qquad [\text{HAc}]_o = \{[(\text{HAc})_2]/K_A\}^{0.5}$$
$$\text{-----------} \text{-} \| \text{-----------} \qquad [\text{HAc}]_o = K_D[\text{HAc}]_w$$
$$\text{WATER} \quad \text{HAc} \rightleftharpoons \text{H}^+ + \text{Ac}^- \qquad\qquad [\text{HAc}]_w = [\text{H}^+][\text{Ac}^-]/K_D$$

Let us call $[\text{HAc}]_o = x, [\text{Ac}^-] = y$, and $[(\text{HAc})_2] = z$. Because $K_N = 1$, we have that $[\text{HAc}]_w$ also equals x. This means that $x = [\text{H}^+]y/1.25\cdot 10^{-5}$, and since $[\text{H}^+] = 10^{-\text{pH}} = 1.26\cdot 10^{-5}$, we obtain $y \approx x$. The total concentration in the water phase is thus $2x$. In the oil, $x^2 = z/K_A$, and thus $z = 10^4 x^2$. The total concentration of the fatty acid in the oil is now $x + 2\cdot 10^4 x^2$; the factor 2 is needed because z refers to dimers. Taking into account that $\varphi = 0.2$, we obtain for the total amount of fatty acid in a liter emulsion

$$1.6x + 0.2x + 0.4\cdot 10^4\, x^2 = 0.01$$

which yields $x = 0.00137$ mole per liter. We now calculate that the total concentration in the oil $(x + 2\cdot 10^4 x^2)$ amounts to 0.0389, and since the concentration of nondimerized acid is 0.00137, the apparent activity coefficient $= 0.035$. The amount of fatty acid in the oil phase is 0.0078 moles, i.e., 78% of the total. Note that in this case the overall partitioning ratio as well as the apparent activity coefficient in the oil depend on the concentration.

Note We have tacitly assumed that all activity coefficients equal unity. This may be reasonable, except for Ac$^-$ (see Section 2.3); taking its activity coefficient into account would cause the total concentration in the water phase to be somewhat higher. (We do not need an activity coefficient for H$^+$, since pH is a measure of activity, not concentration.)

Note A certain amount of the fatty acid may adsorb on the oil–water interface. If the oil droplets are small, this may be an appreciable amount. If the adsorption isotherm is known, the effect can be calculated.

Note If the aqueous phase contains protein, it may well be that some fatty acid becomes associated with the protein, whereby the activity coefficient is lowered and the concentration in the water will thus be increased.

2.3 ELECTROLYTE SOLUTIONS

Ionizable substances, like salts, acids, bases, and polyelectrolytes, partly dissociate into ions when dissolved in water. As a consequence, the osmolarity will be higher than the molarity (see Section 2.2.4). More important, ionic species generally are reactive because of their electric charge, and the charge generally is shielded to a certain extent by the presence of ions of opposite charge, called counterions. This implies that the activity coefficient may be greatly diminished if the concentration of counterions is high.

We will briefly recall some basic facts about dissociation and its consequences, before discussing the magnitude and the importance of ion activity coefficients.

2.3.1 Dissociation

A simple salt like NaCl will in water dissociate according to $NaCl \rightleftharpoons Na^+ + Cl^-$. We now have for the *dissociation constant*

$$K_D = \frac{a(Na^+) \cdot a(Cl^-)}{a(NaCl)} = \frac{[Na^+] \cdot [Cl^-]}{[NaCl]} \times \frac{\gamma_+(Na^+) \cdot \gamma_-(Cl^-)}{\gamma_0(NaCl)} \qquad (2.21)$$

The first right-hand expression is written in activities, and this quotient gives the *intrinsic* dissociation constant. The second right-hand expression is made up of two factors, a quotient of (molar or molal) concentrations that may be called the *stoichiometric* dissociation constant, and a quotient of activity coefficients. All dissociation constants, association constants $(K_A = 1/K_D)$, and solubility products in reference books are intrinsic constants. They apply to concentrations only if the solution is extremely dilute for all ionic species. In other cases, one has to know the activity coefficients. γ_0, i.e., γ for a nonionic species, will mostly be close to unity, but γ_+ and γ_- will generally be < 1, the more so for a higher ion concentration. One may define the *free ion activity coefficient* of NaCl as

$$\gamma_{\pm}(\text{NaCl}) \equiv [\gamma_{+}(\text{Na}^{+}) \cdot \gamma_{-}(\text{Cl}^{-})]^{1/2} \qquad (2.22)$$

Note that the subscripts to the activity coefficient here have the following meanings:

$\pm$ of the salt in its dissociated form
0 of the salt in its undissociated form
$+$ of the cation
$-$ of the anion

The solubility of a salt is given as the *solubility product* K_s, which is to be compared with the activity product K_a. For NaCl this is given by

$$K_a = a(\text{Na}^{+}) \cdot a(\text{Cl}^{-}) = \{[\text{Na}^{+}] \cdot [\text{Cl}^{-}]\} \times \{\gamma_{+}(\text{Na}^{+}) \cdot \gamma_{-}(\text{Cl}^{-})\} \qquad (2.23)$$

The activity product of the ions in solution thus cannot be higher than K_s. To be sure, for a very soluble salt like NaCl, it often would make little sense to use the solubility product, because of the large nonideality at the relevant concentration; but it is a very useful concept for salts of lower solubility, especially if the activities of the cations and anions are not equal, as will often be the case in foods.

For the dissociation of an acid ($\text{HAc} \rightleftharpoons \text{H}^{+} + \text{Ac}^{-}$) we can write, recalling that $\text{pH} = -\log a(\text{H}^{+})$ and that $pK_a = -\log K_D$,

$$\text{pH} - pK_a = \log[\text{Ac}^{-}] + \log \gamma_{-} - \log[\text{HAc}] \qquad (2.24)$$

The pK_a (subscript a for acid) is the pH at which the dissociation is exactly 50% ($[\text{Ac}^{-}] = [\text{HAc}]$), provided that $\gamma_{-} = 1$; however, the latter mostly is < 1, implying that the stoichiometric pK_a is smaller than the intrinsic one, say by 0.1 or 0.2 units. Strong acids have a low pK_a (large K_D), often < 1; fatty acids, for instance, are weak acids and have a $pK_a \approx 4.7$. At a pH one unit higher, the acid will be dissociated for 91% according to (2.24), at a pH one unit lower for 9%, at 2 units lower for 1%, at 3 units lower for 0.1%, etc. This is illustrated in Figure 2.9a.

Di- and triprotic acids have two and three pK_a's, respectively. For instance, the intrinsic pK_a values of citric acid, $\text{CH}_2(\text{COOH})-\text{CH}(\text{COOH}) - \text{CH}_2(\text{COOH})$, are 3.1, 4.7, and 5.4. The three acid groups are almost identical, and the differences in pK are due to the increased charge if more than one group is dissociated. When increasing the pH, the second proton to be dissociated must be removed against the electric potential of two, rather than one negative charge. This needs additional free energy and will thus happen at a higher pH, where the driving force for dissociation is greater. Figure 2.9b gives a titration curve for citric acid; here the degree of neutralization rather than dissociation is given, and only when using a very strong base for the titration and after extrapolation to zero ionic strength would the two be identical. Figure 2.9b also gives a titration curve for phosphoric acid; this has intrinsic pK_a values of 2.1, 7.2, and about 12.7. These differences are far larger than in the case of citric acid, because the three acid groups are essentially different; the neutralization of the various groups can now be distinguished on the titration curve.

For bases comparable relations hold.

The relations involving activity coefficients are slightly more complicated for ions of a valence higher than 1. For instance, for $CaCl_2 \rightleftharpoons Ca^{2+} + 2\,Cl^-$ the dissociation constant is given by

$$K_D = \frac{a(Ca^{2+})a^2(Cl^-)}{a(CaCl_2)} = \frac{[Ca^{2+}] \cdot [Cl^-]^2}{[CaCl_2]} \times \frac{\gamma_+(Ca^{2+}) \cdot \gamma_-^2(Cl^-)}{\gamma_0(CaCl_2)} \quad (2.25)$$

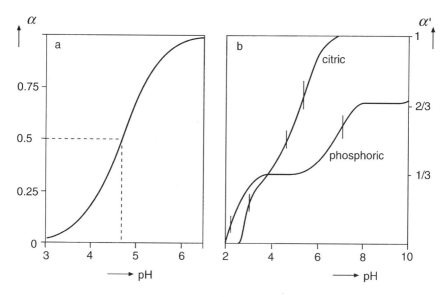

FIGURE 2.9 Dissociation of acids as function of pH. (a) Calculated degree of dissociation (α) of a fatty acid (intrinsic $pK_a = 4.7$) at very small ionic strength. (b) Titration curves, i.e., experimentally determined degree of neutralization (α') by KOH of citric acid and phosphoric acid; the intrinsic pK_a values are indicated by vertical dashes.

K_D is now in $mol^2 \cdot L^{-2}$ rather than moles per liter. Further,

$$\gamma_\pm(CaCl_2) = [\gamma_+(Ca^{2+}) \cdot \gamma_-^2(Cl^-)]^{1/3}$$

or, more generally, for a cation C and an anion A making a salt C_xA_y,

$$\gamma_\pm(C_xA_y) = [\gamma_+^x(C) \cdot \gamma_-^y(A)]^{1/(x+y)} \qquad (2.26)$$

It is generally difficult to determine the ion activity coefficients, and one commonly makes shift with values calculated with semiempirical equations: see Section 2.3.2. Anyway, γ decreases with an increase in total ionic strength. This means that adding any other electrolyte will decrease the activity coefficients, causing the solubility and the dissociation to increase. For instance, if KNO_3 is added to a solution of $CaCl_2$, this affects the latter's dissociation equilibrium. γ_+ and γ_- decrease and γ_0 remains at 1, and—since the intrinsic dissociation constant remains unaltered—it thus follows from (2.25) that the concentrations of Ca^{2+} and Cl^- increase and that of the undissociated salt decreases.

It should finally be remarked that dissociation constants generally depend, and often strongly depend, on temperature. The dissociation may either increase or decrease with temperature, and there are no general rules. The same holds for solubility products.

Note In principle, the ion activity coefficient of a salt $(\gamma_\pm)$ can be determined, but not those of individual ions $(\gamma_-$ and $\gamma_+)$, because their concentrations cannot be varied independently. Nevertheless, the activity coefficients of individual ions are very useful, and as mentioned, one tries to calculate them from theory. Ion-selective electrodes measure chemical potentials (which depend on activities, not concentrations), but the standard potential is unknown. For the measurement of pH, which is the negative logarithm of the hydrogen ion activity, one has therefore arbitrarily chosen a reference potential for a certain buffer, which potential is of course as close to the real one as theory permits it to be calculated.

2.3.2 Debye–Hückel Theory

The ion activity coefficients depend on a great number of factors, but for low ionic strength, electric shielding is by far the main factor. On this basis the Debye–Hückel "limiting law" has been derived. As depicted in Figure 2.10, an ion in solution is, on average, surrounded by more counterions (opposite charge) than coions (same charge), thereby to some extent shielding the charge of the ion. Entropy is highest if all ions are completely distributed at random, but the attractive electric energy between ions of opposite charge tries to arrange the ions in a regular lattice. The attraction is stronger if the ions are on average closer to each other, and consequently the higher the ion concentration, the stronger the shielding (see Figure 2.10). Here all ions contribute, and use is therefore made of the total *ionic strength*, defined as

$$I \equiv \frac{1}{2} \sum_i m_i z_i^2 \qquad (2.27)^*$$

where m denotes molarity and z the valence of the ions. Note that the square of the valence is needed. This implies that a 0.01 molar solution of $CaCl_2$ has an ionic strength of $(0.01 \cdot 2^2 + 2 \cdot 0.01)/2 = 0.03$ molar. Most aqueous foods have an ionic strength between 1 and 100 millimolar (see also Figure 6.8b).

In the theory, the size of the ion also is involved; taking an average value, the ion activity coefficient of a dilute salt solution is roughly given by

$$\gamma_\pm \approx \exp\left[-\frac{42 \cdot 10^5}{(\varepsilon T)^{3/2}}|z_+ z_-|\sqrt{I}\right] \qquad (2.28a)$$

where ε is the relative dielectric constant. For a higher ionic strength, higher valences, and a lower dielectric constant, the activity coefficient is thus smaller. (A lower temperature gives a higher dielectric constant, and this

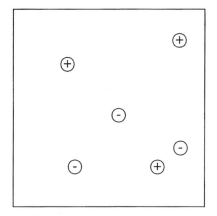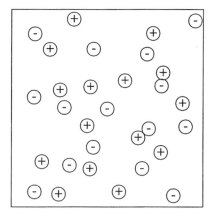

FIGURE 2.10 Electric shielding of ions by ions of opposite charge in a dilute and a more concentrated solution. Highly schematic.

causes the temperature influence to be fairly small.) For water at 20°C, $\varepsilon = 80$, and the equation becomes

$$\gamma_\pm \approx \exp(-1.17|z_+ z_-|\sqrt{I}) \qquad (2.28b)$$

The valence has a large effect; comparing, for instance, a 0.01 molar solution of NaCl with one of $CaSO_4$, the activity constants are calculated as 0.89 and 0.31, respectively (assuming dissociation to be complete).

Figure 2.11 gives the total ion activity coefficients of a few salt solutions, and the calculated values are given for comparison. It is seen that the agreement is good up to an ionic strength of about 0.05. For higher values, the equation gives activity coefficients that are too low. Note that Eq. (2.28) does not discriminate between different ions. In practice, fairly

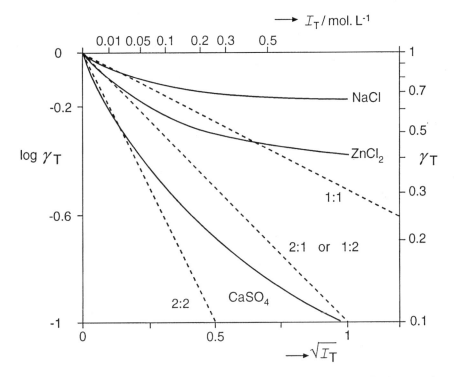

FIGURE 2.11 Total ion activity coefficients (γ_T) as a function of total ionic strength (I_T) of NaCl, $ZnCl_2$, and $CuSO_4$. The broken lines are according to Eq. (2.28) for salts with ions of various valences (indicated).

small differences are found for most inorganic ions, except for the hydrogen ion, which has a higher activity coefficient than predicted by Eq. (2.28).

2.3.3 Not Very Dilute Solutions

The situation becomes far more complicated for higher concentrations, say $I > 0.03$ molar, which still is fairly dilute, i.e., a few times 0.1% for many salts. The most important aspect may be that the **association of ions** into *ion pairs*, i.e., undissociated salt molecules (or ions, e.g., $CaCl^+$), becomes significant. It is often assumed that salts completely dissociate into ions, unless the concentration is very high. This assumption is generally not true, and it would lead to considerable error in many foods. To be sure, most ion pairs are very short lived, but at any time a certain proportion of the ions is in the associated form.

Consider the association of a cation C and an anion A according to $C^{z+} + A^{z-} \rightleftharpoons CA$ where the ion pair CA may be neutral or not, according to the values of $z+$ and $z-$. We may write for the association constant

$$K_A = \frac{a(CA)}{a(C^{z+}) \cdot a(A^{z-})} = \frac{[CA]}{[C^{z+}] \cdot [A^{z-}]} \times \frac{\gamma(CA)}{\gamma(C) \cdot \gamma(A)} \qquad (2.29)$$

where all γ denote free ion activity coefficients, except possibly (CA), which ≈ 1 if CA is neutral. Although K_A varies among ions, in first approximation it is governed by the valence of the ions, as given in Table 2.3. Taking the values given there, we obtain for the proportion dissociated of a salt of valence

1:1	0.01 molar	dissociation $= 0.99$
	0.1 molar	dissociation $= 0.95$
2:2	0.01 molar	dissociation $= 0.33$
	0.1 molar	dissociation $= 0.22$

TABLE 2.3 Order of Magnitude of Intrinsic Association Constants in $(L \cdot mol^{-1})$ for Ions of Various Valences (z)

		z_c	
$|z_A|$	1	2	3
1	1	10	50
2	10	700	10^5

See Eq. (2.29).

and it is seen that especially for valences higher than 1 the effect is great. In calculating the above figures, one has to know γ_+, which means that one has to know the dissociation to obtain the ionic strength, which is needed to calculate the ion activity coefficients. This can be done by first taking an assumed activity coefficient and then reiterating the calculation until the coefficient is in agreement with the dissociation.

This reveals an important problem. Most **ion activity coefficients** given in the literature, including those of Figure 2.11, are *total ion activity coefficients*, i.e., they relate to the total salt concentration; moreover, the ionic strength is calculated *as if* the dissociation of the salt were complete. This is a suitable method if there is only one salt in solution, or if there are only monovalent ions and the concentration is very small. But in all other cases, which implies in almost all foods, where we have several ions, some of which have valences larger than unity, the method does not work. One should rather take the ions only, thus taking the association into account, for calculating the ionic strength as well as the activity coefficients. The latter then are called *free ion activity coefficients*. Another complication is that for concentrated solutions, where this may also mean that there are other solutes than ionizable ones (say, sugars), the concentration unit of moles per liter is not suitable any longer. It is much better to use *molality*, i.e., moles per kg water.

Taking the above considerations into account, it turns out that reasonable agreement between theory and results is obtained if we express the free ion activity coefficient of an ion of valence $|z|$ as

$$\gamma_{+,-} = \exp(-0.8z^2\sqrt{I}) \tag{2.30}$$

where I is in moles per kg water. Some results are given in Table 2.4.

TABLE 2.4 Approximate Free Ion Activity Coefficients[a]

	I			
$|z|$	0.003	0.01	0.03	0.1
1	0.96	0.92	0.87	(0.78)
2	0.84	0.73	0.57	(0.36)
3	0.67	0.49	0.29	(0.10)

Ionic strength is in moles per kg water
[a] Calculated according to Eq. (2.30).

Although the results may be only approximate, they are very much better than those obtained when ignoring association, or even worse, when ignoring activity coefficients. For ionic strengths above 0.1 molal, however, a more refined treatment would be needed to obtain reasonable results. At very high concentrations the activity coefficients tend to increase again (possibly even to > 1), owing to the mechanisms under 2, "High concentration," in Section 2.2.5.

Question 1

What is the value of the ionic strength of a 0.025 molar solution of calcium oxalate?

Answer

Denoting calcium oxalate as CaOx, it will dissociate into Ca^{2+} and Ox^{2-}. Assuming complete dissociation, this results in an ionic strength [see Eq. (2.27)] of $(0.025 \times 4 + 0.025 \times 4)/2 = 0.1$ molar. The activity coefficients would then be [Eq. (3.30)]: $\gamma_+ = \gamma_- \approx \exp(-0.8 \cdot 2^2 \cdot \sqrt{0.1}) = 0.36$. However, association of Ca^{2+} and Ox^{2-} will occur. Assuming the fraction dissociated to be α, Eq. (2.21) can be written as

$$\frac{(\alpha \cdot m \times \gamma)^2}{(1 - \alpha)m} = \frac{1}{K_A}$$

According to Table 2.3, $K_A \approx 700$ L·mol^{-1}. Moreover, $m = 0.025$ mol·L^{-1} and γ was calculated at 0.36, which then yields $\alpha = 0.48$. This would mean that only about half of the salt is in the dissociated form. That implies that the ionic strength is far smaller than supposed, so that the ion activity coefficient is higher than calculated and the calculation of α thus was incorrect. An iterative calculation, inserting adapted values of γ or α until agreement is reached between both parameters, would be needed. This yields about $\alpha = 0.37$, leading to $I \approx 37$ mmolar.

Question 2

Bovine blood serum contains about 10 mg Ca and 50 mg citrate per 100 g. The solubility product of $Ca_3Cit_2 = 2.3 \cdot 10^{-18}$ mol^5·kg^{-5}. Does this not imply that calcium citrate is far supersaturated in the blood?

Answer

The simplest solution appears to be to calculate the molar concentration, which is for Ca (molar mass 40) 2.5 and for citrate (89) 5.6 mmol per liter. Then the ion product is

$$(2.5 \cdot 10^{-3})^3 \cdot (5.6 \cdot 10^{-3})^2 = 4.9 \cdot 10^{-13}$$

which is 210,000 times the solubility product. (If we had equal concentrations of Ca and citrate, we could say in such a case that the supersaturation would be by a factor of $210,000^{1/5} = 11.6$.) We have, however, made several errors: (a) The blood serum contains about 90% water by weight. This implies that the millimolal concentrations become 2.8 and 6.3, respectively. (b) We should ascertain that the citric acid can fully dissociate. Since blood has a $pH \approx 7.2$ and the $pK_a(3)$ of citric acid is 5.2, this is virtually the case. (c) We should take the ion activity coefficients into account. The ionic strength of blood serum is about 0.14 molal. By use of Eq. (2.30) we calculate for ions of valence 1, 2, and 3 activity coefficients of 0.74, 0.30, and 0.07, respectively. We can now calculate an ion activity product of

$$(0.0025 \cdot 0.30)^3 (0.0056 \cdot 0.07)^2 = 6.4 \cdot 10^{-17}$$

Since we know which still is by a factor of 28 higher than the solubility product. (d) However, we also have to take ion association into account. The most important association is likely to be $Ca^{2+} + Cit^{3-} \rightleftharpoons CaCit^-$ and taking from Table 2.3 that $K_A \approx 10^5$ we obtain by use of Eq. (2.29)

$$\frac{[CaCit^-] \cdot \gamma_-(z=1)}{[Ca^{2+}] \cdot \gamma_+(z=2) \times [Cit^{3-}] \cdot \gamma_-(z=3)}$$

Since we know the γ's and we also know that $[CaCit^-] + [Ca^{2+}] = 0.0028$ and that $[CaCit^-] + [Cit^{3-}] = 0.0063$, we can solve the equations and obtain $[CaCit^-] = 0.00261$, implying that there are almost no Ca^{2+} ions left. We obtain for the activity product

$$(0.0028 - 0.00261)^3 \cdot 0.14^3 \times (0.0063 - 0.00261)^2 \cdot 0.07^2 = 10^{-21}$$

which is far below the intrinsic solubility product.

Note   Because of the fairly high ionic strength, Eq. (2.30) is not accurate anymore, and the free ion activity coefficients are likely to be somewhat higher than calculated. On the other hand, other associations undoubtedly occur, for instance of citrate with K^+, which is abundant in blood, thereby further decreasing the free ion concentrations.

2.4 RECAPITULATION

Thermodynamics describes the (changes in) energy (or enthalpy H) and in entropy (S) of a system; entropy is a measure of disorder. These parameters are combined in the free or **Gibbs energy** $G = H - TS$. Absolute values of these parameters cannot be given, but the magnitude of changes in them can often be established. Every system tends to change in the direction of the lowest free energy, for instance by evening out of concentration (increase in entropy) or by reaction between components (decrease in enthalpy). If it has attained such a state, it is stable; if not, it is unstable. However, thermodynamics tells us nothing about rates of change, and some systems can be metastable or change extremely slowly.

A substance in solution has a **chemical potential**, which is the partial molar free energy of the substance, which determines its reactivity. At constant pressure and temperature, reactivity is given by the thermodynamic activity of the substance; for a so-called ideal system, this equals the mole fraction. Most food systems are nonideal, and then activity equals mole fraction times an **activity coefficient**, which may markedly deviate from unity. In many dilute solutions, the solute behaves as if the system were ideal. For such ideally dilute systems, simple relations exist for the solubility of substances, partitioning over phases, and the so-called colligative properties (lowering of vapor pressure, boiling point elevation, freezing point depression, osmotic pressure).

At high concentrations of a (neutral) solute, the activity coefficient is generally greater than unity, often appreciably. The activity coefficient can be markedly below unity if the substance is subject to self-association or to association with (adsorption onto) other substances.

For **ionizable substances**, the activity coefficient is generally smaller than unity, the more so for a higher total ionic strength, due to screening of positive charges by negative ones and vice versa; the coefficient is also smaller for ions of higher valence. For fairly small ionic strength (up to about 0.1 molar), a simple theory predicts the value of the activity coefficients. The smaller the activity coefficient, the higher the solubility of the substance and the stronger its degree of dissociation. This means that addition of a different salt (e.g., NaCl) to a solution (e.g., of calcium phosphate) will increase the degree of dissociation and the solubility of the latter. It should be realized that salts of multivalent ions are not nearly completely dissociated unless the ionic strength is very small. The relations (especially the state of association) of multicomponent salt solutions are intricate.

BIBLIOGRAPHY

For a general treatment of chemical thermodynamics and the properties of solutions, see any of the textbooks on physical chemistry mentioned in Chapter 1. Several advanced texts on thermodynamics exist, but these mostly go into great detail. A good review for chemists still is

D. H. Everett. An Introduction to the Study of Chemical Thermodynamics. Longman, London, 2nd ed. 1971.

A review of partition equilibria and their consequences in foods is

B. L. Wedzicha. Distribution of low-molecular-weight food additives in dispersed systems. In: E. Dickinson, G. Stainsby, eds. Advances in Food Emulsions and Foams. Elsevier, London, 1988, pp. 329–371.

An in-depth treatment of ionic solutions and ion activity coefficients is

R. M. Pytkowicz, ed. Activity Coefficients in Electrolyte Solutions, Vols. 1 and 2. CRC Press, Boca Raton, FL, 1979.

Especially useful is the chapter by

K. S. Johnson, R. M. Pytkowicz. Ion association and activity coefficients in multicomponent solutions, Vol. 2, pp. 1–62.

A useful general discussion of salts in foods, largely on other aspects than discussed in this chapter, is in

D. D. Miller. Minerals. Chapter 9 in O. R. Fennema, ed. Food Chemistry, 3rd ed. Marcel Dekker, New York, 1996, pp. 617–649.

3

Bonds and Interaction Forces

Atoms, groups of atoms, ions, molecules, macromolecules, and particles always are subject to forces between them. These interaction forces may cause chemical reactions to occur, i.e., cause the formation of other molecular species, but they are also responsible for the existence of condensed phases (solids and liquids), for adherence of a liquid to a solid surface, or for aggregation of particles in a liquid. In short, all structures form because of interaction forces. Generally, formation of a structure causes a decrease in entropy, and this may counteract the tendency of formation, depending on its magnitude compared to that of the energy involved.

There are several, rather different, types of interaction forces, although all of them are ultimately due to the electromagnetic force. This force can thus become manifest in various ways. The interactions greatly differ in specificity: what group or molecule will interact with what other group etc.? For instance, they all decrease in magnitude with interparticle distance, but the relation between energy and distance may vary widely. One generally considers the energy needed to bring two particles (or molecules, etc.) from infinite distance to close proximity. Since there is always more than one type of force acting, this energy (U) may be negative or positive, depending on the interparticle distance (h), for instance as depicted in Figure 3.1. The force (F) generally is the derivative of the energy $(F = -\mathrm{d}U/\mathrm{d}h)$ and, as illustrated in the figure, the net force is thus zero where the energy is at minimum; here we have a stable configuration.

Table 3.1 gives an overview of the various types of forces. The first one mentioned cannot lead to bond formation, since it is always repulsive. If two atoms approach closely, their electron clouds start to overlap, and this causes a repulsion that increases very steeply with decreasing distance; it is

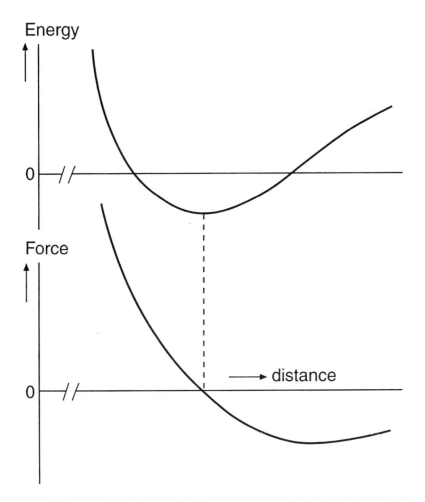

FIGURE 3.1 Hypothetical example of the net energy needed to bring two molecules or particles from infinite separation to a certain mutual distance, and of the corresponding interaction force.

TABLE 3.1 Overview of Possible Interaction Forces Between Atoms, Molecules, and Particles

Type of interaction	Energy involved, $(k_B T)$	Working range (nm)	Dependence of energy on distance[a]	Fixed direction	Attraction or repulsion	Additive
Hard-core	Very large	0.1	r^{-12}	No	R	
Covalent	200	0.2	Complicated	Yes	A	No
Coulomb	200[b]	20[c]	r^{-1}	No	A or R	Yes
Charge-dipole	Up to 50[b]	0.3[c]	r^{-4}	No	A	No
van der Waals	1	1–20	$r^{-6} - h^{-1}$	No	A	Yes
Hydrogen bond	10	0.2	r^{-2}	Yes[d]	A	No
Solvation	"Weak"	Up to 2	$\exp(-h/x)$	No	R (or A)	No
Hydrophobic[c]	5[e]	Up to 2	$\exp(-h/x)$	No	A	[f]

Quantities given are examples.

[a] r is from the center of the atom, h from the surface.
[b] In air or vacuum.
[c] In aqueous solutions.
[d] Some freedom in direction, e.g., 20 degrees.
[e] Free energy.
[f] Proportional to hydrophobic surface area.

therefore called *hard-core repulsion*. If it is ions that approach each other, the interaction is often called Born repulsion.

3.1 TYPES OF BONDS

The next five rows in Table 3.1 relate to what may be properly called bonds. They vary widely in strength, also within one type. The strength is commonly given in terms of energy, expressed in J per mole of bonds. (Note that the expression of a strength in units of energy is quite unlike that for macroscopic systems, where strength commonly refers to the force per unit cross-sectional area needed to cause breaking, i.e., in $N \cdot m^{-2}$.) In general, only covalent bonds and some ion–ion bonds may be strong enough to give "permanent" single bonds. Permanent here means that the atoms or groups bonded stay in this configuration for ordinary times (at least several seconds). If the bond energy is small, thermal motion of the atoms tends to break the bonds within a very short time. The average kinetic (i.e., thermal) energy of a molecule is of the order $k_B T$; this is further discussed in Section 4.3.1. Therefore it is often useful to give interaction energies relative to $k_B T$, as is done in Table 3.1; at room temperature $1 \text{ kJ} \cdot \text{mol}^{-1} \approx 0.4 k_B T$. If bond energy $\ll k_B T$, bonds will not be formed; if it is $\gg k_B T$, permanent bonds will be formed.

A great number of weak bonds acting on the same molecules or ensembles of molecules may also cause permanent bonding. Examples are van der Waals bonds holding molecules in crystals and various weak bonds keeping globular proteins in a compact conformation or keeping particles flocculated.

Covalent Bonds. Covalent bonds exist if some electrons participate in the orbitals of more than one atom. These bonds are highly specific and are extensively discussed in texts on organic chemistry. Here it may suffice to remark that covalent bonds may be very strong (mostly 150–900 $\text{kJ} \cdot \text{mol}^{-1}$) and act over a very short distance in a very restricted range of directions.

Coulomb Forces. These are also called electrostatic forces, ion–ion bonds (if attractive), or charge–charge interactions. They always occur between charged particles, be they ions, protein molecules, or colloidal particles, and they are also quite strong. From Coulomb's law we have for two charges z_1 and z_2,

$$F \propto \frac{z_1 z_2}{\varepsilon r^2} \tag{3.1}$$

where ε is the *relative dielectric constant* of the medium. Since the force (F) is proportional to the product of the charges, it is negative (attractive) if the charges are of opposite sign and positive (repulsive) if of the same sign. The force is inversely proportional to the square of the distance between the centers of the charges (r); this implies that the attractive or repulsive energy is inversely proportional to the distance, and it can therefore act over a relatively long range. It is not direction specific, and it is additive. The latter implies that, if we consider the electrostatic interaction between two molecules or particles carrying more than one charged group, we must sum the interactions between any group on the one molecule with all those on the other molecule.

The Coulomb force also is inversely proportional to the dielectric constant, and since the latter varies widely among materials (e.g., solvents), so do the bond energies. For water, the relative dielectric constant $\varepsilon \approx 80$ at $20°C$. This implies that in water the energies involved are smaller by a factor 80 than most tabulated values, which apply to vacuum ($\varepsilon = 1$). In water, the force may be significant up to a range of about 20 nm. In air ($\varepsilon \approx 1$) or oil ($\varepsilon \approx 3$), the force can act over a longer range. It should be taken into account, however, that the presence of electric charges is generally due to dissociation of ionogenic molecules or groups, and in media of low ε such dissociation may not or hardly occur. This is because the ion–ion bonds themselves are so very strong at low dielectric constant.

Note that ε is the relative dielectric constant, i.e., relative to the absolute dielectric constant (also called dielectric permittivity) of a vacuum, ε_0, which equals $8.854 \cdot 10^{-12}\,C \cdot V^{-1} \cdot m^{-1}$.

Charge-Dipole. Several uncharged molecules bear permanent dipoles, i.e., the geometric centers of the positive charge(s) and the negative charge(s) do not coincide. Such molecules therefore have a dipole moment and they are called *polar*. The dipole moment is the product of charge (expressed in coulombs) and distance between charges; it is mostly given in Debye units (D), where $1D = 3.34 \cdot 10^{-30}\,C \cdot m$. Water is the prime example of a small polar molecule (dipole moment $= 1.85\,D$); see Figure 3.2a. This polarity is the origin of the high dielectric constant of water and it also leads to fairly strong bonds between ions and water molecules. Charge-dipole interactions are always attractive, since the dipole is free to orient in such a way that the positive "end" of the molecule is close to a cation, etc. Ions in water, for instance, are accompanied by a few water molecules. Like ion–ion bonds, charge-dipole interactions are inversely proportional to the dielectric constant of the medium.

van der Waals Forces. These are ubiquitous: they act between all molecules and are always attractive (see Section 12.2.1 for an apparent exception). They may be due to three somewhat different interactions:

1. Dipole–dipole: dipoles on average orient themselves so that the positive end of one molecule is close to the negative one of another, etc.
2. Dipole–Induced Dipole: a dipole always induces a slight unevenness in the charge distribution—i.e., a dipole—in a nonpolar molecule, thereby causing attraction.
3. Induced Dipole–Induced Dipole: even an atom is at any moment a weak dipole, due to the oscillatory motion of its electrons, although on average its dipole moment is zero. Fluctuating dipoles arise and those of neighboring atoms or molecules affect each other so that always a net attraction results. The resulting forces are called *London* or *dispersion forces*, and they act always between all atoms. The other two types need polar groups to be present, which can only exist in molecules. In most pure compounds, the dispersion forces are predominant, but not, for instance, in water, with its strong dipole moment.

The van der Waals forces rapidly decay with distance between molecules, the interaction energy being proportional to r^{-6}. Since the forces

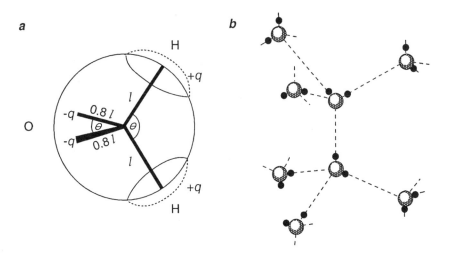

FIGURE 3.2 Structure of water. (a) Model of a water molecule. The distance between the nuclei of O and H is $l \approx 0.1$ nm, the net charges q are 0.24 times the charge of an electron, and the bond angles θ are 109°. After Israelachvilli (see Bibliography). (b) Example of how water molecules form H-bonds with one another; schematic and not to scale. (After O. R. Fennema. Food Chemistry, 3d ed. Marcel Dekker, New York, 1996 (Chapter 2).

are additive, however, the decay between larger bodies may be much weaker. This is discussed in Section 12.2.1.

Hydrogen Bonds. These bonds form between a covalently bound hydrogen atom and an electronegative group like $= O$ or $\equiv N$. In a water molecule, for instance, as shown in Figure 3.2a, the two H atoms each provide a local positive region (since its only electron is exclusively in the orbital with the oxygen) and the O atom provides two negative ones. Hence a net attraction between H and O of different molecules results, if their mutual orientation allows this. The so formed hydrogen bond is to some extent like a covalent one, especially in the sense that it is direction dependent: a small deviation from the optimum orientation results in appreciable weakening of the bond. An H bond is clearly weaker than a covalent one, and it will often be short-lived. On the other hand, H bonds are much stronger than van der Waals interactions.

Water. Extensive hydrogen bonding occurs in water, about as depicted in Figure 3.2b; the resulting configuration is somewhat comparable to that in ice. The molecules try to make as many H bonds as possible, without losing too much entropy. This results in a fluctuating network of bonded molecules: although there are many H bonds at any one time, the bonds continually break and form again, though often in a different configuration. This implies that water, though a liquid, has some ordering, i.e., a structure.

The extensive hydrogen bonding gives water some of its specific properties. Compared to other compounds consisting of small molecules, it has high melting and boiling temperatures; the enthalpies of fusion and of vaporization are high; also the surface tension is high. Moreover, the temperature dependence of several properties is exceptional, such as the well-known maximum in density at 4°C. Some values of water properties are in the Appendices 8 and 9.

Several other molecules can make H bonds with one another, but if they are dissolved in water, H bonds between solute and water are preferentially formed in most cases; the various H bonds are of about the same strength, and association of solute molecules would lead to loss of entropy. This implies that strong hydrogen bonding especially occurs in an apolar solvent, for instance between the carboxyl groups of fatty acids in oil, as discussed in Section 2.2.5, point 4.

For sake of completeness, it may be added that *frictional* forces can act between molecules or between molecules and particles. These arise owing to *external* forces that cause flow, sedimentation, electrophoresis, etc.

3.2 SOLVATION

As mentioned above, net attractive interactions occur between all pairs of molecules (excluding ions for the moment), except at very small distances. In a pure liquid, all these interactions are on average the same, and it makes no difference whether the one or the other molecule is close to a third one. In a solvent (1) with a solute (2), however, some different interactions occur. In the simplest case, where we consider spherical molecules of the same size, the solvent–solute interaction is governed by

$$U_{net} = U_{12} - \frac{1}{2}U_{11} - \frac{1}{2}U_{22} \qquad (3.2)$$

where U_{12} stands for the attractive energy between solvent and solute molecules etc.; hence all U's are negative. If $U_{net} < 0$, *solvation* occurs, i.e., the solute molecules are preferentially surrounded by solvent molecules rather than by other solute molecules, as illustrated in Figure 3.3a. In such a *good solvent*, two solute molecules are on average farther away from each other than they would be if there were no net attraction ($U_{net} = 0$). This results then in a *repulsive* force between the solute molecules, that may be felt over a range of at most a few solvent molecules. It generally implies that the activity coefficient of the solute is smaller than unity (see Section 2.2.1); hence the solute is well soluble.

Solvation should not be interpreted in terms of permanent binding of solute by solvent: a solvent molecule is merely longer near a solute molecule than it would be if $U_{net} = 0$, for instance 10^{-9} s instead of 10^{-12} s. It may be

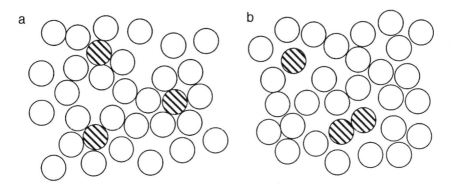

FIGURE 3.3 Solvation. (a) Solute molecules (hatched) are preferentially surrounded by solvent molecules. (b) Solvent molecules tend to stay away from solute molecules (negative solvation).

added that entropic effects can be involved in the interactions if the solvent has some structure (as in water), and one then speaks of contact entropy. In such a case, one cannot express the solute–solvent interaction as a bond energy, but one may—in principle—calculate an interaction free energy.

If $U_{net} > 0$, we have *negative solvation* or, in other words, a *poor solvent* for the solute considered. Now the solute molecules are preferentially near to each other, rather than near to solvent molecules: Figure 3.3b. This generally implies that the solute has a high activity coefficient and poor solubility.

Solvation repulsion may also act between segments of one polymer molecule (Section 6.2.1) or between colloidal particles that have groups at their surface that become solvated (Section 12.4). Negative solvation leads to attraction between polymer segments or between particles.

Hydration. If the solvent is water, solvation is called hydration. It is an intricate phenomenon, since water is such an intricate, not fully understood liquid. Hydration nearly always involves considerable change in entropy, since anything altering the fluctuating network of hydrogen bonds alters entropy. Four kinds of solute molecules or groups may be conveniently distinguished:

1. Ions or ionic groups. Due to the ion–dipole interactions mentioned, small ions tend to be strongly hydrated. Ions move, by diffusion or in an electric field, as if they were accompanied by a number of water molecules. Again, this does not imply that these water molecules are permanently bound: they interchange with other water molecules. Ion hydration is stronger for a smaller ion and a higher valence; cations tend to be more strongly solvated than anions of the same size and valence. The attraction between a proton and a water molecule is so strong that hydronium ions (H_3O^+) occur in water, leaving very few free protons. The hydronium ion is, in turn, hydrated.

   It may further be noted that the formation of ion pairs (e.g., $Na^+ + Cl^- \rightarrow NaCl$, or $-COO^- + H^+ \rightarrow -COOH$) requires desolvation ("dehydration"); especially if the ions or ionic groups involved are small, the increase in free energy involved can be appreciable. In other words, hydration then strongly promotes dissociation of ionizable species.

2. Groups with a strong dipole moment also become hydrated. A case in point is the peptide bonds in proteins.

3. Other somewhat polar groups, such as -OH groups, that can make H bonds with water. Substances with several -OH groups, like sugars, often are said to be hydrophilic. Nevertheless, hydration is

mostly weak in this case. It appears that water molecules adjacent to an -OH group may have either a somewhat shorter or a somewhat longer residence time than water molecules in the bulk, according to the conformation of the OH group in relation to the rest of the solute molecule.

4. Nonpolar or hydrophobic groups. The water molecules cannot make H bonds with these groups. Bringing an apolar molecule or group in water then leads to some breaking of H bonds, which will cause an increase in enthalpy. However, the system tries to make as many H bonds as possible; this leads to a locally altered water structure and thereby to a decrease in entropy. Anyway, the free energy is increased, which implies negative solvation. Similar changes presumably occur at the surface of larger molecules and particles.

The Hydrophobic Effect. If two hydrophobic molecules or groups in water come close together, negative solvation is diminished, which implies a decrease in free energy. This works as if an attractive force is acting between these groups, and this is called *hydrophobic bonding*. Such bonds especially act between aliphatic chains or between aromatic groups. They are largely responsible for the micellization of amphiphilic molecules in water and for the formation of vesicles and membranes of lipid bilayers. They are also important for the conformation of globular proteins (Section 7.2.1). For a large hydrophobic group, the bond free energy is about proportional to the surface area involved, and equals about $4 k_B T (10 \, \text{kJ} \cdot \text{mol}^{-1})$ per nm^2.

The explanation of the hydrophobic effect and the resulting hydrophobic bonding is still a matter of some dispute. For instance, attraction due to dispersion forces may provide a considerable part of the interaction free energy of a hydrophobic bond, varying with the chemical constitution of the groups involved.

The explanation of the *temperature dependence* of hydrophobic bonding is especially intricate and controversial. By and large, at low temperature (near $0°C$), ΔH for bond formation is positive; ΔS is relatively large and positive. The result is a relatively small negative ΔG, i.e., bonds are formed. This would all be in agreement with an overriding effect of water entropy. Above a certain temperature, however, ΔS starts to decrease, but hydrophobic bonding nevertheless increases in strength, because $-\Delta H$ also increases. A hypothetical result is depicted in Figure 3.4, merely to illustrate trends. It may be concluded, that hydrophobic bonding strongly increases with temperature, especially in the range from 0 to about $60°C$. It may further be noted that we have here a clear exception to the

kJ·mol⁻¹

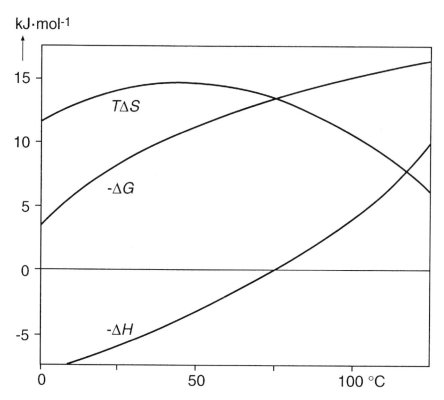

FIGURE 3.4 Assumed relation between the free energy of formation of hydrophobic bonds $(-\Delta G)$ and temperature. Also the entropic $(T\Delta S)$ and enthalpic (ΔH) contributions to ΔG are given. The relations greatly depend on the chemical constitution of the apolar groups involved.

"rule" that changes in enthalpy and entropy are roughly independent of temperature.

Anyway, the hydrophobic effect exists by virtue of the extensive hydrogen bonding in water. It is therefore of importance that various solutes, unless present in quite small concentrations, significantly affect water structure and thereby the hydrophobic effect and its consequences. It concerns alcohols, sugars, well soluble organic molecules like urea, and most salts. These phenomena are poorly understood.

The Hofmeister Series. The specific effects of salts are often arranged in the lyotropic or Hofmeister series, or rather two series, one for anions and one for cations. For some ions of importance in food science these series are approximately as follows:

$$SO_4^{2-} > HPO_4^{2-} > \text{acetate}^- > \text{citrate}^- > Cl^- > NO_3^- > I^- > SCN^-$$

$$NH_4^+ > K^+ > Na^+ > Li^+ > Mg^{2+} > Ca^{2+} > \text{guanidinium}^+$$

The anions on the left-hand side are strongly hydrated, cause a "hydrophilic" water structure around them, and increase the surface tension of water; they tend to stay away from apolar surfaces, which can be interpreted as negative solvation of apolar molecules or groups. Consequently, they enhance hydrophobic bonding and decrease the solubility of apolar substances ("salting out"). The anions on the right-hand side are less hydrated and are not so greatly different from apolar solutes in their effect on water, making the local water structure more "hydrophobic." They tend to increase the solubility of apolar substances ("salting in") and have a fairly small effect on surface tension. The explanation of the series for cations is somewhat less clear, and the observed order is not always the same. The effects of anions and cations are roughly additive, although the anions seem to be dominant. LiCl and NaCl are mostly about neutral with respect to solubility of apolar substances. A solution of $(NH_4)_2SO_4$ is strongly salting out.

It should be noted that these are *specific effects* of ions that are independent of their valence. This is different from the general effects of ions as discussed in Section 2.3, which primarily depend on total ionic strength—i.e., on ion concentration and valence—and which become manifest at far smaller salt concentration.

Note The reader should realize that the term hydration also tends to be used fairly indiscriminately for a number of other phenomena, like the taking up of moisture by a dry material. This is further discussed in Chapter 8.

3.3 RECAPITULATION

Forces acting between molecules or groups not only determine chemical reactivity but also affect several other phenomena, such as formation of condensed phases and aggregation of colloidal particles. Several kinds of forces occur, greatly differing in strength, effective range, effective direction, and additivity. Strength of Coulombic type bonds is inversely proportional to dielectric constant. Bond strength is measured in terms of energy, but in some cases entropy is involved and free energy should be used.

Attractive interaction between solvent and solute molecules causes solvation, as a result of which solute molecules repel each other; this decreases the activity coefficient of the solute and increases its solubility. In other cases, negative solvation occurs, leading to net attraction between solute molecules. Hydration, i.e., solvation by water molecules, is an intricate phenomenon, due to liquid water being strongly hydrogen-bonded. Negative hydration of apolar groups leads to the formation of hydrophobic bonds between these groups. Hydrophobic bonding strongly increases with temperature, being weak or absent below 0°C.

Some solutes, if present in significant concentration, affect water structure and thereby hydrophobic bonding. In this respect salts are arranged in a lyotropic series of decreasing hydration and increasing tendency to enhance solubility of apolar substances.

BIBLIOGRAPHY

We refer again to the textbooks on physical chemistry mentioned in Chapter 1. A clear description is also given by

J. N. Israelachvilli. Intermolecular and Surface Forces, 2nd ed. Academic Press. London, 1992.

A thorough and extensive discussion is found in Chapters 4 and 5 of

J. Lyklema. Fundamentals of Interface and Colloid Science, Vol. 1. Fundamentals. Academic Press, London, 1991.

Theory and consequences of hydrophobic interactions are extensively and clearly discussed in

C. Tanford. The Hydrophobic Effect, 2nd ed. John Wiley, New York, 1980.

4

Reaction Kinetics

Chemical kinetics is generally discussed with respect to reactions between molecules (or ions or radicals) in a gas phase or in a very dilute solution. In foods, we often have other situations. The system never is gaseous, it is rarely very dilute, and it may have more than one phase containing reactants. Changes may occur within molecules, especially macromolecules. Reactions may be between particles, causing, for instance, their aggregation. Numerous other changes may occur, such as phase transitions, leading to a change in rheological properties, color, or other perceptible property. In nearly all such cases we are greatly interested in the rate at which these processes occur. This we cannot derive from the bond energies involved or from other thermodynamic considerations: these may tell us what the driving force is, but in general the rate results from a driving force divided by a resistance, and the resistance may be very large or highly variable.

In this chapter, we will recall some basic aspects of chemical reaction kinetics in solution, starting from an oversimplified point of view and gradually bringing in more complications. We will not discuss theory aimed at explaining reaction rates on a molecular level (molecular reaction dynamics). Other rate processes will be discussed in Chapters 5 and 13.

4.1 REACTION ORDER

Before coming to factors determining reaction rates, it is useful to review the manner in which concentrations depend on time.

The *reaction rate* is usually given as the change in concentration c, i.e., as either $+$ or $-$ dc/dt. According to the units of c, it may be expressed in $mol \cdot L^{-1} \cdot s^{-1}$ (the most common way), $mol \cdot kg^{-1} \cdot s^{-1}$, number $\cdot m^{-3} \cdot s^{-1}$, etc.

For a **zero-order** reaction, the rate remains constant: see Table 4.1. Approximately zero-order reactions occur, for instance, if small quantities of a substance, say one causing an off-flavor, are slowly formed from a very large reservoir of a parent component.

For a **first-order** reaction of the type $A \rightarrow B$ or $A \rightarrow B + C$, we have

$$-\frac{d[A]}{dt} = k[A] \tag{4.2a}$$

where [A] stands for the molar concentration of A and k is the *rate constant*, which in this case is in s^{-1}. k varies with temperature and pressure, but it is generally assumed to be constant otherwise, i.e., independent of concentration; this is often (nearly) true, but not always. Integrating the equation, and introducing the initial concentration of A, we obtain

$$\ln\left(\frac{[A]}{[A]_0}\right) = -kt \tag{4.2b}$$

or

$$[A] = [A]_0 \exp(-kt) = [A]_0 \exp\left(\frac{-t}{\tau}\right) \tag{4.2c}^*$$

where τ is the *relaxation time*. If we plot the log of the concentration versus time, we thus obtain a straight line. The relations are illustrated in Figure 4.1. If, for example, 10% of A is left after D s, this means that 1% is left after $2D$ s, 0.1% after $3D$ s, and so on; D, which equals $2.3/k$, is called the decimal reduction time and is mostly used by microbiologists. The killing of microorganisms and the inactivation of enzymes at high temperature often follow first-order kinetics, at least approximately. Also bacterial growth in the so-called exponential phase follows first order kinetics, but now the sign in Eq. (4.2a) is positive.

The **relaxation time** is mostly used by physical chemists and is the time needed for a certain change to occur over $(1 - 1/e) \approx 0.63$ of its maximum value, after a specified change in conditions has been applied, say a change in temperature or in pH. It is said then that the system relaxes toward a new

TABLE 4.1 Overview of Rate Equations

Order	Rate equation	Integrated form	$t_{0.5}$	Units of k	
0	$-\dfrac{d[A]}{dt} = k$	$[A]_0 - [A] = kt$	$\dfrac{[A]_0}{2k}$	$mol \cdot L^{-1} \cdot s^{-1}$	(4.1)
1	$-\dfrac{d[A]}{dt} = k[A]$	$[A] = [A]_0 \exp(-kt)$	$\dfrac{\ln 2}{k}$	s^{-1}	(4.2)
2	$-\dfrac{d[A]}{dt} = k[A]^2$	$\dfrac{1}{[A]} - \dfrac{1}{[A]_0} = kt$	$\dfrac{1}{[A]_0 k}$	$L \cdot mol^{-1} \cdot s^{-1}$	(4.3a)
2*	$-\dfrac{d[A]}{dt} = k[A][B]$	$\dfrac{1}{[B]_0 - [A]_0} \ln \dfrac{[B][A]_0}{[A][B]_0} = kt$	—	$L \cdot mol^{-1} \cdot s^{-1}$	(4.3b)
n	$-\dfrac{d[A]}{dt} = k[A]^n$	$[A]^{1-n} - [A]_0^{1-n} = (n-1)kt$	—	$L^{n-1} \cdot mol^{1-n} \cdot s^{-1}$	(4.4)

Second-order reactions of two kinds are given: for the reaction $2A \rightarrow A_2$ and for $A + B \rightarrow AB$ (2*). In the latter case different kinds of half time can be defined. n can be any positive number except unity.

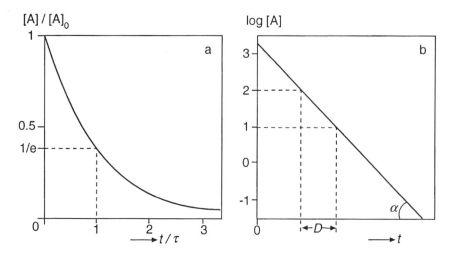

FIGURE 4.1 First-order reaction of the type A → B. (a) Concentration relative to the original one as a function of time (t) over relaxation time (τ). (b) Example of log concentration versus time; $\tan \alpha = k \log e = 0.434k$. D is the decimal reduction time.

equilibrium situation. To be sure, the relaxation time has physical significance only for a first-order reaction, and many changes do not follow such a relation, for instance because more than one relaxation mechanism acts.

For a **second-order** reaction of the type A + B→AB we have

$$-\frac{d[A]}{dt} = -\frac{d[B]}{dt} = \frac{d[AB]}{dt} = k \cdot [A] \cdot [B] \qquad (4.3b)$$

where k is in L·mol^{-1}·s^{-1}. For the case that $[A] = [B]$, a special case of which is a dimerization reaction (i.e., A = B), integration yields

$$\frac{1}{[A]} - \frac{1}{[A]_0} = kt \qquad (4.3a)$$

which implies that a plot of 1/[A] versus time yields a straight line. Notice that also the relative rate depends on concentration. A measure for the (inverse of the) rate is the half time $t_{0.5} = 1/k[A]_0$, i.e., the time needed for half of the reaction to become complete. Many reactions in foods approximately follow second-order kinetics. Further information is given in Table 4.1, also for the more complicated case of A + B→AB.

The reaction order is an empirical concept. Its value has to be determined, since it cannot readily be derived from the stoichiometry, often only apparent, of the reaction: see Section 4.4. Also the rate constant is an empirical quantity to be experimentally determined, and it depends on the reaction order. The combined knowledge of the order, the rate constant, and the initial concentration of the reactant(s) allows calculation of changes occurring. An example is prediction of the extent to which a certain component is formed or degraded during long storage of a food. As is illustrated in Figure 4.2, it may need very precise determination of the time dependent concentration of a reaction product to establish the order, as long as the reaction has not proceeded very far.

Another difference among the rates of reactions of various order is, of course, the dependence on the concentration of the reactant(s). As seen in Table 4.1, the rate is independent of concentration for zero-order, proportional to concentration for first-order, and proportional to concentration squared for second-order reactions.

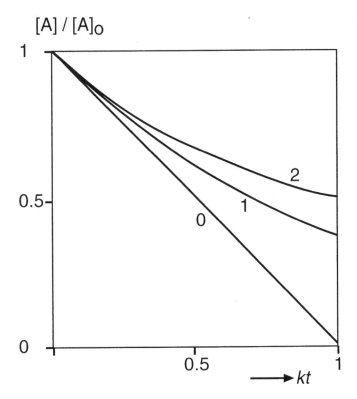

FIGURE 4.2 The change in concentration of reactant A with time for reactions of order 0, 1, and 2. k is the rate constant; t is time.

Note Up till now, we have used concentrations in the rate equations, whereas we have learned in Chapter 2 that activities should be used. In many cases the difference is unimportant, not because it is small, but because we generally have no way of predicting the rate constant from first principles. Whether a reaction is slow because of a low activity coefficient or because of a low rate constant then would be a mere academic question. Nevertheless, there may be situations where it is important to know about activity coefficients, for instance when comparing the same reaction in different media or when the activity coefficient of one of the reactants varies in a different manner with conditions than that of another reactant.

Question

A food company pasteurizes a beverage for 15 s at 70°C in a heat exchanger, and it is then aseptically packaged in 1 liter cartons. The product occasionally spoils, and it is established that a bacterium is responsible. It may be present in concentrations up to 2 per mL. Its first-order rate constant for thermal death at 70°C is determined at $0.7\,s^{-1}$. How long should the pasteurization time be to ensure absence of spoilage?

Answer

A guarantee for absence of spoilage can never be given, since there always will be a probability that one bacterium is present in a package, and since it can grow out to high numbers, this would eventually mean spoilage. Presume that a risk of 1 in 10^5 is taken. That means that the average bacterial count after pasteurization should be 10^{-5} per liter or 10^{-8} per ml. The original count was 2. Applying Eq. (4.2b), we thus have $\ln(10^{-8}/2) = -0.7t$, which directly gives $t = 27.3\,s$, clearly longer than the 15 s previously applied.

Notes
1. Most microbiologists would use a decimal reduction time rather than a rate constant. Here $D = 2.3/0.7 = 3.3\,s$, and reduction by a factor of $2 \cdot 10^8$ means $\log(2 \cdot 10^8) = 8.3$ decimal reductions; hence a heating time of $8.3 \times 3.3 = 27.3\,s$.
2. In practice, there are many uncertainties: (a) The rate constant for thermal destruction of bacteria always varies, even within one population, and those bacteria being most heat resistant have of course the greatest chance of survival; in other words, a curve as given in Figure 4.1b need not be linear and will often curve upwards at long t. (b) The effective heating conditions in a laboratory test may be

different from those in a heat exchanger, for instance the times needed for heating up and for cooling may differ. (c) Occasionally a greater contamination may occur.

This implies that a producer will mostly remain on the safe side, and heat for e.g. 35 s. However, in many products longer heating impairs quality, for instance flavor, and a compromise may be needed.

4.2 CHEMICAL EQUILIBRIUM

In principle, any chemical reaction is **reversible**. Thus if we have $A \rightarrow B$, we also have $B \rightarrow A$. The first reaction may have a rate constant k_1, the reverse one k_{-1}. We therefore have for the rate at which A is transformed

$$-\frac{d[A]}{dt} = k_1[A] - k_{-1}[B] = \left(k_1 - k_{-1}\frac{[B]}{[A]}\right)[A] \tag{4.5a}$$

If we start with A only, it will be changed into B, but the apparent rate constant (the factor between parentheses) will become ever smaller, since [B] increases. When the right hand side of [4.5a] has become zero, i.e., at infinite time, equilibrium is obtained, and it follows that the equilibrium constant, $K \equiv [B]_\infty/[A]_\infty$, equals k_1/k_{-1}. Equation (4.5a) yields upon integration and some rearrangement

$$\frac{[A] - [A]_\infty}{[A]_0 - [A]_\infty} = \exp(-kt) \tag{4.5b}$$

where the rate constant $k = k_1 + k_{-1}$. A good example of such a reaction is the "mutaroration" of reducing sugars like glucose and lactose, i.e., the transmutation of the α anomer into the β anomer and vice versa.

Referring to Section 2.2, we observe that at equilibrium the chemical potentials of A and B must be equal. This leads to

$$\mu_A^\ominus + RT \ln a_A = \mu_B^\ominus + RT \ln a_B$$

and, consequently, to

$$\frac{a_A}{a_B} = \exp\left[-\frac{\mu_A^\ominus - \mu_B^\ominus}{RT}\right] = \exp\left(-\frac{\Delta G^\circ}{RT}\right) \tag{4.6}$$

where, of course, $a_A/a_B = 1/K$. ΔG° is the standard free energy (per mole) for the transition of A to B. Since $\Delta G = \Delta H - T\Delta S$, we have two contributions. $-\Delta H^\circ \approx U^\circ$, the net molar bond energy, values of which are tabulated in reference books. Generally, ΔS° is made up of two terms. The first is due to changes in solvation; its quantity is rarely of the order of magnitude of a covalent bond energy. The second is the mixing entropy, which changes for many reactions, e.g., when $A + B \rightarrow AB$; it is quite generally given by

$$S_{mix} = nR \sum_i \ln x_i \tag{4.7}$$

where n is the total number of moles in the mixture and x stands for mole fraction. Sample calculations show that the change in mixing entropy per mole of reactant in a dilute system is mostly smaller than $10R$, i.e., $|T\Delta S^\circ_{mix}|$ would be $< 10RT$. All this means that for formation of a covalent bond, $|\Delta G^\circ|$ often is not greatly different from U°. Taking the latter at $100\,RT$ (see Table 3.1) and assuming ΔG° to be no less than $90RT$, Eq. (4.6) yields an equilibrium ratio $a_A/a_B \approx e^{-90} \approx 10^{-39}$. This truly means that the reaction would be completed: there is nothing of component A left, taking into account that Avogadro's number is "only" $6 \cdot 10^{23}$.

In several other cases, significant quantities of both molecules A and B occur at equilibrium, because the net change in free energy is far smaller. Often, besides formation of a covalent bond, another one has to be broken, and the two terms (having different signs) may almost cancel. This is presumably the case in the mutarotation reactions mentioned, where the equilibrium constant generally is of the order of unity. Another case is a small bond energy, for instance due to van der Waals attraction. In these situations, entropy changes may play a considerable part, and a mixture of components may have a particular composition because it is in equilibrium; the composition then is *thermodynamically controlled*. Good examples also are the salt association equilibria described in Section 2.3; here the Coulomb energy for bond formation in water is of the order of a few times RT. In other situations, a mixture may have a certain composition that is far removed from equilibrium, but it nearly remains so because the reactions leading to equilibrium are very slow; the composition then is said to be *kinetically controlled*.

Another situation may be that of a **steady state**. The observation that the concentration of a reactant remains constant does not necessarily imply that it is in equilibrium, nor that it does not react at all. Consider, for example, a reaction scheme of the type $A \rightarrow B \rightarrow C$, with consecutive rate constants k_1 and k_2. If [A] is quite large and k_1 is not very small, molecules B will soon be formed, at a rate $k_1[A]$, assuming the reaction to be first order. B is in turn converted into C at a rate $k_2[B]$. Unless $k_2 \gg k_1$, this will lead to an increasing concentration of B until $k_2[B] = k_1[A]$, at which stage the formation and disappearance of B occur at equal rates. Because [A] is very large, it may take a long time before a significant relative decrease in [A] occurs. Consequently, we have an approximate steady state with respect to the concentration of B. Of course, this does not last indefinitely, since ultimately all A will be consumed. This means that it may be of importance to distinguish between an equilibrium state and an (approximate) steady state, since predictions about what would occur in the long run are clearly different for the two cases.

A true steady state can be attained if, for example, the system is confined in a reaction vessel where a solution of A is continuously added to the system while some of the product is continuously removed at the same volume flow rate. Such steady states are by no means exceptional and occur often in living cells or chemical reactors. A steady state then lasts as long as the reaction conditions, including rates of inflow of reactant(s) and outflow of product(s), are kept constant. Also for other rate processes, e.g. involving mass or heat transfer, steady states are often achieved.

Reaction Heat. A reaction can only proceed if $\Delta G < 0$. In relation to Eq. (4.6), it was mentioned that for most reactions $-\Delta H$ is larger than $T\Delta S$. This implies that during the reaction heat is produced (the amount of reaction heat can be measured by calorimetry). The reaction then is said to be *exothermic* and *enthalpy driven*. There are also *endothermic* reactions, where heat is consumed; in other words, $\Delta H > 0$. Because ΔG must be negative for the reaction to proceed, this implies that $T\Delta S > \Delta H$, and the reaction is said to be *entropy driven*.

4.3 RATE THEORIES

Virtually no food is in thermodynamic equilibrium. We all know that food is a source of energy and that, for example, one gram of carbohydrate yields on oxidation in the body about 17 kJ (about 4 kcal). Assuming the elementary reaction to be

$$CH_2O + O_2 \rightarrow CO_2 + H_2O$$

we calculate the molar bond energy difference at about $30 \times 17 = 500 \, kJ \cdot mol^{-1} \approx 200 \, RT$. Inserting this value as ΔG in Eq. (4.6), we come up with the immense figure of 10^{87} for the equilibrium constant. Now this calculation is not too precise, and several refinements must be made, but that does not materially alter the result. So the driving force for oxidation by the O_2 in air of, say, plain sugar is very large; nevertheless plain sugar appears to be stable almost indefinitely.

This clearly asks for an explanation, and an attempt will be made below. To that end it is useful first to say something about the kinetic energy of molecules.

4.3.1 The Maxwell–Boltzmann Distribution

All atoms, molecules, or particles constantly move (velocity v), except at absolute zero temperature. They thus have a translational kinetic energy, according to basic mechanics given by $(1/2)\,mv^2$, where m is mass. (In addition, they have rotational and vibrational kinetic energy, but these do not concern us at this moment.) It can be shown that for all particles, whether small or large, the average kinetic energy $\langle U \rangle$ depends on temperature only, according to

$$\langle U \rangle = \left(\frac{1}{2}\right) m\langle v^2 \rangle = \left(\frac{3}{2}\right) k_B T \tag{4.8}$$

where Boltzmann's constant k_B is a fundamental constant of nature, its magnitude being $1.38 \cdot 10^{-23}\,\mathrm{J \cdot K^{-1}}$. Incidentally, Eq. (4.8) also can be seen as the definition of temperature; notice that temperature thus can only refer to a fairly large ensemble of molecules or atoms, since it is defined in terms of an average.

As mentioned, there is a spread in kinetic energy, and the **energy distribution** of the molecules or particles is at any temperature given by

$$d \ln N = \frac{dN}{N} = 2\left(\frac{U'}{\pi}\right)^{0.5} \exp(-U')\,dU' \tag{4.9}$$

which is known as the Maxwell–Boltzmann distribution; it can also be given as a velocity distribution. N stands for the number of molecules etc., and $U' = U/k_B T$. The distribution is depicted in Figure 4.3a. It can, of course, be recalculated in terms of U for any T, and then it is seen that the distribution becomes wider and flatter for a higher temperature: Figure 4.3b.

The essential point is that a certain proportion of the molecules has a kinetic energy above a specified level (say, 10^{-20} J), which proportion is given by the area under the curve in Figure 4.3b above the specified level relative to the total area. For a given temperature, this proportion is the same for any kind of molecule (or atom or particle). For a higher temperature the proportion is higher. If a higher energy level is specified, the proportion is smaller.

Notes For small species, the translational kinetic energies are very significant, but for large ones they may be (very) small compared to

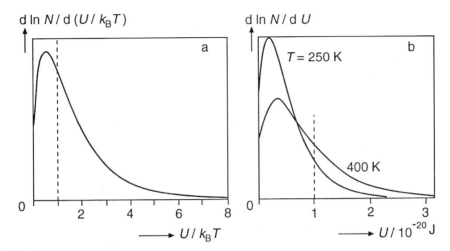

FIGURE 4.3 The Maxwell–Boltzmann distribution of the translational kinetic energy U of molecules or other particles. N is number of molecules, k_B is Boltzmann's constant, and T is temperature (K). (a) Normalized distribution. (b) Distribution for two temperatures.

any mechanical external energy applied to them, say due to streaming of a liquid. This is illustrated when we calculate the average velocity of particles from (4.8) and their mass; some results at 300 K are:

Hydrogen atom   2700 $m \cdot s^{-1}$
Protein molecule, e.g., 15
Emulsion droplet, e.g., 0.001
Billiard ball, about 10^{-10}

An emulsion droplet can readily attain a velocity relative to the liquid of greater than 1 mm per second due to stirring and a very slight touch may move a billiard ball at a speed of $10^{-1} m \cdot s^{-1}$.

Although the velocities are high for small species, they do not say anything about the distances that the molecules travel. For water molecules at 0°C the average velocity is $614 m \cdot s^{-1}$, and this is equally true in ice, liquid water, and water vapor. But in ice the molecules vibrate over a very small distance only (order of 10^{-10} m); in water they also vibrate, but they move with respect to each other as well, thereby traveling over some distance (Brownian motion); in vapor (or air) the molecules travel over far larger distances in the same time. See further Section 5.2.

4.3.2 Activation Energy

The first useful theory of reaction rates was due to *Arrhenius*, and it is easiest to envisage for a bimolecular reaction. It is assumed that the molecules have to overcome an energy barrier before they can react. This is depicted in Figure 4.4. The energy barrier per mole is called the *activation energy*, symbol E_a. As mentioned, the average translational kinetic energy of a molecule is $(3/2)\,k_B T$, and the average kinetic energy involved in a collision of two molecules is given by 2 times $1/3$ of that value, i.e. $k_B T$; the factor $1/3$ arises because the molecules move in 3 dimensions and when they collide this happens in one dimension. The collision may now provide the activation energy needed for the molecules to react. From Eq. (4.9), the proportion of collisions of which the energy is higher than a given value U^* can be derived to equal $\exp(-U^*/k_B T)$. By changing from molecules to moles and by putting $U^* N_{AV} = E_a$, it follows that the rate constant would be

$$k = A\,\exp\!\left(\frac{-E_a}{RT}\right) \qquad\qquad (4.10)^*$$

where A often is called the *frequency factor*. It is also called the

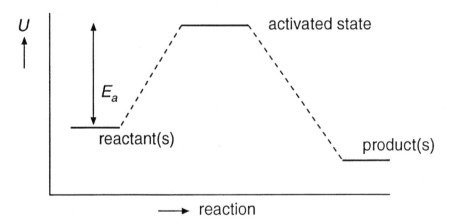

FIGURE 4.4 Arrhenius theory. Illustration of the energy state of the reactants, the reactants in the activated state, and the reaction product(s). E_a represents the activation energy.

preexponential factor and denoted as k_∞ or k_0, i.e., the value that k would attain at infinite temperature or zero activation energy, respectively.

Temperature Dependence. Eq. (4.10) predicts that log k is proportional to $1/T$, and this is indeed very often observed, especially for reactions of small molecules, involving breaking and formation of covalent bonds. The Arrhenius theory can thus be said to be very successful, and as a semiempirical relation Eq. (4.10) is indeed useful in many cases.

Chemists generally express the temperature dependence of a reaction in Q_{10}, i.e., the factor by which a reaction is faster if one increases the temperature by 10K. Bacteriologists use the Z value, i.e., the temperature increase (in K) needed to increase the reaction rate by a factor of 10. These parameters naturally depend on temperature, even if the activation energy is constant. We have

$$Q_{10} = \exp\left(\frac{10E_a}{RT^2}\right)$$

and

$$Z = 2.3\frac{RT^2}{E_a}$$

This implies that errors are made by assuming these parameters to be constant. It is fairly common to plot the log of the time t' needed to obtain a certain effect (e.g., 90% inactivation of an enzyme, or the emergence of a given quantity of a compound), not against $1/T$ but against T, and within a small temperature range an almost straight line is obtained. Extrapolation of such plots beyond the temperature range studied may cause considerable error, especially for large E_a, since it assumes in fact that E_a/T^2 is constant, which is very unlikely.

Note The success of the Arrhenius theory has often induced workers to apply it to other phenomena. Several physical properties of a system tend to depend on temperature like an Arrhenius relation, but this does not necessarily mean that we can assign an activation energy to the phenomenon. A case in point is the fluidity, i.e., the reciprocal of the viscosity, since there is no such thing as an activation energy for fluid motion (a true fluid moves if only the slightest force is applied). In this case, the Arrhenius type relation directly derives from the Maxwell–Boltzmann distribution. The same holds true for diffusion.

4.3.3 Absolute Rate Theory

The Arrhenius theory cannot readily account for the mechanism in monomolecular reactions, and the so called frequency factor is in fact not more than a fitting factor. Moreover, it considers an activation energy only, whereas there is sufficient reason to believe that there may be a positive or negative activation entropy as well. One tries to overcome these deficiencies in the theory of the *activated complex*, also called absolute rate theory; it is largely due to Eyring. Here it is assumed that the molecule or molecules to react attain an activated or transition state of higher free energy (denoted by superscript $\ddagger$). The activated state is induced by collision with other molecules, be they reactants or solvent. An effective collision will not merely enhance the kinetic energy of a molecule but will also cause distortion of bonds, hence a local increase of bond energy. The activated complex formed is very short-lived, and it is in equilibrium with the reactants. Hence we can apply Eq. (4.6) with B being the activated complex. The latter spontaneously decomposes into the reaction products, at a rate derived from the theory of statistical mechanics. The result, for a first-order reaction, is

$$k = \frac{k_B T}{h_p} \exp\left(-\frac{\Delta G^{\ddagger}}{RT}\right) = \frac{k_B T}{h_p} \exp\left(-\frac{\Delta H^{\ddagger}}{RT}\right) \exp\left(\frac{\Delta S^{\ddagger}}{R}\right) \qquad (4.11)$$

where h_p is Planck's constant ($6.626 \cdot 10^{-34}$ J $\cdot$ s) and $\Delta G^{\ddagger}$ is the standard molar activation free energy, etc. The frequency factor $k_B T/h_p \approx 6 \cdot 10^{12}$ s^{-1} at room temperature.

The Eyring theory is generally considered to be fairly rigid and it has been shown to be in good agreement with results for several reactions. Nevertheless, in many cases, especially when we are interested in the temperature dependence, the Arrhenius theory may suffice. For reactions in solution, the *activation enthalpy* $\Delta H^{\ddagger} = E_a - RT$. If E_a is not very small, $RT \ll E_a$ and the temperature dependence is virtually the same in both equations. The relative change in T in the frequency factor ($k_B T/h_p$) then is small compared to that in the factor $\exp(-E_a/RT)$, and the difference between E_a and $\Delta H^{\ddagger}$ is also small.

At this stage it may be good to emphasize that a reaction rate constant is *not* determined by the difference in standard free energy between the reactants and the products to be formed, but by the standard *activation free energy.* The great advantage of the Eyring theory is the introduction of the **activation entropy**, which may be considerable in many cases. Consider, for instance, the heat denaturation of a protein, the essential step of which is a change from the native (globular) conformation to an unfolded state. This may lead to its inactivation if the protein is an enzyme or other biologically active agent. An example is given in Figure 4.5 for the enzyme alkaline phosphatase (EC 3.1.3.1). The activation enthalpy equals about $450 \, kJ \cdot mol^{-1}$; applying Eq. (4.11), while taking only this contribution to $\Delta G^{\ddagger}$ into account, would lead to a presumed rate constant of about $10^{-55} \, s^{-1}$ at 75°C, whereas in fact the reaction proceeds fairly fast at that temperature. The activation entropy is, however, large and positive, presumably owing to the unfolding of the protein leading to a greatly increased conformational

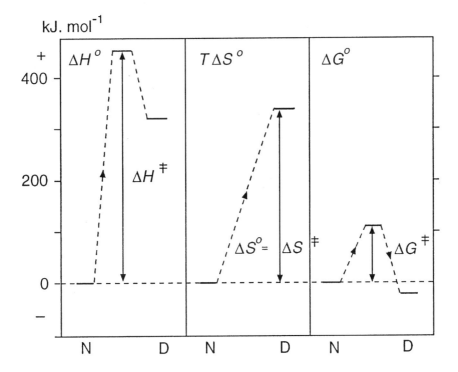

FIGURE 4.5 Examples of the molar enthalpy (H), entropy (S) times temperature (T), and free energy (G) of the enzyme alkaline phosphatase in the native (N) and denatured (D) state; the intermediate state refers to the "activated complex." Results at 340K, derived from kinetic data on inactivation of the enzyme.

entropy (this is further discussed in Section 7.2). Consequently, the reaction can proceed despite the very large activation enthalpy, which, on the other hand, causes the reaction rate to be extremely dependent on temperature.

According to Table 3.1, breaking of a covalent bond would typically take $200\,k_B T$ at room temperature, equivalent to $500\,\text{kJ}\cdot\text{mol}^{-1}$. If breaking of such a bond would be prerequisite for a given reaction to proceed, the activation energy would roughly equal that quantity. In such a case, any compensation by an increase in entropy (conformational or contact entropy) would be small, and the activation free energy would be at least $450\,\text{kJ}\cdot\text{mol}^{-1}$. That implies that such a reaction would never occur, unless the temperature is extremely high. In the present case, we calculate $k \approx 10^{-68}\,\text{s}^{-1}$ at room temperature and $k \approx 10^{-5}\,\text{s}^{-1}$ at 1000°C. In other words, "simple" chemical reactions (i.e., those that do not involve a large activation entropy) that occur at ordinary temperatures, at a perceptible rate, must have a fairly small activation enthalpy, and their rate cannot be strongly temperature dependent. For most of them Q_{10} is 2 to 3, whereas it ranges from 10 to 150 for protein denaturation. Some examples are given in Table 4.2.

This difference has important practical implications for the food technologist. Many foods are heat-treated to ensure microbiological safety and to enhance keeping quality. In nearly all cases, the desired properties are the result of heat inactivation of enzymes: enzymes may themselves cause spoilage, but they are also essential for microbial (and all other) life, implying that irreversible inactivation of some of their enzymes kills the

TABLE 4.2 Typical Examples of the Temperature Dependency of Reactions

Type of reaction	$\Delta H^{\ddagger}(\text{kJ}\cdot\text{mol}^{-1})^{a}$	Q_{10} at 100°C
Many chemical reactions	80–125	2–3
Many enzymatic reactions	40–60	1.4–1.7
Hydrolysis, e.g.	60	1.7
Lipid autoxidation	40–100	1.4–2.4
Maillard reactions	100–180	~2.4
Protein denaturation	150–600	4–200
Killing of microbes	200–600	6–175
Killing of bacterial spores	250–330	9–17

[a] In most cases an apparent average activation enthalpy, since it concerns a number of consecutive reactions.

microbes (as well as microbial spores). Inactivation kinetics of enzymes is often determined by the unfolding kinetics of globular proteins, hence a very strong dependence of its rate—and of the thermal death rate of microorganisms—on temperature. Some results are shown in Figure 4.6. The slope of the log rate constant against $1/T$ greatly differs between phosphatase inactivation or spore killing and Maillard reaction. (We will discuss the curve for plasmin in Section 4.4.) Most chemical reactions, like the Maillard one, are undesirable, whereas killing of microorganisms is needed. By applying a high temperature for a short time, one may ensure the latter while minimizing the former.

It may finally be noted that it is often implicitly assumed that $\Delta H^{\ddagger}$ and $\Delta S^{\ddagger}$ do not depend on temperature. This may not be true for reactions involving changes in hydrophobic interactions; cf. Section 3.2.

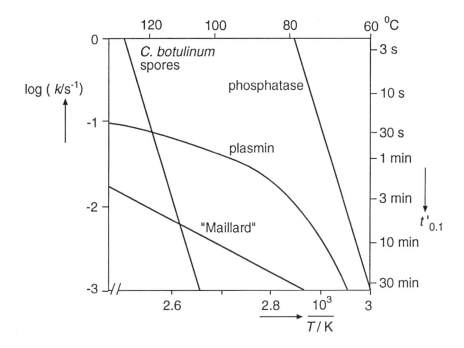

FIGURE 4.6 Dependence of (pseudo) first-order reaction rate constants (k) on temperature (T). Approximate examples for heat inactivation of alkaline phosphatase and plasmin, for killing of *Clostridium botulinum* spores, and for the formation of a certain small amount of Maillard products. $t'_{0.1}$ is the time needed for the reaction to proceed for 0.1 times the final value (not for the Maillard reaction).

Question

In a study on heat inactivation of a peroxidase enzyme (EC 1.11.1.7), it was found that 5 min heating at 72°C left 55%, and 5 s at 80°C left 2.7%, of the enzyme activity. Calculate the molar activation enthalpy and entropy of the reaction.

Answer

To do the calculation, we have to make some assumptions: (a) The heat inactivation is controlled by the heat denaturation. (b) The latter is a first-order reaction; this should, of course, be checked by determining residual activities for various heating times. (c) The reaction rate can be derived from absolute rate theory.

We then have from Eq. (4.2b) that at 72°C (i.e., 345 K) $\ln([A]/[A]_0) = \ln 0.55 = -300 \cdot k$; hence $k = 0.0020 \, \text{s}^{-1}$; similarly at 80°C (353 K) $k = 0.72 \, \text{s}^{-1}$. We can now write (4.11) in logarithmic form as

$$\ln k = \ln(8.7 \cdot 10^{12}) + \frac{\Delta S^{\ddagger}}{R} - \frac{\Delta H^{\ddagger}}{RT}$$

and filling in the values for k and T we obtain two equations with two unknowns. The result is $\Delta H^{\ddagger} = 745 \, \text{kJ} \cdot \text{mol}^{-1}$ and $\Delta S^{\ddagger} = 1860 \, \text{J} \cdot \text{mol}^{-1} \cdot \text{K}^{-1}$.

4.3.4 Diffusion-Controlled Reactions

Equation (4.11) applies to a first-order reaction. For a *second-order reaction* it becomes

$$k_{E,2} = \frac{k_B T}{h_P} \cdot \frac{RT}{p} \cdot \exp\left(-\frac{\Delta G^{\ddagger}}{RT}\right) \tag{4.12}$$

Now the rate constant is in $\text{m}^3 \cdot \text{mol}^{-1} \cdot \text{s}^{-1}$, and to obtain it in $\text{L} \cdot \text{mol}^{-1} \cdot \text{s}^{-1}$, it has to be multiplied by 10^3. At ambient temperature, the factor RT/p then is about 25 liters per mole $= 0.025 \, \text{m}^3 \cdot \text{mol}^{-1}$. p has to be taken as the standard pressure, i.e., $10^5 \, \text{Pa}$, at which the standard molar energies are given.

There is, however, a problem for a second-order reaction: it is implicitly understood that the rate at which the molecules—say, A and B—encounter each other is not limiting the reaction rate. In other words, the system is considered to be ideally mixed (i.e., as if it were stirred at an infinite rate). This is not always reasonable, and we need to consider the encounter frequency. The basic theory is due to *Smoluchowski*: it gives the encounter rate of diffusing particles or molecules, assuming that every encounter leads to reaction. For the reaction $A + B \rightarrow AB$ the result is

$$-\frac{d[A]}{dt} = -\frac{d[B]}{dt} = \frac{d[AB]}{dt} = 4\pi(D_A + D_B)(r_A + r_B)N_A N_B \qquad (4.13)$$

where D = diffusion coefficient, r = collision radius of the molecule, and N = number of molecules per unit volume. If we have a reaction of the type $A + A \rightarrow A_2$, we have to divide the rate by a factor of 2, since we otherwise would count every encounter twice; this also applies to any derived equation. The rate constant is the rate divided by $N_A N_B$ and, converting from molecules per m^3 to moles per liter, we obtain

$$k_S = 4\pi(D_A + D_B)(r_A + r_B)10^3 N_{AV} \qquad (4.14)$$

The Stokes–Einstein relation gives, for the diffusion coefficient of spherical particles,

$$D = \frac{kT}{6\pi\eta r} \qquad (4.15)$$

where η is the viscosity of the solvent and r is now the hydrodynamic radius (see Section 5.3 for more on the relations between D, η, and r). For perfect spheres, collision radius and hydrodynamic radius are equal. Assuming, moreover, $r_A \approx r_B$, we obtain by inserting (4.15) into (4.14)

$$k_S \approx 8 \cdot 10^3 \frac{RT}{3\eta} \qquad (4.16)$$

Comparing now Eqs. (4.12) (Eyring) and (4.16) for dilute aqueous solutions at room temperature, where $\eta \approx 10^{-3}$ Pa $\cdot$ s, and taking $\Delta G^{\ddagger} = 0$ (because it is assumed in the Smoluchowski treatment that every encounter leads to reaction), we obtain rate constants of $1.5 \cdot 10^{14}$ and $6.6 \cdot 10^9$ L $\cdot$ mol$^{-1} \cdot$ s^{-1}, respectively. The encounter rate would thus be very much smaller than the rate according to the Eyring theory, implying that the reaction is diffusion controlled (albeit still quite fast, unless the reactant concentrations are very small). However, zero activation free energy is in many cases unrealistic. Taking, for instance, $\Delta G^{\ddagger} = 25$ kJ $\cdot$ mol^{-1}, which is a fairly small value, we obtain from Eq. (4.12) a rate constant of $7 \cdot 10^9$, i.e., about the same as from (4.16). For an activation free energy of 50 kJ $\cdot$ mol^{-1}, we obtain $3 \cdot 10^6$ L $\cdot$ mol$^{-1} \cdot$ s^{-1}, and now the reaction rate presumably is controlled by activated complex formation. On the other hand, if the viscosity were much higher, and a value of 10^3 Pa $\cdot$ s is not exceptional in a low-moisture food, Eq. (4.16) would yield $7 \cdot 10^3$, and then the reaction may still be diffusion controlled.

4.3.5 The Bodenstein Approximation

We may conclude that in most aqueous solutions, simple bimolecular reactions will proceed according to Eq. (4.12). For bimolecular reactions in a system containing very little solvent, however, the reaction rate will be mostly diffusion controlled, implying that the rate is inversely proportional to the viscosity. In low-moisture foods, the situation may often be intermediate. It appears logical to combine Equations (4.12) and (4.16). Since we must essentially add the times needed for encountering and for the reaction itself, the resultant rate constant would then follow from

$$\frac{1}{k} \approx \frac{1}{k_S} + \frac{1}{k_{E,2}} \tag{4.17}$$

which is a form of the so-called *Bodenstein approximation*. It is called an approximation because a simple addition of the reciprocal rates is mathematically not quite correct. The equation clearly shows, however, that if either k_S or $k_{E,2}$ is much smaller than the other, the other is the effective rate constant.

The **temperature dependence** of k according to Eq. (4.17) needs some consideration. According to the Eyring equation, it is largely determined by the factor $\exp(-\Delta H^{\ddagger}/RT)$, as discussed. Equation (4.16) shows that k_S is proportional to T/η, where η is, for instance, inversely proportional to T. This then would lead to k_S being about proportional to T^2. However, Eq. (4.15), on which this relation is based, is by no means valid for systems that are not homogeneous solutions. One has to use the effective diffusion coefficient (D) directly, as is discussed in Section 5.3.2. For low-moisture systems, D may strongly depend on temperature, the more so for larger molecules; see also Figure 8.9. Altogether, in most situations where Eq. (4.17) would be more or less applicable, there is no simple expression for dk/dT. This implies that the temperature dependence of a reaction in such systems must be experimentally determined over the full temperature range of interest. Another consequence is that an analysis in which $\Delta H^{\ddagger}$ and $\Delta S^{\ddagger}$ are derived for inactivation of an enzyme, as given in Section 4.3.3, would be questionable for low-moisture foods.

Some Complications. Apart from Eq. (4.17) being approximate, it is only valid if Eqs. (4.12) and (4.16) are correct. Several objections can be made and we will just mention the main ones.

The absolute rate theory implicitly assumes that the addition of a given amount of free energy ($\Delta G^{\ddagger}$) would be sufficient for a reaction event to occur. In fact, there is another condition, which is that the reactants must collide in a specific way, because their mutual orientation will determine whether reaction can occur. This may decrease k by one or two orders of magnitude as compared to the predicted value. However, for reactions in solution another effect plays a part. When two reactants encounter each other, they stay close to each other for a relatively long time. According to Einstein, the mutual displacement Δ of two molecules by diffusion over a time t is given by $\langle \Delta \rangle^2 = 6Dt$. In water, the diffusion coefficient $D \approx 1.5 \cdot 10^{-9}\,\text{m}^2 \cdot \text{s}^{-1}$. Assuming the molecules to be in each other's interaction sphere if their separation distance $< 0.3\,\text{nm}$, and putting this value equal to Δ, we arrive at a time $t \approx 10^{-11}\,\text{s}$ before they would diffuse away from each other. This may seem a very short period, but it would allow the molecules to collide, say, 50 times for every encounter. This effect seems often to roughly compensate for the orientation effect mentioned. Nevertheless, Eq. (4.12) may be fairly uncertain.

Smoluchowski's equation (4.13) is reasonably rigid, but in the derivation of (4.16) several assumptions were made. Putting the hydrodynamic radii of the molecules equal to their collision radii is especially questionable. As discussed above, Eq. (4.15) would only apply in homogeneous dilute systems.

Strictly speaking, Smoluchowski's equation only applies if each encounter leads to reaction. If the chance of a reaction event is small, as is the case for large $\Delta G^{\ddagger}$, the concentration gradient of reactant B near a molecule of reactant A will be smaller, and this will upset the relations on which Eq. (4.13) is based.

Altogether, and unfortunately, Eq. (4.17) cannot be seen as more than indicating trends.

4.4 FURTHER COMPLICATIONS

In practice, numerous complications are encountered, concerning the order of reactions, their rate, and their temperature dependence. Some aspects important for the food scientist or technologist will be mentioned, if only to warn against pitfalls in handling or interpretation of kinetic results.

Order and Molecularity. If we have a very simple reaction, like $H + H \rightarrow H_2$ in the gas phase, this is a true bimolecular reaction and also the order is two. Such a simple correspondence between molecularity and order is, however, the exception rather than the rule. For example, a reaction involving water as a reactant in a dilute aqueous solution, for instance, the hydrolysis of an ester,

$$R-CO-OCH_2-R' + H_2O \rightarrow R-COOH + R'-CH_2OH$$

may effectively be first order, although it is a bimolecular reaction. This is because the concentration of water (or the water activity, rather) does not significantly alter during the reaction. Another point is that even a fairly simple reaction may involve a number of elementary reactions or steps, and one of them may be effectively rate determining, thereby also determining the order. Consequently, the order has to be established from experimental results, and it may be very difficult to elucidate the elementary reactions involved. In many cases, it turns out that the order is not an integer number; or the order may change in the course of the reaction; or it may be different with respect to different reactants; or the order with respect to concentration may be different from that with respect to time.

A simple (and probably oversimplified) example will be discussed, the inactivation of an enzyme by heat treatment. At a high temperature, the protein molecule will unfold, but if nothing else happens, it will probably refold after cooling and thereby regain its enzyme activity. This means that the unfolded molecule must undergo a reaction that prevents it from refolding into its native conformation. In the simplest situation we thus have $N \rightarrow U \rightarrow I$, where N is the native, U the unfolded, and I the inactivated state. The second reaction will mostly involve other molecules, but we will assume here that both the first and the second step are first-order reactions. We then have

$$-\frac{d[N]}{dt} = k_1 \cdot [N]$$

and

$$\frac{d[I]}{dt} = k_2 \cdot [U]$$

Simple though this set of differential equations may seem, it has no simple solution. If $k_1 \gg k_2$, we have effectively that all N has been converted into U and $d[I] \approx -d[U]$, which leads to a simple solution [i.e., Eq. (4.2)]. If, on the other hand, $k_1 \ll k_2$, any U that is formed will almost immediately be converted into I, and $[U] \approx 0$, again leading to a simple first-order reaction, now with rate constant k_1.

For the reactions just considered, k_1 is likely to be far more temperature dependent than k_2. An example is shown in Figure 4.6 for the thermal inactivation of plasmin, a proteolytic enzyme. At low temperature, the unfolding reaction N→U is very slow, and the second reaction, albeit slow, is much faster than the unfolding. Consequently, the overall reaction rate is determined by the unfolding, the rate constant of which strongly depends on temperature (see Section 4.3.3). At high temperature, unfolding is extremely fast, and the rate will be determined by U→I, which is far less temperature dependent. At intermediate temperatures, the rate constant given is a pseudo-first-order rate constant, and more elaborate kinetic studies would be needed to describe properly the overall reaction.

From this example, we can draw a few general conclusions. The first is about **uncoupling**. At relatively high temperature, a high concentration of U is formed, whereas at lower temperatures this is not so. In the more general case of A→B→C, the ratio of the concentrations of the products B and C formed will vary with temperature, if the consecutive reactions exhibit different temperature dependencies. An example is potatoes becoming sweet when storing them near 0°C. Broadly speaking, potatoes exhibit two consecutive reactions: the hydrolysis of starch, leading to the formation of sugar, and the conversion of sugar into CO_2 and H_2O by respiration. At room temperature, the latter reaction is the fastest, leaving the sugar concentration low. At low temperature, both reactions are slowed down, but the respiration more so than the hydrolysis, and sugar accumulates. Numerous other examples could be given.

The second point concerns **changing reaction order**. The order may change in the course of the reaction, for instance because one of the reactants or intermediates becomes consumed, thereby leading to a different mix of products and another reaction step dominating the order. This may make it difficult to predict the extent of a reaction after various reaction times from only a few analytical data, the more so since the relation may vary with temperature.

This brings us to the third point, i.e., there is **not a single activation energy** for the temperature dependence of a sequence of reactions. Again, the curve for inactivation of plasmin (EC 3.4.21.7) in Figure 4.6 is a good example, since such a situation is fairly general, although mostly not as extreme. It may also be noted that the accuracy of the Arrhenius relation becomes questionable if the activation energy is small: the rate equation has T in the preexponential factor, and E_a differs by RT from the activation enthalpy. A small apparent activation energy may result if the first of two consecutive elementary reactions is in fact reversible.

In foods, we often have what may be called **reaction cascades**, i.e., a whole series of reactions, partly consecutive, partly parallel, with bifurcations and with more than one reaction pathway leading to the same product. Examples are nonenzymatic browning or Maillard reactions, as well as several changes occurring during heat treatment. Chain reactions may be involved as well, as in the formation of hydroperoxides during the autoxidation of fats:

$$\begin{aligned} &\rightarrow ROO^{\bullet} + RH \rightarrow ROOH + R^{\bullet} \\ &\quad R^{\bullet} + O_2 \rightarrow ROO^{\bullet} \end{aligned}$$

Here $^{\bullet}$ denotes a radical.

Simple reaction kinetics will never suffice, but one can try to elucidate the various elementary reactions and then set up all the rate equations (differential equations) and solve the whole set numerically. In fact, this often is the only way to determine the reaction scheme with any confidence, since only quantitative agreement between calculated and observed product concentrations as a function of time guarantees its correctness. It is far beyond the scope of this book to discuss even a simple example.

Catalysis. Many reactions are catalyzed, i.e., increased in rate, by a compound in solution (homogeneous catalysis) or a group at the surface of a particle (heterogeneous catalysis), where the catalyst is not consumed itself. Examples are various hydrolyzing reactions, like the ester hydrolysis mentioned above, that are catalyzed by H^+ as well as OH^- ions. In such a case the reaction rate greatly depends on pH, though the ions themselves do not appear as reactants in the overall reaction scheme. Ubiquitous in natural foods are enzyme-catalyzed reactions. The simplest case leads to Michaelis–Menten kinetics, but several complications may arise.

Negative catalysis may also occur, since several compounds are known that *inhibit* reactions. For example, some cations, notably Cu^{2+}, catalyze the autoxidation of lipids; chelating agents like citrate may greatly lower the activity of divalent cations, thereby decreasing the oxidation rate. Inhibition of enzymes is frequently observed.

Compartmentalization. Be it of one or more reactants or of a catalyst, compartmentalization often occurs in foods. A simple case is oxidation, e.g., of unsaturated lipids in several foods, where the oxygen needed for the reaction is present in low concentration and has to diffuse from another compartment, viz., the air above the food, to the oxidizable components. This will naturally slow down the reaction.

Monoglycerides dissolved in oil droplets can hardly react with water to be hydrolyzed into glycerol and fatty acids, and certainly not when the reaction has to be catalyzed by an enzyme (an esterase), which is in the water phase. Possibly, the reaction proceeds at the interface between oil and water, which usually means that it will be slow; but if a suitable enzyme adsorbs onto the said interface, the reaction may in fact be quite fast. If one of the reactants is immobilized at a particle surface, this will in general slow down the rate, if only because the diffusion coefficient for that reactant is effectively zero.

In plant and animal cells, many enzymes are compartmentalized, and several are also immobilized, greatly slowing down reactions. After the cells have been mechanically damaged, some reactions may proceed fast. A well-known example is the rapid enzymatic browning of apple tissue after the apple has been cut; here, the cutting allows an enzyme, polyphenoloxidase, to reach its substrate, mainly chlorogenic acid.

Slowness of Reactions. In foods, we are often concerned with reactions that are very much slower than the reactions studied in the chemical or biochemical laboratory. For instance, the maturation of products like hard cheese, wine, and chutney may take years, and the maturation is ultimately due to chemical reactions, often enzyme catalyzed. Slow quality loss or deterioration of foods is ubiquitous. An example is loss of vitamins; when we accept a loss of 10% per year and assume it to be due to a first order reaction, this would imply that its rate constant would be given by $\ln 0.9 = -k \times 365 \times 24 \times 3600$, or $k \approx 3.3 \cdot 10^{-9} \, \text{s}^{-1}$, a very small rate constant. We may also consider loss of available lysine due to Maillard reactions. Taking a very simplified view, this is ultimately due to a second-order reaction between a reducing sugar and the exposed lysine residues of protein. Assume that we have 10% glucose (say in orange juice) and that 1% of it would be in the reducing, i.e., open-chain, form, this gives a molarity of about 0.005. Assume 1% protein, 7% of the residues of which are lysine, of which 80% would be reactive, this gives a molarity of about 0.0047, to which must be added some free lysine, say 0.0003 molar. If we accept a lysine loss of 1% in 3 months, i.e., $6 \cdot 10^{-5}$ molar per 3 months, we have

$$-\frac{d[L]}{dt} = \frac{6 \cdot 10^{-5}}{91 \times 24 \times 3600} = 8 \cdot 10^{-12} \, \text{s}^{-1} = k \times 0.005 \times 0.005$$

yielding $k \approx 3 \cdot 10^{-7} \, \text{L} \cdot \text{mol}^{-1} \cdot \text{s}^{-1}$. Such a low rate constant would imply a fairly large activation free energy, i.e., $136 \, \text{kJ} \cdot \text{mol}^{-1}$ according to Eq. (4.12).

Note Assuming $\Delta S^{\ddagger}$ to be negligible, this also yields $\Delta H^{\ddagger} \approx 136 \text{ kJ} \cdot \text{mol}^{-1}$. Cf. Table 4.2.

These examples may suffice to illustrate that we may have very slow reactions in foods. Here a brief summary is given of the possible causes for slow reactions.

1. The (effective) concentration of reactants is small, which is especially important in second-order reactions. It may be due to small total concentrations, to compartmentalization or immobilization, or to a complicated cascade of reactions with several "side-tracks" that consume reactants for other reactions.
2. The activity coefficient(s) may be small. See Section 2.2.5.
3. The activation free energy is large. If this means that also the activation enthalpy is large, as will often be the case, the reaction rate is strongly temperature dependent.
4. A suitable catalyst is missing or unavailable, or inhibiting substances are present.
5. Diffusion is very slow, because of a very high viscosity. This is often the case in low-moisture products. See also Section 8.4.2.

4.5 RECAPITULATION

Most foods are not in thermodynamic equilibrium. For several possible reactions, the reaction free energy is large and negative, suggesting that the reaction would be fully completed in a very short time, while nevertheless the reaction proceeds slowly. The composition of a reaction mixture then is not thermodynamically controlled but kinetically. In some cases, a steady state rather than an equilibrium state is attained, and it may be useful to distinguish these.

Reaction kinetics, or at least the mathematical formulas describing it, depend on the *order of the reaction*. Orders of 0, 1, and 2 are mostly considered. Reaction order, however, is an empirical number, mostly not equal to the molecularity of the reaction. It has to be determined experimentally, and noninteger values may be observed; moreover, order may change in the course of the reaction or may vary with conditions such as temperature.

The classical *rate theory* due to Arrhenius proceeds on the Maxwell–Boltzmann distribution of the velocity, and thereby the kinetic energy, of molecules or particles; their average kinetic energy equals $(3/2)k_B T$. If two molecules collide with a kinetic energy larger than an *activation energy E_a* for a reaction between them to proceed, they are assumed to react. The proportion of collisions of sufficient energy increases with T, and the reaction rate would be proportional to $\exp(-E_a/RT)$. This is a useful relation to describe temperature dependence, but it is insufficient in other respects; for instance, it is difficult to fit into reactions of order (molecularity) unity. The theory of the *activated complex* proceeds on the basis of an activation free energy, i.e., including an entropy term. It often allows us to make quantitative predictions of the reaction rate. The denaturation of proteins, which is mostly at the root of, and rate determining for, the inactivation of enzymes and the killing of microorganisms, is extremely temperature dependent, because of the very large activation enthalpy ($\approx E_a$); the also very large positive activation entropy then causes the reaction to proceed at a measurable rate. Most "simple" chemical reactions occurring in foods have a much weaker temperature dependency.

Chemical reaction kinetics proceeds on the (often implicit) assumption that the reaction mixture is ideally mixed, and does not consider the time needed for reacting species to encounter each other by diffusion. The encounter rate follows from the theory of *Smoluchowski*. It turns out that most reactions in fairly dilute solutions follow "chemical" kinetics, but that reactions in low-moisture foods may be *diffusion controlled*. In the Bodenstein approximation, the Smoluchowski theory is combined with a limitation caused by an activation free energy. Unfortunately, the theory contains several uncertainties and unwarranted presumptions.

Several further *complications* may arise. A number of consecutive reactions is very common in foods, and then kinetics may become very complicated. There may be a whole cascade of reactions and moreover some components formed may react in various ways, causing the reaction scheme to be branched. In such cases, uncoupling often occurs, i.e., the reaction mixture obtained (relative proportion of reaction products) depends on conditions like temperature. Several reactions can be catalyzed, notably by enzymes, and enzyme activity strongly depends on conditions like temperature and pH. Inhibitors, e.g., of enzymes may further complicate matters. In many foods, reactants, catalysts, or inhibitors are compartmentalized, which often causes a decrease in reaction rate.

BIBLIOGRAPHY

The textbooks mentioned in Chapter 1 all give information on chemical kinetics. Especially the text by P. W. Atkins discusses the topic in detail. Specialized books are

K. J. Laidler. Chemical Kinetics. Harper and Row, New York, 1987.

and

H. Maskill. The Physical Basis of Organic Chemistry. Oxford Univ. Press, 1985.

A series of articles on kinetic studies and modeling are in
Food Technol. 34(2) (1980) 51–88.

including work on experimental procedures for determining kinetic data, interpretation, and modeling, all specifically applied to foods (heat treatment and quality loss).
    A discussion of kinetics in relation to heat treatment that is somewhat more
rigorous is
M. A. J. S. van Boekel, P. Walstra. Use of Kinetics in Studying Heat-Induced
    Changes in Foods, In: P. F. Fox, ed. Heat-Induced Changes in Milk.
    International Dairy Federation, Brussels, 1995, pp. 22–50.

Kinetics of enzyme-catalyzed reactions are treated in all textbooks on biochemistry.
A clear and authoritative discussion is in
J. R. Whitaker. Enzymes, 3rd ed. In O. R. Fennema, ed., Food Chemistry, Dekker,
    New York, 1996, pp. 431–530.

A discussion of the derivation of kinetics for a sequence of reactions is by
W. E. Stewart, M. Caracotsias, J. P. Sørensen. Parameter estimation from
    multiresponse data. *Am. Inst. Chem. Engs. J.* 38 (1992) 641–650.

A similar treatment for foods, especially discussing how to cope with statistical
uncertainties, is in
M. A. J. S. van Boekel. Statistical aspects of kinetic modelling for food science
    problems. *J. Food Sci.* 61 (1996) 477–485, 489.

5

Transport Phenomena

The transport of momentum, heat and mass, or in simpler terms the phenomena of flow, convection, heat conduction, and diffusion, are primarily studied by process engineers and in some aspects also by rheologists. Important though these topics are for the food technologist, they are not the subject of this book. However, some basic concepts are needed in various chapters; this includes aspects of rheology and hydrodynamics. Furthermore, transport phenomena inside solidlike foods often are rather intricate, and this subject is also introduced in this chapter.

5.1 FLOW AND VISCOSITY

Rheologists study what happens with a system, be it an amount of fluid or a piece of solid material, when work (= mechanical energy) is applied to it. Remembering that work equals force times distance, we come to the definition:

> *Rheology* is the study of the relations between the *force* acting on a material, its concomitant *deformation*, and the *time scale* involved.

To keep the relations simple, rheologists tend to use **stress** rather than force; it is defined as the force divided by the area over which it is acting (S.I.

unit $N \cdot m^{-2} = Pa$). The stress can be normal, i.e., in a direction perpendicular to the plane on which it acts, or tangential, i.e., acting in the direction of the plane. Of course, intermediate situations also occur; the stress vector can then be resolved into a tangential and a normal component.

Deformation implies change(s) in the distance between two points in the material. **Strain** is used rather than deformation; it is defined as the relative deformation, i.e., the change in distance (e.g., length of a specimen) divided by the original distance; it is thus a dimensionless quantity. It may be noted in passing that the words stress and strain have almost identical meanings in everyday language, but that they apply to fundamentally different variables in rheology.

The notion of **time scale** needs some elaboration. It is generally defined as the characteristic time needed for an event to occur, e.g., a reaction between two colliding molecules, the rotation of a particle in a flow field, or the transformation of some dough into a loaf of bread. In rheology, the characteristic parameter is the *strain rate*, i.e., the time derivative of the strain. This will be further discussed below. The strain rate is expressed in reciprocal seconds, and the characteristic time scale during deformation is the reciprocal of the strain rate, rather than the duration of the experiment. In many systems, the relation between stress and strain is dependent on the strain rate.

A main problem in doing rheological work is that in most situations stress and strain vary from place to place. Moreover, the strain rate often varies during the deformation. One generally tries to do experiments in such a way that well-defined conditions apply throughout the test piece, thereby establishing true material properties, i.e., results that do not depend on the size or shape of the test piece.

Section 5.1 discusses, besides some basic notions, the rheology of liquids and liquidlike systems, i.e., those systems that exhibit flow. Solidlike systems are discussed in Section 17.1. This all concerns bulk rheology. Surface rheology is discussed in Section 10.8.

5.1.1 Flow

If a stress, however slight, is applied to a fluid, it will flow. A fluid may be a gas or a liquid, and we will primarily consider liquids. The flow may be laminar or turbulent. The latter is chaotic, implying that a volume element may at any moment move in any direction, though the average flow is in one direction. In laminar flow, the streamlines, i.e., the trajectories of small volume elements, exhibit a smooth and regular pattern.

Laminar Flows. These exist in several types, depending on the geometrical constraints. Some examples are shown in Figure 5.1. An important characteristic of laminar flow is the *velocity gradient* Ψ, defined in the figure. Figure 5.1a shows pure *rotational flow* (circular streamlines); here, a volume element in the center will only rotate and not be displaced ("translated") nor deformed. The rotation rate equals the velocity gradient. Figure 5.1c shows hyperbolic streamlines, an example of *elongational flow* (also called extensional flow). A volume element in the center will be deformed by elongation, as depicted in Figure 5.2. The velocity gradient is in the direction of the flow. There is no rotation. Figure 5.1b depicts what is called *simple shear flow* (straight streamlines), although it is not such a

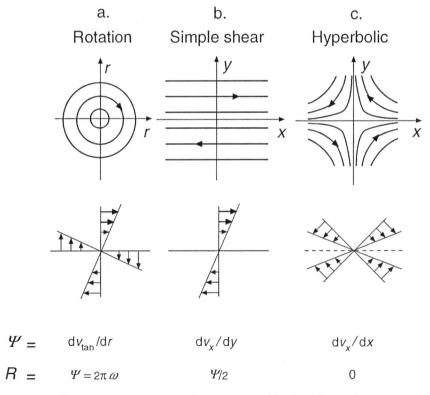

a. Rotation	b. Simple shear	c. Hyperbolic
$\Psi = \quad dv_{tan}/dr$	dv_x/dy	dv_x/dx
$R = \quad \Psi = 2\pi\omega$	$\Psi/2$	0

FIGURE 5.1 Cross sections through three types of laminar flow. The upper row gives the streamlines, the second row the velocity profiles. The flows are two-dimensional, implying that the patterns do not change in the z-direction (perpendicular to the plane of the figure). Ψ = velocity gradient, v = linear flow velocity, R = rotation rate, ω = rotation frequency.

simple flow type. Layers of liquid appear to slide over each other. A volume element is sheared, as depicted in Figure 5.2, but it is also rotated (its diagonal rotates). Simple shear is a rotating flow, and the rotation rate is half the velocity gradient. The gradient is in a direction perpendicular to the direction of flow.

Most people envisage simple shear when laminar flow is considered, but it is not the most common type of flow. To begin with, intermediate types generally occur. An example is the flow type in Figure 5.1a. In the center it is purely rotating, but moving away from the center of rotation, i.e., at increasing r, the flow becomes ever more similar to simple shear. Also when going from either axis outwards in Figure 5.1c, an increasing amount of shear flow is introduced.

Moreover, most flows are not two-dimensional as in Figure 5.1, where the flow pattern does not change in the z-direction. Consider Figure 5.1c and rotate it around one of the axes: it then represents *axisymmetric flow*. By rotation around the x-axis, *uniaxial elongational flow* in the x-direction results; a practical example is flow through a constriction in a tube. By rotation around the y-axis, *biaxial elongational flow* in the x,z-plane results; a practical example is squeezing flow between two closely approaching

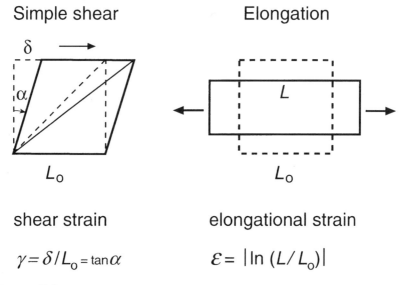

Simple shear   **Elongation**

shear strain   elongational strain

$$\gamma = \delta / L_0 = \tan\alpha \qquad\qquad \varepsilon = |\ln (L / L_0)|$$

FIGURE 5.2 Explanation of the type of strain occurring in a volume element subjected to simple shear (shear strain) or to elongation (strain expressed as Hencky strain).

plates. Some kind of elongational flow is always involved when a liquid is accelerated or decelerated.

Axisymmetric simple shear flow occurs in a straight cylindrical tube; this so-called *Poiseuille flow* is depicted in Figure 5.4a, further on. The figure also illustrates another point. The liquid velocity v equals zero at the wall of the tube and is at maximum in the center, whereas the velocity gradient Ψ is zero in the center and is at maximum at the wall. This is more or less the case in many kinds of flow. The flow velocity at the wall of a vessel always equals zero, at least for a Newtonian liquid (explained below).

Long lasting simple shear flow of constant shear rate is often approximated by Couette flow. The liquid is between two concentric cylinders, one of which is rotating. If the ratio between the radii of the inner and outer cylinders is close to unity, Ψ is nearly constant throughout the gap.

Figure 5.2 illustrates the **strains** resulting from shear and elongational flows. The elongational strain can be expressed in various ways and the so-called *engineering strain*, i.e., $(L - L_0)/L_0$, is often used. The disadvantage is that it gives the strain with respect to the original length, not with respect to the length at the moment of measurement. The latter is to be preferred, and this so-called *Hencky strain* or natural strain is given in Figure 5.2. During flow, the strain alters, and we need to know the *strain rate*, for instance $d\gamma/dt$ or $d\varepsilon/dt$. These strain rates are both equal to the velocity gradient Ψ. (Unless stated otherwise, we will use the symbol Ψ both for the velocity gradient and for the strain rate.)

During flow of a fluid, the relation holds

$$\sigma = \eta\Psi = \frac{\eta d\gamma}{dt} \qquad (5.1)^*$$

where the part after the second equals sign only applies to simple shear flow. The factor η is called the **viscosity** or more precisely the dynamic shear viscosity; the S.I. unit is $N \cdot m^{-2} \cdot s = Pa \cdot s$. Viscosity is a measure of the extent to which a fluid resists flow (what Newton called the "lack of slipperiness"). For a so-called Newtonian liquid, η is independent of the velocity gradient. Pure liquids and solutions of small molecules virtually always show Newtonian behavior, i.e., the velocity gradient is proportional to the stress. The value of the viscosity depends, however, on the type of flow, and its value for elongational flow (η_{el}) is always higher than that for simple shear flow (η_{ss} or simply η). The relation is given by

$$\eta_{el} = Tr \cdot \eta \qquad (5.2)$$

where Tr stands for the dimensionless Trouton ratio. For Newtonian liquids we have: in two-dimensional hyperbolic flow Tr = 2; in uniaxial axisymmetric flow Tr = 3; in biaxial axisymmetric flow Tr = 6.

Any flow exerts a frictional or **viscous stress** $\eta\Psi$ [see Eq. (5.1)] onto the wall of the vessel in which the liquid is flowing or onto particles in the liquid. This has several consequences, the simplest one being that particles move with the liquid; some others are illustrated in Figure 5.3, which applies to *simple shear flow*. A solid sphere rotates, as mentioned. Solid anisometric particles also rotate, but not at a completely constant rate. An elongated particle will rotate slower when it is oriented in the direction of flow than when in a perpendicular direction. This means that such particles show *on average* a certain preference for orientation in the direction of flow, although they keep rotating.

Note This can give rise to so-called flow birefringence; see Section 9.1, under Optical anisotropy.

A liquid sphere can become elongated, if some conditions are fulfilled (see Section 11.3.2). As depicted, it obtains an orientation of about 45° to the direction of flow, and the liquid *inside* the particle rotates. Much the same holds for a (random) polymer coil: it is also elongated, and its rotational motion now implies that the coil is compressed and extended (as indicated in the figure) periodically. It may be noted that this is a good example of the difference between the time scale of an event and the time that an experiment lasts. The time scale of compression/extension is simply $2/\Psi$ (i.e., 1 ms if the shear rate is 2000 s^{-1}), however long the shearing lasts.

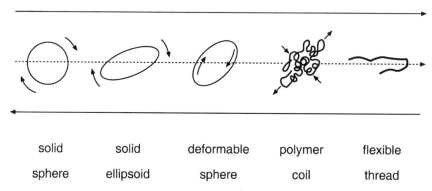

| solid | solid | deformable | polymer | flexible |
| sphere | ellipsoid | sphere | coil | thread |

FIGURE 5.3 Motion of particles in simple shear flow. The arrows indicate the direction of flow relative to the particles. In the central plane the flow velocity is zero or, in other words, the coordinate system moves with the geometric center of the particle. See text.

A long flexible thread tends to align in the direction of flow but periodically folds (as depicted) and stretches again.

In *elongational flow* of the same velocity gradient, the viscous stresses are greater: see Eq. (5.2). Anisometric solid particles become aligned in the direction of flow; flexible particles become extended, i.e., anisometric, and aligned. Particles that are close to each other become separated from each other, while they can stay together in simple shear flow: a doublet of spheres then continues rotating as one dumbbell-shaped particle.

Turbulent Flow. In Figure 5.4, flow through a tube is depicted. In Figure 5.4a, the flow is laminar (simple shear, Poiseuille flow) and the velocity profile is parabolic. If the flow velocity is increased, a flow profile like that in 5.4b may develop. The streamlines become wavy and eddies develop. This implies that the flow becomes more chaotic and is called turbulent. Near the wall the flow is still laminar, at an increased velocity gradient as compared to the situation in 5.4a. For still higher velocity, the flow becomes increasingly chaotic, and the thickness of the laminar layer near the wall decreases, as depicted in 5.4c. The flow profile drawn now gives the *time average* of the flow velocity at various distances from the center. Any volume element is subject to rapid fluctuations both in velocity and in direction of flow. The average flow profile is almost block-shaped, except for a high velocity gradient near the wall.

Eddies (vortices, whorls) are thus superimposed on the average flow direction, and this causes a strong mixing effect. The size of the largest eddies is of the order of the smallest dimension of the vessel (e.g., the pipe diameter). These large eddies transfer their kinetic energy to smaller ones, which transfer it to still smaller eddies, etc. With decreasing eddy size, the

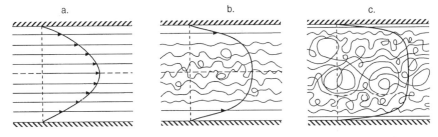

FIGURE 5.4 Streamlines of flow through a tube. In (a) the flow is laminar (Poiseuille flow), in (b) and (c) flow is turbulent. The average flow velocity increases from (a) to (c). In (a) a velocity profile is given, in (b) and (c), average velocity profiles. See text.

local flow velocity is decreasing, but less than proportionally, which implies that the local velocity gradient Ψ is increasing. Flow of a liquid means dissipation of kinetic energy, and the *energy dissipation rate* (in $J \cdot m^{-3} \cdot s^{-1}$), also known as the *power density* (in $W \cdot m^{-3}$), is simply given by

$$\text{power density} = \eta \Psi^2 \tag{5.3}$$

where Ψ is the *local* velocity gradient. This implies that small eddies dissipate more kinetic energy per unit volume than do large ones, and below a certain size all energy is dissipated into heat. This also means that in turbulent flow more energy is dissipated than in laminar flow, which means, in turn, that it costs more energy to produce flow. It is thus *as if* the viscosity of the liquid were larger in turbulent than in laminar flow.

Turbulent flow thus generates rapid local velocity fluctuations. According to the law of conservation of energy (Section 2.1), the sum of the kinetic energy of the flowing liquid and its potential energy must remain the same. The potential energy per unit volume is simply given by the pressure p (unit $Pa = N/m^2 = J/m^3$) and the kinetic energy by $(1/2)mv^2$ (where m is mass). This leads to the *Bernoulli equation*

$$p + (1/2)\rho v^2 = \text{constant} \tag{5.4}^*$$

where ρ is mass density (kg/m^3). Consequently, a high liquid velocity implies a low pressure, and turbulent flow thus generates rapid local pressure fluctuations. These cause *inertial forces* to act on any particle present near an eddy. For low-viscosity liquids like water, the stresses caused by these inertial forces tend to be much higher than the frictional forces caused by laminar flow. (These aspects are further discussed in Section 11.3.3.)

What are the conditions for flow to become turbulent? This depends on the preponderance of inertial stresses—proportional to ρv^2—over frictional or viscous stresses. The latter are equal to $\eta \Psi$ in laminar flow; Ψ is proportional to v/L, where L is a characteristic length perpendicular to the direction of flow. The ratio is proportional to the dimensionless *Reynolds number*, given by

$$\text{Re} \equiv \frac{L \bar{v} \rho}{\eta} \tag{5.5}$$

Here $\bar{v}$ is the average flow velocity, i.e., the volume flow rate (flux) divided by the area of the cross section of the flow channel. The characteristic length is, for instance, a pipe diameter. If now Re is larger than a critical value Re_{cr}, turbulence will set in. Table 5.1 gives the Reynolds number equation for

TABLE **5.1** Reynolds Numbers (Re) for Various Flow Geometries

Flow geometry	Re =	Value of Re_{cr}
In a cylindrical pipe of diameter D	$D\bar{v}\rho/\eta$	2300
Between flat plates at separation distance δ	$2\delta\bar{v}\rho/\eta$	~ 2000
Film (thickness δ) flowing over sloping flat plate	$4\delta\bar{v}\rho/\eta$	~ 10
Flow around a sphere[a] of diameter d	$dv\rho/\eta$	~ 1

[a] Here v is the velocity of the sphere relative to the liquid.

some flow geometries, as well as values for Re_{cr}. The flow around a sphere relates, for instance, to a sedimenting particle, where turbulence will develop in its wake for $Re > 1$.

It may finally be noted that a geometrical constraint that induces elongational flow if $Re < Re_{cr}$ (e.g., a sudden constriction in a tube) tends locally to depress turbulence if $Re > Re_{cr}$. Turbulence is also depressed by the presence of large-molar mass polymers or by a high concentration of dispersed particles in the liquid.

Question 1

Show for simple shear flow that the velocity gradient dv/dy equals the shear rate $d\gamma/dt$.

Answer

Consider Figure 5.2. Since the velocity at the bottom of the volume element equals zero, the velocity gradient dv/dy is given by the velocity at the top divided by L_0, hence by $(d\delta/dt)/L_0 = d(\delta/L_0)/dt = d\gamma/dt$.

Question 2

In a Couette apparatus, i.e., between two concentric cylinders, of which the outer one of radius 10 cm is rotating at 180 revolutions per minute and the inner stationary one has a radius of 9.5 cm; some triglyceride oil is present. The temperature is 20°C. How long will it take before the temperature is raised to 21°C, assuming that no heat loss from the oil to the environment occurs? Tip: Some useful data are in Table 9.2.

Answer

Laminar flow in a Couette apparatus has a practically constant velocity gradient, which makes calculation much easier. To check whether the flow is laminar, the Reynolds number has to be calculated; it is roughly given by Table 5.1 for flat plates. Because one of the "plates" is moving, the maximum velocity v should be taken; it is 2π times radius times angular velocity, i.e., $2\pi \cdot 0.1 \cdot 180/60 = 1.88 \, \text{m} \cdot \text{s}^{-1}$. Since $\delta = 0.005 \, \text{m}$, $\rho = 920 \, \text{kg} \cdot \text{m}^{-3}$, and $\eta = 0.075 \, \text{Pa} \cdot \text{s}$, we obtain Re $= 220$, clearly below the critical value. The velocity gradient $\Psi = v/\delta = 375 \, \text{s}^{-1}$. The energy dissipation rate equals $\eta \Psi^2 = 1400 \text{J} \cdot \text{m}^{-3} \cdot \text{s}^{-1}$. The specific heat c_p of triglyceride oil equals 2.1 $\text{kJ} \cdot \text{kg}^{-1} \cdot \text{K}^{-1}$ or $1.93 \cdot 10^6 \, \text{J} \cdot \text{m}^{-3} \cdot \text{K}^{-1}$. The quotient $\eta \Psi^2 / c_p$ now yields a temperature increase rate of $(1/1380) \, \text{K} \cdot \text{s}^{-1}$. Hence it would take $1380/60 = 23 \, \text{min}$ to raise the temperature by 1 K.

5.1.2 Viscosity

Molecules in a fluid undergo continuous Brownian or heat motion and thus have kinetic energy (Section 4.3.1). When the fluid flows, they have some additional kinetic energy and—owing to the velocity gradient—this energy varies from place to place. Envisaging simple shear flow, adjacent layers have a different velocity. During such flow, some molecules will move by Brownian motion from one layer to another one, which means to one with another velocity; such a molecule thus is accelerated or decelerated. This implies that a (small) part of the kinetic energy related to the flow is lost and converted into heat. This is the classical explanation for the viscosity of gases. The theory predicts that the viscosity of a gas increases with temperature (see Table 5.2) and is virtually independent of pressure. Neither of these two predictions is true for liquids. The explanation of viscosity in a liquid involves other factors, and has much to do with the limited free volume between the molecules: it is difficult for them to move past the other ones and this difficulty is enhanced in the presence of a velocity gradient. This then would mean that the viscosity is far greater in liquids than in gases and decreases with increasing temperature (higher $T \rightarrow$ lower density $\rightarrow$ more free space between molecules).

Table 5.2 gives some examples. It is seen that for homologous compounds, the viscosity increases with molecular size, in accordance with simple theory. It is also seen that there is a considerable variation among various types of molecules. This is related to the attractive interaction forces between molecules, and the existence of hydrogen bonds in water and alcohols is often held responsible for the relatively high viscosity of these compounds. However, the molecular explanation of viscosity is intricate.

TABLE 5.2 Viscosity of Some Fluids and Solutions

Material	Temperature (°C)	η mPa·s	$(\eta/\rho)^a$ mm^2/s
Gases			
Dry air	20	0.018	15
	70	0.021	20
Liquids			
Water	0	1.79	1.79
	20	1.00	1.00
	40	0.65	0.66
	100	0.28	0.29
Diethyl ether	20	0.23	0.32
Ethanol	20	1.20	1.53
Glycerol	20	1760	1400
n-Pentane	20	0.24	0.38
n-Decane	20	0.92	1.26
n-Hexadecane	20	3.34	4.32
Triglyceride oil	10	125	135
	20	75	82
	40	33	37
	90	8	9
Aqueous solutions, 20%			
Ethanol	20	2.14	2.21
Glycerol	20	1.73	1.66
Glucose	20	1.90	1.76
Sucrose	20	1.94	1.80
KCl	20	1.01	0.89
NaCl	20	1.55	1.35
Calculated[b]	20	1.82	

[a] This is called the kinematic viscosity.
[b] According to Eq. (5.11), for $\varphi = 0.2$, $\varphi_{max} = 0.65$ and $[\eta] = 2.5$.

Table 5.2 also gives examples of the viscosity of some aqueous solutions. It is seen that most solutes increase the viscosity, although the increase is quite small for KCl. A 20% aqueous ethanol solution is seen to have a distinctly higher viscosity than each of the pure liquids; this must be related to the contraction—hence a decrease in free volume—occurring upon mixing of the two liquids. Also for solutions it is useful to have a look at the effects of dispersed particles (or molecules).

Dispersions.*　When particles are added to a liquid, the viscosity is increased. Near a particle the flow is disturbed, which causes the velocity gradient Ψ to be locally increased. Because the energy dissipation rate due to flow equals $\eta\Psi^2$, more energy is dissipated, which becomes manifest as an increased macroscopic viscosity. The "microscopic" viscosity, as sensed by the particles, remains that of the solvent (pure liquid) η_s. For very dilute dispersions of solid spherical particles, Einstein derived

$$\eta = \eta_s(1 + 2.5\varphi) \tag{5.6}$$

a very simple relation. Note that only the volume fraction φ of the particles, not their size, affects viscosity (provided that the size is significantly larger than that of the solvent molecules). If the volume fraction becomes larger than about 0.01, the flow disturbances caused by the particles start to overlap. Consequently, the viscosity increases more with φ than predicted by Eq. (5.6). This is illustrated in Figure 5.5. The particles themselves now also sense a greater stress, because the local velocity gradient is increased.

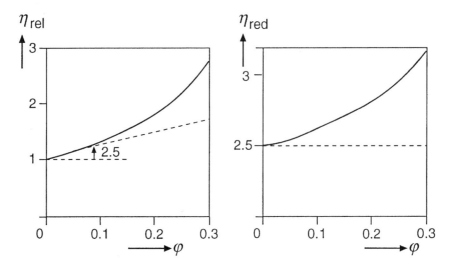

FIGURE 5.5　Example of the effect of concentration on viscosity. The relative (η_{rel}) and the reduced viscosity (η_{red}) of dispersions of spherical particles are given as a function of volume fraction (φ).

---

* It may be helpful for understanding the following part of the present section to consult first some parts of Chapter 9, especially where it concerns particles.

Various quantities are used in relation to the viscosity of dispersions:

relative viscosity: $\quad \eta_{rel} \equiv \dfrac{\eta}{\eta_s}$ $\hspace{4cm}$ (5.7)

specific viscosity: $\quad \eta_{sp} \equiv \dfrac{\eta - \eta_s}{\eta_s} = \eta_{rel} - 1$ $\hspace{2.5cm}$ (5.8)

reduced viscosity: $\quad \eta_{red} \equiv \dfrac{1}{c}\dfrac{\eta - \eta_s}{\eta_s} = \dfrac{1}{c}\eta_{sp}$ $\hspace{2cm}$ (5.9)

intrinsic viscosity: $\quad [\eta] \equiv \dfrac{1}{\eta_s}\left(\dfrac{d\eta}{dc}\right)_{c=0} = \lim_{c \to 0} \eta_{red}$ $\hspace{1.5cm}$ (5.10)

where c means concentration. Notice that for $c = \varphi$, insertion of Eq. (5.6) into (5.10) yields for the intrinsic viscosity $[\eta] = 2.5$. For other systems, other values for $[\eta]$ are observed (see below); $[\eta]$ is a measure of the capacity of a substance to increase viscosity. Often, the concentration of the substance is given as, for instance, $kg \cdot m^{-3}$, rather than volume fraction, implying that η_{red} and $[\eta]$ are not dimensionless but are expressed in reciprocal concentration units.

Concentrated Dispersions. For the viscosity of not very dilute systems, the Krieger–Dougherty equation is often useful. It reads

$$\eta = \eta_s\left(1 - \frac{\varphi}{\varphi_{max}}\right)^{-[\eta]\varphi_{max}} \hspace{3cm} (5.11)$$

Here φ_{max} is the maximum volume fraction (packing density) that the dispersed particles can have. At that value the viscosity becomes infinite (no flow possible). For random packing of monodisperse spheres, $\varphi_{max} \approx 0.65$. For polydisperse systems, its value can be appreciably higher. Note that now particle size becomes a variable, though its spread (e.g., relative standard deviation) rather than its average is determinant.

Equation (5.11) is rigorous for hard spheres in the absence of colloidal interaction forces, where $[\eta] = 2.5$; in the limit of $\varphi \to 0$, it equals the Einstein equation (5.6). Some calculated results are shown in Figure 5.6. An important aspect is that a given small increase in φ gives only a limited increase in viscosity if φ is relatively small, but if it is close to φ_{max}, the increase in η is large. For dispersions of other kinds of particles, the Krieger–Dougherty equation is not quite exact, but it remains useful, provided that $[\eta]$ is experimentally determined (since it can generally not be precisely predicted). However, for deformable particles Eq. (5.11) predicts values that are markedly too high if φ is fairly close to φ_{max}; for rigid anisometric particles η is underestimated at high φ (see Section 17.4).

Intrinsic Viscosity. Table 5.2 gives some values for the viscosity of 20% solutions; it is seen that for some neutral solutes the outcome does not greatly differ from what is calculated by Eq. (5.11) with $[\eta] = 2.5$. This suggests that the Einstein equation (5.6) can reasonably well apply to particles of molecular size. However, the agreement is far from perfect, and several factors may cause $[\eta]$ to deviate markedly from the value 2.5, generally being larger. Some variables are following.

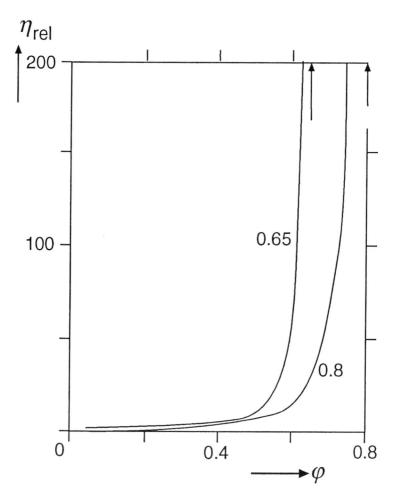

FIGURE 5.6 Relative viscosity (η_{rel}) as a function of volume fraction of spherical particles (φ) according to Eq. (5.11), for two values of φ_{max}, indicated near the curves.

1. **Particle shape**. In emulsions the particles tend to be almost perfectly spherical, but in other food dispersions spherical particles are the exception rather than the rule. In shear flow particles rotate (see Figure 5.3). This means that they sweep out a bigger volume during rotation than does a sphere of the same net volume. In other words, the effective volume fraction of particles is increased. This is more strongly so for prolate (cigar-shaped) ellipsoids or rods, than for oblate (disc-shaped) ellipsoids or platelets. The effect increases with increasing anisometry. Particles with an irregular or dented surface also exhibit a difference between effective and net volume.

2. **Colloidal interaction forces between particles**. If particles repel each other, the viscosity is always increased, except at very small φ; this means that η is generally affected but $[\eta]$ is not. Weak mutual attraction also tends to enhance η somewhat, but if the attraction is strong enough to cause aggregation of particles, $[\eta]$ can be greatly enhanced: the aggregates tend to enclose a lot of solvent, which means that effective φ is greatly increased.

3. **Swelling**. Figure 5.3 schematically depicts a coiled polymer molecule and illustrates that such a "particle" encloses a lot of solvent (although the enclosed solvent is not completely immobilized: see Section 6.2.2). This is comparable to the entrapment of solvent in aggregates mentioned above. Protein molecules always contain some water, i.e., are "swollen."

4. **Particle size**. For anisometric particles, their size has an effect. Small particles show rotational diffusion, and this is more rapid for a smaller particle. This affects the average orientation of the particles, hence the increase in viscosity due to anisometry. Smaller particles would thus give a higher viscosity. Also Factor 2 can come into play: the smaller the particles at a given value of φ, the smaller the interparticle distance and the larger the effect of repulsive interaction can be. On the other hand, attractive forces tend to have a smaller effect on smaller particles.

   The magnitude of most of the effects mentioned also depends on flow type. In elongational flow, for instance, rodlike particles may obtain a parallel orientation in the direction of flow, which tends to decrease the viscosity increase caused by the particles. In turbulent flow, the relations are more complicated.

Strain Rate Thinning. For a Newtonian liquid, η is independent of the magnitude of the stress, or of the velocity gradient, applied. In many cases, however, this is not the case, and the viscosity depends on the velocity gradient (or strain rate) applied, or on the stress applied. The ratio of stress over strain rate then is called the *apparent viscosity*, symbol η_a. In most cases, η_a decreases with increasing value of σ or Ψ, as illustrated in Figure 5.7. Such liquids are commonly called shear thinning, but *strain rate thinning* is a better term. First, the dependency of viscosity on the velocity gradient is not restricted to shear flow; it also occurs in elongational flows. Second, a material may become more resistant to deformation as the strain on it increases—a phenomenon called strain hardening—while at the same time the value of η_a decreases with increasing Ψ value (at the same strain), which means that the material is strain rate thinning. Some doughs and batters show such a combination of properties.

Several types of dispersions show strain rate thinning, and a quantitative explanation is not easily given. We will briefly consider two cases. The first one concerns shear flow. As discussed (above, Factor 4), anisometric particles show rotational diffusion and thereby increase viscosity. This effect will be smaller for a higher shear rate: when the shear-induced rotation is much faster than the diffusional rotation, the latter will have no effect anymore. The shear rate thinning effect is completely reversible. Something comparable happens in polymer solutions (Section 6.2.2).

Another situation arises when attractive forces cause particle aggregation (see Factor 2). A shear or elongational stress may now break down the aggregates, the more so for a higher strain rate. This then will lead to a decrease in effective volume fraction, hence to a decrease in η_a.

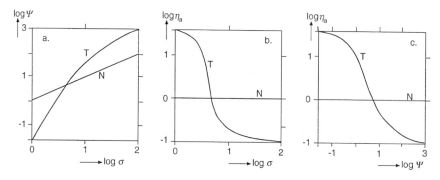

FIGURE 5.7 Examples of the relations between strain rate (Ψ, in s^{-1}), stress (σ, in Pa), and (apparent) viscosity (η_a, in Pa·s) in flow of a Newtonian liquid (N) and one showing fairly strong strain rate thinning (T); logarithmic scales. Note that most of the decrease in η_a occurs over about one decade in σ, and over about three decades in Ψ.

Thixotropy. In most systems the latter effect is not directly reversible: at constant strain rate η_a tends to decrease with time, and it gradually increases again when flow is stopped (aggregates then are formed again). Such behavior is called thixotropy. Many thick liquid foods are thixotropic; a good example is tomato ketchup. For such systems, the apparent viscosity observed will depend not only on the shear rate applied but also on the time during which it is sheared. Moreover, agitation applied to the sample before measurement may have markedly decreased the apparent viscosity, especially when measured at low strain rate.

Question

Consider (a) an oil-in-water emulsion of monodisperse droplets, $\varphi_{oil} = 0.6$; and (b) the same emulsion but at $\varphi_{oil} = 0.5$, to which an amount of a globular protein has been added at $\varphi_{protein} = 0.2$. Which system would have the higher viscosity?

Answer

For these systems we can presumably apply Eq. (5.11). In (a) we have $\varphi = 0.6$; since the particles are spheres, $[\eta] = 2.5$; and because the spheres are monodisperse, φ_{max} would be about 0.65. This yields $\eta_{rel} = 65$. In (b) φ is higher, i.e., 0.7, but φ_{max} will also be higher. Even if the oil droplets touch one another, the very much smaller protein molecules can fit in the gaps between the drops. Presumably, φ_{max} will be about 0.85. Also $[\eta]$ may be taken at 2.5, since most globular proteins are fairly spherical particles. Equation (5.11) then yields $\eta_{rel} \approx 40$; for $\varphi_{max} = 0.82$, we obtain $\eta_{rel} = 51$. Consequently, despite the higher volume fraction in (b), its viscosity would be lower.

> *Note* The validity of Eq. (5.11) may be questioned for such a strongly bimodal particle size distribution. Another approach is first to calculate the viscosity of the protein solution (with $\varphi = 0.4$ and $\varphi_{max} = 0.65$), which yields $\eta_{rel} = 4.7$. Subsequently, this value is taken for η_s in Eq. (5.11) for the emulsion (with $\varphi = 0.5$ and $\varphi_{max} = 0.65$), which also yields $\eta_{rel} = 51$.

5.1.3 Viscoelasticity

Figure 5.8 illustrates what can happen when an amount of material is put under a given stress for some time. Envisage, for instance, a cheese cube onto which a weight is placed and after some time removed. In (a) the stress–time relation applied is shown, and in (b) we see the response of a *purely elastic* material. The material is instantaneously deformed upon applying the stress and it instantaneously returns to its original shape.

> *Note* Actually, instantaneous deformation cannot occur; in practice the deformation rate corresponds to the sound velocity in the material, often of the order of a $km \cdot s^{-1}$.

The material thus has a perfect "memory" for its original shape. The energy applied to achieve deformation is not dissipated but stored: upon removal of the stress, the stored energy is recovered. The ratio of stress over strain is called *modulus*.

In Figure 5.8c we see what happens with a *purely viscous* material, i.e., a Newtonian liquid. As soon as a stress is applied, it starts to flow, and after removal of the stress, flow stops, and the deformation attained remains. The liquid has no memory for its original shape, and the energy applied to cause flow is dissipated into heat.

In Figure 5.8d an intermediate behavior, called *viscoelastic,* is depicted; such a relation is often called a creep curve, and the time-dependent value of the strain over the stress applied is called creep compliance. On application of the stress, the material at first deforms elastically, i.e., "instantaneously," but then it starts to deform with time. After some time the material thus exhibits flow; for some materials, the strain can even linearly increase with time (as depicted). When the stress is released, the material instantaneously loses some of it deformation (which is called elastic recovery), and then the deformation decreases ever slower (delayed elasticity), until a constant value is obtained. Part of the deformation is thus permanent and viscous. The material has some memory of its original shape but tends to "forget" more of it as time passes.

It should be noted that viscoelastic behavior varies widely among materials. The magnitudes of the instantaneous elastic modulus, the apparent viscosity and the elastic recovery, and especially the time scales involved, vary widely. Some viscoelastic substances, like cheese, seem on the face of it to be solids, but they are observed to flow over longer time scales. Others, like egg white, appear to be liquids but show elasticity on closer inspection. Closer inspection may involve sudden acceleration of the liquid, for example by rapidly giving a turn to the beaker containing it. If the liquid then shows oscillatory behavior, best seen in the motion of a few enclosed air bubbles or small particles, it also has some elastic property.

Such a **viscoelastic** or **memory liquid** is another example of a non-Newtonian liquid. Nearly all viscoelastic liquids are also strain rate thinning, but not all strain rate thinning liquids show significant elasticity. Deformation can, of course, be in shear or elongation, etc. However, for viscoelastic liquids, the Trouton ratios [see Eq. (5.2)] are higher, often much

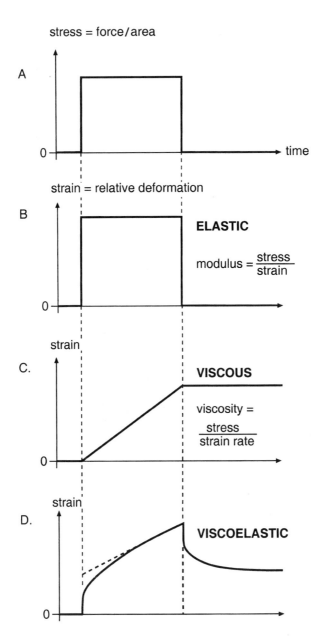

FIGURE 5.8 Various kinds of behavior of a material under stress. (a) Stress applied as a function of time. (b) Resulting strain as a function of time for a purely elastic material. (c) Same, for a Newtonian liquid. (d) Same, for a viscoelastic material.

higher, than those obtained for Newtonian liquids. This implies that it is difficult to reach a high elongation rate during flow of a non-Newtonian liquid with strongly elastic behavior.

Elasticity often stems from the resistance of the bonds in a material to extension or bending. Deformation will thus increase bond energy (see Figure 3.1). Another cause is that the conformational entropy of a material will decrease upon deformation: this occurs especially in polymeric systems and it is further discussed in Chapter 6. In either case, the material will return to its original state upon release of stress—i.e., behave in a purely elastic manner—provided that no bonds have been broken. In a viscoelastic material, part of the bonds break upon deformation. (A purely viscous material has no permanent bonds between the structural elements.)

Dynamic Measurements. Viscoelastic materials thus show an elastic and a viscous response upon application of a stress or a strain. To separate these effects, so-called dynamic measurements are often performed: the sample is put, for instance, between coaxial cylinders, and one of the cylinders is made to oscillate at a frequency ω. Stress and strain then also oscillate at the same frequency. In Figure 5.9, a shear strain (γ) is applied, and it is seen to vary in a sinusoidal manner. If the material is purely elastic, the resulting shear stress (σ) is always proportional to the strain, and the ratio σ_{el}/γ is called the elastic or *storage shear modulus* G' ("storage" because the mechanical energy applied is stored). If the material is a Newtonian liquid, σ is proportional to the strain rate $d\gamma/dt$. Hence the stress is out of phase with the strain by an amount $\pi/2$. The ratio $\sigma_{vis,max}/\gamma_{max}$ is called the viscous or *loss modulus* G'' ("loss" because the mechanical energy applied is lost, i.e., dissipated into heat). The subscripts max denote the highest values of these parameters during a cycle, and these have to be taken because stress and strain are out of phase. The relation with the apparent viscosity is that $\eta_a = G''/\omega$.

For viscoelastic materials, the response is as in Figure 5.9d. Here $\sigma_{ve} = \sigma_{el} + \sigma_{vis}$. A *complex shear modulus* $\tilde{G}$ can be derived, and the following relations hold:

$$\tilde{G} = G' - iG'' \tag{5.12a}$$

$$|\tilde{G}| = \sqrt{(G')^2 + (G'')^2} \tag{5.12b}$$

$$\tan \delta = \frac{G''}{G'} \tag{5.12c}$$

where $i = \sqrt{(-1)}$, $|\tilde{G}|$ is the absolute value of $\tilde{G}$, and δ is the loss angle (see Figure 5.9). The absolute value of the modulus measures total stress over

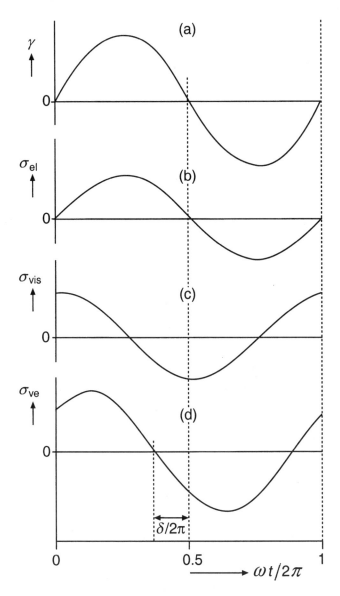

FIGURE 5.9 Illustration of dynamic (oscillatory) measurement of rheological properties. In (a) the applied shear strain (γ) is shown as a function of time t; ω is the oscillation frequency. In (b–d) the resulting shear stress σ is given for an elastic, a viscous, and a viscoelastic response. δ is the phase or loss angle. See text.

total strain. The *loss tangent* tan δ is a measure of the nature of the material. For very small tan δ, the material is solidlike ("elastic" behavior); for large tan δ, it is more liquidlike (more "viscous" behavior).

> *Note* If the material consists of a solid matrix interspersed with a continuous liquid—as is the case for most gels—deformation will always lead to flow of the liquid with respect to the matrix. This causes frictional energy dissipation, hence a finite value of G'' and a finite tan δ. Nevertheless, the system may have a perfect memory, i.e., a response like that in Figure 5.8b. In other words, such a matrix may be called viscoelastic without showing any flow. To be sure, many gels do show some flow upon applying a stress.

The values of G', G'', and tan δ all tend to depend on ω. The most common situation is that G' and G'' increase in magnitude with increasing frequency, though at different rates. Viscoelastic behavior depends on the time scale of deformation, but the relations vary widely among materials. This will be further discussed below.

Relaxation Time. Figure 5.10 illustrates what may happen in stress relaxation experiments. A material is somehow deformed until a given strain is obtained and then kept at that strain (a). In (b) the response of the stress is given. For a Newtonian liquid, the stress will instantaneously go to zero. For a purely elastic solid, the stress will remain constant. For a viscoelastic material, the stress will gradually relax. The figure illustrates the simplest case, where

$$\sigma = \sigma_0 e^{-t/\tau} \qquad\qquad (5.13)^*$$

Here τ is the relaxation time (defined as the time needed for the stress to relax to $1/e \approx 0.37$ of its initial value). In most materials the stress relaxation follows a different course, since there may be a number (a distribution) of relaxation times. Nevertheless, the relaxation time, even if it merely concerns an order of magnitude, is a useful parameter.

Actually, all materials have relaxation times, but these vary tremendously in magnitude. The value is directly related to the proportion of the bonds in the material that spontaneously break per unit time. If this rate is large, τ is small. In liquids, all bonds between molecules break spontaneously, but they have a finite time scale, even if it is only about 10^{-12} s, as in water. This implies that at shorter time scales water would behave as a solid. The bonds in elastic materials may be very long lasting. Solid rocks can have $\tau \approx 10^{14}$ s (a few million years), implying that a rock (e.g., a mountain) will exhibit flow at such time scales. It thus depends on the time scale of observation what we observe or, more precisely, on the Deborah number:

$$De \equiv \frac{\text{relaxation time}}{\text{observation time}} \qquad (5.14)$$

Note This dimensionless ratio is named after the Old Testament prophetess Deborah, who said "the mountains flow before the Lord"; Judges 5:5.

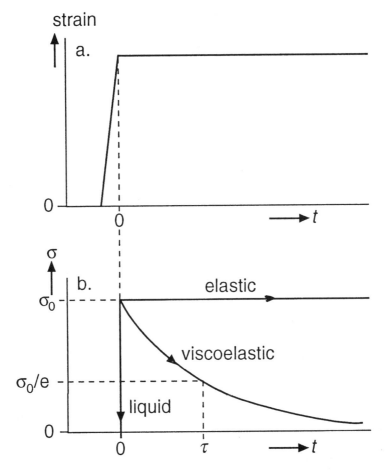

FIGURE 5.10 Stress relaxation. (a) Strain applied to a material as a function of time t. (b) Resulting stress σ for various materials; τ = relaxation time.

If De $\ll$ 1 we call the material a liquid; and at De $\gg$ 1 the material appears solid. It is only if De is of order unity that we observe viscoelastic behavior. Variation of the frequency (ω) in dynamic tests thus amounts to varying the Deborah number.

Nonlinearity. Many rheological measurements, e.g., virtually all dynamic measurements, are carried out in the so-called *linear region*. This means that the strain is proportional to the stress. In practice, however (e.g., during processing), large deformations are often applied, and in most viscoelastic materials the linear region is quite small, i.e., the proportionality of stress and strain is lost at a small value of the strain, say, between 10^{-4} and 0.02. Figure 5.11 gives some examples of the velocity gradient Ψ—be it due to shear or to elongation—resulting from applying increasing stresses (σ). Curve (a) relates to a Newtonian liquid and curve (b) to a strain rate thinning liquid.

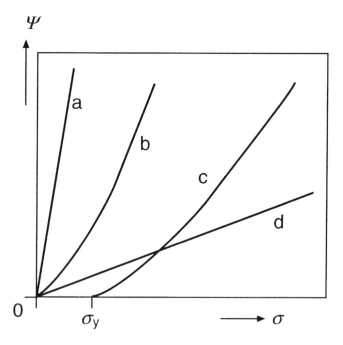

FIGURE 5.11   Examples of the relation between velocity gradient Ψ and stress σ for liquidlike systems. σ_y is yield stress. See text.

Yield Stress. Curve (c) is characteristic for many viscoelastic materials. At small stress, the material behaves elastically, as shown in Figure 5.8b. Above a given stress, however, it starts to flow, and its behavior is like that in Figure 5.8d. This critical stress is called the *yield stress*. Beyond τ_y, the liquid is strain rate thinning. The magnitude of the yield stress varies widely, and values from 10^{-5} Pa (some fruit juices) to over 10^5 Pa (some butters) have been observed. If $\tau_y < 10$ Pa, the yielding behavior tends to go unnoticed at casual observation. (Can you explain this?) Thus the existence of a yield stress does not mean that a high stress is needed to cause a certain flow rate. A high-viscosity Newtonian liquid, like curve (d), may have a much higher apparent viscosity, at least at high σ.

The yield stress is not a clear-cut material property, since its magnitude generally depends on time scale, often markedly so. If the applied stress lasts longer, a lower yield stress is commonly observed. In other words, the Deborah number determines whether or not yielding occurs at a given stress.

A material with a finite yield stress that is made to flow, say in a pipe, may show quite irregular behavior. Near the wall, the velocity gradient, hence the stress, tends to be greatest. This may then imply that near the wall the stress is larger than the yield stress, but further to the center of the tube it is not. In this way, plug flow arises: a high velocity gradient near the wall and a (very) small one in most of the tube. In extreme cases, one speaks of *slip*: the material does not flow, except in a very thin layer near the wall. This can readily occur in several kinds of rheometers. In such a case the results of the test will then merely characterize the strongly altered material near the wall rather than the whole specimen.

Question 1

To check whether an egg is boiled, you can put it on a plate and then set it spinning. If boiled, the egg goes on spinning for a fairly long time; if unboiled, it soon stops. Can you explain this? Can you roughly calculate the time needed for the unboiled egg to stop spinning?

Answer

The inside of an unboiled egg is liquid. In a spinning egg the moving shell will transfer its momentum to the liquid, and a velocity gradient is formed. This means that viscous energy dissipation occurs, thereby decreasing the kinetic energy of the spinning egg. A (hard-)boiled egg is virtually an elastic solid, and a velocity gradient cannot form; the egg merely loses kinetic energy by friction with the plate and the air.

      An order of magnitude calculation is as follows. The kinetic energy is of order mv^2 (Section 4.3.1), where m is the mass of the egg and v the (average) spinning velocity. The energy dissipation rate would equal $\eta_a \Psi^2$ per unit volume [Eq. (5.3)]. For one egg this then becomes $m \cdot \eta_a \Psi^2 / \rho$. The time needed for the kinetic energy to dissipate would be given by kinetic energy over energy dissipation rate. Taking into account that $\Psi \approx v/r$, where r is the effective radius of the egg, we obtain $t \approx \rho r^2 / \eta_a$. Taking in S.I. units $\rho = 10^3$, $r = 0.015$, and $\eta_a = 0.1$ (i.e., 100 times the value for water), the result is 2 s, roughly as observed.

Question 2

Consider a semihard cheese that has the shape of a flat cylinder, height 10 cm. It is put on a shelf and after one month it has gradually sagged to a height of 9 cm; the shape now is roughly like a flat truncated cone. What can you conclude about the rheological properties of the cheese?

Answer

Due to gravitation, the cheese is subject to a stress of $\rho g h$, where ρ = mass density, g the acceleration due to gravity, and h the height of the cheese above the position considered. Sagging means flow, and the cheese must thus have been at a stress above the yield stress it may have. Since the shape of the sagged cheeses is fairly regular, flow must have occurred even close to—say one cm below—the top surface of the cheese. This would correspond to a stress of $10^3 \times 10 \times 0.01 = 100$ Pa. The yield stress must thus be below that value. (Actually, an unequivocal yield stress has never established for semihard cheese.)

      The flow of the cheese is largely elongational. The vertical strain after a month would be about 0.1, and the time needed to achieve this is 1 month $\approx 25 \cdot 10^5$ s, leading to a strain rate (Ψ) of about $4 \cdot 10^{-8}$ s^{-1}. The average stress (σ) in the cheese will have been about $10^3 \times 10 \times 0.05 = 500$ Pa. Consequently, the apparent elongational viscosity (σ/Ψ) would have been of the order of 10^{10} Pa $\cdot$ s.

5.2 DIFFUSION

Diffusion is caused by the thermal motion of molecules (and small particles), which is briefly discussed in Section 4.3.1. The molecules (or particles) can rotate and translate. This section will be restricted to translational diffusion in liquids.

5.2.1 Brownian Motion

When observing a dilute dispersion of small particles (order of 1 μm) under the microscope, one observes—as was originally described by Brown—that the particles display an erratic motion. If there is no convection, every particle makes a random walk, implying that it very frequently alters direction and speed of motion; the change in direction is completely random, that in speed within certain bounds (cf. Section 4.3.1). This Brownian motion is illustrated in Figure 5.12. The figure gives the projection on a plane of the positions of a particle at regular time intervals, connected by straight lines. The positions at 10 times shorter time intervals are also given, and it is seen that the straight lines on the left-hand figure turn into pathways that have an appearance like the total trajectory at left, though at a smaller scale. The average pattern of the Brownian motion is thus independent of the length of the time step considered, unless the latter is extremely short. Actually, the particles may change position, say, 10^8 times per second.

The motion of the particles is due to the heat motion of the solvent molecules, which collide incessantly with a particle. It has become clear that the molecules themselves follow just the same Brownian or diffusional motion as the visible particles, albeit in still shorter time steps. On average—

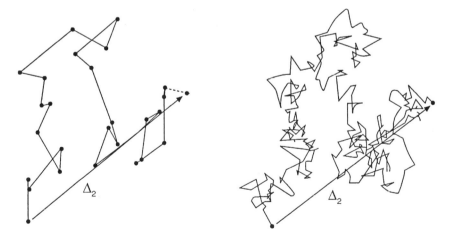

FIGURE 5.12 Brownian motion. Projection on a plane of the trajectory of a gamboge particle, taking observations at constant time intervals. The right-hand side shows the same trajectory, but the time interval is 10 times shorter than at the left-hand side. After observations by Perrin. See text.

i.e., the mean of the distances covered by a large number of identical molecules—the linear displacement x of a molecule is zero (if no flow occurs), but the average of x^2 is finite, and Einstein derived that the *root-mean-square distance* covered in a given direction, Δ_1, will follow the relation

$$\Delta_1 \equiv \ <x^2>^{1/2} = \sqrt{2Dt} \tag{5.15}$$

where t is the time—although with significant statistical variation. It may be noted that the equation considers the absolute value of the distance in a *given direction*, i.e., the projection of the real distance covered on a straight line of given orientation (one-dimensional). The projection on a given plane (two-dimensional), as in Figure 5.12, yields $\Delta_2 = \sqrt{2} \cdot \Delta_1$, and the full distance covered in three dimensions $\Delta_3 = \sqrt{3} \cdot \Delta_1$.

The proportionality constant D is called the **diffusion coefficient** (S.I. unit $m^2 \cdot s^{-1}$). Einstein also derived that $D = k_B T / f$, and taking Stokes's expression for the friction factor f for spheres, the relation becomes

$$D = \frac{k_B T}{6\pi\eta_s r} \tag{5.16}$$

where η_s is the viscosity of the solvent and r is the radius of the molecule or particle. For nonspherical species we need the hydrodynamic radius, which mostly must be experimentally determined.

It follows from Eqs. (5.15) and (5.16) that a big molecule or particle will travel over smaller distances—i.e., diffuse at a slower rate—than a small one, although the pattern of the motion is just the same. For visible particles, Eq. (5.15) can be verified by microscopic observation, and it is found to be exact. Some values for the diffusion coefficient and for root-mean-square distances traveled are given in Table 5.3.

Question

Can you calculate from values given in this chapter the hydrodynamic radius of a sucrose molecule?

Answer

0.46 nm.

TABLE 5.3 Diffusion Coefficients (D) of Some Molecules and Particles in Water at Room Temperature and Times Needed for These Species to Diffuse over Various Distances in a Given Direction, Δ_1

Species	D $m^2 \cdot s^{-1}$	Diffusion time for $\Delta_1 =$			
		10 nm	1 μm	0.1 mm	1 cm
Water	$1.7 \cdot 10^{-9}$	0.03 μs	0.3 ms	3 s	8 h
Sucrose	$4.7 \cdot 10^{-10}$	0.1 μs	1 ms	11 s	30 h
Serum albumin	$6.1 \cdot 10^{-11}$	0.8 μs	8 ms	82 s	10 d
Emulsion droplet[a]	$4.2 \cdot 10^{-13}$	0.1 ms	1 s	3 h	4 y

[a] Diameter 1 μm.

5.2.2 Mass Diffusion

So far, we have tentatively assumed that all molecules and particles are randomly distributed throughout the volume available. If there are two (or more) substances present, say solute and solvent, concentration differences can occur for a number of reasons. If so, the heat motion causes the molecules to attain a (more) random distribution, i.e., the concentration differences will eventually disappear, except over very small distances. The process is thus entropy driven. The rate at which it occurs is generally described by Fick's laws. Fick postulated in his first law that the **diffusional transport rate** is proportional to the concentration gradient according to

$$\frac{dm}{dt} = -DA\left(\frac{\partial c}{\partial x}\right)_t \tag{5.17}$$

where dm is the amount of solute transported in the direction of x through the area A of a cross section perpendicular to x. The amount m can be given in any unit of substance, and the concentration c must be taken in the same units per unit volume. From Eq. (5.17) Fick's second law can be derived, which gives the change in c with time at any place as a function of the local concentration gradient

$$\left(\frac{\partial c}{\partial t}\right)_x = D\left(\frac{\partial^2 c}{\partial x^2}\right)_t \tag{5.18}$$

Total mass transport and concentration profiles as a function of time can be obtained from these differential equations. The solution greatly depends on the boundary conditions, i.e., the geometrical constraints. For the fairly simple case of diffusion through an infinite plane surface, on one side of which a constant concentration c_1 is maintained, whereas at the other side initially $c = 0$, the amount of mass transported is given by

$$m = 2Ac_1\sqrt{\frac{Dt}{\pi}} \tag{5.19}$$

The concentration as a function of the distance x from the surface then is given by

$$c(x) = c_1(1 - \mathrm{erf}\, y) = c_1\left[1 - \frac{2}{\sqrt{\pi}}\int_0^y \exp(-z^2)\mathrm{d}z\right]$$
$$y = \frac{x}{2\sqrt{Dt}} \tag{5.20}$$

where z is an integration variable; erf y, the error function, is tabulated in the Appendix J. Some results are illustrated in Figure 5.13, which also gives the distance x' over which the original concentration difference is precisely halved. From Eq. (5.20) it follows that the relation with the time needed for this to occur, the *halving time* $t_{0.5}$, is given by

$$(x')^2 \approx Dt_{0.5} \tag{5.21}^*$$

This is a very useful equation. Although it is often only approximately correct, because the boundary conditions are not fully met, it gives the order of magnitude of the time scales or distance scales one has to reckon with. For many molecules in water, D is of order $10^{-10}\,\mathrm{m^2 \cdot s^{-1}}$. This yields for $x' = 1\,\mu\mathrm{m}$ a halving time of 10 ms; for 1 mm it is 10^4 s or about 17 min, and for 1 m it is 10^{10} s or about 300 years. When a cube of sugar is put into a cup of tea, and if convective transport of dissolving sugar would not occur, it would take on the order of several times $t_{0.5}$, i.e., several months, before the sugar is more or less evenly distributed.

In practice, however, convection will of course occur, due to currents arising from temperature differences, if not due to stirring, and convection greatly enhances mixing rate. In Section 5.1.1 flow is considered. It follows from Figure 5.1b that mass transport (and transport of heat, for that matter) in a direction perpendicular to the flow rate will not be enhanced by simple shear flow. Elongational flow is more effective to achieve mixing. A look at Figure 5.4 will make it clear that turbulent flow is superior in speeding up mixing; the thickness of the laminar boundary layer will often be a limiting factor for the transport rate perpendicular to the overall flow direction. Pure diffusional transport can be observed by putting the sugar cube mentioned in a weak polysaccharide gel, which does not allow convection, while hardly slowing down diffusion (see Section 5.3.2).

On the other hand, diffusion of a substance into an emulsion droplet of a few fm in radius would take far less than a second. Diffusion thus proceeds very fast at very small distances and takes a very long time at long distances; it cannot be expressed as a linear rate, in $m \cdot s^{-1}$. If a substance has to diffuse into a lump of material, say salt in a loaf of cheese by immersing it in brine, similar rules hold. In the beginning of the process, the quantity of salt taken up is proportional to the square root of time, according to Eq. (5.19): if it takes one day to obtain a total salt content of 1% in the cheese, it will take four days to obtain 2%.

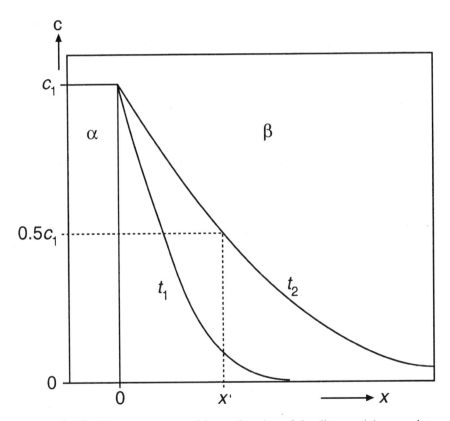

FIGURE 5.13 The concentration (c) as a function of the distance (x) as a solute diffuses from a liquid α into a material β, given for two times after diffusion started, where $t_2 \approx 6 \times t_1$. The amount of α is considered infinite and the liquid α remains ideally mixed.

Some Complications. Above it is implicitly assumed that the diffusion coefficient in the Stokes–Einstein relation (5.16) equals that in Fickian diffusion. This is an oversimplification:

1. In Eq. (5.17) and derived equations, activities rather than concentrations (c) should be used. If the activity coefficients (γ) differ significantly from unity, this can be accounted for by multiplying the diffusion coefficient by a factor $(1 + \mathrm{d} \ln \gamma / \mathrm{d} \ln c)$.

2. Individual ions (say, Ca^{2+}) cannot diffuse independently since they must be accompanied by counterions (say, SO_4^{2-} or $2Cl^-$), to keep the solution electroneutral. (The distance between an ion and its counterion will be on the order of the Debye length $1/\kappa$: see Section 6.3.2.) Moreover, neutral species (say, $CaSO_4$) will also be transported if they show a concentration gradient. The diffusion coefficient of an ionizable component thus is a kind of average of those of the species involved.

3. Equation (5.15) concerns self-diffusion, and mass transport involves mutual diffusion: if the solute diffuses in one direction, then the solvent does so in the opposite one. A kind of average diffusion coefficient must be taken, and only for low solute concentrations is it about equal to the self-diffusion coefficient of the solute, taking the viscosity of the solvent. This is primarily because solute concentration will generally affect the viscosity of the solution; in most cases it is higher than that of the solvent.

4. Mutual diffusion may go along with a change in volume, since many solute–solvent mixtures have a different volume (mostly a smaller volume) than the sum of that of both components. This implies that the frame of reference moves; for instance, the original interface between two layers of different concentration moves, and this means that some transport by flow occurs also.

Question 1

A way to study diffusion rate in liquids not hindered by convection is to separate the liquids into two large stirred compartments by means of a disk of sintered glass. Assume that the one component contains a 10% sucrose solution and the other initially is water. Both compartments are stirred. The sintered glass disk is 3 cm in radius and 3 mm in thickness; the void volume (i.e., available for the liquid) fraction in the disk is 0.25. How much time would it roughly take for 1 gram of sucrose to diffuse through the disk?

Answer

In first approximation, Eq. (5.17) would be directly applicable. From Table 5.3 we derive that for sucrose $D = 4.7 \cdot 10^{-10} \, m^2 \cdot s^{-1}$. A would equal $0.25 \cdot \pi r^2$—where the factor 0.25 derives from the limited void fraction—which makes $0.0007 \, m^2$. The concentration gradient can be taken as the concentration difference $(100 \, kg \cdot m^{-3})$ over the thickness of the disk (0.003 m). This yields for dm/dt about $11 \cdot 10^{-9} \, kg \cdot s^{-1}$. To transport 0.001 kg then takes about 90,000 s, or about one day.

> *Notes*   Inspection of Figure 2.3 shows that the sucrose activity coefficient equals unity in a 10% solution. Nevertheless, correction is needed. In the first place, the diffusion coefficient will be smaller than assumed, because of the increased viscosity of the liquid. From Table 5.2 it can be estimated that this would amount to a factor of about 1.4 for 10% sucrose. In the second place, the concentration gradient will be smaller than assumed, because the channels in the porous glass will be tortuous. Altogether the transport may take almost two days rather than one. See also Figure 5.17.

Question 2

The same setup as in Question 1 is used to study the diffusion rate of sodium dodecyl sulfate. In one experiment, the first compartment contains an 8 millimolar solution, the other one water. A diffusion coefficient of about $5 \cdot 10^{-10} \, m^2 \cdot s^{-1}$ is observed. In a control experiment, one compartment contains 16 millimolar and the other one 8 millimolar SDS. Now the diffusion coefficient turns out to be smaller by a factor of about 5, despite the concentration gradient being the same. How is this to be explained?

Answer

Inspection of Figure 2.8 shows that 8 millimolar SDS would equal the critical micellization concentration. In the first experiment, the concentration gradient of the SDS molecules (or ions, rather) would thus have been as assumed. In the second experiment, both compartments would have contained the same concentration of free SDS, the concentration gradient being merely be due to SDS in micellar form. The micelles clearly have a larger hydrodynamic radius than free SDS species, hence the slower diffusion.

> *Note*   There are some other complications, such as ion activity coefficients being < 1.

5.2.3 Heat Transfer

If a temperature gradient exists in a liquid—and this is also valid for a solid—heat motion of the molecules will cause the temperature to become equal throughout; after all, temperature is proportional to the average kinetic energy of the molecules (Section 4.3.1), and the molecules collide with each other, thereby transferring momentum and thus smoothing out temperature differences. This **diffusion of heat** (or conduction) proceeds in the same way as diffusion of mass. The diffusion coefficient for heat D_H is generally called the thermal diffusivity. Equations (5.18), (5.20), and (5.21) remain valid, replacing c by T. For the transport of heat, the quantity of heat per unit volume $(\rho c_p T)$ has to be used instead of concentration (quantity of mass per unit volume). Thereby, Fick's first equation is transformed into the *Fourier equation*

$$\frac{dq}{dt} = -D_H A \left(\frac{\partial(\rho c_p T)}{\partial x} \right)_t = -\lambda A \left(\frac{\partial T}{\partial x} \right)_t$$
$$\lambda = D_H \rho c_p \qquad\qquad\qquad (5.22)$$

where q is the amount of heat (J), λ the thermal conductivity $(W \cdot K^{-1} \cdot m^{-1})$, ρ the mass density $(kg \cdot m^{-3})$, and c_p the specific heat at constant pressure $(J \cdot kg^{-1} \cdot K^{-1})$. The diffusion equations for heat also apply to solids, since momentum transfer between molecules in a solid is about as frequent as in liquids.

D_H equals about $10^{-7} \, m^2 \cdot s^{-1}$; for most food components this is correct within a factor of two. For crystalline material the value tends to be higher than for a liquid, and ice even has $D_H \approx 10^{-6} \, m^2 \cdot s^{-1}$ (most metals are in the range 10^{-5}–10^{-4}). Since the mass diffusion coefficient nearly always is $< 10^{-9} \, m^2 \cdot s^{-1}$, the diffusion of heat is at least 100 times as fast as the diffusion of mass. Nevertheless, it is still slow at distances longer than a few mm.

Heat can also be transferred by other mechanisms. The most common way is by **convection**, where cold and hot masses of liquid are mixed, so that heat diffusion is merely needed at very small distances; the rate of transfer is thereby greatly enhanced. Heat can be transported by **radiation** from a hot to a cold surface, where both surfaces are separated by a gas phase or by vacuum. We will not discuss the theory but merely mention that the rate of transfer is proportional to $(T_1^4 - T_2^4)$, where T_1 and T_2 are the (absolute) temperatures of both surfaces.

Another mechanism is heat transfer by **distillation**. It is illustrated by the heat transport into a loaf of bread during baking. Applying Eq. (5.21) with $x' = 4 \, cm$ and $D_H = 10^{-7} \, m^2 \cdot s^{-1}$, gives a halving time for heat diffusion to the center of the loaf of about 4 hours, far too long. The viscosity of the dough is far too high to allow significant convection. However, the dough contains a high volume fraction (about 0.8) of fairly large gas cells, for the most part containing CO_2, and in these cells convection is possible. The value of c_p of the gas is too small to allow transport of sufficient heat, but water will evaporate at the hot side of a gas cell (consuming heat) and condense at the cool side (giving up heat). In this way, sufficient heat can be transported. It implies that also mass is transported; indeed, the water content in the center of freshly baked bread is higher by about 4% water than that in the outer layer. (This difference is in fact one of the quality marks of fresh bread; it disappears—by distillation and diffusion—when keeping the bread for a day or so in a plastic bag.)

Question

It has been observed by polarized light microscopy that some emulsion droplets (diameter about $5\,\mu m$), containing a triglyceride oil that can partly crystallize at room temperature, have crystalline fat in their outer layer. This has been ascribed to the cooling of the droplets occurring from the outside, so that crystallization would start there. Is this a reasonable explanation?

Answer

No. Applying Eq. (5.21) with $x' = 1\,\mu m$, we obtain a halving time for the temperature difference of 10^{-5} s. This implies that it would need a cooling rate of order 10^5 $K \cdot s^{-1}$ to achieve a temperature difference of only 1 K. This rate is widely outside the range attainable in everyday life. Even if such fast cooling could be achieved by special apparatus, the temperature gradient would not be from the outside to the inside of a droplet, since that would imply that D_H is much smaller in oil than in water, whereas the two values are about equal.

5.3 TRANSPORT IN COMPOSITE MATERIALS

A liquid food of not very high viscosity can be stirred to speed up transport of heat or mass. Even if it contains dispersed particles, these mostly are small enough to allow rapid diffusion in or out of them (cf. Table 5.3). Many foods, however, are solidlike, and there are even some that contain a lot of water (cucumbers, for example, contain about 97% water); transport generally is by diffusion and in some cases by—greatly hindered—flow. Some examples are

During several processing operations, like drying, extraction, and soaking, diffusion is the rate determining step.

When a food is kept, it may lose substances by diffusion or leaking, such as water or flavor components, or it may take up substances from the environment, e.g., from packaging material.

When a food is kept, the concentration of solutes, which may at first be uneven, becomes (slowly) evened out. This may be of considerable importance for the eating quality if it concerns flavor components: see the introduction to Chapter 9. Likewise, color substances may become evenly distributed—e.g., moving from the fruit to the surrounding yoghurt; and so may water—e.g., from the inside of the bread to its crust, by which the crust loses its crispness.

These processes often are slow, and transport rates may be difficult to predict. Below, a few more or less idealized cases will be considered.

5.3.1 Flow Rates

Through a porous material, liquid may flow, albeit often sluggishly. It is useful to consider the material as a solid matrix, containing several capillary channels or pores. In practice, the pores are always narrow enough and liquid velocity is slow enough to ensure that flow is laminar. The rate of flow as caused by a pressure gradient through a cylindrical capillary is given by the law of Hagen–Poiseuille, which can be written as

$$\bar{v} \equiv \frac{Q}{\pi r^2} = -\frac{\triangle p}{\triangle x} \cdot \frac{r^2}{8\eta} \tag{5.23}$$

where Q is the volume flow rate $(m^3 \cdot s^{-1})$, r the radius of the capillary, p the pressure, and x the distance along the capillary. The velocity profile is illustrated in Figure 5.4a. Equation (5.23) is commonly applied in the determination of viscosity in a capillary viscometer, where the pressure difference mostly is caused by the weight of a column of the liquid.

For flow through a porous material, the superficial flow velocity is given by **Darcy's law**,

$$\bar{v} \equiv \frac{Q}{A} = -\frac{B}{\eta} \cdot \frac{\triangle p}{\triangle x} \tag{5.24}$$

where A is the cross-sectional area (perpendicular to x) through which the liquid flows and B is the permeability. The pressure difference $\triangle p$ acting on the liquid may be due, for instance, to

Gravity, as in the percolation of hot water through ground coffee

An external mechanical pressure, as in ultrafiltration

Capillary suction, as in tea being taken up into a sugar cube (see Section 10.6.3)

An osmotic pressure difference, as over the semipermeable membrane in the apparatus of Figure 2.4 before equilibrium has been reached

An endogenous tendency of the material to shrink, as in the syneresis of renneted milk

The **permeability** may be considered as a material constant, provided that the Reynolds number Re $\ll 1$, which is nearly always the case (approximately, Re $= \bar{v} \cdot \sqrt{B} \cdot \rho / \eta$). By comparing Eq. (5.24) with (5.23), it follows that B (unit m^2) is in first approximation proportional to the square of the diameter of the pores in the material and to the surface fraction of pores in a cross section of the material. In most real materials, the pore diameter shows considerable spread, and Eq. (5.23) shows that Q is about proportional to r^4; hence by far most of the liquid will pass through the widest pores. Moreover, the pores tend to be irregular in shape and cross section, they are tortuous and bifurcate, and some may have a dead end. The permeability may even be anisotropic, i.e., be different in different directions (see Section 9.1).

Accordingly, it is not easy to predict B from the structure of the material. Numerous relations have been proposed, and the one used most is the Kozeny–Carman equation

$$B \approx \frac{(1 - \varphi)^3}{5 A_{sp}^2} = \frac{(1 - \varphi)^3 d^2}{180 \varphi^2} \tag{5.25}$$

where φ is the volume fraction of matrix material and A_{sp} the specific surface area of the matrix (in m^2 per m^3 of the whole material). The part after the second equals sign is only valid if the matrix consists of an aggregate of spheres of equal diameter d. The equation was derived for powders, and it has quite limited validity, but it serves to illustrate two points. The permeability very strongly decreases if the volume fraction increases (or the void volume decreases), especially for small φ; and it decreases if the specific surface area increases, i.e., as the structural elements in the composite material become smaller. Some very approximate magnitudes of B are

Ground coffee	10^{-8} m^2
Renneted milk gel	10^{-12}
Polysaccharide gel	10^{-17}

The very low permeability of several gels is the main reason why they hold water tenaciously.

For very narrow pores, say 10 nm, the permeability depends on molecular size. This is comparable to, but not the same as, the hindered diffusion discussed in the following section.

Question

A company stores margarine in 250 g packages in a cold room, temperature 5°C. The packages are stacked to a height of 1 m. They can be stored for some weeks without problems. One day, the cooling does not work properly and the temperature is 15°C. Nevertheless, it is decided to store the margarine in the room. After a day it is observed that in the packages near the bottom of the stack oiling off has occurred: the inside of the wrappers shows free oil, about a ml per package. Can you roughly estimate the permeability of the crystal network in the margarine? Why would no oiling off be observed at 5°C? You may consider margarine to consist of a space-filling network of aggregated fat crystals, interspersed with a continuous oil phase.

Answer

Assuming that the surface area of a package is 200 cm^2, 1 ml of oil after one day would correspond to a superficial oil velocity of 1/200 cm per day, or about $5 \cdot 10^{-10}$ m$\cdot$s^{-1}. According to Table 5.2, the viscosity of the oil would be about 0.1 Pa$\cdot$s. The pressure exerted by the stack equals $\rho \, g \, h$, which gives for $h = 1$ m about 10^4 Pa. The distance over which the oil has to flow would on average be about 2 cm. Inserting these values in Eq. (5.24) yields a permeability of 10^{-16} m^2. This is indeed the order of magnitude experimentally determined for such fats.

A variable in Eq. (5.24) that is significantly different at 5°C is the oil viscosity; it would be higher by a factor of about 2.8. Moreover, the margarine contains more solid fat at a lower temperature, which may decrease the value of B by a factor of 2.5. Altogether, the flow rate may be about 7 times smaller, which would imply that also at 5°C oiling off would occur after a week. However, for oil separation to occur, the crystal network has to yield: a completely rigid network would carry all the stress (here at most 10 kPa) applied to it, leaving no possibility for shrinkage and thereby for oil separation. Presumably, this means that owing to the higher solid fat content at 5°C, the yield stress is above 10 kPa over a time scale of weeks, whereas it is below that value at 15°C over a time scale of hours. This implies that (slight) sagging must occur in the lowest packages.

5.3.2 Effective Diffusion Coefficients

Even for such a simple case as the diffusion of a small-molecule solute, say sugar, in a gel made of a dilute polymer solution, say gelatin or pectin, the Stokes–Einstein relation (5.16) cannot be applied. This is because the macroscopic viscosity of the system greatly differs from the microscopic viscosity as "sensed" by the diffusing molecules: they just diffuse around the strands of the gel. The effect is illustrated in Figure 5.14, and it is seen that the discrepancy may be by several orders of magnitude, even for quite low concentrations of matrix material. To be sure, the viscosity of a gel is an ambiguous property; it concerns in fact an apparent viscosity determined at a stress larger than the yield stress of the gel. But the example serves to illustrate the point: the effective viscosity cannot be obtained from macroscopic measurements. Nevertheless, diffusion in a gel is slower than in pure solvent, the more so for a higher concentration of matrix material.

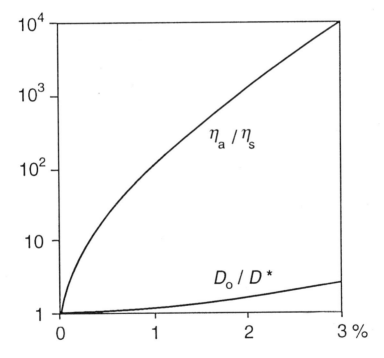

FIGURE 5.14 Example of the effect of polysaccharide concentration (%) on the apparent viscosity of the polysaccharide–water mixture (η_a) and of the effective diffusion coefficient (D^*) of a solute in the mixture. D_0 is the diffusion coefficient in the absence of polymer, η_s the viscosity of the solvent.

Numerous theories have been developed for diffusion in compound systems, and they all depend, of course, on the structure of the matrix; in other words, they are model dependent. In Section 5.3.3 a very simple model of macroscopic regions of different diffusivity will be given. Here, a microscopic approach is taken, where part of the material, the matrix, is inaccessible to the diffusing species; the matrix has pores filled with solvent or solution, through which diffusion can occur. Quite generally, the following factors affecting the **effective diffusion coefficient** D^* can be distinguished:

1. If the *volume fraction* of the matrix is φ, only a fraction $(1 - \varphi)$ of the material is available to the solute. This does not necessarily affect D^*, if the concentration of solute is taken in the solvent, and if this is also accounted for in the boundary conditions used to solve Fick's equations. However, φ is often unknown, nor readily determined. For one thing, the effective φ may be larger than the nominal value, because some of the solvent is not available to the solute (see Section 8.3). In such cases, what will be experimentally observed from the mass transport rate is a smaller D^*.

2. *Tortuosity*: the diffusing molecules have to travel around the obstacles formed by the matrix, thereby increasing the effective path length, hence decreasing D^*. For a low value of φ the effect is small. At high φ, say more than 0.6, the correction factor would be of order $(\pi/2)^{-2} \approx 0.4$.

3. *Constriction*: if the pore radius r_p is not much larger than the radius of the diffusing solute molecule r_m, the molecule frequently collides with the pore wall whereby its diffusion is impeded, the more so for r_m/r_p closer to unity. It is difficult to predict the magnitude of this effect. It can be roughly estimated by the semiempirical Renkin equation, which applies to diffusion of reasonably spherical molecules (or particles) in a straight cylindrical capillary:

$$\frac{D(\lambda)}{D_0} = (1 - \lambda)\,(1 - 2\lambda + 2\lambda^3 - \lambda^5)$$

$$\lambda = \frac{r_m}{r_p} \tag{5.26}$$

The result is given in Figure 5.15, and it is seen that the effect is very strong. Also more sophisticated theories, in which the pores considered are more realistic, show that the ratio λ often is the most important factor limiting diffusion rate in porous materials.

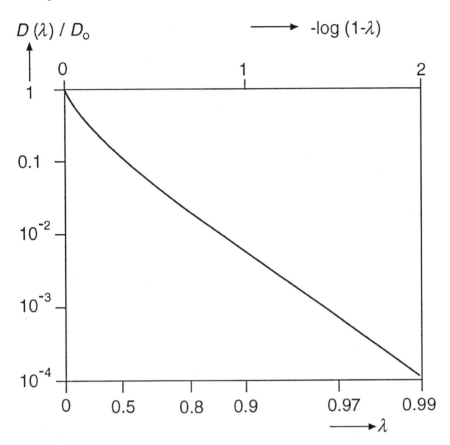

FIGURE 5.15 Relative diffusion coefficient $D(\lambda)/D_0$ as a function of $\lambda = r_m/r_p$ for diffusion of spherical molecules of radius r_m through a cylindrical capillary of radius r_p; calculated according to Eq. (5.26).

4. *Electrostriction.* Ions diffusing (or being subject to flow, for that matter) through a matrix that carries an electrical charge, will be subject to an additional retarding effect, caused by electrostatic attraction. The strength of this effect depends on the ratio of the average distance between ion and charged surface over the Debye length (thickness of the electric double layer; Section 6.3.2). Consequently, the retarding effect will be stronger for narrower pores and for a lower ionic strength.

In most foods of high or intermediate water content, D^* is not very much smaller than D_0, rarely by more than a factor 10. For example, for diffusion of salt in water $D \approx 10^{-9}$ m$^2 \cdot$s^{-1}, whereas in meat D^* is about half that value, and in hard cheeses about 0.2 times D. D^* of sucrose in most fruits is about 10^{-10} m$^2 \cdot$s^{-1}. For drier foods, the differences become larger. All factors mentioned above give a stronger effect for a higher φ value. Figure 5.16a shows some results on D^* for water in foods of variable water content, and it is seen that the effects are large; some studies on drying indicate an even stronger reduction of D^* at very low water content. It should also be considered that the activity coefficient of water may be materially decreased at very low water content; hence a smaller activity gradient, hence slower diffusion. Constriction, i.e., Factor 3 mentioned above, implies that the diffusion of larger molecules will be hindered more than that of smaller ones. This is illustrated by results shown in Figure 5.16b. See also Section 8.4.1.

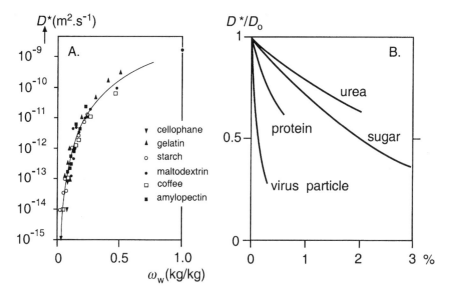

FIGURE 5.16 Examples of hindered diffusion. (a) The effective diffusion coefficient (D^*) of water in various materials as a function of their mass fraction of water (ω_w). (Adapted from Bruin and Luyben; see Bibliography.) (b) The effective diffusion coefficient (D^*) of some molecules and a virus in polymer gels of various concentrations (%). (Very approximate results, obtained from Muhr and Blanshard; see Bibliography.)

Common sense suggests, and Eq. (5.26) predicts, that $D^* = 0$ if $r_m \geqslant r_p$. This is, however, not the case. Even in a crystalline solid, diffusivity is finite, though very small, e.g., $10^{-22}\,\mathrm{m}^2 \cdot \mathrm{s}^{-1}$. (Over a distance of 1 nm $t_{0.5}$ would then equal about 3 hours, and over 1 mm 300 million years.) At the scale of molecules, pores do not have a fixed size: even immobilized structures exhibit Brownian motion and this leads to fluctuating pore sizes, occasionally letting even a fairly large molecule pass. This is quite obvious in polymer gels, and something similar happens in dry materials.

All these observations imply that the prediction of diffusivity in composite materials is far from easy and incompletely understood. Specific interactions on a molecular scale between solute and matrix can also affect diffusivity.

Partial Osmosis. If constriction is significantly stronger for larger molecules—say, of a solute—than for small ones—say, of the solvent—the diffusional fluxes are not anymore equal and opposite, as is the case in unhindered diffusion (barring volume change upon mixing). This results in partial osmosis. For example, if a piece of fruit is put in a concentrated sugar solution, water diffuses faster out than sugar diffuses in, leading to a decrease in volume. A prerequisite then is that the piece of fruit can shrink, which means a compression of the cellular structure (see Section 17.5).

This is a way of "drying" the fruit, called *osmotic dehydration*. It is, however, not pure osmosis, since the fruit is not impervious to sugar: it only diffuses slower, say by a factor of five, than water. If the fruit is left in the sugar solution, it will eventually obtain the same sugar content, relative to water, as in the said solution. The fruit then becomes candied besides dried, which is commonly applied to dates. Several other examples could be given; for instance, when salting cheese by immersing it in brine, the loss of water from the cheese is generally more than twice the uptake of salt.

Question

In the research department of a food company, the diffusion of salt into meat products is studied. Large pieces are immersed in concentrated brine (salt concentration c_1), and the salt uptake per unit surface area is determined as a function of brining time (t). As expected, it is proportional to $\sqrt{t}$. By means of Eq. (5.19), the effective diffusion coefficient D^* is estimated. For lean pork a value of $2.2 \cdot 10^{-10}\,\mathrm{m}^2 \cdot \mathrm{s}^{-1}$ results, for back fat (untrimmed bacon) only $10^{-11}\,\mathrm{m}^2 \cdot \mathrm{s}^{-1}$. As a check it is determined in a separate experiment what the salt content is at a distance of 1 cm from the outside after 5 days of brining. By using Eq. (5.21), it is expected that in the lean pork c (expressed per kg water) will equal nearly $0.5c_1$, whereas it would be negligible for the back fat. However, for the meat the value is somewhat above $0.5c_1$, and for the fat about half that value, i.e., far from negligible. Can you explain these results?

Answer

Above, the factors affecting the magnitude of D^* have been discussed. Factor 1 is that part of the volume of the material is not available for transport: salt dissolves in water and not in the dry matter of the meat. This then implies that the determined values of D^* are too small. If the volume fraction of water equals $1 - \varphi$, then $D^* \approx D_{\text{experimental}}/(1 - \varphi)$. Using this relation, calculate the expected values of c/c_1 at 1 cm after 5 days brining, assuming for lean pork $\varphi = 0.4$ and for back fat $\varphi = 0.9$. The results are 0.56 and 0.28, respectively.

5.3.3 Diffusion Through a Thin Layer

Different regions in a food may have different effective diffusion coefficients for a solute. Moreover, the solubility of the solute may differ. A simple example will be discussed.

Figure 5.17 depicts a situation in which a solute diffuses from left to right because of a difference in concentration in two compartments of phase α, which are separated by a (thin) layer of a phase β. For sake of simplicity, it is assumed that within each entire compartment the concentration is everywhere the same. The solubility is not the same in α and β, and the distribution coefficient or partition ratio (see Section 2.2.2) $K_D = c_\beta/c_\alpha$. Solution of Eq. (5.17) now yields the steady-state equation

$$Q = \frac{D_\beta^* A K_D \triangle c}{x_2 - x_1} \tag{5.27}$$

where Q is the mass flux (amount of mass transported per unit time) through an area A. The hindrance of the diffusional transport by the layer of β is thus stronger for a thicker layer, a smaller value of D^* in the layer, and a lower value of K_D—i.e., a lower solubility of the solute in β. In the example of Figure 5.17 the distribution coefficient is about 0.5, but far smaller values can occur. For instance, in a layer of fat (i.e., a dispersion of fat crystals in oil) between two aqueous compartments, K_D may be as small as 0.002. Hence such a layer will greatly impede the diffusion of hydrophilic components from the one to the other compartment; this is commonly applied in some kinds of pastry to slow down transport of flavor or color substances.

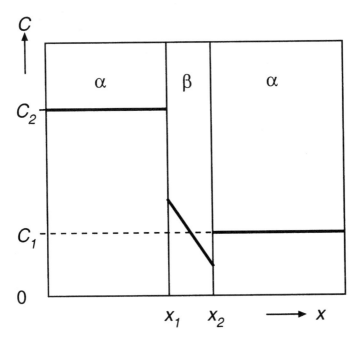

FIGURE 5.17 Example of the concentration (c) of a component as a function of distance (x) as it diffuses through a layer of another phase (β). See text.

Many packaging materials act in much the same way. Another application of Eq. (5.27) has been the development of "edible films," used to separate compartments or to enrobe foods so that water loss or exchange of flavor components is considerably diminished. In practice, the situation may be more complicated. The concentrations on either side of the film do not stay constant, the geometry may be different, there may be more layers of different properties, etc.

Eq. (5.27) shows that the diffusional resistance becomes very small if the layer is very thin. An adsorption layer on the interface between phases, which mostly is of the order of a nm in thickness, thus causes a negligible resistance to diffusion. However, highly condensed lipid bilayers, as are present in cell membranes, may provide a significant diffusive resistance to large hydrophilic molecules, although the membranes are quite thin; water itself can pass relatively fast. In a natural tissue there are numerous membranes in parallel; hence, the total resistance to diffusion may be substantial. In a living cell, mechanisms for active transport of specific substances over a membrane against a concentration gradient exist; thereby, passage of these substances through a tissue will be greatly hindered.

5.4 RECAPITULATION

Rheologists study the relations between stress (force per unit area) and strain (relative deformation) of a material, generally as a function of time scale or rate of strain. For elastic solids the modulus, i.e., stress over strain, is a characteristic parameter; for pure liquids it is the viscosity, i.e., the ratio of stress over strain rate. An elastic solid regains its original shape after the stress is released and the mechanical energy used to deform it is regained; a pure liquid retains the shape attained and the mechanical energy is dissipated into heat.

Flow.   Liquids thus flow when a stress is applied. The flow is laminar if the streamlines are straight lines or smooth curves. The flow is characterized by its velocity gradient (which equals the strain rate) and the type of flow. Flow type can e.g. be rotational or elongational. Best known is simple shear flow (parallel streamlines), which has an elongational and a rotational component. In elongational flow, the velocity gradient is in the direction of flow, in simple shear, normal to the direction of flow. The elongational viscosity is larger than the common or shear viscosity. The flow exerts a frictional stress on particles or other objects; this stress equals viscosity times velocity gradient. The stress can cause particles to rotate and to become deformed.

In turbulent flow, the streamlines are erratic and eddies occur. This leads to a strong mixing effect and to a greater energy dissipation rate. Flow is turbulent if the Reynolds number (proportional to flow velocity, channel dimension, and the inverse of viscosity) is greater than a critical value. Because of Bernoulli's law, which states that the sum of pressure and kinetic energy is constant, the strong velocity fluctuations in turbulent flow induce pressure fluctuations. These cause, in turn, inertial forces to act on any particles present.

Viscosity.   The viscosity of liquids varies among molecular configurations and generally increases with molar mass; it decreases with increasing temperature. The addition of particles or of a solute leads to an increased viscosity, the more so for a higher volume fraction. The increase is generally (almost) independent of particle size, but it does depend on particle shape and on the inclusion of solvent by the particles, particle aggregates, or polymer molecules, since these variables affect the effective volume fraction. The extent to which a substance can increase viscosity is expressed in the intrinsic viscosity (relative increase of viscosity per unit solute, extrapolated to zero concentration). The maximum volume fraction possible for the particles present also affects viscosity, as the latter becomes infinite at the said maximum.

In a Newtonian liquid the viscosity does not depend on the stress or the strain rate applied, but many liquids are non-Newtonian. Many liquid dispersions are strain rate thinning, i.e., the "apparent" viscosity decreases with increasing strain rate. Some of these dispersions are also thixotropic: the apparent viscosity decreases during flow at constant strain rate and slowly increases again after flow has stopped.

Viscoelasticity. Some strain rate thinning liquids, especially polymer solutions, are also viscoelastic. This means that after exerting a stress on the liquid, it deforms at first elastically and then starts to flow; upon release of the stress it regains part of the original shape.

If a given deformation is applied to a viscoelastic material, the stress slowly relaxes; the characteristic time for this is called the relaxation time. The Deborah number (De) is defined as the ratio of this relaxation time over the observation time. For a solid De is very large, for a liquid very small, and for a viscoelastic material of order unity. It thus depends on the time scale of observation whether we call a material solid or liquid. Several foods appear to be solid at casual observation, but show flow during longer observation.

Another phenomenon is that a material may turn out to have solid properties if a small stress is applied, but starts to flow above a certain stress, called the yield stress. Its magnitude varies widely among foods, some apparently true liquids having a small yield stress and some apparently rigid solids flowing above a large stress. Its magnitude also depends on the time scale, being smaller for a longer observation time.

Viscoelastic materials are often studied by means of dynamic, i.e., oscillatory, rheological tests. These yield a complex shear modulus, the resultant of a (real) elastic or storage modulus and a (imaginary) viscous or loss modulus. The ratio of loss over storage modulus is called the loss tangent; the higher it is, the more liquidlike the material is. The loss tangent generally depends on the time scale of the deformation, i.e., on the oscillation frequency.

Diffusion. Molecules show heat motion, and this has some consequences: (a) particles in a liquid undergo Brownian motion, in fact the equivalent of molecular motion at a slower rate; (b) if concentration gradients exist, these are evened out by diffusion, and (c) the same applies to temperature gradients. The rate of these processes depends on the diffusion coefficient, which is inversely proportional to viscosity and to molecule or particle radius. Heat motion is a random process, governed by statistical laws. This causes (the root-mean-square value of) the transport distance to be proportional to the square root of time. Consequently, diffusion is a rapid process at very small distances and very slow at large distances. In practice, the rate of mass transport is commonly enhanced by mixing. Diffusional transport can generally be calculated by integration of Fick's laws. Also the amount of mass transported by diffusion into a lump of material is in first approximation proportional to the square root of time.

Heat.   Heat can also be transported by diffusion, also in solids. The diffusion coefficient then is called thermal diffusivity; it has a fairly constant value that is much larger than that of mass diffusion coefficents. This means that temperature evens out much faster than concentration. To calculate the transport of the amount of heat, the diffusion coefficient in Fick's laws must be replaced by the thermal conductivity. Under various conditions, heat can also be transported by mixing, by radiation, and by distillation.

Composite Materials.   Many solid foods can be considered as solid matrixes, interspersed with a continuous liquid phase. Transport through such a material may be greatly hindered. Flow of the liquid through the matrix under the influence of a pressure gradient is proportional to a material constant called permeability, which is about proportional to pore diameter squared and pore volume fraction.

Transport of mass by diffusion is also hindered. Even in a liquid solution, the effective diffusion coefficient D^* of a solute may be smaller than the diffusion coefficient of a single molecule D, for a number of reasons. On the other hand, D is inversely proportional to liquid viscosity, but this concerns a microscopic viscosity, as sensed by molecule or small particle; the macroscopic viscosity as determined in a viscometer can be much higher. Nevertheless, in materials containing fairly little liquid, D^* may become quite small. Some of the factors involved are constriction (if pore diameter is not much larger than molecule size); tortuosity of the pores (increasing the effective path length); and electrostriction (for ions in a matrix carrying electric charges).

The effect of molecule size on constriction also implies that large molecules will be hindered to a greater extent than small ones. One consequence of this is partial osmosis. If a solute molecule (A) is much larger than the solvent molecules (mostly water), contact of the system with a concentrated solution of A will lead to diffusion of water out of the system and a much slower transport of A inwards. Provided that the material can comply with the concomitant decrease in volume, this means that it will become dehydrated.

The diffusional resistance of a thin sheet of material is primarily governed by the solubility of the diffusing species in the material of the sheet and on its thickness. Very thin sheets, like adsorption layers, tend to cause negligible resistence to most solutes.

BIBLIOGRAPHY

The classical engineering text and reference book on transport phenomena is the thorough treatment by

R. B. Bird, W. E. Stewart, E. N. Lightfoot. Transport Phenomena, 2nd ed. John Wiley, New York, 2002.

A comprehensive handbook giving both theory and tabulated data is

D. W. Green, J. O. Mahoney, R. H. Perry, eds. Perry's Chemical Engineer's Handbook, 7th ed. McGraw-Hill, New York, 1997.

or any older edition.

Especially treating foods is

D. R. Heldman, D. B. Lund. Handbook of Food Engineering. Marcel Dekker, New York, 1992.

Chapter 5 by R. P. Singh treats heat transfer, and Chapter 7 by B. Hallström, mass transfer.

Several books on food rheology exist. A clear discussion of the principles of rheology and of methods of measurement, with an emphasis on liquidlike systems, is

H. A. Barnes, J. F. Hutton, K. Walters. An Introduction to Rheology. Elsevier, Amsterdam, 1989.

Most texts on physical chemistry discuss diffusion. A comprehensive description of diffusion and related phenomena is also given in

K. J. Mysels. Introduction to Colloid Science. Interscience, New York, 1964.

Solutions of the diffusion equations for a wide variety of boundary conditions are given in

J. Crank. The Mathematics of Diffusion, 2nd ed. Clarendon, Oxford, 1990.

The Darcy equation, its application and related phenomena are thoroughly discussed in

A. E. Scheidegger. The Physics of Flow through Porous Media. Oxford Univ. Press, London, 1960.

Diffusion in gels and similar materials is reviewed by

A. H. Muhr, J. M. V. Blanshard. Diffusion in gels. Polymer 23(7, suppl.), (1982) 1012–1026.

A review on drying, including factors affecting diffusion coefficients, is

S. Bruin, K. Luyben. Drying of food materials: a review of recent developments. In: A. S. Mujumbar, ed. Advances in Drying, Vol. 1. Hemisphere, Washington, 1980, p. 155.

Some articles on osmotic dehydration are in

G. V. Barbosa-Cánovas, J. Welti-Chanes, eds. Food Preservation by Moisture Control. Technomic, Lancaster, PA, 1995.

Properties of "edible films" made to retard diffusional transport in foods are discussed in

J. M. Krochta, E. A. Baldwin, M. O. Nisperos-Carriedo, eds. Edible Coatings and Films to Improve Food Quality. Technomic, Lancaster, PA, 1994.

Especially the introductory chapter by I. G. Donhowe and O. R. Fennema provides useful understanding.

6

Polymers

Almost all foods contain macromolecules and almost all of these macromolecules are polymers. Polymers have specific properties, warranting a separate treatment; some aspects of them are discussed in this chapter.

6.1 INTRODUCTION

A polymer molecule in its simplest form is a linear chain of covalently bonded identical monomers. A very simple synthetic polymer is poly-ethylene, which is obtained by linear polymerization of ethylene, CH_2CH_2, yielding $[-CH_2-]_n$, i.e., a very long paraffin chain. Another simple example is poly(oxyethylene), $[-O-CH_2-CH_2-]_n$. n is the degree of polymerization, and it can be very high, up to about a million. The monomers may have one or more reactive side groups.

Such molecules have very specific properties, due to their size and flexibility. A linear molecule with $n = 10^4$, built of monomers with a molar mass of 30 Da, will be taken as an example. The molecule is thus very large, molar mass 300 kDa. Its length-to-diameter ratio is like that of a 10 m string of about 1 mm thickness. If it were tightly coiled up in a sphere, it would assume a diameter of about 10 nm. In other words, it would be of colloidal rather than molecular size, and many polymer molecules are even far larger.

Nevertheless, a polymer molecule cannot be treated as a simple, compact colloidal particle. Because at all monomer–monomer bonds rotation about the bond angle is possible, the molecule can assume numerous different *conformations*. (This leaves the *configuration* of the molecule intact, since a change in configuration involves breakage or formation of covalent bonds.) A polymer molecule in solution, the common situation in foods, would unwind to a considerable extent, leading to even larger dimensions, about 100 nm for the present example. This is because of Brownian motion. Solvent molecules collide with the individual monomer segments, causing a continuous change in conformation of the polymer. Assuming three degrees of freedom for each monomer–monomer bond (i.e., three essentially different orientations), a conformational entropy of $R\ln3^{10,000} \approx 10^5\,\text{J}\cdot\text{mol}^{-1}\cdot\text{K}^{-1}$ can be calculated from Eq. (2.1). By comparison, Eq. (4.7) would yield an entropy of (translational) mixing of the polymer with water of the order of $10^2\,\text{J}\cdot\text{K}^{-1}$ per mole of polymer, for a typical dilute solution. The conformational entropy in solution is thus of overriding importance.

Altogether, polymer solutions show essential differences with normal solutions as well as with colloidal dispersions and need a special treatment. Highly concentrated polymer systems behave differently, again, and tend to form amorphous solids. Typical concentrated synthetic polymers are plastics and rubbers.

Until now, identical linear *homopolymers* were considered, but such molecules are rather exceptional. Polymers can be *heterogeneous* or more complicated in various ways:

> Most polymers, especially synthetic ones, vary in *degree of polymerization* and thus in molar mass. Moreover, long polymer molecules will generally contain a few irregularities, since polymerization without error is almost impossible.
>
> *Heteropolymers* are (purposely) built of more than one type of monomer. Some are copolymers, i.e., constituted of two different monomers (A and B). The latter can be arranged in various ways: evenly (e.g., alternating),
>
> —A—B—A—B—A—B—A—B—A—B—A—B—A—B—A—B—A—
>
> at random,
>
> —B—A—A—A—B—A—B—B—A—B—B—B—A—A—A—A—B—
>
> or in fairly long blocks built of one monomer.
>
> —A—A—A—A—A—A—A—A—A—A—B—B—B—B—B—B—B—

If several different monomers are involved, the molecules can be far more complicated.

Branched polymers occur, and various modes of branching and subbranching have been observed.

Polyelectrolytes contain monomers that are charged. They are virtually always heteropolymers as well, since only part of the monomers will have a charge. A polyelectrolyte may be a polyacid, a polybase, or a polyampholyte (acid and basic groups).

Natural polymers come in a far wider range of composition and properties than synthetic ones. Natural polymers are for the most part of three main types, all of which occur in foods.

Nucleic acids, i.e., DNA and RNA. These are linear heteropolyelectrolytes (four types of monomers) of very large degree of polymerization (especially DNA). Their biological function is the transfer of information. In foods, their concentrations mostly are too small to affect physicochemical properties, and they will not be further discussed.

Proteins, i.e., linear heteropolyelectrolytes, built of amino acids, i.e., 20 different monomers of highly different configuration and reactivity. The degree of polymerization typically is about 10^3. They may have any of several biological functions.

Polysaccharides, i.e., linear or branched heteropolymers of sugars and derived components; several are polyelectrolytes. The degree of polymerization is mostly 10^3 to 10^4. The main biological functions are "nutritional" (primarily starch in plants, glycogen in animals) and "building material" (in plants). The latter are called *structural polysaccharides*, which occur in a great variety of types and mostly form mixed and highly complex structures, especially in cell walls.

Several polysaccharides and proteins can be isolated to be applied in manufactured foods; examples are given in Table 6.1. It should be realized that such preparations may significantly vary in properties and also in purity. Moreover, several natural polymers are *chemically modified* to obtain altered ("improved") properties. For instance, cellulose is insoluble in water, but by carboxymethylation of part of the $-CH_2OH$ groups, the material becomes well soluble (and charged, since free carboxyl groups are formed). Cellulose can also be methylated, whereby it becomes soluble, despite the presence of hydrophobic groups. Starch is modified in various ways, e.g., by cross-linking.

Natural or modified polymers are often used as *thickening* or *gelling* agents, especially polysaccharides. Figure 6.1 gives examples of the effect of polymer concentration on apparent viscosity. Clearly, polysaccharides can be very effective thickeners. Polymers may be used to obtain a certain

TABLE 6.1 Properties of Aqueous Solutions of Some Polymers

Name	Monomers or building blocks	n	M (Da)	Stiffness		Intrinsic viscosity	
				b (nm)	b/L	K (ml/g)	a
Synthetic							
Poly(oxyethylene)	$-O-CH_2-CH_2-$	$200-2 \cdot 10^5$	10^4-10^7	1.6	4	0.01	0.8
Polyglycine	$-NH_2-CH_2-CO-$			0.7	2		
Poly(acrylic acid)	$-CH_2-CH(COOH)-$	$100-3 \cdot 10^4$	$10^4-2 \cdot 10^6$	3.9	15	0.013	0.9
Neutral Polysaccharides							
Amylose[a]	$1 \rightarrow 4$ Linked α-D-glucose	$10^4-6 \cdot 10^5$	$2 \cdot 10^5-10^7$	2.2	5	0.115	0.5
Dextran (linear)	$1 \rightarrow 6, 1 \rightarrow 3$ and $1 \rightarrow 4$ linked α-D-glucose	10–1000	$2 \cdot 10^3-2 \cdot 10^5$	2.5	6	0.1	0.5
Locust bean gum	α-D-Mannose $+25\%$ α-D-galactose side groups	250–2500	10^5-10^5	8	13	0.008	0.79
Acid Polysaccharides							
CMC	$1 \rightarrow 4$ Linked β-D-glucose with $-OCH_2COOH$ on C6	400–5000	10^5-10^6	8	16	0.01	0.91
Alginate	β-D-Mannuronic acid and α-L-glucuronic acid	600–17,000	$10^5-3 \cdot 10^6$	15	40	0.0005	1.15
κ-Carrageenan	β-D-Galactose-4-sulfate-$(1 \rightarrow 4)$-3, 6-anh.α-D galactose.	50–4000	10^4-10^6	20	20	0.01	0.9
Pectin	1-4 Linked α-galacturonic acid, partly methylated	200–600	$4 \cdot 10^4-10^5$	15	30	0.02	0.8
Xanthan	D-Glucose, D-mannose and D-glucuronic acid[b]	300–10,000	$3 \cdot 10^5-10^7$	60	100	0.002	1.07
Proteins							
β-Casein	Amino acids	209	24000	2.5	7	—	—
Gelatin[c]	Amino acids	300–10,000	$3 \cdot 10^4-10^6$		0.003	0.8	

[a] In 0.33 molar KCl[b] Has very large side groups.

[c] Solution properties highly dependent on gelatin type and temperature.

Approximate results from various sources. Conditions: Ionic strength ≈ 0.01 molar, neutral pH, room temperature. K and a are parameters in the Mark–Houwink equation (6.5).

consistency (Section 17.2.2) or to provide stability by arresting or slowing down the diffusion of molecules or the leaking out of liquid (Section 5.2).

Another important function of polymers is to *stabilize dispersions.* Polymers in solution can slow down sedimentation and aggregation of particles, or even prevent these phenomena if they form a gel (Section 13.3). Many polymers, especially proteins, adsorb onto fluid or solid particles, thereby greatly affecting colloidal interactions between the particles, hence their stability. This will be discussed in Sections 10.3.2 and 12.3.1, using basic aspects given in the present chapter. Polymers in solution can also *cause aggregation* of dispersed particles, i.e., instability (Section 12.3.3).

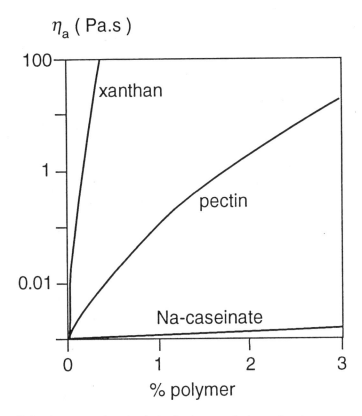

FIGURE 6.1 Apparent viscosity (η_a) of polymer solutions of various concentration (% w/w). Shear rate zero (extrapolated) for xanthan, about $100\,\mathrm{s}^{-1}$ for the other solutions. (Approximate results after various sources.)

In this chapter, the properties of *"hydrocolloids,"* i.e., water-soluble polymers, are primarily discussed, in relation to the functions mentioned in the previous paragraphs. It concerns mainly polysaccharides and gelatin. The emphasis will be on fairly dilute solutions, but concentrated systems are also discussed. Use will be made of polymer science to explain general principles, but it makes little sense to give much quantitative theory, since the polymers involved generally are too heterogeneous to follow the rules derived for the more simple synthetic polymers for which the theory has been derived. Proteins have very specific properties and will for the most part be discussed in Chapter 7.

6.2 VERY DILUTE SOLUTIONS

In this section, uncharged polymers in very small concentrations will be considered. This means that the behavior of a polymer molecule will not be affected by the presence of other polymer molecules, but only by the solvent.

6.2.1 Conformation

Ideal Polymer Chain.   The simplest model of a polymer molecule is a linear chain of n segments, each of a length L, where each segment is free to assume any orientation with respect to its neighbors, and where all orientations have equal probability. The molecule as a whole then has a conformation that can be described by a random walk through space: all steps have the same length L but can be in any direction. This is very much like the path that a diffusing molecule or particle follows in time, and it can be described by the same statistics. An example of such a statistical or random chain is depicted in Figure 6.2. If we take the *average conformation*, i.e., the average over the conformation of the same molecule at various moments or over the conformation of various molecules at the same time, the distribution of chain segments over space is Gaussian (Fig. 6.5, later on, gives examples). If the number of segments is much larger than in Figure 6.2, also each individual molecule has at any moment a more or less Gaussian segment distribution. The end-to-end distance of the chain r (see Figure 6.2) is on average zero, since r actually is a vector that can assume any orientation. Theory shows that for large n the root-mean-square distance r_m is given by

$$r_m \equiv \langle r^2 \rangle^{0.5} = Ln^{0.5} \tag{6.1}$$

r_m is also called the Flory radius. The *radius of gyration* r_g, which is defined as the root-mean-square distance of all segments with respect to the center of mass, is now given by

$$r_g = \frac{r_m}{\sqrt{6}} = 0.41Ln^{0.5} \tag{6.2}$$

The volume that the random coil occupies, including entrapped solvent, can be defined in various ways, but it will anyway be proportional to r_g^3. Because r_g is proportional to $\sqrt{n}$ and n is proportional to the molar mass M of the (hypothetical) macromolecule, the said volume will be proportional to $M^{1.5}$. Since the mass of polymer present in each coil is proportional to M, the specific volume—i.e., the volume occupied per unit mass of polymer—is proportional to $M^{0.5}$.

This is an important conclusion. It implies that the volume occupied by a certain mass quantity of polymer increases with increasing molar mass or molecular length, other things being equal; in other words, the coils are

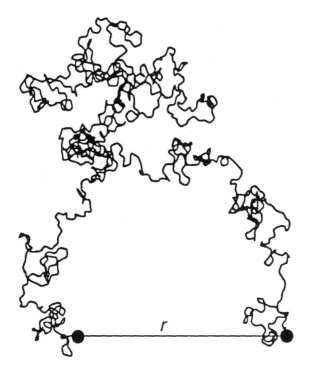

FIGURE 6.2 Example of a projection of a calculated random coil of 250 segments. r is the end-to-end distance. (Adapted from L. R. G. Treloar. The Physics of Rubber Elasticity. Clarendon, Oxford, 1975.)

more tenuous (rarefied, expanded) for a larger M or n. It also implies that the viscosity of a polymer solution of a given concentration increases with increasing M (see Section 6.2.2). This is illustrated by a simple experience in everyday life. When eating a soup that has been thickened by starch, we can generally observe that the viscosity of the soup decreases during eating. This because the we contaminate the soup via the spoon with saliva. Saliva contains amylase, an enzyme that hydrolyzes the large starch molecules into smaller fragments. This means that the average M will gradually decrease, with a concomitant effect on viscosity.

Complications. The above qualitative statements are generally true, but Eq. (6.1) is mostly not obeyed precisely, because the theory is an oversimplification. The main complications are

1. **Stiffness of chain**. An actual polymer chain will always be less flexible than the statistical chain considered above, because the bond angle between monomers is fixed. Figure 6.3 demonstrates that the position of a monomer–monomer bond with respect to the next one can only be at any point of a circle, rather than at any point of a sphere of radius L. Bulky side groups of the monomer may further restrict the freedom of orientation. Moreover, in the case of polyelectrolytes, electrical repulsion between chain segments may limit flexibility; this is discussed in Section 6.3.2. Nevertheless, the same laws may hold for the polymer conformation, by taking into account that the position of a bond with respect to another that is several monomers away can still be random. This is illustrated in Figure 6.4. A long chain can always

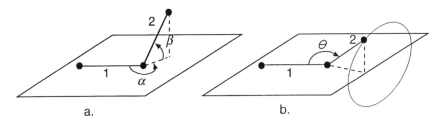

a.   b.

FIGURE 6.3 Orientational possibilities of two segments of a polymer, where a segment is depicted by its axis. Segment 1 is fixed in the horizontal plane. (a) Ideal chain; the angles α and β can assume any value. (b) More realistic chain; the bond angle θ is fixed (about 109 degrees for a C—C—C bond) and the end of segment 2 must be on the circle depicted.

be described as a random chain of n' "statistical chain elements" of average length b, where $n'b = nL$. Any statistical chain element thus contains b/L monomers, and the ratio b/L is a measure of the stiffness of the chain.

Note Some theories use the concept of persistence length q, where effectively $2q = b$.

Eq. (6.1) now reads

$$r_m = b\sqrt{n'} \tag{6.3}$$

This would imply that the specific volume occupied by the polymer becomes larger than predicted by (6.1), by a factor $(b/L)^{1.5}$. Experimental values for b/L rarely are below 4 and can be much larger; Table 6.1 gives some examples.

It is also seen in Table 6.1 that the number of statistical chain elements nL/b may be fairly small. If n' is smaller than about 25, Eq. (6.3) is not valid any more because the average distribution of the segments is not gaussian any more. Instead, the molecule assumes an elongated form, and r_m will be larger than predicted by Eq. (6.3). The extreme is a stiff rodlike molecule of length nL.

Some linear homopolymers tend to form a regular helix. A case in point is amylose, which tends to form a helix in aqueous solutions. Nevertheless, also in this case Eq. (6.2) may remain

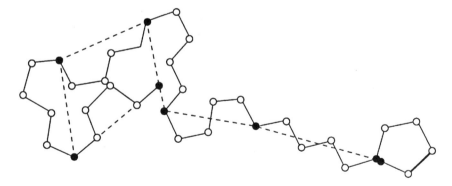

FIGURE 6.4 The effect of the stiffness of a polymer chain on its conformational freedom. This is illustrated for a two-dimensional case, with a fixed, obtuse bond angle, implying two possible conformations at each bond. Although over a distance of two or three segments, the position cannot vary at random, this is possible over a distance of, say, 5 segments, as illustrated. The broken lines would then indicate the statistical chain elements.

valid, since the helix now is a flexible chain, albeit shorter and of larger diameter.

2. **Excluded volume and solvent quality**. Up till here, the volume taken up by the polymer itself, i.e., n times the volume of a monomer, has been neglected. In other words, such an ideal random chain has no volume, which would imply that two different segments can occupy the same place in the solvent at the same time. This is, of course, physically impossible, which is why the statistics of a real chain are different from those of a random walk (diffusion). Instead of this, a self-avoiding random walk should be considered, and the average conformation then is different, r_m being proportional to n' to the power 0.6, rather than 0.5. This means that the molecule is more expanded than an ideal chain.

Another complication is that segments of a polymer molecule may show a net interaction, either attractive or repulsive, when close to each other. The tendency to interact is often expressed in Flory's *solvent–segment interaction parameter* χ. In Section 3.2 the interaction energy U between solvent and solute molecules is discussed. χ is a dimensionless number proportional to $-U_{net}$, when adapting Eq. (3.2) (i.e., $U_{net} = 2U_{12} - U_{11} - U_{22}$) to interactions between solute molecules (1) and polymer segments (2). If $\chi = 0$, i.e., there is no net solvent–segment interaction, the conformation of a polymer molecule follows the self-avoiding random walk just mentioned. In this case, the quality of the solvent for the polymer is considered good. For a higher χ, the solvent has poorer quality. For $\chi = 0.5$ the solvent is said to be an *ideal or theta solvent*: the net attraction between segments and the effect of finite segment volume just compensate each other. This means that the polymer molecule behaves as if it had no volume, so that its conformation will be like that of an ideal chain of n' segments. Examples of the segment distributions for a polymer in a good and an ideal solvent are in Figure 6.5.

Note Actually, the solvent–segment interaction may also involve an entropic contribution called contact entropy, especially when the solvent is water; see Section 3.2.

The solvent–polymer interaction can also be described by other parameters, and in the following the *excluded volume parameter* β will be used. It is the factor by which the segment volume must be multiplied in order to describe the conformation of the polymer. It follows that $\beta = 1 - 2\chi$. For an ideal solvent, $\beta = 0$, and if $\beta = 1$, the conformation of the molecule follows a self-avoiding random walk.

Most water-soluble polymers have $0 \leqslant \beta \leqslant 1 (0.5 \geqslant \chi \geqslant 0)$. This implies that for a linear homopolymer

$$r_m \approx r_g \sqrt{6} \approx b(n')^\nu \qquad n' > \sim 25 \qquad\qquad (6.4)^*$$

where ν varies between 0.5 and 0.6. Addition of small-molecule solutes (salts, sugars, alcohols, etc.) in high concentrations may significantly affect β. For β values only slightly smaller than zero, most polymers are insoluble. Notice that an "ideal" solvent actually is a fairly poor solvent. Solubility is discussed in Section 6.5.1.

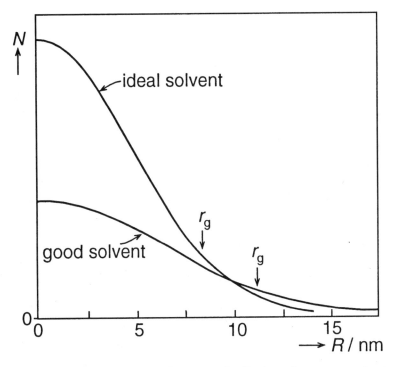

FIGURE 6.5 Calculated average frequency distribution of monomers about the center of mass of a polyethylene molecule of 2000 CH_2 monomers, for $b = 0.46$ nm. N is the number of monomers per unit volume per unit of R (arbitrary scale). R is the distance from the center of mass. Examples for an ideal solvent ($\beta = 0$) and a good solvent ($\beta = 1$). r_g is the radius of gyration. (Adapted from Tanford; see Bibliography.)

3. **Polydispersity**. As mentioned, most polymers show a range in degree of polymerization or molar mass. This means that suitable averages should be taken. The theory is fairly involved and will not be discussed here.

4. **Branched polymers**. Some polymers are highly branched, and amylopectin is a prime example. In such a case, the volume taken up by a polymer molecule will be much smaller than that of a linear molecule of the same number of the same segments. The exponent in the relation between r_m or r_g and n' will generally be smaller than 0.5. For amylopectin, exponents of 0.41 to 0.43 have been observed. Calculation of the conformation of branched polymer molecules is generally not possible.

5. **Heteropolymers**. Several complications may arise. The stiffness and the bulkiness of the chain may be different in different stretches. The same may be true for the quality of the solvent, which implies that the excluded volume parameter of the polymer is a kind of average value. Specific interactions may occur between different side groups. Simple theory is not available.

Moreover, polymers may be *polyelectrolytes*, which are discussed in Section 6.3. The electric charge generally causes the chains to be stiffer.

Polysaccharides. Many natural or modified natural polymers exhibit most or all of the complications mentioned. This is especially true for many polysaccharides, which tend to be fairly stiff molecules; see Table 6.1. The β values are generally small, often < 0.1. The chemical constitution varies considerably, some polysaccharides (e.g., xanthan) having very large side groups on the primary chain. Besides causing steric hindrance, which also makes the chain stiffer, side groups may exhibit (weak) mutual attraction. Some of these polymers tend to form helices in solution. Altogether, polysaccharides vary widely in such properties as solubility, tendency to form a gel, and extent of expansion. When characterizing the latter by the hydrodynamic voluminosity, i.e., the hydrodynamic volume of a polymer molecule per unit dry mass of polymer, values ranging from 10 to more than 10^3 ml/g have been observed for polysaccharides of $M = 10^6$ Da. Notice the extremely large values possible (largely due to the stiffness mentioned), which would even be larger for higher M.

The complications mentioned imply that the average conformation can mostly not be calculated from first principles. However, many of the parameters mentioned can be determined, such as molar mass and composition. The radius of gyration r_g can be obtained from light scattering experiments, β from osmometry. If molecules of varying (average) molar mass are available, v and n', and thus the length of a statistical chain element b, can be established. These data can then be used to predict behavior. To take an example, the hydrodynamic radius r_h of an unbranched polymer molecule is to a good approximation given by

$$r_h = \frac{2}{3} r_g \tag{6.5}$$

This allows calculation of its diffusion coefficient according to Eq. (5.10), or of its sedimentation rate in a centrifugal field. It is also of interest in relation to the effect of the polymer on viscosity, which is discussed in the following section.

Question

Consider a solution of amylose in 0.33 molar KCl. In this condition, the excluded volume parameter $\beta = 0$. Assume $M = 10^6$ Da. What would be the diffusion coefficient of an amylose molecule in a very dilute solution? And what amylose concentration would be needed for the amylose molecules plus entrapped solvent to fill the whole volume?

Answer

Amylose is essentially a linear chain of anhydroglucose monomers. Molar mass of the monomer thus equals that of glucose minus water, i.e., 162 Da. This means that $n = 10^6/162 = 6173$ monomers. From tabulated bond lengths and angles it can be calculated that the monomer length $L \approx 0.5$ nm. Table 6.1 gives $b/L \approx 5$, leading to $b \approx 2.5$ nm and $n' = nL/b \approx 1235$. To find r_g, Eq. (6.4) can be applied with $v = 0.5$ (because the solvent is ideal), yielding 36 nm. According to Eq. (6.5), the hydrodynamic radius would be 2/3 times r_g, yielding 24 nm. Applying Eq. (5.16) with $T = 300$ K and $\eta = 10^{-3}$ Pa·s yields a diffusion coefficient $D \approx 9 \cdot 10^{-12}$ m^2·s^{-1}, which is by a factor 75 smaller than D for glucose. It may be noticed that the simple assumption of a molecular radius proportional to $M^{1/3}$, as applies for compact molecules, would lead to a diffusion coefficient smaller than that of glucose by a factor $(10^6/180)^{1/3} \approx 18$, rather than 75.

To calculate the volume occupied, we need to have a reasonable value for the effective radius. Figure 6.5 shows that it would be roughly 1.5 times r_g for an ideal solvent, i.e., 54 nm. The volume of a molecule is $(4/3)\pi r^3$, leading to $6.6 \cdot 10^{-22}$ m^3. This implies that $15 \cdot 10^{20}$ molecules can be packed in a m^3 and dividing by N_{AV} leads to 0.0025 moles. Multiplying by M gives 2.5 kg·m^{-3}, or about 0.25% w/w. A very small concentration would thus be sufficient to "fill" the whole system.

It should be noticed that the results obtained would greatly depend on conditions, such as M, b/L, and β. For instance, if $M = 10^5$, we would have $r_g \approx 11\,\text{nm}$, $D \approx 29 \cdot 10^{-12}\,\text{m}^2 \cdot \text{s}^{-1}$, and the concentration needed would be about 9% (try to check these results).

6.2.2 Viscosity

In Section 5.1.2 the effect of solute molecules and particles on viscosity is briefly discussed. It follows that the intrinsic viscosity $[\eta]$ is a measure of the extent to which a certain solute can increase viscosity. (Remember that $[\eta]$ equals specific viscosity $(\eta/\eta_s - 1)$ divided by concentration for infinitesimally small concentration.) According to the Einstein equation (5.6) the specific viscosity of a dispersion of spheres is 2.5φ, where φ is volume fraction. This means that $[\eta] = 2.5\varphi/c$ for $c \to 0$, where c is concentration in units of mass per unit volume. For a very dilute polymer solution the effective volume fraction can be given as the number of molecules per unit volume N times $(4/3)\pi r_h^3$, where r_h is the hydrodynamic radius; see Eq. (6.5). Furthermore, $N = c \cdot M/N_{AV}$. For the amylose mentioned in the question just discussed, $r_h \approx 25\,\text{nm}$ and $M = 10^6\,\text{Da}$. It follows that $[\eta]$ would equal

$$2.5 \times \left(\frac{4}{3}\right)\pi(25 \cdot 10^{-9})^3 \times \frac{N_{AV}}{10^6} = 10^{-4}\,\text{m}^3/\text{g}$$

or 1 dl/g. The value observed is 1.15 dl/g, close to the calculated result. It can be concluded that the hydrodynamic volume of the macromolecule (with interstitial solvent) is far smaller than the volume calculated from the approximate outer radius of the polymer coil, in this case about 55 nm, leading to a difference in $[\eta]$ by a factor $(55/25)^3 \approx 11$. This implies that a random coil molecule does not move with all its interstitial solvent, but that it is partly permeated by solvent. (It is also said that the polymer coils are partly draining.) On the other hand, the volume fraction, and thereby the intrinsic viscosity, would be far smaller, here by a factor of about 70, if the molecule did not contain interstitial solvent.

In the calculations just made, Eq. (6.4) was applied (with $v = 0.5$). For most polymers, this is not allowed. The theory for the relation between viscosity and polymer conformation is rather intricate and not fully worked out, and one mostly makes shift with the semiempirical *Mark–Houwink* relation,

$$[\eta]_0 = K\langle M\rangle^a = K\langle M\rangle^{3v-1} \tag{6.6}$$

where M is expressed in daltons. The relation between the exponents a and v follows from the following proportionalities. Intrinsic viscosity is proportional to the hydrodynamic volume of the polymer molecules, hence $[\eta] \propto Nr_g^3/nN = r_g^3/n$. From Eq. (6.4) we have $r_g \propto (n')^v$ and for one kind of polymer we also have that $n \propto n' \propto M$. Consequently, $[\eta] \propto M^{3v-1}$.

Polymers vary widely in the value of K, which can in general not be rigorously derived from theory. It may further be noticed that the range $v = 0.5$ to 0.6 corresponds to $a = 0.5$ to 0.8. This range applies only for very long linear polymers in fairly good solvents. In practice, a values of 0.5 (for amylose or dextran) to almost unity are observed. Some very stiff and charged polysaccharides can even exhibit an exponent > 1. Branched polymers generally have $a < 0.5$; for amylopectin (highly branched) it equals about 0.3. Some data are given in Table 6.1. Anyway, $[\eta]$, and thereby viscosity, greatly depends on molar mass of the polymer, the more so for a higher value of a.

Strain Rate Dependence.   The outline given above is, however, an oversimplification, because it has been implicitly assumed that the flow would not affect orientation or conformation of the polymer coil. This is not true. It is always observed that η, and thus $[\eta]$, is affected by the shear rate applied. As discussed in Section 5.1.1, various types of flow can occur and, more generally, we should say strain rate or velocity gradient, rather than shear rate. However, we will restrict the discussion here to simple shear flow.

An example is shown in Figure 6.6, lower curve. At very low shear rate, the solution shows Newtonian behavior (no dependence of η on shear rate), and this is also the case at very high shear rate, but in the intermediate range a marked *strain rate thinning* is observed. The viscosity is thus an apparent one (η_a), depending on shear rate (or shear stress). It is common practice to give the (extrapolated) intrinsic viscosity at zero shear rate, hence the symbol $[\eta]_0$ in Eq. (6.6). The dependence of η on shear rate may have two causes.

First, all molecules or particles rotate in a shear flow, but if they are not precisely spherical, the flow causes an *orientation* aligned in the direction of the flow to last for a longer time than orientation perpendicular to the flow (Section 5.1.1). This implies that average flow disturbance, and thereby viscosity, is smaller. The alignment depends on the rate of rotary diffusion of the particles in relation to the magnitude of the shear rate. The rotary

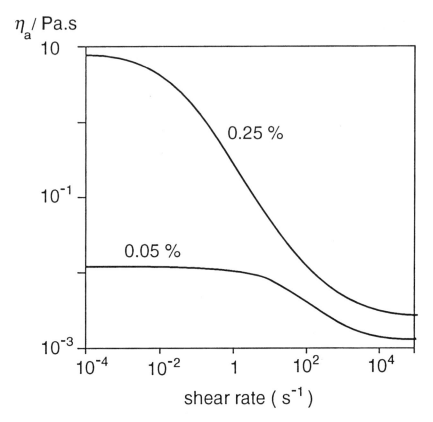

FIGURE 6.6 Dependence of the apparent viscosity η_a of xanthan solutions (concentration indicated) on shear rate; low ionic strength. (Approximate results from various sources.)

diffusion time of a solid sphere of radius r is given by

$$\tau_{\text{rot}} = \frac{4\pi r^3 \eta_s}{k_B T} \tag{6.7}$$

where η_s is the viscosity of the solvent. The relation also applies to spherical random coils, with $r \approx r_g$. For elongated particles, the rotation time is longer than predicted by Eq. (6.7). If $\tau_{\text{rot}} \gg 1/$shear rate, orientation will readily occur. If shear rate is very small, the rotary Brownian motion will be overriding, no orientation will occur, and the viscosity will be higher. The extent of the viscosity difference depends on the deviation from sphericity of the molecule or particle. Because of the random variation in conformation, some of the polymer molecules will always be nonspherical. Larger deviations occur with more elongated molecules, i.e., with some very stiff polymers, which generally implies that the exponent a in Eq. (6.6) is relatively large. The value (more precisely, the range of values) of the shear rate where the transition occurs is inversely proportional to τ_{rot}. The latter is longer for a higher radius of gyration of the polymer molecule. This means that the transition from high to low apparent viscosity occurs at smaller shear rates for a larger molar mass, a better solvent quality, and a greater stiffness of the polymer molecule.

Second, *deformation* of a particle like a random coil polymer molecule may occur in a shear field. As illustrated in Figure 5.3, the particle becomes elongated in one direction and compressed in another one. Since the particle also rotates, this effectively means that the particle is repeatedly compressed and elongated, which goes along with solvent locally being expelled from it and locally being taken up. This causes additional energy dissipation and thus an increased viscosity. The polymer molecule has a natural *relaxation time* for deformation: if it is deformed by an external stress and then the stress is released, it takes a time τ_{def} for the deformation to be diminished to $1/e$ of its original value.

Note Actually, there may be a spectrum of relaxation times, implying that τ_{def} would be an average value.

If $\tau_{def} \gg 1/\text{shear rate}$, the molecule cannot deform during flow, implying that it keeps its roughly spherical shape, and the viscosity remains relatively small. At very small shear rate, the molecule can fully deform twice during every rotation, and viscosity is relatively large. The theory for the relaxation time is not fully worked out, but it may be stated that the relation for τ_{def} is of the same form as for τ_{rot} given in (6.7). The same variables thus apply for the deformation mechanism, which presumably has a larger effect on the shear rate dependence of the viscosity than the orientation mechanism, especially for large molecules.

Till here, only the interaction between one molecule and the solvent has been considered. Unless polymer concentration is extremely small, mutual interaction between polymer molecules will further affect viscosity and its shear rate dependence. This is discussed in Section 6.4.3.

Note The reader may wonder whether the values of the apparent viscosity at such extreme shear rates as 10^{-4} or $10^4 \, \text{s}^{-1}$ are of any practical significance. It will be seen, however, in Chapters 11 and 13, that such velocity gradients do indeed occur and affect, for instance, the physical stability of dispersions.

Question

For a solution of (a certain type of) locust bean gum in water at room temperature, the Mark–Houwink parameters are $K = 8 \cdot 10^{-4}\,\mathrm{m^3 \cdot kg^{-1}}$ and $a = 0.79$. What would be the zero shear rate viscosity of a 0.01% solution, assuming that M equals (a) 10^5 or (b) 10^6 Da? ($\eta_s = 1\,\mathrm{mPa \cdot s}$.) Are the results obtained by applying Eq. (6.6) reliable?

Answer

Equation (6.6) directly gives for (a), and (b) that $[\eta]_0 = 7.1$ and $44\,\mathrm{m^3 \cdot kg^{-1}}$, respectively. Taking Eqs. (5.9) and (5.10), and assuming $d\eta/dc$ to be independent of c, it is derived (check this) that

$$\eta = \eta_s(1 + c[\eta]_0)$$

at very small shear rate. 0.01% corresponds to $c \approx 0.1\,\mathrm{kg \cdot m^{-3}}$, and it follows that for (a), and (b) the viscosity would be 1.7 and 5.4 mPa·s, respectively.

    To check whether these results are reasonable, it is useful to invoke the Einstein equation (5.6)

$$\eta = \eta_s(1 + 2.5\varphi)$$

and compare it with the relation above; it follows that φ would equal $[\eta]c/2.5$. This gives for sample (a) $\varphi = 0.28$ and for (b) 1.76. The latter value is clearly impossible, and the derivation given thus is invalid; the actual viscosity would be very much higher than calculated above. Actually, also 0.28 is much too high a φ value to ensure that $d\eta/dc$ is independent of c (the viscosity actually would be about 2.4 mPa·s in this case).

6.3 POLYELECTROLYTES

In this section some specific aspects of polymers bearing electric charge will be discussed. The reader is referred to Section 2.3, Electrolyte Solutions, for basic aspects.

6.3.1 Description

A polymer containing electrically charged groups is called a polyelectrolyte or a macroion. Three types can be distinguished:

Polybases or cationic polymers; these mostly contain —NH$_2$ or =NH groups that can be protonated at sufficiently low pH (mostly 7–10). They are unimportant in foods.

Polyacids or anionic polymers; the most common charged groups are carboxyl groups (—COO$^-$), which are often present in glucuronic acid or comparable residues, as in gum arabic, pectins, alginates, and xanthan gum. Most carrageenans contain —O—SO$_3^-$ groups.

Polyampholytes, which contain both positively charged (basic) and negatively charged (acidic) groups. Well-known are the proteins, DNA, and RNA.

The *valence*, i.e., the number of charges per molecule (z), naturally depends on the degree of polymerization (n) and on the type of polyelectrolyte. In some, all monomers have a charge, as in poly(acrylic acid): (—CH(COO$^-$)—CH$_2$—)$_n$. More often, only part of the monomers is charged. Some polysaccharides, notably pectins, contain methylated carboxyl groups —COOCH$_3$; often, part of these groups are or become hydrolysed, leading to —COO$^-$ + CH$_3$OH.

Ionization. The number of charges depends, of course, on the *degree of ionization*, hence on pH. The sulfate groups mentioned are virtually always ionized. Carboxyl groups, however, have a pK_a value of about 4.7, or about 3.4 for uronic acid groups (primary carboxyl group on a sugar ring), implying that their ionization can vary greatly over the pH range occurring in foods. At low pH, where the hydrogen ion activity is high, they are fully protonated, having no charge; at high pH they are ionized, having negative charge.

The situation is more complicated than for a simple monovalent acid. It is useful to recall Eq. (2.24) given in Section 2.3.1 on dissociation of an acid HAc:

$$pH - pK_a = \log[Ac^-] - \log[HAc] + \log \gamma_-$$

where γ_- is the ion activity coefficient of Ac$^-$. Calling the fraction of the molecules that is dissociated α, the equation can be rewritten as

$$\log \frac{1-\alpha}{\alpha} = pK_a - pH + \log \gamma_- \tag{6.8}$$

Figure 6.7a gives titration curves of a sample of poly(methacrylic acid). We will first consider the curve for low ionic strength, where γ_- is not much smaller than unity. It is seen that Eq. (6.8) with $pK_a = 4.7$ is not nearly obeyed: the curve is shifted to higher pH (by almost 2 units), and the slope is far too small (spanning nearly four pH units rather than two).

The main cause for the discrepancy is that the molecule has *many ionizable groups*, rather than one. If it is partly charged, any additional proton to be dissociated has to be removed against the electric potential caused by several neighboring $-COO^-$ groups, rather than just one. See Section 2.3.1, especially Figure 2.9b. The extent to which the presence of many ionizable groups affects the titration curve depends on the immediate chemical environment of these groups, notably on the distance between them. In practice, one often finds a so-called Henderson–Hasselbalch relation,

$$\log\frac{1-\alpha}{\alpha} = \frac{pK_{av} - pH}{f} \tag{6.9}$$

where α is now the fraction ionized (or neutralized) averaged over all groups

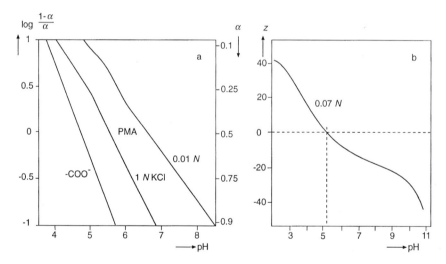

FIGURE 6.7 Titration curves. (a) The degree of ionization α as a function of pH for a single carboxyl group (calculated) and for solutions of poly(methacrylic acid) in 0.01 and 1 normal KCl. (b) The (average) valence z of β-lactoglobulin as a function of pH. (After various sources.)

at the prevailing ionic strength; pK_{av} is the average pK_a of all groups and f is a constant > 1.

It is seen in Figure 6.7a that the curve does not precisely follow Eq. (6.9); this is due to the PMA undergoing a conformational transition at a certain pH, which leads to a different distance between charged groups. It is generally observed that the conformation of a polyelectrolyte (somewhat) affects its titration curve.

The latter is part of the explanation of the third point to be mentioned about Figure 6.7a, viz., the dependence of the titration curve on *ionic strength*. Both pK_{av} and f are smaller at higher ionic strength. This is partly clear from Eq. (6.8); at higher ionic strength, the activity coefficient γ_- is smaller, the more so for a higher α, according to the Debye–Hückel theory (Section 2.3.2). Moreover, the shielding of charges is stronger at higher ionic strength, causing the removal of a proton from a charged molecule to be easier, as the proton senses a smaller attractive electrostatic potential.

Altogether, it is mostly impossible to calculate the titration curve from the molecular structure of the polyelectrolyte, although the explanations given are useful in a semiquantitative sense. In practice, one just determines a titration curve. An example is in Figure 6.7b for a protein, i.e., a polyampholyte. Such a molecule has an isoelectric pH, i.e., a pH at which the net charge is zero. For a polyacid (or a polybase), net zero charge means no charge, but this is not so for a polyampholyte, which will have several positive as well as (an equal number of) negative charges at its isoelectric pH.

Note Even the net charge of most individual molecules will not be zero, since the charge distribution shows statistical variation.

This has some consequences for the behavior of these molecules, which will be discussed further on. Quite in general, the presence of charged groups causes greater heterogeneity of the polymer, and polyelectrolytes are virtually never true homopolymers.

The presence of charges on a polymer in solution has several important **consequences**:

1. The solution must be *electrically neutral* (unless a large external electrostatic potential gradient is applied). If the molar polyelectrolyte concentration is m, and the average valence is z, there must be mz counterions (of unit valence) in solution, for instance Na^+ ions for a polyacid.
2. The presence of counterions causes the *Donnan effect*, discussed in Section 6.3.3.

3. The polyelectrolyte itself *contributes to the ionic strength*. Recall that $I = (1/2)\Sigma(m_i z_i^2)$. For a high polymer m is very small, but $|z|$ may be very high. For example, a 10^{-5} molar concentration of a polyelectrolyte with $z = -100$ (which is not exceptional) would yield an ionic strength of at least $(1/2)(100 \times 10^{-5} \times 1^2 + 10^{-5} \times 100^2) = 0.05$ molar; the first term between parentheses is due to the counterions, the second to the polyelectrolyte. Although the Debye–Hückel theory as discussed in Section 2.3.2 cannot be precisely applied, the high ionic strength of a polyelectrolyte solution does have large effects on several properties. Some examples are given later on.

4. The electric charge on the molecule strongly affects its *conformation*, causing it to be more expanded than a neutral polymer. The extent of expansion can be very high if z is high, but it markedly decreases with increasing ionic strength. The change in conformation affects several properties, for instance the viscosity of the solution; see Section 6.3.2. Other properties affected are diffusion coefficient, sedimentation rate (in an ultracentrifuge), light scattering, the second virial coefficient (Section 6.4.1), and the chain overlap concentration (Section 6.4.2).

5. Polyelectrolyte molecules of like charge *repel each other*. Unless the concentration is very small, this may affect the distribution of molecules over the available space, hence light scattering and the second virial coefficient. It also causes the conformation of the molecules to be somewhat less expanded at a higher concentration.

6. If polyelectrolytes *adsorb* onto (uncharged) particles, this gives the particles an electric charge, causing interparticle repulsion. This is discussed in Chapter 12.

7. If an electric potential gradient is applied in the solution, polyelectrolyte molecules will move in the direction of the electrode of opposite charge. This is called *electrophoresis*. If the polyelectrolyte is immobilized, the solvent will move in the electric field, a process called electroosmosis. These principles are applied in several laboratory techniques but will not be discussed here.

8. *Solubility* may strongly depend on pH and ionic strength. This is especially important for proteins and is discussed in Section 7.3.

9. Specific interactions may be caused by the presence of polyvalent counterions or oppositely charged polyelectrolytes.

6.3.2 Conformation and Viscosity

Electric charges on a polymer greatly affect its conformation in aqueous solution, because like charges repel each other. This depends, of course, on the number of charges per molecule (z), hence on pH, and also on **ionic strength**. Electric shielding of ions by other ions is discussed in Section 2.3.2, and charged groups are shielded in the same way. The repulsive force due to electric charges is proportional to the square of the electric potential that they generate. If the potential equals ψ_0 at the surface of the charge, shielding causes it to decrease with distance (h) from the charge according to

$$\psi = \psi_0 e^{-\kappa h} \tag{6.10a}$$

which applies for fairly low $|\psi_0|$ (say, < 50 mV) and a flat geometry. Around a charged sphere of radius r the relation is

$$\psi = \psi_0 \frac{r}{r+h} e^{-\kappa h} \tag{6.10b}$$

In water at room temperature, κ is given by

$$\kappa \approx 3.2\sqrt{I} \tag{6.10c}*$$

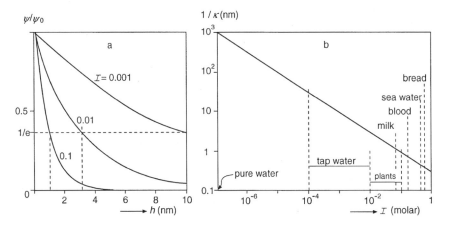

FIGURE 6.8 The electric double layer. (a) The electrostatic potential ψ relative to its value at $h = 0$, as a function of distance h from a flat charged surface in solutions of various ionic strength I (molar); see text. (b) The thickness of the electric double layer $1/\kappa$ as a function of ionic strength in aqueous solutions at room temperature; approximate values of I in (the aqueous part of) some materials are indicated.

if κ is in nm^{-1} and the ionic strength I in mol·L^{-1}. It follows from Eq. (6.10a) that ψ is reduced to ψ_0/e (≈ 0.37 times ψ_0) at a distance h that equals $1/\kappa$. $1/\kappa$ is called the *thickness of the electric double layer* or *Debye length*. It is a measure for the distance over which electrostatic interactions are significant. In Figure 6.8a, the dependence of the electric potential on distance is illustrated for some values of the ionic strength. It is seen that at high I, $1/\kappa$ is very small, comparable to the size of a very small molecule. Figure 6.8b shows the dependence of $1/k$ on I. To give an idea of the values of I to be encountered in foods, examples are given. Some fabricated foods, especially when pickled or salted, have much higher ionic strengths, leading to very small values of $1/k$ and very weak electrostatic interaction.

Figure 6.9 schematically depicts the **conformation** of a polyelectrolyte (a polyacid) at various ionic strength. At high I (small $1/k$) the negative charges can only sense each other if they are very close. This implies that the conformation is not greatly different from that of a similar but uncharged polymer. As the ionic strength decreases, the molecule becomes more expanded, because the charges sense each other over a longer distance. The

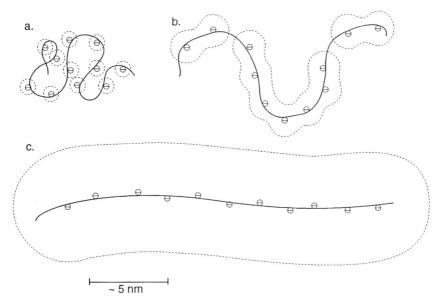

FIGURE 6.9 Examples of the conformation of a polyacid molecule at various ionic strengths, which would roughly be 400, 80, and 7 mmolar for (a), (b), and (c). The dotted lines are at a distance of about $1/\kappa$ of the charges. Only ions that are part of the polymer are indicated. Highly schematic.

radius of gyration thus is relatively large. If the molecule is not very long, it will attain an almost rodlike conformation at very low ionic strength. The conformation will also depend on the *linear charge density*. This is inversely proportional to the mutual distance b_{ch} between charged groups, where $\langle b_{ch} \rangle = n\,L/|z|$. The extent of expansion of the polyelectrolyte molecule increases with decrease of the product $\kappa \times b_{ch}$.

Quantitative theory for the volume occupied by a polyelectrolyte molecule is available, but it will not be discussed here; there are too many complications for most of the natural polyelectrolytes. Nevertheless, the semiquantitative reasoning given provides important understanding. Quite in general, the presence of charges makes the molecule much *stiffer*; cf. Section 6.2.1, complication 1. Moreover, it causes the volume exclusion parameter β to be higher (cf. complication 2), or in other words, the solvent quality is effectively enhanced. The negatively charged polysaccharides, such as carrageenans, alginates, and xanthan, are all rather stiff and expanded molecules (unless the ionic strength is high), the power a in Eq. (6.6) generally being close to unity.

Viscosity. The effects of charge (degree of dissociation) and ionic strength on expansion (radius of gyration) of polyelectrolytes are, of course, reflected in the extent by which they increase viscosity. Figure 6.10 gives an example. It is seen that a higher degree of dissociation and a lower salt concentration both yield a higher intrinsic viscosity, as expected. However, it is also seen that at large α, $[\eta]$ decreases again, at least at low salt concentration. This may be explained partly by the contribution of the polyelectrolyte to the ionic strength. At high z, i.e., high α, this contribution is considerable, as was discussed in Section 6.3.1, Consequence 3. This means that an increase in α leads to an appreciable increase in ionic strength, and at low salt concentration this is sufficient to decrease the radius of gyration of the molecule and thereby $[\eta]$. If the ionic strength is kept constant at high z, the anomaly is much smaller. Figure 6.11 further illustrates the effect of ionic strength on intrinsic viscosity of a polyacid. Notice that the exponent a markedly increases with decreasing ionic strength, leading to very high intrinsic viscosities.

It can be concluded that charged polysaccharides can be very effective thickening agents, but that the viscosity of such solutions strongly decreases with increasing ionic strength I. At very high salt concentration, charged polysaccharides behave virtually like neutral polymers.

The kind of counterions present may also affect polyelectrolyte conformation. At very high salt concentration, part of the ionized groups on the polymer will become neutralized (due to ion pair formation), causing a more compact conformation. This depends somewhat on the kind of

counterion. The association of counterions generally decreases in the order
$K^+ > Na^+ > Li^+$, following the Hofmeister series (Section 3.2), but divalent
cations (Mg^{2+}, Ca^{2+}) are more effective. At high concentration, divalent
cations can also form salt bridges between acid groups. Most of these
bridges are intramolecular bonds, thereby causing a further reduction of the
volume occupied by the polymer. At high polymer concentration, bridging
by divalent counterions may cause gelation.

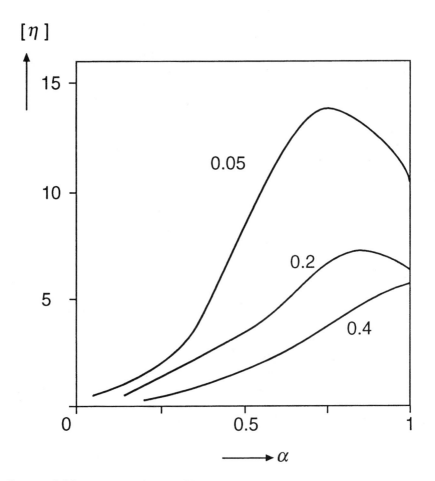

FIGURE 6.10 Intrinsic viscosity $[\eta]$ (in dL/g) of aqueous solutions of partially
esterified poly(methacrylic acid) as a function of degree of ionization α, at various
concentrations of NaCl (mmolar, indicated). (From results by J. T. C. Böhm. *Comm.*
Agric. Univ. Wageningen 74(5) (1974).)

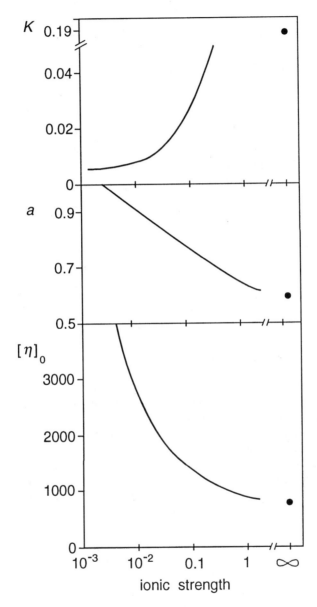

FIGURE 6.11 Viscosity parameters of solutions of carboxymethyl cellulose (Na salt) at various ionic strengths (molar). Mark–Houwink parameters K (ml/g) and a, and intrinsic viscosity $[\eta]_0$ (ml/g) for a molar mass of 10^6 Da. (From results by W. Brown and D. Henley. *Makromol. Chemie* 79 (1964) 68.)

The behavior of **polyampholytes** is in some respects different. At a pH far from its isoelectric point, a polyampholyte behaves roughly like a polyacid (high pH) or a polybase (low pH), but near the isoelectric pH the molecule has about equal numbers of positive and negative charges, and these attract each other, thereby causing a less expanded conformation. (Moreover, the solubility may be much decreased: Section 7.3.) The effect of ionic strength on conformation will then be opposite to that of a simple polyelectrolyte: at lower ionic strength, the oppositely charged groups attract each other over a greater distance, leading to a more compact conformation. In other words, near the isoelectric pH, salt screens (diminishes) attraction, far away from this pH, it screens repulsion.

> *Note* As mentioned already, the exposé given in this section is an oversimplification. The most important point may be that at a high (linear) charge density, i.e., at small $nL/|z|$, the Debye–Hückel theory on the screening of electric charge is no longer valid. At some distance from the polyelectrolyte chain, the potential then is nearly insensitive to (changes in) the charge density; the deviation from the Debye–Hückel theory is especially great at low ionic strength. This factor also contributes to the downward trend of the curves in Figure 6.10 at high α.

6.3.3   The Donnan Effect

Principles. The condition of electroneutrality implies that a polyelectrolyte molecule in solution is accompanied by counterions (small ions of opposite charge). Unless a strong external electrical potential is applied, the polyelectrolyte cannot be separated from its counterions. To be sure, the counterions freely diffuse toward and away from the polyelectrolyte molecule, but this only applies to individual ions. Each polyelectrolyte molecule is always accompanied by the same number of counterions, apart from statistical fluctuations around the average.

The polyelectrolyte also affects the distribution of coions (small ions having a charge of the same sign). This is known as the Donnan effect, which may be illustrated by envisaging a system with two compartments that are separated by a semipermeable membrane. This membrane does not allow passage of polymers but is permeable for small ions. Assume that one compartment initially contains the polyelectrolyte (PE^{z-}) and sufficient counterions (Na^+) to neutralize the charge, while the other compartment initially contains NaCl. At equilibrium, the activity of NaCl (which means the free ion activity $a_{\pm}$) must be equal in both compartments. This implies that Cl^- will diffuse towards the other compartment, and the condition of electroneutrality forces an equal transport of Na^+. This is depicted below, where m means molar concentration. The vertical line depicts the membrane.

Ions	Na^+	PE^{z-}	Cl^-	Na^+	Cl^-
Initial concentration	m_1	m_1/z	0	m_2	m_2
Equilibrium concentration	$m_1 + x$	m_1/z	x	$m_2 - x$	$m_2 - x$

The equilibrium condition is that the ion activity product of NaCl

$$a(Na^+) \times a(Cl^-) = [Na^+] \times [Cl^-] \times (\gamma_\pm)^2$$

is equal on both sides of the membrane. (See Section 2.3.1 for the theory of electrolyte solutions.) Assuming dissociation of the salt to be complete and $\gamma_\pm$ to be equal on both sides, the condition would become

$$(m_1 + x)x = (m_2 - x)^2 \quad \text{or} \quad x = \frac{m_2^2}{m_1 + 2m_2}$$

Calculated examples are in Table 6.2. The first conclusion to be drawn is that the charge of the polyelectrolyte is compensated in two ways, i.e., by an excess of counterions and by a depletion of coions. It is further seen that the absolute (x) and relative amounts of NaCl (x/m_2) transferred increase with increasing salt concentration. The ratio of concentrations of Na^+ on either side of the membrane is closer to unity for higher salt concentration,

TABLE 6.2 The Donnan Effect[a]

m_2	x	x/m_2	$(m_1 + x)/(m_2 - x)$	ΔO
0.3	0.056	0.19	4.33	0.20
1	0.33	0.33	2.00	0.34
3	1.29	0.43	1.33	0.44
10	4.76	0.48	1.10	0.49
100	49.75	0.50	1.01	0.51

[a] Calculations of the change in concentration x (millimolar) of coions and counterions between a compartment (1) containing 0.1% of a polyelectrolyte of $M = 10^5$ Da (i.e., 0.01 mmolar) and valence $z = -100$. Compartment 1 initially contained $m_1 = 1$ mmolar Na^+; compartment 2 m_2 mmolar NaCl. $(m_1 + x)/(m_2 - x)$ is the ratio of the equilibrium concentration of Na in compartment 1 to that in 2; the concentration ratio of the Cl^- ions is the reciprocal of the values in this column. ΔO is the difference in osmolality (millimolar) between the two compartments.

and this also holds for the Cl^- ions. For $m_2 \gg m_1$, the salt composition is nearly the same in both compartments.

If a mixture of salts is present, the concentration ratio of all univalent cations will be equal to that of Na^+, provided that the total concentration of univalent cations is the same. This thus also applies to H^+. Taking the upper row of Table 6.2, where the ratio is 4.33, it follows that the pH would be lower by $\log 4.33 = 0.64$ units in the compartment containing the polyelectrolyte. Such pH differences can indeed be measured. For a ratio of 2, the pH difference would be $\log 2 = 0.30$, etc. Increasing ionic strength thus leads to smaller differences in salt composition and in pH. For a polybase, the pH would be higher in the compartment with polymer. For coions, the concentration ratio is the inverse of that for counterions.

It also follows that the **osmotic pressure** will be higher in the compartment containing the polyelectrolyte, and water will be drawn in until the osmotic pressure is equal on either side, provided that this would not provoke a hydrostatic pressure difference. Accurate prediction of the osmotic pressure is intricate, because of large nonideality, and we will only consider the osmolality. As long as the volumes remain unchanged, a difference in osmolality of

$$\frac{m_1}{z} + \frac{m_1 m_2}{m_1 + 2m_2}$$

is produced. Some results are in Table 6.2. It is seen that the difference is far larger than that due to the polymer only, which would have been $m_1/z = 0.01$ mmolal in the present case. This implies that a simple measurement of osmotic pressure to determine the molar mass of a polymer would yield highly erroneous results for polyelectrolytes. Nevertheless, reliable determinations can be made, but the theory is intricate, and painstaking experimentation is needed.

Polyelectrolyte Conformation. Up till here, we have considered two compartments separated by a semipermeable membrane. However, this is not essential to the existence of the Donnan effect. In Figure 6.9 (especially 6.9c) a polyelectrolyte molecule is depicted with a volume around it that contains an excess of counterions and that is depleted of coions. Actually, there is no sharp boundary surface involved, since the difference in concentration between counterions and coions gradually decreases with distance from the polyelectrolyte molecule. This is reflected in the gradual decrease in electric potential, as depicted in Figure 6.8. Since there are no restrictions to volume changes, solvent (water) will be drawn to the region close to the polyelectrolyte molecule, to even out the osmotic pressure between that region and the salt solution farther away. It may be noted that this provides an alternative explanation for the expansion of a polyelectrolyte molecule at low ionic strength, as discussed in Section 6.3.2. In fact, treatment of polyelectrolyte conformation in terms of osmotic pressure leads to the same results as that based on electrostatic repulsion, discussed earlier. The two theories are equivalent.

Consequences. Due to the effect discussed in the previous paragraph, it is difficult to remove specific ions from a polyelectrolyte. Of course, the condition of electroneutrality implies that counterions must always be present, in proportion to the valence of the polymer. They can be exchanged by other ions, say Ca^{2+} by K^+. However, several "washings" are needed: at high ionic strength because the ion concentration is high, at low ionic strength because the volume containing counterions is large. This means that usually several washings of the polyelectrolyte with a salt solution of other composition would be needed.

Another consequence of the Donnan effect is that it is difficult, if not impossible, to calculate ionic strength and composition of a solution of polyelectrolytes and salts, especially if the polyelectrolyte concentration is high. One should try to separate a portion of the salt solution from the mixture, without applying a substantial chemical potential difference. This can be done, for instance, by ultrafiltration. The ultrafiltrate then contains, ideally, no polyelectrolyte, but all the salts. It can be chemically analyzed, and from the result the ionic composition can be calculated (Section 2.3.3).

Note In the derivations given in this section we have assumed complete dissociation of salt and identical ion activity coefficients $(\gamma_\pm)$ in both compartments. Complete dissociation may not occur at high ionic strength; see Section 2.3.3. The second conditions will not be met at very low ionic strength (I). The activity coefficients will be different owing to the presence of the polyelectrolyte, which has a very low mass concentration but a very high valence. Taking the example in Table 6.2 for $m_2 = 1$ mmolar, Eq. (2.27) yields $I = 51$ and 0.67 mmolar for compartments (1) and (2), respectively. Equation (2.30) then gives for $(\gamma_\pm)^2$ values of 0.7 and 0.96, respectively. These complications materially affect the results. The presence of small ions of higher valence further complicates the theory.

6.4 MORE CONCENTRATED SOLUTIONS

So far, we essentially considered interactions between one polymer molecule and the solvent. In this section, mutual interactions between polymer molecules come into play. Nevertheless, the solutions remain fairly dilute, i.e., at most a small percentage of polymer.

6.4.1 Nonideality

Nonideality of solutions is discussed in Section 2.2.5. It can be expressed as the deviation of the colligative properties from that of an ideal, i.e., very dilute, solution. Here we will consider the virial expansion of osmotic pressure. Equation (2.18) can conveniently be written for a neutral and flexible polymer as

$$\Pi = RT \left(\frac{1}{V_p} \varphi + \frac{\beta}{2 V_s} \varphi^2 + \frac{1}{3 V_s} \varphi^3 + \cdots \right) \tag{6.11}$$

where V_p is the molar volume of the polymer (in $m^3 \cdot mol^{-1}$), V_s the molar volume of the solvent, φ the net volume fraction of polymer present (i.e., concentration in kg per m^3 divided by the mass density of the polymer), and β is the excluded volume parameter defined in Section 6.2.1. By determination of osmotic pressure over a range of concentrations, the number-average molar mass and β can be derived by use of Eq. (6.11).

Some calculated results are in Table 6.3. It is seen that the nonideality (i.e., the magnitude of the second and third virial terms in comparison to the first virial term) may be very large, especially for large n: a large n implies a small first virial term, and the second and third terms are independent of n. If the polymer behaves like an ideal random chain ($\beta = 0$), the second virial term equals zero. For $\beta < 0$, it is even negative, but then the solubility of the polymer is quite small (Section 6.5.1).

Even far stronger nonideality may occur for **polyelectrolytes**, especially at low ionic strength. The chain is much more expanded (high β). Moreover, the condition of electroneutrality implies that the polymer is accompanied by additional ions. The second virial term then approximately equals

$$\frac{z^2}{4 m_c V_p^2} \varphi^2 \tag{6.11a}$$

where z is the valence of the polymer and m_c is the concentration of (monovalent) counterions (in $mol \cdot m^{-3}$). Results are also given in Table 6.3. It is seen that for $\varphi = 0.003$ and $m_c = 1$, the second virial term gives rise to an osmotic pressure over 3 bar ($126 \, RT \approx 310 \, kPa \approx 3.1 \, bar$). These results

TABLE 6.3 Osmolality of Polymer Solutions[a]

φ	0.003	0.01	0.03
1st virial term			
$n = 100$	0.75	2.50	7.5
$n = 1000$	0.075	0.25	0.75
2nd virial term			
$\beta = 0$	0	0	0
$\beta = 0.5$	0.125	1.39	12.5
2nd virial term			
$m_c = 1$	126	1400	12,600
$m_c = 10$	12.6	140	1260
3rd virial term	0.005	0.02	0.5

[a] Given are the virial terms $(\mathrm{mol \cdot m^{-3}})$ calculated according to Eq. (6.11). The osmotic pressure in Pa thus equals $RT \approx 2500$ times the sum of the virial terms. The second virial term is given for a neutral polymer (main variable β) and for a polyelectrolyte (main variable m_c). Molar volume of solvent $V_s = 18 \cdot 10^{-6}$, of polymer $V_p = 4 \cdot 10^{-5} \, nm^3 \cdot \mathrm{mol^{-1}}$. $n =$ number of monomers per molecule. For the polyelectrolyte the valence is $0.3n$, and $m_c =$ concentration of counterions $(\mathrm{mol \cdot m^{-3}})$.

imply that determination of the molar mass of a polyelectrolyte via osmotic pressure should be done at very high ionic strength and requires extrapolation to very small φ values. (It may be noted that some other complications arise, but these can be taken into account.)

6.4.2 Chain Overlap

When the concentration of polymer in solution is increased, the coiled molecules effectively fill the whole volume at a certain critical concentration. This was briefly discussed in the Question in Section 6.2.1. At still higher concentration, chain overlap occurs; in other words, the molecules interpenetrate, becoming mutually entangled. This is illustrated in Figure 6.12 for a neutral polymer.

Polymer scientists distinguish various **regimes**, as shown in Figure 6.13, which gives an example of a state diagram; actually, it represents quite a small part of the diagram, albeit the part that is the most relevant at the moment. The variables are the (net) volume fraction of polymer φ and the excluded volume parameter β (or the solvent-segment interaction parameter χ). In the domain or regime called dilute, the relations discussed in Section 6.2 hold; going from A via B ("ideal") to C, the molecules decrease in extent of expansion. Both in the "semidilute" and the concentrated regime, there is chain overlap, but the relations between properties of the solution and polymer characteristics are different. In the domain labeled phase separation, the polymer is not fully soluble: see Section 6.5.1.

Two of the boundaries between the regimes are roughly given by $\beta = \pm \varphi$: see Figure 6.13. The critical point for phase separation is approximately

$$\varphi_{cr} = \frac{1}{1 + q^{0.5}} \tag{6.12a}$$

$$\beta_{cr} = 1 - (1 + q^{-0.5})^2 \tag{6.12b}$$

where $q =$ the net volume of a polymer molecule divided by the volume of a solvent molecule. It follows that q is proportional to n, and for most polysaccharides in water $q \approx 0.04 \, M$, which mostly is $500-10^5$. Some results are given in Table 6.4. If n is high, β_{cr} is seen to be very close to zero and φ_{cr} is quite small. φ_{cr} also marks the boundary between the dilute and the concentrated regime, but the range in β over which this boundary extends, i.e., about from $-\beta_{cr}$ to β_{cr}, is very small for large q: see Table 6.4. Many food polymers in water have β-values above 0.01, which means that they would pass from the dilute to the "semidilute" to the concentrated regime, when increasing their concentration (φ).

Actually, we have used the term "semidilute" in a loose sense here (hence the quotation marks); it denotes neither dilute, nor concentrated. Polymer scientists distinguish a semidilute and a marginal regime. Scaling

FIGURE 6.12 Schematic examples of well soluble polymer molecules in solution at increasing concentrations. (Modified from P. G. de Gennes. Scaling Concepts in Polymer Physics. Cornell Univ. Press, 1979.)

laws have been derived for the boundaries between the regimes in the state diagram and for some characteristic properties. We will not discuss those in detail, since (a) the numerical constants are unknown and (b) the laws do not really apply to most food polymers because of their heterogeneity: chain length (n) varies between molecules and solvent quality (β) and stiffness

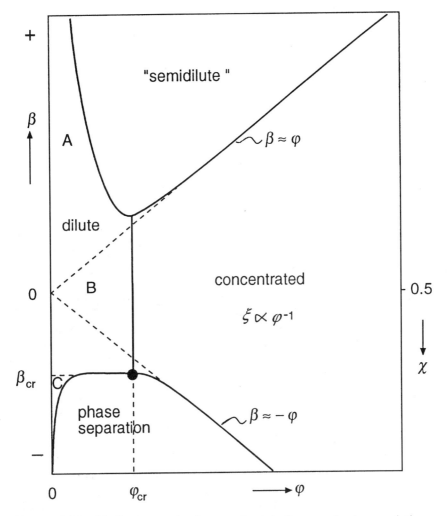

FIGURE 6.13 Idealized example of a part of a state diagram of polymer solutions. $\varphi =$ net polymer volume fraction, $\beta =$ excluded volume parameter, and $\xi =$ correlation length. The critical point for phase separation is denoted by ●. See text.

TABLE **6.4** Examples of the Critical Volume
Fraction φ_{cr} and the Critical Volume Expansion
Factor β_{cr} as a Function of the Molecular Volume
Ratio Polymer/Solvent q, According to Eq. (6.12)

q	φ_{cr}	β_{cr}
10^2	0.091	-0.21
10^3	0.031	-0.06
10^4	0.010	-0.02
10^5	0.003	-0.006

(b/L) vary along the chain. Nevertheless, it may be useful to have some idea
about the concentration needed for chain overlap; it can roughly be given by

$$\varphi_{ov} \approx (\beta n)^{-1} \tag{6.13}$$

which will roughly hold if the polymers are fairly stiff and β is not much
larger than 0.1, which is true for most polysaccharides. Clearly, chain
overlap is already reached for very small concentrations, often well below
1%.

The **correlation length** ξ is a parameter that quantifies the distance over
which fluctuations in a system are correlated. In a dilute polymer solution it
roughly equals the radius of gyration. For conditions of chain overlap, ξ can
be defined as the diameter of the "blobs" illustrated in Figure 6.14. The
value of ξ decreases with increasing stiffness, β and φ, in a manner
depending on the regime. In practice, the most important variable
determining the correlation length is thus the concentration. The degree
of polymerization n has no effect: if in a "semidilute" or a concentrated
solution some polymer molecules would be split into shorter ones, or if some
polymer molecules would be joined to form longer ones, the properties of
the solution would not or hardly change. Of course, the value of n is an
important variable in determining whether chain overlap occurs [Eqs.
(6.12a) and (6.13)], but once in a nondilute regime, it further has little effect
on macroscopic properties.

In the **"semidilute" regime**, the molecules cannot distribute themselves
at random over the volume. The polymer concentration fluctuates with a
wavelength equal to the correlation length. The system can be seen as a kind
of network with mesh size comparable to ξ. The "network" continuously
changes conformation due to Brownian motion. Over distances along the
polymer chain $< \xi$, which implies short time scales for molecular motion,
polymer sub-chains behave as in a dilute solution; interactions between two

adjacent sub-chains are the same whether the two belong to the same molecule or not. For distances $> \xi$, i.e., at longer time scales, behavior is different; the diffusivity of a whole polymer molecule is greatly reduced. Generally, macroscopic properties of the solution do not depend on the radius of gyration of the polymer, but on the correlation length. The osmotic pressure, for instance, is now approximately given by kT/ξ^3. This means that every "blob" in the system behaves as if it were one molecule, insofar as the colligative properties are considered. Notice that Eq. (6.11) is not valid outside the dilute regime.

In the **concentrated regime**, the concentration of the polymer is fairly even. Nevertheless, a correlation length can be defined, and it is proportional to $1/\varphi$. For most food polymers, this regime will be reached at fairly high polymer concentrations, say above 3%, unless β is very small.

For **polyelectrolytes**, the stiffness (b/L) is mostly larger than for the same polymer uncharged. This would mean that the correlation length is larger under most conditions. However, the relations determining the boundaries in the state diagram and the correlation length in the various regimes have not been well established. For charged food polysaccharides in the nondilute regimes, the correlation length will probably increase with increasing charge and decreasing ionic strength.

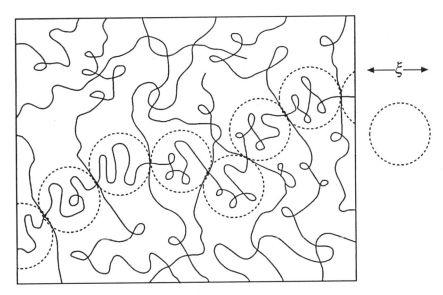

FIGURE 6.14 Explanation of the concept of correlation length ξ in a semidilute or concentrated polymer solution. (From P. G. de Gennes. Scaling Concepts in Polymer Physics. Cornell Univ. Press, 1979.)

6.4.3 Viscosity

Polymeric thickening agents used in foods typically are well soluble polysaccharides, with an excluded volume parameter β clearly above zero. This implies that especially the dilute and semidilute regimes are often of importance. Viscosity of very dilute solutions has been discussed in Sections 6.2.2 and 6.3.2. For higher concentrations, the reduced viscosity (η_{sp}/c) is higher, as is true for any system (see Figure 5.5), but for polymer solutions the viscosity increases far stronger with concentration as soon as the chain overlap concentration is reached.

It has been shown (and made plausible from theory) that a general relation exists between relative or specific viscosity at very low shear rate and dimensionless concentration of polymer $c[\eta]_0$. This relation is shown in Figure 6.15 as a log–log plot. The *critical concentration for chain overlap* c^* would equal about $4/[\eta]_0$; at that concentration, $(\eta_{sp})_0 \approx 10$ or $\eta \approx 11\eta_s$ (at very low shear rate). c^* is mostly expressed in g/100 mL. It may not exactly correspond to the critical volume fraction φ_{ov} given by Eq. (6.13). Most workers use c^*, as determined from the break in a log–log plot of specific viscosity versus concentration.

> *Note* For several polymers, there is a limited concentration region where the slope gradually changes from small to large and one takes as the critical concentration that where the steepest slope begins, often denoted c^{**}. Possibly, the concentrated regime then has been reached.

For $c < c^*$, the slope is 1.2–1.4, i.e., not unity, which would be expected for dilute solutions. A reason for the discrepancy may be that most polysaccharides are so stiff that (part of) the molecules are already somewhat rodlike; they may then already hinder each other in a shear flow at concentrations well below φ_{ov}. Above c^*, the slope is close to 3.3 for most polysaccharides. Galactomannans (guar and locust bean gums) show a steeper slope, about 4.4. This is supposed to be due to specific attractive interactions between side groups on the polymer chains. Some further data are in Table 6.5. Remember that these relations only apply at extremely small shear rates. At higher shear rates, c^* remains about the same, but the slope above c^* is far smaller, e.g., about 2. A steep slope of $\log (\eta/\eta_s - 1)_0$ versus $\log c$ also means a very strong dependence of viscosity on molar mass M of the polymer. In the "semidilute" regime, $(\eta/\eta_s - 1)$ is approximately proportional to $M^{3.4}$, at constant mass concentration. This relation is far from exact, but the strong dependence of η on M, much stronger than in the dilute regime, is unmistakable.

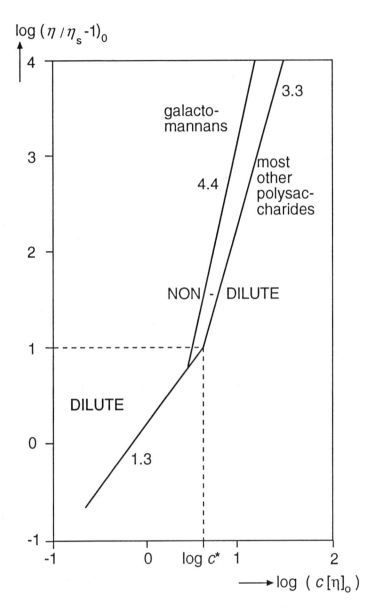

FIGURE 6.15 Viscosity of polysaccharide solutions. Specific viscosity at very low shear rate $(\eta/\eta_s - 1)_0$ versus dimensionless concentration $(c[\eta]_0)$; $[\eta]_0$ is the zero shear rate intrinsic viscosity. (From results by E. R. Morris et al. *Carbohydr. Polym.* 1 (1981) 5.

TABLE 6.5 Viscosity of Polysaccharide Solutions[a]

Polysaccharide	c^* g/100 mL	slope
Xanthan[e]	0.1	3.9
Na-alginate[e]	0.2	3.3
Locust bean gum	0.2	4.4
Guar gum	0.25	4.4
Pectin[e]	0.3	3.3
λ-carrageenan[e]	0.4	3.3
Dextran (linear)	2.5	3.3

[a] The critical concentration c^* for chain overlap and the slope of the relation $\log(\eta_s)_0$ versus $\log c$ for $c > c^*$. Actually, results vary among samples and, for charged polysaccharides, with ionic strength. The latter is fairly high (about 0.1 molar) for the data presented.
[e] polyelectrolyte.

Part of the data presented in Figure 6.15 and Table 6.5 concern polyelectrolytes, but they were obtained at fairly high ionic strength. It may be recalled that c^* would be much smaller at low ionic strength. Also the slope for $c > c^*$ in plots, as in Figure 6.15, tends to be somewhat higher at low I. Altogether, the viscosity obtained at a certain polyelectrolyte concentration is far higher at low than at high ionic strength.

For $c > c^*$, the viscosity is strongly **strain rate thinning**, far stronger than in the dilute regime. This is illustrated in Figure 6.6 for xanthan solutions. The lower concentration is just below, the higher one well above c^*. It is seen that the apparent viscosity decreases by 3.5 orders of magnitude over the range of shear rates applied. The explanation for this decrease is somewhat related to the effects discussed in Section 6.2.2, but the mechanism is a different one. Above the chain overlap concentration, the polymer chains exhibit *entanglements*; Figure 6.16 illustrates this. The smaller the correlation length, which implies the higher the polymer concentration, the greater the number of entanglements per unit volume. When applying a shear rate to the solutions, the shearing stress causes disentangling. This is why the viscosity is so high: the disentangling requires a relatively large amount of energy. If the polymer molecules become fully disentangled, the viscosity will be very much smaller. However, Brownian motion of the polymer chains causes new entanglements to form. The time available for the formation of entanglements during shearing is roughly the reciprocal of the shear rate. At very low shear rate, as many entanglements are formed as are loosened per unit time, and the viscosity is Newtonian,

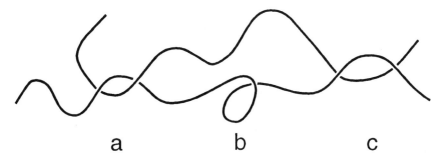

FIGURE 6.16 Entanglements. Depicted are two polymer chains. At (a) they form an entanglement, but not at (b) or (c).

i.e., independent of shear rate, and high. At very high shear rate, the time allowed for entanglements to form is very short, and the molecules are (almost) fully disentangled, yielding a relatively low viscosity. The relation between apparent viscosity η^* and shear rate depends on polymer type and concentration and on solvent quality.

Question 1

Consider an aqueous solution of carrageenan, at a concentration above the chain overlap concentration, and compare it with a solution of the same mass concentration, but with a carrageenan of larger molar mass. Would the following properties be larger, smaller or the same?
    Apparent viscosity
    Dependence of apparent viscosity on concentration
    Extent of shear rate thinning
    Correlation length
What would happen if to the carrageenan solution an equal volume were added of
    Water
    Ethanol
    NaCl solution, final concentration 0.1 molar
    HCl solution, final concentration 1 molar

Question 2

What is the zero shear rate viscosity of 0.25% aqueous solutions of

Dextran
Na-alginate at high ionic strength
Locust bean gum
Xanthan at high ionic strength
Xanthan at low ionic strength

What would the viscosities be at a shear rate of $10\,s^{-1}$?

Answer

Table 6.5 gives the critical concentration for chain overlap c^* and the slope of the log–log relation for $c > c^*$ for systems 1–4. With the help of Figure 6.15 or a newly drawn graph, the specific viscosity at zero shear rate can be read off. Assuming the viscosity of the solvent, water, to be $1\,mPa \cdot s^{-1}$, the viscosity directly follows. For example, for Na-alginate, $c^* = 0.2\,g/dL$. Since $c^*[\eta]_0 = 4$, $[\eta]_0 = 20$ and $c[\eta]_0 = 5$. The logarithm of $5 = 0.7$, and we read from Figure 6.15 that $\log(\eta_{sp}) = 1.33$ or $\eta_{sp} = 21$, leading to $\eta = 22\,mPa \cdot s$, all at zero shear rate. There are some complications. For dextran, $c < c^*$, and the lower slope curve should be used. For locust bean gum, Figure 6.15 shows a $c^*[\eta]_0$ value of 2.7. For xanthan, the slope is different, i.e., 3.9. The results are in the Table.

Polysaccharide	c^* g/dl	$c^*[\eta]_0$	$c[\eta]_0$	η_0 mPa·s	$\eta_{a,10}$ mPa·s
Dextran	2.5	4	0.4	1.5	1.2
Na-alginate, high I	0.2	4	5	22	10
Locust bean gum	0.2	2.7	3.4	17	8
Xanthan, high I	0.1	4	10	300	20
Xanthan, low I	0.05	4?	20	10^4	50

Those for xanthan, low I (about 1 mmolar), have been obtained by using the earlier given value of $c^* \approx 0.05\,g/dL$. The value for η_0 obtained agrees well with the result shown in Figure 6.6. It can also be read from that graph that $\eta_{a,10}$, i.e., the apparent viscosity at a shear rate of $10\,s^{-1}$, is about $50\,Pa \cdot s$. For the other systems, obtaining values for $\eta_{a,10}$ is almost guesswork, and such results are given in the table. It may be noticed that the range of viscosities is very much wider for η_0 (a factor 7000) than for $\eta_{a,10}$ (a factor 40).

6.4.4 Gelation

In a polymer solution above the chain overlap concentration, *entanglements* occur (Fig. 6.16). If such a solution is deformed, part of the polymer chain sections between entanglements will become stretched, by which their conformational entropy decreases. Each chain section tends to regain its original conformation, which implies that the material behaves in an elastic manner. If the deformation is very slow, the entangled chains can slide along each other while continuously seeking a conformation of largest entropy, and the main result of the entanglements is an increase in viscosity (Section 6.4.3). If the deformation is fast, considerable stretching of chain sections will occur, and the elastic effect will be strong. We now have a *viscoelastic* or memory fluid: see Section 5.1.3. If *permanent cross-links* are formed between entangled chains, a *gel* is obtained, as is discussed in Section 17.2.2.

6.5 PHASE SEPARATION

Polymers have, like all solid materials, a limited solubility, but if the concentration becomes greater than the solubility, separation into liquid phases occurs, not precipitation. The factors governing this are briefly discussed below.

6.5.1 One Solute

Going back to Section 6.4.2, Figure 6.13 shows a (hypothetical) result for the solubility of a homopolymer. One relevant relation is given in the figure, the others are in Eq. (6.12), with some calculated examples in Table 6.4. It follows that the important variables are the excluded volume parameter β and the polymer–solvent molecular volume ratio q (proportional to the degree of polymerization n). It is also seen that the critical value of β for solubility is close to zero for high molar mass polymers.

   To explain these observations, it may be remembered that the solubility of a substance is reached if its molar free energy is equal in the undissolved and the dissolved states. This can be expressed as $-\Delta G_{mix} = \Delta_{d \to u} G = \Delta H - T \Delta S = 0$, and since entropy generally decreases when a solute goes from the dissolved to the undissolved state ($\Delta S < 0$), ΔH must be negative (net attraction between solute molecules). For small spherical molecules, the enthalpy change precisely equals T times the mixing entropy at the solubility limit. For a polymer the situation is more complicated. The enthalpy term is directly related to β, and for $\beta < 0$, there is net attraction between polymer segments. The translational mixing entropy is very small; it goes to zero when q becomes very large, since the molar concentration then is negligible. Consequently, β_{cr} will be very close to zero for high q, as observed. However, on precipitation of the polymer, it would lose *conformational entropy*, by a very large amount (see Section 6.1). This implies that precipitation will not occur, unless β is very small (near or below -1). Instead, phase separation occurs: the solution separates into one with a low and one with a high concentration of polymer. In both phases, the conformational entropy still is large. The highly concentrated, and thereby viscoelastic, polymer solution is often called a *coacervate*.

The phenomena involved are further illustrated in Figure 6.17 for simple homopolymers, calculated according to Flory–Huggins theory. It gives phase diagrams in the $\beta - \varphi$ domain, for some values of q. An example of phase separation is depicted for a polymer–solvent mixture of $q = 100$, $\varphi = 0.1$, and $\beta = -0.525$. What will occur at equilibrium is a separation into phases A and B, containing a volume fraction of polymer of 10^{-4} and 0.5, respectively. The ratio of the volumes of A over B is given by $(0.5 - 0.1)/(0.1 - 10^{-4}) = 4$, i.e., the (very) dilute part makes up 80%, the concentrated part 20%. It is also seen that at larger, i.e., more "normal," values of q, the low-concentration phase is extremely dilute, almost pure water. The few polymer molecules in the dilute phase are called collapsed coils, meaning that they have a relatively small radius of gyration. The exponent ν in Eq. (6.4) is < 0.5, approaching the minimum value of $1/3$.

If β is small (highly negative), the high concentration phase is very concentrated. In such a phase, i.e., in the regime called "concentrated" in Figure 6.13, other relations hold than in the dilute or "semidilute" regimes. The correlation length is inversely proportional to φ. The viscosity is not proportional to that of the solvent, but the system behaves like a polymer melt, containing water as a plasticizer. The viscosity is extremely large for low water content (small $1 - \varphi$).

Figure 6.17 shows two kinds of curves, "*binodals*" and one example of a "*spinodal*." The binodal is the curve for $\Delta G_{\text{mix}} = 0$, and below that curve the solution is supersaturated. This does not necessarily mean that phase separation occurs. The spinodal curve, given by $(\partial^2 \Delta G_{\text{mix}}/\partial 2\varphi)_{T,p} = 0$, bounds a regime in which phase separation is spontaneous. Here, any thermal fluctuation in the system will lead to the formation of regions of different composition, which means phase separation. In the regime between the binodal and the spinodal curves, phase separation depends on a mechanism of nucleation and growth of a phase. This is discussed in Section 14.2.3; see especially Figure 14.8. The system is thus metastable, and it may take a long time before phase separation occurs, especially if the initial φ is high (very high viscosity).

According to theory, the solubility will increase if temperature is increased. The larger value of $-T \Delta_{d \rightarrow u} S$ means that ΔH has to be more negative for phase separation to occur. There are, however, several exceptions to this rule, for instance because ΔH may depend on temperature. For many polysaccharides, the temperature effect seems to be fairly small.

This brings us to the question about the applicability of the Flory–Huggins theory for food polymers. For polyelectrolytes, the theory is invalid, unless ionic strength is very high. In Section 7.3 the solubility of proteins will be discussed. Very few polysaccharides are simple homo-

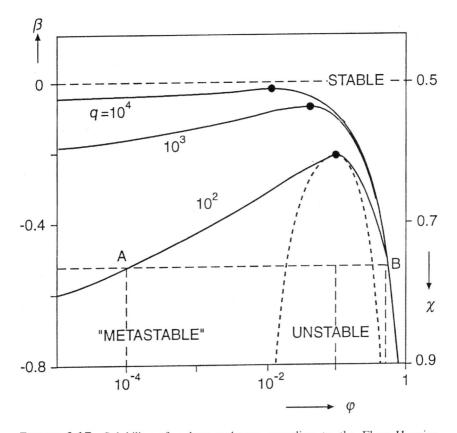

FIGURE 6.17 Solubility of a homopolymer according to the Flory–Huggins theory. Variables are the excluded volume parameter β (or the polymer–solvent interaction parameter χ), the net volume fraction of polymer φ, and the polymer-to-solvent molecular volume ratio q. Solid lines denote binodal, the broken line spinodal decomposition. Critical points for decomposition (phase separation) are denoted by •. See text.

polymers, and several specific interactions between groups on the polymer may play a part, implying that precise prediction of phase separation is generally not possible. Nevertheless, the trends observed generally agree with theory. When adding ethanol to a polymer solution, whereby β is decreased, most polysaccharides will indeed show spontaneous separation into a highly concentrated viscoelastic phase and a very dilute one.

That even small changes in *chemical structure* can have an enormous effect on solubility is illustrated by the difference between amylose and cellulose. Both are 1→4 linked linear chains of glucose, but as shown in Figure 6.18, amylose is a polymer of α-glucose and cellulose of β-glucose. This implies that amylose cannot form a straight chain, but that cellulose can. The latter molecules can become perfectly aligned, forming stacks that are almost true crystals, held together by Van der Waals attraction and hydrogen bonds. Accordingly, cellulose is completely insoluble in water. A similar structure cannot occur for amylose, which is to some extent soluble in water. Actually, even amylose can form a kind of microcrystalline regions, since the linear molecules can readily form regular helices, which then can become stacked. Dextrans are chemically almost identical to amylose, but have some branching of the polymer chain, which prevents this kind of stacking. Consequently, most dextrans are well soluble in water.

Note The term phase separation is often used indiscriminately when separation into layers is observed. In this section true phase separation is considered. Although both phases are aqueous solutions, there is a phase boundary between them, exhibiting an interfacial tension, albeit small, mostly $< 0.01\,\mathrm{mPa \cdot s}$.

6.5.2 Polymer Mixtures

If a solution contains two polymers at high concentration, phase separation generally occurs, especially if the polymers have a high molar mass. Phase separation may be of two kinds. It is illustrating to consider Eq. (2.19), which gives a virial expansion (of the osmotic pressure) for a mixture of two solutes, 2 and 3, 1 denoting the solvent. The so-called mutual second virial coefficient B_{23} now determines what will happen. If $B_{23} > 0$, the two polymers tend to stay away from each other; they are preferentially surrounded by identical molecules (or by solvent). This may lead to a separation in a phase rich in polymer 2 (and poor in 3) and one rich in 3 (and poor in 2). Figure 6.19a gives a hypothetical phase diagram. The polymers show *segregative* phase separation and are said to be *incompatible*. If, on the other hand $B_{23} > 0$, the two polymers attract each other. This leads to *associative* phase separation in a phase rich in both polymers, called a *complex coacervate*, and a phase depleted of polymer. A hypothetical phase diagram is in Figure 6.19b. Also for $B_{23} \approx 0$, complex coacervation will generally occur at high polymer concentration.

   Incompatibility. For two polymers that do not have side groups causing mutual attraction, incompatibility mostly occurs, and the concentration needed for phase separation is smaller as the molar mass of the polymers is higher. A good example is the phase separation occurring

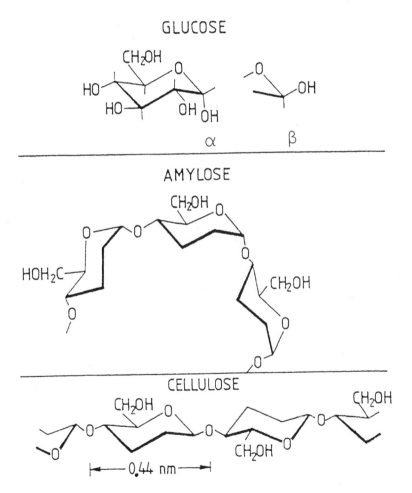

FIGURE 6.18 A glucose molecule in α- and β-configuration, and formation of 1–4 glycosidic polymers from those. The α-form produces amylose, the β-form cellulose.

between amylose and amylopectine in fairly dilute (gelatinized) starch solutions. The solvent quality also affects phase separation; for instance, incompatibility is generally less at large sugar concentrations.

Figure 6.19a gives a hypothetical example. The tie lines indicate how the separation will be. A mixture of composition A will separate into phases of composition B and C; the ratio of the volumes of the solutions of composition A to B is as the ratio of the distances AC/AB. The longer the tie line, the stronger the incompatibility. The dot gives the critical point, i.e., the composition at which the tie line vanishes. The heavy line giving the solubility is a binodal. It is mostly not possible to calculate the phase diagram from the properties of the two polymers.

In the concentration range between the binodal and the spinodal (not shown), separation may be very slow. Even spinodal decomposition may take long to become visible, because the system often is concentrated and very viscous, implying slow diffusion of the polymers; a concentrated phase may even tend to gel. Mostly, one of the phases forms droplets, and the system may be called a *water-in-water emulsion*. Which phase becomes the continuous one depends on the concentration ratio of both polymers. If c_3/c_2 is larger than the ratio at the critical point, the *continuous phase* tends to become the one rich in c_2, and vice versa. The interfacial tension between

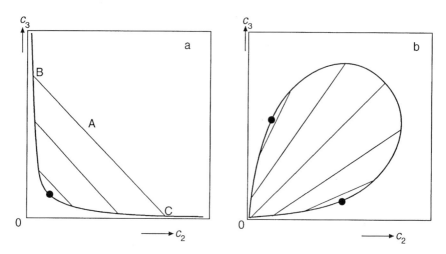

FIGURE 6.19 Idealized cases of phase separation in aqueous mixtures of two polymers, concentrations c_2 and c_3. (a) Segregative phase separation or incompatibility. (b) Associative phase separation or complex coacervation. The heavy lines denote the binodal (solubility limit), the thin ones are tie lines. The dots indicate critical points.

the phases is very small, 10^{-7} to $10^{-4}\,\mathrm{N\cdot m^{-1}}$, being higher for greater incompatibility.

An important case of incompatibility is formed by mixtures of **proteins** and neutral polysaccharides. Examples are in Figure 6.20. It is seen that the phase diagrams may be very asymmetric, the protein concentrations needed for phase separation being much larger than the polysaccharide concentration. The asymmetry is stronger for globular proteins than for more or less unfolded molecules like gelatin or casein. Proteins are polyelectrolytes, and if the pH is not close to the isoelectric point and the ionic strength is low, phase separation does not occur. This is due to the Donnan effect (Section 6.3.3). The presence of counterions implies that phase separation would go along with separation of salt ions, causing considerable loss of mixing entropy. This, of course, counteracts any decrease in enthalpy due to phase separation. As seen in Table 6.2, the relative difference in salt concentration between the "compartments" decreases with ionic strength, and it almost vanishes at 0.1 molar. It is indeed observed that phase separation only occurs at ionic strengths of 0.1 molar or higher, unless the pH is near the isoelectric point of the protein. In the latter case a low salt content promotes phase separation, presumably because the solubility of the protein strongly

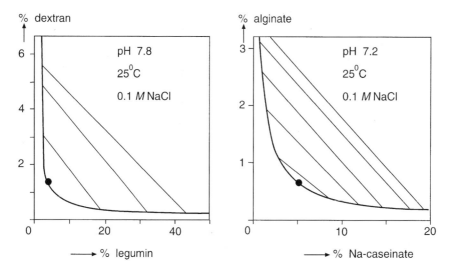

FIGURE 6.20 Examples of incompatibility of proteins and polysaccharides. (After results by V. B. Tolstoguzov.)

increases with increasing ionic strength (see Section 7.3). Two polyelectrolytes of equal charge density may show segregative phase separation irrespective of ionic strength, since separation would not cause separation of counterions.

The incompatibility may have various **consequences**. It may be a nuisance when a homogeneous liquid is desired, since it may lead to slow separation into layers. If one of the phases is high in protein, any aggregation of the protein during heating will proceed faster. Moreover, at higher temperature the incompatibility of proteins and polysaccharides is generally greater, and the lower viscosity allows faster phase separation to occur. On the other hand, the phenomenon can be made useful in concentrating one of the polymers; it can be seen as a kind of membraneless ultrafiltration. Careful optimization of conditions can yield a practicable process. Water-in-water emulsions can be useful stages in manufacturing foods, for instance when either of the phases is made to gel. The droplets can be very easily deformed, because of their very small interfacial tension, and this can be employed in making threadlike particles that can then be made to gel ("spinneretless spinning").

Complex Coacervation occurs if the two types of polymers have side groups that are mutually attractive. A prime example is a mixture of a protein below its isoelectric pH (positive groups) and an acid polyelectrolyte (negative groups), the first known system being an acid solution of gelatin with gum arabic. Figure 6.19b gives an idealized phase diagram. As in Figure 6.19a, the tie lines give the composition and the volumes of the two phases obtained.

Coacervation of two polyelectrolytes occurs especially at low ionic strength I, for instance below 0.2 molar. At higher I, the charges are sensed at very small distances only; see Figure 6.8. Some proteins may also exhibit complex coacervation with a polyacid at a pH above their isoelectric point. An example is given by caseinate with κ-carrageenan, where some positive groups on the caseinate are responsible for the attractive interaction, although the protein has more negative groups. In such a case, coacervation only occurs at intermediate I: if I is very small, the negative and positive groups on the protein molecule cannot be sensed separately, whereas this can occur at higher I (cf. Figure 6.9).

Actually, attractive interactions between two types of polymers can become manifest in various ways. Besides coacervates, small soluble complexes can be formed. If the interactions are weak, a homogeneous weak gel may form. If the interactions are strong, coprecipation of both polymers may occur; this can be applied to separate proteins from solutions, for instance with carboxymethyl cellulose. The relations governing which of these phenomena will occur are not fully understood.

6.6 STARCH

In this section a particular food polymer will be discussed in some detail. Starch is a very important nutrient, providing roughly half of the edible energy used by mankind. Moreover, large quantities are isolated and used as functional ingredients, starch being the most important thickening agent applied in manufacture and preparation of food.

In its native form, starch occurs as a very concentrated polymer mass, and discussion of its properties may provide some understanding of such systems in general. On the other hand, native starch granules provide an example of a highly specific structure. Another example of a concentrated system of polymers is the cell wall of plants. A cell wall contains several different polysaccharides, and its structure is even far more intricate and specific than that of starch. It will not be discussed here.

6.6.1 Description

Starch is a polymer of α-D-glucose. It occurs in most higher plants species, in the form of roughly spherical granules, ranging from 2 to 100 μm in diameter. The granules consist of about 77% starch and about 1% other dry matter (lipids, proteins, minerals), the remainder of the mass being water. Starch granules often are surrounded by large quantities of water, but they do not dissolve, in accordance with their physiological function. Starch provides energy when needed, which may be a long time after its synthesis, for example during the germination of a seed. It is then hydrolyzed by enzymes (amylases) to yield sugars, mostly maltose, that are well soluble and can be metabolized.

Starch consists of two components. The one is **amylose**, a linear polymer of $1 \rightarrow 4$ linked anhydroglucose monomers; see Figure 6.18. The degree of polymerization n ranges from several hundred to about 10^4. Amylose mostly makes up 20 to 30% of starch. Amylose in solution readily forms helices of a pitch of about 6 monomers and a diameter of about 1 nm. These helices can accommodate hydrophobic molecules in their central cavity, especially fatty acid chains; fairly stable compounds are formed. Iodine can be included in a similar way, and the compound has an intense blue color, which allows identification of amylose and determination of its concentration.

Amylopectin, the main component, has $n = 10^5 – 10^8$, but it should be realized that the size of very large polymer molecules ($n > 10^6$) is always uncertain, since the solubilization procedure needed for determination readily causes breaking of very long molecular chains. Amylopectin is highly branched, 4–5% of the monomers also having a $1 \rightarrow 6$ linkage. Its structure is still a matter of some debate; a fairly well-established model is in Figure 6.21. The average length of the branches is a few tens of monomers, although some are smaller and some larger. The figure shows that most of the branches take part in only one "cluster." The longer ones span two or even three clusters, thereby providing connections between these.

When viewed with a polarizing microscope, with crossed polarizer and analyzer, native starch granules show marked **birefringence**, which means that they can be seen.

Note A birefringent material has two or even three optical axes, which causes the refractive index to vary with the direction of the wave vector of the polarized light.

Consequently, the state of polarization of the light passing through the material is generally changed, and part of the emerging light can pass the analyzer; a nonbirefringent material remains dark. Birefringence is caused by a regular ordering of molecules over distances of about half a wavelength or more. This does not necessarily imply that the material is crystalline, since stacks of polymer molecules with a roughly parallel orientation also show strong birefringence. Birefringence can be positive or negative with respect to the direction of the wave vector, according to the direction at which the

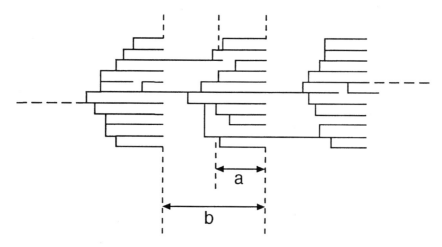

FIGURE 6.21 Schematic illustration of the structure of a small part of an amylopectin molecule. The lines denote stretches of linear $1 \rightarrow 4$ linked anhydroglucose units, the branching points are $1 \rightarrow 6$ linkages. The molecule grows from left to right. *a* denotes the approximate thickness of a microcrystalline region, *b* the repeat distance of such regions. (Redrawn from S. Hizukuri. *Carbohydr. Res.* 147 (1986) 342.)

refractive index is higher. The sign observed in a starch granule indicates that the molecules have a roughly radial orientation. This fits with the growth of a granule being in the radial direction, starting at the so-called hilum, which becomes its center.

Whether a material is crystalline or not can be established by x-ray diffraction. X-rays have a very small wavelength, of the order of 0.1 nm, which implies that individual atoms may cause scattering. If the atoms (or small molecules) occur at regular distances, sharp diffraction maxima occur, and the crystal structure can be derived from the diffraction pattern. Native starch indeed shows a distinct diffraction pattern. It can be concluded from these and other data that 20 to 45% of the starch is in a crystalline form. Since this concerns almost exclusively amylopectin, this material would be crystalline for 30 to 50%. A starch granule thus has **crystalline regions**, the remainder being amorphous. The crystalline regions are small (order of 10 nm) and are called (*micro*) *crystallites*.

The starch structure is similar to that of concentrated synthetic polymers below a given temperature. Above this temperature we have a polymer melt, and upon cooling, microcrystallites are formed, in which parts of the long molecules have a parallel orientation, as illustrated in Figure 6.22. The driving force is the lowering of contact enthalpy occurring

FIGURE 6.22 Highly schematic picture of microcrystalline regions in a mass of a linear polymer at very high concentration (little or no solvent). The reader should remember that a two-dimensional picture cannot give a fully realistic representation of a three-dimensional structure in a case like this.

upon close packing. As soon as polymer molecules take part in such crystallites, the remainder of each molecule loses much of its freedom to reorientate, the more so if more of the material is crystalline. As becomes clear from the figure, any further crystallization would require considerable reorientation, also of the chain sections present in crystallites. This means that crystallization virtually stops at a certain fraction crystalline, not because thermodynamic equilibrium has been reached, but because crystallization has become infinitely slow due to these geometric constraints. The crystallites are not true three-dimensional crystals: there are no crystal faces in directions roughly perpendicular to the polymer chains. Also in other directions, the crystallites mostly have no sharp boundaries. This imperfectness implies, for instance, that there is no sharp melting temperature, but a melting range of at least several K (cf. Figure 6.25, further on).

The similarity between crystallization in starch granules and in synthetic polymers is, however, limited. The latter only show crystallite formation if they are linear polymers, whereas in starch it concerns the highly branched amylopectin. The crystalline structure in starch is formed during starch synthesis and is largely irreversible. If the structure is disrupted by melting (see Section 6.6.2), it does not reform on cooling.

The structure is intricate. Crystallites occur in two types of chain packing, designated A (in most cereals) and B (in potato starch). Both contain water of crystallization, 10 and 20% by mass, respectively. The "clusters" depicted in Figure 6.21 are predominantly crystalline. However, the molecular chains are not straight (this is geometrically impossible: see Figure 6.18) but form double helices, about 1 nm in diameter. Figure 6.21 may suggest that crystalline layers (thickness a) alternate with other, amorphous layers, but the structure is more complicated. The crystalline material is in highly curved strips that form large helices: see Figure 6.23. In potato starch such a helix has an outer diameter of 18 nm and an inner diameter of 8 nm; the strips are thus 5 nm wide and have a thickness $a \approx 4$ nm. The pitch of the helices, which equals b, is about 10 nm. In the cavity of each helix and in the space between the strips, amylose and noncrystalline amylopectin is found; presumably, the amylose is predominantly in the cavity.

Native starch granules are thus very stiff or rigid particles. They have crystalline regions, which will contain about half of the water present, and the remainder of the starch and water forms a glass. The glassy state is discussed in Section 16.1.

It may finally be mentioned that starch shows considerable variation in properties, among granules of one source and among plant species and cultivars. This concerns average granule size and size distribution; granule shape; concentrations of nonstarch components; amylose/amylopectin ratio; molecular structure of amylopectin; type of molecular ordering in the crystallites (type A or B; a third type, C, is also mentioned, but it probably is a mixture of A and B); proportion crystalline; large scale structure of the crystalline material, for instance the distances *a* and *b*. Some cultivars make no or very little amylose, especially "waxy" maize. The amylopectin in potato starch contains phosphate groups, linked by esterification.

6.6.2 Gelatinization

Starch granules can be isolated from various plant materials, washed and air-dried to obtain a powder. When the powder is put in cold water, the granules take up a little water, but that is about all that happens. Native

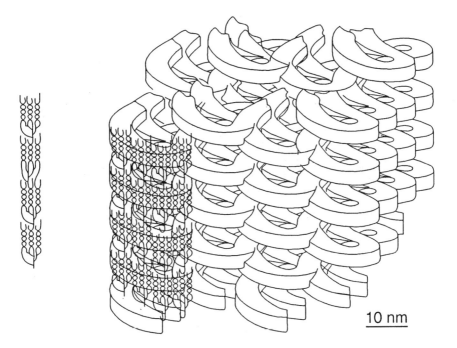

10 nm

FIGURE 6.23 Schematic model of the crystal morphology in potato starch. At left, the building blocks of double molecular helices are shown. At right, shape and stacking of the large helices is depicted. The large helices are packed in a tetragonal array, and neighboring helices partly interlock as they are shifted by half the pitch with respect to each other. (Adapted from a figure by G. T. Oostergetel and E. F. J. van Bruggen. *Carbohydr. Polym.* 21 (1993) 7.)

starch is physically and chemically inert, and it shows little digestion in the human gut. To change it into a functional product, it is generally heated in an excess of water. This causes what is called *gelatinization*, which involves a number of changes, including water uptake.

The most conspicuous change may be **melting** of the crystallites. When a suspension of starch granules in water is heated, their birefringence disappears. Examples are in Figure 6.24a; in each separate granule, the temperature range over which birefringence disappears is significantly more narrow. The melting is also observed with x-ray diffraction and calorimetry. The melting enthalpy ranges from 12 to 22 J per g dry starch, i.e., 40–50 J per g crystalline material. This is much less than the melting enthalpy of crystalline sugars, which is generally of the order of $400 \, \mathrm{J} \cdot \mathrm{g}^{-1}$. This indicates, again, that the degree of ordering in starch crystallites is far less perfect than in a sugar crystal.

The melting temperature depends on water content, as depicted in Figure 6.24b. This phenomenon is comparable to the melting point depression commonly observed for impure solid materials (e.g., imperfect crystals). Flory has derived an equation for the melting temperature T_m as a function of the volume fraction of polymer φ in concentrated polymer–

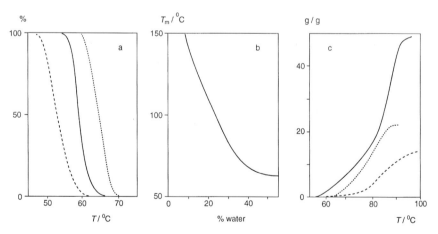

FIGURE 6.24 Gelatinization of starch from wheat (- - -), potato (——), and maize (corn) ($\cdots$). (a) Percentage of starch granules showing birefringence as a function of temperature T during heating up. (b) Melting temperature T_m as a function of water content (% w/w). (c) Water uptake in g per g dry starch during heating in an excess of water as a function of temperature. (Approximate results from various sources.)

solvent mixtures:

$$\frac{1}{T_m(\varphi)} - \frac{1}{T'_m} = \frac{R}{\Delta H_m} q[(1 - \varphi) - \chi(1 - \varphi)^2] \tag{6.14}$$

where T'_m is the melting temperature of the pure polymer, ΔH_m the melting enthalpy per mol of monomer, q the volume ratio of monomer to solvent molecule, and χ the solvent–segment interaction parameter (which equals $(1 - \beta)/2$; see Section 6.2.1). For water contents below 30%, which is about the range for which Eq. (6.14) would be valid, a relation as depicted in Figure 6.24b can be reasonably predicted, if χ is assumed to be 0.5 or even somewhat larger ($\beta \leqslant 0$). (Incidentally, this implies, again, that starch, like many polysaccharides, is not very hydrophilic.) Exact agreement with Eq. (6.14) is not to be expected, since the crystalline ordering in starch is irreversible (see, however, Section 6.6.3). The extrapolated value for melting of pure starch would be about 250°C; it cannot be experimentally determined.

There is, however, an important complication. Data as presented in Figure 6.24a have been obtained by slow heating of starch granules in an excess of water. During this process, the granules can take up water, which they tend to do with increasing temperature. During melting, the water content will be close to the equilibrium content at that temperature, which is the plateau value obtained at water contents above 50 or 60% (Figure 6.24b). If starch–water mixtures of lower water content (about 35–60%) are heated, two melting temperatures may be observed, the plateau value and a higher one. Some results obtained by differential scanning calorimetry are shown in Figure 6.25. The explanation is presumably that part of the crystallites have been in a position to take up sufficient water to melt at a low temperature, whereby little water is left to be taken up by the remaining crystallites. It should be realized that the diffusion rate of water in a concentrated starch system can be very small (see Figure 5.16a). It is also seen in Figure 6.25 that the melting occurs over a wider temperature range at low water content, indicating that the crystallites are less perfect. Heating in combination with water uptake may thus alter the crystallites before they melt.

In the **gelatinization process**, four stages can be distinguished. They occur when an aqueous starch suspension is heated while stirring.

1. Below the melting point, the *granules swell* somewhat by taking up water, increasing in volume by 30–40%. Some of the amylose leaches from the granules.
2. *Melting* is discussed above. During melting, more amylose leaches. Melting does not necessarily imply that all supermolecular

structure disappears. Considerable entanglement of amylopectin chains will persist, and possibly even part of the large helix structure will remain.

3. At higher temperature, *swelling becomes extensive*: see Figure 6.24c. Virtually all *amylose leaves the granules*. This actually is a phase separation between amylose and amylopectin, as is discussed in Section 6.5.2. Highly swollen though often irregularly

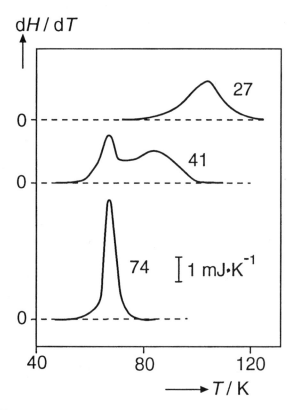

FIGURE 6.25 Differential scanning calorimetry (DSC) as applied to potato starch–water mixtures; % water (w/w) indicated near the curves. In DSC, the differential amount of heat (or more precisely, enthalpy, H) needed to increase the temperature $(\mathrm{d}H/\mathrm{d}T)$ is registered as a function of temperature (T), and any melting (or similar transition) causes a peak in the heat uptake. The peak area is proportional to the melting enthalpy. In the figure, the specific heat capacity of the material has been subtracted, providing horizontal base lines. (After results by J. W. Donovan. *Biopolymers* 18 (1979) 263.)

shaped granules are still observed, but they should not be envisaged as having a wall or membrane: the amylopectin is held together by covalent bonds and by entanglements. The swelling causes a very great increase in apparent viscosity, especially at fairly high starch content, since then the volume will be fully taken up by swollen granules. The extent of swelling greatly varies among starch types, as illustrated in Figure 6.24c. For potato starch, even far stronger swelling has been reported, up to 1 kg of water per g of starch. This will only occur at very low ionic strength; the phosphate groups present will be ionized (unless the pH is low), causing considerable electric repulsion between chains (cf. Section 6.2.2).

4. At still higher temperature, the swollen *granules are broken up* into far smaller fragments, especially during vigorous stirring. This may be partly due to disentangling, partly to breaking of amylopectin molecules. It causes the apparent viscosity to decrease again; the explanation is disputed.

The various stages may overlap, and the amount of swelling in stage 2 as compared to that in stage 3 markedly varies among starch types; compare wheat and maize starch in Figure 6.24. If starch granules have been damaged (broken), as for instance during the grinding of wheat, some water may be taken up at earlier stages. Gelatinization depends on several conditions: amount of water present, temperature regime, intensity of stirring, presence of other substances. For example, high concentrations of some sugars cause the melting temperature to increase. If the pH is low, say < 3, heating may cause marked hydrolysis of the starch, leading to a far lower apparent viscosity.

In practice, gelatinization is applied in fairly dilute solutions to achieve **binding** (thickening), as in soups or gravies. Actually, gelatinized starch is not a very efficient thickening agent; the concentration needed to obtain a certain increase in viscosity is about 10 times higher than for several other polysaccharides. The main reason is the highly branched character of amylopectin, giving rise to a small value (about 0.25) for the exponent in the Mark–Houwink equation (6.5). During the cooking of vegetables or the baking of bread, gelatinization occurs as well, but stage 4 is generally not reached, partly because the starch solution is not stirred. Moreover, swelling may be restricted by the amount of water available. In bread, the ratio of water to starch is about unity, implying that the granules cannot nearly attain full swelling; nevertheless, they have lost most of their crystallinity. In even dryer products, as in some biscuits, stage 2 is not reached and the granules are much like native ones.

Note The reader may be cautioned that the term gelatinization temperature is used rather indiscriminately. The method used to obtain it may have been loss of birefringence, loss of x-ray diffraction pattern, differential scanning calorimetry (where the result depends on the scanning rate), or even increase of apparent viscosity in a particular instrument (e.g., an "amylograph"). These methods may give quite different results, depending in a different way on water content, especially the last one. Moreover, cultivars of one species may vary in gelatinization temperature.

6.6.3   Retrogradation

When keeping a gelatinized starch solution or paste at ambient temperature, physical changes are observed, which are generally lumped under the name retrogradation. These changes can be of various types: a solution becoming turbid, precipitation of part of the starch, gel formation, or a once formed gel becoming stiffer and more brittle. It greatly depends on conditions, especially starch/water ratio and temperature, what change will occur. These changes have the same origin: ordering of molecules or parts of molecules occurs, and such a change can be quantified by calorimetry. Upon heating, energy is consumed in disordering the structure over a certain fairly narrow temperature range, and this ΔH may be considered as a melting enthalpy. ΔH increases with time; see e.g. Figure 6.26.

Dilute Systems.   We will first consider what will happen in an *amylose* solution. Actually, amylose is very poorly soluble in water at room temperature (although it is well soluble in some salt solutions, notable KCl; cf. Table 6.1). It readily forms helices in water, of which at least part are double helices. These helices tend to align, forming parallel stacks that may be considered *microcrystallites*. X-ray diffraction shows the chain packing to be similar to that of B type crystallites in native starch. As much as 70% of the amylose may become crystalline. It depends on amylose concentration what the consequences will be. An amylose solution at 65°C has a chain overlap concentration c^* (see Section 6.4.2) of about 1.5%. If a more dilute solution is cooled, *precipitation* of amylose will occur.

In more **concentrated** amylose solutions, a *gel* will be formed. The gel stiffens during keeping, although the correlation between the increase in ΔH and the increase in stiffness is far from perfect: Figure 6.26. Much the same happens in a gelatinized *starch* solution. Here, amylose and amylopectin have phase separated, and the swollen granules contain virtually no amylose. This implies that the amount of water available for amylose is reduced, and at starch concentrations of 2–4% (depending on starch type), the chain overlap concentration for amylose will be reached. Above this concentration, cooling will lead to the formation of an *amylose gel*, enclosing swollen granules.

Gelatinized starch systems thus show retrogradation, as can be measured by calorimetry. However, the change in ΔH proceeds for a far longer time than in an amylose gel—compare Figures 6.26 and 6.27—suggesting that also amylopectine exhibits retrogradation. The latter is corroborated by studying amylose-free systems. Starch crystallites are observed, albeit in far smaller proportion than in pure amylose. The crystallites are about 10 nm in size, and their structure is to some extent comparable to the B type crystallites in native starch.

Reversibility. Nevertheless, the supermolecular structure present in native starch granules, like the one depicted in Figure 6.23, does not reappear upon cooling after the starch has been gelatinized. This is borne out by the observation that the melting enthalpies involved, about 2 to 10 J

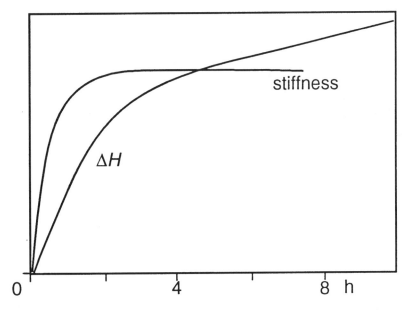

FIGURE 6.26 Retrogradation of a 7% amylose solution. The solution was cooled from 90 to 26°C and then kept for the time indicated. Stiffness (as shear modulus) and melting enthalpy (ΔH) were monitored; arbitrary scales. (After results by M. J. Miles et al. *Carbohydr. Res.* 135 (1985) 271.

per g of starch, are much smaller than those for native starch (12–$22 \ J \cdot g^{-1}$). The crystallinity obtained after gelatinization and cooling is, however, reversible: it disappears at high temperature—which is applied to alleviate the staling of bread—and reforms, albeit slowly, after cooling. The theory for crystallization of concentrated polymer systems is roughly applicable.

The **rate of retrogradation** strongly depends on *temperature*: see Figure 6.27a. At a temperature that is further below the melting point, the "supersaturation" is greater. Although supersaturation is not a well-defined concept in polymer systems, it is clear that the driving force for crystallization is greater. For a starch/water ratio of unity, the melting temperature is about 75°C. The staling of bread, which is primarily due to retrogradation, will thus proceed faster at 5°C (refrigerator) than at 25°C (ambient). In a freezer staling rate is very much slower; see Section 16.3 for the explanation.

Retrogradation rate strongly depends on *water content* (Figure 6.27b), and the relation is similar to that with temperature. According to Eq. (6.14), the "supersaturation" will be greater for a higher volume fraction of starch. On the other hand, at very low water content the mobility of the polymer chains will be very small, which will reduce the rate at which crystallites are formed. For 20% water at room temperature the mobility is effectively zero and no retrogradation occurs.

Among other variables is *starch type* (Figure 6.27c); it may be noticed that, especially during the first few hours after gelatinization, the retrogradation rate may be very different. Gelatinization conditions and

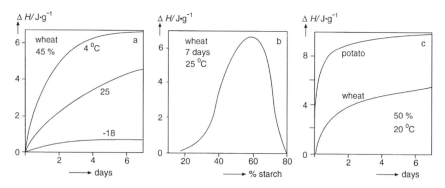

FIGURE 6.27 Retrogradation of gelatinized starch. Effects of temperature (a), starch concentration (b), and starch type (c) on melting enthalpy (ΔH). Starch type and concentration, and storage temperature and time indicated. (Approximate results from various sources.)

the presence of other components also play a role. It is often assumed that lipids may have considerable effect. As mentioned, some lipids (especially those having an acyl chain of 12 to 16 carbon atoms) form inclusion complexes with amylose. This does somewhat affect the properties of fairly concentrated starch gels, but it has never been shown to a have a significant effect on retrogradation as such.

Quite in general, great care should be taken in directly relating retrogradation, for instance as measured by an increase of melting enthalpy, to physical changes occurring. Mechanical properties, whether evaluated sensorily or by rheological measurement, mostly depend, in a very involved manner, on the number and the properties of such structural elements as microcrystallites. Moreover, modified starch is often used; for instance, chemical cross-linking causes a decrease in gelation temperature and in the extent of retrogradation.

6.7 RECAPITULATION

Polymers are macromolecules that consist of long chains, for the most part linear, of identical or similar units called monomers. They derive special properties from their molecular structure, of which the possibility of assuming very many conformations is the most striking one.

Conformation. In solution, the average conformation tends to be much expanded, causing the molecules to occupy relatively large volumes. For simple polymers, theory is available to predict properties. The most important variables are molecular mass or size, the stiffness of the chain, and the solvent quality, for instance as expressed in the solvent exclusion parameter β. The greater these parameters, the more expanded is the polymer chain in solution. This implies that the relative viscosity of a polymer solution (relative to that of the solvent) can be very high and increases with increasing molar mass, for the same mass concentration. The viscosity is an apparent one, since it strongly decreases with increasing strain rate.

Most **natural polymers** are greatly heterogeneous, having specific structures, and polymer theory can only be used in a broad, semiquantitative sense. For most polysaccharides and for some proteins (gelatin, casein) the theory nevertheless is useful. For most polysaccharides, the solvent quality of water is rather poor, but many of them nevertheless have a quite expanded conformation, because they are relatively stiff molecules (starch being an exception). For globular proteins other theory is needed. Proteins and several of the polysaccharides are polyelectrolytes.

Polyelectrolytes, i.e., polymers containing charged groups (negative or positive or both), show some specific relations. Due to the repulsive effect of the charges if of the same sign, the polymer chain is relatively stiff. This implies a very expanded conformation, whereby polyelectrolytes produce very high viscosities. This greatly depends on the charge density, which in turn may depend on pH, and on ionic strength. At a high ionic strength, shielding by counterions causes the distance over which the repulsive effect is sensed to be far smaller, leading to a less expanded conformation. At high ionic strength, say above 0.2 molar, the difference with neutral polymers mostly tends to be small.

Polyelectrolytes also show the **Donnan effect**. Charged macromolecules always are accompanied by counterions, i.e., small ions of opposite charge, in order to make the solution electroneutral. In the presence of salt, also the distribution of coions (small ions of the same charge) is affected, and relatively more so for a lower salt concentration. These phenomena are most readily observed when the polymer solution is contained within a semipermeable membrane, where the osmotic pressure is greatly affected by the Donnan effect. However, the effect also occurs for a polyelectrolyte in solution; this implies, for instance, that it will be difficult to remove all ions of a certain species, even if exchanged for other ions of the same charge sign.

Concentrated polymer solutions show strong nonideality. This is, for instance, observed in the osmotic pressure being very much higher than would follow from the molar concentration. The main variables are the β value and the volume fraction of polymer, and for polyelectrolytes also charge and ionic strength.

The **solubility** of neutral polymers depends primarily on molecular size and β. For long polymers, a β value just below zero leads already to very poor solubility. Generally, the polymer does not precipitate, but phase separation occurs into a highly concentrated solution (a coacervate) and a very dilute one. Quite in general, a number of "regimes" can be distinguished for polymer–solvent mixtures, depending on β value and concentration. Besides the dilute, there are semidilute and concentrated regimes.

Outside the dilute regime, **chain overlap** occurs, which implies that polymer molecules are mutually entangled. This greatly increases viscosity, as well as the dependence of viscosity on concentration and the extent of strain rate thinning. Moreover, the solution shows elastic besides viscous behavior, and if intermolecular cross-links are formed, a gel is obtained. The chain overlap concentration decreases with increasing molecular size, β value, and stiffness. For overlapping chains, the solution is characterized by a correlation length, which does not depend on molecular size, and which is generally smaller for a higher concentration, greater stiffness, and higher β value.

Phase separation in mixtures of polymers especially occurs at high concentrations and for large molar mass. The separation can be of two kinds. If the two polymers show mutual affinity, associative phase separation or complex coacervation occurs, i.e., a separation into a solution high in both polymers (a complex coacervate) and a very dilute solution. In most other cases, segregative phase separation or incompatibility is observed, i.e., separation in a phase rich in polymer A but poor in polymer B, and vice versa.

Starch. Starch is a polyglucan mixture, containing the linear amylose (about 25%) and the very large, heavily branched amylopectin. Native starch granules are virtually insoluble in water, mainly because part of the amylopectin is crystalline. The crystals consist of stacks of double helices, which stacks are arranged in a complex supermolecular structure.

Starch can be **gelatinized** by heating in excess water, which implies melting of the crystallites and extensive swelling; amylose and amylopectin become phase separated. Highly viscous solutions or pastes are obtained, although starch is not a very efficient thickening agent as compared to most other polysaccharides, mainly because of the strong branching of amylopectin.

On cooling and keeping such systems, **retrogradation** occurs, which originates from partial crystallization of the starch and can cause physical changes like precipitation, gel formation, or gel stiffening (e.g., staling of bread). Amylose shows much more and quicker crystallization than does amylopectin. The original supermolecular structure is not regained. Retrogradation itself is reversible: it can be undone by heating and starts again upon recooling. The retrogradation rate increases with decreasing temperature and increasing water content. The crystallization in gelatinized starch is similar to the partial crystallization in concentrated linear polymer systems.

BIBLIOGRAPHY

The classical treatment of the chemistry and physics of polymers is very thorough, but not easy to read; it is

P. J. Flory. Principles of Polymer Chemistry. Cornell Univ. Press, Ithaca, 1953.

A somewhat easier book, more up to date and laying stress on concepts and principles, is

P. G. de Gennes. Scaling Concepts in Polymer Physics. Cornell Univ. Press, Ithaca, 1979.

A brief and lucid discussion of the properties of polymers in solution is in the introductory chapter of

G. J. Fleer, M. A. Cohen Stuart, J. M. H. M. Scheutjens, T. Cosgrove, B. Vincent. Polymers at Interfaces. Chapman and Hall, London, 1993.

A still very useful book, that also discusses many natural polymers (including proteins), as well as experimental methods, is

C. Tanford. Physical Chemistry of Macromolecules. John Wiley, New York, 1961.

Donnan equilibria are treated in most texts on physical chemistry.

Phase separation in polymer solutions is extensively discussed in

P.-A. Albertsson. Partition of Cell Particles and Macromolecules, 2nd ed. Almqvist and Wiksell, Stockholm, 1971.

The theory and practice of food polymers are amply discussed in

S. E. Hill, D. A. Ledward, J. R. Mitchell, eds. Functional Properties of Food Macromolecules, 2nd ed. Aspen, Gaithersburg, MD, 1998.

A detailed treatment of the molecular structure of several polysaccharides and of the viscosity of polysaccharide solutions is in

R. Lapasin, S. Pricl. Rheology of Industrial Polysaccharides: Theory and Applications. Blackie, Glasgow, 1995.

Much practical information on a range of natural and modified polysaccharides, including starch, is given by

A. M. Stephen, ed. Food Polysaccharides and Their Applications. Marcel Dekker, New York, 1995.

and also by

G. O. Phillips, P. A. Williams. Handbook of Hydrocolloids. Woodhead, Cambridge, 2000.

Many aspects of starch are treated in

T. Galliard, ed. Starch: Properties and Potential. John Wiley, Chichester, 1987.

7

Proteins

Proteins are polymers, more specifically polyelectrolytes, which are discussed in Chapter 6. However, proteins were hardly considered in that chapter because they are highly specific and intricate molecules. They are built of 20 different monomers, with side groups of different reactivity. Proteins evolved to fulfil a wide range of highly specific physiological functions, and each protein has a specific composition and conformation. Every protein species is unique; the number of species occurring in nature is presumably far over 10^{10}. Chemical reactivity is at least as important as physical chemistry for protein properties in general and for many problems related to proteins in foods. Despite these qualifications, some important physicochemical rules can be derived, and this is the subject of this chapter.

Proteins play many roles in foods, the most important one being nutritional. Another one is flavor binding, and so is flavor formation during processing and storage. The activities of the proteins called enzymes are of obvious importance. Two main groups of physicochemical functional properties of proteins can be distinguished:

1. Their role in formation and stabilizing of emulsions and foams, and in stabilizing suspensions. These functions depend on the propensity of proteins to adsorb at most interfaces. These aspects are discussed in Chapters 10–13.

2. The formation of more or less ordered macroscopic structures, especially gels. This is governed by the tendency of many proteins to aggregate or to form intermolecular cross-links on heating, change of pH, etc. Gels are discussed in Section 17.2.

In the present chapter, protein solubility, conformation, and conformational stability are the main subjects. See Chapter 8 for water relations.

7.1 DESCRIPTION

The chemistry of proteins is covered in most texts on food chemistry and, of course, on biochemistry. For the convenience of the reader, a brief review is given in this section.

7.1.1 Amino Acids

Proteins are linear polymers of α–L amino acids linked by the formation of peptide bonds:

$$NH_2-CHR-COOH + NH_2-CHR^*-COOH \rightarrow$$
$$NH_2-CHR-C0-NH-CHR^*-COOH + H_2O$$

The degree of polymerization n ranges from 50 to several 100. R denotes a side group; basically 20 different ones exist in nature and are given in Table 7.1, together with some properties. According to the side group, they can be categorized as

> Aliphatic: Ala, Val, Leu, Ile
> Aromatic: Phe, Tyr, Trp, His
> Charged: $Asp^-, Glu^-, Lys^+, Arg^+, His^+, (Cys^-)$
> Somewhat polar, uncharged: Asn, Gln, Ser, Thr, (Tyr)
> Sulfur containing: Cys, Met

This leaves Gly, which has a mere —H as a side group, and proline, which is not a primary amino acid.

> *Note* Proline is commonly called an imino acid, although it does not contain an imino group, C=NH, but a secondary amino group, C—NH—C.

Note that some belong to more than one category.

TABLE 7.1 Properties of Amino Acid Residues

Name of acid	Symbols[a]		Side chain	Reactive group	pK[b]	Φ^c	Φ^d
Glycine	Gly	G	—H			0	0
Alanine	Ala	A	—CH$_3$			2.1	3.6
Valine	Val	V	—CH(CH$_3$)CH$_3$			6.3	13.0
Leucine	Leu	L	—CH$_2$CH(CH$_3$)CH$_3$			7.5	16.6
Isoleucine	Ile	I	—CH(CH$_3$)CH$_2$CH$_3$			7.5	16.6
Serine	Ser	S	—CH$_2$OH	Hydroxyl			
Threonine	Thr	T	—CHOHCH$_3$	Hydroxyl		1.7	< 0
Aspartic acid	Asp	D	—CH$_2$CO$_2^-$	Carboxyl	4.0		
Asparagine	Asn	N	—CH$_2$CONH$_2$	Amide			
Glutamic acid	Glu	E	—CH$_2$CH$_2$CO$_2^-$	Carboxyl	4.5		
Glutamine	Gln	Q	—CH$_2$CH$_2$CONH$_2$	Amide			
Lysine	Lys	K	—(CH$_2$)$_4$NH$_3^+$	ε-Amino	10.6		
Arginine	Arg	R	—(CH$_2$)$_3$NHC(NH$_2$)$_2^+$	Guanidine	12.0		
Cysteine	Cys	C	—CH$_2$SH	Thiol	8.5	4.2	1.4
Methionine	Met	M	—CH$_2$CH$_2$SCH$_3$	Thio-ether		5.4	5.9
Phenylalanine	Phe	F	—CH$_2$—C$_6$H$_5$	Phenyl		10.4	8.5
Tyrosine	Tyr	Y	—CH$_2$—C$_6$H$_4$—OH	Phenol	~10	9.0	< 0

TABLE 7.1 Continued

Name of acid	Symbols[a]		Side chain	Reactive group	pK[b]	ϕ^c	ϕ^d
Tryptophan	Trp	W		Indole		14.2	5.8
Histidine	His	H		Imidazole	6.4	2.1	< 0
Proline	Pro	P	(Note e)			5.9	.

[a] Three-letter and one-letter symbols are given. [b] pH of 50% ionization at zero ionic strength in an unfolded peptide chain.
[c] Hydrophobicity of side group (kJ per mole of residues) from relative solubility of amino acids in water and ethanol or dioxane.
[d] Same, from relative solubility of side chain analogues in water and in cyclohexane.
[e] Secondary amino acid:

Other side groups occur when the protein is *conjugated*, implying that other groups become covalently bonded to some amino acid residues. This may concern phosphorylation (mostly of Ser or Thr); glycosylation, which takes many forms; hydroxylation (mostly of Pro or Lys); as well as attachment of some other groups. One thus commonly speaks of the conjugated phosphoproteins and glycoproteins; the latter may contain a wide variety of glucide groups, from one per molecule to several times 10% by mass. Metalloproteins contain one or more, bivalent or trivalent, cations, that are tightly bound, but not by covalent bonds; in other words, they are not part of the protein, though the biological function of the protein generally depends on their presence. A still weaker association is involved in the formation of lipoproteins. Most of these are not proteins but complexes of several protein molecules and several lipid molecules, held together by weak forces (van der Waals, hydrophobic, etc.).

The side groups determine chemical reactivity, i.e., the possibility of forming covalent bonds. Also much of the physicochemical behavior follows from the nature of the side groups. Some can be involved in hydrogen bonding, either as a hydrogen donor ($-OH$ and $=NH$) or as an acceptor ($=O$, $-O-$, $=N-$ and $-S-$). The bulkiness greatly varies, group molar mass ranging from 1 (Gly) to 130 (Trp).

Some groups can be ionized, i.e., carry an *electric charge*. A proton can dissociate from a carboxyl group above a certain pH, giving a negative charge; and several other groups become protonated below a certain pH, giving positive charges. Table 7.1 gives pK values ($pK = pH$ of 50% dissociation; see Section 2.3.1). Some ionizable groups can be added to those in Table 7.1. Several proteins are glycosylated, and some of these glucides contain carboxyl groups. Other proteins are phosphorylated, especially the caseins, which contain phosphoserine residues (R is $-CH_2O-PO_3H_2$). Two protons can dissociate from a phospho group, giving pK values of about 1.5 and 6.5.

Proteins are thus *polyelectrolytes*; see Section 6.3.1 for a general discussion, especially on titration curves (Figure 6.7). It may be clear from that discussion that a titration curve cannot be obtained by merely adding the titrations of the separate ionizable groups. This is the more so for most proteins, where the ionizable groups often are quite close to each other, and where adjacent groups can be of the same or of opposite charge; the pK of a group can readily be shifted by a full unit. This also means that the *isoelectric* pH can vary with conditions, e.g., ionic strength. Many proteins used in the food industry have an isoelectric pH not far from 5, implying that they are negatively charged at neutral pH.

Another important property of side groups is their solvation by water, because this greatly affects protein conformation (see below) and solubility

(Section 7.3). It is to be expected that the charged groups can be strongly hydrated (Section 3.2) and that the aliphatic groups and the aromatic ones will be hydrophobic (except His owing to its charge and possibly Tyr owing to its —OH).

As an estimate for the tendency to form hydrophobic bonds, one often assigns a *hydrophobicity* to the side groups. One way of finding such values is by determining the solubility of the amino acid in water and in an organic solvent, say ethanol. The Gibbs free energy of transfer of the amino acid from ethanol to water is then given by

$$\Delta_{E \to W} G = RT \ln \left(\frac{c_{sat,E}}{c_{sat,W}} \right) \tag{7.1}$$

expressed in kJ per mole. Here c_{sat} is the solubility of the amino acid per unit volume. Because the peptide bond is hydrophilic, the free energy of transfer of glycine (side group —H) is subtracted, to obtain a value that is presumed to be characteristic for the side group of the amino acid in a peptide chain. In other words,

$$\Phi = \Delta_{E \to W} G \text{ (amino acid)} - \Delta_{E \to W} G \text{ (glycine)} \tag{7.2}$$

where Φ is the hydrophobicity in kJ per mole of residue, mostly according to the scale of Tanford–Bigelow, which is also used to calculate an average hydrophobicity (over all residues) of a protein. For hydrophobic residues $\Phi > 0$. There are, however, several other scales of hydrophobicity, by taking another organic solvent (or taking the vapor phase instead), or by determination on side group analogues (e.g., methane for alanine), all putting $\Phi = 0$ for glycine. Unfortunately, the scales differ considerably. Values according to two scales are given in Table 7.1. There is general consensus on Leu, Ile, Val, Phe, and Trp being hydrophobic (and also Ala, but with a small Φ), whereas no agreement exists about Tyr, Pro, Met, or Cys.

7.1.2 Primary Structure

The composition of proteins is variable. The percentage of a certain amino acid may readily vary by a factor of 10 among a number of proteins. Some amino acids do not occur in some proteins, notably Cys. The composition largely determines the nutritional quality and to a considerable extent the chemical reactivity of a protein. Most other properties depend on the *primary structure*, i.e., the sequence of amino acid residues, because the primary structure determines the higher structures (see below), which in turn determine properties like conformational stability and solubility. Nevertheless, it is mostly not possible to predict the higher structures from the primary structure. One reason is that the number of possible primary structures is, for all purposes, infinite. As an example we take proteins of 100 residues (most proteins are far larger), of which $20^{100} \approx 10^{130}$ different primary structures can exist! Even if one assumes that many interchanges of amino acids would not materially alter the properties of the protein, taking the effective number of different residues to be as small as six, this leads to $6^{100} \approx 10^{78}$ different species. Assuming that of each of those one molecule existed, their total mass would be the number times the average residue molar mass $(\approx 0.12\,\text{kg} \cdot \text{mol}^{-1})$ over Avogadro's number, i.e., $\approx 10^{53}\,\text{kg}$, which is greater than the presumed total mass of the universe.

The actual variation in primary structure really is very great. Some proteins have a fairly regular structure, like collagen, where the greater part of the amino acid sequence consists of repeats of Gly-Pro-Pro or Gly-Pro-Lys, of which the third residue may be hydroxylated. By far most proteins have much more intricate primary structures. Some of these are illustrated in Figure 7.1; note the difference in structural heterogeneity.

Figure 7.2 shows the configuration and dimensions of a *peptide unit*. The *peptide bond* has some special features, since the electron distribution over the O, C, and N of the bond is intermediate between that of the two structures

This causes the peptide bond to be flat: rotation about the CO—NH axis is not possible. The bond is in the trans configuration, which is far more stable than the cis one. The peptide bond also has a significant dipole moment of 3.5 Debye units; this implies that it tends to be hydrated. The H of the —NH group can act as a donor, the =O as a hydrogen acceptor in forming hydrogen bonds.

Figure 7.2 gives an idea about the *flexibility* of the peptide unit; rotation about the ψ and φ angles is possible. This rotation is by no means unlimited, because the side groups sterically prevent several conformations, according to their size and shape. Even in the case of polyglycine (side group —H), some bond angles are clearly preferred. The peptide chain would have about 4 degrees of (conformational) freedom per peptide unit, though this may vary with the side groups involved. An unfolded peptide chain is much more flexible than most polysaccharide chains.

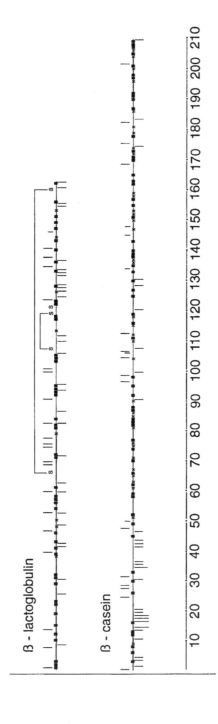

FIGURE 7.1 Illustration of some aspects of the primary structure of two proteins, β-lactoglobulin (genetic variant B) and β-casein (variant A2). The straight line denotes the amino acid sequence, from the N-terminal end on the left-hand side to the C-terminal end. Solid squares on the line denote hydrophobic residues (Leu, Ile, Val, Phe, and Trp), a cross (×) indicates a Pro residue, and the vertical lines denote charge at a pH of about 6.6 (positive upwards, negative downwards); the longer lines denote serine-phosphate ($z \approx +0.5$). Cysteine residues are denoted by S, and —S—S— bridges by connecting lines.

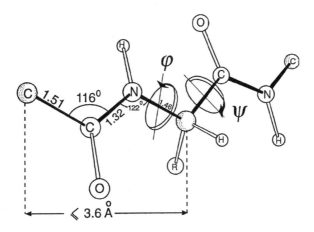

FIGURE 7.2 Bond angles and lengths (Å) and rotational freedom in the peptide unit.

Despite the flexibility of the peptide chain, most proteins do not assume random conformations in solution. Contrariwise, the conformation mostly is highly ordered, and some higher levels of structural organization are distinguished, i.e., secondary, tertiary, and quaternary structures.

7.1.3 Secondary Structure

This concerns fairly regular arrangements of adjacent amino acid residues. Several types exist, but the most common ones are α-helices and β-strands.

In the right-handed **α-helix**, the peptide chain forms a helix (like a cork screw) with the side groups on the outside, where each turn takes 3.6 residues (18 residues making 5 turns); the translation of the helix is 0.15 nm per residue (i.e., a pitch of 0.54 nm per turn), compared to 0.36 nm per residue for a stretched chain (Figure 7.2). The helical conformation is stabilized by H-bonds, between the O of peptide bond i and the NH of peptide bond $i + 4$. Moreover, enhanced van der Waals attraction is involved. The possibility for the latter to occur varies among amino acid residues, which means that not all of them readily partake in an α-helix. Ala, Glu, Phe, His, Ile, Leu, Met, Gln, Val, and Trp have strong tendencies to form helices, whereas Pro, owing to its cyclic structure, is a "helix breaker."

The formation of an α-helix is a clear example of a *cooperative transition*. Although each of the bonds involved is weak, at most a few times $k_B T$, the collective bond energy of the whole structure may be sufficient to stabilize it. This is comparable to the formation of a molecular crystal from a solution, where bond energies often are 1 or 2 $k_B T$ per molecule. Doublets of these molecules would be very short-lived, but a crystal of sufficient size is stable, mainly because each molecule now is involved in a number of bonds, e.g., six. A prerequisite, both in a crystal and in a helix, is a very good fit of the bonds. This is the case in an α-helix, where the H-bonds are almost perfectly aligned. The cooperativity principle implies that an a-helix cannot be very short, as is indeed observed.

In the **β-strand** the peptide chain is almost fully extended, although slightly twisted (the translation equals about 0.34 nm per residue rather than 0.36 nm when stretched). A single β-strand is unstable, but several of them can be aligned to form *β-sheets*, which can either be parallel or, more commonly, antiparallel. Note that the strands in one sheet need not be from nearby regions in the primary structure. In the sheets, several H-bonds are formed, again stabilizing the conformation by the cooperativity principle. The antiparallel β-sheet seems to be somewhat more stable than the parallel one. A single β-strand can also be stabilized by other structures, e.g., by alignment with an α-helix.

The presence and abundance of the various secondary structure elements in a protein in solution can in principle be determined by spectroscopic techniques. This also relates to some smaller scale structure, like reverse turns. However, definitive results on secondary structure can only be obtained by determination of the complete conformation.

7.1.4 Tertiary Structure

By means of x-ray diffraction of crystalline protein and of NMR spectroscopy of the molecules in solution, the complete three-dimensional structure of a protein can be established. Many proteins show an intricate, tightly folded structure, which includes secondary structure elements. Generally, hydrophilic amino acid side groups are predominantly at the surface, and hydrophobic ones in the core of the structure. The driving forces for folding are discussed in Section 7.2.1. The role of water is essential and it may be stated that

polypeptide chain + water = protein

It depends on the proportions of hydrophobic and hydrophilic residues, as well as on the length of the peptide chain, what the overall structure can be. This is illustrated in Figure 7.3. Assuming a hydrophilic outer layer of one peptide chain, i.e., about 0.5 nm in thickness, a larger protein molecule can accommodate a greater proportion of hydrophobic residues in its core. A method of excluding hydrophilic residues from the core, if they make up a high proportion of the residues, is the formation of an elongated shape (Figure 7.3c). On the basis of the outer shape, proteins can be classified into three groups: globular, fibrous, and disordered.

In a **globular protein** the peptide chain is tightly folded into a roughly spherical shape. The secondary structures (α-helices and β-sheets) roughly span the diameter, and they are often linked by reverse turns. The latter are sharp bends in the peptide chain, to some extent stabilized by hydrogen bonds. The charged amino acid residues Asp, Glu, Arg, and Lys are predominantly at the surface, and the charge density is generally between 0 and 2.5 charges per nm^2, depending on pH. Val, Leu, Ile, Phe, Ala, Gly, and Cys are mostly inside. Globular proteins always have a fairly high average hydrophobicity, over 4 kJ per mole of residues, and contain relatively few Pro residues.

However, these statements are by no means universal rules. Globular proteins, which make up by far the greatest proportion (99%?) of protein species, evolved to fulfil specific functions, which are mostly related to the capacity to bind specific molecules or groups, mostly called "ligands." This applies to enzymes, transport proteins, antibodies (immunoglobulins), etc. Their tertiary structure thus is highly specific, and the primary structure needed mostly does not allow complete segregation of hydrophobic and hydrophilic residues. Consequently, the outside often contains several hydrophobic residues, whereas some hydrophilic ones are in the core. Nevertheless, the geometrical constraints mentioned above imply that a protein of more than, say, 250 residues cannot assume a globular

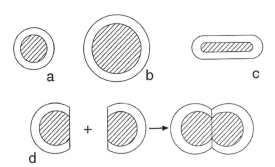

FIGURE 7.3 Illustration of the effect of size and shape on the proportion taken up by the hydrophobic core (hatched) of a protein, taking the hydrophilic outer layer to be of constant thickness (e.g., 0.5 nm). In (a) the proportion is 0.20, in (b) 0.50, and in (c) 0.12. In (d) molecules with a hydrophobic patch on the surface are illustrated, which tend to associate into a quaternary structure. Highly schematic.

conformation, unless it is very hydrophobic. Often, such large molecules contain a number of separate, globular *domains* of 100 or more residues. The domains generally are connected by a single short peptide strand.

The conformation of globular proteins is often stabilized by covalent bonds that are due to *posttranslational modifications*, i.e., changes occurring during protein synthesis, but after the primary structure has formed. These include the conjugations mentioned before, especially various glycosylations. Another modification is oxidation of the —SH groups of two cysteines, forming an intramolecular —S—S— bridge. Figure 7.1 shows that two —S—S— bridges and one free thiol group occur in β-lactoglobulin (a fairly small globular protein). Many large globular proteins contain a great number of sulfur bridges.

For small-molecule globular proteins and for globular domains, the following *size relations* roughly apply. From determinations on several of such proteins, the volume in nm^3 equals on average $1.27M$, where molar mass M is in kDa. This implies that the density of the protein then is about $1300\,kg \cdot m^{-3}$, which is very high, comparable to the density of a crystal, and leaving very little free space. (The interior mostly does contain some water molecules, i.e., a few % by mass, which may be considered part of the protein structure.) The surface area that can make contact with water is about $9.3 \times M^{2/3}nm^2$. This area is on average about 1.64 times the surface area of a sphere of the same volume. A globular protein or domain is thus not a perfect sphere, having an uneven surface.

It may finally be noted that the conformation of a globular protein is *not rigid*: (a) Limited conformational changes often occur upon a change in conditions, such as the binding of a ligand. (b) Vibration of molecular segments occurs even in the core, allowing exchange of small species, e.g., protons, albeit slowly. In fact, a protein molecule is rather dynamic, and its conformation represents an average. (c) As will be seen in the next section, there will always be a proportion of the molecules, however small, in a partly or fully unfolded state. (d) Side groups at the surface may have considerable conformational freedom and can be reactive.

For *fibrous proteins* very different relations hold. Most of them are "structural proteins." This means that they act as construction materials, as in silk (fibroin), tendon (collagen), blood clots (fibrin), or muscle (myosin). These molecules have a very elongated structure, most of them containing a high proportion of β-sheet, and they often have a fairly regular primary structure. The average hydrophobicity is low, $< 3.5\,kJ \cdot mol^{-1}$, and their molar mass generally is very large. The large size is mostly due to —S—S— bridges and other cross-links between peptide chains, thereby producing fibers. These posttranslational modifications also serve to stabilize the secondary structure.

7.1.5 Other Aspects

Some proteins do not fit in the scheme of globular versus fibrous, and they may be called **disordered**. Such proteins have a range of conformations rather than one. The group includes some proteins of which the natural function is nutritional, such as the caseins and some storage proteins, e.g., the glutelins in wheat grains. Caseins have a primary structure that prevents tight folding of the peptide chain, partly due to the many proline residues; see Figure 7.1. Also several denatured globular proteins and gelatin (derived from collagen) can be considered disordered. Although the greatest number of protein species occurring in nature is globular and the greatest mass probably fibrous, disordered proteins are rather important in foods. Casein and gelatin are often applied because of their functional properties, and globular proteins often are denatured during processing.

Disordered does not mean that no secondary or other regular structure exists, but that the conformation is (much) closer to a random coil than in the other proteins. Only gelatin above $30°C$ and β-casein at about $5°C$ have been shown to behave like random coil polyelectrolytes. Glutelins have a very high molar mass and are highly branched, undoubtedly due to —S—S— bridges linking the originally formed smaller peptides.

The linking of protein molecules to form larger units is termed **quaternary structure**. The term is especially used to denote association by noncovalent bonds. Globular proteins ("monomers") then form dimers or trimers or larger aggregates, but in a specific orientation. The monomers may be identical, which is the more common case, or not. The bonds mostly are hydrophobic, although salt bridges may also be involved. In general, globular proteins that have several apolar side groups at their surface tend to form a quaternary structure. The driving force is, of course, minimizing contact between water and apolar groups; see Figure 7.3d. The structures can often be dissociated by altering the pH (generally if farther away from the isoelectric point), or the temperature (either to high or to low values), or the ionic strength (generally if lower).

Homogeneity and Purity. Unlike polysaccharides, proteins are synthesized in a precisely prescribed way, i.e., as encoded in the DNA. This would mean that all molecules of a certain protein species are exactly equal. There are, however, various disturbing factors.

(a) For all proteins studied, *genetic variants* exist that differ in one or more amino acids. Unless the protein is obtained from one organism that produces only one variant, it will be a mixture.

(b) *Post-translational modifications* occur in the cell producing the protein. Of these, proteolytic removal of nonfunctional parts and —S—S— bridge formation generally are identical for all molecules, but glycosylation, phosphorylation, and hydroxylation may induce significant variability.

(c) Changes may occur during isolation, storage, and processing. Many modifications can occur, such as deamidation, changing Asn into Asp and Gln into Glu; —S—S— bridge reshuffling; partial proteolysis; several cross-linking reactions (especially at high temperature); and partial or full denaturation (see Section 7.2).

It should further be realized that virtually all protein preparations used in practice are mixtures of several protein species and generally contain many other substances.

Question

In a publication by Miller et al. (J. Mol. Biol. 196 (1987) 641) concerning a series of one-domain globular proteins, molar mass 5–35 kDa, the surface area of the various groups on the peptide chains was calculated and it was established which part of that surface is in contact with water and which part buried in the interior. Of the average surface area in contact with water 51% would be nonpolar, 24% polar noncharged, and 19% charged; for the interior these figures were 58, 39, and 4%, respectively. This appears to disagree with the generally accepted idea that hydrophobic residues are predominantly in the interior and the hydrophilic ones at the outside. Can you think of factors that may explain this discrepancy? Take the size relations of protein molecules into account and also Eq. (7.2).

Answer

Three factors would contribute to the results obtained by Miller et al. (a) In these fairly small molecules, the core cannot accommodate all of the apolar residues. Taking the size relations given, $M = 20$ kDa would lead to a volume of $1.27 \times 20 = 25.4 \, \text{nm}^3$, i.e., a radius of 1.8 nm. Assuming the outer layer to have an average thickness of 0.5 nm (see Figure 7.3), this leads to the core comprising $(1.3/1.8)^3 = 0.38$ of the volume. However, the molecules are not nearly perfect spheres and the outer surface area would not be $4\pi \times 1.8^2 = 41 \, \text{nm}^2$, but about $9.3 \times M^{2/3} = 68 \, \text{nm}^2$, causing a greater proportion of the molecular surface to be exposed. The authors found indeed that the part of the molecule that is not exposed to the solvent was only 15–32% for the proteins studied. (b) If amino acid residues are in the interior, so must be their peptide bonds. The so-called hydrophobicity [Eq. (7.2)] relates to the side groups. However, all peptide bonds are clearly hydrophilic, because of their dipole. A large proportion of the polar noncharged surface in the interior did indeed comprise peptide bonds. (c) The terms polar and nonpolar do not say much about the quantitative aspects of the solvation of these groups by water. Note that the charged, i.e., very hydrophilic, groups were very predominantly at the outside. Also, by far the greater part of the "true" hydrophobic residues (Leu, Ile, Val, Phe, Trp) were observed to be in the core. Nevertheless, a large proportion of the surface of the molecules in contact with the solvent consisted of, albeit not very strongly, hydrophobic groups. This appears to be generally true for globular proteins of fairly small size.

7.2 CONFORMATIONAL STABILITY AND DENATURATION

In this section only globular proteins will be considered. Their tightly folded native conformation (designated N) may change into a more or less unfolded conformation (U). This change may be called denaturation, but the conformation change may be reversible, and several authors reserve the word denaturation for irreversible unfolding, or for the loss of a specific activity.

The loss of the native conformation generally has several important consequences:

1. Loss of *biological activity*, notably enzyme activity, because binding of molecules or groups is always involved, which depends in a precise manner on the conformation of at least part of the protein molecule.

2. Decreased *solubility*, because more hydrophobic groups become exposed to water. Also the surface activity may be altered.

3. Increased *reactivity* with other compounds or among groups on the protein itself, because reactive side groups become exposed. In general, most of the charged side groups are already exposed in the native state, such as the ε-amino groups of lysine, which are involved in the Maillard reaction. Some other groups are mostly buried in the native state, notably the thiol group of cysteine, which often becomes quite reactive when exposed.

4. Increased susceptibility to *attack by proteolytic enzymes*, because peptide bonds inside a tightly-coiled protein molecule are not accessible to these enzymes.

5. Increased *hydrodynamic size* and its consequences; see Chapter 6.

7.2.1 Thermodynamic Considerations

We will first consider fairly ideal cases, implying globular protein molecules containing one domain, and conditions that do not cause appreciable changes other than unfolding. Figure 7.4 gives examples of the change from the native to the unfolded conformation (and vice versa), induced in various ways. The change can be monitored from the change in hydrodynamic properties (e.g., viscosity) or in spectroscopic properties (e.g., optical rotation) and expressed as the fraction changed; it is generally observed that various methods give (almost) identical results. It is also observed that the changes are reversible and occur quickly (time scale of the order of a second). This all means a dynamic equilibrium of two reactions: unfolding, N→U, reaction constant k_U; and (re)folding, N→D, reaction constant k_F. We may rewrite Eq. [4.6] as

$$\frac{\Delta_{N \to U} G}{RT} = \frac{\Delta_{N \to U} H}{RT} - \frac{\Delta_{N \to U} S}{R} = \ln\left(\frac{[N]}{[U]}\right) = \ln K \qquad (7.3)$$

where $K = k_F / k_U$ is the equilibrium constant. An example of ΔG as a function of temperature is shown in Figure 7.5. It is seen that there are two temperatures at which $\Delta G = 0$, implying $[N] = [U]$; these are called the *denaturation temperatures*. ΔG may be considered a measure of the

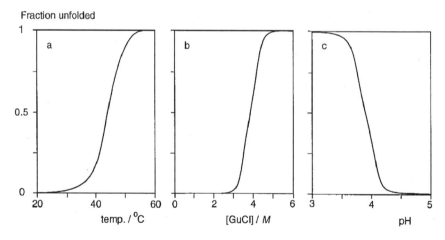

FIGURE 7.4 Transition of proteins from the native to the unfolded state or vice versa. (a) Ribonuclease at pH 3.15, as a function of temperature. (b) Lysozyme as a function of guanidinium chloride concentration. (c) Nuclease A as a function of pH.

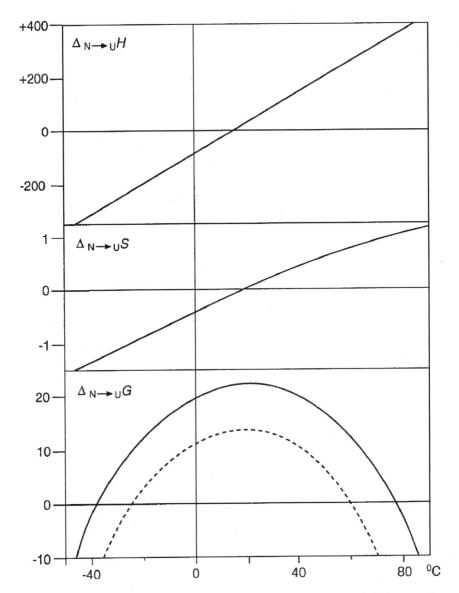

FIGURE 7.5 Approximate example of the stability of a fairly small globular protein as a function of temperature. Change from the native to the unfolded state in enthalpy (ΔH in kJ·mol^{-1}), entropy (ΔS in kJ·mol^{-1}·K^{-1}) and in Gibbs energy (ΔG in kJ·mol^{-1}). The broken line (only for ΔG) applies to a pH farther away from the isoelectric point. The temperatures at which $\Delta G = 0$ are called denaturation temperatures.

conformational stability, and it is seen to be fairly small, in this case about $25 \, \text{kJ} \cdot \text{mol}^{-1}$ at its maximum. Equation (7.3) then yields $\ln K \approx 10$, implying that a fraction $e^{-10} \approx 4 \cdot 10^{-5}$ of the molecules would be in the unfolded state. For $\Delta G = 10 \, \text{kJ} \cdot \text{mol}^{-1}$, it would amount to about 2%.

According to Eq. (7.3), a van't Hoff plot, i.e., of $R \ln K$ versus $1/T$, would yield from its slope ΔH, and from its intercept ΔS. There is, however, a difficulty, since ΔH and ΔS generally depend on temperature in the case of protein unfolding, as illustrated in Figure 7.5. ΔH can also be determined by DSC (differential scanning calorimetry: see Figure 6.25 for an explanation). The example given in Figure 7.6a shows a sharp peak in heat uptake, almost like a melting transition. When results from a van't Hoff plot and DSC can be compared, good agreement is mostly observed for single-domain proteins (within 5% or so). From such results and the observed denaturation temperatures, and with some interpolation, values for ΔH and ΔS can be derived as a function of temperature, to obtain curves like those in Figure 7.5.

Figure 7.4 shows that the transition N→U occurs over a very small range of the variable applied. This is typical for a *cooperative transition*, where several bonds are broken (or formed) simultaneously. In other words, the molecule would either be in the native or in a (nearly) fully unfolded state. Also the narrowness of the peaks on DSC plots (Figure 7.6) points to a cooperative transition. For the unfolding at high temperature, the

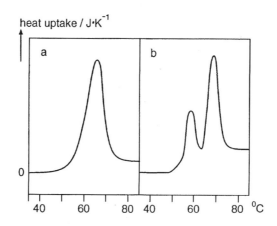

FIGURE 7.6 DSC diagrams (increasing temperature) showing denaturation of proteins; initial heat uptake ("base line") has been shifted to zero. (a) Lysozyme (pH 2.5). (b) Transferrin.

abruptness directly follows from the strong temperature dependence of ΔG, which is largely due to ΔH being very large (see Figure 7.5). Unfolding due to a decrease in pH must be due to a change in ionization of side groups. For a simple ion, however, a change in ionization from 0.1 to 0.9 occurs over a pH range of 2 units (see Figure 2.9a), whereas Figure 7.4c shows a transition over only 0.5 pH units. The explanation must be that the protein involved has some buried His groups in its native conformation and if the molecule unfolds, all of these can become ionized. This then happens at pH 3.8 rather than at the pK of 6.4 (Table 7.1), where the driving force for unfolding presumably is too small.

Several workers have tried to calculate the various **contributions to the free energy** involved in stabilizing the native conformation. The subject is, however, still somewhat controversial, at least partly because of differences in terminology. We will now consider what terms are involved in the hypothetical stability equation and how large they may be for a small molecule or domain of about 100 residues. For most terms, reliable estimates cannot be made.

The first four terms are negative, hence **promote the native conformation**:

1. *Hydrogen bonds.* It should be realized that all of the groups potentially involved can also make H-bonds with water, and the latter mostly are stronger. This would mean that no internal H-bonds are formed. However, the predominantly apolar interior of a globular protein molecule may have a dielectric constant as low as 5, as compared to 80 in water. That would cause the internal H-bonds to be much stronger, having a net bond energy of about $16 \, kJ \cdot mol^{-1}$. Assuming 50 H-bonds per molecule, this would lead to a term in $\Delta_{U \to N}G$ of roughly $-800 \, kJ \cdot mol^{-1}$.

2. *Hydrophobic interactions.* It has long been accepted that these provide a major stabilizing force, but at present many workers assume them to play a fairly small, although not a negligible, part. However, since changes in enthalpy as well as entropy are involved, and since it is difficult to separate hydrophobic interaction from other forces, the author feels that a definite answer has not yet been given. Anyway, the presence of hydrophobic residues is essential for obtaining the native conformation, since they allow H-bonds to form in an apolar environment; in that sense, the $-800 \, kJ \cdot mol^{-1}$ mentioned in item 1 does include hydrophobic interaction.

3. *Van der Waals attraction.* Of course, this also occurs between groups and water molecules, but the net effect is likely to be stabilizing. Van der Waals attraction cannot be fully separated from hydrophobic interaction. Moreover, some dipole–dipole interaction may be involved. It concerns an enthalpic contribution.

4. *Salt bridges.* This concerns only a few bonds; they may on average contribute by an amount of about $-20\,\text{kJ}\cdot\text{mol}^{-1}$ to ΔH.

Some other terms are positive, hence **promote the unfolded state**:

5. *Conformational entropy.* In the unfolded state each peptide unit would contribute about 4 degrees of freedom (Ω), but many side groups would also have more conformational freedom. Assuming the increase in Ω to be 6 to 8 per residue, this would lead to a $\Delta S = R\ln\Omega^{100}$ between 1500 and $1700\,\text{J}\cdot\text{mol}^{-1}\cdot\text{K}^{-1}$. Altogether, $\Delta_{U\to N}G$ would be about $500\,\text{kJ}\cdot\text{mol}^{-1}$.

6. *Hydration* of polar side groups and peptide bonds. This must provide a large term in ΔG, including enthalpic and entropic contributions. It cannot be separated from the interactions mentioned in items 1 and 2.

7. *Bending and stretching* of covalent bonds. Although such deviations from the state of lowest energy may be minor, it is inconceivable that all the bonds in a tightly folded protein molecule would remain completely undistorted. Since about a thousand bonds are present, the contribution to ΔH may be significant.

8. Mutual *repulsion of charged groups* on the surface. This is an enthalpic term, that may be appreciable if the pH is far away from the isoelectric point.

Despite the uncertainties in the magnitude of the various terms, it is obvious that the thermodynamic stability of a globular protein results from the difference between two large terms. For a 100 residue protein, these terms are about $10^3\,\text{kJ}\cdot\text{mol}^{-1}$ or more, whereas their difference would only be about $25\,\text{kJ}\cdot\text{mol}^{-1}$. Observed values of $\Delta_{U\to N}G$ for globular proteins under physiological conditions range for the most part between -20 and $-65\,\text{kJ}\cdot\text{mol}^{-1}$, far less than the strength of one covalent bond (about $-500\,\text{kJ}\cdot\text{mol}^{-1}$). This implies that even small changes in conditions may lead to unfolding. At high temperature, the increase in conformational entropy of the protein in the unfolded state will generally be overriding: the mentioned ΔS of about $1.6\,\text{kJ}\cdot\text{mol}^{-1}\cdot\text{K}^{-1}$ (item 5, above) leads to a change in $\Delta(TS)$, hence in ΔG, by about $50\,\text{kJ}\cdot\text{mol}^{-1}$ for a 30K increase in temperature. At very low temperature, hydrophobic interactions become very small or may even become repulsive (see Figure 3.4). At extreme pH values, electrostatic repulsion (item 8, above) may be sufficient to cause unfolding. Changes in solvent quality may significantly affect solvation free energy.

The stability is thus due to a very subtle balance. Small changes in the primary structure may leave the conformation virtually the same, but it is also possible that replacement of only one residue by another one leads to an unstable molecule. Globular proteins evolved to be stable under physiological conditions, and stability under other conditions may not be needed. Although more stable conformations would in principle be possible, it may be that such conformations cannot be reached because of the intricacies of the folding process. It is by no means certain that the native conformation of a protein is the one of lowest free energy possible; the conformation may just represent a local minimum.

There are some mechanisms that *enhance conformational stability*:

Formation of —S—S— bridges between Cys residues adjacent in the folded state; these would anyway prevent complete unfolding.

Other posttranslational modifications, notably glycosylation of specific residues, presumably after folding of the peptide chain.

Formation of structures on a larger scale than that of a domain (e.g., "domain pairing" or "subunit docking").

Binding of ligands mostly alters the conformation somewhat, but then generally stabilizes it.

Complications. The discussion given above is to some extent an oversimplification. The following complications may arise.

1. The protein consists of *more than one domain* and these can unfold independently of each other. A fairly simple example is in Figure 7.6b. In other cases, deconvolution of the DSC diagram may be less straightforward. In some proteins, two (or three) domains are so closely similar that they unfold (almost) simultaneously. In such a case, ΔH obtained from a DSC diagram would be about twice (or three times) the value derived from a van't Hoff plot. Serum albumin (molar mass 66 kDa), for example, consists of three similar domains of about 190 residues. Mutual interaction of domain unfolding may occur, but in most cases each domain unfolds just like a small globular protein.

2. There may be *intermediate stages* between the native and the unfolded state. In some cases a transient and not very stable "prefolded state" is observed, which appears to be similar to that mentioned next.

3. Some proteins exhibit at some pH values, for instance, about three units below the isoelectric pH, an intermediate conformation designated *"molten globule state"*. Such a state appears to be characterized by

Being not fixed, the molecules showing a population of states;

A hydrodynamic size somewhat larger than in the native state, but far smaller than in the unfolded conformation;

A large amount of secondary structure, but any tertiary structure appears to be fluctuating and many residues are in contact with water;

A conformational entropy that is almost the same as that in the unfolded state, which seems difficult to reconcile with the abundance of secondary structure;

Transitions with the native state are slow and cooperative, those with the unfolded state fast and noncooperative.

Characteristics and significance of the molten globule state are still a matter of debate.

4. In many cases, *refolding* to the native state is *incomplete* or not precise. For many proteins, the peptide chain produced in the cell assumes its native folded state with the aid of specific helper proteins, called chaperonins. These are missing when a protein has been denatured and is allowed to refold in practice. Several proteins refold into a near-to-native state, which then may or may not slowly change into the native conformation.

In conclusion, proteins are *individuals*. Although the same factors determine the conformation and the conformational stability, the net result is highly variable. Moreover, they react in a different way on differences in environmental conditions, which is further illustrated below.

7.2.2 Denaturation

Because of the relatively small stability of most globular proteins, they can be induced to unfold in many ways. If the conditions causing unfolding (high temperature, extreme pH, etc.) are removed again, the ultimate result will depend on the extent to which the peptide chain will regain its native conformation and possibly on the rate of this refolding. Reactions occurring when the protein is in the unfolded state may partially prevent this.

Prevention of refolding to the native state can be caused by several reactions, but this has been studied insufficiently. The following ones may be involved.

1. *Aggregation* of the unfolded molecules. Upon unfolding, many hydrophobic residues become exposed, which generally leads to a substantially decreased solubility. This would undoubtedly cause refolding to be slower, but it is uncertain whether it would be prevented. It may well be that intermolecular cross-linking is involved (points 3 and 5), since this may occur much faster when the molecules are already aggregated due to noncovalent bonding.

2. *Trans* $\longrightarrow$ *cis* change of the configuration of peptide bonds. The trans conformation is very stable, except for a peptide bond involving the N of proline

which does not have a double-bond character, implying that rotation about the —CO—N $<$ bond is possible; trans and cis conformations have about equal stability in this case. The change involves an activation enthalpy of about $85\,kJ \cdot mol^{-1}$. The relaxation time thus is about 20 min at $0°C$ and 1 s at $70°C$ $(Q_{10} \approx 3)$. This implies that these peptide bonds will have about equal amounts of trans and cis at high temperature in an unfolded chain; rapid cooling may then readily cause locking of cis bonds. In such a case, refolding would not be in the native conformation, but slow renaturation may occur after cooling. After all, some protein molecules, though it may be a very small number, will be in the unfolded state, even at the optimum temperature for stability.

3. *Reshuffling* (scrambling) *of sulfur bridges.* This occurs if the molecule is unfolded and contains a free —SH, of which at least a little is present in its ionized form, i.e. at pH $>$ about 6. The reaction occurring is schematically

$$R_a—S—S—R_b + R_c—S^- \rightleftarrows R_a—S—S—R_c + R_b—S^-$$

Interchange of disulfides can also occur among different protein molecules, producing molecular aggregates. These reactions occur faster at higher temperature and higher pH. At very high pH, —S—S— bridges are in fact weak bonds that readily break.

4. *Deamidation.* At high temperatures, the residue Asn may show the reaction

$$—CONH_2 + H_2O \longrightarrow —COOH + NH_3$$

thereby forming Asp. It may also happen with Gln, giving Glu, though to a far smaller extent. These reactions occur faster at a lower pH. The changes would generally lead to a nonnative conformation upon refolding.

5. At very high temperatures various *cross-linking* reactions between side groups are possible. Moreover, peptide bonds may even be hydrolysed, depending on pH.

6. The absence of *chaperonins*, as discussed above.

Denaturing Agents. As mentioned, several agents or conditions can cause denaturation. They may be categorized as follows:

1. *High or low temperature.* Conformational stability as a function of temperature is discussed in the previous section; see especially Figure 7.5. By and large, unfolding at low temperature occurs because hydrophobic bonds then are very weak or even repulsive. At high temperature, the increased effect of the conformational entropy becomes overriding. At high temperature, irreversible changes in protein configuration may well occur,

preventing return to the native state. This is far less likely at low temperature. Kinetic aspects are discussed in Section 7.2.3.

2. *Extreme pH*. Most proteins are the most stable near their isoelectric pH. Figure 7.5 gives an example of the stability curve (ΔG versus T) at two different pH values. The decreased stability at extreme pH values must be ascribed to electrostatic repulsion between groups of like charge and to the impossibility of forming internal salt bridges.

Figure 7.7a shows some relations between pH and denaturation temperature, which give a clear example of the way in which destabilizing agents enhance each other's effects. Varying pH is thus a method of causing denaturation at a fairly low temperature, where irreversible changes are less likely to occur, especially at low pH. At high pH, sulfur bridges tend to break, as mentioned.

3. *Solvent quality*. Various solutes added in high concentrations affect solvent quality and thereby solubility and conformation of macromolecules; see Sections 3.2 and 6.2.1. Solutes may thus affect conformational stability. Relations are not straightforward for proteins, because they have polar as well as apolar groups, that may be affected in opposite manner. For *salts* (ions), the Hofmeister series (Section 3.2) is mostly obeyed. Examples are in Figure 7.8a. It is seen that the very hydrophilic ions at the beginning of the series, i.e., ammonium and sulfate, stabilize the conformation, whereas those at the other end, guanidinium and thiocya-

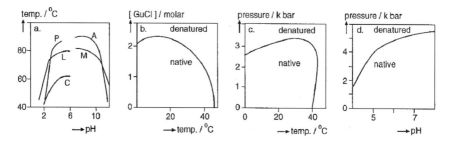

FIGURE 7.7 Combined effects of two variables on conformational stability of globular proteins. (a) Denaturation (unfolding) temperature as a function of pH for papain (P), lysozyme (L), cytochrome C (C), parvalbumin (A), and myoglobin (M). (b) Effect of concentration of guanidinium chloride concentration and temperature on conformation of lysozyme at pH 1.7. (c) Effect of pressure (1 kbar $= 10^8$ Pa) and temperature on conformation of chymotrypsinogen. (d) Effect of pressure and pH on conformation of myoglobin (20°C).

nate, have a strong destabilizing effect, acting even at room temperature (Figure 7.7b).

Many *neutral* solutes also have a distinct effect: e.g., Figure 7.8b. Well known is the destabilizing effect of urea at high concentration, similar to that of guanidinium chloride. Urea makes strong H-bonds with water, undoubtedly altering water structure, but the explanation of its denaturing effect is not quite clear. It appears to bind to peptide bonds, thereby dehydrating them. Solutes like ethanol, that are far less polar than water but are nevertheless readily soluble in water, tend to be destabilizing. On the one hand they strengthen H-bonds and salt bridges (because of the lower dielectric constant), but on the other they strongly weaken hydrophobic

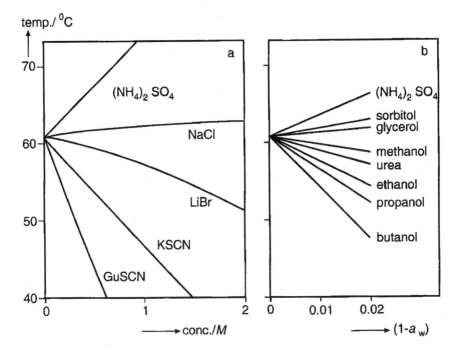

FIGURE 7.8 Effect of concentration of (a) various salts (Gu = guanidinium) and (b) various organic solutes on the denaturation (unfolding) temperature of ribonuclease A. In (b) water activity a_w is used as the independent variable, rather than molar concentration; $(1 - a_w)$ is about proportional to the mole fraction of solute. (After various sources, mainly von Hippel and Wong, J. Biol. Chem. 240 (1965) 3909.)

bonds. The overall result mostly is loss of the native conformation, but not formation of a highly unfolded state. Several sugars and polyols tend to stabilize the native conformation, when present at high concentrations. This is similar to the increased stability usually observed in systems with a very low water content (Section 8.4.2).

4. *Specific reagents.* Here solutes are meant that destabilize conformation at low concentration. A prominent example is reagents that *reduce a sulfur bridge* to —SH groups, such as mercaptoethanol and dithiothreitol. Note that this implies a change in configuration (primary structure). The reduction leads to further unfolding of a denatured protein, as observed by changes in hydrodynamic size. For example, serum albumin has an intrinsic viscosity of 3.7 ml·g^{-1} in its native form, 16.6 in 8 molar urea, and 43.2 in 8 molar urea after breaking of all sulfur bridges.

Various *detergents* denature proteins at concentrations in the range of 1 to 10 millimolar. SDS (sodium dodecylsulfate) is often used, but detergents with a more apolar chain are even more effective. The apolar part strongly binds to hydrophobic regions in the protein molecule, thereby disrupting the native structure, but it appears that the ionic group is also essential. Detergent-denatured proteins tend to fully regain their native conformation after removal of the detergent by dialysis.

5. *High pressure.* Very high pressure treatments are used in food processing for several purposes, for instance to kill microorganisms, while very little chemical reactions occur. The main mechanism is that high pressures cause denaturation, or at least unfolding, of globular proteins. The unfolding occurs over a narrow pressure interval, indicating a cooperative transition between two states. The pressure needed greatly depends on temperature (Figure 7.7c) and pH (7.7d).

The effect of pressure on a chemical equilibrium follows the *principle of le Chatelier*: an increase in temperature shifts the equilibrium in the direction of highest enthalpy for an endothermic reaction, an increase in pressure in that of smallest volume. The relation giving the temperature effect is obtained by differentiating Eq. (7.3) at constant pressure, resulting in

$$R\left[\frac{\partial \ln K}{\partial(1/T)}\right]_p = \Delta_{N \to U}H \tag{7.4a}$$

where $\Delta H > 0$ for $T > T_{opt}$ (Figure 7.5). For the pressure effect a similar relation holds

$$RT\left(\frac{\partial \ln K}{\partial p}\right)_T = -\Delta_{N \to U}V \tag{7.4b}$$

where K is the equilibrium value of $[N]/[U]$ and V is total molar volume of the system. Since a very high pressure is needed to cause unfolding of a protein (often much more than 1000 bar), the concomitant volume change must be small. Hydrophobic bond formation mostly goes along with $\Delta V > 0$, and high pressure would thus lead to breaking of these bonds, implying destabilization of the native conformation. On the other hand, formation of most electrostatic bonds, H-bonds, and van der Waals interactions involve a negative ΔV, and high pressures would be stabilizing. A full explanation of the effect of pressure on protein stability is lacking. Figure 7.7c shows that under some conditions, relatively moderate pressures slightly stabilize (slightly increase the unfolding temperature). Very high pressures always cause unfolding.

For some proteins, for instance ovalbumin, high pressure may cause irreversible *aggregation*. Whether it occurs may depend on the rate of pressure increase. Moderately high pressure generally causes dissociation of most quaternary structures. This is not surprising, since the association is generally due to hydrophobic interactions.

6. *Adsorption.* Adsorption phenomena are discussed in Sections 10.2 and 10.3. Proteins are surface active, which implies that they lower the interfacial free energy upon adsorption. For instance, protein adsorption at an air–water interface lowers surface tension by about $30\,\text{mN}\cdot\text{m}^{-1}$, which equals $0.03\,\text{J}\cdot\text{m}^{-2}$. We will here consider the effect on protein conformation.

Proteins adsorb onto almost all surfaces, whether air–water, oil–water, or solid–water. There is only one exception: the adsorbent is a solid that is hydrophilic and charged, and the protein has a charge of the same sign as the solid and is a "hard" protein. The latter implies that the protein has a relatively stable globular conformation, i.e., a fairly high $\Delta_{N \to U}G$. "Soft" proteins also adsorb at hydrophilic solid surfaces, even of the same charge. Adsorption may thus primarily involve electrostatic attraction, in which case protein conformation is not greatly affected. However, other solids, oil, and air provide hydrophobic surfaces, where the main driving force for adsorption generally is hydrophobic interaction. Since most apolar residues are buried in the core of a globular protein, adsorption generally involves a marked change in conformation. This is borne out by results of spectroscopic studies, which show a change in secondary and loss of tertiary structure. DSC applied to an adsorbed protein generally shows a denaturation peak that is smaller (or even negligible), and that occurs at a lower temperature as compared to the protein in solution.

Adsorption of *enzymes* generally leads to loss of enzyme activity, whether measured in the adsorbed state or after desorption. By and large, the activity loss is greater under conditions (temperature, pH, etc.) where conformational stability is less, and if the adsorbent is more hydrophobic, presumably because the driving force for conformational change is greater. It is sometimes observed that adsorption from a more dilute enzyme solution leads to more inactivation. The explanation may be that at low concentration adsorption is slow, allowing adsorbed molecules to expand laterally, which implies conformational change. If adsorption is fast, a densely packed adsorbed layer is rapidly formed, which would prevent lateral expansion. In agreement with this, it has been observed that some proteins do not greatly change conformation when merely adsorbing onto an air–water interface, but when the air–water surface is expanded, for instance by deforming an air bubble, considerable change occurs. Beating air into a protein solution can therefore cause denaturation.

It has further been observed for several enzymes that adsorption onto an oil–water interface causes complete inactivation, whereas only partial inactivation may occur due to adsorption onto an air–water surface. The reason may be that hydrophobic segments of the molecule can penetrate into an oil phase, but not into air. This would be because the net attractive energy between these segments and oil can be greater than that between segments and water, whereas the attractive energy between any group of a protein and air will be virtually zero. This must cause a greater driving force for loss of native configuration at the oil–water interface. A fairly stable enzyme like lysozyme, which can regain activity after various unfolding treatments at low temperature, does not regain it after adsorption onto oil droplets, even at its isoelectric pH. This leads to the important conclusion that more than one unfolded state can exist, and that some of these states permit return to the native state, whereas others do not.

7. *Shear stress.* It has been observed that some enzymes under some conditions (such as the presence of specific solutes) show inactivation when the solution is subjected to simple shear flow for a considerable time, especially when the temperature is not much below the denaturation temperature. The extent of inactivation then is proportional to the product of shear rate and treatment time. Most workers agree that the shearing stresses applied (for instance 1 Pa at a shear rate of $1000 \, \text{s}^{-1}$ and a viscosity of $1 \, \text{mPa} \cdot \text{s}$) are far too small to affect protein conformation. In some cases, denaturation at the air–water interface may have occurred, but in other cases this possibility has been ruled out. There is no generally accepted explanation of the effect.

Question

Can you explain why bringing an aqueous solution of a globular protein into a shear field, shear rate $10^4 \, s^{-1}$, would not cause its unfolding, whereas adsorption onto the air surface would? Assume a protein molecule of $5 \times 5 \times 5 \, nm^3$ in size.

Answer

Water has a viscosity of $1 \, mPa \cdot s$ and a shear rate of $10^4 \, s^{-1}$, thus would cause a shear stress of $10 \, Pa$ $(N \cdot m^{-2})$. The stress acts on a surface of area $5 \times 5 \, nm^2$, exerting a maximum tensile force of $10 \times 25 \cdot 10^{-18} = 25 \cdot 10^{-17} \, N$ on a protein molecule. The force acts over a distance of $5 \, nm$. For every rotation of the molecule, an amount of mechanical energy of $5 \cdot 10^{-9} \times 25 \cdot 10^{-17} = 1.25 \cdot 10^{-24} \, J$ $(N \cdot m)$ is acting on it, which equals about 0.0003 times $k_B T$. We have seen that the conformational stability of a globular protein is about $25 \, kJ \cdot mol^{-1}$ or more, corresponding to about 10 times $k_B T$ for a single molecule. This is more than $3 \cdot 10^4$ the mechanical energy applied to the molecule during a rotation in the shear field.

We have also seen that the change in free energy upon adsorption of a protein on the water surface amounts to about $0.03 \, J \cdot m^{-2}$. This corresponds to a free energy of $0.03 \times 25 \cdot 10^{-18} = 7.5 \cdot 10^{-19} \, J$ per adsorbed molecule, or about 185 times $k_B T$. Roughly that amount of energy would be available to act on a protein molecule on adsorption, and this is much larger than the value of the conformational stability.

7.2.3 Denaturation Kinetics

This section is essentially restricted to heat denaturation of globular proteins. This is a very important subject for food technologists. It is the basis of the inactivation of enzymes. Since microorganisms depend for their metabolism on enzymes (including those that act as transport regulators in the cell membrane), killing of microbes is governed by protein denaturation. Aggregation of globular proteins often occurs after denaturation, as in heat-set protein gelation. Knowledge of the kinetics of these processes is essential, especially for optimizing the time–temperature combination for eliminating undesired microorganisms and enzymes, while minimizing quality loss due to heat treatment. The reader is advised to study Chapter 4 first, Sections 4.3.3 and 4.4 in particular. There it is explained that the rate of a reaction does not depend on the difference in free energy between the states before and after, i.e., in the present case on $\Delta_{N \to U} G$, but on the activation free energy $\Delta G^{\ddagger}$.

Unfolding Kinetics. Figure 4.5 gives an example (the enzyme alkaline phosphatase) of kinetic parameters for protein denaturation. It is seen that the activation enthalpy $\Delta H^{\ddagger}$ would be very large, in accordance with the cooperative transition from native to unfolded state. Although the bonds involved are weak, they are numerous, and most of them have to break simultaneously. The value of $\Delta H^{\ddagger}$ was obtained using Eq. (4.11), by taking the temperature dependence of the denaturation reaction (enzyme inactivation, in this case); it is larger than $\Delta_{N \to U} H$, but of the same order of magnitude. The large $\Delta H^{\ddagger}$ implies a very strong dependence of the reaction rate on temperature, as illustrated in Figure 7.9. The large positive $\Delta S^{\ddagger}$ compensates for the large $\Delta H^{\ddagger}$, causing a fairly small $\Delta G^{\ddagger}$ and a fairly fast reaction rate.

Many proteins follow this trend. Some results are given in Figure 7.9. It seems logical to assume that the molar $\Delta H^{\ddagger}$ and $\Delta S^{\ddagger}$ values are roughly proportional to molar mass (M). The number of bonds to be broken, and the corresponding bond enthalpies and contact entropies, will be about proportional to molecular size. Also the increase in conformational entropy of the peptide chain upon unfolding will be about proportional to the number of amino acid residues n, and thereby to M. (Remember that $S_{conf} = R \ln \Omega$ (per mole), where the number of degrees of freedom Ω is the number per residue, presumably about 7, to the power n. This would yield $\Delta S = Rn \ln 7$.) A rough correlation is indeed observed between the activation enthalpy and entropy and molar mass, but it is far from perfect. This may be due to denaturation occurring separately in separate domains in the same molecule (especially for large M), and to variation in the number of —S—S— bridges or other conformational details.

The reasoning just given implies another oversimplification. Figure 7.5 shows that $\Delta_{N \to U} H$ and $\Delta_{N \to U} S$ are not constant but depend markedly on temperature. It may be argued that this relates to differences in H and S between the native and the unfolded states and not to the activation enthalpy and entropy, but it is very unlikely that the latter two behave quite differently from the former. The "observed" $\Delta H^{\ddagger}$ is thus an apparent activation enthalpy.

Does this imply that the idea of an activation free energy etc. would not apply? Figure 7.10a gives an example of the first-order rate constant for unfolding (a fully reversible reaction in the present case). It is seen to be about $1 \, s^{-1}$ at T_{den}, 67°C. By applying Eq. (4.11) we arrive at $\Delta G^{\ddagger} = 84 \, kJ \cdot mol^{-1}$, definitely higher than the free energy difference between the two states, which is, by definition, zero at this temperature. Consequently, $\Delta H^{\ddagger}$ will be higher than $\Delta_{N \to U} H$. (It seems unlikely that $\Delta S^{\ddagger}$ is greatly different from $\Delta_{N \to U} S$.)

Irreversibility. The reasoning given above is also an oversimplification in another sense. Most of the results discussed concern loss of biological (e.g., enzyme) activity, and the activities then are determined after cooling to room temperature. It thus concerns an irreversible change, rather than reversible unfolding. The simplest case would be

$$N \underset{k_f}{\overset{k_u}{\rightleftharpoons}} U \overset{k_i}{\longrightarrow} I$$

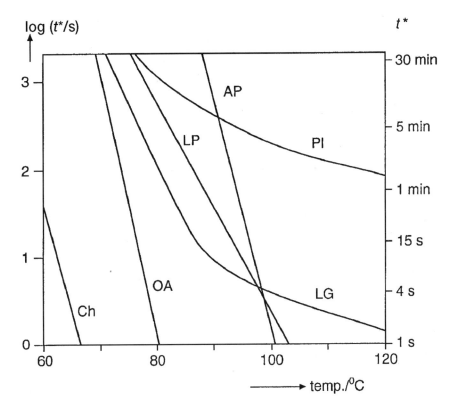

FIGURE 7.9 Heating times needed (t^*) for inactivation (reduction of activity to about 1%) of the enzymes chymosin (Ch), lipoxygenase (LP), acid phosphatase (AP), and plasmin (Pl); and for 30% of ovalbumin (OA) and β-lactoglobulin (LG) to become insoluble. Because of the narrow temperature intervals involved, also plots versus T, rather than $1/T$, can be approximately linear.

where I stands for "inactive" or "irreversibly changed" form. If, moreover, all reactions were first order, the rate of change observed would be given by

$$v = \frac{d[I]}{dt} = k_i[U] \qquad [U] = f([N]_{t=0}, k_u, k_f, t) \tag{7.5}$$

If the rate constant for refolding k_f is zero, the second step is not needed, and the unfolding step is rate determining; in this case Eq. (7.5) takes the form $v \approx k_u[N]$, as described above.

More realistically, all U will refold on cooling (see, e.g., Figure 7.10a), and what is effectively observed is the loss of N and U. The rate equation thus is

$$v = -\frac{d([N] + [U])}{dt} = k_{obs} \times ([N] + [U]) \tag{7.6}$$

Combination with (7.5) and invoking the equilibrium constant $K = [U]/[N]$, as in Eq. (7.3), yields for the observed rate constant

$$k_{obs} = \frac{k_i}{1 + 1/K} \tag{7.7}$$

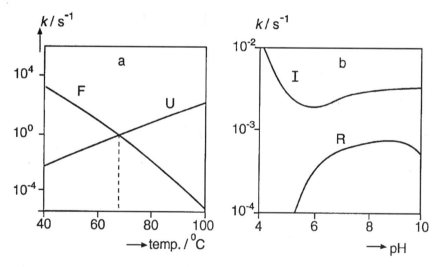

FIGURE 7.10 Refolding of denatured proteins. (a) Rate constants (k) for unfolding (U) and refolding (F) of lysozyme at various temperatures. (b) Rate constants for inactivation (I) at 76°C and for reactivation (R) at 35°C of peroxidase at various pH.

For $T \gg T_{den}, K \gg 1$, and $k_{obs} \approx k_i$, and the latter probably has a weak temperature dependence as compared to k_u. At $T \ll T_{den}, K \ll 1$, and $k_{obs} \approx k_i K$. Now the temperature dependence will be very strong: K strongly depends on T (see, e.g., Figure 7.4a), and although this is far less so for k_i, the dependence of the reaction rate on temperature is as the product of both variables.

It will now depend on the magnitude of k_i at T_{den} (where $K = 1$) what is observed, taking into account that the values of k_{obs} would range between about $3 \cdot 10^{-4}$ and $1\,s^{-1}$ in practice (heating times between 1 s and 1 h). If k_i is very small, denaturation is negligible. If, for instance, $k_i = 0.1\,s^{-1}$ at $T_{den}, k_{obs} = 0.05\,s^{-1}$; in such a case a strong temperature dependence will be observed at low T, gradually changing to a weak dependence at high T. Several enzymes show such behavior, and an example is given by plasmin (Figures 4.6 and 7.9). If k_i is large, say $10^3\,s^{-1}$, only the steep relation between reaction rate and temperature is experimentally accessible, and $k_{obs} = k_i K$.

The last case is the most common one for enzyme inactivation and it is nearly always observed for the killing of microorganisms. The latter is to be explained by the fact that the killing of a microbe will depend on irreversible inactivation of the most unstable of its essential enzymes, and that would be one following the inactivation pathway discussed here, since the other situations imply a smaller k_i and would therefore imply greater heat stability.

Note. On the other hand, microbes of one species, or even of one strain, show some physiological variation in heat stability, which may possibly affect curves like those in Figure 7.9.

Complications. In several cases, other relations are observed, and only for some of these have fitting explanations been found. From such studies, supplemented with some reasoning, the following complications may be derived. It will often be difficult to distinguish between them. We will not discuss the various types of kinetics observed or derived from inactivation models.

1. *Intermediate steps* in reaching the unfolded state. One example is dissociation of quaternary structure. This is not uncommon, since several enzymes exhibit such structures, which often are essential for enzyme activity. Another possibility is unfolding in steps, as mentioned in Section 7.2.1. Such complications can in principle cause so-called grace-period behavior (Figure 7.11a) as well as the opposite, i.e., decelerating inactivation. The latter may be a gradual decrease in rate, or there may be two fairly distinct first order stages, as shown in Figure 711.b.

2. The reaction U→I may not be first order, but *second order*. A case in point is aggregation of unfolded protein, which is rather common during heating of not very dilute protein solutions. The extent of "denaturation" then is generally estimated from the increase in turbidity or from the proportion of protein having become insoluble. Often, the overall reaction order so observed is between one and two, and it may change in the course of the reaction.

3. The change U→I may involve *more than one reaction*, either simultaneously or in sequence. For instance, a state I_1 may still allow slow refolding (where one may think of the trans–cis equilibrium of peptide bonds involving proline), whereas a state I_2 does not allow refolding.

4. An unfolded globular protein generally is very susceptible to *proteolytic cleavage*, if a suitable protease is present in the active state. Such a situation can readily produce inactivation curves like those in Figure 7.11b, where the fast inactivation between 45 and 55°C would be due to proteolysis, until also the protease has attained the unfolded, i.e., inactive, state. One example is the autodigestion at intermediate temperatures shown by several proteolytic enzymes; an example is in Figure 7.11c.

5. The rates of *heating and cooling* may significantly affect the shape of the inactivation curves, especially in the case just mentioned. Another cause may be *slow reactivation* (refolding) as exemplified in Figure 7.10b; in such cases, also the time elapsed between cooling and estimation of residual activity plays a part.

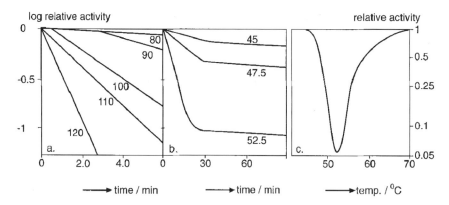

FIGURE 7.11 Heat inactivation of some enzymes, expressed as activity divided by initial activity. (a) A bacterial protease as a function of heating time at various heating temperatures (°C, indicated). (b) Luciferase, same variables. (c) A bacterial protease heated during 30 min as a function of heating temperature.

It may finally be recalled that conditions during heating, such as pH and ionic composition, greatly influence denaturation rate. In some liquids, the pH changes significantly during prolonged heating. Another variable may be the presence of a suitable ligand, especially a substrate for an enzyme. Many enzymes are far more heat stable in the presence of substrate.

Question

The low-temperature inactivation of a proteolytic enzyme presented in Figure 7.11c was observed to proceed faster when the enzyme preparation had been further purified. What would be the explanation?

Answer

As mentioned, the inactivation was due to autodigestion. Molecules still in the native, i.e., enzymatically active, conformation can split unfolded ones. We thus have

$$N \underset{\leftarrow}{\rightarrow} U$$

$$N + U \underset{\leftarrow}{\rightarrow} NU \longrightarrow N + U_1 + U_2$$

where U_1 and U_2 are the enzymatically inactive fission products. However, besides $N + U \underset{\leftarrow}{\rightarrow} NU$, the competing association

$$N + P \underset{\leftarrow}{\rightarrow} NP$$

will occur, where P means any other protein that can be attacked by the enzyme. This would slow down the splitting of U, the more so for a larger [P]. Since purification of an enzyme preparation essentially means removal of other proteins, it will diminish competition and hence increase splitting of U.

7.3 SOLUBILITY

The solubility of a component is defined as its concentration in a saturated solution c_{sat}, i.e., in a solution in contact and in equilibrium with crystals or a fluid phase of the component. Such a situation can often not be realized for a protein, because crystals cannot be obtained, and determination of the solubility then is somewhat questionable. Nevertheless, fairly consistent results can generally be obtained.

The solubility of proteins in aqueous media is widely variable, from virtually zero to about 35% by volume. Both protein composition and conformation, and the environment, i.e., solvent composition and temperature, cause the variation. There is no good quantitative theory based on first principles: proteins are too intricate. Nevertheless, some general rules can be given and the effect of some environmental variables can be predicted semiquantitatively.

The solubility of a globular protein is closely related to its *surface properties*, i.e., on the groups that are in contact with the solvent. The free energy involved is primarily due to hydrophobic and electrostatic interactions. The greater the proportion of apolar groups on the surface of a protein, the poorer its solubility in water, whereas a larger proportion of charged groups enhances solubility. It may thus be useful broadly to classify proteins as hydrophobic or hydrophilic, meaning the surface properties. As was already mentioned, a fairly large surface hydrophobicity often leads to the association of polypeptide chains into a specific quaternary structure. These oligomeric proteins then may have good solubility, since many apolar groups have been shielded from the solvent (as illustrated in Figure 7.3d).

Size. The larger the protein molecules, the smaller the decrease in molar translational entropy upon precipitation. Hence the larger the decrease in free energy and the smaller their solubility. This especially applies to *globular* proteins, which have a relatively small conformational entropy. *Disordered* proteins behave somewhat differently at a concentration above the solubility: they tend to form a "coacervate," i.e., a highly concentrated aqueous phase, rather than a precipitate (see Section 6.5.1). It then is quite difficult to determine the magnitude of the solubility.

Many "*structural*" proteins, i.e., those that provide mechanical properties to a system, have a very poor solubility. This is at least partly due to very large size. The water insoluble glutelins of wheat flour consist of peptides that are cross-linked by —S—S— bridges into very large molecules (they are, moreover, rather hydrophobic). Most structural proteins are fibrous. Collagen, the main component of tendon and also abundant in skin, cartilage, and bone, largely consists of triple helices of long peptides. The helices are closely packed to form fibrils, in which they are covalently bonded to each other. Such a material is completely insoluble. When collagen is boiled in water, many of the covalent bonds are broken (including some peptide bonds) and the helices unfold. In this way gelatin is obtained, and gelatin is well soluble (at least at temperatures above 30°C), because it has very few hydrophobic amino acid residues. (See Section 17.2.2 for more about gelatin as a gelling agent.)

Salting In. The effect of *electrostatic interactions* on solubility can be derived from the Debye–Hückel theory, which is briefly discussed in Section 2.3.2. It concerns electrostatic repulsion, not bond formation. A compact globular protein can be considered as a large ion, and the theory gives the activity coefficient $\gamma_\pm$ of ions. The solubility of an ionic component should be expressed as its activity at saturation; in other words, a_{sat} is fixed at any temperature. Since $a = c \times \gamma$, a decrease of $\gamma_\pm$ then leads to an increase in solubility in terms of concentration (c). A relation for $\gamma_\pm$ is given in Eq. (2.28), but that is an oversimplification in the sense that all ions were considered small and of roughly equal size. A protein molecule is relatively large, and then a better approximation is given by

$$\gamma_\pm \approx \exp\left[\frac{-0.37z^2\kappa}{1 + \kappa R}\right] \tag{7.8}$$

which holds for water at room temperature. Here, z is valence, R the protein radius in nm, and κ the Debye parameter (1/thickness of the electric double layer) in nm^{-1}. κ depends on total ionic strength I (molar), according to Eq. (6.9b): $\kappa \approx 3.2\sqrt{I}$. A protein being a macroion, z^2 can be very large. Even at the isoelectric pH it will not be zero, since it has to be averaged over all molecules. Then $<z>$ is zero, but $<z^2>$ is not, because of the stochastic variation in charge; the root-mean-square value of z will often be about 2 at the isoelectric pH. Some calculated examples are in Table 7.2. They are not reliable for the highest ionic strength or for the highest valence. Nevertheless, they clearly show that the activity coefficient will strongly decrease, and thereby the solubility strongly increase, for a pH farther away from the isoelectric point (larger $|z|$) and for a higher ionic strength. This is

TABLE 7.2 Values of the Free Ion Activity Coefficient

$<z^2>^{1/2}$	ionic strength/millimolar		
	2	20	200
2	0.86	0.75	0.60
4	0.55	0.32	0.20
8	0.09	0.011	0.002

Calculated according to Eq. (7.8) for various ionic strengths of the solution and valence (z) of the macroion considered; its ionic radius was taken as 3 nm.

indeed what is observed for fairly hydrophilic globular proteins; an example is in Figure 7.12. The increase in solubility by increasing ionic strength is called salting in.

A complication is that it may take a long time before equilibrium is reached. For instance, β-lactoglobulin is quite soluble at neutral pH, even at low ionic strength, as indicated by Figure 7.12. If a concentrated solution of pH 7 is brought to pH 5.2, however, it may take some days before visible protein aggregation occurs.

Equation (7.8) does not hold for proteins of a more or less unfolded conformation: the protein may then form a coacervate rather than a precipitate (Section 6.5.1). However, the trends would be the same. Another point is that extreme pH values may lead to unfolding, even at room temperature. The molecules then generally have such a high charge as to be well soluble, despite the exposure of apolar groups. Most proteins used in the food industry are soluble at pH > 9. An additional cause would be disruption of —S—S— bridges at such a pH.

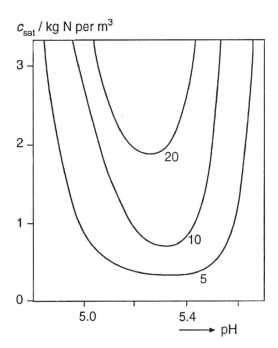

FIGURE 7.12 Salting in. Solubility (c_{sat}) of β-lactoglobulin as a function of pH for various NaCl concentrations (indicated, millimolar).

Salting Out. Many proteins are fairly hydrophobic, several seed storage proteins, for example, such as those from soya beans. Most of these are insoluble in water, especially at their isoelectric pH. Their solubility is primarily determined by *solvent quality*. The chemical potential of a protein in solution is to a large extent determined by the solvation free energy of the groups in contact with solvent, which is positive for apolar groups, and about zero or negative for most other groups. In first approximation its value per unit apolar surface area (A_{ap}) may be equated to the surface tension of the solvent, since air also is "apolar." By and large, solutes that lower the solubility of proteins also increase the surface tension of the aqueous phase. The surface tension mostly increases linearly with solute concentration by an amount σ per mole of solute. The increase in surface free energy would then be proportional to A_{ap} times σ. If the solute is a salt, its concentration can be expressed as ionic strength. As discussed above [Eq. (7.8)], ionic strength also affects solubility in another way. Combining the two relations, and further simplifying (7.8), would roughly lead to

$$\ln\left(\frac{c_{sat}}{c^*}\right) \approx C_1 z^2 \sqrt{I} - C_2 \sigma I \tag{7.9}$$

where c_{sat} is the molar solubility, c^* the hypothetical solubility for $I = 0$, and C_1 and C_2 are constants. Like c^*, they depend on the type of protein, C_2 being proportional to the molar hydrophobic surface area A_{ap}. σ is a characteristic of the salt present, and all constants depend on temperature.

It follows that for small I the first term is overriding (salting in), and for large I the second one (salting out). This is illustrated in Figure 7.13a. The trends predicted are observed, but the agreement between theory and observation is not perfect. One reason is that a solution of zero ionic strength cannot be obtained (except at the isoelectric pH), because the protein is charged and contains counterions; especially the protein itself contributes to I (see Section 6.3.1, item 3). This implies that c^* cannot be determined with accuracy. Moreover, Eq. (7.9) overestimates the salting-in effect for high z. Another complication is that at high ionic strength, say above 0.3 molar, ionized groups at the protein start to bind, i.e., form ion pairs with, counterions, thereby lowering the electric charge. Nevertheless, the salting-in effect is about the same for different salts (if the ionic strength rather than the molar concentration is used), whereas salting out greatly depends on type of salt (Figure 7.13b), as predicted. The linear relation between $\log(c_{sat}/c^*)$ predicted by Eq. (7.8) at high I is also observed (Figure 7.13c).

The effectiveness of salts to reduce solubility greatly varies and generally follows the *Hofmeister series* (see Section 3.2). The salts that increase the conformational stability of globular proteins (Section 7.2.2) generally decrease their solubility; compare Figures 7.8a and 7.12b. However, the agreement is not perfect. The greater effect of sodium sulfate compared to the ammonium salt, for example, does not agree with their position in the Hofmeister series. $(NH_4)_2SO_4$ is often used for salting out proteins, as it is very effective and also very soluble, so that high concentrations can be reached. Since the salt enhances conformational stability, the precipitation of the protein is reversible: diluting the system with water or dialyzing the salt away causes the protein to dissolve again in an undenatured state. Ions at the end of the Hofmeister series, such as chlorides, do not cause salting out of hydrophilic proteins (Figure 7.13b). In accordance with this, H_2SO_4 is generally much more effective than HCl in realizing isoelectric precipitation of proteins. Proteins vary greatly in their susceptibility to become salted out, as shown in Figure 7.13c. Hydrophobic proteins can even be salted out by high concentrations of NaCl.

Other Aspects. Several *nonionic solutes* also affect solubility. Most sugars increase the surface tension of water somewhat, and tend to reduce protein solubility. Alcohols strongly decrease surface tension and tend to increase markedly the solubility of hydrophobic proteins. On the other hand, they decrease the dielectric constant, thereby enhancing electrostatic bonding. This implies that alcohols tend to decrease solubility at higher concentrations, especially of hydrophilic proteins. The same holds for most

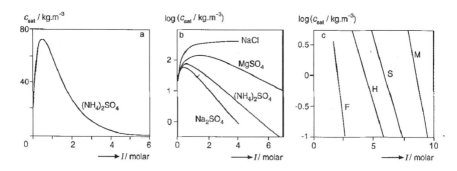

FIGURE 7.13 Salting out: solubility of proteins (c_{sat}) as a function of ionic strength (I). (a and b) Hemoglobin; salts indicated. (c) fibrinogen (F), hemoglobin (H), serum albumin (S), and myoglobin (M); salt $(NH_4)_2SO_4$.

other organic solvents. The presence of other polymers may lead to phase separation (Section 6.5.2), which might be interpreted as a decrease of solubility.

The dependence of solubility on *temperature* varies. We will consider here the range of zero to about 50°C; at still higher temperature unfolding may occur. For hydrophilic proteins, the solubility may increase with temperature, by up to 4% per K. For more hydrophobic proteins, solubility decreases with increasing temperature, by up to 10% per K. This is in accordance with the strong temperature dependence of hydrophobic bonds in the range considered (Fig. 3.4). Low temperature may also cause dissociation of quaternary structures.

When considering the effect of *denaturation* on the solubility of globular proteins, two cases should be distinguished. The first is denaturation by such agents as detergents, urea, or guanidinium salts. As long as these compounds are present, the solubility is enhanced. This stands to reason, since the denaturation (unfolding) occurs because the free energy is lower for an increased solvent–solute contact. The second is irreversible denaturation, especially as caused by heat treatment. The increased exposure of apolar groups now allows many intermolecular hydrophobic bonds to be formed, i.e., cause aggregation. Some denatured proteins are virtually insoluble. This will greatly depend, however, on pH and ionic strength. Figure 7.14a gives an example for whey protein. This is a mixture, for the most part consisting of globular proteins, isoelectric pHs around 5, that are subject to irreversible heat denaturation. Here turbidity was used as a measure for aggregation, hence for "insolubility."

This brings us to a final remark about the solubility of *protein preparations*, i.e., the more or less crude mixtures as applied in the food industry. Solubility is an essential criterion for most functional applications. The tests applied to assess this quality generally involve mixing of a given amount of the material with a given amount of a specified solvent, usually a buffer. Vigorous mixing then is followed by centrifuging at specified conditions. The amount of protein or nitrogen in the supernatant is determined and compared to the total amount present. The result is so many "percent soluble." Some results are shown in Figure 7.14b.

Some remarks should be made about these tests. First, the meaning of the word solubility is fundamentally different from the definition used by physical chemists, given at the beginning of this section. Suppose that a "solubility" of 50% is observed. If this were a true solubility, doubling of the amount of solvent would lead to 100%. For the protein preparation, doubling the amount of solvent may well leave the result at 50%; in other words, half of the material would be well soluble, and the other half not at all. In most cases, however, the situation will be somewhere in between. This is illustrated in Figure 7.14c. Here Line 1 indicates a relation that would be observed for a single pure protein of limited solubility. Line 2 indicates a relation for a protein mixture of which 10% is very soluble, the rest being fully insoluble. It may be clear that the results of these tests are not unequivocal quantities.

A second remark is that the test result may depend on conditions, such as the manner and intensity of stirring and of centrifuging. Stirring may disrupt large protein aggregates, but it may also cause copious beating in of air, and some proteins become denatured and aggregated upon adsorption onto air bubbles. Small aggregates of proteins may escape centrifugal sedimentation, whereas large aggregates may not. Time and temperature during the test may determine to what extent an equilibrium situation is reached.

Finally, the result may depend on conditions during manufacture (especially when these have caused denaturation), and the presence of other components in the product (e.g., phenolic compounds), as discussed above. Moreover, some of the other components may contain nitrogen (10% nonprotein N is not exceptional), which will be reckoned as soluble protein, if only N content is determined.

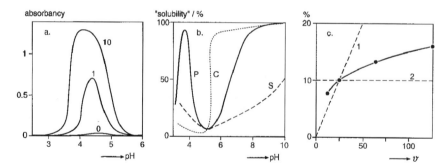

FIGURE 7.14 "Solubility" of protein preparations as a function of pH. (a) Turbidity (expressed as absorbancy) of solutions of a whey protein isolate, heated at 70°C for various times (indicated, minutes). (From results by S. Damodaran, see Bibliography). (b) Solubility (percentage of protein in supernatant after centrifuging) of various protein products: sodium caseinate (C), peanut (P), and soya (S) proteins. (Approximate results after various sources.) (c) Solubility (as in b) of the protein in a potato juice extract (pH = 7.0, $I = 0.2$ molar) as a function of solvent volume (v, in ml). See text for lines 1 and 2. (After results by G. A. van Koningsveld. Ph.D. thesis, Wageningen University, 2001.)

Questions

1. The globular protein β-lactoglobulin generally is a mixture of two genetic variants, A and B. One difference is that A has Asp at position 64, whereas B has Gly. How could this affect the solubility profile as given in Figure 7.12?

2. Figure 7.1 shows that β-lactoglobulin and β-casein have an equal proportion of hydrophobic residues. Nevertheless, for a not too small ionic strength, β-lactoglobulin is well soluble at its isoelectric pH, whereas β-casein is not at all. How is this to be explained?

3. Figure 7.15 gives examples of the association of β-casein in aqueous solutions at various temperature and ionic strength, as determined by light scattering. Can you qualitatively explain the shape of the curves and the differences caused by the variables applied?

Answers

1. The substitution of Asp for Gly means an additional positive charge (at neutral pH). This would cause a somewhat lower isoelectric pH, according to Figure 6.7b, by about 0.1 pH unit. The average charge of the mixture would thus hardly vary over this range, making the "peak" of minimum solubility somewhat broader.

2. At the isoelectric pH, the solubility of a protein is essentially determined by the hydrophobicity of its surface. β-lactoglobulin is a globular protein, which implies that its strongly hydrophobic residues are largely buried in the core or

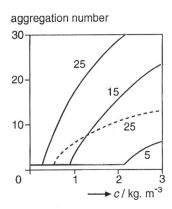

FIGURE 7.15 Association of β-caseinate (expressed as the average number of molecules in an aggregate) as a function of protein concentration at various temperatures (indicated, °C) at pH 7.0. Ionic strength 0.2 molar (full lines) or 0.05 molar (broken line).

shielded from the solvent due to association (it occurs as a dimer). β-casein is a disordered protein, having many hydrophobic residues exposed to the aqueous phase. This would cause a negative solvation free energy, implying very small solubility.

3. Figure 7.1 shows that the part of the molecule comprising the first 47 residues differs greatly from the rest (162 residues). It follows that

First part: 53% charged, $z = -15$,   22% hydrophobic
Other 162:   14% charged, $z = -2.5$,   31% hydrophobic

The molecule thus has a hydrophilic "head" and a long and flexible, rather hydrophobic, "tail." It resembles a huge soap molecule. Such molecules tend to form micelles above a fairly well-defined concentration (Figure 2.8), in accordance with the shape of the curves in Figure 7.15. In such aggregates, the polar heads are at the outside, and the more apolar tails at the inside. The driving force for micellization is hydrophobic interaction between these tails, and this interaction strongly decreases with decreasing temperature in the range considered (Figure 3.4). At smaller ionic strength, mutual repulsion between the negatively charged heads will be sensed over a longer distance, which would counteract the tendency to form micelles. This is in qualitative agreement with the trends observed.

7.4   RECAPITULATION

Description.   Proteins are polyelectrolytes, but they occur in such a bewildering variety of composition, structure, and properties that physicochemical polymer theory is of limited use for understanding them. The properties ultimately depend on the primary structure of a protein, i.e., what amino acid residues occur and in what sequence. The 20 amino acid building blocks differ in several respects. The most important general properties may be their charge, which determines the charge of the protein as a function of pH; and the hydrophobicity, which is of prime importance for conformation and solubility.

The *conformation* is the total three-dimensional folding of the peptide chain. Some levels of structure can be distinguished. The secondary structure involves fairly regular orderings of amino acid residues, especially the α-helix and the β-sheet. These are strongly hydrogen bonded, primarily via $=$O and NH of the peptide bonds. The tertiary structure involves the further folding of the peptide chain, including the secondary structure elements. Three types of tertiary structure can be distinguished, viz., (a) globular, which implies a tightly compacted chain forming a roughly spherical mass, with most of the very hydrophobic residues at the inside and nearly all charged residues at the outside; (b) fibrous, which implies elongated structures of mostly not very hydrophobic peptide chains; and (c) disordered, which structure somewhat resembles a random coil. The structures are often stabilized by posttranslational modifications, which may include glycosylation of some residues, and formation of intramolecular cross-links in the form of —S—S— bridges between Cys residues. Many large protein molecules form separate globular domains of 100–200 residues. Quaternary structure involves association of protein molecules into larger entities of specific order.

Conformational Stability. The compact conformation of globular proteins is due to a great number of weak intramolecular bonds. If the molecule unfolds, which leads to a greatly increased conformational entropy, this is generally via a cooperative transition, which implies that intermediate conformational states do not or hardly occur. The stability, defined as the free energy difference between the folded and unfolded states, is fairly small. It is, however, the sum of two very large terms, one promoting and the other opposing unfolding. This means that small changes in conditions can already lead to unfolding. The bonds involved are for the greater part H-bonds, but these can only be strong in an apolar environment, implying that the presence of hydrophobic residues is essential in obtaining a folded, i.e., globular, conformation. Unfolding generally occurs at high or very low temperature, at extreme pH, at very high pressure, and upon adsorption onto hydrophobic surfaces (solid, oil, or air). Several solutes may cause unfolding due to altering the solvent quality, such as salts that are "high" in the Hofmeister series. Other solutes have specific effects, such as the breakage of —S—S— bridges.

Denaturation of a globular protein may be equated to the unfolding of its peptide chain; it can also be related to the effects it has, such as loss of biological (e.g., enzyme) activity, or aggregation. If these changes are to be permanent, refolding of the peptide chain into its native conformation should be prevented. Several reactions, which especially occur at high temperature or high pH, can cause changes in configuration that do prevent refolding. The kinetics of denaturation, particularly of heat denaturation, is of great practical importance. In general, the kinetics is intricate, though in many cases the denaturation rate is controlled by the unfolding reaction. Then, the reaction is first-order and has a very steep temperature dependence. The latter is due to the very large activation enthalpy (numerous bonds have to be broken simultaneously). The very large entropy change (e.g., the increase in conformational entropy) causes the reaction nevertheless to proceed at a reasonable rate at moderate temperatures (mostly 50–80°C). Several complications can cause more complicated denaturation kinetics.

Solubility. The solubility of a protein is primarily determined by the preponderance of charged groups and of hydrophobic groups that are exposed to the solvent. Considering a protein as a macroion, the Debye–Hückel theory on the ion activity coefficient can be applied. It predicts that a smaller net charge and a lower ionic strength in the solvent lead to a smaller solubility. This is indeed observed, and most proteins are virtually insoluble at their isoelectric pH in the absence of salt. Adding some salt then may cause dissolution: i.e., "salting in." High concentrations of salt may lead to "salting out," but this is very salt specific. The salts, like many other solutes, affect solvent quality. The greater the number of hydrophobic groups that are in contact with the solvent, the smaller the solubility, and this effect is enhanced by lowering solvent quality. Unfolding of a globular protein thus greatly lowers its solubility, as more hydrophobic groups become exposed. It often leads to aggregation and possibly gel formation.

The "solubility" as determined by practical tests on crude protein preparations used in industry, is a different property. It may be fairly small because the solubility of all the proteins present is fairly small, but also— and, more generally—because some of them are virtually insoluble, whereas others are well-soluble.

BIBLIOGRAPHY

Some textbooks on physical chemistry treat some of the aspects discussed here, e.g., Chang (see Chapter 1). Most texts on biochemistry give much information on protein structure, function, reactivity, and other properties. For food scientists the following introduction can be recommended:

S. Damodaran. Amino acids, peptides and proteins. In: O. R. Fennema, ed. Food Chemistry, 3rd ed. Marcel Dekker, New York, 1996, Chapter 6.

Comprehensive and clear is the monograph

T. E. Creighton. Proteins: Structures and Molecular Properties, 2nd ed. Freeman, New York, 1993.

Conformational stability is extensively discussed in two reviews by

P. L. Privalov. Stability of proteins. Adv. Protein Chem. 33 (1979) 167–241 and 35 (1982) 1–104.

A monograph on denaturation is

S. Lapanje. Physicochemical Aspects of Protein Denaturation. John Wiley, New York, 1978.

Kinetics of heat inactivation of enzymes is discussed by

R. W. Lencki, J. Arul, R. J. Neufeld. Biotechnol. Bioeng. 40 (1992) 1421–1426, 1427–1434.

Molecular mechanisms involved in heat inactivation are treated by

T. J. Ahern, A. M. Klibanov. Meth. Biochem. Anal. 33 (1988) 91–127.

Effects of high pressure on globular proteins are reviewed by

V. B. Galazka, D. A. Ledward. In: S. E. Hill, D. A. Ledward, J. R. Mitchell, eds. Functional Properties of Food Macromolecules, 2nd ed. Aspen, Gaithersburg, MD, 1998, Chapter 7.

8

Water Relations

The water content of foods varies widely, and several properties of foods greatly depend on water content. This concerns, among other things, rates of changes—notably various kinds of deterioration—rheological properties, and hygroscopicity. The relations are far from simple, partly because water is not a simple liquid. Most of the intricacy relates to the many constituents in food, each of which may interact with water in a different way.

The reader is advised to consult Sections 2.2 (especially 2.2.1) and 3.2 before starting with this chapter.

8.1 WATER ACTIVITY

Many of the relations mentioned become simpler if water activity (a_w) rather than water content is considered. Following Eq. (2.6), the chemical potential μ_w of water in a solution is given by

$$\mu_w = \mu_w^\circ + RT \ln a_w \tag{8.1}$$

where μ_w° is the chemical potential of pure water. a_w is expressed as mole fraction and varies between 0 and 1. The chemical reactivity of the water, as in a reaction formula, is exactly proportional to a_w, rather than to water concentration. Water activity is often a much better indicator for water-dependent food properties (such as its stability) than gravimetric water content, in the first place because several constituents may be inert with respect to water. This is easily understood when we realize that for an ideal solution the water activity equals the mole fraction of water, i.e.,

$$a_w(\text{ideal}) = x_w \equiv \frac{m_w}{m_w + \Sigma m_{s,i}} \tag{8.2}^*$$

where m is molar concentration of water (subscript w) and of solutes (subscript s). It is thus the number concentration of solute molecules that counts. Very large molecules (say, of starch) have very little effect on x_w (which equals $1 - x_w$), and water-insoluble components (fats, substances in solid particles) not at all.

> *Note*   Equation (8.2) often is called Raoult's law. Strictly speaking, however, Raoult's law only applies to lowering of vapor pressure by a solute. Cf. Eq. (8.6).

Figure 8.1 gives some examples of the water activity versus mole fraction of water for some simple solutions. It is seen that Eq. (8.2) is poorly obeyed, except for sucrose at very small x_s. Deviations from ideality are discussed in Section 2.2.5. They can be expressed in a virial expansion, as given in Eq. (2.18) for the osmotic pressure Π. The relation between the two properties is

$$-\ln a_w = \frac{\Pi M_w}{RT\rho_w} \approx 1.8 \cdot 10^{-5} \frac{\Pi}{RT} \tag{8.3}$$

(in SI units) where M is molar mass (kDa) and ρ mass density. For $a_w > 0.9$, $-\ln a_w \approx 1 - a_w$. The quantity $\Pi/RT\rho_w$ equals the *osmolality* of the solution (for an ideal solution the number of moles of molecules, ions, and particles per kg of water). For a single solute, an approximate relation follows from the Flory–Huggins theory (see Chapter 6), and the combination of Eqs. (6.10) and (8.3), with some rearrangement, leads to

$$\ln a_w \approx -x_s\left(1 + \frac{1}{2}\beta q x_s + \frac{1}{3}q^2 x_s^2 + \cdots\right) \tag{8.4}$$

where x_s is the mole fraction of solute, q is the volume ratio of a solute molecule over a water molecule, and β is the solute–solvent interaction parameter defined in Section 6.4.1. The relation would hold if the volume fraction of solute is not too high, say < 0.3.

Important **causes of nonideality** are

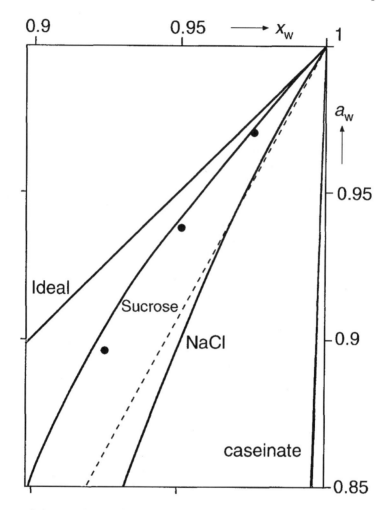

FIGURE 8.1   Water activity (a_w) versus mole fraction of water (x_w) for aqueous solutions of sucrose, NaCl, and Na-caseinate. Also given are the ideal relation $(a_w = x_w)$, the relation for NaCl assuming complete dissociation (broken line), and some points (•) for sucrose according to Eq. (8.4).

1.   *Dissociation* of the solute. For NaCl, m_s should be replaced by $m_{Na} + m_{Cl}$, assuming complete dissociation. The broken line in Figure 8.1 has been drawn on this assumption. It is seen that dissociation explains by far the greater part of the nonideality for NaCl, but not all of it.

2.   *Solute molecule size* or, more precisely, the parameter q in Eq. (8.4). Figure 8.1 shows a few points calculated for $\beta = 1$, which implies no

net solvent–solute interaction, and $q = 12$, the approximate value for sucrose in water. It is seen that the correction is somewhat overestimated in this case.

3. *Solvent–solute interaction.* If there is net attraction or association between solute molecules and water molecules ("hydration"), $\beta > 1$. If the association is strong, i.e., the solvent quality is very good, it is *as if* some water molecules were removed. This then is *as if* the mole fraction of water were decreased. The resulting decrease of a_w is only appreciable if x_s is fairly large. For instance, if $x_s = 0.02$, and thus a_w (ideal) = 0.98, "removal" of 1 mol water per mol solute would yield $a_w = (0.98 - 0.02)/(0.02 + 0.96) = 0.9796$, which is hardly different. On the other hand, if $x_s = 0.3$, and a_w (ideal) = 0.70, the same "removal" would result in $a_w = 0.4/0.7 = 0.57$, which is a substantial decrease.

If the solvent quality is poor, $\beta < 1$, and solute molecules tend to associate with each other. This is *as if* there were fewer solute molecules, hence a higher x_w, hence a higher a_w. The latter effect occurs in the sucrose solution, partly compensating for the volume exclusion effect; the results reasonably fit Eq. (8.4) with $\beta = 0.64$. This then implies that sucrose is (slightly) hydrophobic.

Figure 8.1. also gives an approximate curve for solutions of caseinate, i.e., a *protein.* Here all nonidealities mentioned are involved. Casein has an isoelectric pH of about 4.6 and at neutral pH it is thus negatively charged, having about 14 net charges per molecule. This would mean that the (univalent) counterions present increase the molar concentration by a factor of 15. (Actually, most protein preparations contain more salt than strictly needed for electroneutrality.) Casein has a large molar mass (about 23 kDa), implying that the molar volume ratio solute/solvent $q \approx 900$. This implies a large nonideality, as is the case for all polymers; see Section 6.4. Solvent–solute interactions also play a part, but it is difficult to find out in what manner, since for some groups on the protein (especially ionized groups) water is a good solvent, whereas for hydrophobic groups solvent quality is poor. It is known that the caseinate molecules tend to associate, forming clusters of, say, 15 molecules at about 1% concentration. Presumably, the overall solvent quality therefore is such as to (slightly) increase a_w.

Most foods show marked nonideality, and calculation of a_w from the composition is generally not feasible. For a mixed solution, calculation may be done according to the so-called Ross equation,

$$a_w = a_{w,1} \times a_{w,2} \times a_{w,3} \times \cdots \tag{8.5}$$

where $a_{w,i}$ means the water activity as determined for solute i at the same molar ratio of i to water as in the mixture. Equation (8.5) can be derived from the Gibbs–Duhem relation [e.g., Eq. (2.13)], assuming interactions (i.e., net attraction or repulsion) between different solute molecules to be negligible. It gives fairly good results for a_w down to about 0.8, the relative error in $(1 - a_w)$ generally being $< 10\%$. At lower water activity, the situation becomes very complicated. To mention just one point, the dissociation of ionogenic groups is suppressed, the more so at a lower a_w, thereby altering (decreasing) hydration of these groups.

   Determination of water activity thus is generally needed. This can be done by bringing the food in equilibrium with the air above it. At equilibrium, the a_w values in the aqueous phases of the food and in air are equal. Since moist air shows virtually ideal behavior at room temperature and pressure, we simply have

$$a_{w,f} = a_{w,v} = \frac{p_v}{p_{v,sat}} \tag{8.6}^*$$

where subscript f stands for food and v for vapor; $p_{v,\,sat}$ is the vapor pressure of pure water at the temperature of measurement. The water activity of the food thus equals the relative humidity (expressed as a fraction) of the air above it, provided that equilibrium has been reached.

   Examples of the water activity of several foods, in relation to water content, are shown in Figure 8.2. For some foods a range is given, in other cases just one example. It is seen that most points fall within a band having a width corresponding to a variation in $(1 - a_w)$ by a factor of about three. Two kinds of exceptions are observed. The one is exemplified by the point labeled "brine." This represents a type of seasoning that is almost saturated with NaCl. Because of the small molar mass of the molecules and ions dissolved, the effective x_w is relatively small. Foods of a high fat content form the other exception. Skim milk and cream have exactly the same aqueous phase, and thereby the same a_w, despite the large differences in water content. Margarine has an even lower content of aqueous phase, and an extreme is cooking oil, which may have a water content of about 0.15% at room temperature, but virtually no substances that can dissolve in water; hence, a_w will approach unity.

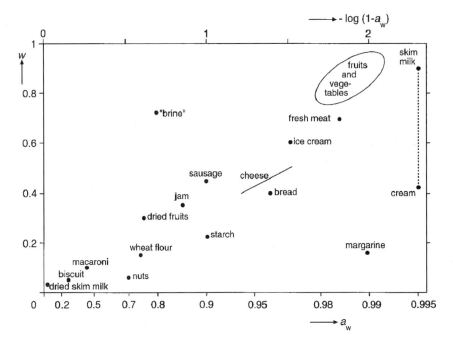

FIGURE 8.2 Approximate water activity (a_w) of some foods versus mass fraction of water (w).

Question 1

In what direction will the water activity of a liquid food change (higher, lower, or virtually unaltered) if the following changes occur, keeping other factors constant?

1. Addition of sodium chloride
2. Addition of native starch (granules)
3. Heating the food with starch and cooling again to the original temperature without loss of water
4. Enzymatic hydrolysis of the protein present
5. Emulsifying oil into it
6. Freezing part of the water

What is the explanation in each case?

Question 2

Calculate the water activity of a 25% (w/w) solution of glucose in water, assuming ideal behavior. What would be the effect of solute size on the result, assuming the volume ratio glucose/water to be 6? Assuming that the observed $a_w = 0.960$, then what would you conclude?

Answer

The molar mass of glucose $= 180\,g/mol$, that of water 18. We thus have in a kg of solution $250/180 = 1.39\,mol$ glucose and $750/18 = 41.67\,mol$ water, and application of Eq. (8.2) yields $a_w = 0.968$. If no net interactions occur between solute and solvent molecules, Eq. (8.4) with $\beta = 1$ can be applied to calculate the effect of solute size:

$$a_w = \exp[-0.032(1 + 1/2 \times 6 \times 0.032 + \cdots)] = 0.965.$$

The nonideality would thus be small.

If indeed a_w were 0.960, that would imply net attraction between glucose and water, i.e., hydration; Eq. (8.4) would then yield that the solute–solvent interaction parameter $\beta = 2.87$, an unlikely high figure. Another way to calculate roughly the hydration is as follows. We can write the a_w value of $0.965 = 27.57/(27.57 + 1)$, where 1 stands for 1 mole of glucose. Assume now that one glucose molecule "removes" x molecules of water from the solution. For the "observed" a_w we would then have

$$\frac{27.57 - x}{28.57 - x} = 0.960$$

yielding $x = 3.6$ molecules of water "removed" per molecule of glucose. This too is a very unlikely high figure. In other words, the "observed" a_w would not be correct. Indeed, exact determination has shown $a_w = 0.9666$, implying that β would be < 1.

8.2 SORPTION ISOTHERMS

Physical chemists distinguish between adsorption and absorption. *Adsorption* is a surface phenomenon. Consider a solid or liquid phase (the *adsorbent*), in contact with another, fluid, phase. Molecules present in the fluid phase may now adsorb onto the interface between the phases, i.e., form a (usually monomolecular) layer of *adsorbate*. This is discussed in more detail in Section 10.2. The amount adsorbed is governed by the activity of the adsorbate. For any combination of adsorbate, adsorbent, and temperature, an adsorption isotherm can be determined, i.e., a curve that gives the equilibrium relation between the amount adsorbed per unit surface area, and the activity of the adsorbate. Powdered solid materials in contact with moist air can thus adsorb water. The amount adsorbed increases with increasing relative humidity (a_w) of the air. Note that the amount adsorbed per kg adsorbent would be proportional to the adsorbent's specific surface area, i.e., to the fineness of the powder.

Liquid or amorphous materials may (also) show *absorption*. Here, the absorbate can dissolve in the absorbent, or it can be seen as adsorbing onto the surface of a great number of fine pores in the absorbent. Anyway, the amount absorbed would be proportional to the mass of absorbent, other things being equal. In most dry foods, it is unclear whether the mechanism is adsorption or absorption; in liquid foods, it is always the latter. It is rarely observed in a dry food that the equilibrium amount of water taken up depends on the specific surface area of the food. Generally, the term "sorption" is used, leaving the mechanism involved out of consideration.

It is customary to construct *sorption isotherms* or vapor pressure isotherms of foods, where water content (either as mass fraction or as mass of water per unit mass of solids) is plotted against a_w. Figure 8.3a gives an example of such an isotherm; here, the material is a powder below 10 and a liquid above 30% water.

Figure 8.3a gives the whole range of water contents, but if the water content is plotted on a linear scale, any differences, which are of greatest importance for low water contents, are not shown in sufficient detail. Consequently, one mostly plots only the part below $a_w \approx 0.9$. Examples are given in Figure 8.3b, and considerable variation among foods is seen. The

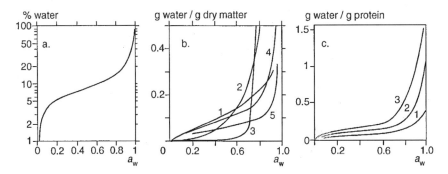

FIGURE 8.3 Water vapor pressure or water sorption isotherms of foods. Given are water content versus water activity (a_w). (a) skim milk (powder). (b) Various foods: meat (1), apple (2), boiled sweet (3), skim milk (4), and peanuts (5). (c) Caseinate systems (water content expressed as g per g dry protein); pure caseinate (1), curd or renneted milk (2), and cheese (3).

relations predominantly depend on food composition, and the basic factors governing them are given in Section 8.1. Figure 8.3c shows, for example, that the presence of small molecules or ions has considerable effect on the isotherm, in agreement with Eq. (8.2); curd contains several solutes (altogether about 4 mmol per g protein), and cheese contains added salt as well (about 5 mmol per g protein).

For true adsorption, several equations for adsorption isotherms have been derived, based on various theories. Such equations often are applied to water sorption isotherms of foods as well. However, one cannot speak of adsorption in the case of most foods, as mentioned above, because there is no (or a very limited) phase surface onto which water can adsorb. Moreover, most foods contain numerous components; even if phase surfaces were present they must be very inhomogeneous. In the author's opinion, it therefore makes little sense to use such equations. Only for relatively simple and homogeneous systems, like pure starch granules, can some theories be more or less applicable, but not for real foods. Mathematical fitting of experimental data may be useful for practical purposes, and since the equations generally have three or four adjustable parameters, a reasonable fit can often be obtained. But one cannot attribute physical significance to the parameters derived in this way, such as a "monolayer water content."

Another approach is to proceed from the chemical composition. This may work for some fairly dry foods, although it implicitly assumes an absorption mechanism, which is that certain chemical groups "bind" certain amounts of water, and by determining the concentration of these groups, the water sorption can be calculated. It concerns especially ionized groups (a few water molecules per group) and dipoles, such as a peptide bond (<1 water molecule per group). This method works reasonably well for proteins around $a_w = 0.5$.

Hysteresis. A sorption isotherm is in principle determined by placing a small sample of the food of known water content in air of a given humidity and temperature and then determining the weight of the sample after various times. After the weight does not change any more, which often takes several days, the equilibrium water content is considered to be reached. By doing this experiment at a range of air humidities (a_w), an isotherm is obtained. One can do this with samples that are successively brought to a higher or to a lower water content, and the curves so obtained are usually not identical. There is hysteresis between the "desorption" and the "adsorption" isotherm, as illustrated in Figure 8.4a. This means that *thermodynamic equilibrium is not obtained*, at least at low water content. The consequence then is that the water activity is undefined, since a_w is by definition an equilibrium property. This would mean that the scale of the x-axis in graphs like those in Figure 8.3 does not represent the water activity of the food, at least at low a_w. In an actual food sample of low water content, one may imagine that at any spot a certain a_w prevails, but it cannot be determined and it would probably vary with place and time.

Many foods show even stronger hysteresis than in Figure 8.4. It is further seen in Figure 8.4c that intermediate sorption curves are obtained when one starts drying or wetting somewhere in between, especially when starting at fairly low a_w; other materials show similar, though not precisely the same, behavior. The explanation of the hysteresis is far from clear. It can be stated that the drying involved in determining the isotherm alters the physical state of the food in such a way that it cannot relax again to the original state after taking up water, or only very sluggishly. Both curves may be considered to represent a metastable state, in the sense that the values do not change over the time scale of interest. This is true enough, but gives little further understanding. In the Note at the end of Section 8.3, a possible mechanism (closing of pores in the material) is mentioned.

Another factor would be that equilibrium is not reached because the rate of diffusion of water through the sample is too small. Figure 5.9 gives, as an example, a diffusion coefficient $D \approx 10^{-15} \, \text{m}^2 \cdot \text{s}^{-1}$ at about 2.5% water, and as water content becomes smaller, D becomes ever smaller. Applying the simple Eq. (5.13) ($x'^2 = D \cdot t_{0.5}$, with $x' =$ diffusion distance—

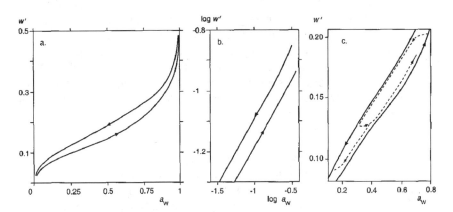

FIGURE 8.4 Sorption isotherms—w' in g per g dry starch versus water activity a_w—of native potato starch, obtained when decreasing water content (desorption) and when increasing it (absorption). (a) Linear scales. (b) Part of the same data on a log/log scale. (c) Part of the same data, also showing intermediate curves (broken). (After results by C. van den Berg, Ph.D. thesis, Wageningen University, 1981.)

i.e., about particle size—and $t_{0.5}$ = time needed to halve a difference in concentration over a distance x'), we obtain for water transport

$$D = 10^{-15}, \quad x' = 0.1 \text{ mm} \rightarrow \quad t_{0.5} = 10^7 \text{ s} \approx 4 \text{ months}$$

$$D = 10^{-15}, \quad x' = 1 \, \mu m \quad \rightarrow \quad t_{0.5} = 10^3 \text{ s} \approx 15 \text{ minutes}$$

$$D = 10^{-18}, \quad x' = 1 \, \mu m \quad \rightarrow \quad t_{0.5} = 10^6 \text{ s} \approx 12 \text{ days}$$

Actually, zero water content will never be reached. The sample should be very finely divided to obtain even reasonable sorption curves at low water content. For instance, a curve for a boiled sweet as in Figure 8.3b can only be obtained if the material is finely ground.

It is seen in Figure 8.4a that both curves converge at high a_w, as is to be expected, and they also seem to do so at low a_w. However, the latter is not really the case. In Figure 8.4b the same relation is shown on a log–log scale for low water contents, and it is seen that the relative difference does not decrease with decreasing a_w. One often sees desorption isotherms drawn through the origin, but this is in fact misleading. That point is never reached on dehydration at room temperature, and the lowest water content reached may be a few percent (see, e.g., Figure 8.6b, below).

How then is the point of zero water content obtained? This can be done by drying at high temperature, often at about 100°C, where the diffusion coefficient generally is some orders of magnitude higher than at room temperature (see also Figure 8.9b, later on). Moreover, the driving force for water removal (the difference in chemical potential of water between sample and air) then is greater, the more so when drying under vacuum. On the other hand, prolonged keeping at high temperature, mostly for several hours, may cause chemical reactions (e.g., involving uptake of oxygen) or vaporization of other substances than water, making determination of the water content somewhat unreliable. Consequently, it may be preferable to dry the sample under vacuum at a somewhat lower temperature.

Sorption Enthalpy. When removing water from a product, heat is consumed. This is because a lower water content goes along with a lower water activity, and the water has to be removed against a water activity gradient or, in other words, against an increasing osmotic pressure. The sorption heat or enthalpy ΔH_s generally increases as a_w decreases (see Figure 8.5a), which would imply that removal of water becomes ever more difficult in the course of drying. However, ΔH_s mostly is small: it is rarely over 20 kJ per mole of water, and its average (integrated) value over the whole drying range is $0.2 - 2 \text{ kJ} \cdot \text{mol}^{-1}$. This is far smaller than the enthalpy of

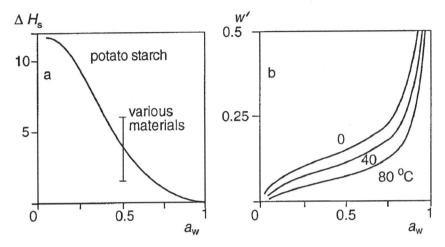

FIGURE 8.5 (a) Sorption enthalpy ΔH_s (kJ $\cdot$ mol^{-1}) as a function of water activity a_w, for potato starch. Also the range obtained at $a_w = 0.5$ for various materials is given. (b) Desorption isotherms—w' in g water per g dry matter versus water activity—of (dried) potato at three temperatures (indicated).

evaporation of water, which is 43 kJ $\cdot$ mol^{-1} at 40°C. Difficulty of removal of the last bit of water is thus not due to strong "binding" but to very slow diffusion, as discussed above. Note that the sorption enthalpy has always to be supplied, whether water is removed while remaining liquid (as in reversed osmosis), or by evaporation. In the latter case, also the enthalpy of evaporation has to be supplied.

Because of the finite ΔH_s, the water activity of a product increases with increasing temperature (at a given water content), relatively more so for a higher ΔH_s, which implies a lower a_w. This is expressed in the relation of Clausius–Clapeyron:

$$\frac{\mathrm{d}(\ln a_w)}{\mathrm{d}(1/T)} = \frac{-\Delta H_s}{R} \tag{8.7}$$

Figure 8.5b gives an example of the temperature dependence of a_w. It is seen that a given water content is reached at a higher water activity for a higher temperature. This is an additional reason why drying is easier at a high temperature.

Hygroscopicity. A dry material will take up water from the surrounding air if the latter has a higher a_w (relative humidity) than the material. The rate of water uptake will be faster for a greater difference in a_w, a larger specific surface area of the material, and a higher diffusion coefficient of water in the material. If water uptake occurs readily and if it leads, moreover, to a perceptible change in properties of the material, the latter is called hygroscopic. The changes may be of various kinds. Biscuits and comparable foods lose their crispness upon water uptake; this is discussed in Section 16.1.2. Most foods that primarily consist of polymers become soft. Dry foods that contain a lot of water-soluble components tend to become tacky upon water uptake: a concentrated sugar solution *is* very tacky. If the food is a powder, the tackiness readily causes caking of the powder particles. Further water uptake may cause deliquescence.

Hygroscopicity is reflected in the shape of the sorption isotherm. If it has a small slope (small dw/da_w), the material is little hygroscopic; if the slope is large, it may be strongly so. Looking at Figure 8.3b, we may conclude that material 2 (dried apple) probably is quite hygroscopic for most relative humidities; boiled sweets (material 3) will be little hygroscopic for relative humidity < 0.6, and greatly so if humidity is > 0.7. Knowledge of

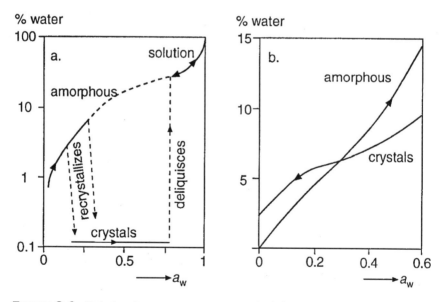

FIGURE 8.6 Relation between water content (weight %) and water activity for various situations. (a) Sucrose: in solution, amorphous and crystalline; recrystallization takes a very long time (years) at low water content (say 3%) and is quite fast at values over 6%. (b) Dried skim milk, with amorphous or crystalline lactose; see text.

the sorption isotherm is thus of importance for deciding under what humidity of the air a food material should be stored.

In many foods, the situation is more complicated. Often, the material, or an important component, may occur either in crystalline or in amorphous form. Figure 8.6a gives examples for sucrose: crystalline, amorphous, and in solution. Crystalline sucrose, if pure, cannot contain water and any water present is adsorbed on the crystal faces, and this is a very small amount. Amorphous sucrose can take up water, because this means in fact dilution of an extremely concentrated solution. If sufficient water has been taken up, the diffusion coefficient of sucrose has become large enough for it to crystallize. This then leads to the release of water. The a_w and the rate at which these changes occur greatly depend on temperature.

Figure 8.6b shows sorption isotherms for dried skim milk, of which about 50% is lactose. The lactose can remain amorphous when the milk is spray-dried. It can also be made with crystalline lactose, and the normal crystalline form is a monohydrate. The water of crystallization is not available as a solvent, which means that a_w is smaller than for amorphous sugar at the same water content, provided the latter is fairly small. At higher water content, the curves cross, and amorphous sugar may crystallize, taking up part of the water.

It may further be noted, that water content, and thereby a_w, may greatly vary from place to place in a food during water uptake (as well as during drying).

Question 1

Consider a powder recently obtained by spray drying a liquid food. At room temperature a water activity of 0.3 is observed. How does a_w alter if the following changes are applied?

(a) Increase the temperature to 80°C. (b) Keep at that temperature for some days in completely dry air, then cool again to room temperature. (c) Then put the powder in moist air at room temperature until it has the same weight (hence, the same water content) as originally. (d) Keep this powder in a closed container at room temperature for a month.

Answer

(a) According to the relation of Clausius–Clapeyron [Eq. (8.7)] a_w will increase; see also Figure 8.5b. (b) This treatment implies that water has been lost, hence $a_w < 0.3$. (c) In the original powder, the water activity would correspond to that on a desorption isotherm, in this case on an adsorption isotherm. According to Figure 8.4a, now a_w will be > 0.3. (d) The treatment described in (c) cannot have led to equilibrium. In a spray-dried product, the particles are for the most part about 0.05 mm in diameter or larger. Given the small diffusion coefficient at the prevailing water contents, it will take at least some weeks before a homogeneous distribution of water throughout the powder particles is obtained. This implies that the outside of the powder particles would contain more water than the inside. Keeping the powder will thus lead to a lower water content in the outer layer. Since the water activity as observed will roughly correspond with that at the outside, its value will decrease upon storage.

Note The results would be rather different if a considerable part of the material could crystallize, say in step (c).

Question 2

Suppose that you make a dough from 1 kg wheat flour, 600 g water and 15 g salt. The flour contains 14% water and 65% starch. Assume that the dough can be considered as a two-phase system: (a) containing starch and part of the water and (b) the remaining components, in which gluten will take up most of the dry matter. How can you establish the water content of both phases?

Answer

First, determine the water activity of the system; suppose that the answer is 0.975. Then find out what the water content is of starch at that a_w value; Figure 8.4a shows that it will be about 0.4 g water per g dry starch. This then would mean that the starch phase contains $650 \times 0.4 = 280$ g water, or 30%. This leaves for the rest $600 + 0.14 \times 1000 - 280 = 460$ g water, leading to a water content of 67%. It is thus clear that the "gluten phase" contains far more water than the "starch phase."

Notes

1. Actually, Figure 8.4a relates to potato starch, not wheat starch, but the difference will be small.

2. To check the calculation, the water activity of a gluten system of comparable composition should be found or estimated.

3. In doing this exercise, you will see that the sorption isotherms should be exactly known and that even at the high vales of a_w involved, hysteresis of the isotherms cannot be fully neglected.

8.3 "WATER BINDING"

The quotation marks in the heading of this section signify that water binding is a questionable term. In the author's view, the concept has caused more confusion than understanding. We all know that many solid foods hold water. It is also observed that fairly dry foods often are quite stable to deterioration, although these foods may contain a significant amount of water. This has led to the notion that at least part of the water is bound. It is, however, uncertain what that means.

In Section 3.2, solvation, including *hydration*, is briefly discussed, and it is clear that solute molecules and ions, or chemical groups on macromolecules and surfaces, can be hydrated. This means that one or a few water molecules reside longer (e.g., for 10^{-10} s) at a given site than they would in pure water (about 10^{-13} s). Especially charged groups, and to a lesser extent groups with a dipole, are hydrated, but not most hydroxyl groups. One may now speak of bound water, but it is often uncertain what the amount would be. Another case of water binding is water of hydration in crystals or in polymer crystallites; here the residence time of the water molecules may be much longer, up to years. Globular protein molecules contain some buried "structural" water molecules, and their residence time may be long, say several minutes; however, it only concerns very little water, some grams per 100 g protein.

There are several relations between water content and some macroscopic property that have been interpreted by *assuming* that part of the water is of a special category, often equated with "bound water." For instance,

1. *Nonreactive water*. At low water content, the reactivity of water is decreased, but reactivity is precisely given by the water activity. In other words, all the water has a smaller reactivity. Of greater importance, almost all reactions are slower in a concentrated system than in a dilute solution, including reactions not involving water. As is discussed in Section 8.4, this has other causes, especially small diffusivity. The term "nonreactive water" thus makes no sense.

2. *Nonsolvent water*. When a solution can be separated from a food, for instance by centrifugation or by ultrafiltration, it is often observed that the mass ratio of a given solute S to water in that solution is larger than the ratio in the whole food. This can be interpreted as part of the water that is left behind (in the pellet or in the retentate) being not available as a solvent. This may be bound water, but the greater part of it may not. It is generally observed that the amount of nonsolvent water depends on the nature of the solute, and it generally increases with increasing molar mass of solute. This phenomenon can be ascribed to *negative adsorption* of the solute, for

instance with respect to the "surface" of a macromolecule. If there is no net attraction or repulsion between the surface and the solute, the latter stays away (is sterically excluded) from the surface. This is illustrated in Figure 8.7a. When R_w is the radius of a water molecule and R_s that of a solute molecule, a layer of thickness R_w is devoid of water, and one of thickness R_s devoid of solute. In first approximation, the amount of nonsolvent water then is given by

$$w_{ns} = \frac{1}{2}(R_S - R_W)\rho A \tag{8.8}$$

in kg water per kg polymer, where ρ = mass density of water and A the specific surface area ($m^2 \cdot kg^{-1}$) of the polymer. The factor 1/2 derives from the concentration profile that develops due to Brownian motion of solute; the factor depends somewhat on conditions of separation. Figure 8.7b gives some results for nonsolvent water for various solutes (most of them sugars) with respect to milk protein. Equation (8.8) is not precisely obeyed, but this is only to be expected, since it is assumed in the equation that the protein surface is flat and smooth and that the solute molecules are spherical, and neither will be the case. The value of w_{ns} corresponding to R_w may be considered to be true hydration water, about 0.15 g per g protein in the present case.

The amount of nonsolvent water for a certain solute in a food can be considerable, especially if the solute is a large molecule, the concentration of polymer in the food is large, and the specific surface area of the polymer is large. Most of the nonsolvent water is, of course, freely exchangeable with bulk water. Nevertheless, negative adsorption of a solute causes a_w to be decreased, since the effective concentration of solute is increased.

Note For a solute that shows positive, i.e., real, adsorption onto a polymer present, the amount of nonsolvent water may be negative. In such a case a_w is increased.

3. *Nonfreezing water.* When a food is cooled to slightly below 0°C, not all water will freeze, owing to the freezing point depression caused by the solutes. This is discussed in Section 15.3. But even at a low temperature, say −30°C, part of the water will not freeze, and this is often considered to be bound water. However, at low temperatures, where most of the water is frozen and where the viscosity of the remaining solution becomes extremely high, diffusivity of water is so small that freezing becomes infinitely slow. This is further discussed in Section 16.2. The quantity of nonfreezing water can depend on the conditions during freezing.

4. *Immobilized water.* This is meant to be water that does not leak out of a solidlike food. It is present in closed cells, in open pores in a solid matrix (like a sponge), or between chains of coiled polymers. Binding sites for water need not be present for the water to be held, and "bound water" clearly is a misnomer. Better names are held, trapped, or imbibed water. Its amount can be large: in several gels, one g of polymer can readily hold 100 g of water. Actually, it is generally an aqueous solution that is held, rather than pure water.

What is the mechanism by which water is held? A simple explanation is that *water fills space*. A coiled polymer molecule has a certain equilibrium conformation (Section 6.2.1), and there are water molecules in the spaces between polymer segments. Removal of water means shrinking of the coil, which costs free energy, because it implies a decrease in conformational entropy. Something similar is true for a polymer network or gel (Section 6.4.4) and for other gellike structures (Section 17.2): deformation of the network needs a force and thereby energy. This implies that the water in the network has a decreased activity, but the decrease is extremely small. Equation (8.3) gives the relation between a_w and osmotic pressure. The same relation would hold for the mechanical pressure needed to remove water. From (8.3), the change in a_w equals the change in pressure multiplied by

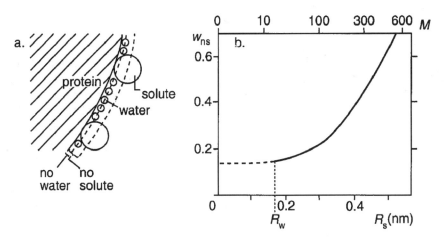

FIGURE 8.7 Negative adsorption (steric exclusion) of a solute S from the surface of a protein molecule or particle. (a) Schematic explanation. (b) Relation between nonsolvent water w_{ns} and molecular radius R_s of the solute (mostly sugars) for micellar caseinate; a scale of molar mass (M, in Da) is also given. R_w is the radius of a water molecule.

$1.8 \cdot 10^{-5}/RT \approx 7 \cdot 10^{-9}$ per Pa at room temperature. If we apply an external pressure of 1 bar (10^5 Pa), it would be equivalent to a decrease in a_w by 0.0007. Such a small change cannot even be detected by most methods.

In a situation where the food is a porous water-filled matrix in contact with air, a contact line matrix-water-air exists. Assuming the pores to be cylindrical, a capillary pressure can be calculated (see Section 10.5.2), and it is higher for a smaller pore diameter. For rigid pores with a radius of 1 μm in a matrix that is completely wetted by water, a pressure of 1.4 bar would be obtained, corresponding to a lowering of a_w by 0.001. In practice, the lowering of a_w would be less, because the conditions mentioned would not be completely fulfilled.

The quantity of imbibed water in a gellike system would vary with all factors that affect the equilibrium state of swelling of the gel, such as concentration of cross-links and solvent quality and, in the case of polyelectrolytes (proteins), pH and ionic strength. Lowering of solvent quality may be seen as a decrease in hydration of the polymer. Since it would also cause shrinkage of the gel, it is tempting to explain the decrease in the amount of water held as a decrease in the amount of water "bound." However, it concerns very different amounts of water. This is illustrated in Figure 8.8. It is seen that the amount of water held by the protein varies between 2 and 7 g/g, whereas the water associated with polar groups (predominantly the peptide bonds) of a protein rarely is > 0.2 g/g and does not change significantly with pH.

Methods. The various cases discussed above would all provide a means of determination of what has been considered to be bound water. Moreover, other methods have been applied, such as deriving the amount of "monolayer water" from an assumed relation between water content and a_w (Section 8.2); or from decreased "water mobility" as deduced from NMR spectroscopy. The methods give widely varying results. Gelatin, for instance, yields values ranging from 0.2 to 100 g per g protein, the latter value representing the amount of water held in the gel. Even when excluding methods determining held water, the results obtained may vary by more than an order of magnitude.

It is therefore advisable not to use the concept of bound water. It would be much better to be specific and speak of held water, nonsolvent water, or nonfreezing water, according to the method of determination or according to the relevance for the effect considered. After all, the phenomena discussed are real and often of importance.

Note The presence of very narrow and rigid, but noncylindrical, pores in a material could in principle explain hysteresis between

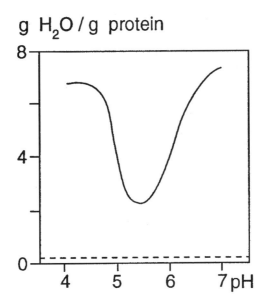

FIGURE 8.8 Amount of water held by myofibrillar muscle protein as a function of pH at low ionic strength. The broken line indicates bound water.

desorption and adsorption curves, as has been shown for rigid porous materials, say minerals. However, such hysteresis is generally observed in the region of high a_w, quite unlike that shown in Figure 8.4. In most foods the pores would not be rigid, and the negative capillary pressure would cause them to become closed, strongly altering the sorption isotherms. Possible relations between the presence of non rigid pores and sorption hysteresis are currently far from clear.

Question

To a solution of sucrose in water, mole fraction 0.05, 5% of a protein is added. The protein dissolves, it has a molar mass of 20 kDa, and it contains just sufficient Na^+ to neutralize the net 10 negative charges per molecule. To what extent will the addition of protein change the water activity of the solution?

Answer

The sucrose solution has an a_w of 0.94 according to Figure 8.1. A mole fraction of 0.05 implies per 100 moles $5 \times 342 = 1710$ g of sucrose and $95 \times 18 = 1710$ g water, i.e., a mass fraction of 0.5. Hence 50 g of protein per kg solution corresponds to 100 g per kg water, or $(100/20 =)$ 5 mol protein per $(1000/0.018 =)$ 55,555 mol water. Since 10 mol Na^+ is present per mol protein, the total solute amounts to 55 mol, yielding a mole fraction of water in the protein solution of 0.999. However, a protein is a very large molecule and we should consider the volume ratio solute/solvent q. Assuming the mass density of the protein in water to be $1400 \, kg \cdot m^{-3}$, we derive $q \approx 800$, and insertion into Eq. (8.4) with $\beta = 1$ yields $a_w = 0.9986$. By applying the Ross equation (8.5), a_w of the mixture would be $0.94 \times 0.9986 = 0.9387$. Another nonideality must be nonsolvent water for sucrose. Figure 8.7b would yield for a molecule of 342 Da a quantity of about 0.4 g per g of protein. Assuming this value to hold for the protein involved (which is by no means certain), we arrive at $0.4 \times 50 = 20$ g nonsolvent water per 500 g, or 4% of the water. The quantity $(1 - a_w)$ should then be multiplied by 1.04. The water activity would then become $(1 - 1.04 \times 0.06) \times 0.9986 = 0.9363$. This would mean that $(1 - a_w)$ would be increased by only about 6% due to the addition of a fairly large quantity of protein, and the greater part of this effect would be due to nonsolvent water.

8.4 REACTION RATES AND WATER CONTENT

Most, though not all, changes or reactions occurring in foods proceed slower at a smaller water content. Often, water activity is considered to be the key variable, but the situation may be far more complicated. Unfortunately, reliable quantitative theory is not available. We therefore can only give some general considerations and examples on physical changes, chemical reactions, and microbial growth.

8.4.1 Physical Changes

Important examples are loss or uptake of water, gases, and volatiles, and crystallization, often resulting in changes in mechanical and some physical properties. An example of crystallization is given for sucrose in Figure 8.6a. At 6% water, it takes about 2 days before sucrose crystallization becomes manifest; at 3% water, it takes about 500 times longer. Water content thus has a very large influence. This is via its effect on *molecular mobility*, i.e., the diffusion coefficient of the molecules considered. This was discussed to some extent in Chapter 5, especially Section 5.3.2 and Figure 5.16a. The smaller the mass fraction of water w, the smaller the effective diffusion coefficient D_{eff}. The smaller w, the steeper this dependence and the more difficult it is to make quantitative predictions. At very small w, the system may be in a glassy state; this is further discussed in Section 16.1. Besides water content, other factors play a part, for instance,

1. *Composition of the dry matter.* Figure 5.16a suggests that for several materials roughly one curve is found, but closer inspection shows differences by about an order of magnitude in D_{eff} for the same w. If the dry matter contains a water soluble substance of fairly small molar mass, diffusivity can be markedly greater than in its absence, at least at small w. If glycerol is added to an aqueous system of not too low water content, say $w > 0.5$, it will cause the diffusion coefficient to decrease, since glycerol has a higher viscosity than water. In a very dry food, however, say for $w < 0.1$, the presence of glycerol will generally increase the diffusion coefficient, since its presence means a greater proportion of liquid. The presence of sugars, like glucose or sucrose, may also enhance diffusivity of water at very small w. These substances then act as "plasticizers" (see Section 16.1).

2. *Molecular size.* Point 1 concerns the properties, including molecular size, of the materials that form most of the mass of the system, for convenience called the matrix. Here we consider the size of diffusing molecules, which may be present in small quantities only. According to the Stokes–Einstein relation [Eq. (5.16)], the diffusion coefficient of a molecule is inversely proportional to its radius. At small w this dependence is much stronger, since it now concerns diffusion through narrow pores in the matrix (see Figure 5.15). Semiquantitative examples are given in Figure 8.9a, where the upper curve relates to water. It is seen that at small w the differences become very large. This phenomenon explains retention of volatiles during drying of foods (e.g., coffee essence): most volatiles concerned have a distinctly larger molar mass than water. Most gases, on the other hand, will diffuse relatively fast.

To be sure, interaction forces between the diffusing molecules and the material of the matrix would also play a part, but this has received little study.

3. *Temperature.* Figure 8.9b gives some examples, and it seen that the effect again is very strong at small w, involving several orders of magnitude. This temperature effect is all that allows reasonably fast drying of several materials, or determination of dry matter content of most foods by oven drying.

4. *Physical inhomogeneity* of the system. Some authors have reported that diffusivities in systems of small w were markedly greater than expected on the basis of some theory, or when extrapolating from higher w values. Such a discrepancy may well be due to physical inhomogeneity of the matrix. Especially at very small w, tiny cracks may develop in the matrix, allowing much faster transport of small molecules. This phenomenon

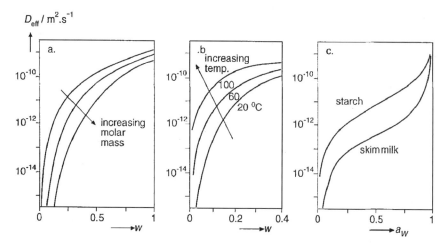

FIGURE 8.9 Effective diffusion coefficient (D_{eff}) of molecules in systems of various water contents. (a) Diffusivity of solutes of various molar mass as a function of mass fraction water (w). (b) Diffusivity of water at some temperatures as a function of w. (c) Diffusivity of water in two systems as a function of water activity. Very approximate, only to illustrate trends.

greatly upsets the interpretation of results on diffusivity, since it is not well known what factors cause crack formation; it greatly depends on the composition of the matrix.

8.4.2 Chemical Reactions

Examples of the dependence of the rate of chemical reactions, including an enzyme-catalyzed one, are in Figure 8.10. The reader should realize that relative rates are given, i.e., relative to the maximum rate shown in the graph. The latter rate may be much slower than the maximum possible rate (as in c and e). Moreover, the absolute rates of the various reactions considered may differ by orders of magnitude.

Most authors plot relative reaction rates against water activity, as in Figure 8.10, but this is not always practical, and Figure 8.11, below, gives rates against mass fraction of water w. These figures show that the relations can vary widely. Important factors affecting reaction rates are

1. *Water activity*. If water is a reactant, the rate will decrease with decreasing a_w. For instance, if $a_w = 0.5$, the rate would be slower by a factor of 2 than at $a_w \to 1$, if nothing else changes. In practice, the rate may be smaller by an order of magnitude or more; consequently, other factors must

then be more important. Nevertheless, reactions involving water, like the various kinds of enzyme-catalyzed hydrolysis, will all go to zero rate as $a_w \to 0$, irrespective of other considerations. The rate of a reaction in which water is produced should increase as a_w decreases, but the author is not aware of unequivocal proof for such a situation in a food.

    2. *Diffusivity*. The rate of a bimolecular reaction may decrease if the diffusion rate of the reactants decreases. In Section 8.4.1 diffusivity is discussed, but we need to consider two other aspects. First, even if D_{eff} is a

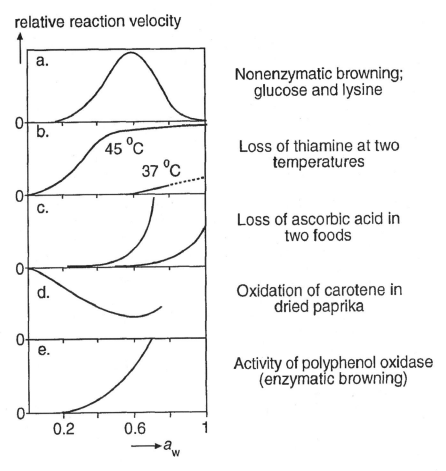

FIGURE 8.10 Relative rate of some reactions occurring in foods plotted against the water activity (a_w) at which the food was stored.

function of w only, the relation between a_w and w may vary considerably. This then means that the relation between diffusivity and a_w varies among foods. Figure 8.9c shows that the difference can be considerable, even for intermediate moisture foods. The second point was discussed in Section 4.3.3. The rate constant of a chemical reaction is generally determined by its activation free energy ($\Delta G^{\ddagger}$), and this also holds for a bimolecular reaction [Eq. (4.12)]. But if $\Delta G^{\ddagger}$ is very small, or D_{eff} is very small, the reaction may be diffusion controlled [Eq. (4.14)]. It depends on the reactions involved and on the further composition of the system at what (low) water content diffusivity becomes rate limiting. To give a very rough indication, the reactions in a food are likely to be diffusion controlled if D_{eff} for water is $< 10^{-14}\,\text{m}^2\cdot\text{s}^{-1}$, unless all of the reacting molecules are very small.

In many intermediate-moisture foods, most chemical reactions will not be diffusion limited. In low-moisture foods (e.g., $a_w < 0.2$), however, the most important cause of chemical stability will generally be small diffusivity. This would apply to all examples in Figure 8.10, except for oxidation of carotene. A relation like that in Figure 8.10d is generally observed for oxidation reactions, where water is not a reactant, and even at $a_w \approx 0$ perceptible diffusion of O_2 may occur; see further point 6, below.

3. *Concentration of reactants.* For a bimolecular reaction in aqueous solution, the concentration of reactants increases with decreasing water content, and the reaction rate would then be proportional to concentration squared. This is undoubtedly the main cause for the increase in reaction rate with decreasing a_w shown in Figure 8.10a; at still lower a_w, other factors are overriding. Also the composition of the reaction mixture may alter when lowering water content, for instance because a component partly crystallizes or becomes dissolved in oil (if present).

4. *Activity coefficients* of the reactants generally alter upon water removal. This was discussed in Section 2.2.5. For ionic species, the activity coefficient will decrease; for neutral ones it will generally increase, and the effect can be very large at very low a_w. These changes presumably explain part of the variation in the relation between reaction rate and a_w. Especially for unimolecular reactions, like the denaturation of proteins, variation in factors 2 and 3, above, would have no effect, but activity coefficients may change. This is equivalent to stating that the solvent quality is altered. Figure 8.11 gives some examples of—what are effectively—heat denaturation rates, and they show great variation. It is even possible that the rate increases with decreasing w, as for lipoxygenase, despite the notion that protein denaturation goes along with increased hydration (Section 7.2). However, even at $w = 0.15$, water activity would probably be > 0.6 at $72°C$ in the system studied.

It is generally observed, for heat inactivation of enzymes or killing of bacteria, that at a smaller w both the (apparent) activation enthalpy $\Delta H^{\ddagger}$ and the activation entropy $\Delta S^{\ddagger}$ decrease. This implies that the temperature dependence of the reaction becomes less; generally, the relative change in $\Delta S^{\ddagger}$ is greater than that in $\Delta H^{\ddagger}$, causing the reaction rate constant to decrease, but this is not always so. These relations are poorly understood, but they are of great practical importance for the stability of enzymes and microbes during drying.

    5. *Enzyme activity* (e.g., Figure 8.10e) may change (decrease) because the concentration of components affecting protein conformation increases, especially if it affects the active site of the enzyme; this may involve ionic strength, pH, and solvent quality. It is difficult to separate this effect from that of decreased diffusivity.

    6. *Catalysts and inhibitors* also change in concentration with water content. Presumably, Figure 8.10d provides an example. Here water would be an inhibitor of one or more of the reaction steps in oxidation. In such a

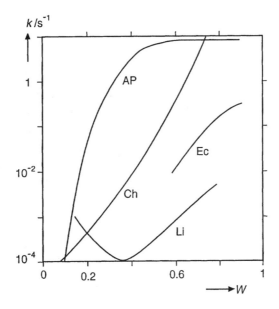

FIGURE 8.11 First-order reaction rate constants k for heat inactivation, plotted against mass fraction of water w. AP: alkaline phosphatase, in skim milk, 80°C. Ec: killing of *Eschericia coli*, in skim milk, 63°C. Ch: chymosin, in whey, 80°C. Li: lipoxygenase, in sucrose/calcium alginate, 72°C.

case, water activity would be the determinant variable. Another example may be the curve for chymosin in Figure 8.11. Here, k is already greatly decreased at w values where neither a_w nor D_{eff} can have changed appreciably, and it must be assumed that concentration of a stabilizing component has occurred. In many cases, water removal causes a significant change in pH.

It must unfortunately be *concluded* that the relations between reaction rate and water content of foods vary widely and that our understanding of this is incomplete. This means that one often has to rely on experimental determination.

8.4.3 Microbial Growth

It is well known that microorganisms cannot proliferate in dry foods and thus cannot cause spoilage. It depends, of course, on how dry the food is, and Figure 8.12 gives some examples of growth rate as a function of water

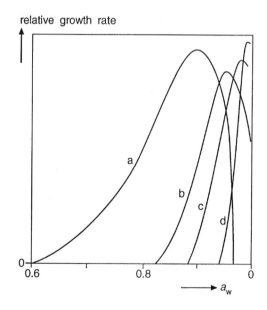

FIGURE 8.12   Relative growth rate (arbitrary scale) of some microorganisms as a function of water activity (a_w). (a) A xerophilic mold, *Xeromyces bisporus.* (b) A common mold, *Aspergillus flavus.* (c) A yeast, *Saccharomyces cerevisiae.* (d) A bacterium, *Salmonella* sp.

activity. Microorganisms vary widely in the water activity that they can tolerate. Generally, the lowest a_w for growth is for

Bacteria 0.98–0.9, halophilic ones down to ~ 0.75
Yeasts about 0.9, osmophilic ones down to ~ 0.6
Molds 0.92–0.8, xerophilic ones down to ~ 0.65

Osmotic Pressure. The lowest a_w for growth is generally much higher than the lowest one for most chemical reactions. This points to a very different inhibitory mechanism. The simplest explanation is that a low a_w means a high osmotic pressure (Π), and that the organism cannot tolerate this. If Π is high, it tends to draw water from the cell, thereby concentrating the cell contents and damaging the metabolic system. The organism tries to keep water in the cell, and it has several mechanisms for this, but they all fail at very high Π. From Eq. (8.3) we can calculate the relation

$$\frac{\Pi}{Pa} \approx 135 \cdot 10^6 (1 - a_w) \tag{8.9}$$

at room temperature and for not too low a_w. This implies that already at $a_w = 0.92, \Pi > 100$ bar. It may also be noted that a high Π need not kill the organisms, and activity of bacterial enzymes has often been observed. Bacterial spores can survive at very high Π, though not germinate.

There are, however, several *complications*. The terms halophilic and osmophilic already suggest that different substances have different effects. Halophilic bacteria can tolerate a high concentration of salts, but not of sugars, and it is the other way round for osmophilic yeasts. Table 8.1 gives some examples of the lowest a_w tolerated, when caused by various components. It is seen that the variation is considerable, by a factor of 5 in $(1 - a_w)$. This implies that the components have specific effects. Ethanol and glycerol cannot be kept out of the cell by most organisms, and thus do not cause an osmotic pressure difference over the cell membrane.

TABLE 8.1 Minimum Water Activity for Growth of Two Bacteria, Where the Water Activity Is Lowered by Addition of Various Components

Component	*Staphylococcus aureus*	*Lactococcus lactis*
Ethanol	0.973	—
Polyethylene glycol	0.927	—
Glycerol	0.89	0.924
Sucrose	0.87	0.949
Sodium chloride	0.86	0.965

Presumably, ethanol then is more damaging for the cytoplasm than glycerol. (It may be noted that $a_w = 0.973$ caused by ethanol corresponds to 7% (w/w) of ethanol in water.) But also substances that can be kept out of the cell, like polyethylene glycol and sucrose, show a significant difference. Table 8.1 also shows considerable difference between the two bacteria, even in the order of inhibitory activity of the components.

Moreover, other conditions determine the lowest a_w tolerated. Figure 8.13 illustrates how two inhibiting components can reinforce each other's effect. How this quantitatively works out greatly depends on the components and the organism, and the curves can be markedly asymmetric. Nevertheless, the trend illustrated in the figure is often obeyed. One may also determine contour lines for equal rate of growth. Other conditions, notably temperature and the kinds and concentrations of nutrients present, also affect the result. Actually, one could in principle "construct" n-dimensional graphs, in which the combined effect of all factors inhibiting and promoting growth for a certain organism would be given. The mutual enhancement of inhibiting factors is at the basis of "mild conservation,"

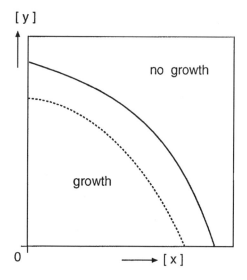

FIGURE 8.13 Growth of a microorganism as a function of the concentration of two components X and Y, say, a sugar and an acid. The full contour line marks the boundary between growth and no growth at optimum conditions (say, optimum temperature), the broken one at some suboptimum condition. Only meant to illustrate trends.

where an attempt is made to keep foods fresh without intense heat treatment and without adding large concentrations of bactericidal components.

8.5 RECAPITULATION

Water Activity. The reactivity of water in a food is precisely given by its water activity, which is mostly expressed as a fraction, thus ranging from 0 to 1. In a dilute and ideal solution, a_w equals the mole fraction of water, but in most foods there are several nonidealities, and it may be very difficult to predict a_w from composition. This means that it has to be determined, which can be done by measuring the relative vapor pressure of air in equilibrium with the food.

Many food properties correlate better with water activity then with water content, especially if the food contains substances that are more or less indifferent to water, like crystals or oil droplets. This does not mean that relations between a property and a_w are the same for all foods, not even in a relative sense. Some properties, notably rheological ones, generally correlate better with water content than with a_w.

Sorption. The relation between a_w and mass fraction of water w, at constant temperature, is called a *sorption isotherm*. It is a useful relation, giving information about hygroscopicity and about drying conditions to be applied. For almost all foods, it does not reflect true water adsorption, since much water is in a (concentrated) solution rather than adsorbed onto a well-defined surface. It mostly takes a very long time to determine a sorption isotherm, especially at the low a_w end, presumably because the diffusivity of water then becomes very small. The only way to obtain (near) zero water content generally is drying at high temperature, where diffusivity is much greater. Moreover, it is commonly observed that going to progressively lower a_w values ("desorption") gives a sorption curve that differs significantly from one obtained by progressively increasing a_w ("adsorption"). This hysteresis indicates that no true equilibrium is reached, implying that water activity is actually undefined. Nevertheless, sorption isotherms can be of great practical use.

Hydration. Hydration or true binding of water to food components generally involves only small quantities of water. Only polar groups, and to a lesser extent dipoles, can "bind" water molecules, i.e., immobilize them for a short time. It is often stated that bound water does not react, but water reactivity is just proportional to a_w. Bound water would not be available as a solvent, but nonsolvent water is primarily due to negative adsorption of a solute onto particles or macromolecules and greatly depends on the nature, especially on its molecular size, of the solute. Bound water would not freeze, but some water remains unfrozen when its diffusivity has become virtually zero, so that its crystallization would take infinite time. And held or imbibed water is certainly not bound, but just mechanically entrapped. In other words, water fills space. Water binding as a general term therefore is confusing.

Reaction Rates. Dry foods are generally more stable than those with a higher w. This can be due to various mechanisms, and the relations are generally not well understood. Most physical changes depend on diffusivity, which becomes progressively smaller with smaller w. Most chemical reactions proceed slower for a smaller w. If water is a reactant, a lower a_w means a slower reaction. At fairly low w, reactions may become diffusion–limited, which generally is the ultimate cause of stability in very dry foods. Activity coefficients may be greater at lower w, and removal of water may increase the concentration of reaction inhibitors or catalysts. It is possible that a reaction proceeds faster at a smaller w, at least over a certain range. For a bimolecular reaction this may be due to a higher concentration of reactants; moreover, water itself may be an inhibitor.

Microorganisms generally do not grow at low a_w, although great variation is observed among various organisms. The prime cause for inhibition is a high osmotic pressure. However, it is generally observed that different components at the same a_w, which also means at the same osmotic pressure, markedly differ in inhibitory capacity. This means that one cannot state that a certain a_w is the minimum one at which a given organism can grow, since it would depend on the presence and concentration of many components.

BIBLIOGRAPHY

A somewhat different discussion, also involving glass transitions, and giving more practical information, is
O. R. Fennema. Water and ice. In: O. R. Fennema, ed. Food Chemistry, 3rd ed. Marcel Dekker, New York, 1996, Chapter 2.

Basic aspects of water activity are discussed by
T. M. Herrington, F. C. Vernier. In: S. T. Becket, ed. Physico-Chemical Aspects of Food Processing. Blackie, London, 1995, Chapter 1.

A series of symposium reports, containing many interesting articles, comprises e.g.
L. B. Rockland, G. F. Stewart, eds. Water Activity: Influences on Food Quality. Academic Press, 1981.

D. Simatos, J. L. Multon, eds. Properties of Water in Foods. Nijhoff. 1985.

H. Levine, L. Slade, eds. Water Relationships in Foods, Plenum Press, 1991.

G. V. Barbosa-Cánovas, Weltí-Chanes, eds. Food Preservation by Moisture Control: Fundamentals and Applications. Technomic, Lancaster, 1995.

The role of water in relation to microbial growth is treated by

G. W. Gould. Drying, raised osmotic pressure and low water activity. In: G. W. Gould, ed. Mechanisms of action of food preservation procedures. Elsevier, London, 1989, pp. 97–118.

9

Dispersed Systems

Most foods are dispersed systems; in other words, they are physically heterogeneous. This means that their properties are not fully given by their chemical composition. For a homogeneous liquid, like apple juice or cooking oil, it is often reasonable to assume that thermodynamic equilibrium exists. In such a case, full knowledge of chemical composition will, in principle, give all properties, including reactivity, of the system, for given external conditions (temperature, pressure, etc.). A heterogeneous system has structural elements, and making these mostly costs energy. The system is therefore not in thermodynamic equilibrium; it can be manufactured in various ways, leading to a variety of structures. The properties of the system are determined by its structure and by the chemical composition of its structural elements.

Difference in structure generally implies differences in properties. Take, for example, ice cream. It is made (by freezing and agitation) of ice cream mix and air, and the two systems are very different, as is illustrated in Figure 9.1. Nevertheless, they have exactly the same chemical composition. Also the properties are very different, as we all know: when we let ice cream melt, we obtain a product of very different appearance, consistency, and eating qualities. However, melted ice cream is not quite the same as ice cream mix as, for instance, larger fat globules or clumps of them will be

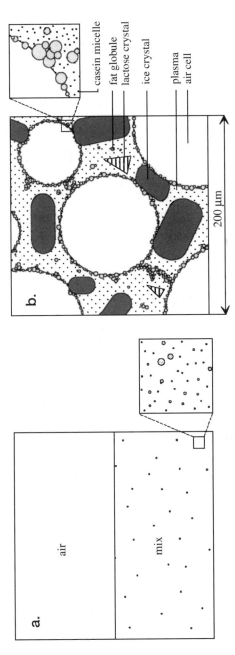

FIGURE 9.1 Example of the structure of (a) ice cream mix + air, and (b) the ice cream made of it (at about −5°C). The fat globules contain two phases: triglyceride oil and crystals. Highly schematic.

present. This illustrates the other point made: structure and properties depend on the history (process steps applied, storage conditions, temperature history) of the product.

Besides those mentioned, some other properties may greatly depend on structure. Several kinds of structure give the food a certain consistency, and this may greatly reduce transport rates in the system; these are discussed in Chapter 5. In many foods, the various chemical components are fully or partly compartmentalized: in cells, tissue fragments, emulsion droplets, etc. This implies that reactions between those components may be greatly hindered; this is mentioned in Section 4.4. The compartmentalization may also apply to *flavor* substances, and it will slow down their release during eating. If the compartments are fairly large it can lead to fluctuation in flavor release during eating, thereby enhancing flavor, because fluctuation can offset adaptation of the senses to flavor stimuli.

Note the response of the senses to a continuous stimulus generally decreases in time; this is called adaptation.

Thus a compartmentalized food generally tastes quite different from the same food that has been homogenized.

If a system is physically heterogeneous it can also be *physically unstable*. Several kinds of change may occur during storage, which may be perceived as a change in consistency or color, or as a separation into layers. Moreover, during processing or usage, changes in the dispersed state may occur. These may be desirable—as in the whipping of cream—or undesirable—as in overwhipping of cream, where butter granules are formed.

Most of the ensuing part of this book deals with dispersed systems. These generally have one or more kinds of interface, often making up a considerable surface area. This means that *surface phenomena* are of paramount importance, and they are discussed in Chapter 10. *Colloidal interaction* forces between structural elements are also essential, as they determine rheological properties and physical stability; these forces are the subject of Chapter 12. The various kinds of physical instability are treated in Chapter 13, and the nucleation phenomena involved in phase transitions in Chapter 14. Specific dispersed systems are discussed in Chapters 11 and 17. The present chapter explains important concepts and discusses geometrical aspects.

9.1 STRUCTURE

Structure can be defined as the distribution over space of the components in a system. This is a purely geometrical concept, as it just concerns angles and distances. The physical building blocks of such a system may be called *structural elements*, i.e., regions that are bounded by a closed surface, where at least some of the properties within such a region are different from those in the rest of the system. Structural elements can be particles, such as air bubbles, oil droplets, crystals, starch granules, cells, etc. If these particles are separate from each other, the system is called a *dispersion*. Figure 9.2 illustrates a dispersion with various structural elements. Also the continuous material surrounding the particles in a dispersion is a structural element. A structural element can be heterogeneous itself, containing further structural elements. Think of starch granules in a cell of a potato, or of fat crystals in the oil droplets of an emulsion (Fig. 9.2, E). Structural elements can also be (nearly) space filling, like parenchyma cells in the soft tissue of a fruit or myofibrils in a muscle.

In many cases, internal *interaction forces* act between structural elements. By internal we mean in this context that the forces have their origin in the (properties of the) materials making up the structural elements. This excludes external forces, e.g., caused by gravity, by flow, or by an electric field. The interaction forces can be attractive or repulsive, and the

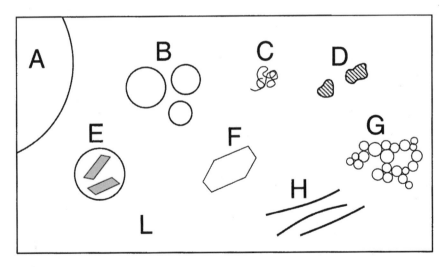

FIGURE 9.2 A liquid dispersion with various structural elements. (A) Gas bubble. (B) Emulsion droplets. (C) Polymer molecule. (D) Solid particles (amorphous). (E) Oil droplet with fat crystals. (F) Crystal. (G) Floc or aggregate of particles. (H) Fibers. L, Continuous phase. Highly schematic and not to scale (for instance, F is likely to be orders of magnitude larger than C).

net result always depends on the distance over which they are acting. This is discussed in Chapter 3 and is further worked out in Chapter 12. Net attractive interaction between particles can cause their aggregation into a floc (Fig. 9.2, G). Such a floc is also a structural element and the interaction forces determine (part of) the structure.

> *Note* This is a very common situation, and some authors include these forces in the description of structure.

Geometrical structure can, in principle, be seen: by eye or by means of a microscope. Interaction forces can only be derived from mechanical experiments. It is essential to know about the forces to understand the structure and most properties of the system.

Structure should not be confused with attributes like consistency or texture. *Consistency* is defined by rheologists as the resistance of a material against permanent deformation, characterized by the relation between deformation and external force working on it. Its magnitude is determined by structure and interaction forces of the material. The original meaning of *texture* is the perceivable (visible or tactile) inhomogeneity of a surface, be it an outer surface or a cut one. It is a direct consequence of structure. Currently, food scientists tend to use the word texture as also including consistency (or, more precisely, rheological and fracture properties), especially if it concerns sensory perception. It is indeed often difficult to distinguish between the two in sensory testing.

A **phase** is defined as a part of a system that is homogeneous and bounded by closed surfaces, at which surfaces at least some of the intensive parameters change abruptly (see Section 2.1). This looks much like the definition of a structural element, but there are differences. First, a structural element need not be physically homogeneous, whereas a phase is.

> *Note* This does not imply that a phase consists of one chemical component: most phases in foods contain many components, like all aqueous solutions encountered.

Second, the distance over which intensive parameters change is always on the order of a few molecules, and the criterion of abruptness of the phase boundary implies that the smallest dimension of a structural element must be several times that of a molecule, for the element to constitute a phase. A phase boundary is a true interface, which contains an amount of interfacial free energy, and onto which other, surface active, molecules can become adsorbed.

Whether structural elements constitute true phases is an important issue, as it marks the difference between what colloid scientists call *lyophilic* (= solvent loving) or reversible, and *lyophobic* (= solvent hating) or irreversible, systems. It may further be noted that dispersions are called *colloids* if the particles are larger than most molecules and too small to be visible, i.e., a size range of about 10^{-8} to 10^{-5} m.

A **lyophilic** heterogeneous system is—or, more precisely, can be—in thermodynamic equilibrium. It does not cost energy to make it: it forms spontaneously on mixing the components. Important examples are

Macromolecular solutions, especially of polymers. These molecules can be so large that they should be considered as particles. It may, however, depend on the property considered whether such a system is homogeneous or heterogeneous. A dilute polymer solution can be treated as homogeneous when considering its dielectric properties or the diffusion of salt through it. But it can scatter light, like other dispersions, and in explaining its viscosity the heterogeneity is essential. Even the viscosity of a dilute sucrose solution can be reasonably explained by Einstein's equation (5.6) for a dispersion, although the system is homogeneous in almost all respects.

Association colloids are formed by fairly small molecules that associate spontaneously into larger structures. A clear example is the formation of micelles, i.e., roughly spherical particles of about 5 nm diameter, by amphiphilic molecules like soaps: see Figure 2.8. At high concentrations, such molecules can in principle form a range of structures, called *mesomorphic or liquid crystalline phases*, which are briefly discussed in Section 10.3.1. To be sure, the whole system is called a mesomorphic phase, not its structural elements.

The particles in lyophilic dispersions cannot be considered to constitute a phase: they have no sharp boundary.

Lyophobic systems, on the other hand, contain particles that do make up a phase. It costs energy to make them and they never form spontaneously. A lyophobic dispersion always has a *continuous phase*, which means that one can envisage a molecule moving from one end of the system to the other in any direction, without ever leaving that phase. The particles make up the *disperse(d) phase*. According to the state of the two phases, five types of dispersions can be distinguished, as given in Table 9.1. Gas–gas dispersions do not occur: gases are fully miscible. Most liquids encountered in foods are also fully miscible, but triacylglycerol oils and aqueous solutions are not; this leaves two emulsion types, oil-in-water and water-in-oil. Most foams are gas bubbles in an aqueous solution. The dispersed phase in a suspension can consist of crystals or of amorphous particles. Fogs, aerosols, and smokes will not be discussed in this book, and powders hardly.

The properties of the **continuous phase** determine many properties of the system. If the continuous phase is a liquid, it determines (a) what substances can be dissolved in the system; (b) it greatly affects the

TABLE 9.1   Various Types of Lyophobic Dispersions

Dispersed phase	Continuous phase	Dispersion type
Gas	Liquid	Foam
Liquid	Gas	Fog, aerosol
Liquid	Liquid	Emulsion
Solid	Gas	Smoke, powder
Solid	Liquid	Suspension, sol

interaction forces between the particles; and (c) the possibility of loss of substances by evaporation, etc. An oil-in-water emulsion, such as a fairly concentrated cream, differs greatly in properties from a water-in-oil emulsion, such as a low fat spread, although both systems may have nearly the same chemical composition and contain droplets of about the same size. Table 9.2 gives some important properties of materials that can make up the continuous phase of foods.

The continuous phase can be a *solid* or have some characteristics of a solid. This implies that the structural elements are immobilized, which considerably enhances physical stability of the system. When making such a dispersion, the continuous phase is always liquid, but it can solidify afterwards, e.g., by lowering the temperature or by evaporating the solvent. The liquid can become crystallized, form a glass (Section 16.1), or turn into a gel. Especially the last named situation is frequently encountered in foods. Also the "solvent," generally an aqueous solution, in the continuous phase then is more or less immobilized (Section 5.3). If the gel is a classical polymer gel (Section 17.2.2), the polymer molecules provide a continuous network, but do *not* make up a continuous phase: the polymer strands cannot be seen as a structure *in* which other molecules can be present and diffuse.

The structure of **food systems** can be more complicated. This is discussed in greater detail in Chapter 17. A system can have two continuous phases (in theory even a greater number). A good example is bread crumb, where both the gas phase and the "solid" *matrix* are continuous: one can blow air through a slice of bread. The matrix is, by and large, a continuous gluten phase containing partially gelatinized starch granules. Such a *bicontinuous* structure of a solid and a fluid phase is called a *sponge*. Cheese is a dispersion of oil droplets (the droplets also contain fat crystals) in a proteinaceous continuous phase. However, a fairly hard and well matured cheese often exhibits oiling off if the ambient temperature is not too low, which proves that also the oil phase has become continuous.

TABLE 9.2 Approximate Values of Some Physical Constants of Oil (i.e., a Liquid Mixture of Triacylglycerols), Water, a Saturated Sucrose Solution (About 66% Sugar by Mass in Water), and Air. All at about 20°C (Except ΔH_f)

Property	Symbol	Unit	Oil	Water	Sugar solution	Air
Mass density	ρ	$kg \cdot m^{-3}$	920	990	1320	1.2
Refractive index	n_D	—	1.45	1.333	1.451	1.000
Specific heat	c_p	$kJ \cdot kg^{-1} \cdot K^{-1}$	2.1	4.2	2.8	1.0
Heat of fusion	ΔH_f	$kJ \cdot kg^{-1}$	150–200	313	—	—
Heat conductivity	λ	$mW \cdot m^{-1} \cdot K^{-1}$	160	580	270	0.26
Viscosity	η	$mPa \cdot s$	70	1.00	120	0.018
Surface tension	γ	$mN \cdot m^{-1}$	40	73	78	—
Dielectric constant	ε	relative	3	80	35	1
Solubility[a] in						
water			0	∞	∞	+
ethanol			+	∞	+	+
hexane			∞	trace	trace	+
Vapor pressure	p_v	Pa	<1	2300	1990	—

[a] 0 insoluble, + soluble, ∞ miscible in all proportions.

Figure 9.1b gives an example of a complicated structure. Ice cream can contain seven or more different structural elements, making up six phases: the continuous phase (an aqueous solution), air, ice, oil, crystalline fat, lactose. It may be noted that the system depicted is bicontinuous in that the clumped fat globules also form a continuous network. During long storage at low temperature, part of the air phase can become continuous. At a lower temperature, a greater part of the water freezes and a continuous ice phase can be formed; moreover freezing causes less room being available between the various particles.

This brings us to the subject of **packing**: what is the maximum volume fraction (φ_{max}) of particles that a dispersion can contain? For hard, smooth, monodisperse spheres, it is in theory for a cubic arrangement 0.52, for a hexagonal arrangement, the closest packing possible, the value is 0.74. In practice, one often observes for solid spheres that $\varphi_{max} \approx 0.6$, because friction between spheres hinders attaining their closest packing. For emulsions, where the drops are relatively smooth, a common value is 0.71. If the particles are

More polydisperse, φ_{max} is larger, because small particles can fill the holes between large ones.

More anisometric or have a rougher surface, φ_{max} is smaller, mainly because these factors cause more friction between particles, making it difficult for them to rearrange into a denser packing.

Aggregated, φ_{max} is smaller, often very much so. In most aggregates, much of the continuous phase is entrapped between the primary particles: see Figure 9.2, G.

More deformable, φ_{max} is larger, and values up to 0.99 have been observed in foams, where the large bubbles can readily be deformed.

How much material would be needed to make a *continuous network*, i.e., a structure that encloses all of the continuous phase? This can readily be calculated for a regular geometry. Assume anisometric particles, more in particular cylinders of length L and diameter d. When making a cubical configuration of these particles, i.e., each edge of a cube is one particle, we obtain

$L/d =$	5	10	20	50
φ needed $\approx$	0.094	0.024	0.006	0.001

It is seen that very little material can suffice, provided, of course, that attractive interaction forces keep the network together. In practice the situation is more complicated; some systems are discussed in Chapter 17.

Order. For very small volume fraction φ, the particles in a dispersion can be distributed at random. For higher φ, there is always some *order*. This is illustrated in Figure 9.3a and 9.3b. Around a sphere of radius R a volume of $(4/3)\pi(2R)^3$ is not available for other spheres of the same radius (taking the center of a sphere as representing its position). This implies that random distribution is not possible. The higher φ, the less random the distribution is, and at the closest packing (φ_{max}), monodisperse spheres would show perfect order.

If the particles are *anisometric*, which means that they have different dimensions in different directions, fairly close packing always leads to *anisotropy* of the system, which means that some properties depend on the direction considered. This is illustrated in Figure 9.3c. Anisotropy is also possible for low volume systems, or for spherical particles: it all depends on their arrangement. A good example of visible anisotropy is in bread, where the gas cells are generally not spherical, but elongated. During the oven rising, when the cells become much larger, the dough is confined by the mold, and expansion of the cells in a horizontal direction is limited. When cutting a slice of bread in the normal way, this is clearly observed, but when the loaf is cut parallel to the bottom, the cells look spherical. Muscle tissue is very anisotropic, and especially the mechanical strength greatly depends on the direction of the force applied. The same holds true for many plant organs, like stems.

Anisotropy can be manifest in several physical properties and at various scales, for instance,

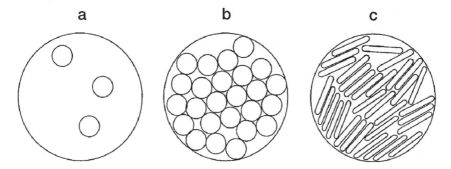

FIGURE 9.3 Order induced when particles are present at high volume fraction in a confined space. Illustrated for two-dimensional systems.

Optical anisotropy. The best known example is that where the refractive index depends on direction. Such a material is said to be *birefringent*: when linearly polarized light passes through it, its velocity depends on direction (on the angle between the plane through the light beam and the polarization plane). Interference of the emerging rays then causes the light to become elliptically polarized. Such a material generally appears bright when viewed by a polarizing microscope. The origin is at molecular scale. Most crystals are birefringent (e.g., all sugar crystals) and fibrous molecules in a (partly) parallel orientation also. Good examples are plant cell walls and native starch granules.

Permeability (see Section 5.3.1). Transport of liquid through the material can be greatly dependent on direction. The prime example is wood, which derives from vascular tissue that consists of long tubular cells (trachea), needed for transport of water. Some natural food materials also show this dependency. It finds its origin at a scale of, say, a micrometer. Anisotropy with respect to *diffusion* of small molecules is rarely encountered in foods. For instance, the rate of diffusion of salt into muscle tissue is not direction dependent (along or across the fibers).

Mechanical properties, i.e., rheological and fracture phenomena. This anisometry is widespread among natural foods and among fabricated foods made by spinning or extrusion. It is mostly caused by relatively large structural elements (micrometer to millimeter scale).

The most complicated structures are found in natural foods, but even a superficial discussion would take too much space. Just one example is given in Figure 9.4. Even from this simplified picture, it is obvious that an apple has numerous structural elements and that these can be arranged in a *hierarchical order*. At every level, specific aspects are of importance. To obtain a full understanding of the properties of the system, the structure and the interactions between structural elements would have to be studied at several length scales, from molecular to macroscopic. This brings us to the next section.

Question

You have a so-called low-fat spread, i.e., a system containing about equal volumes of triglycerides and an aqueous phase. The system is not a liquid, but it is spreadable. Can you think of a very simple way to find out whether it is oil or water continuous?

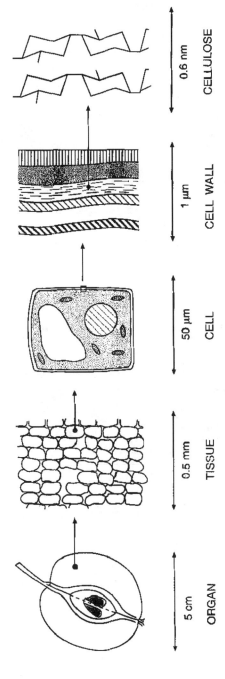

FIGURE 9.4 Structure of an apple or, more precisely, of some small parts of an apple, at various scales. Highly simplified and schematic.

9.2 IMPORTANCE OF SCALE

For homogeneous systems, knowledge of properties and phenomena on a molecular or *microscopic* scale often suffices to understand, or even predict, what happens on a *macroscopic* scale. For dispersed systems, it is generally necessary, or at least useful, to invoke structural elements on an intermediate or *mesoscopic* scale. A glance at Figure 9.5 shows, however, that the linear mesoscopic scale may span a wide range, say, by four orders of magnitude, and the volume of a cell may be large enough to contain 10^{15} (small) molecules.

The size of the structural elements often determines or greatly affects several properties of a dispersed system. This concerns changes during processing and storage as well as static properties of the finished product. Below, some examples of dependencies on scale will be given, for the most part illustrated for simple dispersions.

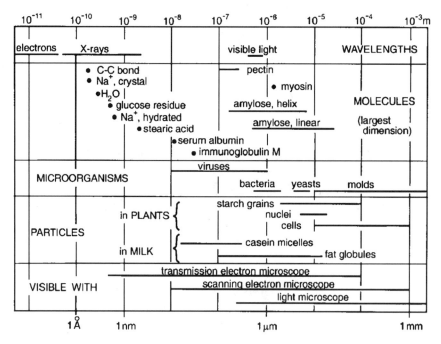

FIGURE 9.5 Length scales. Approximate values.

9.2.1 Geometric Aspects

The **number** of particles present per unit volume (N) is, for a given volume fraction (φ), inversely proportional to particle diameter (d) cubed. For monodisperse spherical particles, $\varphi = N\pi d^3/6$. Examples are in Table 9.3.

The **specific surface area** (A), i.e., the surface area per unit volume of dispersion, is proportional to φ and inversely proportional to d. For monodisperse spherical particles, $A = 6\varphi/d$. The amount of material needed to cover particles is proportional to A, and this amount can be appreciable if the particles are small. Examples are in Table 9.3.

Question

Which of the values given in the table is in fact impossible?

The **distance** between particles (x) can be defined in various ways. Assuming a regular cubic arrangement of particles, the smallest distance between adjacent ones is given by

$$x = N^{-1/3} - d = d\left(\frac{0.81}{\varphi^{1/3}} - 1\right) \tag{9.1 a}$$

where the second part is valid for spheres only. The average free distance over which a sphere can be moved before it touches another one is, assuming the distribution of the spheres throughout the volume to be random, given

TABLE **9.3** Number and Surface Area of Spherical Particles of Various Diameters in a Dispersion at a Volume Fraction of 0.5 (the volume occupied by surface layers around the spheres is also given)

Diameter of spheres, μm	0.1	1	10
Number per ml	10^{15}	10^{12}	10^9
Surface area, m^2 per ml	30	3	0.3
Surface layer of 2 nm, % v/v	6.2	0.6	0.06
Same, of 10 nm thickness	36	3.1	0.30

by

$$x = \frac{d}{6} \left(\frac{1}{\varphi} - \frac{1}{\varphi_{\max}} \right) \tag{9.1 b}$$

where $\varphi_{\max}$ equals, for instance, 0.7. Results for both equations are in Figure 9.6, and it is seen that the dependence on φ is considerable and that especially at low φ the difference between the two is considerable.

Pore Size.   If aggregated particles make up a continuous phase, and the voids are filled with a continuous fluid phase, an important characteristic is the (average) pore size, i.e., the width of the channels between particles. This applies to powders and to particle gels. The pore size greatly affects

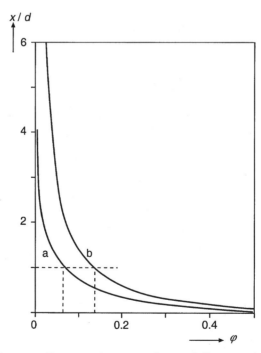

FIGURE 9.6   Average distance x between spheres of diameter d as a function of volume fraction φ, according to Eqs. (9.1a) and (9.1b).

transport rates through the system (Section 5.3) and capillary phenomena (Section 10.6). The pore size is governed by the same variables as is interparticle distance, although the relations are more complicated.

Diffusion Times. Brownian motion of molecules and particles is discussed in Section 5.2. The root-mean-square displacement of a particle is inversely proportional to the square root of its diameter. Examples are given in Table 9.4. The diffusion time for heat or matter into or out of a particle of diameter d is of the order of $d^2/10D$ where D is the diffusion coefficient. All this means that the length scale of a structural element, and the time scale needed for events to occur with or in such a structural element, generally are correlated. Such correlations are positive, but mostly not linear.

Separability. The smaller the particles in a dispersion, the more difficult it is to separate them from the continuous phase. This is illustrated by the approximate pore size in various separation membranes:

paper filter	20 μm	1 kPa
microfiltration	1 μm	5 kPa
ultrafiltration	10 nm	30 kPa
nanofiltration	1 nm	1 MPa

The smaller the pores, the higher the pressure to be applied to achieve substantial flow through the membrane, as indicated. See also under "sedimentation," below.

9.2.2 Forces

Internal forces. A fluid particle exhibits an internal pressure due to surface forces, the Laplace pressure, which is inversely proportional to its diameter (Section 10.5.1). This means that the particle resists deformation, the more so the smaller it is. The colloidal forces that may act between particles, keeping them aggregated, often are about proportional to d (Chapter 12). This means that the stress (force over area) involved would, again, be inversely proportional to d.

External forces acting on a particle generally increase steeply with d. For instance, the viscous stress exerted by a flowing liquid is given by $\eta \Psi$, where η is viscosity and Ψ velocity gradient. This means that the viscous force (stress times area) acting on a particle or aggregate is proportional to d^2. Comparison with internal forces then leads to the conclusion that for small particles, external stresses are unlikely to overcome internal forces, while this may be easy for large particles. Gels made of particles generally are firmer for smaller particles.

Sedimentation rate is the result of opposite forces acting on a particle, gravitational and frictional, leading to a proportionality to d^2 (Section 13.3). Again, it is more difficult to separate small particles from the continuous phase than large particles. Table 9.4 gives examples; notice that for small molecules Brownian motion tends to be much faster than sedimentation, whereas the opposite is true for large particles.

Coalescence of emulsion droplets tends to occur more readily for larger droplets (Section 13.4).

Altogether, dispersions of small particles tend to be more stable, often to a considerable extent, than those of large particles.

9.2.3 Optical Properties

Hardly ever do we see light directly emanating from a light source; it is nearly always scattered. *Scattering* of light, which can be due to reflection, refraction, or diffraction, occurs at sites where the refractive index changes, for instance at a phase boundary; see Figure 9.7. This means that we can see such a boundary.

Refraction. The *refractive index* n of a homogeneous material equals the ratio of the wavelength of the light in vacuum over that in the material. The value of n also depends on the wavelength of the light λ; it generally decreases with increasing λ. The refractive index is commonly given as n_D, i.e., at $\lambda = 589$ nm (the sodium D line). Table 9.2 gives some values. n decreases with increasing temperature.

TABLE 9.4 Motion of Spherical Particles of Various Sizes in Water

Diameter (μm)	0.001	0.01	0.1	1	10
Brownian motion	1200	390	120	39	12
Sedimentation by gravity		0.02	2	200	2×10^4
Same, centrifuge at $1000\,g$	0.2	20	2000		
Same, ultracentrifuge at $10^5\,g$	20	2000			

Room temperature. Root-mean-square displacement ($< x^2 >^{0.5}$) in μm by brownian motion over one hour. Sedimentation rate in μm per hour, assuming the particles to differ in density from water by $100\,\text{kg} \cdot \text{m}^{-3}$

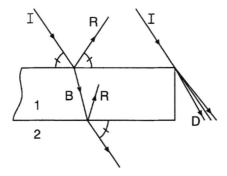

FIGURE 9.7 Illustration of reflection, refraction, and diffraction of light that is incident upon a particle of transparent material (1) of a refractive index that is higher than that of the surrounding medium (2). I is incident light, R, reflected light, B broken or refracted light, and D diffracted light. Part of the light passing near the edge of a particle shows diffraction, and the angle by which it is diffracted is appreciable only if the particle is not very large compared to the wavelength.

In most cases, we are interested in the relative refractive index m, i.e., the ratio of the refractive indices of the materials on either side of a phase boundary. The higher the value of $|m - 1|$, the stronger the refraction of a light beam at that boundary. Whether we can see a boundary depends on the contrast with the environment, which follows from the fraction of light reflected.

Reflection. For perpendicular incidence of light on a plane surface, the fraction reflected is given by $(m - 1)^2/(m + 1)^2$; for $m = 1.5$ this fraction equals 0.04, for $m = 1.1$ it is 0.002, which is very little. At an oil–air boundary m equals about 1.45. For foods containing no air cells and no crystals, m is smaller than 1.1 for most structural elements, and it is often much closer to unity.

For oblique incidence, the reflection is stronger, and moreover refraction occurs, further enhancing contrast. Consequently, oil droplets in water ($m = 1.09$) can readily be observed in a simple light microscope. Figure 9.5 gives the minimum size needed for particles to be visible with various microscopic techniques. However, measures to enhance contrast are often needed if m is close to unity.

Scattering. If the particles are small, which means of the order of the wavelength of light (about 0.5 μm) or smaller, the scattering of light can no longer be separated into reflection, refraction, and diffraction. If the size

d is much smaller than the wavelength (Rayleigh scattering), the amount of light scattered by each particle is about proportional to d^6, λ^{-4} and $(m-1)^2$; since the volume of a particle is proportional to d^3, the total scattering per unit volume (or mass) of particles is proportional to d^3. As the particles become larger, the dependence of scattering on d and on λ becomes weaker, and finally total scattering decreases with increasing d/λ. This is illustrated in Figure 9.8. It is seen that scattering is at maximum for $d \cdot \Delta n/\lambda \approx 0.5$. For visible light (average $\lambda = 0.55\,\mu$m) and $\Delta n = 0.1$, this yields for the optimum d about $3\,\mu$m. We can now understand that the appearance of an emulsion of oil in water (φ being, e.g., 0.03) will depend on droplet size as follows:

$d \approx 0.03\,\mu$m:	grayish, almost transparent
$d \approx 0.3\,\mu$m:	blueish, transmitted light being red
$d \approx 3\,\mu$m:	white
$d \approx 30\,\mu$m:	less white, maybe some color

The smallest particles hardly scatter light. For those of about $0.3\,\mu$m scattering greatly increases with decreasing wavelength. This means that blue light (short λ) is scattered far more intensely than red light (long λ),

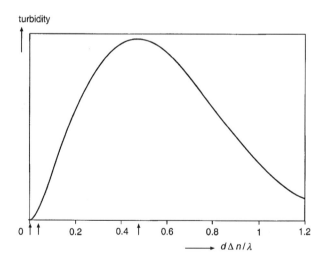

FIGURE 9.8 Turbidity or total scattering per unit mass of particles as a function of particle size d, refractive index difference Δn, and wavelength λ. Approximate results for small Δn. The arrows denote droplets of 0.03, 0.3, and $3\,\mu$m for an average oil-in-water emulsion at $\lambda = 0.55\,\mu$m.

which can thus pass the emulsion almost unscattered. For still larger particles, the dependence of scattering on λ is very small, and the emulsion (or other type of dispersion) appears white, the more so if the particle concentration is larger.

Absorption. The possible color of the last mentioned emulsion needs further explanation. Up till now, we have implicitly assumed that light is only scattered, not absorbed. If absorption occurs, we should use the complex refractive index $\tilde{n} = n - i n'$, where n' determines the adsorption ($i = \sqrt{-1}$). The relations now become more complicated. n' is related to the specific extinction γ according to $\gamma = 4\pi n'/\lambda$. The absorbency as determined in a spectrophotometer equals $0.434 \, \gamma L$, where L is the optical path length. Generally, n' is far more strongly dependent on wavelength than is n, and it gives rise to a fairly narrow absorption peak as a function of λ; consequently, we see the material to be colored if the absorption occurs for visible light. Assume that we have a strong absorbency, being unity for an optical path length of one mm at $\lambda = 0.55 \, \mu m$. We now calculate that $n' = 0.0001$. This means $n' \ll |n - 1|$, and under such conditions scattering tends to predominate over absorption; hence the emulsion still looks white. However, if the drops are large, the scattering per unit mass of oil becomes small (Figure 9.8), and now absorption can be perceived. The yellow color of an oil will then give a somewhat creamy color to the emulsion. Drops of mm size can, of course, be seen as such, the color included.

Diffuse Reflection. Light scattering by homogeneous small particles of fairly simple shape can accurately be calculated. Measurement of scattering by a dispersion can therefore yield information on concentration, size, and possibly shape of the particles. However, this is straightforward only if the dispersion is dilute. For concentrated systems, *multiple scattering* occurs, which means that light scattered by one particle is subsequently scattered by many others. Moreover, *interference* occurs of the light scattered by particles that are close to each other, and some structural elements may absorb light. In most disperse foods, all these phenomena will happen. If light falls onto such a system, some light will be directly reflected, but a much greater part will penetrate and be scattered numerous times at the surfaces of structural elements, before it emerges again; this is called *diffuse reflection*. In the meantime, some of the light will often be absorbed, generally at specific wavelengths, giving rise to color. The phenomena involved generally are far too complicated to calculate the amount and wavelength dependency of the diffusely reflected light. Consequently, diffuse reflection by food systems has to be measured and can hardly be predicted, if at all. If the food contains substances that absorb light at certain wavelengths, the resulting color can roughly be predicted. Furthermore, the rules of thumb on whiteness given above can be of some use to explain whiteness. But otherwise, we have to rely on empiricism.

9.3 PARTICLE SIZE DISTRIBUTIONS

The particles in a dispersion are hardly ever of the same size. Nature may often succeed in making rather monodisperse systems, like protein molecules, cells, or wheat kernels, although wider distributions also occur, e.g., starch granules. Most man-made dispersions have a fairly wide range of sizes and are thus polydisperse or heterodisperse; examples are emulsion droplets, particles obtained by grinding (flour, etc.), and spray-dried milk. Since many properties of a dispersion depend on particle size, as we have seen in the previous section, such properties may also depend on the distribution of sizes: how many particles of each possible size are present? This is the subject of this section.

9.3.1 Description

We start by defining a size variable x. Various definitions can be chosen: x may be particle diameter, molar mass, number of molecules in a particle, particle volume, etc. The *cumulative number distribution* $F(x)$ is now defined as the number of particles with a size smaller than x. Consequently $F(0) = 0$ and $F(\infty) = N =$ the total number of particles in the dispersion. The dimension of $F(x)$ generally is $[L^{-3}]$ (where L stands for length), i.e., number per unit volume, but other definitions can also be taken, for instance number per unit mass. Often, a cumulative distribution is recalculated to a percentage of N, hence putting $F(\infty)$ at 100%.

The *frequency distribution* of the number is now defined as

$$f(x) = F'(x) = \frac{dF(x)}{dx} \tag{9.2}$$

The frequency distribution is thus a differential quantity, and it is given in number per unit volume per unit of x. If x is expressed in units of length, e.g., particle diameter, the dimension of $f(x)$ is $[L^{-4}]$.

So far we have assumed the distribution to be a continuous one, which is nearly always a good approximation, because of the very large number of particles in most dispersions. In practice, a distribution is often split into size classes. We then have N_i particles in size class i and

$$N_i = \int_{x_i - \frac{1}{2}\Delta x}^{x_i + \frac{1}{2}\Delta x} f(x)\, d(x) \tag{9.3}$$

where Δx is the class width and x_i is the value of x characterizing class i (midpoint of class). The approximated frequency is now given by $N_i/\Delta x$, the cumulative distribution by $\Sigma\ N_i$.

Figure 9.9 illustrates this. At the left, the (presumed) measured values are indicated; for convenience, the points are connected by straight lines. The right-hand graph shows the same data in the form of a histogram. Notice that the class width is not the same for every class. A smoothed curve is drawn through the histogram to show the continuous frequency distribution; since only limited information is available (8 points), this curve is to some extent conjectural.

Besides the number distribution, distributions of *mass, volume, surface area* or other characteristics of the particles can be made. In such a case, the number frequency should be multiplied by mass, volume, etc. of the corresponding particles. For instance, if it concerns spheres, where x equals the diameter, the volume frequency distribution is given by $\pi x^3 f(x)/6$. In other words, one can choose both for the abscissa and for the ordinate the kind of variable that is most suitable for the presentation of the results, depending on the problem studied. Plotting volume frequency versus particle diameter is the most common presentation.

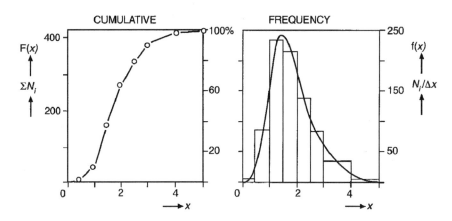

FIGURE 9.9 Example of a particle size distribution, given as a cumulative and as a frequency distribution. The points denote the measured values. A continuous frequency distribution is also shown.

In the next part, we will for sake of simplicity consider spherical particles. The following parameters can be defined:

d = sphere diameter, dimension [L]; in other words $x = d$

$$S_n = \int_0^\infty d^n f(d) \mathrm{d}d \approx \sum_{i=1} N_i d_i^n \tag{9.4}$$

S_n is the nth moment of the distribution, dimension $[L^{n-3}]$. It has no physical meaning, but it is useful as an auxiliary parameter.
$S_0 = N$ = total number of particles, dimension $[L^{-3}]$.

$$d_{ab} = \left(\frac{S_a}{S_b}\right)^{1/(a-b)} \tag{9.5}$$

This is a general equation for an average diameter, and the kind of average depends on the values for a and b; dimension [L].

$$c_n = \left(\frac{S_n S_{n+2}}{S_{n+1}^2} - 1\right)^{1/2} \tag{9.6}$$

c_n is the relative standard deviation or the variation coefficient of the distribution weighted with the nth power of d; it is thus dimensionless.

9.3.2 Characteristics

Figure 9.10 gives an example of a size frequency distribution of considerable width. It would be a reasonable example for a homogenized emulsion, assuming the d scale to be in 10^{-7} m. It is seen that the number frequency can give a quite misleading picture: more than half of the volume of the particles is not even shown in the number distribution. We will use this figure to illustrate some characteristic numbers.

Average. The most important one is usually an average value, often an *average diameter*. As indicated by Eq. (9.5), several types of diameter can be calculated, by choosing values for a and b. Examples are

$$d_{10} = \frac{S_1}{S_0} = \textit{number average} \text{ or mean diameter}$$

$$d_{30} = \left(\frac{S_3}{S_0}\right)^{1/3} = \textit{volume average } \text{diameter}$$

$$d_{32} = \frac{S_3}{S_2} = \textit{volume/surface average}, \text{ or Sauter mean, diameter}$$

$$d_{63} = \left(\frac{S_6}{S_3}\right)^{1/3} = \textit{volume-weighted average volume } \text{diameter}$$

Other characteristic diameters are

> The *modal* diameter, corresponding to the peak of the number frequency distribution, and the modal volume diameter
>
> The *median* diameter d_m, which divides the distribution into halves of equal number, i.e., $F(d_m) = (1/2)\, F(\infty)$. One can also use the median volume diameter.

Figure 9.10 shows that the various averages can differ widely. The sum $a + b$ is called the order of the average, and the higher it is, the larger the average. The common mean, d_{10}, is a poor measure of the center of the

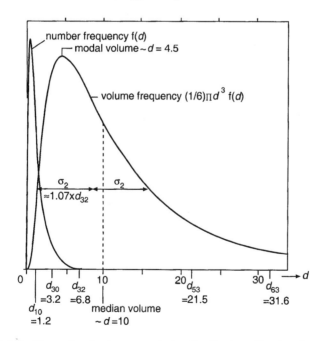

FIGURE 9.10 Example of a log-normal size distribution. Number frequency and volume frequency versus diameter (d) are given, and the various types of characteristic diameters are indicated, as well as the distribution width.

distribution. This type of average should be inserted into Eq. (9.1a) for a polydisperse system. The most commonly used average may be d_{32}, which is generally not far removed from the modal volume diameter. It gives the relation between volume fraction and specific surface area of the particles A (dimension $[L^{-1}]$), according to

$$A = \pi S_2 = \frac{6\varphi}{d_{32}} \tag{9.7}$$

taking into account that $\varphi = \pi S_3/6$. d_{32} also is the appropriate average in Eq. (9.1b) for a polydisperse system, and in relations for pore size or permeability [e.g., Eq. (5.25)].

Width. If the size distribution is very narrow, i.e., almost monodisperse, an average suffices to characterize it, and the various types of average differ only slightly from each other. Otherwise, the *distribution width* is important. It can best be expressed as a relative width, according to Eq. (9.6). Taking $n = 2$, the resulting parameter is c_2, i.e., the standard deviation of the size distribution weighted with the particles' surface area, divided by d_{32} (= the mean of the distribution weighted with the surface area). Values of c_2 are rarely below 0.1 (very narrow) or over 1.3 (very wide); in most fabricated foods the range is between 0.5 and 1.

An absolute width, e.g., expressed as a standard deviation in micrometers, tells very little. Figure 9.10 shows a quite wide distribution ($c_2 \approx 1.07$); its absolute width would be about 0.7 μm, assuming it to represent a homogenized emulsion. A collection of glass marbles ($d \approx 10$ mm) tends to be very monodisperse (e.g., $c_2 = 0.02$), but the standard deviation of the diameters would then be 200 μm, i.e., about 300 times as high.

Shape. If c_2 is not small, say > 0.4, the *shape* of the size distribution may become of importance. Several equations exist for frequency distributions (among which the log-normal one depicted in Figure 9.10 represents several dispersions rather well), but for a small width they all give nearly the same curve (assuming d_{32} to be the same). For larger width, differences in shape become important. For instance, distributions can be more or less skewed, or truncated, or even bimodal.

We will not treat mathematical equations for frequency distributions. It may be added, however, that the often used normal distribution is rarely suitable for particle size distributions. For instance, it allows the presence of particles of negative size; in other words, $f(x) > 0$ for $x \leqslant 0$. A log-normal distribution (in which log x has a normal distribution) gives $f(x) = 0$ for $x = 0$.

9.3.3 Complications

Many particles encountered in practice are not true homogeneous spheres, although emulsion drops and small gas bubbles may be virtually so. Deviations can be of two types.

The particles may be *inhomogeneous*. They may be hollow or have a more intricate internal structure. If the inhomogeneity varies among particles, a particle size distribution is insufficient to characterize the dispersion. For instance, the mass average diameter and the volume average diameter may be markedly different; this is illustrated by several spray-dried powders, where some of the particles have large vacuoles, while others have not. If the oil droplets in an emulsion are coated with a thick layer of protein, the smallest droplets contain far more protein per unit amount of oil than the largest ones, as illustrated in Table 9.3.

The particles may be *anisometric*, i.e., deviate from the spherical form. Moreover, the particles may have a rough surface. The two cases cannot be fully separated because intermediates occur, but a sphere can have a rough surface and a platelet can be smooth. One parameter now is insufficient to characterize a particle. If all particles have approximately congruent shapes (think of a collection of screws of various sizes), it may be possible to use just one size parameter, e.g., length. For irregularly shaped but not very anisometric particles, as found in several powders, one often defines an *equivalent sphere diameter*. This can be defined in various ways, such as

d_v = diameter of a sphere of the same volume
d_s = diameter of a sphere of the same surface area
d_f = diameter of a sphere that sediments at the same rate
d_e = the edge of the smallest square through which the particle can pass, or sieve diameter
d_p = diameter of a circle of the same surface area as the perpendicular projection of the particle on the plane of greatest stability (explained in Section 9.3.4)

To characterize the shape, various parameters are used. The deviation from spherical is often expressed by the following *shape factor*

$$\psi = \left(\frac{d_v}{d_s}\right)^2 = 4.836 \frac{(\text{volume})^{2/3}}{\text{surface area}} \tag{9.8}$$

The smaller the value of ψ, the more anisometric the particle. Table 9.5 gives examples. Surface roughness further decreases the value of ψ. Often, the

TABLE 9.5 Shape Factor ψ [Eq. (9.8)] for Various Particles, Either Calculated or Determined

Calculated	sphere	1
	cube	0.81
	brick (1 : 2 : 4)	0.69
	tetrahedron	0.67
	cylinder, length/diameter $= 1$	0.87
	same 5	0.70
	same 20	0.47
	postcard, e.g.	0.05
Determined	sand: somewhat rounded grains	0.8
	ground lime: irregular shapes	0.65
	gypsum: flaky crystals	0.5
	mica: very thin flakes	0.2

shape factor varies considerably among individual particles. However, if average and range of ψ do not significantly vary among size classes, a graph of frequency versus a suitable equivalent sphere diameter, supplemented by an average shape factor, may be sufficient to characterize the dispersion.

Another example of irregularly shaped particles is given by *aggregates* of emulsion droplets or of small solid particles in a suspension. The aggregates enclose a variable amount of solvent (Fig. 9.2, G), and it is difficult to classify them according to size, and especially according to mass of primary particles. For a powder of true spheres or other particles having a fixed shape, the specific surface area A, e.g., in $m^2 \cdot kg^{-1}$, is proportional to d^{-1}. For aggregates, however, A may be (almost) independent of d, if d is the aggregate diameter, since the primary particles determine the value of A. In practice, several powders show a proportionality of A with d to a power between -1 and zero. For very porous particles, the exponent can also be close to zero.

9.3.4 Determination

Numerous methods exist for the determination of average particle size or size distribution. They will not be treated here. Nevertheless, it may be useful to discuss briefly some general aspects and examples, to warn the reader against some of the pitfalls that may be encountered. Reliable determination of particle size distribution is notoriously difficult, and all methods employed have limitations and are prone to error. When comparing two different methods on the same sample, one should not be surprised when the results differ by a factor of two, and even larger errors may occur. Limitations exist in the particle size range accessible, in the maximum particle concentration, in the method of preparing the sample and thereby in the probability of producing artifacts, etc.

What Is Measured. The methods employed vary widely in underlying principle. Some methods count and size *individual particles*. Best known are the various types of microscopy applied. Figure 9.5 shows the useful size ranges. In principle, there are few other limitations, but the method of making preparations can introduce errors, especially in electron microscopy. Some other methods sense individual particles in a very dilute dispersion flowing through an opening. Particles can be sensed by scattered light, by the change in electrical conductivity when passing through a narrow hole (as in the Coulter counter), etc. If the relation between particle size and signal intensity is known, a size distribution can be determined.

Other methods directly split the sample, or the particles in a sample, into some *size classes*. The paramount example is sieving, but that is only useful for particles that are (a) large, (b) smooth, (c) fairly isometric, and (d) fairly hard (undeformable). Something similar can be achieved by determining sedimentation rate. By application of gravity and centrifugal fields of various intensity, a wide range of particle sizes is accessible.

A wide variety of methods determines a *macroscopic property*, that is subsequently related to particle size. The prime example is scattering of light or other radiation, either static or dynamic. In static scattering (see also Section 9.2), several different methods can be used, but in all of them the time-averaged intensity of the light scattered by the particles is measured in some way. Static scattering can be applied for a wide range of sizes, from polymer molecules to millimeter particles, but each separate method allows a narrower range. Dynamic scattering measures the magnitude of the Doppler shift in wavelength due to the Brownian motion of the particles, and the shift is thus a measure of the diffusion coefficient, and hence of particle size. It is useful for particles up to a size of 1 μm diameter.

Various methods give rise to various *types of size or average*. In Section 9.3.3 some types of equivalent sphere diameter were mentioned. d_s would result from a method where total surface area is determined, e.g., by an adsorption method. d_f is the result of sedimentation analysis and d_e of sieving. Some particle sensing methods, e.g., change in conductivity, yield d_v. For polydisperse spheres, the average diameter obtained would be d_{32} for a surface area–related method, and d_{53} for sedimentation analysis. Scattering methods can yield a wide variety of averages, up to order 9 (d_{63}), according to the method and the particle size range.

Accuracy. Several kinds of uncertainty can arise. *Systematic errors* can readily occur for indirect methods. The signal measured can depend on other factors besides particle size. The relation between the magnitude of the signal and the particle size may not be known with sufficient accuracy; often a linear relation is assumed, but this is not always true. A fairly general problem is that the method underestimates or even does not notice the smallest particles. This means that the average size is overestimated, especially averages involving S_0, such as d_{10} and d_{30}; an estimate of d_{32} may then be far closer to reality, since the smallest particles contribute fairly little to total surface area and even less to total volume.

Even for direct methods, such problems may exist. Several microscopic methods see in fact cross sections or thin slices of the material. Assuming the particles to be spherical, a number of circles is seen, and the problem is to convert their diameter distribution into that of the original spheres; this is known as the "*tomato salad problem.*" Good solutions exist for spheres, but for anisometric particles the problem is far more difficult, especially if the system as a whole is anisotropic. This is treated in texts on stereology. The particles in a dispersion are often allowed to sediment before viewing them; the supporting plane then is commonly the plane of greatest stability, which means that it is the plane for which the distance to the center of mass of the particle is minimal. Consequently, average d_p is always larger than average d_s for anisometric particles. (Can you explain this?) Furthermore, it may be difficult to distinguish between separate particles being close to each other and aggregates of particles, or even some irregularly shaped particles.

Conversion of the raw data to a size distribution especially poses problems for indirect methods. For instance, in scattering methods a range of data (a spectrum) has to be determined to allow the derivation of anything else than an average size, be it a range of wavelengths or a range of scattering angles. If the particle size distribution is known (together with some other characteristics like particle shape and refractive index), it is relatively easy to calculate a spectrum. But the inverse problem, calculating the distribution from a spectrum, is far more difficult, especially because the amount of information and its accuracy are limited. The algorithms involved always involve rounding off and even shortcuts, and may lead to considerable error.

Finally, the *reproducibility* should be taken into account. The sample taken should be representative, and this is often difficult for powders, which are prone to segregation between large and small particles. In dispersions, segregation due to sedimentation or combined aggregation and sedimentation may occur. Sizing will also give rise to random errors, especially when the particles are anisometric. The greatest uncertainty is generally due to random errors in counting or, more precisely, in establishing the number of particles in a size class. This is largely owing to the Poisson statistics of counting. The standard deviation of the number of particles in a certain volume is, for complete random distribution, equal to the square root of the average. This means that the relative standard deviation of the number of particles in a size class i is equal to or larger than $1/\sqrt{N_i}$, if N_i is the number actually counted (i.e., before any multiplication with a dilution factor, etc.). Counting just one particle thus leads to an uncertainty (standard deviation) of over 100%. This becomes especially manifest for very large particles, as can be derived from Figure 9.10. It is seen that very small numbers of large particles can give rise to a large proportion of particle volume (or mass), which means that the large-particle end of a volume distribution is often subject to large errors. Especially if the size distribution is wide (high value of c_2), tens of thousands of particles may have to be counted to obtain reliable results.

Question 1

Butter and margarine are water-in-oil emulsions. These products are inevitably contaminated with some microorganisms, and especially yeasts can cause spoilage. The organisms can only proliferate in the aqueous phase, and cannot move from one drop to another. If the number of yeasts present is thus very much smaller than the number of drops, the fraction of the aqueous phase that is contaminated may be too small to give perceptible spoilage.

In a certain margarine, the volume fraction of aqueous phase is 0.2. The number of water drops is counted by microscopy and is estimated at 10^9 per ml of product. The aqueous phase of the freshly made product is separated (by melting and centrifuging) and the count of yeasts is determined at 5×10^5 per ml, which means 10^5 per ml product. It is concluded that $10^5/10^9$ or 0.01% of the aqueous phase is contaminated, which would give negligible spoilage. Is this conclusion correct?

Answer

No. The reasoning followed above would imply that the fraction of the aqueous phase that is contaminated is proportional to the number average volume of the drops: $S_3/S_0 = (d_{30})^3$. However, the chance that a droplet is contaminated with a yeast cell is proportional to its volume. Consequently, the drop volume distribution should be weighted with d^3, which then means that the volume fraction contaminated is proportional to S_6/S_3, or to $(d_{63})^3$. Since the droplet size distribution of the products concerned tends to be very wide, the difference will be large. For $S_0 = 10^9 \, \text{cm}^{-3}$ and $\varphi = 0.2$, we derive $d_{30} = 7.3 \, \mu\text{m}$. We have no way of determining d_{63}, but assuming a size distribution similar to the one depicted in Figure 9.10, we see that d_{63} would be about 10 times d_{30}. The volume ratio would then be 10^3, implying that

10^3 times $0.01\% = 10\%$ of the aqueous phase would have been contaminated with yeasts, which is not negligible.

This is only a crude reasoning. Moreover, it is not quite correct, since for such a large contamination, the chances that a large droplet becomes contaminated with more than one yeast cell is not negligible. Nevertheless, it illustrates the importance of using the appropriate type of average, as well as the usefulness of obtaining a reasonably narrow droplet size distribution in butter and margarine.

Question 2

The following raw size classification of the droplets in an o/w emulsion is given:

Size class i (µm)	N_i (in $10^4 \, \mu m^3$)
< 0.5	7
$0.5 - 1$	20
$1 - 1.5$	51
$1.5 - 2$	47
$2 - 2.5$	33
$2.5 - 3$	18
$3 - 4$	18
$4 - 5$	8
> 5	3

Calculate

The number frequency distribution
The volume frequency distribution
d_{10} ($1.97 \, \mu m$)
d_{30} ($2.49 \, \mu m$)
d_{32} ($3.11 \, \mu m$)
The volume fraction of oil (0.17)
The specific surface area ($0.32 \, \mu m^{-1} = 0.32 \, m^2/ml$)

Approximate results are given in brackets. Why approximate?

9.4 RECAPITULATION

Structure. Virtually all foods have a complicated composition, and the great majority of those also have a distinct *physical structure*, adding to the complexity. In other words, most foods are dispersed systems, in which two or more kinds of structural elements can be distinguished. Most of the aspects covered in the following chapters concern dispersed systems, and this chapter defines concepts and introduces some general aspects.

A wide range of highly involved structures occurs, especially in natural foods, though manufactured foods can also have an intricate structure. In the simplest case, we have a dispersion of particles, say emulsion droplets, in a continuous liquid. Another category is that of "soft solids," in which many structural elements are bonded to each other. Interaction forces between structural elements have an often overriding effect on the properties of dispersed systems, and these forces could even be considered as being part of the structure. Geometrical structure can in principle be seen through a microscope, but not forces, of course.

The existence of a physical structure has several important consequences. It determines a number of *physical properties,* notably mechanical ones: viscosity, elasticity, consistency, fracture properties. Because of the physical inhomogeneity, some kinds of *physical instability* can occur, often manifest as a kind of segregation, and these are generally undesired. Chemical components are to some extent compartmentalized, which may affect their mutual reactivity, hence the *chemical stability* of a food. The compartmentalization also tends to affect *flavor.* The result of some process operations can greatly depend on the physical structure of the system, and several processes are intended to produce or to alter structure.

It is important whether the various structural elements can be said to constitute *phases* or not. If not, we generally have a so-called *lyophilic* system, which is in principle in equilibrium. In most foods, structural elements do constitute phases, which implies that they have phase boundaries in which free energy is accumulated. This means an excess of free energy, hence a *lyophobic* system; it costs energy to make it, and it is inherently unstable. The properties of such foods thus depend on the manufacturing and storage history and, for natural foods, on growth conditions.

Scale. Besides knowledge of the composition and the material properties of the structural elements, their *size* is important. The length scales involved span a wide range, from molecular to visible, i.e., about six orders of magnitude. Especially natural foods may show a hierarchy of scales, all of which would need study for a full understanding of properties. Several static properties, such as visual appearance and consistency, greatly depend on the size of the structural elements, and so may the rates of change occurring in foods. In general, length scale and time scale—the time needed for a change in structure or composition to occur—are correlated. Dispersions with small particles are in most instances far more stable than those with large ones. External forces applied to a system have stronger effects on large particles than on small ones, which has direct consequences for their separability.

Size Distributions. Since size is important, so is its *distribution*. Size distributions can be presented in various ways, e.g., cumulative or as a frequency (which is the derivative with respect to size of the cumulative distribution); as a number or as a volume distribution; versus diameter or (molar) mass, etc. Various *types of averages* can be defined and calculated, and their values can differ by more than an order of magnitude if the size distribution is relatively wide. It depends on the problem involved what type of average should be taken. Distribution width can be defined as standard deviation over average; for most food dispersions, it ranges between 0.2 and 1.2. If the width is considerable, the shape of the distribution may vary significantly. For anisometric particles, characterization of size poses additional problems.

Determination of size distributions can be done by several methods, each having its limitations and pitfalls. Accurate determination is notoriously difficult. Systematic and random errors are involved. Several methods are indirect ones, determining some macroscopic property, for instance, a light scattering spectrum. The conversion of these data to a size distribution is generally difficult and may lead to considerable error.

BIBLIOGRAPHY

Several aspects briefly touched on in this chapter are far more elaborately discussed in texts on colloid science. A somewhat outdated but still very interesting book is

K. J. Mysels. Introduction to Colloid Science. Interscience, New York, 1965.

Far more up-to-date is

R. J. Hunter. Foundations of Colloid Science, Vol. 1. Clarendon, Oxford, 1987.

of which Chapter 3 gives an extensive discussion on particle size and shape, including methods of determination.

Information on the physical structure of several foods and on some of its implications is in

J. M. Aguilera and F. W. Stanley. Microstructural Principles of Food Processing and Engineering, 2nd ed. Aspen, Gaithersburg, MD, 1999.

Diffuse reflection of foods is discussed by

F. M. Clydesdale. Color measurement. In: D. W. Gruenwedel and J. R. Whitaker, eds. Food Analysis, Vol. 1, Physical Characterization. Marcel Dekker, New York, 1984, Chapter 3.

Several chapters on microscopic and other methods of determining the physical structure of foods are in

E. Dickinson, ed. New Physico-Chemical Techniques for the Characterization of Complex Food Systems. Chapman Hall, London, 1995.

Several texts and reference books exist about microscopic techniques, including manuals by manufacturers of microscopes. A thorough discussion of the quantitative relations between what is observed in a two-dimensional cut or slice and the real three-dimensional structure is by

E. R. Weibel. Stereological Methods, Vols. 1 and 2. Academic Press, London, 1979, 1980.

Especially Chapter 2 in Volume I gives a clear and useful introduction.

A fairly simple but very useful and clear introduction on size distributions is by

J. D. Stockham and E. G. Fochtman. Particle Size Analysis. Ann Arbor Science, 1977.

Especially Chapters 1, 2, and 11 are recommended. More elaborate and giving much about methods is

T. Allen. Particle Size Measurements, 5th ed. Two volumes. Chapman Hall, London, 1997.

It is especially meant for analysis of powders, but Volume 1 gives much general and useful information.

10

Surface Phenomena

As we have seen in the previous chapter, most foods are dispersed systems, and many of the structural elements constitute separate phases. This means that there are phase boundaries or interfaces, and the presence of such interfaces has several important consequences. Substances can adsorb onto the interfaces, and if the interfacial area is large, as is often the case, the adsorbed amounts can be considerable. The adsorption can strongly affect colloidal interaction forces between structural elements, i.e., forces acting perpendicular to the interface; this is discussed in Chapter 12. Other forces act in the direction of the interface, and these are treated in the present chapter. Altogether, surface phenomena are of considerable importance during processing and for the physical properties, including stability, of most foods.

Strictly speaking, the word surface is reserved for an interface between a condensed phase (solid or liquid) and a gas phase (mostly air). In practice, the words surface and interface are often used indiscriminately.

10.1 SURFACE TENSION

It is common experience that fluid systems consisting of two phases try to minimize their interfacial area. For instance, if the one phase is present as a blob, it tends to assume a spherical shape, which is the smallest surface area possible for a given volume. An obvious example is a rain drop in air. If two of such drops collide, they generally coalesce into one bigger drop, thereby lowering total surface area. This is also commonly observed for fairly large oil drops in (pure) water. Since any system tries to minimize its free energy, it follows that at an interface between two phases free energy is accumulated. This is called *surface or interfacial free energy*. For a homogeneous interface, it is logical to assume that the amount of surface free energy is proportional to surface or interfacial area. Consequently, the surface (or interface) is characterized by its *specific surface free energy*. It can be expressed in units of energy per unit area, i.e., $J \cdot m^{-2}$ in the SI system.

In Figure 10.1a a metal frame is depicted in which a piece of string is fastened. By dipping the frame in a soap solution, a film can be formed between frame and string. (Such a film cannot be made of pure water, as will be explained in Section 10.7.) Figure 10.1b illustrates that this film too tries to minimize its area. By pulling on the film, as depicted in c, its area can be enlarged. The film thus exerts a tension on its boundaries, and *this tension acts in the direction of the film surfaces.* It is called the *surface tension*, and it is expressed in units of force per unit length, i.e., in $N \cdot m^{-1}$ in SI units. Notice that it concerns a two-dimensional tension; in three dimensions, tension (or pressure) is expressed in newtons per square meter. Since $1 \text{ J} = 1 \text{ N} \cdot \text{m}$, the surface tension has the same dimension as the specific surface free energy. In fact, these two parameters have identical values (provided

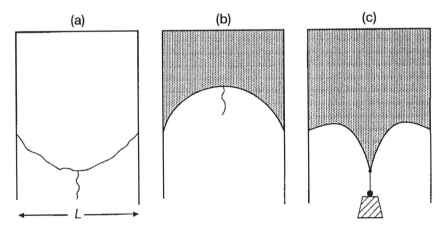

FIGURE 10.1 Illustration of surface tension. See text. The presence of a soap film is depicted by gray. The weight in c is of order 1 gram.

that it concerns a pure liquid) and are different manifestations of the same phenomenon: work has to be done to enlarge the surface area. For surface or interfacial tension, the symbol γ is used. In Figure 10.1, the force by which the film pulls at the horizontal bar of the frame is γ times L times two, since the film has two surfaces.

The phenomenon just mentioned allows *measurement* of the surface tension, as illustrated in Figure 10.2a. One measures the net downward force

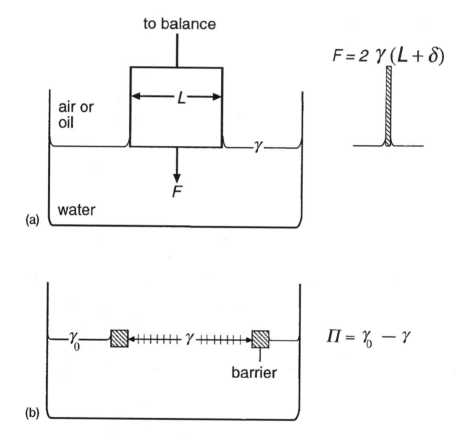

FIGURE 10.2 Surface tension and surface pressure. (a) Measurement of surface or interfacial tension by means of a Wilhelmy plate (width L, thickness δ). The plate is attached to a sensitive balance. (b) Illustration of surface pressure (Π) caused by surfactant molecules (depicted by vertical dashes). Between the barriers the surface tension is lowered and a net two-dimensional pressure acts on the barriers.

by which the liquid pulls at the plate and divides it by the perimeter of the plate in a plane parallel to the liquid surface. A prerequisite is that the plate is fully wetted by the (lower) liquid, since otherwise the interfacial force on the plate does not act in the downward direction. To obtain the net force, the weight of the plate has to be taken into account. The apparatus of Figure 10.2a can be used to measure surface tension, i.e., the tension acting at a surface between a liquid and air (or another gas), or interfacial tension, i.e., the tension between two liquids. In the latter case, one has to correct for the buoyancy of the plate caused by the upper liquid.

Over what *range* can surface tension forces be sensed? Figure 10.3 shows a water drop hanging on a horizontal solid surface. For the sake of simplicity, it is assumed that the drop is a half sphere and that its surface thus meets the solid at a right angle. If such a drop is very large, it would fall off, but a small drop can be kept suspended by surface tension. How large can the radius of the drop be before it would fall off? At a level just below the solid, the surface tension γ pulls the drop upwards. The total upward force then equals circumference $(2\pi R)$ times γ. The downwards force is given by the volume of the half sphere $((2/3)\pi R^3)$ times mass density $(\rho \approx 10^3$ $kg \cdot m^{-3})$ times gravitational acceleration $(g \approx 9.8 \, m \cdot s^{-2})$. According to Table 10.1, water has $\gamma = 0.072 \, N \cdot m^{-1}$, and we calculate for the maximum possible radius $R = 0.0047 \, m$. This then means that surface forces can affect the shape of a system over several millimeters.

For a *molecular explanation* of surface tension, we refer to Chapter 3, where it is stated that attractive forces act between all molecules, i.e., the van der Waals forces; for some molecules, also other attractive forces act. Consider a horizontal surface between a liquid, say oil, and air. Oil molecules present in the surface sense the attractive forces due to the oil molecules below, and hardly any attractive forces due to the air molecules above, because there are so few air molecules per unit volume. This does not mean that the molecules in the surface are subject to a net downward force, since that would imply that these molecules immediately move downwards;

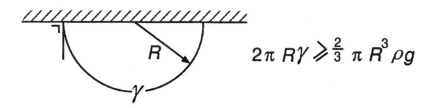

$$2\pi R\gamma \gtrless \frac{2}{3}\pi R^3 \rho g$$

FIGURE 10.3 Drop of surface tension γ hanging on a horizontal solid surface.

TABLE 10.1A Some Values of Surface and Interfacial Tensions—Liquids

Material	γ against air[a]	γ against water
Water 0°C	76	·
25°C	72	·
60°C	66	·
100°C	59	·
0.02 molar Na-dodecylate	43	0
Protein solutions	≈ 50	0
Saturated NaCl solution	82	0
Diethyl ether	20	0
Ethanol	22	0
Benzene	29	35
Paraffin oil	30	50
Triacylglycerol oil	35	30
Mercury	486	415

[a]Saturated with vapour of the material

TABLE 10.1B Some Values of Surface and Interfacial Tensions Between Solids and Liquids (Rough Estimates)

Solid	Liquid	γ
Ice, 0°C	Water	25
Sucrose	Saturated sucrose solution	5
Triacylglycerol crystal	Triacylglycerol oil	4
Triacylglycerol crystal	Water	31

Approximate data in $mN \cdot m^{-1}$ at room temperature, unless mentioned otherwise.

on average, the net force equals zero. This is because of the hard core repulsion between molecules. As illustrated in Figure 3.1, molecules attain on average a mutual distance where the net interaction energy is at minimum, which implies that the net force equals zero. Nevertheless, it costs energy to move oil molecules from the oil towards the surface, and this is what occurs when the surface area is enlarged.

Table 10.1 gives examples of the surface tension of liquids and of interfacial tensions. The values are fairly small for the organic liquids shown; for these, the attractive energy is predominantly due to van der Waals forces. For water, γ is higher, owing to the extensive hydrogen bonding between water molecules (Section 3.2). For mercury, γ is very high;

here the very strong metallic bonds are responsible. The interfacial tension γ_{12} between two condensed phases 1 and 2 is generally smaller than the sum of the surface tensions of both ($\gamma_1 + \gamma_2$); see Table 10.1. This is because the value of the surface tension is relatively high owing to the near absence of molecules in air, whereas at an interface between condensed phases, molecules at the interface are subject to attractive interactions with molecules of the other phase. For pure liquids, surface and interfacial tension always decrease with increasing temperature; an example is in Table 10.1.

Equilibrium. From a *thermodynamic* point of view, interfacial tension is an equilibrium parameter. When enlarging an interface at a high velocity, equilibrium distribution and orientation of the molecules in the interface cannot be directly attained, and in order to measure γ, the rate of change in interfacial area should be slow and reversible. Nevertheless, when enlarging a liquid surface at conditions that do not allow the establishment of equilibrium, a force can be measured, hence a surface or interfacial tension can be derived, which differs from the equilibrium value. It may be a transient value, but it is also possible that a constant surface tension is measured; it then concerns a steady state. In other words, from a *mechanical* point of view, interfacial tension need not be an equilibrium value.

Also a *solid* has a surface tension, but when creating a new solid surface, it may take a very long time before the molecules near the surface have attained an equilibrium distribution and orientation. The same holds true for the solid–liquid interface. Moreover, since a solid surface cannot be enlarged (without grinding etc.), γ cannot be measured. All the same, there is a surface free energy and it becomes manifest in other phenomena, partly to be discussed later. This allows making rough estimates of γ, and such values are given in Table 10.1B.

In this chapter, we will primarily consider fluid interfaces.

10.2 ADSORPTION

The presence of solutes in a liquid may affect the surface tension. Examples are in Table 10.1, and the matter is further illustrated in Figure 10.4. It is seen that the solute may cause γ to increase or to decrease. An example of a special case is the dependence of γ on concentration for Na-dodecyl sulfate: a very small concentration suffices to cause a large decrease in γ, whereas a further increase in concentration has very little effect. This is due to the solute preferentially accumulating at the surface, and after a fully packed monolayer has been obtained (roughly speaking), no further accumulation takes place. Solutes showing this kind of behavior are called *surfactants*.

The accumulation of a compound at a surface or an interface is called *adsorption*. Adsorption is a very common phenomenon and can occur at all solids or liquids in contact with a gaseous or a liquid phase. The compound adsorbing is called the *adsorbate*, the material onto which it adsorbs is the *adsorbent*. The following notation will be used, either as such or in subscripts: S = solid, A = air (or a gaseous phase), W = water (or an aqueous solution), and O = oil.

The adsorbent can be a solid or a liquid. The adsorbate is dissolved in a liquid or is (present in) a gas. Adsorption on an A–S interface concerns, for instance, adsorption of water from moist air on a solid; see Section 8.2. Also other volatiles can adsorb from air, e.g., flavor compounds. At the liquid–solid interface, the solid generally is in contact with solvent as well as adsorbate, i.e., solute molecules. For a liquid–liquid interface (generally O–W) the adsorbate may be soluble in both liquids. The molecules adsorbed

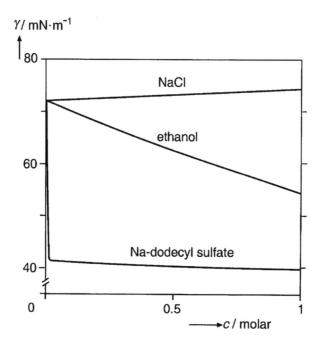

FIGURE 10.4 Effect of the concentration c of various solutes on the surface tension γ of water.

often stick out partly in the one, partly in the other phase. This is not possible at solid interfaces.

At a fluid interface (gas–liquid or liquid–liquid) interfacial tension can be measured, and adsorption leads to lowering of γ (Figure 10.4). The extent by which γ is decreased is called the *surface pressure*, defined as

$$\Pi \equiv \gamma_0 - \gamma \qquad\qquad (10.1)^*$$

Π is, like γ, expressed in $N \cdot m^{-1}$. If the surface containing adsorbate is confined between barriers, as is illustrated in Figure 10.2b, the surface pressure appears to become manifest as a force acting on the barrier, which equals Π times the length of the barrier. Nevertheless, the force is not due to an autonomous pressure but to a difference in the tensile (two-dimensional) stresses acting on either side of the barrier. A surface tension always "pulls," never "pushes."

10.2.1 The Gibbs Adsorption Equation

Adsorption occurs because it lowers the free energy of the system. According to Gibbs, the chemical potential of the adsorbate is at equilibrium equal in the solution and at the surface. He further postulated an infinitely thin *dividing plane* between the two phases and then derived the equation

$$d\Pi = -d\gamma = RT\Gamma\,d\ln a \qquad\qquad (10.2)^*$$

where R and T have their normal meanings. Γ is the *surface excess concentration* (in moles per square meter) of the adsorbate, usually abbreviated to surface excess. a is the *thermodynamic activity of the adsorbate* in the solution. Note that it does not matter in what units a is expressed, since $d \ln a = (1/a)\,da$. The equation is valid, and exact, for one solute at equilibrium. It is especially useful if γ can be measured, i.e., for fluid interfaces.

The *surface excess* can be defined in various ways. Actually, there is no true dividing plane, but rather an A–W interface that is not sharp, since molecules have a finite size and moreover exhibit Brownian motion. Hence the "interface" extends over a layer of some molecular diameters. In the derivation of Eq. (10.2), the position of the dividing plane has been chosen so that the surface excess of the solvent is zero. In Figure 10.5 the concentration of the solute is depicted as a function of the distance from the dividing plane (z). In Figure 10.5a, there is no adsorption: the two hatched areas on either side of the dividing plane are equal. (Because of the definition

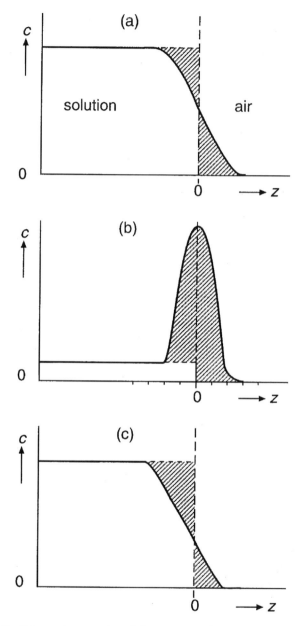

FIGURE 10.5 Schematic examples of the concentration c of a solute in water as a function of the distance z in a direction perpendicular to the air–water surface. The vertical broken line gives the position of the dividing plane. In (b) a scale in nm is given as an example.

of the dividing plane, the same curve holds in this case for the concentration of the water, but at a different vertical scale.) In 10.5b, there *is* adsorption, and the adsorbed amount per unit area, i.e., the surface excess, is represented by the hatched area under the curve. Notice that in case a some adsorbate is present at the interface, but there is no excess, as there is in case b. In Figure 10.5c, *negative adsorption* is depicted: the solute stays away from the interface. According to Eq. (10.2), the surface tension must then increase with increasing concentration, for instance as in Figure 10.4, curve for NaCl. In case a, γ is not altered, and in case b γ is decreased.

Surfactant Concentration. The parameter a needs some elaboration. In a very dilute system, a may equal the concentration of the adsorbate (if expressed in the same units), but that is not always true, as discussed in Section 2.2. Even if it is true, it concerns the concentration *in the solution*, not in the total system. This means that the concentration adsorbed, which equals Γ times the specific surface area of the adsorbent, has to be subtracted from the total concentration.

In Figure 10.6a, examples are given of the relation between γ and log concentration for some surfactants. Assuming for the moment that the activity of the solute equals its concentration, the slopes of these curves would be proportional to the surface excess Γ. The steeper the slope, the higher Γ, implying that Γ increases with increasing surfactant concentration. This is illustrated by the corresponding *adsorption isotherms* given in Figure 10.6b. It is seen that for a considerable range in c, the slope, and thereby Γ, is almost constant; this is further discussed in Section 10.2.3. It is also seen that the curves in Figure 10.6a show a *sharp break* and that for c beyond the break, γ is virtually constant. The latter seems to imply that Γ then is virtually zero. This is clearly impossible: it would mean that with increasing concentration of surfactant its adsorption sharply drops. The explanation is that above the break the thermodynamic activity of the surfactant a remains virtually constant. The break roughly occurs at the *critical micellization concentration* (CMC), above which next to all additional surfactant molecules go into micelles; for some systems, the curve stops at the solubility limit of the surfactant, for instance for Na-stearate (C18) in Figure 10.6. Micellization and the CMC are further discussed in Section 10.3.1.

Figure 10.6 relates to Na soaps of varying chain length, and it is seen that the longer the chain, the lower the CMC. This means that soaps with a longer chain length are—other things being equal—more surface active: less is needed to result in a certain Γ. The concept of *surface activity* is often loosely used as referring to the lowering of γ: the component giving the lowest γ then would be the most surface active one. This is, however, an ambiguous criterion, because different surfactants give different values for the lowest γ attainable. Consequently, it is better to relate surface activity to the concentration at which Π has half its maximum value. Two of such concentrations are indicated in the figure, and it follows that their

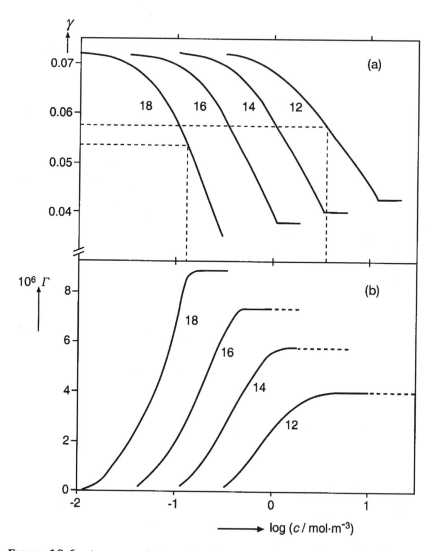

FIGURE 10.6 Aqueous solutions of sodium soaps of normal fatty acids of various chain length (number of C-atoms indicated). (a) surface tension γ ($N \cdot m^{-1}$) against concentration c in solution. (b) Calculated surface excess Γ ($mol \cdot m^{-2}$) against concentration.

magnitudes decrease by a factor 28 when going from 12 to 18 C atoms. (As a parameter to characterize the surface activity, the surfactant concentration at which Γ reaches a given proportion [say, 1/2] of its maximum value can also be taken.)

It may finally be noted that the surface free energy is no longer equal to the surface tension if the latter is altered by the presence of a surfactant. A full explanation is intricate, but it has to do with the decrease in mixing entropy (translational and conformational) of the solution upon depletion of surfactant due to its adsorption. For surfactants like globular proteins, the difference is generally small, since the high molar mass of the protein limits the change in translational mixing entropy, and the change in conformational entropy of the solution will also be small.

Question

Consider a saturated aqueous solution of Na-stearate (C18). 10% pure oil is added to it and the mixture is emulsified so that droplets of average size $d_{32} = 1\,\mu m$ are assumed to be formed. What proportion of the surfactant would then become adsorbed onto the oil droplets? Make the same calculation for Na-myristate (C14) at a concentration that equals its CMC. Assume Figure 10.6 to be valid.

Answer

From the figure, the solubility of Na-stearate is slightly below $10^{-0.5}$ or about $0.3\,\text{mol}\cdot\text{m}^{-3}$. The specific oil surface area would equal $6/d_{32} = 6\cdot 10^{6}\,\text{m}^2$ per m^3 of oil. For $1\,\text{m}^3$ of water, the area A then is $6\cdot 10^{5}\,\text{m}^2$. At the solubility limit, $\Gamma \approx 9\,\mu\text{mol}\cdot\text{m}^{-2}$, and $A\cdot\Gamma$ would amount to about $5.4\,\text{mol}\cdot\text{m}^{-3}$. However, this corresponds to 18 times the amount of Na-stearate present, which is clearly impossible. It is even unlikely that an emulsion can be made.

Doing the same calculation for Na-myristate, we would obtain $3.20\,\text{mol}\cdot\text{m}^{-3}$ for the CMC and $3.48\,\text{mol}\cdot\text{m}^{-3}$ for $A\cdot\Gamma$, which is above the CMC, though not greatly so. What now will happen is that Γ will be smaller than its plateau value, so that the concentration left in solution (CMC $- A\cdot\Gamma$) will produce precisely that Γ value according to Figure 10.6b. This value is about $4.4\,\mu\text{mol}\cdot\text{m}^{-2}$. (Check this calculation.)

10.2.2 Adsorption Isotherms

Adsorption isotherms provide important knowledge, for instance, if the amount of surfactant needed to make an emulsion has to be established, as illustrated in the Question just discussed. Information about Γ is also needed when the stability of a dispersion has to be studied or explained, since several forms of instability greatly depend on the Γ value of the adsorbed surfactant.

For adsorption at a **solid surface**, it is often envisaged that the surface contains a finite number of identical specific adsorption sites, i.e., atomic groups that can each bind an adsorbate molecule. Further assuming (a) the binding to be reversible (which generally implies that it is not due to formation of a covalent bond) and (b) that the occupation of binding sites does not affect the affinity of the surfactant for neighboring sites, the following simple adsorption isotherm was derived by Langmuir:

$$\Theta = \frac{\Gamma}{\Gamma_\infty} = \frac{c/c_{0.5}}{1 + c/c_{0.5}} = \frac{c'}{1 + c'} \tag{10.3}$$

where Θ is the proportion of adsorption sites occupied and $c_{0.5}$ is the concentration in solution at which $\Theta = 0.5$; $1/c_{0.5}$ is thus a measure of the surface activity of the adsorbate. The normalized Langmuir isotherm is shown in Figure 10.7. At very small c, the adsorbed amount is proportional to c; the concentration in solution has to be relatively large to achieve a high saturation of the adsorption sites, e.g., $c' \approx 100$ for $\Theta = 0.99$. However, the assumptions on which the Langmuir equation is based are not nearly always met, especially not at high Θ.

For **fluid surfaces** and one (pure) surfactant, we can apply the Gibbs equation to construct an adsorption isotherm, provided that the relation between γ and surfactant activity is precisely known. In Figure 10.6b such results are shown. Strikingly, a constant surface excess is reached at concentrations markedly below the CMC. This is due to the (virtually) constant slope of γ versus log c in this region. Strictly speaking, Γ cannot be constant for increasing surfactant activity. However, surfactant concentration rather than activity has been used in making Figure 10.6b, and it is very likely that the activity coefficient of the surfactant will decrease with increasing c near the CMC (see, e.g., Figure 2.7c). In other words, $-d\gamma/d \ln a$ would actually still increase, leading to a (slightly) increasing Γ value.

The Langmuir equation may be applicable for adsorption on fluid surfaces under "ideal" conditions. The parameter Θ should then be interpreted as being equal to Γ/Γ_∞, where Γ_∞ is the maximum surface excess attainable. Comparison with an actual adsorption isotherm (taken from Figure 10.6) in Figure 10.7 shows marked differences. These must be due to deviations from ideality. The most important causes of *nonideality* are (a) a difference in molecular size between surfactant and solvent, and (b) net interaction energy between surfactant molecules when in the adsorption layer, which can vary greatly among surfactants.

The adsorption isotherms for surfactants at the A–W and the O–W interface need not be the same. This is primarily because the net interaction energy mentioned will often be different for these interfaces. It may even make a difference whether the oil is a paraffin or a triacylglycerol mixture. Another condition affecting the isotherm is temperature. By and large, the surface excess will be smaller for a higher temperature at the same activity of the surfactant; the difference can be considerable (cf. Figure 8.5b).

The **surface activity** of a solute, i.e., the reciprocal of $c_{0.5}$, is higher if the free energy needed to remove a surfactant molecule from the interface is larger. For an aqueous solution, this is the case if the hydrophobic part of the molecule (e.g., an aliphatic chain) is larger relative to the hydrophilic part (e.g., a carboxyl group). This explains the trend shown by the surfactants in Figure 10.6; it may also be noted that in such cases (homologous surfactants) the surface activity is closely related to the solubility. For the same ratio hydrophobic/hydrophilic, the free energy for

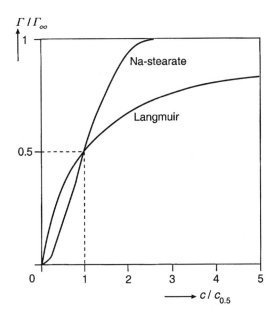

FIGURE 10.7 Adsorption isotherms. Surface excess Γ over Γ_∞ (i.e., Γ at surface saturation) versus concentration in solution c over $c_{0.5}$ (i.e., c for $\Gamma/\Gamma_\infty = 0.5$). Langmuir isotherm and calculated isotherm for Na-stearate at the A–W surface.

removal will be larger for a larger molecule. This is the main reason why polymeric substances can be so very surface active (see Section 10.3.2).

Up till here, we have implicitly assumed that adsorption is restricted to a monolayer. Actually, *multilayer adsorption* can occur, especially if the concentration of surfactant is approaching its solubility. Adsorption of a second layer tends to go along with very little increase in surface pressure; in other words, the interaction forces leading to this adsorption often are weak.

Finally, a few words about **negative adsorption**. Figure 10.4 shows that the increase in γ for NaCl is proportional to c. This is quite generally observed for negative adsorption. Hence $d\gamma/dc$ is constant. Since $(1/c)\,dc = d \ln c$, this means that $d\gamma/d \ln c$ is proportional to c. According to Eq. (10.2), $d\gamma/d \ln c$ is proportional to $-\Gamma$. Hence a very simple "adsorption isotherm" results, $-\Gamma$ being proportional to c. This can be interpreted as the presence of a solvent layer of constant thickness adjacent to the interface that is devoid of solute. The thickness of the layer follows from the slope of γ versus c, and it turns out to be of the order of 1 nm, i.e., of molecular dimension. Negative adsorption can thus be interpreted as being due to steric exclusion of the solute by the interface; see also Figure 8.7.

10.2.3   Surface Equations of State

This concerns the relation between surface pressure Π (or surface tension γ) and surface excess Γ. In the simplest case, the relation is given by the equation

$$\Pi = \frac{n}{A} RT = \Gamma RT \tag{10.4}$$

where n is the number of moles and A the interfacial area. It may be no surprise that Eq. (10.4) is only valid for very small values of Γ. At higher Γ, repulsive and attractive forces between surfactant molecules will for a considerable part determine Π. It may be noted that Eq. (10.4) is very similar to Eq. (2.17) for the osmotic pressure of an ideally dilute solution, which can be written as $\Pi_{osm} = (n/V)RT$. The surface pressure may indeed be considered as a two-dimensional analogue of the osmotic pressure.

Examples of observed surface equations of state are in Figure 10.8.

Note   Actually, they are merely relations of state, since no equation is given. However, the word equation is commonly used.

It is seen that the deviation from ideality varies widely. Three aspects of nonideality will be discussed.

1. If the *surface fraction* covered θ is appreciable, large deviations from Eq. (10.4) occur. The discrepancy is especially large for large molecules, because the value of Γ (to be expressed in moles per m^2) remains very small. In first approximation, the correction term would be $(1 - \theta)^{-2}$, leading to

$$\Pi = \frac{\Gamma RT}{(1 - \theta)^2}$$

$$\theta = \pi r^2 N_{AV} \Gamma$$

(10.5)

where r is the radius of a surfactant molecule in the plane of the interface. This equation is especially useful for globular proteins as surfactants. As an example the relation for lysozyme, a hard globular protein of 14.6 kDa, is given in Figure 10.8, and it appears that the correction for finite surface fraction is overriding. (The value calculated for r equals about 1.6 nm, in good agreement with the radius of the molecule in solution, about 1.7 nm.)

2. There often is a *net interaction energy* between the surfactant molecules in the interface, and if the interaction is attractive, Π will be smaller (at the same Γ); if it is repulsive, Π will be higher. The former appears to be the case for Na-stearate (Figure 10.8). For most small-molecule surfactants, the attractive interaction appears to be smaller at the

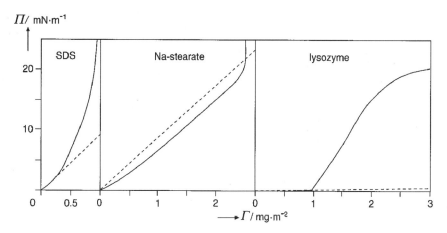

FIGURE 10.8 Surface equations of state, i.e., the relation between surface pressure (Π) and surface excess (Γ) for Na-dodecyl sulfate (SDS), Na-stearate (C18), and lysozyme at the A–W interface. Also the relations according to Eq. (10.4) are given (broken lines).

O–W than at the A–W interface, the reason that Π tends to be higher at the O–W interface. See Figure 10.14 for proteins.

3. At high Π, the relation becomes intricate. Many small-molecule surfactants tend to give near vertical curves, for Na-stearate up till $\Pi = 37 \, \text{mN} \cdot \text{m}^{-1}$ at an A–W interface.

When increasing the **temperature**, Eq. (10.4) would predict that Π increases with T. However, a temperature increase will generally cause a decrease in Γ, as mentioned in Section 10.2.2, hence a decrease of Π. Moreover, the interaction energy between molecules in the monolayer may change. The net result cannot be easily predicted. Nevertheless, γ tends to decrease with increasing temperature, because γ_0 will decrease.

10.3 SURFACTANTS

Surfactants can fulfil a wide variety of functions. Some of the most important can be categorized as follows.

1. They are essential in the formation of foams and emulsions (Chapter 11). Some surfactants can also be used to destabilize emulsions or foams.

2. Because they adsorb onto particles, they can alter the interaction forces between these (Chapter 12).

3. By adsorption onto interfaces, they can greatly alter wetting properties (Section 10.6).

4. Small-molecule surfactants can give rise to a series of association colloids, including micelles, mesomorphic structures, vesicles, and microemulsions.

5. They can promote "dissolution" of substances, either by uptake in micelles of the surfactant (Section 10.3.1) or by adsorption onto small particles.

6. Some small-molecule surfactants interact with proteins in solution and cause partial unfolding of the latter.

A very important function is detergency, which is generally due to a combination of functions 3 and 5.

The suitability of a surfactant for the various functions is greatly dependent on its chemical structure. A crucial difference is that between the following two types.

> Fairly small amphiphilic molecules, sometimes called soaps (although strictly speaking a soap is a salt of a fatty acid) or emulsifiers (although they are used for a number of other purposes as well). In this chapter we will use the word "amphiphile" for short. The molar mass is generally < 1 kDa. They are discussed in Section 10.3.1.

Only a limited number are used in foods. These include polar lipids, especially monoacylglycerides and phospholipids.

Macromolecules, generally linear polymers of molar mass > 10 kDa. The chemical composition varies widely. They are discussed in Section 10.3.2. Proteins make up by far the most important category used in foods.

These two types differ widely in most properties. Moreover, they can interact with each other in various ways; this is discussed in Section 10.3.3.

10.3.1 Small-Molecule Surfactants

These surfactants generally consist of an aliphatic chain ("tail") to which a polar "head" group is attached. The aliphatic part would readily dissolve in oil (if separated from the head group); the head group would readily dissolve in water. This is why these substances preferentially go to an O/W interface: the total free energy then is smallest. They also adsorb onto an A/W interface, because then the aliphatic chain is not surrounded by water molecules, which also causes a decrease in free energy, though less than for adsorption on an O/W interface. According to the nature of the head group, the surfactants are classified as *nonionic* (neutral), *anionic* (negatively charged head group in water, unless the pH is quite low) and *cationic* (positively charged). Examples are given in Table 10.2. Phospholipids are special: they have two aliphatic chains, and the head group of most types is *zwitterionic*, which means that it contains a positive as well as a negative charge.

An important characteristic of small-molecule water-soluble surfactants is their tendency to form **association colloids** in water. Some examples are given in Figure 10.9. Phospholipids readily form *bilayers*, the basic structure of all cell membranes. A *vesicle* is a closed bilayer. The primary driving force for association is the hydrophobic effect (see Section 3.2). Close packing of the hydrophobic tails greatly diminishes their contact with water. This lowers free energy, despite the resultant decrease in mixing entropy. The presence of the polar heads counteracts the association, since they repel each other by electrostatic repulsion or by hydration. Consequently, high surfactant concentrations are needed for most types of association colloids to form, except for micelles and some vesicles. A bewildering variety of *mesomorphic phases* can be formed, according to chemical structure of the surfactant, temperature, water content, and other variables. One fairly common type, a *lamellar phase*, is depicted in Figure 10.9b; it is typically formed at a water content of, say, 50%.

TABLE 10.2 Some Small-Molecule Surfactants and Their Hydrophile–Lipophile Balance (HLB) Values

Type	Example of surfactant	HLB value
Nonionics		
Aliphatic alcohols	Hexadecanol	1
Monoacylglycerols	Glycerol mono stearate	3.8
Monoacylglycerol esters	Lactoyl monopalmitate	8
Spans	Sorbitan mono stearate	4.7
	Sorbitan mono oleate	7
	Sorbitan mono laurate	8.6
Tweens	Poly(oxyethylene) sorbitan mono oleate	16
Anionics		
Soaps	Na oleate	18
Lactic acid esters	Na stearoyl-2-lactoyl lactate	21
Teepol[a]	Na dodecyl sulfate	40
Cationics[a]	Palmityl trimethyl ammonium bromide	large
Phospholipids (zwitterionic)	Lecithin	∼9

[a]Not used in foods but as detergents.

A small part of a phase diagram is given in Figure 10.10a. An important characteristic is the temperature at which the apparent solubility of the surfactant markedly increases, called the *Kraft point*, T^*. Above that temperature micelle formation can occur. Below T^*, the molecules can form α-*crystals*, which have a very small solubility in water. The Kraft temperature increases with increasing chain length and can be as high as 60°C for some surfactants. Mesomorphic phases only form above the Kraft temperature. Figure 10.9b illustrates that at lower temperature a so-called α-*gel* can be formed. An α-gel is not in thermodynamic equilibrium; it tends to change into a dispersion of crystals in water.

The latter change can be quite slow if the surfactant is not pure, and virtually all technical preparations constitute a mixture. Chain length, chain saturation, and type of head group can vary. This is of considerable practical importance. For example, when a fairly pure preparation of saturated monoglycerides is added to a liquid food at, say, 60°C, it will readily disperse, forming micelles. On cooling to below the Kraft point, however, crystals will form, and very little monoglyceride is left in solution. If then oil is added and the mixture is agitated to form an emulsion, the monoglyceride will barely or not at all reach the O–W interface and is thus inactive. If a less pure preparation is used, crystal formation can be delayed

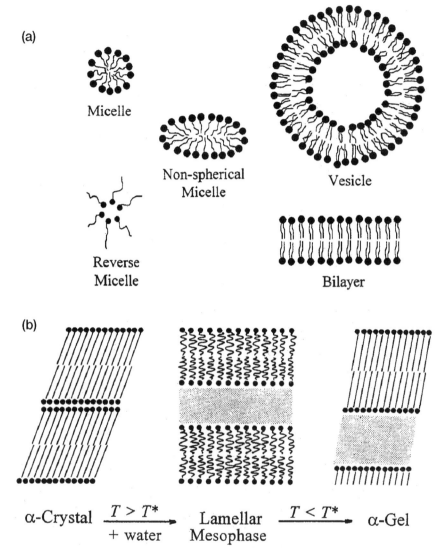

FIGURE 10.9 Highly schematic examples of some association colloids. (a) Micelles and bilayers. (From E. Dickinson, D. J. McClements. Advances in Food Colloids. Blackie, 1995.) (b) Crystal, lamellar, and gel structures of simple surfactant water mixtures; T is temperature, T^* Kraft temperature. (Modified from a figure by N. J. Krog.)

for quite long times, and the monoglycerides do adsorb at interfaces. Once adsorbed, they remain so, even at low temperature.

The region in the phase diagram denoted "crystals + water" often contains other structures as well. For phospholipids, which are generally mixtures and are very poorly soluble in water, vesicles and "*liposomes*", i.e., fragments of liquid crystalline phases (or possibly of α-gel), have been observed.

Chain Crystallization. At the O–W interface it is often observed— especially for surfactants containing a saturated aliphatic chain—that lowering of the temperature leads to a marked decrease in interfacial tension. A fairly sharp critical temperature can be noted, which is called the chain crystallization temperature T_c. Above T_c the chains of the adsorbed surfactant are presumed to be flexible, i.e., as in a liquid. Below T_c the chains would be rigid, as in a crystal. Chain crystallization can only occur if the Γ value is high. Upon heating the system, heat is taken up at T_c and the amount per mole is of the same order of magnitude as, though somewhat smaller than, the molar melting enthalpy of α-crystals of the surfactant. T_c is lower by 20 to 50 K than the α-crystal melting temperature.

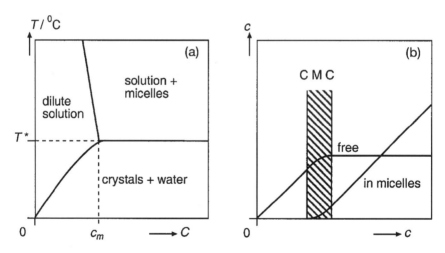

FIGURE 10.10 Micellization. (a) Simplified phase diagram of a surfactant and water. T is temperature, T^* Kraft temperature, c is molar concentration, c_m is CMC. As an example, T^* may be 20°C, c_m 0.3 mol·m^{-3} (or about 0.01% w/w). (b) Concentration of surfactant in free or in micellar form as a function of total concentration c.

The difference decreases as the concentration of the surfactant, hence the value of Γ, increases. The plateau values of Γ differ greatly below and above T_c, being, e.g., 6 and 3 $\mu mol \cdot m^{-2}$, respectively. Consequently, the interfacial tensions also differ, being, e.g., 7 and 20 $mN \cdot m^{-1}$, respectively. Making an emulsion at, say, 40°C with an excess of glycerol monopalmitate, and then cooling it to 5°C, may thus lead to a considerable reduction in γ, although it will take some time before equilibrium adsorption has been attained.

Micelle formation is briefly discussed in Section 2.2.5, item 4; see especially Figure 2.8. Soap micelles typically contain 50 to 100 molecules, and the radius is roughly 2 nm (about the length of a surfactant molecule). The core of a micelle contains a little water, at most one molecule per surfactant molecule. The size and shape of the micelles closely depend on the molecular configuration of the surfactant. Micelles are dynamic structures. They are not precisely spherical, and surfactant molecules move in and out. Characteristic times for these processes are a matter of debate, but they seem to be of the order of 10 μs. Presumably, a micelle can disappear in 10–100 ms upon dilution.

Ideally, micellization occurs above a sharply defined concentration, the *critical micellization concentration* or CMC. Surfactant molecules added above the CMC tend to go into the micelles, as is illustrated in Figure 10.10b. This means that the thermodynamic activity of the surfactant does not increase above the CMC, and neither do the resultant colligative properties, such as osmotic pressure. Most ionic surfactants give a sharper transition from solution to micelles than many nonionic, although even the former show a transition zone rather than a transition point, as illustrated in Figure 10.10b. Tweens, for instance, which are mixtures in that aliphatic chain length and the number of oxyethylene groups are variable, do not show a clear CMC, although micellelike structures are present at high concentration. In a solution containing different surfactant molecules, mixed micelles readily form.

The CMC is smaller for a longer chain length, as measured by the number of carbon atoms n; see, for instance, Figure 10.6. The CMC tends to decrease somewhat with increasing temperature (see Figure 10.10a).

For ionic surfactants, the *ionic strength* has a large effect. Figure 10.11 shows that with increasing NaCl concentration, the CMC of Na dodecyl sulfate (SDS) greatly decreases and the surface activity increases. This is because the negative charge of the head groups is shielded to a greater extent for a higher ionic strength, so that the mutual repulsion of these groups acts over a smaller distance. This implies that a denser packing is possible (higher Γ), hence a lower γ at the same bulk concentration of SDS. Explaining the phenomenon in other words, the thermodynamic activity of SDS increases when adding NaCl. From the discussion in Section 2.3, it follows that the activity is given by $\gamma_\pm (c_+ \times c_-)^{0.5}$, where $\gamma_\pm$ is the ion activity coefficient, c_+ is the molar concentration of the cation (Na) and c_- that of the anion (dodecyl sulfate). Since the molar concentration of SDS is very small, addition of NaCl will greatly increase the ion concentration product; on the other hand, the higher ionic strength diminishes the ion activity coefficient [Eq. (2.28b)], but the former effect is predominant.

> *Note* In principle, the Gibbs adsorption equation as given [Eq. (10.2)] should be modified for ionic adsorbates. A well dissociated surfactant like SDS can be considered as consisting of two adsorbing species: Na^+ and dodecyl sulfate$^-$. Since electroneutrality must be maintained, the counterion Na^+ can be said to

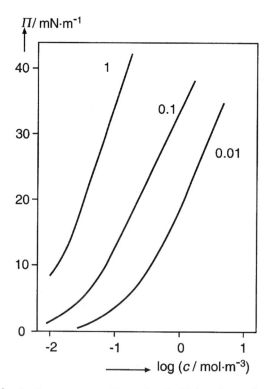

FIGURE 10.11 Surface pressure Π at the A–W interface of aqueous sodium dodecyl sulfate solutions as a function of concentration c, for various molar quantities of NaCl added (indicated near the curves). The curves end at the CMCs. (From data by E. Matijevic, B.A. Pethica. Trans. Faraday Soc. 58 (1958) 1382.)

"adsorb," albeit at some distance from the interface. However, the theory is intricate and there is no complete consensus. Since the potential error is generally small, except at very low ionic strength, we will not discuss the point.

Another variable of a surfactant is its **HLB value**, which is a measure of the balance of hydrophobic and hydrophilic parts of the molecule. Some values are given in Table 10.2. If HLB is large, the substance is well soluble in water; if small, it is well soluble in oil. For HLB $\approx$ 7, the solubility is about equal in both phases and generally not very high. The HLB value is smaller for a longer or a more saturated aliphatic chain (compare the various Spans in Table 10.2), and it is larger for a more polar or bigger polar head group. Ionic surfactants always have HLB values > 7. A head group with poly(oxyethylene) chains also causes a high HLB value, although it markedly decreases with increasing temperature. This is because the oxyethylene groups become dehydrated at high temperature.

Surfactants also differ in the lowest *interfacial tension* they can give. For most amphiphiles, final γ at the A–W interface ranges between 35 and 42 mN $\cdot$ m^{-1}. At the triglyceride oil–water interface the variation is relatively larger, because γ_0 is smaller and Π is, in first approximation, the same as for an A–W interface. Typical results are between 3 and 5 mN $\cdot$ m^{-1}, but smaller values can be obtained for some mixtures of surfactants. An equimolar mixture of a Span and a Tween, for instance, can give a higher surface excess than either of them, and this then leads to a lower γ. Addition of salt tends to give a lower γ for ionic surfactants, as mentioned above; a similar but smaller effect has been observed for several nonionics.

An important function of amphiphiles is **detergency**. Most detergents are ionic surfactants, because these readily form micelles. Micelles can accommodate hydrophobic molecules in their interior, and this is what happens during washing processes. They can thus "solubilize" oil, though to a limited extent; the amount is proportional to and of the same order as the concentration of micelles. In such a way, oil-soluble vitamins or flavor substances can be dispersed in water. Something similar can happen in an oil that contains surfactants, where reverse micelles can form; see Figure 10.9a. These contain in general fewer surfactant molecules than do "regular" micelles. The water inside can "solubilize" water-soluble substances, such as proteins, in the oil. The latter has been used to achieve enzyme action on substances dissolved in oil.

10.3.2 Polymers

In this section, we will consider only water-soluble polymers adsorbing at homogeneous surfaces, A–W, O–W, or S–W. The **conformation** in which polymers adsorb will be considered first.

Adsorption of *homopolymers* is possible but not very common: most of them are either insufficiently surface active or hardly soluble. The conformation of an adsorbed homopolymer molecule will be roughly as depicted in Figure 10.12a. Of course, the chains sticking out into the aqueous phase show considerable random variation in conformation due to Brownian motion. *Copolymers* that contain both hydrophobic segments and (usually a greater number of) hydrophilic segments are very suitable surfactants. They would adsorb roughly as depicted in Figure 10.12b, although the conformation will greatly depend on the distribution of the hydrophobic segments over the chain.

Most *polysaccharides* used in foods are predominantly hydrophilic and not surface active. Some polysaccharides, however, notably gum arabic, contain minor protein moieties, and do adsorb onto O–W (and presumably A–W) interfaces. By chemical modification, hydrophobic groups can be introduced. The best known examples are cellulose ethers, such as methyl cellulose and hydroxypropyl cellulose, which substances are well soluble in water (at least below 40°C) and strongly surface active.

The polymeric surfactants of choice in foods are **proteins**. The polypeptide backbone is fairly polar, but several side groups are hydrophobic (see Section 7.1). Protein adsorption is briefly discussed in Section 7.2.2, subheading "Adsorption". All proteins are surface active and adsorb at O–W and A–W surfaces. Globular proteins often retain a fairly compact form, although conformational changes do occur: see Figure 10.12c. Nonglobular proteins, such as gelatins and caseins, tend to adsorb in a way comparable to Figure 10.12b. For β-casein the (average) conformation on adsorption is fairly well known: see Figure 10.12d. The picture agrees well with the primary structure of β-casein (see Figure 7.1): a very hydrophilic N-terminal part, and a long tail containing several hydrophobic side groups.

The **surface activity** of a protein and an amphiphile are compared in Figure 10.13. It is seen that the protein is much more surface active. The molar bulk concentration needed to reach Γ_∞ differs by 4 orders of magnitude. If the mass concentration is plotted, the curves are closer, but the difference is still by more than two orders of magnitude. The main cause is the larger molar mass of the protein. It implies that the free energy of adsorption per molecule (roughly equal to $\Pi/\Gamma N_{Av}$) is very much larger than that of the amphiphile; for the protein it would be about 60 times $k_B T$,

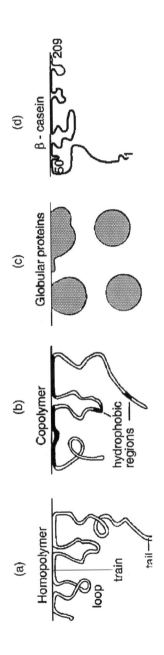

FIGURE 10.12 Mode of adsorption of various polymers from an aqueous solution. Very approximate. See text. In (c) some nonadsorbed molecules are shown for comparison. In (d) the numbers refer to the peptide sequence.

and for the amphiphile of the order of $1\,k_BT$. Consequently, the affinity of the protein for the surface is far greater than that of the amphiphile.

On the other hand, the value of Π reached is clearly larger for the amphiphile, provided that the surfactant concentration is high enough. The explanation must be that for a polymer, and hence a protein, a very dense packing of surfactant material at the interface cannot be reached. This is in accordance with the observation that for most polymers and for most amphiphiles the surface excesses expressed in unit mass are roughly the same: a few $mg \cdot m^{-2}$, despite the "thickness" of the polymer layer being clearly higher for a polymer (say, 10 nm) than for an amphiphile (about 2.5 nm).

For some proteins at some conditions, *multilayer adsorption* can occur, as indicated by a dotted curve for Γ/Γ_∞; a second layer is very weakly

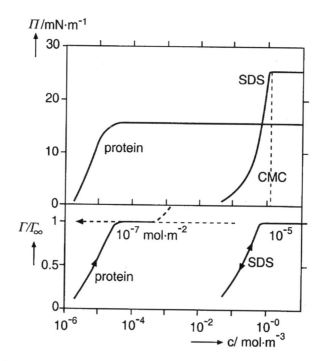

FIGURE 10.13   Surface pressure (Π) and surface excess (Γ; the plateau value Γ_∞ is indicated near the curves) at the triglyceride oil–water interface as a function of the concentration in solution (c) for a protein (β-casein) and an amphiphile (SDS). The value of γ without surfactant $\approx 30\ mN \cdot m^{-1}$.

adsorbed. Apart from this phenomenon, high surface loads can be obtained by adsorption of protein aggregates (e.g., casein micelles) rather than free molecules; by formation of a gel layer at the interface (e.g., of gelatin at low temperature); or by covalent intermolecular cross-linking (e.g., formation of —S—S— bonds between β-lactoglobulin molecules). Generally, lateral interaction forces act between globular protein molecules in an adsorption layer, and these forces markedly strengthen with time.

Reversibility of Adsorption. Apparently, the data in Figure 10.13 imply that the Gibbs equation (10.2) does not hold for the protein. As we have seen, it is valid for the amphiphile. However, the slopes $d\Pi/d \ln c$ given in the figure differ only by a factor 2 between the two surfactants, whereas the values of Γ_∞ differ by two orders of magnitude. The explanation is not fully clear. Application of the Gibbs equation to polymers is anyway questionable, because it is generally not known what the relation is between concentration (c) and activity (a) of the surfactant. Moreover, proteins and other polymers are virtually always mixtures.

As indicated in Figure 10.13 by an arrow on the Γ–c curve, it appears as if lowering the solution concentration does not lead to *desorption* of protein, which would also defy the Gibbs equation. This observation is based on washing experiments, where an O–W emulsion is diluted with solvent and then concentrated again by centrifuging; repeating this a few times may be expected to remove all protein, but it does not. This is mainly because a very low value of c can hardly be reached. Assume that after dilution of the emulsion we would have $c = 0$, $\varphi = 0.01$, $d_{32} = 0.6\,\mu$m, and $\Gamma_\infty = 10^{-7}\,\text{mol}\cdot\text{m}^{-2}$; the total concentration of protein in the emulsion then is $0.01\,\text{mol}\cdot\text{m}^{-3}$. Consequently, only 0.3% of the adsorbed protein would have to desorb to reach $3\cdot10^{-5}\,\text{mol}\cdot\text{m}^{-3}$ in solution, which is roughly the equilibrium concentration. A decrease in Γ by 0.3% cannot, of course, be determined. It would take of the order of 50 washings to achieve significant desorption. Moreover, desorption tends to be very slow. One reason is the very large decrease of free energy per molecule upon adsorption, here about 60 times $k_B T$, which means that the activation free energy for desorption is very large. Moreover, the concentration difference between the solution adjacent to the interface and further away from it cannot be larger than about $3\cdot10^{-5}\,\text{mol}\cdot\text{m}^{-3}$, and this would lead to very slow diffusion away from the interface [cf. Section 10.4, Eq. (10.6)]. In other words, desorption would be extremely slow.

Nevertheless, desorption can occur over fairly short time scales, e.g., 15 min. It has been shown, for instance by using radio-labeled molecules, that flexible large polymer molecules do exchange between bulk and interface. Another indication is that two molecules of about equal surface activity, like α_{s1}-casein and β-casein, can displace each other within about 15 min when the one is present at the surface of emulsion droplets and the other is added to the emulsion afterwards. It appears likely that a flexible polymer is not desorbed at once, but that one segment (or a short train of segments) at the interface is displaced by one of another molecule at the time, etc. This would not be possible for globular proteins adsorbed as depicted in Figure 10.12c; indeed, the evidence for fairly rapid exchange or mutual displacement of these proteins is less convincing. Desorption rate is further discussed in Section 10.4.

The surface equation of state of proteins is highly nonideal, as mentioned before: see, e.g., Figure 10.8, lysozyme. Equation (10.5) indicates that the expansion of a molecule in the interface (larger radius) would lead to a larger value of θ and hence to a higher value of Π. It also makes a difference onto what surface the protein is adsorbed, as illustrated in Figure 10.14. Presumably, the diversity is due to differences in the amount of surface area occupied by each protein molecule in the interface, which will, in turn, depend on the state of unfolding of the peptide chain. For instance,

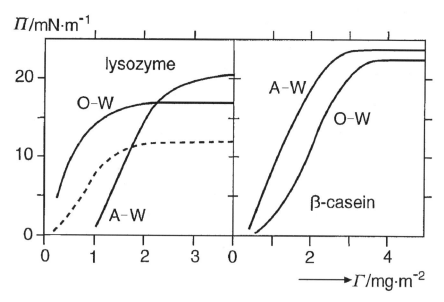

FIGURE 10.14 Relation between surface pressure (Π) and surface excess (Γ) for two proteins at the A–W and O–W interfaces. The broken line is for predenatured lysozyme at the A–W interface. (From results by D. Graham, M. Phillips. J. Colloid Interf. Sci. 70 (1979) 403, 415.)

it is fairly clear from Figure 10.14 that casein molecules expand much more at the A–W interface than lysozyme does. It is, however, also obvious that different proteins do not react in the same manner on a change in interface type; the author is not aware of a reasonable explanation based on protein structure. It is fairly certain that globular proteins tend to become more strongly denatured on adsorption at an O–W interface than at an A–W; this follows, e.g., from studies on the loss of activity of adsorbed enzymes. Heat denaturation of a protein before adsorption also leads to a change in the Π–Γ relation; see Figure 10.14, lysozyme. This may (partly) be due to the aggregation that often occurs upon heat denaturation. Furthermore, the surface equations of state of proteins also depend on conditions like pH, ionic strength, temperature, etc., but few unequivocal results are available.

Typical results for the plateau value of Γ at the O–W interface of globular proteins range from 2 to $4 \, \text{mg} \cdot \text{m}^{-2}$, the higher values corresponding to larger molecules. For nonglobular proteins (gelatin, caseins), values between 3 and 5 are generally observed. As a rule of thumb, Γ tends to be slightly smaller at an A–W interface, and smaller still at a S–W interface, but there are exceptions. For aggregated proteins, e.g., as caused by heat denaturation, far higher Γ values can be obtained, often 10–$15 \, \text{mg} \cdot \text{m}^{-2}$ or even higher.

Question

In a study on O–W emulsions made with β-casein, a plateau value of $\Gamma = 4 \, \text{mg} \cdot \text{m}^{-2}$ was observed. What would be the number of amino acid residues per casein molecule at the interface? You may assume that fully unfolded peptide chains at the interface would give a plateau value of $\Gamma = 1 \, \text{mg} \cdot \text{m}^{-2}$.

Answer

For a surfactant of molar mass $M \, \text{g} \cdot \text{mol}^{-1}$, a surface load of $\Gamma \, \text{g} \cdot \text{m}^{-2}$ corresponds to $\Gamma \cdot N_{AV}/M$ molecules per m^2. The area occupied per molecule is thus $M/\Gamma \cdot N_{AV} \, \text{m}^2$. Applying this to unfolded peptide chains and taking for M the average value of an amino acid residue, i.e., 115 (see Section 7.1), we calculate that the surface area taken up by one residue will equal $0.115/6 \cdot 10^{23} \, \text{m}^2 \approx 0.2 \, \text{nm}^2$. For β-casein (see Figure 7.1), $M = 24,000 \, \text{g} \cdot \text{mol}^{-1}$ and $\Gamma = 4 \cdot 10^{-3} \, \text{g} \cdot \text{m}^{-2}$, which yields a surface area per molecule of $10 \, \text{nm}^2$. Dividing by $0.2 \, \text{nm}^2$ results in a figure of about 50 for the number of residues at the interface. Comparing this with the total number of residues for β-casein of 209, it means that about a quarter of the residues would be directly involved in the adsorption. It should be realized that this is a rough figure, because it is not known what fraction of the interface is covered by adsorbate in the two cases.

10.3.3 Mixtures

In this section mixtures of amphiphiles and proteins are considered. As seen in Figure 10.13, an amphiphile tends to give a lower interfacial tension than a protein does, if the concentration of the amphiphile is high enough. Suppose that an aqueous solution contains both protein and amphiphile in about equal mass concentrations. If the surfactant concentrations are on the order of the CMC of the amphiphile, one would expect the amphiphile to predominate in the interface (A–W or O–W), possibly even displacing all of the protein, because that would give the lowest free energy in the system. At concentrations far below the CMC of the amphiphile, the protein would dominate in the interface. Such phenomena do indeed occur.

Figure 10.15 illustrates what happens if an emulsion made with a protein (β-casein) is subjected to increasing concentrations of an **ionic**

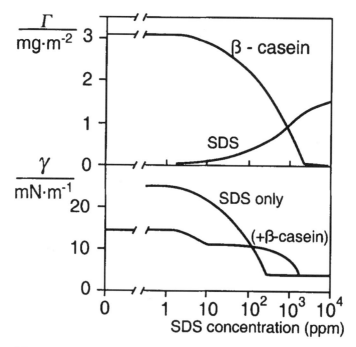

FIGURE 10.15 Surface excess (Γ) in an O–W emulsion and interfacial tension (γ) at the O–W interface for β-casein as a function of the concentration of Na-dodecyl sulfate (SDS) present. γ is also given for SDS only. (From results by J. A. de Feijter et al. Colloids Surfaces 27 (1987) 243.)

amphiphile (SDS); for β-lactoglobulin and SDS similar results were obtained. At high amphiphile concentration the protein is indeed completely displaced; at very low amphiphile concentration the interface only contains protein. The situation is, however, more complicated. The SDS concentration has to be higher than its CMC (by a factor of about 5) to displace all protein; and at intermediate concentrations (10–100 ppm SDS), the interfacial tension is smaller for the mixture of surfactants than for either surfactant alone. This points to association between protein and amphiphile, both in solution (thereby decreasing the activity of the amphiphile) and in the interface (thereby increasing total Γ and decreasing γ).

SDS is an anionic surfactant, and most ionic amphiphiles do indeed show attractive interactions with proteins, although this will depend on pH, ionic strength, etc. These interactions range from strong binding of a small number of amphiphiles (one or two molecules per protein molecule) to formation of large mixed aggregates of amphiphiles and proteins. Such aggregates can possibly adsorb. Other situations may occur, like two-dimensional phase separation between amphiphile and protein, or (weak) adsorption of protein on top of a layer of amphiphiles.

Most **nonionic surfactants** (like Tweens) interact weakly with proteins, if at all. A nonionic therefore tends merely to displace proteins, if its concentration is high enough. At lower concentration, two-dimensional phase separation occurs: generally, islands of the amphiphile are formed in a network of protein. Of course, the plateau value of γ produced by the amphiphile must be significantly smaller than that of the protein. This may explain why most monoglycerides do not or only partly displace proteins from an O–W interface, unless the temperature is below the chain crystallization temperature of the monoglyceride. The latter situation allows much closer packing of monoglyceride at the interface and hence a lower value of γ, hence displacement of proteins. This is, e.g., observed in ice-cream mix, where fat globules are still covered with protein, despite the presence of, say, glycerol monopalmitate, but lose the protein at low temperature (e.g., 5°C).

10.4 TIME EFFECTS

A surfactant can be transported to a certain site at an interface by adsorption from solution and by lateral transport in the interface. In all cases it leads to an even distribution of surfactant over the interface and to a lowering of interfacial tension. These processes take time.

For the case of adsorption of a surfactant, its transport toward the interface often proceeds by **diffusion**. For a surfactant concentration in solution c and a final surface excess Γ_∞, the amount of surfactant needed to give an adsorbed monolayer could be provided by a layer of solution adjacent to the interface of thickness $\delta = \Gamma_\infty/c$. Assuming Γ_∞ to be about $3\,\text{mg}\cdot\text{m}^{-2}$ and a surfactant concentration c of $3\cdot10^6\,\text{mg}\cdot\text{m}^{-3}$ (0.3%), we obtain $\delta = 1\,\mu\text{m}$, i.e., a very thin layer would suffice. Applying Eq. (5.21), we would arrive at a halving time for adsorption $t_{0.5} = \delta^2/D$, where D is the diffusion coefficient of the surfactant in the solution. The conditions underlying Eq. (5.21) are not fully met, however, and a more elaborate analysis (accounting for the development of a concentration gradient and the decreasing possibility for adsorption with increasing values of Γ/Γ_∞) leads to

$$t_{\text{ads}} \approx 10\,\frac{\Gamma_\infty^2}{c^2 D} \tag{10.6}$$

where t_{ads} is the time needed to obtain almost complete adsorption ($\Gamma/\Gamma_\infty \approx 0.97$).

For small-molecule surfactants, D would be of the order of $3\cdot10^{-10}\,\text{m}^2\cdot\text{s}^{-1}$. For the above-mentioned quantities, Eq. (10.6) then predicts an adsorption time of about 30 ms, a time too short to measure Γ (or even γ with common methods). The surfactant concentration can, however, be far smaller. Figure 10.6 shows that Na-stearate, a very surface active amphiphile, gives for $c = 0.13\,\text{mol}\cdot\text{m}^{-3}$ a surface excess $\Gamma = 9\cdot10^{-6}\,\text{mol}\cdot\text{m}^{-2}$. This would lead to $t_{\text{ads}} \approx 10\,(9\cdot10^{-6}/0.13)^2/3\cdot10^{-10} = 160\,\text{s}$. For a still lower concentration, adsorption would also occur, but Γ_∞ would then be smaller, and the calculated adsorption times are of comparable magnitude. In other words, for small-molecule surfactants, diffusion times are always short, from milliseconds to a few minutes.

This need not be true for macromolecular surfactants, since these are much more surface active than amphiphiles (see Figure 10.13), and the diffusion coefficients are also smaller, say, by a factor of 4. From Figure 10.13 we derive that β-casein at a concentration as small as $900\,\text{mg}\cdot\text{m}^{-3}$ can provide a surface excess of $2.5\,\text{mg}\cdot\text{m}^{-2}$ at the A–W interface. This then would lead to $t_{\text{ads}} \approx 10(2.5/900)^2/7\cdot10^{-11} \approx 10^6\,\text{s}$ or about 12 days. Very long times can indeed be observed.

If the surface equation of state is known for the surfactant adsorbing, its combination with a diffusion equation would yield the dependence of γ on time. It is, however, often very difficult to predict the evolution of $\gamma(t)$. Following are some **complications**:

1. **Convection.** Transport of surfactant to the interface may be by convection rather than diffusion. Convection is likely to occur if the distance over which the surfactant has to be transported (about equal to Γ_∞/c) is not

very small, and it can significantly speed up adsorption. During such processes as foam and emulsion formation, very intensive convection is deliberately induced and adsorption times as short as a microsecond may result in some cases (Section 11.3.1).

2. **"Consumption"** of surfactant may occur upon adsorption if it tends to dissolve in the other phase, which may be a slow process.

3. **Micelles.** It is still a matter of debate whether the dissociation of surfactant molecules from micelles would lead to retardation of adsorption. It appears very unlikely that the time scale involved would be longer than a second. On the other hand, some surfactants have a very small solubility in water (e.g., phospholipids), and they are often present in small solid lumps or as vesicles. In such a case, it may take a very long time before surfactant molecules have reached the interface.

4. **Adsorption barrier.** A free energy barrier for adsorption of a surfactant would cause a decrease in adsorption rate, and several kinds of such barriers have been postulated. Apparently, electrostatic repulsion can indeed cause a decrease in adsorption rate. This will occur if the adsorbing species is highly charged (e.g., a protein at a pH far removed from its isoelectric point), and moreover ionic strength is low (so that $1/\kappa$ is larger than the distance between the surfactant molecules at the interface).

5. **Mixtures.** Nearly all preparations of surfactants used in practice are mixtures, and the components vary in surface activity and in the lowest γ they can produce. Often, some components that can produce a very small γ are present in minor quantities, and these eventually tend to predominate in the interface after they have finally reached it, displacing other surfactants. For soluble small-molecule surfactants, the time scale will mostly be short, as explained above, whereas for a mixture of polymeric surfactants (say, proteins) it may take a long time before a "final" value of γ is reached.

6. **Change of conformation.** Polymeric surfactants, especially proteins, may undergo changes in conformation that lead to a decrease in interfacial tension. Such changes may take a long time. For flexible proteins, such as β-casein, time scales up to 10 s may be involved. For globular proteins, conformational changes upon adsorption may take up to 10^3 s.

7. **Partial desorption.** Because a polymer molecule can change its conformation on adsorption, it may unfold and cover a much larger amount of interface than it would when equilibrium between dissolved and adsorbed molecules has been reached. This implies that the interface may be almost fully covered with surfactant in an early stage of adsorption, which will greatly reduce the rate of further adsorption: parts of the adsorbed molecules have to desorb before additional molecules can adsorb, and such partial desorption will be a slow process. These phenomena are illustrated in Figure 10.16 for flexible polymers. Something similar will happen with many

globular proteins; presumably, the changes in surface area covered per molecule are smaller but take a longer time. It should further be noted that very fast adsorption (high polymer concentration, intensive convection) may leave insufficient possibility for unfolding of a polymer (protein) before the surface is covered; after all, unfolding too takes time.

Figure 10.17a gives some examples of the adsorption rate of a protein at various concentrations. Calculation according to Eq. (10.6) leads to values of about 0.6 s, 1 min, 100 min, and 7 days for the decreasing concentrations given. It is clear that the observed adsorption times are much longer. The main reason must be complication 7 just mentioned.

Dynamic Surface Tension. For several kinds of practical problems, the surface tension of a surfactant solution at very short time scales is important. A case in point is foam formation, where the time scales of the relevant processes often are of order 1 or 10 ms. Since determination of γ in static experiments is not possible at such time scales, one often determines what is called dynamic surface tension. Here the surface of a surfactant solution is rapidly expanded, and γ_{dyn} is measured as a function of the expansion rate $\xi = d \ln A/dt$, where A is surface area; the time scale then is taken as the reciprocal of ξ. A surface confined between barriers (as depicted in Figure 10.2b) can be expanded by moving the barriers away from each other. Expansion can also be induced by letting the solution flow over the rim of a vertical cylinder at a certain rate. Generally, a steady state (i.e., constant γ) is obtained: the supply of surfactant from the bulk to the

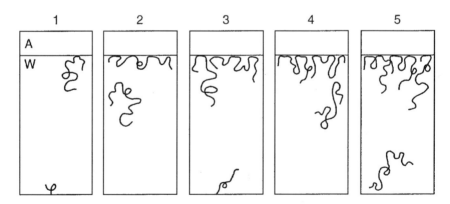

FIGURE 10.16 Illustration of various stages in the adsorption of flexible polymer molecules onto an A–W interface. Highly schematic.

interface then compensates for its dilution by expansion of the surface. Moreover, the conformation of macromolecular surfactants should be constant.

Invariably, γ_{dyn} is larger for a faster expansion rate within a certain range of ξ values. At high ξ, the value of the pure solvent γ_0 is reached, at very low ξ, the equilibrium value. Some examples are in Figure 10.18, and it is seen that, as expected, γ_{dyn} is smaller for a larger surfactant concentration, even if the plateau value of γ is the same. Figure 10.18 relates to A–W interfaces; measurements at O–W interfaces can also be done, although not as easily. Unfortunately, very high expansion rates cannot be obtained, especially not at an O–W interface (because of the high viscosity of the oil).

As mentioned, changes in *protein conformation* readily occur upon adsorption. Results as, for instance, given in Figure 10.14 relate to

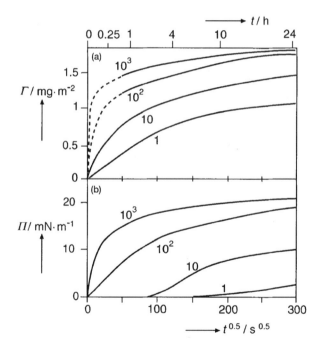

FIGURE 10.17 Adsorption of ovalbumin at the A–W interface; pH = 6.4, ionic strength = 0.02 M. (a) Surface load (Γ). (b) surface pressure (Π) as a function of the square root of adsorption time (t). The numbers near the curves denote the protein concentration in ppm. (From results by J. de Feijter, J. Benjamins. In: Food Emulsions and Foams, ed. E. Dickinson. Roy. Soc. Chem., 1987, p. 74.)

(apparent) equilibrium values of γ, i.e., after no further change in γ is observed. At far shorter time scales the relation between π and Γ may be quite different, primarily because of the conformational changes occurring upon adsorption. This is borne out by some results given in Figure 10.18, which gives γ_{dyn} values as a function of expansion rate. For β-casein, the values at which γ_{dyn} become constant agree well with the time scales calculated from Eq. (10.6): 156 and 25 s for 0.1 and 0.25 g/L, respectively. This means that changes in conformation must have been complete within about 10 s. For β-lactoglobulin, it took far longer to reach the final γ value, although the adsorption time should also have been about 25 s; it appears that the conformational change needed about 10^3 s.

In conclusion, it may take fairly long times for protein conformations at an interface to become constant, be it after adsorption or after a change in physicochemical conditions. The same holds for protein composition at the interface if competition between proteins can occur. Moreover, the

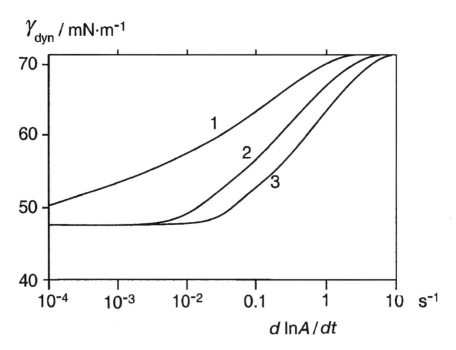

FIGURE 10.18 Dynamic surface tension (γ_{dyn}) of some protein solutions as a function of expansion rate (d ln A/dt). 1, β-lactoglobulin, 0.25 g/L. 2, β-casein, 0.1 g/L. 3, same, 0.25 g/L. (Approximate results by courtesy of H. van Kalsbeek, A. Prins.)

conformation changes of globular proteins upon adsorption may be to some extent irreversible. Consequently, γ, Γ, and surface composition may markedly depend on the history of the adsorbed layer.

Desorption of Proteins. If a spread layer of protein on water is applied between barriers (cf. Figure 10.2b) and the layer is then compressed to a given value of Π that is kept constant, desorption of protein occurs. The protein can readily diffuse away, since the subsurface concentration will be (far) below the value that would be in equilibrium with the surface load. Some results for globular proteins are given in Table 10.3. The times involved are quite long. This is primarily due to the large activation free energy for desorption. Its value equals at least γ/Γ J · mol^{-1}, assuming Γ to be expressed in mol · m^{-2}. For a an average protein, the value would be about 250 kJ · mol^{-1} or about $100 k_B T$ per molecule, which is quite substantial.

It is further seen that for a higher value of Π, which means a greater difference in chemical potential of the protein between interface and "solution," desorption is faster. In practice, unlike the situation to which the table refers, the surface pressure will generally decrease when protein is desorbed; this means that the rate of desorption will become ever slower. The variation in desorption rate among proteins (at the same value of Π) must primarily be due to differences in the activation free energy for desorption. As expected, desorption tends to be slower for a higher molar mass, but the correlation is not perfect. Besides its molar mass, the conformation of a protein at the interface will affect the activation free energy for desorption.

It may finally be mentioned that the desorbed protein may be in a denatured state (see Section 7.2.2).

TABLE 10.3 Characteristic Times (h) needed for Desorption of Some Proteins of Various Molar Mass (M) from the A–W Interface

Protein	M/kDa	Surface pressure/mN.m^{-1}						
		15	20	25	30	35	40	45
Insulin	6	3	0.5					
β-Lactoglobulin	17		8	3	2			
Myoglobulin	17			5	2.5	1		
γ-Globulin	160				18	8	4	
Catalase	230					5	2	1.5

Source: From results by F. MacRitchie. *J. Colloid Interf. Sci.* 105 (1985) 119.

Question

An emulsion technologist wishes to compare two commercial water-soluble small-molecule surfactant preparations, A and B, to be used for making O–W emulsions. The interfacial tension is measured between the aqueous phase and the oil to be used, in the presence of either surfactant. For A it is observed that $\gamma = 6 \, \text{mN} \cdot \text{m}^{-1}$; for B the value is not immediately constant and decreases from about 6 to $2 \, \text{mN} \cdot \text{m}^{-1}$ in two minutes. It is concluded that with preparation B the emulsion droplets will obtain a smaller interfacial tension than with A. Is this conclusion warranted?

Answer

No. The fact that γ decreases with time for surfactant B implies that the latter must contain some component(s) in small concentrations that decrease γ to a lower value than the main component(s) can. Application of Eq. (10.6) with $t_{ads} = 120 \, \text{s}$ and $D = 3 \cdot 10^{-10} \, \text{m}^2 \cdot \text{s}^{-1}$ leads to a thickness of the layer δ that provides the minor surfactant of about 60 μm; in other words, the effective surface-to-volume ratio would be $1/60 \cdot 10^{-6} = 17 \cdot 10^3 \, \text{m}^{-1}$. In the emulsion, however, the ratio of O–W surface area to the volume of the surfactant solution may be far greater. It would be given by $6 \cdot \varphi / d_{32}(1 - \varphi)$; for an assumed oil volume fraction $\varphi = 0.25$ and average droplet size $d_{32} = 1$ μm, this leads to a value of $2 \cdot 10^6 \, \text{m}^{-1}$, i.e., more than 100 times that during the macroscopic measurement of γ. This implies that the concentration of the minor components at the droplet interface would be very small, probably having a negligible effect on γ.

10.5 CURVED INTERFACES

10.5.1 Laplace Pressure

We all know that the pressure inside a bubble is higher than atmospheric. When we blow a soap bubble at the end of a tube and then allow contact with the atmosphere, the air will immediately escape from the bubble: it shrinks and rapidly disappears. This is a manifestation of a more general rule: if the interface between two fluid phases is curved, there always is a pressure difference between the two sides of the interface, the pressure at the concave side being higher than that at the convex side. The difference is called the Laplace pressure p_L.

Figure 10.19 serves to give an explanation for a sphere. At any equator on the sphere, the surface (or interfacial) tension pulls the two "halves" toward each other with a force that equals γ times the circumference. The surface tension thus causes the sphere to shrink (slightly), whereby the substance in the sphere is compressed and the pressure is increased. At equilibrium, the excess inside pressure times the area of the cross section of the sphere provides a force that is equal and opposite to the force caused by the surface tension. Hence, the relation $p_L = 2\gamma/R$.

For a soap bubble of 1 cm diameter and a surface tension of $50\,\text{mN} \cdot \text{m}^{-1}$, p_L will be $2(2 \times 0.05/0.005) = 40\,\text{Pa}$. The factor 2 before the bracket is due to a soap film having two surfaces; here, each surface has (almost) the same radius and the same surface tension. (A "soap bubble" is in fact a very thin spherical shell of a soap solution.) For a gas bubble of the same size in the same soap solution, p_L would be 20 Pa. Note that p_L is greater for a smaller bubble: for one of 1 µm diameter, p_L would amount to $2 \cdot 10^5\,\text{Pa}$, or 2 bar. For an emulsion droplet of that size, where γ is smaller, for instance $10\,\text{mN} \cdot \text{m}^{-1}$, a Laplace pressure of 0.4 bar results.

The given relation for a sphere is a special case of the general **Laplace equation**

$$p_L = p_{\text{concave}} - p_{\text{convex}} = \gamma \left(\frac{1}{R_1} + \frac{1}{R_2} \right) \qquad (10.7)^*$$

The curvature of a surface can at any place be characterized by two principal radii R_1 and R_2. R_1 is found by constructing a plane surface through the normal to the surface at the point considered. The curved surface intersects the plane, resulting in a curve to which a tangent circle is constructed. The plane then is rotated around the normal until the curvature

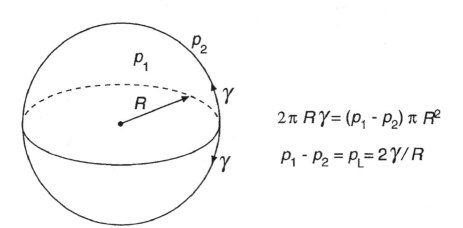

FIGURE 10.19 Derivation of the Laplace pressure for a sphere of radius R, where the sphere's surface is the phase boundary between two fluids, with interfacial tension γ; p means pressure.

is at maximum, i.e., the radius of the tangent circle is at minimum. This is the first principal radius of curvature R_1. R_2 is the radius of the corresponding circle in a plane through the same normal that is at a right angle to the first one. For a sphere $R_1 = R_2 = R$. For a circular cylinder, R_1 is the cylinder radius R_{cyl} and $1/R_2 = 0$; hence, $p_L = \gamma/R_{cyl}$.

Curved surfaces can also be *saddle-shaped*. Figure 10.20 shows an example. Suppose that a surfactant film is made between the two frames. Surface tension causes the film to assume the smallest surface area possible. In the situation depicted, this surface is saddle-shaped. Moreover, the surface has zero curvature. As drawn for the middle cross section of the film, the principal radii of curvature are equal, but of opposite sign, since the tangent circles are at opposite sides of the film (which is, actually, the definition of a saddle-shaped surface). In other words, $p_L = 0$ because $1/R_1 + 1/(-R_2) = 0$. This is true for every part of the film surface.

It may be concluded that a confined film onto which no net external forces act will always form a surface of zero curvature. If formation of such a surface is geometrically impossible—for instance, if in a situation as depicted in Figure 10.20 the two circular frames were much farther apart—a film cannot be made.

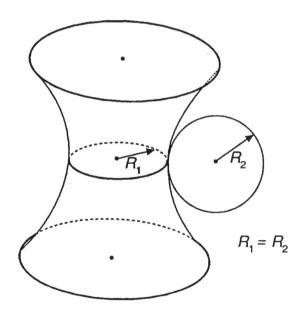

Figure 10.20 Representation of a soap film formed between two parallel circular frames. At a point on the film surface two tangent circles are drawn.

Similarly, a fluid body with a closed surface on which no net external forces act will always adopt a spherical shape. This was already concluded in Section 10.1 on thermodynamic grounds; it can yet also be shown by invoking the Laplace pressure. Figure 10.21 shows a spherical drop that is deformed to give a prolate ellipsoid. Near the ends of the long axis (near a) we have $p_L = 2\gamma/R_a$, whereas near b, $p_L = \gamma/(1/R_{1,b} + 1/R_{2,b})$. The latter value is the smaller one, since both $R_{1,b}$ and $R_{2,b}$ are larger than R_a. Consequently liquid will flow from the pointed ends to the middle of the drop, until a spherical shape is attained. Only for a sphere is the Laplace pressure everywhere the same.

A general conclusion then is that, in a fluid body with a closed surface, the fluid always wants to move from regions with a high curvature and hence high p_L to those of a low p_L. This also holds for a fluid body that is partly confined, as for instance an amount of water in an irregularly-shaped solid body. Only by applying external forces can gradients in p_L be established. The larger the equilibrium value of p_L, the higher the external stress needed to obtain nonequilibrium shapes.

10.5.2 Capillarity

Curved fluid surfaces can give rise to a number of capillary phenomena. It is well known that water rises in a narrow or "capillary" glass tube (if the glass

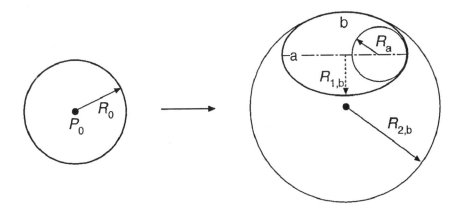

FIGURE 10.21 Illustration of the increase in Laplace pressure when a spherical drop (or bubble) is deformed into a prolate ellipsoid. Cross sections are shown in thick lines; the axis of revolution of the ellipsoid is in the horizontal direction. Two tangent circles to the ellipse are also drawn.

is well cleaned). This is shown in Figure 10.22a. If the diameter of the tube is not more than a few mm, the meniscus in the tube is a nearly perfect half sphere, readily allowing calculation of the Laplace pressure. This pressure difference is compensated for by the weight of the water column, as is

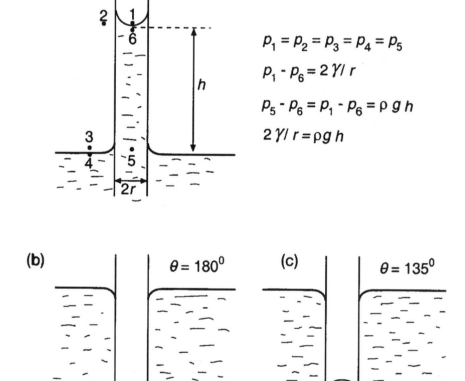

FIGURE 10.22 Capillary phenomena. (a) Rise of liquid in a capillary if the contact angle $\theta = 0$ (perfect wetting). p is pressure, γ is surface tension, ρ is mass density, and $g = 9.81\ \text{m} \cdot \text{s}^{-2}$. (b, c) Effect of contact angle.

explained in the figure. The equality gives for the height of the column

$$h = \frac{2\gamma}{r\rho g} \tag{10.8}$$

For $r = 0.5\,\text{mm}$, $\gamma = 72\,\text{mN} \cdot \text{m}^{-1}$, $\rho = 10^3\,\text{kg} \cdot \text{m}^{-3}$, and $g = 9.8\,\text{m}^2 \cdot \text{s}^{-1}$, this results in $h = 29.4\,\text{mm}$. It provides a method of determining surface tension. The meniscus at the wall of a wider vessel has to be explained in the same way, but the mathematics is complicated, as is the shape of such a meniscus. If the wall of the capillary is not at all wetted by the liquid, capillary depression results, as shown in Figure 20.22b; Eq. (10.8) also holds. We will discuss contact angles (as in Figure 10.22c) and capillary displacement in Section 10.6.

10.5.3 The Kelvin Equation

As discussed, the gas in a small bubble in a liquid is at increased pressure. According to Henry's law, which states that the solubility of a gas in a liquid is proportional to its pressure, its solubility is increased. From the Laplace equation (10.8) we can derive

$$\frac{s(r)}{s_\infty} = \exp\left(\frac{x'}{r}\right)$$
$$x' = \frac{2\gamma V_D}{RT} \tag{10.9}$$

which is known as the Kelvin equation. Here s means solubility, which is thus larger for a sphere of radius r than in the absence of curvature ($r = \infty$). x' is an auxiliary parameter of dimension length, V_D the molar volume of the disperse phase (in $\text{m}^3 \cdot \text{mol}^{-1}$, it equals molar mass over mass density), and RT has its usual meaning.

Unlike the Laplace equation, which can only be applied to fluid surfaces, the Kelvin equation is valid for any phase boundary, also solid–gas, solid–liquid, and even solid–solid. As for the Laplace equation, Eq. (10.9) can also be modified to accommodate nonspherical curved surfaces; then $2/(1/R_1 + 1/R_2)$ should be used instead of r. The values for the solubility should give the thermodynamic activity of the substance involved, for the equation to be generally valid; for ideal or ideally dilute solutions, concentrations can be used. Most gases in water show ideal behavior. It should further be noted that Eq. (10.9) applies for one component; if a particle contains several components, it should be applied to each of them separately. Finally, there are some conditions that may interfere with the applicability of the Kelvin equation; these will be discussed in Sections 13.6 and 14.2.

Some examples are given in Table 10.4, to show the influence of some variables on the results. It is seen that, especially for gases, a considerable increase in solubility can occur for fairly large bubbles. For an air bubble of 1 μm diameter, the solubility ratio would be about 5, and such a bubble would soon dissolve. It is indeed true that such very small gas bubbles are rarely observed. Although Eq. (10.9) does not tell us anything about the rate at which changes due to increased solubility occur, it may be clear that a bubble that starts dissolving will generally do so at an ever faster rate, because its radius decreases, whence the solubility excess increases during the process.

The increase in solub wi decreasing radius has some important consequences.

Nucleation. If a new   hould be formed because a dissolved
substance has beco   rsaturated, this may nevertheless not
occur. Formation   iew phase needs nucleation, i.e., the
spontaneous forma   egions of a few times molecular size in
diameter, say 2 nm.   rose, for instance, the data in Table 10.4
then predict a solu   io of 2.4, and nucleation would thus need
a supersaturatior   at magnitude, otherwise the tiny sucrose

TABLE 10.4 Increase in Solubility of Dispersed Substances Due to Surface Curvature

Disperse phase Continuous phase	Water triglyc. oil	Triglyc. oil water	Air water	Sucrose saturated soln.
r (μm)	0.5	0.5	50	0.01
γ (mN) $\cdot$ m^{-1}	10^a	10^a	40^a	5
V_D (m$^3 \cdot$ mol^{-1})	$1.8 \cdot 10^{-5}$	$7.7 \cdot 10^{-4}$	0.024	$2.2 \cdot 10^{-4}$
x' (nm)	0.14	6	770	0.9
$s(r)/s_\infty$	1.0003	1.012	1.016	1.09

r = radius (of curvature).
γ = interfacial tension.
V_D = molar volume (disperse phase).
x' = characteristic length.
$s(r)$ = solubility of disperse phase if radius is r.
s_∞ = solubility if $1/r = 0$.
[a] surfactant present.
Calculated according to Eq. (10.9) for $T = 300$ K.

regions would immediately dissolve again. Nucleation is discussed in Chapter 14.

Ostwald ripening. Consider two water droplets of different diameters in an oil. The water in the smaller one then has a greater solubility in the oil than the larger one. Consequently, the water content of the oil near the small droplet will be higher than that near the large one, and water will diffuse from the former to the latter. In other words, small droplets will shrink and large droplets grow. Such Ostwald ripening is a very common cause of slow coarsening of dispersions, whether the particles are gaseous, liquid, or solid. It especially occurs in foams. It is discussed in Section 13.6.

Particle shape. Most solid particles tend to be nonspherical, which means that their curvature varies along the surface. This is especially obvious for crystals, where most of the surface is flat while the curvature is very high where two crystal faces meet. This then means that the solubility of the material also varies, and this readily causes local dissolution of material, which is likely to become deposited at sites of small curvature. Table 10.4 shows that for a sucrose crystal a considerable solubility ratio (1.09) is found for $r = 10$ nm. However, where crystal faces meet, the shape would be cylindrical rather than spherical, leading to a solubility ratio of about 1.045. This is certainly sufficient to cause a crystal edge to become rounded in a saturated solution, and if the crystals are very small, they would likely be almost spherical. Indeed, microscopic evidence shows that many crystals of µm size are roughly spherical and that larger crystals often show rounded edges.

Capillary condensation. If the concave side of a curved surface is considered, the radius of curvature in Eq. (10.9) should be taken as negative, implying that the solubility of the material at the convex side of the surface would be locally decreased. This is indeed observed. Consider, for instance, a glass object that has little crevices on its surface. If the surrounding air is saturated with water, this leads to condensation of water in the crevices, because they would have a negative curvature. To explain this further, consider a porous material, assuming for convenience that the pores are cylindrical capillaries of 0.2 µm diameter; the pores contain some water, whereby curved air–water surfaces exist. The Kelvin equation now predicts that the saturation ratio for water in the air near an A–W meniscus is 0.99. For water-saturated air, this will lead to local condensation, and given enough time all capillaries will become filled with water. In practice, the situation is more complicated: the pores are of varying diameter and of irregular shape, and the pore surface often is heterogeneous. Nevertheless the uptake of water from the air can occur even if the solid matter itself cannot dissolve any water, although the pores must be very fine.

Several fine *powders* of such materials exhibit increasing stickiness—which is caused by interparticle attraction—at increasing water activity, often starting above $a_W \approx 0.9$. This is presumably due to capillary condensation, producing tiny water bridges between particles.

Question 1

Figure 10.23a shows two vessels each containing an amount of liquid. Are these stable situations or will the liquids start to move? If so, in what direction?

Answer

Left figure, liquid moves to the left. Right figure, to the right. Explain.

Question 2

Figure 10.23b shows a cylindrical thread of liquid in air that has obtained a shape as indicated. At what place is the Laplace pressure higher, at A or at B? What will happen with the liquid, assuming no external forces to act? Do you know of a situation in which this happens?

Answer

Assume that the original cross-sectional radius of the cylinder $r = 1$ unit. Then the amplitude of the "wave" on the cylinder ≈ 0.4, so the radii R_1 at A and B are 0.6 and 1.4, respectively. At A, $R_2 = -5$, and at B, $R_2 = 5$. Consequently at A, $p_L \approx 1/0.6 - 1/5 = 1.47$, whereas at B, $p_L \approx 1/1.4 + 1/5 = 0.91$. Although one of the principal radii of curvature is negative near A, the pressure at A is higher than that at B; consequently, liquid will flow from A to B. The thread is unstable and will eventually break up into drops. This is what commonly happens when a thin stream of water falls from a faucet, because—according to Rayleigh—any developing varicose wave of wavelength $> 2\pi r$ will cause the Laplace pressure to be higher in the narrow than in the wide parts of the thread.

(a)

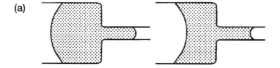

(b)

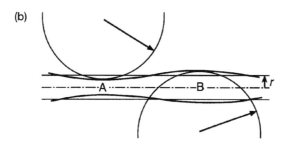

FIGURE 10.23 (a) Representation of an amount of liquid confined in containers with different wetting properties. (b) Cylindrical thread of liquid of original radius r that has undergone an axisymmetric varicose deformation; two radii of curvature indicated.

10.6 CONTACT ANGLES AND WETTING

If three different phases are in pairwise contact, thus giving three different phase boundaries, there is a *contact line* where the three phases meet. Two rather different cases can be distinguished:

> A solid, a liquid, and a fluid phase (gas or liquid); for instance, a water drop on a metal surface in air or in oil.
> Three fluid phases, of which one (and not more than one) is mostly gaseous; for instance, an oil drop on an air–water surface.

10.6.1 Contact Angle

We will first consider Figure 10.24a. At equilibrium there must be a balance of surface forces at any point on the contact line. Since all interfacial tensions act over the same length, this means that the tensions must balance. Considering the forces in the horizontal direction, the equilibrium condition implies that the interfacial tension γ_{AS} equals the sum of γ_{LS} and the projection of γ_{AL} on the solid surface. This leads to the **Young equation**,

$$\gamma_{AS} = \gamma_{LS} + \gamma_{AL} \cos \theta \qquad (10.10)$$

given here for the three phases (A, L, S) in the situation depicted. The characteristic parameter is the contact angle θ, which is conventionally taken in the densest fluid.

For the system solid paraffin, water and air, which is depicted in Figure 10.24c, a value of $\theta \approx 106°$ is measured. $\gamma_{AW} = 72 \, \text{mN} \cdot \text{m}^{-1}$, but the other interfacial tensions cannot be measured. However, $72 \times \cos 106 = -20$, which value must equal the difference $\gamma_{LS} - \gamma_{AS}$; this $20 \, \text{mN} \cdot \text{m}^{-1}$ is the same value as the difference between the corresponding values for liquid paraffin against water and air, as given in Table 10.1. However, the equality is not perfect, and for other systems the discrepancy may be greater.

Different substances give, of course, different contact angles and θ increases when going from Figure 10.24a to b to c. In case a it is said that the solid is preferentially wetted by the liquid, in case c by the air. It is also possible that $(\gamma_{AS} - \gamma_{LS})/\gamma_{AL} > 1$, which would imply $\cos \theta > 1$, which is of

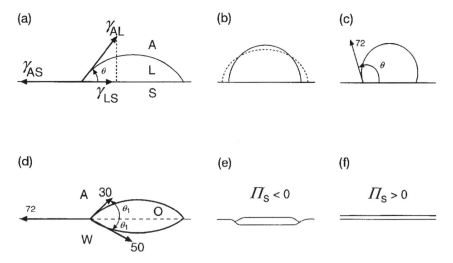

FIGURE 10.24 Contact angles (θ) of liquids at an A–S surface (upper row) and at an A–W surface (lower row). The pictures of the drops are cross sections through the largest diameter. A is air, L is liquid, O is oil, S is solid, W is water (or an aqueous solution). Numbers are interfacial tensions (γ) in $\text{mN} \cdot \text{m}^{-1}$. Π_s is spreading pressure. The scale varies among the panels.

course impossible. In that case, $\theta = 0$ and the solid is said to be fully wetted by the liquid: given enough space, the liquid will spread in a thin layer over the solid. Many liquids, including water and nearly all aqueous solutions, give zero contact angle at clean glass. The other extreme is that $\theta = 180°$; in that case the solid is not wetted at all, and a small drop tends to "float" on the solid, making no contact.

All these configurations are, however, modified if also other forces are acting on the system. In most situations, gravity plays a part, and the shapes given would then only be true if it concerned very small drops (e.g., < 0.1 mm). Figure 10.24b shows by a broken line the effect of gravity for a certain case. The drop becomes flatter and attains a larger diameter, but the contact angle (in this case 90°) remains the same.

In the case depicted in Figure 10.24a, the vertical component of the interfacial tension A–L, which equals $\gamma_{AL} \cdot \sin \theta$, is balanced by elastic reaction forces of the solid. Such forces cannot be exerted by (Newtonian) liquids, and for three fluid phases a *lens* as depicted in Figure 10.24d may be formed; the example given concerns paraffin oil, water, and air. Now two equations for a balance of forces must be met,

$$\gamma_{aw} = \gamma_{AO} \cos \theta_1 + \gamma_{OW} \cos \theta_2$$
$$\gamma_{AO} \sin \theta_1 = \gamma_{OW} \sin \theta_2$$

(10.11)

(*Note*: For convenience, we have taken the contact angle $\theta = \theta_1 + \theta_2$, in the oil phase, rather than in the denser water phase.) The three interfacial tensions can be determined (see data in Table 10.1) and the configuration of the drop at the interface can thus be calculated. The shape of the drop and of the A–W interface near the drop will also be affected by gravity. If θ is fairly small, a large drop will attain a shape as in Figure 10.24e.

Following Eq. (10.11), a two-dimensional *spreading pressure* can be defined,

$$\Pi_s = \gamma_{AW} - (\gamma_{AO} + \gamma_{OW})$$

(10.12)

If $\Pi_s > 0$, spreading of oil over the A–W interface will occur, as depicted in Figure 10.24f. For example, if the oil is a triglyceride oil, the data in Table 10.1 give $\Pi_s = 72 - (30 + 35) = 7$ mN $\cdot$ m^{-1}; hence the oil will spread over the surface. Given sufficient space, the oil layer can spread until it is one molecule thick. The data in Table 10.1 also indicate that paraffin oil will not spread over an A–W surface. Actually, Eq. (10.12) also applies to solid surfaces, but it is generally not so well obeyed (see Section 10.6.2).

For the case that Eq. (10.12) applies and $\Pi_s > 0$, it is easy to see that the total interfacial free energy is smallest if the water surface is covered by an oil layer. Also the solutions of Eqs. (10.10) and (10.11) represent minimum values of the interfacial free energy of the system. The equations given can also be derived by minimizing interfacial free energy.

10.6.2 Wetting

In Figure 10.22a the material of the capillary—say, glass—is completely wetted by the liquid—say, water. In other words, the contact angle equals zero. If the contact angle is finite, the radius of curvature of the meniscus is $> r$ and it will be given by $r/\cos \theta$. Consequently, Eq. (10.8) is modified to

$$h = \frac{2\gamma \cos \theta}{r\rho g} \tag{10.13}$$

This implies that the capillary rise is decreased. If $\theta = 90°$, there is no capillary rise, and for $\theta > 90°$, there is capillary depression. In Figure 20.22c as compared to b, the capillary depression is reduced by a factor $\cos 135/\cos 180 = 0.71$.

Contact angles can be modified by the addition of a **surfactant**, since that alters interfacial tension. This is illustrated in Figure 10.25a, for spherical solid particles (S) at an O–W interface. The contact angle in the aqueous phase decreases as the surfactant concentration increases, and it can even become zero, implying that the particle enters the aqueous phase, being dislodged from the O–W interface. Figure 10.25b gives a kind of state diagram. Straight lines of constant θ go through the origin. At or above the line for $\theta = 0$, the solid would be completely wetted by the aqueous phase; at or below $\theta = 180°$, it is completely wetted by oil; in between, there is partial wetting.

The system to which the data in Figure 10.25 roughly apply is of solid triacylglycerols (saturated, long-chain; β'-crystals), liquid triacylglycerol oil and a solution of Na-lauryl sulfate (SDS). It is seen that, without surfactant, the solid is far better wetted by the oil than by water, as is to be expected. From Table 1, we obtain $\gamma_{OW} = 30\,\text{mN} \cdot \text{m}^{-1}$, and it is generally found that the contact angle for fat crystals at a pure O–W interface $\approx 150°$. This leads to $\gamma_{OW} \cos \theta = -26$ and taking from Table 1 $\gamma_{OS} = 4$ (value derived from crystallization kinetics; see Chapter 15), we obtain $\gamma_{WS} \approx 30\,\text{mN} \cdot \text{m}^{-1}$, i.e., (nearly) the same value as γ_{OW}. If SDS is added, γ_{OW} and θ both decrease, in such a way that the curve in Figure 20.25b is practically straight and of slope -1; this implies that the value of γ_{OS} remains constant. The Young equation (10.10) then predicts that the difference $\gamma_{WS} - \gamma_{OW}$ (which we presumed to be almost zero in the absence of SDS) remains constant. Application of the Gibbs equation (10.2) then leads to the conclusion that SDS must adsorb to the same extent to the triglyceride oil–water interface as to the crystal–water interface, giving the same decrease in interfacial tension.

If the concentration of SDS is about 35 mmolar (i.e., well above the CMC), the solid becomes even completely wetted by the aqueous phase. Although most fat crystals are not nearly spherical, and the geometrical relations thus become more complicated, complete wetting can also be achieved in practice. In the so-called Lanza process fat crystals are separated from a dispersion in oil by washing with an aqueous solution of a suitable amphiphile.

For other systems, the relations may be quite different. Often, the plots of $\gamma_{OW} \cos \theta$ versus γ_{OW} are linear, but of a slope > -1. Moreover, many of such plots go almost through the origin, which implies that the contact angle will remain about the same, unless γ_{OW} becomes very small. Such a situation

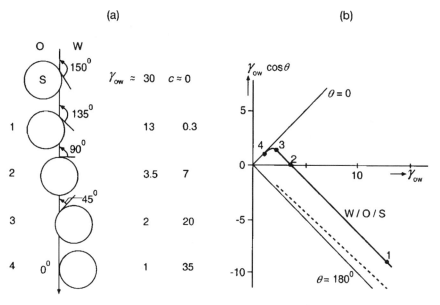

FIGURE 10.25 Effect of the concentration (c, in mol $\cdot$ m^{-3}) of a surfactant on the contact angle (θ) of spherical particles of a solid S at the O–W interface. γ is interfacial tension in mN $\cdot$ m^{-1}. (a) Configuration of the particles and some quantitative data. (b) State diagram of $\gamma_{OW} \cos \theta$ versus γ_{OW} and plot of the data for W–O–S systems. The numbered points correspond to the numbers near the particles at left. The broken line refers to another system, where $\theta \approx 150°$.

is depicted by the broken line in Figure 10.25b; it applies to solid tristearate, water, and a triacylglycerol oil with various oil-soluble surfactants.

Adhesion. Solid particles can thus become lodged in a fluid interface, like fat crystals in an O–W or A–W interface. This is often called *adhesion*. An important parameter is the strength of the adhesion, for instance when considering so-called *Pickering stabilization* of emulsion droplets by small particles (see Section 13.4.2), or flotation (small particles in a liquid can be removed by adsorption onto air bubbles in the liquid and subsequent creaming). The free energy needed to remove a particle from the interface, i.e., for θ going from its equilibrium value to zero, is for spherical particles given by

$$\Delta_{\theta-0}G = \pi r^2 \gamma_{OW}(1 - \cos\theta)^2 \tag{10.14}$$

Taking as an example the particles in Figure 10.25a for $c = 0$, we have $\gamma_{OW} = 0.03\,\mathrm{N\cdot m^{-1}}$, $\theta = 150°$, and assuming a particle radius $r = 1\,\mu m$, we obtain for ΔG $3.3\cdot10^{-13}\,\mathrm{J}$, corresponding to about $8\cdot10^7$ times $k_B T$. This means that the particle will never become dislodged spontaneously (i.e., owing to its Brownian motion). In other words, fat crystals do very strongly adhere to an O–W interface, illustrating again that surface forces can have large effects. Even for a particle as small as 10 nm radius, with $\gamma = 0.002$ and $\theta = 45°$ (Fig. 10.22a, No. 4), ΔG would amount to about 13 times $k_B T$, implying that it would not readily become dislodged ($e^{-13} \approx 2\cdot10^{-6}$). For still smaller particles, we enter the realm of macromolecules, and a molecular treatment as given in Section 10.2 would be needed.

Complications. Several factors affect the contact angles and thereby wetting. The interfacial tensions may differ from the tabulated values and change with time if the liquids involved are to some extent mutually soluble; dissolution may be a slow process. Such changes become more important if surfactants are present. In most cases, the surfactant (mixture) is added to one phase, and upon contact between the phases it may go not only to phase boundaries, which takes some time but also dissolve in one of the other phases, which would probably take a longer time. All these processes may affect contact angle and wetting, the latter especially if the spreading pressure [Eq. (10.12)] is close to zero.

Important complications often arise for the **wetting of solids**, say of water on a solid surface in contact with air or with oil (Fig. 10.24, upper row). These include

The solid *surface is inhomogeneous*. This is virtually always the case if the material contains several components that are not fully mixed, which is very common in foods; consider, e.g., chocolate, biscuits, several sweets. The contact angle will then vary with place. Even if the solid is made up of one chemical substance, the surface may be inhomogeneous: the various faces of a crystal generally differ in surface tension, and various faces may be present at the surface. This will generally be the case at a solid fat surface.

The water added may *dissolve some of the substances*, for instance the sugar crystals at the surface of a chocolate or a biscuit.

The *surface may be rough* (uneven) at a scale clearly larger than molecular sizes. The contact angle with respect to the average plane through the surface may then be very different at various places, even if the true contact angle is the same. This is illustrated in Figure 10.26a. It should be understood that the figure depicts a two-dimensional situation, whereas in three dimensions the relations are far more intricate.

These phenomena give rise to **contact angle hysteresis**, which is illustrated in Figure 10.26b. If a drop of water is applied to a solid and the contact angle is acute $(0 < \theta < 90°)$, the water tends to spread over the surface. In other words, the contact line advances over the solid. If water is removed from the drop, for instance by means of a small syringe, the contact line will recede. On a sloping surface, a drop will often slide downwards, exhibiting an advancing and a receding edge. It is invariably observed that the *advancing* contact angle is larger than the *receding* one. The hysteresis is generally between 1 and 60°; 10–20°C is quite common. The cause must be found in the three complications just mentioned. A satisfactory quantitative theory has not been developed. By adding

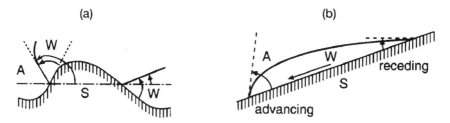

FIGURE 10.26 Contact angles of aqueous solutions (W) on solid surfaces (S) in air (A). (a) Effect of surface roughness. Contact angles given with respect to the true solid surface (inner ones) and to the apparent or average surface (depicted by — · — · — · — ·). (b) Illustration of the difference between an advancing and a receding contact angle.

surfactants, the advancing and receding contact angles are often affected to a different extent; in other words, the hysteresis then depends on surfactant type and concentration.

Question

Two aqueous detergent solutions, A and B, are compared for their ability to remove fat (F) from china (C). Suppose that the interfacial tensions (in $mN \cdot m^{-1}$) are as follows:

A–F = 16	B–F = 12
A–C = 25	B–C = 20
C–F = 35	

1. Which detergent would be superior?
2. Can the data just given be determined? If not, what would you do to find an answer?

Answer

1. By application of Eq. (10.10), we can calculate the contact angle θ, as measured in the detergent phase (D). This leads to $\cos \theta = (\gamma_{FC} - \gamma_{DC})/\gamma_{FD}$. For D = A, it yields $(35 - 25)/16 = 0.625$ or $\theta = 51°$; for D = B we have $(35 - 20)/12 = 1.25$, implying that $\theta = 0$. If the contact angle in the aqueous phase is zero, the detergent will completely wet the china and thus dislodge the fat from it; consequently, B will be the superior detergent. An even simpler way to reach this conclusion is by using Eq. (10.12). How?
2. No, the data cannot be obtained, because interfacial tensions involving a solid surface cannot be determined. What can be measured, however, is the contact angle, and that is a sufficient criterion, as mentioned.

10.6.3   Capillary Displacement

When a porous solid, in which the pores are filled with air, makes contact with a liquid, this will lead to the liquid displacing the air if the contact angle as measured in the liquid is acute or zero. An example is given by a sugar cube brought in contact with tea: the tea is immediately sucked into the pores between the sugar crystals. A liquid can also be displaced by another, immiscible, liquid. An example is a plastic fat—i.e., a continuous network of fat crystals filled with a continuous oil phase—where the oil can be displaced by an SDS solution of sufficient strength; cf. Figure 10.25. If gravity is acting and the displacement is upwards, the displacing liquid will move ever slower until the maximum height given by Eq. (10.13) is reached. In a horizontal direction, there would be no limit to the liquid penetration.

For straight cylindrical pores and in the absence of counteracting forces (like gravity), the *penetration rate* can readily be calculated. According to Poiseuille, the mean linear flow rate v of a liquid of viscosity η in through a pore of radius r is given by

$$v = \frac{r^2}{8\eta} \frac{\Delta p}{L} \tag{10.15}$$

where Δp is the pressure difference acting over a distance (capillary length) L. In the present case Δp is due to the Laplace pressure and is thus given by $(2/r)\gamma \cos \theta$. Insertion into (10.15) then yields

$$v = \frac{r\gamma \cos \theta}{4\eta L} \tag{10.16}$$

To give an example, if the displacing liquid is water ($\gamma = 72\,\mathrm{mN} \cdot \mathrm{m}^{-1}$, $\eta = 1\,\mathrm{mPa} \cdot \mathrm{s}$, $\theta = 0$, and pore radius and length are $1\,\mathrm{mm}$ and $10\,\mathrm{cm}$, respectively, we obtain $v = 18\,\mathrm{cm} \cdot \mathrm{s}^{-1}$, i.e., very fast. For $r = 10\,\mu\mathrm{m}$, $\gamma = 50\,\mathrm{mN} \cdot \mathrm{m}^{-1}$, $\theta = 45°$, $\eta = 10\,\mathrm{mPa} \cdot \mathrm{s}$ and $L = 1\,\mathrm{cm}$, we would have $v = 0.9\,\mathrm{mm} \cdot \mathrm{s}^{-1}$, still appreciable.

However, the pores in a porous solid are virtually never cylindrical. They are tortuous, which increases their effective length; they vary in diameter and shape, which causes the effective resistance to flow to be larger than that given in Eq. (10.15) and the effective Laplace pressure to be smaller than for the average pore radius. Most importantly, the effective contact angle will be significantly larger than the true contact angle A–W–S. This is similar to the situation depicted in Figure 10.27a. In a pore of variable diameter and shape, it may well be that for a true value of $\theta = 45°$, the meniscus of the liquid in the pore tends to become convex (as seen from the air) at some sites rather than concave; this implies that the liquid will not move at all. In many systems, the true contact angle has to be smaller than about 30° for the effective angle to be acute, i.e., for capillary displacement to occur.

Another complication is that equilibrium values of γ and θ are often not reached during displacement, for instance because adsorption of surfactant is too slow.

An important example of capillary displacement concerns the **dispersion of powders** in water. Most powders have particles in the range of 5 to 500 μm. An example is flour, which has to be dispersed in an aqueous liquid for the particles to swell and so obtain a dough. Another example is milk powder, which has to be dispersed in water to achieve its dissolution, thereby reconstituting milk. Most dried soups contain particles that become dispersed and other particles that also dissolve.

Figure 10.27a depicts a heap or lump of powder particles on water. Water should penetrate into it. This is usually aided by stirring, whereby water can penetrate from all sides. If the water does not penetrate fast enough, fairly firm lumps are formed, in which the outside is a gluey layer of partly dissolved or swollen powder particles, whereas the inside is still dry. It is quite difficult to disperse (and dissolve) such lumps.

Following are important phenomena in the dispersion process:

1. The *contact angle* air–water–powder particle should be acute. It can be measured by making a solid and smooth tablet of the powder by applying high pressure, and then let a drop of water fall on it and immediately measure θ; this is an advancing contact angle. It should be smaller than about 30°. As illustrated in Figure 10.27b, even for a fairly small value of θ, the meniscus of the water tends to become flat with increasing penetration between two particles, thereby stopping further penetration.

2. The *size of the pores* between the particles should be large to allow fast penetration. Although Eq. (10.16) only holds for cylindrical pores, the trends predicted apply to the present case. The pores are smaller for smaller powder particles and for a larger spread in particle size. A sample that is a mixture of two powders, one fairly coarse and the other quite fine, is notorious for slow penetration, since the small particles fill up the holes between the large ones.

3. In many powders, the void fraction between particles is large, over 0.5. This means that most of the pores are relatively wide. However, wetting

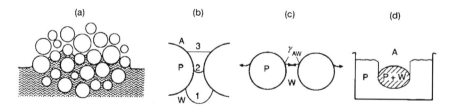

FIGURE 10.27 Dispersion of a powder in water; A = air, W = water, P = powder (particle). (a) Heap of powder on water into which water is penetrating. (b) Penetration of water between two particles at various stages (1–3) at constant contact angle (about 45°). (c) Pulling action (arrows) of the surface tension of water γ_{AW} on powder particles. (d) Situation after a large water drop has fallen on a layer of powder in a cup.

causes *capillary contraction*. As illustrated in Figure 10.27c, the surface tension of the water tends to pull the particles toward each other at a certain stage. The resulting contraction can be considerable. When a drop of water is allowed to fall on a layer of powder, a situation as depicted in Figure 10.27d may occur; it is no exception if the total volume of the powder + water is reduced by a factor of 2. This then causes the pores between the particles to become much narrower, slowing down penetration.

4.   If the powder *particles can swell* on the uptake of water, they will do so, given enough time. This will further decrease pore size, hence penetration rate.

5.   If (material from) the powder *particle dissolves* in water, this causes an increase in viscosity of the penetrating liquid, thereby further slowing down penetration.

A combination of phenomena 2, 3, and 4 or 5 will readily lead to the penetration coming to a standstill and hence to the formation of lumps. To give a powder **instant properties**, i.e., easy dispersability, the powder particles are often *agglomerated* into fairly large units. Such a powder then has large pores, that allow rapid penetration of water, and the agglomerates become readily dispersed, after which they can either dissolve or swell, according to powder type. If needed, the *contact angle can be effectively decreased* by coating the particles with a thin layer of lecithin. (Lecithin is a food grade surfactant that does not immediately dissolve upon contact with water; moreover, most lecithin preparations readily give a thin layer on the particles.) If a given powder cannot be readily dispersed it generally helps to increase temperature (i.e., decrease viscosity) and to apply vigorous stirring.

10.7   INTERFACIAL TENSION GRADIENTS

Figure 10.28a depicts an interface between pure water and a pure oil, where the water is caused to flow parallel to the interface. At the interface, there is a velocity gradient $\Psi = dv_x/dy$. There is thus a *tangential (shear) stress* $\eta_W \Psi$ acting on the interface (η_W is the viscosity of water, about 1 mPa · s). The interface cannot withstand a tangential stress, which implies that the liquid velocity must be continuous across the interface: interface and oil also move. The velocity gradient is not continuous, since $\eta_W \neq \eta_O$, and the shear stress must be continuous; in the picture, $\eta_O \approx 5 \eta_W$. If the upper fluid is air rather than oil, the velocity gradient in air will be very much larger, since $\eta_W \approx 5500 \eta_A$ (see Table 9.2).

In Figure 10.28b the situation is the same except for the interface containing a surfactant. For the moment we will assume that the surfactant is not soluble in either phase. Now the flow will cause surfactant to be swept

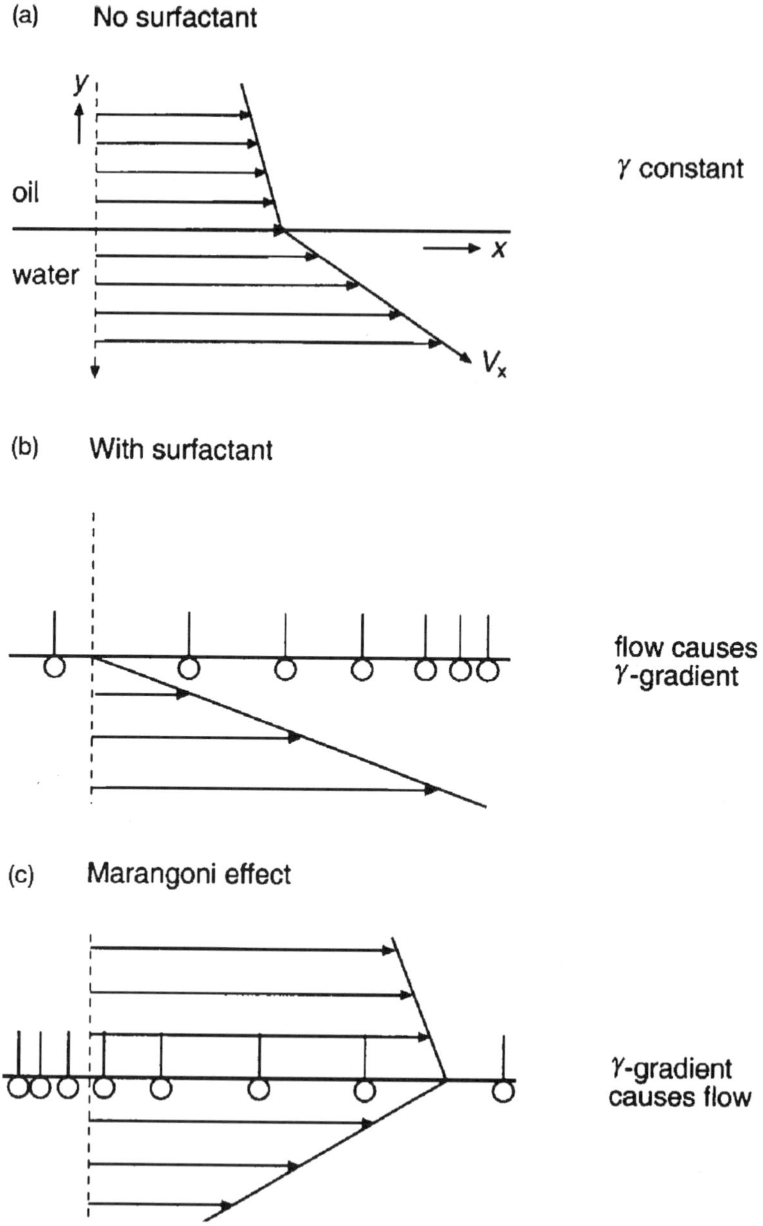
(a) No surfactant
y
oil
water
x
V_x
γ constant
(b) With surfactant
flow causes
γ-gradient
(c) Marangoni effect
γ-gradient
causes flow

FIGURE 10.28 Interfacial tension gradients in relation to flow of the bordering liquids. v = linear velocity, γ = interfacial tension. See text for explanation. Highly schematic.

downstream. This implies that a γ-**gradient** is formed. Such a gradient means that there is a *tangential stress in the interface* of magnitude $d\gamma/dx$ (in $N \cdot m^{-1}/m = Pa$). If this stress is large enough, the surface will be arrested, and thereby any motion of the oil. The equality of stresses is given in the relation

$$\eta_W \left(\frac{dv_x}{dy}\right)_{y=0} = \frac{d\gamma}{dx} \qquad (10.17)$$

For a constant gradient, we can use $\Delta\gamma$ rather than $d\gamma$; its value can at most equal the value of Π. Assuming this to be $0.03 \, N \cdot m^{-1}$, and $\Psi = 10^3 \, s^{-1}$ (i.e., quite a large value), we have $\eta_W \cdot \Psi = 1 \, Pa$, and Eq. (10.17) can be fulfilled for Δx up to $3 \, cm$. This implies that a fluid interface that contains surfactant and that is of mesoscopic size, say $0.01-100 \, \mu m$, would act as a solid wall for nearly all tangential stresses that may occur in practice.

A γ-gradient is thus very effective in withstanding a tangential stress and arresting tangential motion of an interface. Actually, the situation is more complicated. Generally, the surfactant is soluble in at least one of the phases, and exchange between interface and bulk will thus occur. Moreover, Eq. (10.17) is not always fully correct. See further Section 10.8.3.

In Figure 10.28c another situation is depicted. Now surfactant is applied at a certain spot on the interface. The surfactant will immediately spread over the interface in all directions, because that will cause a decrease of interfacial free energy. Hence a γ-gradient is formed, and this will exert a tangential stress on both liquids, causing them to flow. This is called the *Marangoni effect*. For an air–liquid interface, Eq. (10.17) will hold. It should be understood that a γ-gradient generally is a fleeting phenomenon, since it tends to be evened out by surface motion and exchange of surfactant with the bulk.

The Marangoni effect can be induced most easily by adding an insoluble surfactant onto a liquid surface, but the effect is quite general: any γ-gradient causes flow of the adjacent liquid(s) in the direction of increasing γ. A temperature gradient in an interface does also produce a γ-gradient (see, e.g., Table 10.1), even in the absence of surfactant.

The formation of γ-gradients can only be achieved (at constant temperature) by substances that alter interfacial tension. This capacity may be the most important property of surfactants. To illustrate this we give here a few examples.

Foaming and emulsification. The formation of γ-gradients is all that allows the formation of foams and of most emulsions. Consider making a foam. Very soon vertical films (lamellae) of liquid (generally water) between air bubbles will be present. As illustrated in Figure 10.29a, the water will flow downward, and if no surfactant is present, the A–W interfaces cannot carry a stress and the water will flow as if there were no interfaces. In other words, the water falls down like a drop. This then means that the foam immediately, i.e., within seconds, disappears, as is commonly observed. If a surfactant is present, a situation as in Figure 10.29b will generally exist. The downflow of water immediately causes formation of a

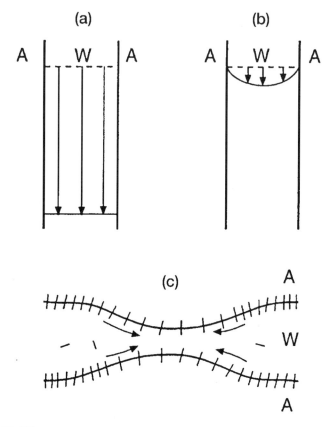

FIGURE 10.29 Foam lamellae. (a) Downflow of water between two air bubbles in the absence of surfactant. (b) Same, in the presence of surfactant. (c) Gibbs mechanism of film stability; surfactant molecules depicted by short lines. The arrows indicate motion of the surface and of the bordering liquid. See text for explanation. Highly schematic.

γ-gradient, hence the A–W interfaces can withstand a tangential stress. Hence the interfaces act like solid walls, so that the downflow of water is very much retarded, the more so as the lamella becomes thinner. Thus the foam will have a lifetime that is orders of magnitude longer than in the absence of surfactant.

Diffusion across an interface. Consider a pond containing pure water. If the air above it is dry, water will evaporate from the surface, especially if the wind is blowing. The air flow will readily be turbulent, so that water vapor can be transported from the pond surface by convection. Now a surfactant is added, enough to produce a monomolecular layer on the pond, and the evaporation rate is markedly reduced. It is often assumed that the surfactant layer provides resistance to evaporation because water cannot readily diffuse through it. However, the layer is very thin (a few nanometers) and can only cause a small resistance to diffusion (see Section 5.3.3). The main explanation of the reduced evaporation must be that the wind over the surface causes a γ-gradient, so that the surface can now withstand a tangential stress; hence a laminar boundary layer of air will be formed near the surface, and the diffusion of water vapor through the boundary layer (which may be about a millimeter thick) will cause a considerable decrease in transport rate.

Film stability. The formation of γ-gradients is all that allows "stable" liquid films to be made. A film of pure water immediately breaks. To be sure, a thin film is never stable in the thermodynamic sense, but its lifetime can be quite long if it contains surfactant. Figure 10.29c illustrates the so-called *Gibbs mechanism* for film stability. If for some reason a thin spot forms in a film, this implies a local increase in surface area, hence a local decrease in surface load, hence a local increase in surface tension, hence motion of the film surfaces in the direction of the thin spot, hence a Marangoni effect, i.e., flow of liquid toward the thin spot, hence a self-stabilizing mechanism. Actually, a more elaborate treatment of film stability is needed (see Section 13.4.1), but the Gibbs mechanism is essential.

Wine tears. The wine in a glass may show the formation of wine tears on the glass wall above the wine surface, which phenomenon is enhanced when the glass is gently rocked. It occurs especially with wine of a fairly high ethanol content. The explanation is illustrated in Figure 10.30. In frame 1 we see the meniscus. Ethanol will evaporate from the thin layer in the meniscus, locally decreasing the ethanol content. This will cause an increase of surface tension (see Figure 10.4). Hence a γ-gradient is formed, hence the Marangoni effect will transport wine upward (frame 2), hence a thicker rim of ethanol-depleted wine is formed (frame 3). This rim is subject to Rayleigh instability (comparable to the phenomenon discussed in Question 2 at the end of Section 10.5), hence "tears" will be formed at regular distance intervals. The tears will grow by further upflow of wine caused by evaporation and become sufficiently heavy to move downwards due to gravity. Other tears will grow, and so on. This process goes on until the ethanol content has become too low.

Spreading Rate. If a gradient in interfacial tension occurs in a liquid interface, because the interface is suddenly expanded or some surfactant is locally applied to the interface, the *interfacial tension will be evened out*, i.e., become the same everywhere at the interface. The *rate* at which this occurs if of considerable importance for the extent of the Marangoni effect. It proceeds as a *longitudinal surface wave*. The linear velocity of the wave on an A–W interface is given by

$$v = 1.2(\eta \rho z)^{-1/3} |\Delta \gamma|^{2/3} \tag{10.18}$$

where z is the distance over which the wave has to travel. The density ρ and viscosity η are those of the aqueous phase. For a wave on an O–W interface, $\eta_W \rho_W$ must be replaced by $(\eta_W^2 \rho_W^2 + \eta_O^2 \rho_O^2)^{0.5}$. This means that in many cases, $\eta \cdot \rho$ of the most viscous phase can be taken, the other one being negligible.

If there is only one surfactant, the wave velocity can be interpreted as "the rate of spreading of surfactant" over the interface. Actually, the word spreading is to some extent misleading, since *it is the interface that moves*, taking the surfactant with it. Putting $v = dx/dt$, integration of Eq. (10.18)

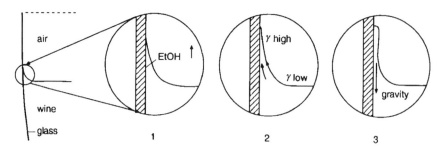

FIGURE 10.30   Formation of wine tears.

gives for the "spreading" time

$$t_{spr} \approx z^{4/3} \eta^{1/3} \rho^{1/3} |\Delta\gamma|^{-2/3} \qquad (10.19)$$

Some sample calculations: for an O–W interface, assuming $\Delta\gamma = 0.02 \, mN \cdot m^{-1}$ and $\eta_O \rho_O = 70 \, kg^2 \cdot m^{-4} \cdot s^{-1}$ we obtain for

$z =$	$1 \, \mu m$	$0.1 \, mm$	$1 \, cm$
$t_{spr} =$	$0.6 \, \mu s$	$0.3 \, ms$	$0.1 \, s$

which shows that the motion can be very fast. For an A–W surface, it would be faster by a factor $70^{1/3} \approx 4$. The values predicted by Eqs. (10.18) and (10.19) agree well with experimental results on A–W interfaces at macroscopic distances.

Figure 10.31 illustrates what would happen if a surface is "instantaneously" enlarged. The surfactant spreads according to Eq. (10.19) over the clean surface created. The surface excess will thus be decreased (by about a factor of 2), and the interfacial tension will be enlarged (see Section 10.2.3). This means that adsorption equilibrium does not exist anymore [Eq. (10.2)], and surfactant will be adsorbed until the original interfacial tension is reached again (assuming the total amount of surfactant present to be in excess). The rate of adsorption will be given by Eq. (10.6).

Two cautioning remarks may be useful. First, Eqs. (10.18) and (10.19) are only valid as long as $\Delta\gamma$ is constant. Its value will often decrease during the process. It is difficult to deduce how large the effect will be, in part because it will depend on the surface equation of state, and the author is unaware of a quantitative treatment of this problem. It may be that spreading times can be as much as a factor of 10 longer than given by Eq. (10.19).

Second, Eq. (10.18) only holds for evening out of the interfacial tension. If more than one surfactant is present, the composition of the adsorbed surfactant layer may differ from place to place, although γ is everywhere the same. Especially for poorly soluble surfactants, evening out of the surface layer composition then has to occur by surface diffusion, which may be quite slow.

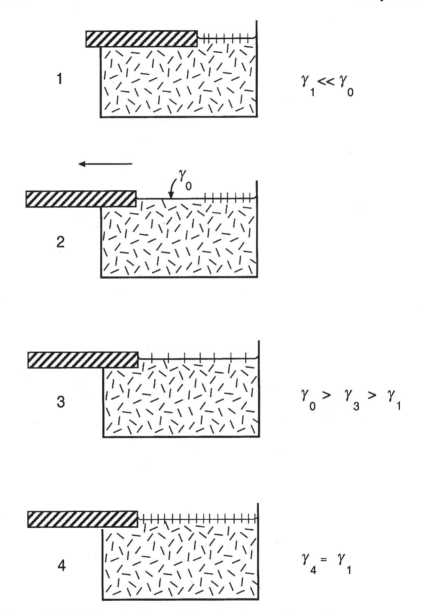

$\gamma_1 \ll \gamma_0$

$\gamma_0 > \gamma_3 > \gamma_1$

$\gamma_4 = \gamma_1$

FIGURE 10.31 Vessel with a solution of surfactant (denoted by short lines), of which the surface is "immediately" enlarged (frame 2). Spreading of surfactant (3) and adsorption (4). γ = surface tension. See text. Highly schematic.

Question

Assume that the experiment illustrated in Figure 10.31 is done with a solution of Na-stearate and with one of Na-myristate. Each solution has a surface tension of $50 \, mN \cdot m^{-1}$. The barrier is moved over 5 cm, thereby increasing the A–W surface area by a factor of two. For which solution is the spreading of surfactant fastest? What is the time needed to restore the initial value of γ for each solution?

Answer

To calculate the spreading time, Eq. (10.19) is needed. The variables z (5 cm), η (1 mPa $\cdot$ s) and ρ (10^3 kg $\cdot$ m^{-3}) are the same for both solutions, and this would also hold for $\Delta\gamma = \gamma_0 - \gamma = 0.072 - 0.05 = 0.22 \, N \cdot m^{-1}$. A spreading time of 0.23 s then follows. Adsorption is needed to restore γ, and the data in Figure 10.6 give at $\gamma = 0.05$, that for C18 $c = 0.16 \, mol \cdot m^{-3}$, and $\Gamma = 8.8 \, \mu mol \cdot m^{-2}$; for C14, these values would be 7.1 and 4.0, respectively. Inserting these data into Eq. (10.6) yields for the adsorption time a value of 100 s for the stearate solution and of 0.01 s for the myristate.

We may conclude that for Na-stearate the spreading is much faster than the adsorption (because of its low bulk concentration). This implies that during adsorption the initial Γ value is half of $8.8 \, \mu mol \cdot m^{-2}$ (since the surface area was doubled) and also half that amount would have to be adsorbed. Equation (10.6) has Γ^2 in the denominator, and replacing Γ by half the final value leads to an adsorption time of 25 rather than 100 s, which is still very much longer than the spreading time. Consequently, we have spreading first, followed by adsorption, as shown in panels 3 and 4 of Figure 10.31.

For Na-myristate, the adsorption is much faster (because of its relatively high bulk concentration) than the spreading, by a factor 23. This means that the value of $|\Delta\gamma|$ decreases rapidly, greatly slowing down the spreading rate. In fact, very little spreading will occur, and the adsorption time is indeed about 0.01 s. In the sequence given in Figure 10.31, panel 3 can thus be omitted.

If the same calculations are done for Na-palmitate, spreading and adsorption cannot be so nicely separated, and calculation of the times involved is far more intricate.

10.8 INTERFACIAL RHEOLOGY

Basic aspects of rheology are discussed in Sections 5.1.1 and 2. This concerns "bulk rheology." Rheological theory can also and usefully be applied to the deformation of fluid interfaces. A main problem is that an interface cannot exist by itself; it is the boundary between two phases and these phases must be deformed with the interface. Surface or interfacial rheology then is concerned with *excess* quantities; in other words, the force needed to deform the bulk materials is somehow subtracted.

As in bulk rheology, various modes of deformation can be applied in interfacial rheology. Some important variants are depicted in Figure 10.32. *Bending* of an interface produces a Laplace pressure [Eq. (10.7)]; the higher the surfactant concentration, the smaller the bending force needed. It may further be noted that a close-packed surfactant layer can fairly strongly resist bending, though only if the radius of curvature is of molecular dimension (order of 1 nm). Bending will not be further discussed in this section.

The other two modes of interfacial rheology only become manifest in the presence of surfactant. An essential difference between deformation in *shear* (Fig. 10.32b) and in *dilatation/compression* (Fig. 10.32c) is that in the former case Γ is constant, whereas in the latter cases the local surface area is enlarged or diminished, whereby Γ varies. This implies that interchange of surfactant between bulk and interface will in most cases occur. It should be added that a change in the area of a surface element does not necessarily imply that the total surface area is changed: expansion at one place can be compensated for by compression elsewhere. An increase of total surface area would mean that the total surface free energy is increased, which needs the application of forces. This is generally not included in surface dilatational rheology.

In practice, interfaces are often subjected to a combination of the deformations mentioned. As in bulk rheology, there are some other variables. First, the response of a material to a force can be elastic or viscous. Elastic response means immediate deformation, where the strain (relative deformation, i.e., $\tan \alpha$ in shear and $\Delta A/A$ in dilatation) is related to the force; on release of the force, the strain immediately becomes zero. In viscous deformation, the force causes flow or, more precisely, a strain rate (d $\tan \alpha/dt$ or d $\ln A/dt$); this occurs as long as the force lasts, and upon release of the force the strain achieved remains. For most systems, the behavior is viscoelastic. Second, deformation can be fast or slow, and time scales between a microsecond and more than a day may be of importance. Third, the relative deformation (strain) applied can be small—i.e., remain close to the equilibrium situation—or be large.

Surface rheology is in two dimensions. The stresses involved are thus given as unit force per unit length, i.e., in $N \cdot m^{-1}$ in SI units (as compared to $N \cdot m^{-2}$ for three dimensions). Surface elastic moduli are expressed in $N \cdot m^{-1}$, and surface viscosities in $N \cdot s \cdot m^{-1}$.

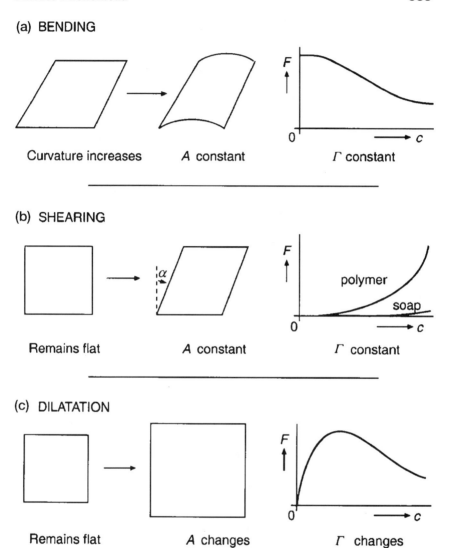

FIGURE 10.32 Interfacial rheology. Schematic illustration of various types of deformation of a square surface element and the approximate result for the force (F) needed for deformation as a function of the concentration of surfactant (c). In (a) the surface element is seen in perspective, in (b) and (c) from above. A = area of surface element, Γ = surface excess.

10.8.1 Surface Shear

Application of a two-dimensional shear stress can be done in various ways, for instance as depicted in Figure 10.33a; here an annular surface or interface is sheared, and the torque on the disc (or, alternatively, on the vessel) is measured. Naturally, also the bordering liquids are sheared, and to obtain true surface parameters, the torque measured in the absence of surfactant has to be subtracted.

Most workers determine *surface shear viscosity* η^{SS}, defined as the (two-dimensional) shear stress over the shear rate. To be sure, most surface layers are viscoelastic and shear rate–thinning, and one thus determines an apparent viscosity η_a^{SS}; often, the surface shear rate applied is of order $0.1 \, \mathrm{s}^{-1}$. The value obtained has been called "film strength," a very misleading term. It is questionable whether a monolayer can be called a film, since this word generally refers to a far thicker layer that has two surfaces. More important, the property measured is not a strength, which would be the stress needed for the adsorbed layer to break or maybe to yield. In fact, also a surface shear modulus E^{SS} can be measured and, for a large strain, yielding or fracture can possibly occur but systematic experiments in that direction appear to be lacking.

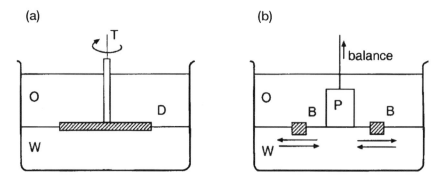

FIGURE 10.33 Principles of interfacial rheological measurements. (a) In shear. A thin disc (D) is at an O–W interface and is made to rotate (or oscillate); the torque on the disc can be measured, e.g., via a torsion wire (T). (b) In expansion/compression. Barriers (B) at an O–W interface are moved, thereby increasing or decreasing the interfacial area between them; the interfacial tension is measured by means of a Wilhelmy plate (P). Both kinds of measurement can also be made at A–W and A–O surfaces.

The prime cause of the surface shear viscosity is friction between surfactant molecules; the cause of surface shear elasticity is attractive forces between those molecules, leading to a more or less continuous two-dimensional network. For a closely packed layer, the effects may be substantial. For layers of small-molecule surfactants, however, the values of η_a^{SS} are generally immeasurably small, about $10^{-5} \, \text{N} \cdot \text{s} \cdot \text{m}^{-1}$ or less. For adsorbed polymers, values between 10^{-3} and $1 \, \text{N} \cdot \text{m} \cdot \text{s}^{-1}$ have been reported.

For example, for Na-caseinate at an A–W surface, $\eta_a^{SS} \approx 2 \cdot 10^{-3} \, \text{N} \cdot \text{s} \cdot \text{m}^{-1}$ has been observed. Other experiments reveal that the thickness of such an adsorbed layer equals about 10 nm. Interpretation of η_a^{SS} as a bulk viscosity of a layer of that thickness would yield $\eta_a = 2 \cdot 10^{-3}/10^{-8} \approx 2 \cdot 10^5 \, \text{N} \cdot \text{s} \cdot \text{m}^{-2}$. Furthermore, $\Gamma \approx 3 \, \text{mg} \cdot \text{m}^{-2}$ is observed for adsorbed Na-caseinate; this yields, for the average concentration of caseinate in a layer of 10 nm, a value of 30% (w/v). A Na-caseinate solution of that concentration does indeed have a bulk viscosity of about $2 \cdot 10^5 \, \text{N} \cdot \text{s} \cdot \text{m}^{-2}$. It thus appears as if surface shear viscosity may be interpreted as the bulk viscosity of a layer of unknown thickness. An adsorbed layer that is thicker (compare, e.g., frames 2 and 5 in Figure 10.16) or denser would thus yield a higher value of η_a^{SS}.

Caseinate is a mixture of fairly flexible polymers. Most proteins are of globular conformation, and their surface properties are not easy to interpret. The values of η_a^{SS} are much higher and tend to increase with the age of the film. It may take a day to obtain a more or less constant value, which is typically 0.1–$0.5 \, \text{N} \cdot \text{s} \cdot \text{m}^{-1}$. However, the surface layer is clearly viscoelastic, and the apparent viscosity obtained will strongly depend on measurement conditions, especially the shear rate. Actually, it cannot always be ruled out that the proteinaceous surface layer is subject to yielding or fracture upon large deformation; this would imply that "slip" occurs in the rheometer, leading to a greatly underestimated viscosity.

The increase in apparent viscosity with time points to slow rearrangements of protein structure and possibly to the formation of intermolecular bonds. A protein like β-lactoglobulin, which contains an —SH group, is known to be subject to —S—S— bond reshuffling, leading to bonds between molecules if these are close to each other. In an adsorbed layer, η_a^{SS} keeps increasing for days, leading to values well over $1 \, \text{N} \cdot \text{s} \cdot \text{m}^{-1}$.

Surface shear viscosity (and modulus) may thus tell us something about conformation and thickness of adsorbed protein layers, and especially about changes with time and with composition. The latter may involve the addition of small quantities of an amphiphile, which tends to greatly reduce η_a^{SS}, or partial displacement of an adsorbed protein by another one. However, a clear and simple theory is not available; combination with the determination of other surface properties is generally needed for a reasonable interpretation of the results.

10.8.2 Surface Dilatation

Figure 10.33b illustrates how to measure surface dilatational parameters. An essential aspect is that the interfacial tension is directly measured. If the surface area A is increased, γ is increased, and vice versa. In principle, one can also measure a two-dimensional stress, but it is far easier to measure γ, which also has the advantage of excluding any effect of the coupling of bulk flow with that of the interface. Preferably, the shape of the surface element remains unchanged upon expansion–compression, to avoid any shearing in the plane of the surface. The change in γ upon a change in A proceeds as a longitudinal wave, the velocity of which is given by Eq. (10.18).

A *surface dilatational modulus* is defined as

$$E^{SD} \equiv \frac{d\gamma}{d\ln A} \qquad (10.20)^*$$

If it concerns a monolayer of an amphiphile that is insoluble in the bordering phases, the modulus is purely elastic (although at strong compression, i.e., large $-\Delta A/A$, the surface layer may collapse), and E^{SD} is constant in time and independent of the dilatation rate. If the surfactant is soluble, exchange of surfactant between interface and bulk occurs, and E^{SD} will be time dependent. This means that also an apparent *surface dilatational viscosity* can be measured:

$$\eta_a^{SD} \equiv \frac{\Delta\gamma}{d\ln A/dt} \qquad (10.21)$$

which tends to be strongly strain rate thinning.

Note A more sophisticated treatment is possible by introducing a complex modulus, as discussed in Section 5.1.3; see Figure 5.9.

Prediction of E^{SD} from measurable parameters is often possible. We will mention two fairly simple cases. In the first one, an interface is bounded by a semi-infinite surfactant solution (the surfactant is not soluble in the other phase), and the transport of surfactant to and from the interface is governed by diffusion. The result is

$$E^{SD} \approx \frac{d\Pi/d\ln\Gamma}{1 + (Dt)^{0.5}dc/d\Gamma} \qquad (10.22)$$

where D = diffusion coefficient, t = time scale of the deformation, and c is bulk surfactant concentration (mol $\cdot$ m^{-3}); Γ should be in mol $\cdot$ m^{-2}. If $t \to$ 0, i.e., very fast deformation, the denominator $\to 1$, and E^{SD} is purely elastic and depends on the surface equation of state only. For finite t, exchange of surfactant with the bulk occurs, which is governed by surfactant concentration c and the adsorption isotherm (which determine dc/dΓ, see Section 10.2.2) and, of course, by D.

   Some results are shown in Figure 10.34. As predicted, E^{SD} decreases with increasing time scale, because of increasing exchange of surfactant with the bulk. At small t, the modulus is larger for a higher surfactant concentration; the explanation is simply that the value of Π is higher, so $\Delta\Pi$ can be larger. At long time scales, the highest modulus is observed for the

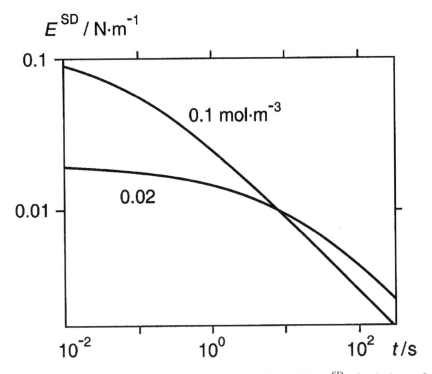

FIGURE 10.34 Values of the surface dilatational modulus E^{SD} of solutions of decanoic acid at two concentrations (indicated) as a function of time scale t. (Adapted from E. H. Lucassen-Reynders. Anionic Surfactants. Surfactant Series Vol. 11. Marcel Dekker, New York, 1981, p. 201.)

lowest value of c, because transport of surfactant by diffusion is slower, as it has to occur over a greater distance. The dependence of E^{SD} on surfactant concentration is, for a constant value of t, like that in Figure 10.35.

The second case concerns a *thin film*. Here the change in Π is governed by the limited amount of surfactant in the bulk liquid in the film, since by far most of the surfactant will be in the adsorbed layers. The time for diffusional transport (normal to the surfaces) is taken to be negligible in the thin film; it would nearly always be < 0.1 s. The modulus then is purely elastic. Rather than the modulus, the **Gibbs elasticity** of the film is given, by

$$E_{G} = 2\,\frac{\mathrm{d}\Pi/\mathrm{d}\ln\varGamma}{1 + (\delta/2)\mathrm{d}c/\mathrm{d}\varGamma} \tag{10.23}$$

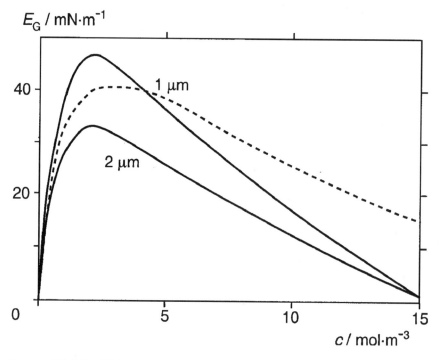

FIGURE 10.35 Gibbs elasticity E_{G} of films made of solutions of SDS of various concentrations c, for two film thicknesses (indicated). The broken line gives an approximate example of results to be expected for a mixture of surfactants. (Adapted from J. Lucassen, in E. H. Lucassen-Reynders (Ed.). Anionic Surfactants. Surfactant Series Vol. 11. Marcel Dekker, New York, 1981, p. 232.)

where δ is the film thickness. The factor of 2 arises because a film has two surfaces.

Some results are given in Figure 10.35. It is seen that E_G goes through a maximum at increasing surfactant concentration, and that the maximum is higher for a thinner film. At very high c values, E_G tends to go to zero. This is because $dc/d\Gamma$ then becomes very large or, in other words, at very high c (above the CMC) the original value of Π can be fully restored. A similar trend is predicted by Eq. (10.22): at fairly long time scales and a high c value, E^{SD} tends to go to zero. In practice, this is rarely observed. The reason must be that most surfactant preparations are mixtures. Some components are present at low concentrations and these can still substantially contribute to the modulus. The broken line in Figure 10.35 gives an example of a relation for a surfactant mixture.

Surface dilatational properties tend to differ considerably between amphiphiles and **polymers**, partly because the surface equation of state is different. From Eqs. (10.22) and (10.23) it follows that if (a) t is short or δ is thin and (b) $dc/d\Gamma$ is small (which is true if Γ is small), $E^{SD} \approx d\Pi/d \ln \Gamma$. Figure 10.36 gives Π versus $\ln \Gamma$, and the slopes of the curves will thus roughly give E^{SD}. It is seen that SDS will give a substantial modulus at far smaller Γ values (in $mg \cdot m^{-2}$) than the proteins do. This is a general trend.

Another difference may be even more important. Polymers can change their conformation in an interface and thereby the interfacial tension. If A is increased, causing a decrease in Π, polymer molecules in the surface may expand and thereby increase the value of Π again, without any additional adsorption occurring. Looking again at Figure 10.36, we see very different curves for β-casein and lysozyme, although these proteins do not differ greatly in molar mass or in hydrophobicity. However, the former can readily unfold on adsorption, and the latter is a compact protein. Thus β-casein would give a higher Π value for the same value of Γ. The determination of curves like those of Figure 10.36 demands quite some time, at least several minutes, whereas the conformational changes of the casein after adsorption will be finished within 10 s (see Section 10.4). This implies that at much shorter time scales, the Π/Γ curve may be quite different for β-casein and be more like that of lysozyme.

This all means that the application of the theory, as given in this section, to polymers is questionable, primarily because of the conformational changes mentioned, and possibly also because of coupling with surface shear effects. Interpretation of results on E^{SD} of adsorbed proteins is still a matter of debate. The moduli are often highly nonlinear, greatly decreasing with increasing ΔA.

Surface dilatational properties are essential in several phenomena of practical importance, because these determine what γ will be at fast deformation and what γ-gradient can be formed. The latter determines the magnitude of the tangential stress that an interface can withstand and the strength of a Marangoni effect. We may mention the importance of E^{SD} for formation of foams and emulsions, stability of thin films, and rate of Ostwald ripening.

10.8.3 Stagnant Surfaces

In Section 10.7 it was explained how flow along an interface can induce a γ-gradient (Fig. 10.28b), and how such a γ-gradient can arrest lateral movement of the interface [Eq. (10.17)]. This is certainly true for an

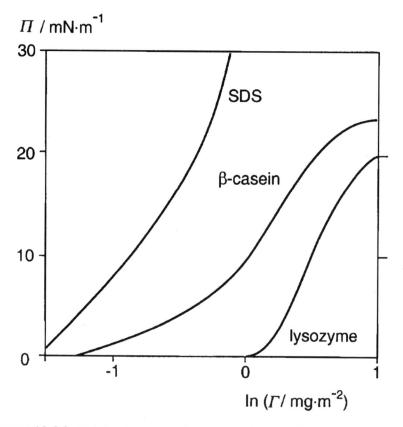

FIGURE 10.36 Relation between surface pressure Π and surface excess Γ at the A–W surface for three surfactants.

insoluble monolayer at the interface: on dilatation the interface reacts in a fully elastic manner. In most cases, however, the surfactant is soluble, and the value of the γ-gradient will depend on the value of the surface dilatational modulus, roughly given by Eq. (10.22). In other words, E^{SD} is time scale dependent. The higher E^{SD}, the more the motion of the interface will be slowed down. The various situations are depicted in Figure 10.37. In the presence of a surfactant, the common situation is the one depicted in frame b, implying that there will nearly always be some, albeit slight, interfacial motion. This then means, for instance, that the downward flow of liquid in a foam lamella will be faster than suggested by Figure 10.29b.

Nevertheless, in some situations—nearly always involving macromolecular adsorbates—truly stagnant surfaces are observed. This would mean that Eq. (10.17) does not hold. Its left-hand side gives the shear stress at the interface, and this cannot be wrong, but the right-hand side $(d\gamma/dx)$ need not be the only tangential stress exerted by the interface. If a liquid flows along a *solid surface*, no γ-gradient is developed, but the elastic reaction force of the

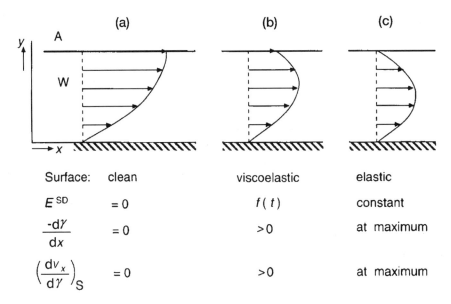

Surface:	clean	viscoelastic	elastic
E^{SD}	$= 0$	$f(t)$	constant
$\dfrac{-d\gamma}{dx}$	$= 0$	> 0	at maximum
$\left(\dfrac{dv_x}{d\gamma}\right)_S$	$= 0$	> 0	at maximum

FIGURE 10.37 Effect of surface dilatational modulus (E_{SD}) on the motion of an A–W surface; the aqueous phase flows over a solid support. $v =$ linear velocity; $t =$ time scale; subscript S means at the surface. Highly schematic. It is assumed that the distance over which the surface can move is small enough to allow Eq. (10.17) to be valid.

solid provides the stress needed to arrest the surface. Something similar must happen for some adsorbates at a *fluid interface*. Probably, a polymeric surface layer cannot be considered as an infinitely thin layer, and it may be that a stress can develop in the layer, i.e., in a direction normal to the surface. The phenomenon is yet insufficiently understood; it is certainly of importance, for instance for the rate of drainage in foams.

10.9 RECAPITULATION

Surface Tension. The presence of an interface between two phases goes along with an excess free energy that is proportional to the interfacial area. For a clean fluid interface the specific interfacial free energy (in $J \cdot m^{-2}$) equals the surface or interfacial tension (in $N \cdot m^{-1}$). This is a two-dimensional tension acting in the direction of the interface, which tries to minimize the interfacial area. The surface tension of a solid cannot be measured.

Adsorption. Some substances tend to adsorb onto an interface, thereby lowering the interfacial tension; the amount by which it is lowered is called the surface pressure. The Gibbs equation gives the relation between three variables: surface pressure, surface excess (i.e., the excess amount of surfactant in the interface per unit area), and concentration—or, more precisely, thermodynamic activity—of the surfactant in solution. This relation only holds for thermodynamic equilibrium, and the interfacial tension in the Gibbs equation is thus an equilibrium property. Nevertheless, also under nonequilibrium conditions, a tension can be measured at a liquid interface.

The relation between surface excess and surfactant concentration (in solution) is called the adsorption isotherm, that between surface excess and lowering of interfacial tension the surface equation of state. Both depend greatly on the type of surfactant and also on the type of interface (e.g., air–water or oil–water). A surfactant is said to be more surface active if the adsorption isotherm is shifted to lower concentrations: in other words, less surfactant is needed to obtain a given surface excess or a given decrease of interfacial tension.

Surfactants. Surfactants come in two main types: small amphiphilic molecules (for short called "amphiphiles") and polymers, among which are proteins. Small-molecule surfactants readily exchange between surface and solution, and a dynamic equilibrium is thus established, in accordance with the presumptions of the Gibbs equation. Most amphiphiles exhibit a critical micellization concentration (CMC), greatly varying among surfactants. Additional surfactant added above the CMC forms micelles; this means that the surfactant activity does not increase any more, and the surface excess at any surface present reaches a plateau value. Another variable is the hydrophile/lipophile balance (HLB value); the higher this value, the better soluble is the surfactant in water and the less in oil. For HLB = 7, the solubility in both phases is about equal.

Polymeric surfactants are generally (far) more surface active, but they give lower surface pressures than most amphiphiles. At the plateau value of the surface excess they are not very tightly packed (most amphiphiles are), but they extend fairly far into the solution. The exchange between solution and interface may be very slow, and the Gibbs equation does not seem to hold. Most amphiphiles can displace polymers from the interface, if present in sufficient concentration, since they give a lower interfacial tension. Mixed surface layers can also be formed.

Time Effects. Surfactants that adsorb are often transported to the interface by diffusion. For most amphiphiles this is a fast process, the times needed ranging from a millisecond to a few minutes. For polymers, it can be much slower. For mixtures of surfactants, changes in surface composition and interfacial tension may take a long time. Several complications can arise, such as very slow adsorption of poorly soluble surfactants (e.g., phospholipids), or a greatly enhanced adsorption rate due to convection. In processes like foam formation, the interfacial tension at short time scales is of importance; to obtain such values, one determines so-called dynamic surface tensions, i.e., values obtained at rapidly expanding surfaces.

Curvature. For a curved liquid interface, the pressure at the concave side is always higher than that at the convex side, by an amount called the Laplace pressure; its value is greater for a smaller radius of curvature and a larger interfacial tension. This has several consequences, such as capillary rise of a liquid in a thin pore, if the material is wetted by the liquid. Another consequence is that the material at the concave side (say in a droplet) has an increased solubility in the surrounding fluid, the more so for a smaller radius of curvature. The relation is given by the Kelvin equation, which also holds for solid materials. The phenomenon is responsible for the supersaturation needed for nucleation of a new phase to occur, for Ostwald ripening (small particles in a dispersion tend to dissolve, whereas the large ones grow), and for capillary condensation in fine pores.

Contact Angles. Where three phases are in contact with each other, the phase boundaries meet at a given contact angle, determined by the three interfacial tensions (Young's equation). The contact angle determines whether and to what extent the wetting of a surface by a liquid occurs;

thereby the cleaning action of detergents, the occurrence of capillary displacement, the wetting by and the dispersion of powders in a liquid; and the adhesion of solid particles to a liquid interface. The contact angle can be substantially affected by the addition of surfactant. The contact angle on a solid surface is generally not an equilibrium property: an advancing angle tends to be larger than a receding one.

It may be added here that the four basic laws of capillarity, i.e., the equations of Gibbs [(10.2)], Laplace [(10.7)], Kelvin [(10.9)] and Young [(10.10)], all describe manifestations of the same phenomenon: the system tries to minimize its interfacial free energy. (Another manifestation is found in the Hamaker equations; see Section 12.2.1.) These laws describe equilibrium situations. Moreover, dynamic surface phenomena are of great importance.

Interfacial Tension Gradient. An interfacial tension gradient can occur in an interface containing surfactant. The tension then varies in the direction of the interface. This creates a two-dimensional stress in that direction. The gradient can be induced by flow of the bordering liquid, and if the gradient is large enough, the interface is arrested (it does not move with the flowing liquid). This is essential in making foam, since it greatly slows down the drainage of liquid from it. A gradient can also be induced by locally applying a surfactant or by locally increasing the interfacial area (by bulging). Then the interface will move to even out the interfacial tension, and this motion will drag bordering liquid with it, the Marangoni effect. The latter is essential for the stability of thin films containing surfactant. Evening of the interfacial tension proceeds as a longitudinal surface wave, which can have a very high velocity.

Interfacial Rheology. The extent to which interfacial tension gradients and the resulting effects occur depends on the surface dilatational modulus. This concerns one type of surface rheology, where a surface element is expanded or compressed without changing its shape, and the change in interfacial tension is measured. The modulus is primarily determined by the exchange of surfactant between interface and bulk and thus is time dependent. In a thin film, it is determined by the limited amount of surfactant present (twice its value is called the Gibbs elasticity). The modulus is zero in the absence of surfactant and goes through a maximum when the concentration is increased. For polymeric surfactants, the modulus is also affected by changes in conformation.

Another kind of interfacial rheology is done in shear: a surface element is sheared without altering its area, and the force needed is measured. It is due to friction or attraction between surfactant molecules. Modulus and viscosity tend to be negligible for amphiphiles but may become appreciable for polymeric surfactants, especially proteins. Interpretation of the results is not easy. A high surface shear modulus probably can cause the interface to act more or less as a solid in the lateral direction, thereby producing a fully stagnant surface if liquid streams along it.

Surfactants fulfil several **functions**, and some important ones are the following:

- Their presence permits the formation of interfacial tension gradients, which may be considered the most essential function. For instance, this is all that allows the formation of foams and emulsions and provides inherent stability to thin films.
- They lower the interfacial tension of liquid interfaces, thereby facilitating bending of the interface, hence deformation and breakup of drops and bubbles.
- They affect contact angles and thereby capillary phenomena such as wetting, adhesion of particles, capillary displacement, and dispersion of powders in a liquid.
- By adsorption onto particles they may greatly affect colloidal interaction forces between those particles. Repulsive forces may provide long-term stability against aggregation; attractive forces may allow the formation of continuous networks.

BIBLIOGRAPHY

A standard text on several aspects discussed in this chapter, including methods of measurement is

A. W. Adamson, A. P. Gast. Physical Chemistry of Surfaces, 6th ed. John Wiley, New York, 1997.

It contains little information on surfactants and very little on dynamic surface properties. An excellent monograph is

E. H. Lucassen-Reynders, ed. Anionic Surfactants: Physical Chemistry of Surfactant Action. Marcel Dekker, New York, 1981.

Especially see Chapters 1 (adsorption), 5 (surface dilatational phenomena) and 6 (dynamic properties of films).

A comprehensive description of many small-molecule surfactants used in foods is given by

N. J. Krog. Food emulsifiers and their chemical and physical properties. In: K. Larsson, S. E. Friberg, eds. Food Emulsions, 2nd ed. Marcel Dekker, New York, 1990.

Also interesting is

E. Dickinson, D. J. McClements. Surfactant micelles in foods. In: Advances in Food Colloids, Blackie, Glasgow, 1995, Chapter 8.

Adsorption of polymers is extensively treated in

G. J. Fleer, M. A. Cohen Stuart, J. M. H. M. Scheutjens, T. Cosgrove, B. Vincent. Polymers at Interfaces. Chapman and Hall, London, 1993.

Adsorption properties of biomolecules (proteins, lipids, polysaccharides), and a range of techniques for studying such properties, are discussed in

A. Baszkin, W. Norde, eds. Physical Chemistry of Biological Interfaces, Marcel Dekker, New York, 2000.

The basic aspects of wetting properties of (milk) powder are made very clear by

A. van Kreveld. Neth. Milk Dairy J. 28 (1974) 23.

A very thorough discussion of dynamic interfacial properties is given by

D. A. Edwards, H. Brenner, D. T. Wasan. Interfacial Transport Processes and Rheology. Butterworth, Boston, 1991.

The determination of some dynamic surface properties is treated by

A. Prins. Dynamic Surface Tension and Dilational Interfacial Properties. In: E. Dickinson, ed. New Physicochemical Techniques for the Characterization of Complex Food Systems. Chapman and Hall, London, 1995.

Stagnant layers are discussed by

A. Prins. Stagnant surface behaviour and its effect on foam and film stability. *Colloids Surfaces A* 149 (1999) 467.

Dynamic interfacial properties of proteins are reviewed by

M. A. Bos, T. van Vliet. Interfacial rheological properties of adsorbed protein layers and surfactants: a review. *Adv. Colloid Interf. Sci.* 91 (2001) 437–471.

11

Formation of Emulsions and Foams

There are two main methods of making dispersions. One is *supersaturation* of the continuous phase with the component that is to become the disperse phase. This involves the processes of nucleation (Chapter 14) and growth (Chapter 15 for solids). The particles can be obtained in a wide range of sizes. The method is applied in food processing, for instance to obtain sugar crystals that are subsequently separated from the liquid, or to make a plastic fat—a dispersion of aggregated fat crystals in oil. The other method involves *dividing a material* into small particles that are then suspended in a liquid. Making suspensions of solid materials is done by dry or wet grinding, which tends to be a very difficult operation if small particles are desired. The theory involves fracture mechanics and shows that it is very difficult to break up small particles into still smaller ones, which is also observed in practice (see Section 17.1.2). An example of the process is the lengthy milling of the (liquid) cocoa mass in chocolate manufacture to obtain very small sugar particles (about $10\,\mu m$).

Emulsions and foams are dispersions of two fluids, which implies that the interface between the phases is deformable. This makes the breakup of one of the materials into small particles far easier, which does not imply that quantitative understanding of the phenomena involved is easy. Some aspects are discussed in this chapter.

Many aerated foods are dispersions of air (e.g., whipped egg white) or carbon dioxide (e.g., a head on beer) in water. Emulsions come in two types: oil-in-water (O–W) and water-in-oil (W–O); in foods, the oil is nearly always a triglyceride oil. O–W emulsions include milk and several milk products, creams, mayonnaise, dressings, and some soups. Very few foods are true W–O emulsions; butter, margarine and most other spreads contain aqueous droplets in a mass of oil and crystals.

In the making of foams and emulsions, hydrodynamic and interfacial phenomena interact. Relevant basic aspects are discussed in Section 5.1 and Chapter 10, respectively.

11.1 INTRODUCTION

To make an emulsion (foam), one needs oil (a gas), water, energy, and surfactant. The energy is needed because the interfacial area between the two phases is enlarged, hence the interfacial free energy of the system increases. The surfactant provides mechanisms to prevent the coalescence of the newly formed drops or bubbles. Moreover it lowers interfacial tension, and hence Laplace pressure [Eq. (10.7)], thereby facilitating breakup of drops or bubbles into smaller ones.

> *Note* The word bubble is sometimes reserved for an air cell in air, i.e., a spherical cell surrounded by a thin film. A cell of air in a liquid would then be called a cavity. For convenience, we will nevertheless call the latter a bubble.

Several **methods** can be used to make emulsions or foams, for instance:

1. **Supersaturation.** This is not applied in food emulsion making but is fairly common in foams. A gas can be dissolved in a liquid under pressure, and then the pressure is released, so that gas bubbles are formed. The gas should be well soluble in water to obtain a substantial volume of bubbles, and carbon dioxide is quite suitable. It is applied in most fizzy beverages. CO_2 can also be formed in situ by fermentation, as in beer and in a yeast dough. Initiation of gas bubble formation is discussed in Section 14.4.

2. **Injection.** Gas or liquid is injected through small openings, for instance in a porous sheet, into the continuous phase. In this way bubbles or drops are directly formed. They are dislodged from the sheet by buoyancy (some foams) or by weak agitation (most emulsions). The method for making emulsions is often called membrane emulsification. The "membrane" generally consists of porous glass or ceramic material.

3. **Agitation.** In agitation (stirring, beating, homogenizing), mechanical energy is transferred from both phases to the interfacial region—which

can lead to the formation of bubbles or drops—and from the continuous phase to these particles—whereby they can be disrupted into smaller ones. It is by far the most common method for making food emulsions and foams. Since all forces have to be transferred via the continuous phase, it is inevitable that a large proportion of the mechanical energy applied is dissipated, i.e., converted into heat.

A special type of agitation is due to *cavitation*. This is the formation of vapor cavities in a liquid by local negative pressures, and the subsequent collapse of these cavities. The latter phenomenon generates shock waves that can disrupt nearby particles. Cavitation can be induced by ultrasonic waves, and ultrasound generators are useful for making emulsions in small quantities.

4. **Chemical energy.** Some surfactant mixtures, during gentle mixing of the emulsion ingredients, can produce instability of an oil–water interface. Instability means that capillary waves form spontaneously, which can lead to droplet formation by a kind of "budding." To the author's knowledge this is not applied in food manufacture: the surfactant concentrations needed are too high to be acceptable in a food.

For the technologist, the prime question when making a foam or an emulsion concerns the physicochemical properties of the product made. This mainly concerns

Volume fraction of dispersed phase φ, since it determines some essential properties, such as rheological ones. For an emulsion, the value desired can mostly be predetermined by the proportions of oil and water in the recipe. This is often different for foams, especially when made by beating: the value of φ obtained then depends on several conditions (see the next section). In foams one often speaks of the overrun, i.e., the percentage increase in volume due to incorporation of gas. The relation is percentage overrun $= 100\varphi/(1 - \varphi)$.

Particle size distribution, $f(d)$; see Section 9.3. It is of considerable importance for the physical stability of the system. Generally, the smaller the droplets or bubbles, the more stable the system.

Surface layer of the particles, i.e., thickness and composition. Again, this greatly affects the stability.

Emulsion type, i.e., O–W or W–O. This determines several properties: see Section 9.1.

Moreover, the *economy* of the process is of importance; thus may include the amount of foam obtained (hence the overrun), the amount of surfactant needed (related to the specific surface area $A = 6\varphi/d_{32}$), and the amount of energy needed.

An important point to be made is that processes occurring during *formation* and changes in the system at rest (*"instability"*) should be distinguished. Conditions or ingredients—such as the surfactant type—that are suitable for making an emulsion or a foam are not necessarily suitable for obtaining a stable system. Several different kinds of instability exist, and they are discussed in Chapter 13. For most emulsions, the formation and the changes occurring in the system made can be studied separately, but this is generally not so for a foam, because a foam is far less stable and it takes more time to make it: formation and subsequent changes may in fact occur simultaneously. This is because of differences in particle size, volume fraction, density of the disperse phase, etc. between foams and emulsions. Some typical values of internal variables are given in Table 11.1. The quantitative differences between an emulsion and a foam can be large enough to obtain systems with very different properties.

Finally, making an emulsion or a foam is a highly complex process, involving several different, though mutually dependent, phenomena in the realm of hydrodynamics and dynamic surface properties. The quantitative relations depend on the composition of the system, the construction of the apparatus used, and the energy input level. We will not consider all these aspects but merely outline the most important principles involved.

TABLE 11.1 Order of Magnitude of Some Quantities in Foams and Emulsions at Room Temperature[a]

Property	Foam	Foam	Emuls. W–O	Emuls. O–W	Units
Drop/bubble diameter	10^3	100	3	0.5	μm
Volume fraction	0.9	0.9	0.1	0.1	—
Drop/bubble number	10^9	10^{12}	10^{16}	10^{18}	m^{-3}
Interfacial area	0.005	0.05	0.2	1.2	$m^2 \cdot ml^{-1}$
Interfacial tension	40	40	6	10	$mN \cdot m^{-1}$
Laplace pressure	10^2	10^3	10^4	10^5	Pa
Solubility D in C	2.1^b	2.1^b	0.15	0^c	% v/v
Density difference D – C	-10^3	-10^3	10^2	-10^2	$kg \cdot m^{-3}$
Viscosity ratio D/C	10^{-4}	10^{-4}	10^{-2}	10^2	—
Typical time scale[d]	10^{-3}	10^{-4}	10^{-5}	10^{-6}	s

[a] Oil phase is a triacylglycerol oil, gas phase is air, D = disperse phase, C = continuous phase.
[b] If the gas phase is CO_2, the solubility is about 100% v/v, but strongly dependent on composition of the aqueous phase, especially pH and salt composition.
[c] Oils often contain minor components that are somewhat soluble in water, but the solubility of the triacylglycerols is generally negligible.
[d] Time needed for separate events during formation, e.g., the deformation time of a drop or bubble.

Question 1

A company wants to make O–W emulsions by means of membrane emulsification. They decide to use a "membrane" with pores of 1 μm diameter (*d*). Can you think of some necessary conditions for making this endeavor a success, mainly by using knowledge gained from Chapter 10? Tip: Make a simple drawing of the situation. You may assume the pores to be cylindrical and at a right angle to the membrane surface.

Answer

First, the material of which the membrane is made should preferentially be wetted by the aqueous phase, since otherwise no drops would be formed at the end of the pores, but an oil layer spreading over the membrane. Ideal would be a zero contact angle as measured in the aqueous phase. Second, the oil should be pressed through the pores, and the pressure should be larger than the Laplace pressure. The latter will be at maximum when the oil is making a half sphere at the end of a pore, and it then equals $4\gamma/d$, which may amount to (see Table 10.1) $4 \times 0.01/10^{-6} = 40$ kPa or 0.4 bar. (Actually, the pressure will have to be substantially larger, to realize a reasonable flow rate of the oil through the membrane.) Third, the pores should be at sufficient distance from each other to avoid contact between the drops emerging from the pores, since otherwise the coalescence of the drops may readily occur. Experience shows that droplet diameter obtained is at least 3 times the pore diameter, which implies that the mutual distance must be at least 3 μm.

Question 2

You are whipping 200 ml of a protein solution with a small kitchen beater. After 2 minutes you have obtained 400 ml of foam. Microscopic observation shows that the bubbles are about 0.2 mm in diameter. Can you make a rough estimate of the fraction of the net energy applied that has been used to create the A–W surface?

Answer

A small kitchen beater has an electric motor consuming, e.g., 150 W. Assuming that the net power uptake then would be 50 W, this would imply a net energy uptake of $50 \times 2 \times 60 = 6000$ W·s or J. The increase in surface area would be $A = 6\varphi/d = 6 \times 0.5/2 \cdot 10^{-4} = 15,000$ m^2/m^3 or 6 m^2 in 400 ml. Assuming $\gamma = 0.05$ N·m^{-1}, the total amount of surface free energy added due to whipping would be $6 \times 0.05 = 0.3$ J. This would equal 0.005% of the net energy input.

> *Note* This example illustrates the very poor efficiency of such a beating method with respect to energy use. In industrial machines for foam or emulsion making a higher efficiency can be obtained, though rarely over 0.3%.

11.2 FOAM FORMATION AND PROPERTIES

When a foam is made in a vessel containing a low-viscosity liquid, either by injection of air bubbles or by beating, phenomena as illustrated in Figure 11.1 will generally occur. As mentioned, foam bubbles are fairly large, and they cream rapidly, at a rate of $1\,\text{mm} \cdot \text{s}^{-1}$ or faster. They thereby form a layer on top of the liquid. As soon as the layer is a few bubble diameters in thickness, buoyancy forces cause the bubbles to deform each other against their Laplace pressure, further increasing the volume fraction of air in the layer. Often, the bubble size distribution becomes fairly monodisperse, because small bubbles disappear by Ostwald ripening [see Section 10.5.3, especially Eq. (10.9)]: small bubbles have a high Laplace pressure, so the air inside has an increased solubility in water, and the air diffuses to larger bubbles nearby. Since the diffusional distance is small, this process can be quite rapid.

11.2.1 Geometry

The phenomena just mentioned lead to formation of a **polyhedral foam**: the shape of the air cells approximates polyhedra. For cells of equal volume, the shape would be about that of a regular dodecahedron (a body bounded by 12 regular pentagons), and the edge q then equals about $0.8r$, where r is the radius of a sphere of equal volume. Actually, the structure is less regular, because of polydispersity. Moreover, close packing of true dodecahedrons is not possible. In a "two-dimensional" foam, say, one layer of bubbles

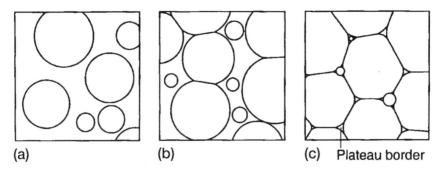

(a)   (b)   (c)   Plateau border

FIGURE 11.1   Subsequent stages, from (a) via (b) to (c), in the formation of a foam after bubbles have been made. The thickness of the films between bubbles is too small to be seen on this scale.

between two parallel glass plates, cells of equal volume can show a regular array of close-packed hexagons, as in a honeycomb.

The thin films between bubbles always meet at angles of 120°, assuming the surface tension to be constant, which is generally the case. For angles of 120° a balance of forces exists, as is illustrated in Figure 11.2a. If the structure is not completely regular—and it never is—at least some of the films must be curved. This is illustrated in Figure 11.2b. Now the Laplace pressure in the central bubble depicted is higher than that in the four surrounding cells, and the central one will disappear by Ostwald ripening. This leaves a configuration of four planes meeting at angles of 90°. Although in this case a balance of forces is also possible, the slightest deviation from 90° causes an unstable configuration, which will immediately lead to rearrangement, as depicted. (In other words, a configuration with 120° angles coincides with the lowest possible surface area, hence a minimum in surface free energy.) In a real foam, such rearrangements occur continuously and follow an intricate pattern. Anyway, the foam becomes coarser.

An important structural element in a foam is the **Plateau border**, i.e., the channel having three cylindrical surfaces that is formed between any three adjacent bubbles. Similar channels are formed where two bubbles meet the wall of the vessel containing the foam. Figure 11.3a shows a cross section, and it follows that the Laplace pressure inside the Plateau border is smaller than that in the adjacent films. This means that liquid is sucked from the films to the Plateau borders, whence it can flow away (drain), because the Plateau borders form a connected network. The curvature in the Plateau border is determined by a balance of forces, Laplace pressure versus

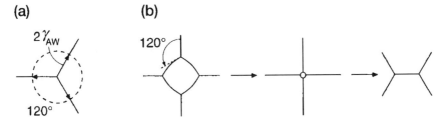

(a) **(b)**

$2\gamma_{AW}$ 120°

120°

FIGURE 11.2 Cross sections through bubble configurations in foams. (a) Balance of forces where three flat films meet. (b) Change of configuration when a small bubble amidst four larger ones disappears.

gravitational pressure. The relation is

$$p_L = \frac{\gamma}{R} = \rho_w g H \tag{11.1}$$

where ρ_w is the density of the aqueous phase (assuming that of air to be negligible) and H is the height above the bottom of the foam. This means that R is smaller at a greater height in the foam (compare Figures 11.3a and b), i.e., the Plateau borders are thinner and the foam contains less water. The bubbles in the bottom layer are practically spherical. Assuming all of the water to be located in the Plateau borders, it can be derived from Eq. (11.1) and the geometry of the system that the volume fraction of air in the foam at a given height H will be given by

$$\varphi \approx 1 - 0.5 \left(\frac{\gamma}{\rho_w g H q} \right)^2 \tag{11.2}$$

where the length of the Plateau borders q equals about 0.4 times the bubble diameter d. The equation is only valid for values of H above which the foam is truly polyhedral. To give an example: for $d = 0.3$ mm and $\gamma = 50$ mN·m^{-1}, and at a height above the liquid of 0.1 m, we obtain $\varphi \approx 0.91$; for $d = 1$ mm, φ would then be as large as 0.99. This would mean that a foam can become very "dry."

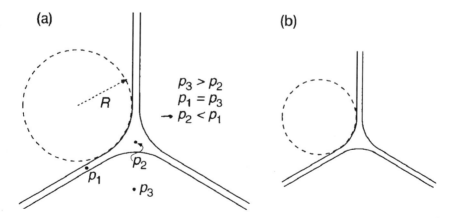

FIGURE 11.3 Cross sections through Plateau borders and films. (a) Illustration of the difference of the Laplace pressure between border and film. (b) Plateau border at a greater height in the foam.

11.2.2 Drainage

It will, of course, take time to obtain a dry foam, since the water has to drain from it, and drainage is greatly hindered by the narrowness of the films and channels in the foam. Drainage theory is intricate and not fully worked out, and we will only consider drainage from a single vertical film due to gravity. The film has a thickness δ, width q, and height h. The gravity force acting on the film is given by the mass of the film times g; i.e., $F_{gr} = \delta q h \rho_w g$. This gives a shear stress on each film surface of $(1/2)\, F/h\, q = (1/2)\, g\delta\rho_w$. The surface can withstand a shear stress due to the formation of an interfacial tension gradient, resulting from the flow of the liquid; this is discussed in Section 10.7 (see especially Figure 10.28b). The reaction stress due to a γ-gradient would equal $\Delta\gamma/h$. The maximum value that $\Delta\gamma = \gamma - \gamma_0$ can reach equals Π. This gives for the maximum height that a film can have to prevent slip, i.e., to prevent the film surfaces from moving with the liquid flowing down, is given by

$$h_{max} = 2\frac{\Pi}{\rho_w g \delta} \tag{11.3}$$

Assuming $\Pi = 0.03\,\text{N}\cdot\text{m}^{-1}$ and $\delta = 0.1\,\text{mm}$, we obtain $h_{max} = 6\,\text{cm}$. This would imply that even for large bubbles and thick films, the γ-gradient can become large enough to prevent slip.

Assuming this to be true, the relation for the volume flow rate $Q\ (\text{m}^3\cdot\text{s}^{-1})$ out of a vertical film is given by $Q = 2\,\rho_w\, gq\delta^3/3\eta$, where η is the viscosity of the continuous phase. Integration yields, for the time it needs for the film to reach a thickness δ,

$$t(\delta) \approx 3\frac{\eta h}{\rho g \delta^2} \tag{11.4}$$

Assuming $h = 0.5\,\text{mm}$, $\eta = 2\,\text{mPa}\cdot\text{s}$, and $\delta = 10\,\mu\text{m}$, the time would be 3 s; for $\delta = 20\,\text{nm}$ we obtain 9 days. This means that a fairly thin film is rapidly obtained, but that it would take a long time for a film to become so thin that colloidal repulsion forces between the two film surfaces become significant (generally at $\delta \approx 20\,\text{nm}$).

These results should be viewed with caution. In the first place, they concern a single film, not a foam, and can therefore only give trends. Secondly, we have assumed the film surfaces to be immobile, which would imply that the surface dilational properties are completely elastic. This is rarely the case, especially if it concerns small-molecule surfactants. Consequently, some slip will occur at the film surfaces, and drainage will be faster than that of Eq. (11.4). As discussed in Section 10.8.3, several protein surfactants can cause the surfaces to be fully stagnant. It is indeed observed that such surfactants cause significantly slower drainage than small-molecule surfactants do, for a foam of the same ρ and bubble size distribution.

Question

In Figure 11.3, two cross sections through a Plateau border with adjacent films are depicted. Assume now that these pictures represent two different situations but at the same height in the foam. Which of the parameters listed below can be the cause of the difference? Reason for each parameter whether it would be higher in (a) or in (b) or has no effect:

> The surface tension
> The density of the liquid
> The age of the foam
> The viscosity of the liquid
> The pressure above the foam (for this parameter, the question has two

different correct answers, depending on what is assumed to be constant).

11.2.3 Overrun

Ideally, all of the liquid in a vessel is incorporated in the foam made by beating or air injection, but this is often not achieved. A layer of liquid remains, with a layer of foam on top. The amount of foam obtained will result from a balance between making the bubbles, drainage, and coalescence. Several variables affect these processes and thereby the resulting overrun. Coalescence of bubbles at the top of the foam with the air above will especially decrease overrun.

Surfactant Concentration.   The first requirement is that sufficient surfactant is present to cover the total air surface produced (i.e., give it a surface load comparable to the plateau value Γ_∞). Assume an overrun of 300%, i.e., $\varphi = 0.75$, and an average bubble diameter $d_{32} = 60\,\mu m$, this implies a specific surface area $A = 6\varphi/d_{32} = 0.05\,m^2$ per ml foam, or $0.2\,m^2$ per ml liquid. Assume, moreover, that $\Gamma_\infty = 3\,mg \cdot m^{-2}$; this corresponds to $0.6\,mg$ of adsorbed surfactant per ml liquid. Since some excess of surfactant is needed to reach full surface coverage (cf. Figure 11.15b), we would need at least about 1 mg of surfactant per ml, or 0.1%. If an overrun of 1000% is desired, with a bubble size of 30 μm, a similar calculation leads to 1% surfactant.

Figure 11.4a gives some examples (obtained in small-scale laboratory equipment). To mention a more practical situation, overrun values of 400–1800% have been obtained when beating 1% whey protein solutions. Another point is that an optimum protein concentration for overrun is often observed. The decrease in overrun upon a further increase of concentration may be due to increased viscosity, hence slower drainage, hence a foam containing more liquid; but this has not been well studied.

Other (possible) effects of surfactant concentration are mentioned below.

Surfactant Type. The surfactant of choice for making food foams is nearly always protein, in order to obtain sufficiently stable foams (see Chapter 13). Figure 11.4 shows the differences between two proteins; in practice, a far wider range of variation is observed. Figure 11.5 shows results of some proteins in relation to dynamic surface tension. It is seen that ovalbumin and lysozyme give virtually no decrease in γ at a high expansion rate, i.e., short time scales; in other words, $\Pi \approx 0$. This then means that γ-gradients cannot form and drainage will be so fast as to prevent foam formation. The differences among the proteins shown cannot be due to a difference in the rate of protein transport to the surface, since the

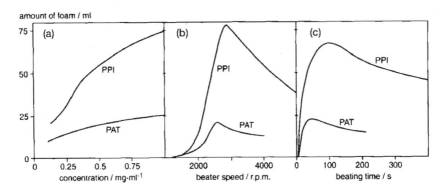

FIGURE 11.4 Amount of foam formed from dilute solutions of a potato protein isolate (PPI) and purified patatin (PAT); pH = 7.0, ionic strength = 0.05 molar. Foam was made on a small scale by beating. (a) Effect of protein concentration. (b) Effect of beater speed (revolutions per minute). (c) Effect of beating time. (Results by courtesy of G. van Koningsveld.)

concentration was equal and all of the proteins are well soluble. The most likely explanation is the rate of (partial) unfolding of the protein after becoming adsorbed; these time scales roughly are 100 s for lysozyme, 0.25 s for β-lactoglobulin and 0.1 s for β-casein. Comparing this with the results in Figure 11.5, it follows that the characteristic time scale during bubble formation would be roughly 0.5 s.

To make a foam with slowly unfolding proteins, it will be advantageous to use a method that involves longer time scales. This is the case for injection (Method 2 in Section 11.1), and it is indeed observed that most globular proteins, even those that are known to have a high conformational stability, give copious foams by injection (although the bubble size tends to be large). Foam can also be made by beating if the protein concentration is much higher, and ovalbumin (the main protein of egg white) provides a good example. Altogether, a substantial excess of protein over the amount needed for A–W surface coverage has to be present in nearly all cases.

Another variable is the effective **molar mass** of the surfactant. Smaller molecules tend to give a higher molar surface load at the same mass concentration in solution, and thereby higher Π-values. Protein hydrolysates often have superior foaming properties, where faster unfolding of smaller peptides may also play a part. (On the other hand, the foam tends to be less stable against Ostwald ripening.) Aggregation of proteins, e.g.,

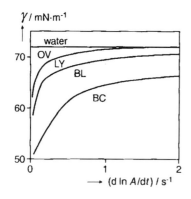

Protein		% overrun	d/mm
OV	ovalbumin	8	0.8
LY	lysozyme	8	0.7
BL	β-lactoglobulin	67	0.1
BC	β-casein	82	0.1

FIGURE 11.5 Foaming properties of some proteins (solution of 0.25 mg per ml) in relation to the dynamic surface tension. γ is surface tension; d ln A/dt is the surface expansion rate; d is the approximate average bubble diameter. (After results by H. van Kalsbeek, A. Prins. In: E. Dickinson, J. M. Rodríguez Patino, eds. Food Emulsions and Foams. Roy. Soc. Chem., Cambridge, 1999, pp. 91–103.)

caused by denaturation, tends to decrease overrun, simply because Γ_∞ then is much higher, implying that more protein is needed to cover a given surface area. Heat denaturation of whey proteins tends to decrease overrun by a factor of about four.

Beating conditions have considerable effect on overrun, but few systematic studies have been done. During beating, large bubbles are formed, and these can be broken up into smaller ones (Section 11.3). Especially at high φ-values, *coalescence of bubbles* can occur. Coalescence is due to rupture of the film between bubbles, and during beating these films are frequently stretched, i.e., increased in area by tensile forces acting on the film. This also means that the film becomes thinner.

The cause of stretching may simply be that a rod of the beater draws an aqueous film through a large bubble. Furthermore, films can be stretched because bubbles are pressed toward each other by pressure fluctuations. These can, again, be caused by the beater. According to the Bernoulli equation (5.4), we have $p + (1/2)\,\rho\,v^2 = $ constant in a liquid; here p is local pressure and v local velocity. At high velocity, the local pressure will thus be small. Assuming that a beater rod moves through the liquid at $5\,\mathrm{m\cdot s^{-1}}$, a pressure change of $-(1/2) \times 10^3 \times 5^2) = -12,500\,\mathrm{Pa}$ will result. A moment later the rod has passed the bubble(s) and the pressure increases again. The bubbles thus periodically decrease and increase in volume (according to Boyle's law: $p \times V = $ constant) and are frequently pressed together. Hence frequent and possibly considerable stretching of films will occur.

As mentioned in Section 10.7, a fairly high value of the *Gibbs elasticity* E_G is needed to keep a film stable by means of the Gibbs mechanism, depicted in Figure 10.29c. Figure 10.35 shows examples of E_G as a function of surfactant concentration c in the liquid. There is an optimum concentration c_{opt} for E_G. If a film is stretched, its thickness will decrease, leading to a higher value of E_G. More important, however, is that the magnitude of c will decrease on stretching, because additional surfactant will become adsorbed onto the enlarged film surfaces (replenishing of surfactant from the bulk of the liquid is negligible in a thin film). As long as the local concentration is above c_{opt}, E_G will increase on stretching, but if c is or becomes $< c_{opt}$, E_G will rapidly decrease, and the film will probably break. Hence there is a critical concentration for breaking the film, probably somewhat below c_{opt}. This has been confirmed by stretching of isolated films.

Figure 11.4b shows some examples of the effect of *beater speed* on foam volume obtained, and the results are in qualitative agreement with the reasoning just given. When increasing the beater velocity, the volume of foam formed at first increases, presumably because more bubbles are formed. Still higher velocities cause a decrease, presumably because pressure fluctuations then become large enough to cause sufficient stretching of films between bubbles to induce coalescence.

Generally, at a smaller surfactant concentration, the maximum overrun is smaller and occurs at a lower beater velocity. This would be because adsorption of surfactant at the newly created A–W interface will cause depletion of surfactant from the bulk, and the concentration of surfactant in the films will become critical at an earlier stage for a smaller initial concentration.

Beating Time. Figure 11.4c illustrates that overrun at first increases during beating—as is only to be expected—but then decreases. Since such "overbeating" is typical for globular proteins as a surfactant, the most likely explanation is surface denaturation, leading to protein aggregation. During beating frequent expansion and compression of film surfaces occurs, and this may readily cause strong unfolding and subsequent aggregation of globular proteins.

11.2.4 Some Properties

The discussion given above concerns **polyhedral** foams, which are formed at volume fractions above that for a close packing of spheres. This critical value of φ equals about 0.7. The average value of φ is rarely above 0.95 for a food foam. A polyhedral foam has a certain rigidity and may be called a gel. The most important rheological parameter is the yield stress (see Section 5.1.3), which should be large enough for the foam to keep its shape under gravity. These and other gel properties are discussed in Chapter 17 (see especially Section 17.4).

If a low-viscosity liquid is beaten to form a foam, it will inevitably become a polyhedral foam. Liquid will always drain from it. For a volume fraction of 0.9, the overrun is 900%, for $\varphi = 0.95$, it is even 1900%; such a foam would make a very fluffy food. Most aerated foods are different. They are **dilute foams** in the sense that the bubbles are separate from each other and they remain spherical. To prevent the bubbles from creaming, the continuous phase should have a yield stress. This can be achieved in several manners:

> By a *gelling polymer*. Gelatin makes a viscous liquid at temperatures above 30°C, and the liquid can then be beaten to form bubbles, which cream very sluggishly. Cooling the system to below 20°C then causes gelling, and the foam is mechanically stabilized.
> The continuous phase can be *solidified*, which occurs in various ice cream–like products. Air is beaten in the liquid while it is frozen,

whereby most of the water is converted into ice; this makes a "solid foam."

Particles that adhere to the A–W interface (see Section 10.6) can make bridges between air bubbles, thereby forming a bubble network. The prime example is whipped cream, where fat globules (partially solid emulsion droplets) cover the air bubbles and also make a continuous network in the aqueous phase; generally, $\varphi \approx 0.5$. Something similar occurs in (high-fat) ice cream: see Figure 9.1.

When beating *egg white*, which contains about 10% protein, part of the protein becomes denatured at the A–W interface, thereby aggregating, and the particles so formed also cover the air bubbles and make a continuous network. The air content can be much higher than in ice cream.

A *heat treatment* can convert some more or less liquid foam systems into a solid foam. A prime example is, again, foam based on egg white—such as foam omelettes and meringues—since at high temperature protein denaturation occurs, which causes gelation. Another well-known case is provided by bread, where baking causes (a) the gas cells in the dough to grow and (b) a fairly stiff gel to be formed of the continuous phase (the dough). Moreover, the foam is converted into a sponge, because most of the thin films between the gas cells break as they become brittle at high temperature. (Question: Why is it necessary that the films break to obtain a good loaf of bread?)

All of these systems can be made so as to have a high yield stress and hence good stand-up properties. A high yield stress also prevents leakage of liquid from the foam.

The most important foam property may be stability against various physical changes. Something on this has been mentioned already, and more will be given in Chapter 13.

11.3 BREAKUP OF DROPS AND BUBBLES

Making bubbles has been discussed in Section 11.2. Making drops is easy: just stir a suitable surfactant solution with some oil and coarse drops are obtained. The difficulty is to make small drops/bubbles. This occurs by the breakup (disruption) of bigger ones. A drop must be strongly deformed to be disrupted, and deformation is counteracted by the Laplace pressure, which increases with increasing deformation (see Figure 10.21) and with decreasing drop size. The ratio between external and internal stress acting on a particle is expressed by a dimensionless Weber number,

$$\text{We} \equiv \frac{\sigma_{\text{ext}}}{(1/2)p_{\text{L}}} = \frac{\sigma_{\text{ext}}d}{2\gamma} \tag{11.5}$$

We has to be of order unity for breakup to occur. Assuming $d = 10^{-6}$ m and $\gamma = 0.01$ N·m^{-1}, σ_{ext} should be about $2 \cdot 10^4$ Pa. Since this implies a pressure difference of that magnitude over a distance d, the local pressure gradient would have to be of order 10^{10} Pa·m^{-1}, a very high value, difficult to achieve by agitation of a liquid.

Consequently, special **machines** are needed to obtain the small emulsion droplets often desired. Numerous types are in use, and we will just mention those that are most common. *Stirrers* are often used, but the intensity of agitation is often too small. Higher intensities can be achieved with rotor–stator type stirrers, as depicted in Figure 11.6a. *Colloid mills* are also rotor–stator machines (Fig. 11.6b), but they have larger dimensions, and the slit between rotor and stator is far more narrow. They are especially used for highly viscous liquids. *High-pressure homogenizers* are pumps that force the liquid through a very narrow slit in the homogenizer valve, depicted in Figure 11.6c; the valve block is pressed onto its seat by a spring. The pressure drop over the valve is, for instance, 20 MPa, and the liquid velocity in the valve is very high. Because of the very small size of the valve slit, a homogenizer cannot be used to make emulsions; it is especially suitable to break up the drops of a coarse emulsion into very small ones.

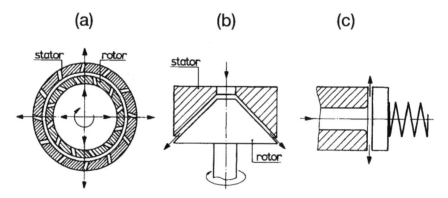

(a)   **(b)**   **(c)**

FIGURE 11.6 Active part of some emulsifying machines. (a) Rotor–stator type stirrer ("ultra-turrax"). (b) Colloid mill. (c) Valve of a high-pressure homogenizer. The slit width in (b) and (c) is greatly exaggerated.

In this section, γ *is assumed constant*. This means either that no surfactant is present (then single drops are considered) or that there is a large excess of surfactant. Furthermore, when speaking about drops, bubbles are also implied. Most of the theory discussed is only valid for small volume fractions.

11.3.1 Regimes

The relations governing the size of the drops obtained depend on hydrodynamic conditions. Accordingly, some different regimes can be distinguished, and in each regime equations can be derived that predict parameters like local stress, resulting drop size, and time scales of various events. The main factors determining what the regime is are the type of force and the flow type. The reader is assumed to be familiar with the material in Section 5.1.1.

The **types of forces** causing disruption of drops can be

Frictional, generally called *viscous*, forces, which are due to the continuous liquid flowing along the drop surface, hence the force acts for the most part *parallel* to the drop surface. (Some authors speak of shear forces.) The local stress would equal viscosity times local velocity gradient $(\eta \cdot \Psi)$.

Inertial forces, due to local pressure fluctuations, which in turn are generally evoked by local velocity fluctuations according to the Bernoulli equation (5.4). They act for the most part *perpendicular* to the drop surface. The local stress would equal $\rho \bar{v} \Delta v$, where $\bar{v}$ is the average liquid velocity and Δv is the local velocity difference.

Flow Types. The main distinction is between:

Laminar flow, occurring if Re $< \sim 2000$. The Reynolds number is given in Eq. (5.5); see also Table 5.1.
Turbulent flow for Re $> \sim 2500$.

These variables give rise to three **regimes** for droplet (or bubble) breakup:

1. *Laminar/viscous* forces (LV), i.e., breakup by viscous forces in a laminar flow field
2. *Turbulent/viscous* forces (TV)
3. *Turbulent/inertial* forces (TI)

Whether in turbulent flow viscous or inertial forces are predominant depends on the *drop Reynolds number* $\mathrm{Re}_{\mathrm{dr}}$ (Table 5.1), where the characteristic length equals drop diameter d, and the velocity is that of the drop relative to the adjacent liquid. The transition occurs for $Re_{dr} \approx 1$. Some particulars about the regimes are given in Table 11.2 and the mathematical expression given are discussed further on.

Which regime can be expected for various systems? This depends in the first place on the size of the apparatus used. If it is very small, Re is small and the regime becomes LV.

Note If the dimension of the active zone is so small as to be comparable to drop or bubble size, as in some laboratory homogenizers, we have a regime that may be called "bounded laminar flow," and different relations hold. It will not be discussed here.

For larger machines the flow is nearly always turbulent for aqueous systems, where the viscosity of the continuous liquid η_C is small. We expect for

Foams, nearly always regime TI, unless the viscosity of the continuous phase η_C is very high. The bubbles tend to be relatively large (virtually always $> 10\,\mu m$), implying $Re_{dr} > 1$.

TABLE 11.2 Various Regimes for Emulsification and Foam Bubble Formation. Valid for Small Volume Fractions

Regime	Laminar, viscous forces, LV	Turbulent, viscous forces, TV	Turbulent, inertial forces, TI
Re, flow	$< \sim 2000$	$> \sim 2500$	$> \sim 2500$
Re, drop	< 1	< 1	$> 1^a$
$\sigma_{ext} \approx$	$\eta_C \Psi$	$\varepsilon^{1/2} \eta_C^{1/2}$	$\varepsilon^{2/3} d^{2/3} \rho^{1/3}$
$d \approx$	$\dfrac{2\gamma We_{cr}}{\eta_C \Psi}$	$\dfrac{\gamma}{\varepsilon^{1/2}\eta_C^{1/2}}$	$\dfrac{\gamma^{3/5}}{\varepsilon^{2/5}\rho^{1/5}}$

^{a}For $d > \eta_C^2/\gamma\rho_C$.

Symbols:

Re = Reynolds number	We = Weber number
d = drop diameter	ρ = mass density
γ = interfacial tension	σ = stress
ε = power density	Ψ = velocity gradient
η = viscosity	

Subscripts:

C = continuous phase	ext = external
cr = critical value for breakup	

O–W emulsions, regime TI if η_C is not much higher than that of water; for a higher viscosity it tends to be regime TV, especially if the resulting d is very small;

W–O emulsions: here, η_C is relatively large, and the regime will often be TV. At small Re—as in a colloid mill—the regime is likely to be LV.

11.3.2 Laminar Flow

Two of the various types of laminar flow, and the effects such flows have on drops, are depicted in Figure 11.7. It also gives the definitions of the velocity gradient.

Simple shear flow causes rotation, and also the liquid inside a drop subjected to shear flow does rotate. The drop is also deformed, and the *relative deformation* (i.e., the strain) is defined as $D \equiv (L - B)/(L + B)$; see Figure 11.7a for definition of L and B. For small Weber numbers, we simply

(a) Simple shear flow

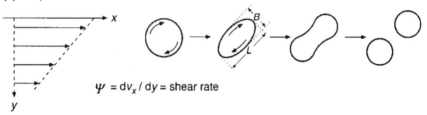

$\Psi = dv_x / dy =$ shear rate

(b) Elongational flow

$\Psi = dv_x / dx =$ elongation rate

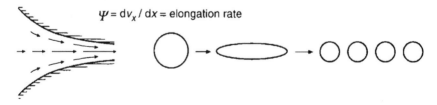

FIGURE 11.7 Two types of laminar flow, and the effect on deformation and breakup of drops at increasing velocity gradient (Ψ). The flow is two-dimensional, i.e., it does not vary in the z-direction. More precisely, the flow type in (b) is "plane hyperbolic flow."

have D = We. For larger values of We, the drop is further deformed as depicted, and it breaks if We > We_{cr}. Values of We_{cr} are given in Figure 11.8, curve for $\alpha = 0$. Inserting the value of We_{cr} in Eq. (11.5) yields for d the value of the largest drop size that can remain unbroken in the flow field applied. The average drop size resulting would be somewhat smaller. Table 11.2 gives the equation for d.

. It is seen that the *viscosity ratio* η_D/η_C has a large effect on We_{cr} in simple shear flow, i.e., on the viscous stress needed to break up a drop of a

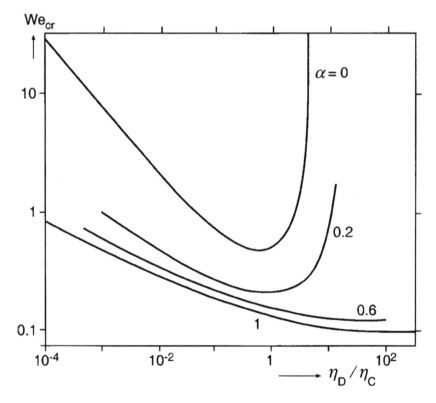

FIGURE 11.8 The effect of the viscosity ratio, drop over continuous phase (η_D/η_C), on the critical Weber number for drop breakup in various types of laminar flow. The parameter α is a measure of the amount of elongation occurring in the flow: for $\alpha = 0$, the flow is simple shear; for $\alpha = 1$, it is purely (plane) hyperbolic.

given size. If $\eta_D > 4\eta_C$, the drop is not disrupted at all. The magnitude of D remains small and the slightly elongated drop starts to rotate (see also Figure 5.3). This is because the time needed for deformation of the drop, which is given by

$$\tau_{def} \approx \frac{\text{drop viscosity}}{\text{external stress}} = \frac{\eta_D}{\eta_C \Psi} \tag{11.6}$$

will be longer than the time needed for half a rotation, given by π/Ψ: $\eta_D/\eta_C\Psi = \pi/\Psi$ yields $\eta_D = \pi\eta_C$, close to the critical value.

Note The part after the second equal sign in Eq. [11.6] only applies in simple shear flow. In plane hyperbolic flow, $2\eta_C$ should be inserted in the denominator: see below.

Also if $\eta_D \ll \eta_C$ breakup is difficult. At $\eta_D/\eta_C = 10^{-4}$, which is about the magnitude in most foams, We_{cr} would be as large as 30. At such a small viscosity ratio, the bubble or drop is deformed into a long thread before breaking. However, for some protein surfactants, the surface layer of the drop can be stagnant (Section 10.8.3) and then the drop can presumably break at a smaller Weber number.

In **elongational flow**, there is no internal rotation in the drop, and it becomes elongated, irrespective of the viscosity ratio. If We is high enough, the drop attains a slender shape (see Figure 11.7b), in which it is subject to Rayleigh instability (see Figure 10.23b) and thereby breaks into a number of small drops. As depicted in Figure 11.8, curve $\alpha = 1$, We_{cr} is always smaller for elongational than for simple shear flow. This is because of the absence of rotational flow inside the drop and because the effective viscosity in elongational flow is larger than the shear viscosity (by a factor of two for a Newtonian liquid in plane hyperbolic flow).

The flow pattern can be **intermediate** between simple shear and pure elongation, and such a situation is very common in stirred vessels and comparable apparatus at $Re < Re_{cr}$. The extent to which elongation contributes to the velocity gradient can be expressed in a simple parameter, here called α, that can vary between 0 and 1. As shown in Figure 11.8, a little elongation suffices to reduce markedly the magnitude of We_{cr} and allows breakup at a higher viscosity ratio.

The results in Figure 11.8 have been obtained for (large) single drops in precisely controlled flow at constant conditions. They agree well with theoretical predictions. Results on average droplet size of emulsions (η_C fairly high) obtained in a colloid mill reasonably agree with theory.

Question

How large a velocity gradient would be needed to make an O–W emulsion with droplets of 1 μm diameter in plane hyperbolic laminar flow, assuming the effective interfacial tension to be $0.01 \, \text{N} \cdot \text{m}^{-1}$? Is the resulting gradient likely to be achievable in practice?

Answer

About $2 \cdot 10^6 \, \text{s}^{-1}$.

11.3.3 Turbulent Flow

Turbulent flow is characterized by the presence of *eddies* (vortices, whirls), which means that the local flow velocity u generally differs from its time-average value $\bar{u}$. The local velocity fluctuates in a chaotic manner, and the average difference between u and $\bar{u}$ equals zero. However, the root-mean-square velocity difference,

$$u' \equiv \langle (u - \bar{u})^2 \rangle^{1/2} \tag{11.7}$$

is finite. Its value generally depends on direction, but for large Re and small length scale, the turbulence can be regarded as being *isotropic*, implying that u' does not significantly depend on direction. This condition is often more or less fulfilled at the scale of drops or bubbles during agitation.

Kolmogorov theory gives relations for the effects of isotropic turbulent flow. These relations are in fact scaling laws, but most constants in the equations are of order unity. The flow shows a spectrum of eddy sizes (l). The largest eddies have the highest value of u'. They transfer their kinetic energy to smaller eddies, which have a smaller u' value but a larger velocity gradient (u'/l). Small eddies have thus a high specific kinetic energy; they are called *energy bearing eddies*, size l_e. The local velocity near such an eddy depends on its size and is given by

$$u'(l_e) \approx \varepsilon^{1/3} l_e^{1/3} \rho^{-1/3} \tag{11.8}$$

Here ρ is the mass density of the liquid, and ε is the *power density*. The latter is defined as the amount of energy dissipated per unit volume of liquid per unit time (i.e., in $\text{J} \cdot \text{m}^{-3} \cdot \text{s}^{-1} = \text{W} \cdot \text{m}^{-3}$).

The eddies have lifetimes given by

$$\tau(l_e) \approx \frac{l_e}{u'(l_e)} = l_e^{2/3} \varepsilon^{-1/3} \rho^{1/3} \tag{11.9}$$

Hence the smaller the eddies, the shorter their lifetime. Moreover, smaller eddies have a higher local power density. Inside a small eddy, Re is small, hence the flow is laminar, and in laminar flow ε equals $\eta \cdot \Psi^2$; local Ψ will equal $u'(l_e)/l_e \approx 1/\tau(l_e)$. Eddies below a certain size cannot be formed, since the local value of ε would be so high that the kinetic energy would be fully dissipated as heat.

Regime TI. The very simple Eq. (11.8) provides the basis for the relations for drop or bubble disruption in regime TI. Because of their abundance and short lifetime, the energy-bearing eddies cause rapid local velocity fluctuations, which according to the Bernoulli equation (5.4) cause pressure fluctuations given by

$$\Delta p(l_e) \approx \rho[u'(l_e)]^2 \approx \varepsilon^{2/3} l_e^{2/3} \rho^{1/3} \tag{11.10}$$

If now Δp near a drop is larger than its Laplace pressure, the drop will be broken up. A more detailed analysis of the eddy size spectrum shows that drop disruption is most effective if $d = l_e$. Hence putting the Laplace pressure equal to Eq. (11.10), and inserting d for l_e, results for d in the maximum size that a drop can have in the turbulent field, because larger ones will be disrupted. The relation is

$$d_{max} \approx \varepsilon^{-2/5} \gamma^{3/5} \rho^{-1/5} \tag{11.11}$$

Disruption probably occurs via the sudden formation of a local protrusion on a drop, which then breaks off. This would mean that fairly wide droplet size distributions result, as is indeed observed: see Figure 11.9a.

In most cases, Eq. (11.11) would also be valid for the average diameter obtained (generally d_{32} is used), albeit with a somewhat smaller proportionality constant. If power input is varied, power density may vary proportionally, which then implies that the average drop size obtained would be proportional to energy input to the power -0.4. In a stirrer, ε is proportional to revolution rate cubed. In a high-pressure homogenizer, the homogenizing pressure p_H can be varied, and the power density is proportional to $p_H^{1.5}$; it follows that average drop size is proportional to $p_H^{-0.6}$. These relations are well obeyed for dilute emulsions with an excess of surfactant. Figure 11.9 gives some practical results. In part b, lower curve, the slope is indeed exactly -0.6.

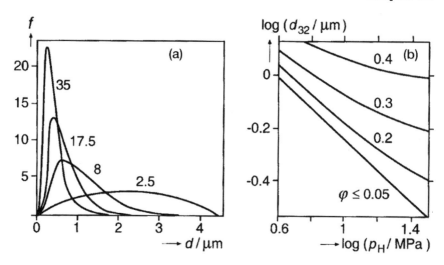

FIGURE 11.9 Examples of the droplet size (distribution) in O–W emulsions resulting from treatment in a high-pressure homogenizer, at various homogenizing pressures p_H. (a) Volume frequency distributions (f) in % of the oil per 0.1 μm class width versus droplet diameter d; p_H indicated in MPa. (b) Average diameter d_{32} as a function of p_H for various oil volume fractions φ (indicated); the aqueous phase was a solution containing about 3% protein.

Regime TV. As mentioned, the relation for regime TI can only hold if $\text{Re}_{dr} > 1$. Re_{dr} can be obtained by combining Eqs. (11.8) and (5.5). Using Eq. (11.11) to eliminate ε, we obtain

$$\text{Re}_{dr} = \frac{\gamma^{1/2}\rho^{1/2}d^{1/2}}{\eta_c} \tag{11.12}$$

This then leads to the condition given in Table 11.2 ($d > \eta_c^2/\gamma\rho$).

If $\text{Re}_{dr} < 1$, the flow near a drop is laminar, and it would be broken up in the regime TV. Now the local flow velocity is needed at the scale of a droplet d [i.e., $u'(d)$], and this scale is (much) smaller than l_e. According to the Kolmogorov theory the velocity is given by

$$u'(d) = \varepsilon^{1/2}d\eta^{-1/2} \quad d \ll l_e \tag{11.13}$$

The shear stress acting on the drop is given by η_C times the velocity gradient $u'(d)/d$. The flow type is probably close to plane hyperbolic flow, and Figure 11.8 shows that in this case We_{cr} will not strongly depend on the viscosity ratio. The relation for droplet size then becomes

$$d_{max} \approx \gamma \varepsilon^{-1/2} \eta_c^{-1/2} \tag{11.14}$$

This is again a very simple relation. It has not been as well tested as Eq. (11.11), to the author's knowledge.

Notice that in the regime TV the viscosity of the viscous phase is a variable governing droplet size, but not in TI. Of course, the magnitude of η_c would determine whether the regime is TI or TV. If the conditions are such that Re_{dr} is close to unity, the relations governing drop size will be between those given in Eqs. (11.11) and (11.14), and viscosity will have some effect, say, $d \propto 1/\eta_C^{1/4}$.

11.3.4 Complications

The discussion given in Section 11.3 is to some extent an oversimplification, not so much because the relations given are approximations, but rather because several conditions must be fulfilled for them to be applicable, since foam and emulsion formation may involve a number of other variables. A few will be mentioned by way of illustration.

> The *viscosity of the disperse phase* η_D is not a variable in any of the equations for the turbulent regimes. Nevertheless, it is often observed that a higher value of η_D leads to a higher average drop size. This is because the time needed for deformation of a drop may be longer than the lifetime of the eddies that would cause its disruption. The latter is given by Eq. (11.9). Table 11.2 gives the stress acting on a droplet, and Eq. (11.6) can then be used to calculate the deformation time. For example, for $\epsilon = 10^{10}$ W $\cdot$ m^{-3}, $d = 1\,\mu m$, and $\eta_D = 0.1$ Pa $\cdot$ s, we would obtain for the deformation time $5\,\mu s$ and for the eddy lifetime $0.5\,\mu s$ (try to check these calculations). This is clearly impossible and the drops can only be broken up by eddies larger than $1\,\mu m$, which have a longer lifetime but provide a smaller stress; hence d_{max} would be larger than $1\,\mu m$.
>
> The value of the *power density may greatly vary among sites* in the apparatus. Near the tip of a stirrer, ε would have a much higher value than further away. It means that the effective volume for droplet disruption is much smaller than the total volume of stirred liquid. This has two consequences. First, part of the mechanical energy is dissipated at a level where it cannot disrupt drops (and is thereby wasted). Second, droplet breakup takes a long time, because

every small volume unit of emulsion has to pass close to the stirrer tip to allow formation of the smallest droplets obtainable.

This brings us to what may be considered the most important point. Formation of small drops (bubbles) involves a number of different processes, roughly illustrated in Figure 11.10. Drops are deformed and may or may not be broken up. Moreover, drops frequently encounter each other and this may or may not lead to their *(re)coalescence*. In the meantime, surfactant adsorbs, which affects the result of the various processes. Each of these processes has its own time scale, depending on a number of variables. All of the processes occur numerous (say, 100) times during stirring, beating, homogenizing, etc., starting with large drops that gradually give rise to smaller ones. Finally, a steady state may be reached, in which disruption and recoalescence balance each other. In practice, such a state is generally not quite reached.

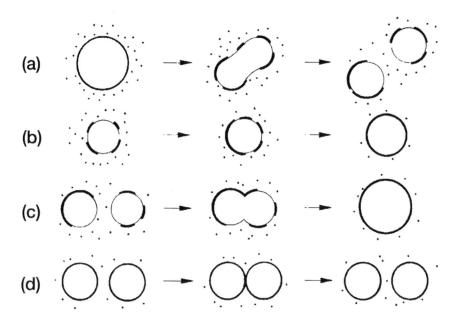

FIGURE 11.10 Various processes occurring during making of small drops (or bubbles). The drops are depicted by thin lines and the surfactant by heavy lines and dots. Highly schematic and not to scale.

Question

Consider again Question 2 at the end of Section 11.1.

1. Can you make a rough estimate of the proportion of the liquid used in the experiment in which the energy is dissipated at the highest power density level, i.e., the level that yields the small bubbles? You may assume that $\gamma = 0.05\,\mathrm{N \cdot m^{-1}}$.

2. Assuming that it is possible to let the beaters turn at twice the original revolution rate, what would then be the bubble size obtained? What conditions then need to be fulfilled?

Answer

1. The overall power density in the liquid was $50\,\mathrm{W}/200\,\mathrm{ml}$, or $25 \cdot 10^4\,\mathrm{W \cdot m^{-3}}$ (the turbulent energy is virtually confined to the aqueous phase, hence $200\,\mathrm{ml}$). Since d obtained is known $(2 \cdot 10^{-4}\,\mathrm{m})$, we can use Eq. (11.11) to calculate the effective power density. This gives $2 \cdot 10^{-4} = 0.05^{3/5} \cdot \varepsilon^{-2/5} \cdot 1000^{-1/5}$, which yields $\varepsilon = 62 \cdot 10^4\,\mathrm{kW \cdot m^{-3}}$. Assuming all of the energy to be dissipated at the highest rate, this then would occur in $25/62$ or 40% of the volume. It is, however, unlikely that the power density distribution would be so uneven. Assuming that in the rest of the liquid the average value of ε amounts to 20% of the peak value, the maximum value of ε would hold in about 25% of the volume.

2. It is remarked above that for a revolving stirrer, ε is proportional to v^3, v being the revolution rate. According to Eq. (11.11), $d \propto \varepsilon^{-0.4} \propto v^{-1.2}$, which then would lead to $d = 200/2^{1.2} = 87\,\mu\mathrm{m}$. Conditions are that surfactant is present in excess (since the specific surface area, and hence the amount of adsorbed surfactant, would be larger by a factor 2.3) and that the given time of beating (2 min) would be long enough to obtain a steady state.

11.4 ROLE OF SURFACTANT

It is often assumed that the role of a surfactant in emulsification is the lowering of γ, thereby facilitating the breakup of drops. Although this is true enough, it does not explain the prime role of the surfactant. Assume that we try to make emulsions (a) of paraffin oil and water, and (b) of triglyceride oil and water. The interfacial tensions O–W are 50 and $30\,\mathrm{mN \cdot m^{-1}}$, respectively (see Table 10.1). To equalize the conditions, we add some surfactant to sample (a) to obtain also here $\gamma = 30\,\mathrm{mN \cdot m^{-1}}$. However, the result will be that the agitation of sample (a) produces an emulsion, whereas this fails for sample (b), despite the equality of γ. The surfactant thus plays at least one other role. This is the *formation of γ-gradients*, as discussed in Sections 11.2.2 and 3 for foam making. The same applies in emulsification.

Different surfactants give different results, as illustrated in Figure 11.11. Plateau values are obtained at high surfactant concentrations. Emulsification was in the regime TI, and according to Eq. (11.11) the droplet size obtained should be proportional to $\gamma^{0.6}$. This roughly agrees with the plateau values of d_{32} obtained, though not precisely. Moreover, the shapes of the curves show clear differences. Some factors causing such differences have been identified, but a full explanation cannot yet be given.

11.4.1 Droplet Breakup

The first question is, What is the **effective interfacial tension** during droplet deformation and breakup? This depends on surface load Γ, which depends on the adsorption rate of the surfactant. Actually, this rate can be very high, since surfactant is not transported by diffusion but by convection. This would imply that Γ can be fairly high. However, every breakup event occurs

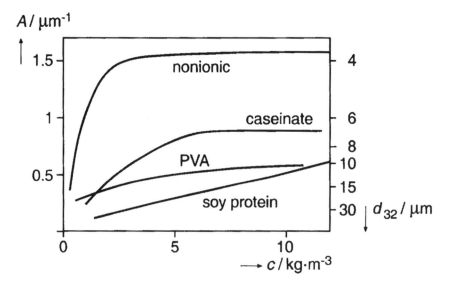

FIGURE 11.11 Specific droplet surface area A (and average droplet size d_{32}) as a function of total surfactant concentration c, obtained at approximately constant emulsification conditions for various surfactants; PVA = poly(vinyl alcohol); also for soy protein a plateau value of A is reached, at about $20\,\mathrm{kg \cdot m^{-3}}$. Approximate plateau values for the interfacial tension γ are 3, 10, and 20 $\mathrm{mN \cdot m^{-1}}$ for the nonionic, caseinate, and PVA, respectively.

very fast (say, in 1 μs), and it involves a temporary increase of the surface area of the drop, maybe by a factor of 2, which would mean halving the value of Γ. Hence at the moment of disruption the value of γ can be significantly higher than the equilibrium value.

Comparing surfactant types, small-molecule amphiphiles tend to give a lower γ-value than proteins: see Figure 10.13. Hence amphiphiles give a smaller average droplet size, as illustrated in Figure 11.11.

Naturally, the concentration of surfactant c becomes smaller during emulsification due to **depletion**: the total interfacial area markedly increases due to the decrease in droplet size, the more so for a higher φ value. Thus more surfactant becomes adsorbed. This also happens during foam formation, but the increase in area is far smaller. A decreased c value gives a longer adsorption time. Consequently, the effective γ value will increase, often strongly, during the emulsification process.

Presence of surfactant at the interface will also directly affect **deformation**. The surfactant allows formation of a γ-gradient. This would affect the deformation mode of a drop, which has indeed been observed. Moreover, enlarging the interfacial area causes γ to increase, as mentioned. This implies that the interfacial free energy increase includes two terms: $\gamma \cdot dA + A \cdot d\gamma$. The first term is due to the deformation of the drop being counteracted by its Laplace pressure; the second is due to surface enlargement being counteracted by the surface dilational modulus E^{SD}. Making use of Eq. (10.20) for E^{SD}, we obtain

$$\gamma \, dA + A \, d\gamma = \gamma \, dA + AE^{SD} \, d\ln A = (\gamma + E^{SD}) \, dA \qquad (11.15)$$

Model experiments have indeed shown some additional resistance to deformation (over that caused by the Laplace pressure). To what extent these phenomena affect the droplet size obtained, for instance by homogenization, is not yet clear.

11.4.2 Prevention of Recoalescence

An essential role of the surfactant is to prevent the newly formed drops from coalescing again. Drops frequently encounter each other during the emulsification process. Recoalescence has been shown to occur in the following type of experiment. Two O–W emulsions are made that have identical properties, except that two oils are used that differ, e.g., in refractive index. These emulsions are then mixed and the mixture is rehomogenized. By comparing the refractive index of the droplets so obtained with that of the original ones, it follows that droplets of mixed oil composition have indeed been formed.

The rate of recoalescence presumably depends on the time lapse between mutual drop encounters τ_{enc} relative to the time needed for adsorption of surfactant τ_{ads}. When two drops come very close to each other before they have acquired sufficient surfactant, they may coalesce, as illustrated in Figure 11.10c. It can be derived from theory that in all regimes,

$$\frac{\tau_{ads}}{\tau_{enc}} \approx \frac{6\pi\Gamma\varphi}{dc} \tag{11.16}$$

From this relation it follows qualitatively that more recoalescence will occur if c is smaller, if φ is larger, and if ε is larger, since this implies that d becomes smaller. This agrees with experimental results on recoalescence and also with results on average droplet size, as shown, for instance, in Figure 11.9b; an increase in φ in that case also implies a stronger decrease in c during emulsification.

> *Note* The picture in Figure 11.10c is to some extent misleading, in that bare patches are shown on the drop surface. In fact, the surfactant will generally be distributed more evenly over the surface, since the spreading rate of surfactant will be quite high. Equation (10.19) predicts spreading times $< 1\,\mu s$ for small drops under most conditions.

Most surfactants provide colloidal **repulsion** to act between droplets, thereby preventing their close approach—hence coalescence—upon encountering each other, and it appears logical to assume that this mechanism also acts during emulsification (see Figure 11.12a). However, the stress by which the drops are driven toward each other can be very large. The magnitude is about the same as that of the stress needed to break up the drops, hence of the order of the Laplace pressure of the drops. For drops of $1\,\mu m$, this means about $2 \cdot 10^4$ Pa. In Chapter 12, colloidal interactions between particles are discussed. It can be derived that the so-called disjoining pressure, the colloidal interaction force per unit area that drives the drops apart, will for $1\,\mu m$ droplets rarely be larger than $2 \cdot 10^2$ Pa if due to electrostatic repulsion. This would thus be far too small to prevent collision and thereby coalescence.

Nevertheless, small-molecule ionic surfactants like sodium dodecyl sulfate, which provide repulsion by electrostatic forces only, can be very efficient in making emulsions with very small droplets, even at low concentrations. If, then, moreover, the ionic strength is high, which will greatly reduce the disjoining pressure, the droplet size distribution obtained is the same as at low ionic strength, although the finished emulsion shows rapid coalescence (and not at low ionic strength). Hence prevention of

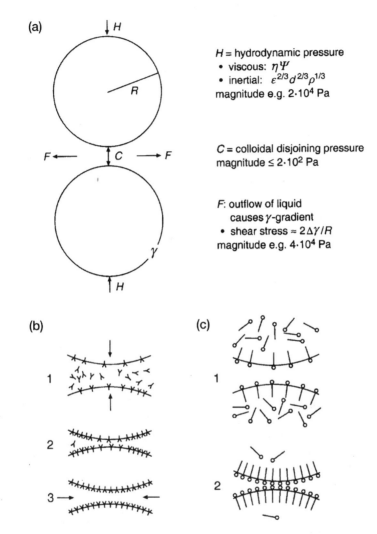

FIGURE 11.12 Schematic illustration of the mechanisms relevant for the prevention of recoalescence of newly formed oil drops during emulsification. (a) Forces involved; the magnitude of the stresses given would apply to $R \approx 0.5\,\mu$m. (b) Formation of a γ-gradient in the gap between approaching drops and subsequent Marangoni effect if the surfactant (Y) is in the continuous phase. (c) Evolution of the distribution of surfactant (φ), and absence of Marangoni effect, if the surfactant is in the disperse phase. Not to scale.

coalescence in a finished emulsion and during emulsification are due to different mechanisms, at least for many surfactants.

Figure 11.12b shows two drops coming close to each other during emulsification; it must be assumed that they are not (yet) fully covered by surfactant. Now **γ-gradients** will be formed as depicted, because (a) the outflow of liquid from the thin film will produce them (see Figure 10.28b); and (b) any additional adsorption of surfactant during the approach will be least where the film is thinnest. The viscous stress that can be balanced by such a γ-gradient will be roughly twice $\Delta\gamma$ over the distance, for which we may take about $d/2$. Assuming that $\Delta\gamma$ can reach a value of $0.01 \, \text{N} \cdot \text{m}^{-1}$, and that $d = 1 \, \mu\text{m}$, the maximum stress would be about $4 \cdot 10^4$ Pa, i.e., on the order of the hydrodynamic pressure acting on the droplet pair. It would certainly be enough to slow down greatly the approach of the droplets, because it would fully prevent slip of the liquid flowing out of the film at the droplet surfaces. It cannot, in itself, prevent collision of the drops, but it would slow down their approach sufficiently during the time that the drops are being pressed together. Equation (11.9) gives the approximate lifetime of an eddy, and it would roughly be a microsecond in the present case. Hence, before the drops are close enough to coalesce, they would probably move away from each other.

It may be noticed that this stabilizing mechanism resembles the one prevailing in foam formation discussed in Section 11.2.3: the essential aspect is the formation of γ-gradients for a long enough time. In foam formation the recoalescence of bubbles occurs also, and γ-gradients play a part in counteracting this. However, the gradients tend to be much smaller than in the case of emulsification. The maximum value of $\Delta\gamma$ may be somewhat larger, because there is more time for the surfactant to adsorb and change conformation, but the effective distance by which it has to be divided is much larger: the bubbles are typically two orders of magnitude larger than emulsion droplets (Table 11.1).

The stabilizing mechanism discussed has been called the **Gibbs–Marangoni effect**. The Marangoni effect would lead to an inflow of continuous phase into the film because of the γ-gradient formed. This will presumably not occur in an early stage, because the hydrodynamic pressure will cause a stronger outflow. As soon as this pressure relaxes, however, the Marangoni effect may become significant, driving the drops apart. The strength of the γ-gradient developing will depend on the *Gibbs elasticity* of the film. Its magnitude will at least equal twice the surface dilatational modulus and be higher for a film of a thickness far smaller than its diameter. Figure 11.13 gives some examples of surface pressure against surface excess, plotted in such a way that the slope of the curve would equal the value of E^{SD} (for not too high Γ). It is seen that SDS gives already an appreciable modulus at a quite small Γ value (expressed in mg·m^{-2}), whereas for β-casein E^{SD} would be very small for $\Gamma < 1$ mg·m^{-2}. Hence prevention of recoalescence would be less. Most small-molecule surfactants are indeed superior in preventing recoalescence as compared to proteins. It may further be noted that Figure 11.13 will even overestimate the value of E^{SD} for β-casein, because the Π value will be smaller for the same Γ value at the very short time scales involved; this is discussed in Section 10.8.2.

An important characteristic of an emulsion is whether oil or water makes up the continuous phase. This is governed by **Bancroft's rule**, which

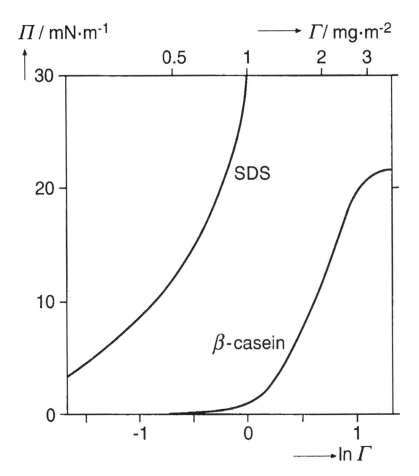

FIGURE 11.13 Surface pressure Π versus ln(surface excess, Γ) at the O–W interface for sodium dodecyl sulfate and β-casein.

states that the phase in which the surfactant is best soluble becomes the continuous one. The explanation presumably is that γ-gradients will not readily form if the surfactant is in the drops, as illustrated in Figure 11.12c. Although a γ-gradient will develop upon fast approach of two drops, it tends to become counteracted by ongoing adsorption of surfactant from the inside of the drop, and then depletion of surfactant does not depend on the thickness of the film between the drops. Hence there will be no Marangoni effect, and the newly formed drops may readily coalesce. The theory for the approach of two droplets has been worked out in more detail than can be given here, but one result is given in Figure 11.14. The approach velocity of two drops is given relative to the velocity in the absence of surfactant. The

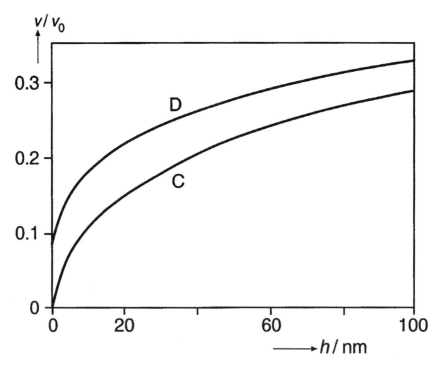

FIGURE 11.14 Linear rate of approach (v) of two drops, relative to the rate in the absence of surfactant (v_0) as a function of the distance between the drop surfaces (h), if the surfactant is in the disperse phase (D) or in the continuous one (C). Calculated for a drop of 1 μm diameter. (After results by I. Ivanov, P. A. Kralchevsky. Proc. 2nd World Congress on Emulsion, Vol. 4, 1997, pp. 145–152.)

conditions are equal, except for the phase in which the (small-molecule) surfactant is present. It is seen that the retardation of the approach is stronger if the surfactant is in the continuous phase and that it then even tends to zero velocity at very small distance. This further underpins Bancroft's rule.

Note that the rule does not speak of the phase in which the surfactant is initially dissolved, but that it concerns their solubility ratio, for small-molecule surfactants given by their HLB value.

Note It concerns solubility in terms of thermodynamic activity, hence surfactant present in micelles should not be included in the solubility ratio.

If the HLB value is not greatly different from 7, the surfactant can probably be dissolved in either phase. After agitation has started, most of the surfactant will readily move to the phase in which it is best soluble. For HLB $\approx$ 7, it will reach about equal concentrations in both phases; this is presumably the reason why such surfactants tend to be poor emulsifiers.

Bancroft's rule also explains why proteins cannot be used as surfactants to make a W–O emulsion: they are insoluble in oil.

11.4.3 Surface Layers of Proteins

Changes occurring in proteins due to adsorption are discussed in Section 7.2.2, topic "Adsorption." In Section 10.3.2 some particulars of proteins as surfactants are given. During emulsion or foam formation with protein(s) as surfactant, however, *no equilibrium adsorption* is attained. In Figure 11.15a an adsorption isotherm for β-casein at the O–W interface (obtained by slow adsorption onto macroscopic interfaces at quiescent conditions) is compared with an apparent adsorption isotherm calculated from emulsions made ($d_{32} = 1\,\mu m$) at various protein concentrations. It is seen that the apparent surface activity of the casein in the emulsions is far smaller, the bulk concentration needed to obtain a plateau value of Γ; being about 3 orders of magnitude larger than for quiescent adsorption. In the emulsion the plateau value of Γ; is higher by about 30%.

For small-molecule surfactants, emulsification leads to equilibrium adsorption. This means that knowledge of the adsorption isotherm, surfactant concentration, and total droplet surface area permits calculation of Γ;. This is not possible for proteins.

Figure 11.15b gives the same result for β-casein, but now plotted as Γ; versus concentration over specific surface area. If emulsions are made at

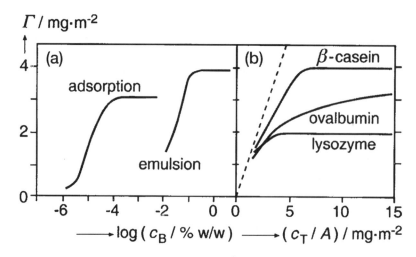

FIGURE 11.15 Surface excess (Γ) of proteins at the O–W interface as a function of protein concentration. (a) Results for β-casein obtained by quiescent adsorption onto a plane interface or by emulsification; c_B is concentration in the bulk (continuous) phase. (b) Results obtained by emulsification for various proteins; c_T is total concentration in the system, and A is the interfacial area produced by emulsification. The broken line would be obtained if all of the protein became adsorbed.

variable oil–water ratio, this way of plotting leads to consistent results (plotting as in Figure 11.15a does not). Results for some other proteins are also given, and the differences cannot be easily explained. It may be noted that during formation of emulsions and foams by agitation, the interface is repeatedly expanded and compressed. This will induce conformational changes ("denaturation") in the adsorbed molecules, some of which changes will be irreversible or very sluggishly reversible. Some molecules will become desorbed and may readsorb in a changed form. Ovalbumin—and other proteins showing a similar relation for Γ—become more or less aggregated during emulsification, especially at high concentration.

Question 1

An emulsion is made of 100 mL triglyceride oil and 400 mL of a 20 millimolar NaCl solution. 15 g of lauric acid (C12) is dissolved in the oil. The mixture is homogenized at a high pressure and the resulting $d_{32} = 0.7\,\mu m$. (a) Why is it beneficial to add NaCl to the aqueous phase? (b) What type of emulsion is formed and why? (c) What is the concentration of lauric acid in the aqueous phase after emulsification? (d) Does the surface excess at the droplets reach the plateau value? (e) What would have been the result if not 15 but 10 mg of lauric acid had been added to the oil and why? Tip: Consult Figure 10.6.

Answers

(b) O–W. (c) $2.04\,g \cdot L^{-1}$. (d) Yes. (e) Larger droplet size.

Question 2

A food company makes an O–W emulsion by using a high-speed stirrer. The aqueous phase consists of a solution of skim milk powder, protein content 4%, ionic strength 0.1 molar. The oil volume fraction is 0.2. A problem is that the emulsion shows significant creaming. Microscopic examination shows drops that are on average about $2\,\mu m$ in diameter, which is considered to be too large. Since the stirrer is already used at its highest intensity, some surfactant is added prior to emulsification, and 0.03% SDS does indeed reduce creaming. To reduce the creaming further, 0.12% SDS is applied, but now creaming is enhanced rather than reduced. Can you explain these observations?

Answer

For a diameter of $2\,\mu m$ and $\varphi = 0.2$, the specific droplet surface area would be about $6 \times 0.2/2 = 0.6\,m^2 \cdot ml^{-1}$. The solution contains about 40 mg of protein per ml, far more than sufficient to cover the droplets (see, e.g., Figure 11.15). Covering the droplets with SDS only would give $\Gamma \approx 1\,mg \cdot m^{-2}$, (see, e.g., Figure 10.8), which would need 0.6 mg SDS per ml (0.06%) + 0.8 times the CMC (about 0.025%), i.e., almost three times the amount added, hence the surface layers of the drops would still consist predominantly of protein. However, we have seen that small concentrations of SDS may be sufficient to reduce markedly the recoalescence during emulsification. Hence a decrease in droplet size, hence reduced creaming. If 0.12% SDS is added, it would almost completely replace the protein at the interface (see Figure 10.15). Although the drops obtained will be smaller, they may now readily coalesce afterwards. This is because SDS stabilizes the emulsion by means of electrostatic repulsion, but the ionic strength of the aqueous phase would be high enough to reduce greatly this repulsion, as mentioned above. Coalescence leads to large drops and hence to enhanced creaming.

11.5 RECAPITULATION

Foam bubbles can be made via supersaturation of a gas (generally CO_2) or by beating in air. Emulsions are generally made by agitation. Emulsion formation can be considered without involving emulsion stability, but foam formation and instability cannot fully be separated, since the time scales for formation and changes due to instability overlap. This is primarily because foam bubbles tend to be larger than emulsion droplets, e.g., 100 versus 1 μm. Important properties of the product determined by the formation process are (a) the drop or bubble size distribution, (b) the composition of the surface layers around drops or bubbles, and (c) in emulsions the type (O–W or W–O).

Foam.   An important variable in foam making is the overrun. It is the resultant of various processes during beating: bubble formation, drainage, and bubble coalescence. Bubbles cream rapidly, forming a close-packed layer from which liquid drains. In this way a polyhedral foam is generally formed. The bubbles are separated by thin films that end in Plateau borders; the latter form a continuous network through which liquid can drain. Drainage causes separation into foam and liquid, and the foam gets an ever higher φ value. This gives the foam a certain firmness. Many aerated foods are bubbly foams, meaning that φ is smaller than about 0.6. Such systems are made more firm by various means, mostly gelation of the continuous phase.

Making emulsions is simple, but making small drops is difficult, because the Laplace pressure, which causes the resistance to deformation and breakup, increases with decreasing drop (or bubble) diameter. To obtain a stable system, small drops are generally desired. The prime occurrence in emulsion formation is thus the breakup of drops into smaller ones. External stresses are needed to overcome the Laplace pressure. Different forces act in different regimes. Viscous forces, which act in a direction parallel to the drop surface, can arise in laminar flow. This can be shear flow or elongational flow, and the latter is more effective in breaking up a drop, especially if the drop has a high viscosity. Breakup can also be caused by inertial forces, which act where local velocity fluctuations cause pressure fluctuations; these forces act normal to the drop surface. Fluctuations can arise in turbulent flow, and they increase with the intensity of agitation, given as the dissipation rate of mechanical energy or power density (in $W \cdot m^{-3}$). Depending on conditions, i.e., drop size and liquid viscosity, a turbulent flow can nevertheless be laminar close to a drop, breaking it by viscous forces.

Roles of the Surfactant. Emulsions and foams cannot be made without surfactant being present. The prime function of the surfactant is the formation of interfacial tension gradients. Thereby drainage of liquid from a film between bubbles or drops is greatly retarded, which is essential in foam formation and in the early stages of emulsification. Moreover, drops can recoalesce after being formed, since two drops are often closely pressed together during the process, and this at a stage when they are not yet fully covered with surfactant. Colloidal repulsion between drops is generally too weak to prevent coalescence during emulsification. However, the formation of a γ-gradient greatly slows down the outflow of liquid from the gap between drops, thereby often preventing their coalescence. The strength of this mechanism depends on the Gibbs elasticity of the film between drops, which varies among surfactants and with surface load. The elasticity is higher and the mechanism more effective if the surfactant is present in the continuous phase rather than in the drop. This explains Bancroft's rule: the continuous phase becomes the one in which the surfactant is best soluble.

The surfactant also lowers the interfacial tension, thereby facilitating droplet breakup. The effective γ value during breakup depends on surfactant type and concentration and on the rate of transport to the drop surface. Approximate equations are available for this rate, and also for the stresses acting on a drop, the drop size resulting from breakup, and the frequency at which the drops encounter each other.

Proteins are not very suitable for making fine emulsions; in other words, it takes more energy to obtain small droplets than with a small-molecule surfactant. This is primarily due to their large molar mass. It causes the effective γ value that they can produce at the O–W interface to be fairly large. Moreover, their molar concentration is small at a given mass concentration, causing the Gibbs elasticity to be relatively small. This means that prevention of recoalescence is less efficient. Proteins are not suitable to make W–O emulsions, as follows from Bancroft's rule: they are insoluble in oil. The adsorption layer of proteins on the droplets obtained by emulsification is not an equilibrium layer, whereas it is for small-molecule surfactants.

BIBLIOGRAPHY

Basic aspects of importance for the topic of the chapter are given in

J. Lucassen. Dynamic properties of free liquid films and foams. In: E. H. Lucassen-Reynders, ed. Anionic Surfactants: Physical Chemistry of Surfactant Action. Marcel Dekker, New York, 1981, pp. 217–265.

A compilation of papers on foam properties and making is

A. J. Wilson, ed. Foams: Physics, Chemistry and Structure. Springer Verlag, London, 1989.

A symposium report edited by H. N Stein is

The Preparation of Dispersions. Chem. Eng. Sci., Vol. 48 (2) 1993.

It includes some articles on emulsification and one on foam formation. Two reviews on emulsion formation are

P. Walstra. Formation of emulsions. In: P. Becher, ed. Encyclopedia of Emulsion Technology. Vol. 1. Basic Aspects. Marcel Dekker, New York, 1983, pp. 58–127.

and

P. Walstra, P. E. A. Smulders. Emulsion formation. In: B. P. Binks, ed. Modern Aspects of Emulsion Science. Roy. Soc. Chem., Cambridge, 1998, pp. 56–99.

12

Colloidal Interactions

In Chapter 3, bonds and interaction forces between species of molecular size were briefly discussed. In this chapter, the same forces come into play when they act between larger structural elements, mostly particles. Moreover, Chapter 3 mostly concerns bond energy, whereas the interaction forces discussed now can be in considerable part due to entropic effects: mixing, conformational, and contact entropy can all play a part.

Colloidal interaction forces act primarily in a direction perpendicular to the particle surface; forces primarily acting in a lateral direction were discussed in Chapter 10. We merely consider "internal forces," which find their origin in the properties of the materials present. This excludes forces due to an external field, such as gravitational, hydrodynamic, and external electric forces; these are involved in some subjects of Chapter 13.

12.1 GENERAL INTRODUCTION

Consider two particles in a liquid, a large distance apart. Colloid scientists try to predict the free energy needed to bring these particles from infinite distance to a close distance between the particles' surfaces h. If that energy is positive, it means that the particles repel each other; if it is negative, they attract each other. For particles of a size of the order of 1 μm, the distance between the surfaces over which interaction forces are significant is nearly always much smaller than the particle size. The interaction free energy is often denoted V; it is often divided by $k_B T$, the kinetic energy involved in the encounter of two particles in Brownian motion. This is done because the interaction curves (V versus h) are often used to predict whether particles that encounter each other by Brownian motion will aggregate or not.

Interaction Curves. Figure 12.1 gives a hypothetical example of such a curve. At very close distance (< 0.5 nm), the interaction free energy is always (large and) positive, due to hard-core repulsion (Section 3.1), but at longer distances the shape of the curves can vary widely. In the drawn curve, a *primary minimum* is shown at A, a *maximum* at B, and a *secondary minimum* at C. Starting at large h, the particles will at some time reach position C. If it is a shallow minimum, the particles can diffuse away from each other; if the minimum is deeper, they may stay aggregated without actually touching each other. If the maximum at B is substantial, the particles will not pass it. The maximum can be seen as a free energy barrier, retarding or even preventing close approach. However, the maximum value of V *cannot* be seen as representing an activation free energy as used in molecular reaction kinetics (Section 4.3). The slowing down of aggregation by an energy barrier will be further discussed in Section 13.2.1. If the particles can come closer, i.e., pass the free energy barrier, they will reach the primary minimum A.

The broken lines in Figure 12.1 give two extremes. In a situation characterized by the upper curve, aggregation will never occur. In other words, the dispersion is stable against aggregation. If the lower curve would apply (no maximum and a deep minimum), particles will reach each other and will stay forever at touching distance, provided that the external conditions (temperature, pH, etc.) are kept constant. In other words, *irreversible aggregation* has occurred. At intermediate situations, slow aggregation may occur, or aggregation may be to some extent reversible.

Importance. Some situations in which colloidal interaction forces play an essential role are the following:

Aggregation of particles, i.e., does it occur and if so at what rate?

The *rheological properties* of particle gels greatly depend on the strength of the interaction forces between the particles (Section 17.2.3).

The *stability of thin films* against rupture (Section 13.4.1).

In understanding the last case, the interaction between two flat plates or surfaces is needed. Assuming the dimensions of the film to be much larger than its thickness h, the interaction free energy V then should be given per unit surface area of the film, i.e., in $J \cdot m^{-2}$. In the case of particles, it is given in joules per particle pair.

In some cases, the force involved is needed rather than the energy. This is simply given by $-dV/dh$ (cf. Figure 3.1). For particle pairs the result is in

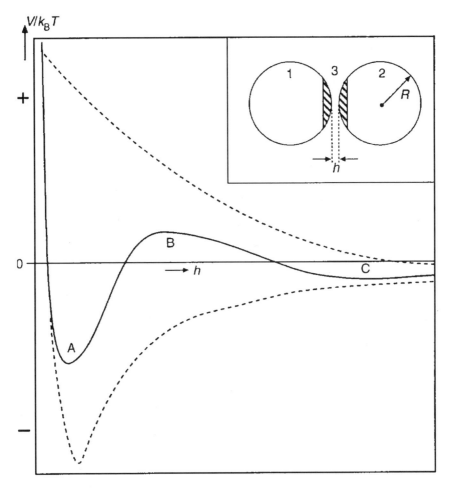

FIGURE 12.1 Hypothetical example (solid curve) of the colloidal interaction free energy V divided by $k_B T$, as a function of interparticle distance h. The insert defines the geometry involved for the case of two spheres of radius R. The broken lines give two other examples of interaction curves.

newtons, for films in pascals $(N \cdot m^{-2})$. The net pressure in a film due to interaction forces is called the *disjoining pressure*.

The nature of the various attractive and repulsive forces that can act is discussed in the rest of this chapter. Because entropic contributions often are involved, we need to consider free energy (more precisely Gibbs energy) rather than internal energy. Especially the repulsive forces tend to depend greatly on the substances adsorbed onto particles or film surfaces, i.e., on surfactant properties and concentration.

12.2 DLVO THEORY

The first useful theory of colloidal interaction forces and colloid stability (against aggregation) was developed independently by Deryagin and Landau and by Verwey and Overbeek. Hence it is called the DLVO theory. It takes into account the combined effects of van der Waals attraction and electrostatic repulsion.

12.2.1 Van der Waals Attraction

Van der Waals forces are briefly discussed in Section 3.1; for interaction between atoms or small molecules, their strength decays with intermolecular distance to the power -6. In the Hamaker–de Boer treatment, two macroscopic bodies are considered, and the van der Waals interaction between each atom in one of the bodies with all of the atoms in the other body are summed (actually a double integration procedure is applied). The result is that the total interaction energy V can be given by the product of a material property, called the Hamaker constant A (expressed in J or units of $k_B T$), and a term depending on the geometry of the system. These relations are relatively simple.

Geometry. For two spheres, radii R_1 and R_2, separated at a distance h between the particle surfaces, and if $h \ll R_1, R_2$, the result is

$$V_{\text{vdW,s}} = -\frac{A\overline{R}}{12\,h}$$
$$\overline{R} = \frac{2R_1 R_2}{R_1 + R_2} \tag{12.1}$$

Hence the repulsive energy is inversely proportional to interparticle distance. For two parallel flat plates of a size $\gg h$ and a thickness larger than about 5 times h, the result is

$$V_{vdW,p} = -\frac{A}{12\pi h^2} \qquad (12.2)$$

where $V_{vdW,p}$ is expressed in $J \cdot m^{-2}$. Also for some other geometries equations have been derived.

Hamaker Constants. Calculation of the magnitude of the Hamaker constant of a material proceeds by quantum mechanics, and we will merely give results. For two bodies (particles) of the same material, separated by vacuum, the Hamaker constant of this material can be inserted in Eqs. (12.1) and (12.2). By convention, the Hamaker constant is called A_{11} for interaction between two bodies of material 1, etc. Table 12.1 illustrates the relations in other cases. Eq. (12.3) is the most general one. If the material of bodies 1 and 2 is the same, Eq. (12.4) results. Note that the Hamaker

TABLE 12.1 Hamaker Constants (A) for Interaction Between Two Particles of Materials 1 and 2, Separated by a Material 3

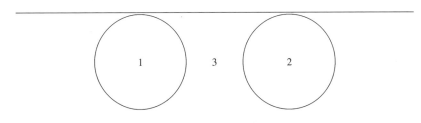

• GENERAL CASE:

$$A_{12(3)} = A_{12} - A_{13} - A_{23} + A_{33} \approx (\sqrt{A_{11}} - \sqrt{A_{33}})(\sqrt{A_{22}} - \sqrt{A_{33}}) \quad (12.3)$$

• IF $1 = 2$:

$$A_{11(3)} = A_{11} - 2A_{13} + A_{33} \approx (\sqrt{A_{11}} - \sqrt{A_{33}})^2 \qquad (12.4)$$

which implies $A_{11(3)} \approx A_{33(1)}$

• IF $3 = $ VACUUM:

$$A_{12(3)} = A_{12} \approx \sqrt{A_{11} \cdot A_{22}} \qquad (12.5)$$

constant of the *medium* between the two bodies concerned is also involved in the interaction energy. If, moreover, the materials 1 and 3 are similar in composition and density, $A_{11(3)}$ tends to be small. If there is no material 3, but vacuum, A_{33} equals zero; for most gases, including air, it is negligible. We then need Eq. (12.5).

The equations in Table 12.1 also give expressions after the $\approx$ sign that are based on the *Berthelot principle*, which states that $A_{12} = (A_{11} \cdot A_{22})^{0.5}$; see Eq. (12.5). The relation is not fully correct, but it serves to illustrate some important points. The first is that the Hamaker constant is always positive if the particles are of the same material [Eq. (12.4)]; this implies that the interaction energy is always negative, i.e., the force between the bodies is always attractive. Secondly, the Hamaker constant will remain the same if the materials of the particles and of the fluid between them are interchanged. In other words, the attraction between two water drops across air will be equal to the attraction between two air bubbles (of the same radius) in water. If the three materials are all different, Eq. (12.3) shows that A can be negative, i.e., the interaction *can* be repulsive. This has been experimentally confirmed, for instance for films of some organic liquids between glass and air [A_{22} in Eq. (12.3) then equals zero, while $A_{11} > A_{33}$; cf. Table 12.2].

Currently, the more elaborate and precise Lifshits theory for the calculation of Hamaker constants is applied rather than the Hamaker–de Boer theory; it yields values that agree well with experiment. Some results

TABLE 12.2 Values of the Hamaker Constant of Various Materials Across Vacuum (A_{11}) and Across Water ($A_{11(3)}$), in 10^{-21} J

Material	A_{11}	$A_{11(3)}$
Water	37	—
Ethanol	42	
n-Pentane	38	3.4
n-Decane	48	4.6
n-Hexadecane	52	5.4
Solid paraffin	71	
Solid stearic acid	50	~3
Polystyrene	65	9
Glass	~90	~13
Bovine serum albumin		~5
Biological cells		~1

(Division by 4.05 gives A relative to $k_B T$ at room temperature.) For the most part calculated by Lifshits theory.

are in Table 12.2. It is seen that the values tend to increase with molar mass for a homologous series and are higher for a crystalline than for a liquid material (compare solid paraffin with n-hexadecane). The values across water tend to be about an order of magnitude smaller than those across air. For most food systems, the values for solids or liquids in liquids are of order $k_B T$. For air–water–air the Hamaker constant is about 10 times larger. Values for proteins in water are not well known; the result for BSA in Table 12.2 refers to protein crystals. In practice, proteins are often swollen, i.e., contain water, and the Hamaker constant across water will then be (markedly) smaller.

Some Complications. As mentioned in Section 3.1, the London dispersion forces generally constitute the most important part of the van der Waals attraction. These forces are due to mutually induced dipoles. Since the dipoles fluctuate very fast, and because the electromagnetic wave involved in establishing the attraction travels at finite speed, a dipole in one particle and its induced counterpart in the other particle will be out of phase if the distance between them is large. This *retardation* becomes appreciable for h values larger than about 5 nm. To be sure, this only applies to the dispersion forces, and not to those dependent on a permanent dipole. However, the latter forces become screened by any electrolyte present between the two particles. Altogether, the Hamaker constant tends to decrease in magnitude with increasing h value. To give an example for water, calculated by Lifshits theory, at 1 nm the effective value of A would be 38, and at 5 nm $31 \cdot 10^{-21}$ J. Nonetheless, van der Waals attraction can be appreciable at relatively large distances (say, up to 100 nm).

Another significant point is the influence of an *adsorbed substance* on the effective value of A. For not very small particles ($R \gg h$) only some of the molecules in each particle contribute to the van der Waals interaction. In the insert in Figure 12.1, the "active part" is indicated by hatching; it roughly concerns layers with a thickness equal to the separation distance h. Assuming adsorption layers to have a thickness of, say, 2 nm, it will be clear that for small h (say, 5 nm) the value of A is largely determined by the adsorbate rather than by the material of the particles. In passing, the magnitude of h tends to be uncertain for particles onto which a surfactant is adsorbed, especially if the latter is a polymer.

Applications. Despite these complications, the Hamaker–de Boer–Lifshits theory is very useful. It is commonly applied in virtually all theories on colloidal interaction. If no other colloidal interaction forces act, it can be used directly. A case in point in foods is triglyceride crystals in triglyceride oil (Hamaker constant of order 0.5 $k_B T$). Here no other substantial forces act except hard-core repulsion. The latter is taken care of by assuming that very strong repulsion prevails below a certain value of h, often taken as 0.5 or 1 nm. [At $h = 0$, Eq. (12.1) would predict the attraction energy to be infinite.] Another example is a water film between large air bubbles, if the surfactant present does not cause repulsion and if h is larger than twice the thickness of the surfactant layer.

> *Note* It would take far too much space to discuss the *Lifshits theory* here. For the interested reader, we give one example of an approximate equation (error $< 5\%$) for the unretarded Hamaker constant of particles (1) in a medium (3). The equation is often useful and it reads
>
> $$A_{11(3)} \approx \frac{3}{4} k_B T \left(\frac{\varepsilon_1 - \varepsilon_3}{\varepsilon_1 + \varepsilon_3} \right)^2 + \frac{3 h_P v_e}{16 \sqrt{2}} \frac{(n_1^2 - n_3^2)^2}{(n_1^2 + n_3^2)^{3/2}}$$
>
> Here ε is the static (zero frequency) relative dielectric constant; h_P is Planck's constant, i.e., $6.626 \cdot 10^{-34}$ J·s; v_e is the main UV adsorption frequency, which equals for most substances involved $2.9 - 3.0 \cdot 10^{15}$ s^{-1}; and n is the refractive index for visible light (generally taken at a wavelength of 589 nm). The first term in the equation is due to dipole–dipole and dipole–induced dipole interactions, and the second term is due to London dispersion forces (unretarded). The first term is always smaller than $(3/4) k_B T$; the second term can be much larger.

Question 1

Consider a flat film of water, thickness $h = 10$ nm, between two large air bubbles. You may assume that at that h value van der Waals attraction is the only interactive force. Calculate the disjoining pressure in the film, and do this also for a similar oil–water–oil film.

Answer

As mentioned in Section 12.1, the disjoining pressure is given by $p_{\text{disj}} = -\mathrm{d} \, V / \mathrm{d} \, h$. Differentiating Eq. (12.2) then yields $p_{\text{disj}} = -A/6 \, \pi h^3$. The disjoining pressure thus is negative, as it should be for an attractive force. According to Table 12.2, the value of A_{11}, i.e., for water across air, equals $37 \cdot 10^{-21}$ J, and the value for air across water is virtually the same. We thus obtain $p_{\text{disj}} = -2000$ Pa, a considerable pressure. The value of the Hamaker constant for oil across water depends on the type of oil and is less well known; assuming $A = 4 \cdot 10^{-21}$ J, we arrive at $p_{disj} \approx -210$ Pa, i.e., much smaller than for air–water–air.

Question 2

When a solution of 1 or 2% tristearate in triglyceride oil, made at about 70°C, is cooled to room temperature, small tristearate crystals form, and these tend to aggregate rapidly. Assuming the crystals to be spheres of about 0.1 μm diameter, what would be the attractive free energy between them? Is this sufficient to cause aggregation? It is also observed that the addition of a little glycerol monolaurate to the system tends to cause disaggregation. How could that be explained?

Answer

As mentioned above, the interaction can be calculated by Eq. (12.1), assuming A to equal 0.5 times k_BT and h, 0.4 nm. Since R is 50 nm, the result would be $-5.2 k_BT$. This is sufficient to cause aggregation, although not high enough to prevent occasional disaggregation. Addition of glycerol monolaurate presumably leads to its adsorption onto the crystal surfaces (it will not in itself crystallize at room temperature). This would increase the closest possible distance between approaching crystals. Assuming that h would then be at least 2 nm, the resulting value for $|V_{vdw}|$ can become at most $1 k_BT$, insufficient to cause much aggregation.

12.2.2 Electrostatic Repulsion

Any aqueous surface or interface carries an electric charge. The surface of pure water is negatively charged, due to a preferential adsorption of OH^- ions. Also in salt solutions, there is a certain preference for some ions, generally anions, to be located at the interface. In many food systems, ionic surfactants (including proteins) are adsorbed at an aqueous interface, thereby giving rise to a considerable charge. Of course, the charge may depend greatly on pH.

Electrostatic Potential. The presence of an excess charge at the interface causes it to have an *electrostatic surface potential* ψ_0. This may cause repulsion between two charged surfaces that are close together. In Section 3.1, Coulombic electrostatic interaction was mentioned [Eq. (3.1)], but the repulsion between charged surfaces separated by an aqueous phase cannot be considered Coulombic. This is because the charge is shielded by ions, as was discussed in Section 2.3.2; see Figure 2.10. Counterions (ions of a charge sign opposite to that at the interface) accumulate near the interface, whereas coions (same charge sign as at the interface) are excluded. This is shown in Figure 12.2a.

In this way, an **electric double layer** is formed: the charge at the surface is compensated for by the excess of counterions and a depletion of coions in the double layer, and electroneutrality is maintained. The distribution of the

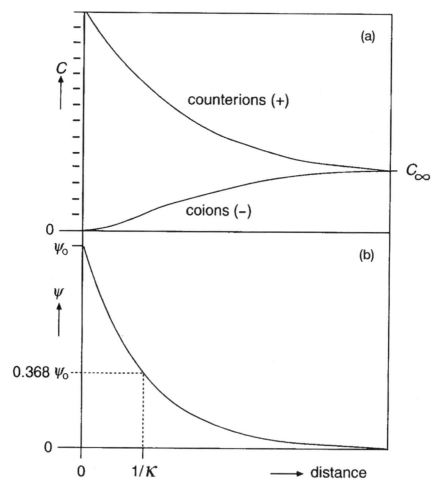

FIGURE 12.2 The electric double layer. (a) The concentration c of counterions and coions near a negatively charged surface; c_∞ is the bulk electrolyte concentration. (b) The decay of the electrostatic potential ψ near that surface. The ("nominal") thickness of the electric double layer $1/\kappa$ is also indicated.

ions is governed by two opposing effects. The attractive energy between ions of opposite charge pulls the counterions to the surface to neutralize the charge (and in a similar manner, the coions are pushed away from the surface). This is counteracted by the loss in mixing entropy that it would cause. The system will strive toward a minimum in free energy, and from this principle the distribution of the ions can be calculated.

Accordingly, the *electrostatic potential will decrease when going away from the interface* (Fig. 12.2b), according to the Debye–Hückel theory discussed in Section 6.3.2. We recall the equation

$$\psi = \psi_0 \exp(-\kappa h) \tag{6.32}$$

where ψ_0 is the surface potential and h the distance from the surface. The shielding parameter κ depends primarily on the total ionic strength I and is given by

$$\kappa^2 = \frac{2IN_{\mathrm{AV}}e^2}{\varepsilon_0 \varepsilon k_{\mathrm{B}} T} \tag{12.6}$$

where e is the charge of the electron, ε_0 the permittivity of vacuum, and ε the relative dielectric constant of the medium. For water at room temperature the equation can be reduced to

$$\frac{1}{\kappa} \approx \frac{0.30}{\sqrt{I}} \approx 0.30[(1/2)\Sigma m_i z_i^2]^{-0.5} \tag{12.6a}$$

where $1/\kappa$ is the Debye length or the nominal *thickness of the electric double layer* (illustrated in Fig. 12.2b) in nm and I is in moles per liter; m is molar concentration and z, valence of an ion. Figure 6.8 illustrates the dependence of ψ on h and values of $1/\kappa$ for various values of I; it also indicates the magnitude of the ionic strength in some foods.

The relation between the surface charge density and the magnitude of the surface potential is far from simple. Consider an adsorption layer of ionic surfactants. The surface excess may be as high as $5 \cdot 10^{-6}$ mol $\cdot$ m^{-2}, corresponding to 3 molecules per nm^2. Following the discussion in Section 6.3.1, we can conclude that the dissociation of, for instance, carboxyl groups may be greatly diminished; see especially Figure 6.7a. The pK of these groups may readily be shifted upwards by two pH units. Sulfate groups have a much lower pK and will still be dissociated at neutral pH. However, the concentration of counterions at the charged interface cannot be as high as would agree with Eq. (6.32). Hence the decrease of ψ with h is different at very small h values, affecting the value of ψ_0. Actually, one has to account for a layer of about the thickness of a hydrated counterion where these ions are closely packed; it is called the *Stern layer*. In practice, however, one may take for ψ_0 the potential at the Stern layer (see below).

The "double layer" discussed should not be interpreted as a static layer, adhering to the particle. Molecules and ions diffuse in and out. When a charged particle is subject to an electrokinetic experiment—which means that it is moving relative to the salt solution in which it is dispersed due to an externally applied electrostatic potential gradient—the *slipping plane* between particle and liquid is close to the geometrical particle surface; for hard and smooth particles it roughly coincides with the outside of the Stern layer. The **zeta potential**, which is the electrical potential determined in such an electrokinetic experiment (usually electrophoresis), is the potential at the slipping plane. It tends to agree with the value of ψ_0 needed to calculate the electrostatic repulsion.

Repulsive Free Energy.   When the surfaces of two particles in an aqueous phase come close to each other, their electric double layers start to *overlap*, as illustrated in Figure 12.3. This will cause a local increase in potential, which implies that work must be applied to bring the particles closer together. From this increase in potential, the repulsive free energy can be calculated. In another approach to the same problem, the increase in osmotic pressure caused by the overlap of the double layers is calculated, which also yields a repulsive free energy. The two methods give identical results. The mathematics of the theory is quite involved, and we will merely give some results for the case that $|\psi_0|$ is not too high, say below 30 mV; this is nearly always true in food systems.

For identical *spheres*, the relation becomes

$$V_{\text{El,s}} = 2\pi\varepsilon_0\varepsilon R\psi_0^2 \ln(1 + e^{-\kappa h}) \approx 4.5 \cdot 10^{-9} R\psi_0^2 \ln(1 + e^{-\kappa h}) \qquad (12.7)$$

where the part after the $\approx$ sign is valid for water at room temperature (in SI units). If the spheres do not have the same radius, this can be accounted for as in Eq. (12.1). There is one other condition for Eq. (12.7) to hold, which is that $\kappa R \gg 1$. In practice, this means that the theory is valid for $h \ll R$.

For the repulsion between infinite *flat plates*, the result is

$$V_{\text{El,p}} = 2\varepsilon_0\varepsilon\kappa\psi_0^2 e^{-\kappa h} \approx 1.4 \cdot 10^{-9} \kappa\psi_0^2 e^{-\kappa h} \qquad (12.8)$$

Again, the part after the $\approx$ sign is valid for water as the medium at room temperature. The results of Eq. (12.7) are in J, those of (12.8) in $J \cdot m^{-2}$.

A different situation arises if the particles are separated by a **dielectric medium** (i.e., not conducting electricity). A case in point is a W–O emulsion, e.g., aqueous droplets in a triglyceride oil. The droplets contain, say, protein, which adsorbs onto the W–O interface, thereby giving rise to a surface charge and thus to a surface potential ψ_0. The oil has a relative dielectric constant of about 3; ionic species can virtually not dissolve in oil, hence $I \approx 0$ and $1/\kappa \to \infty$. Thus there is no electric double layer, and the electrostatic repulsion is almost purely *Coulombic*, which acts over far

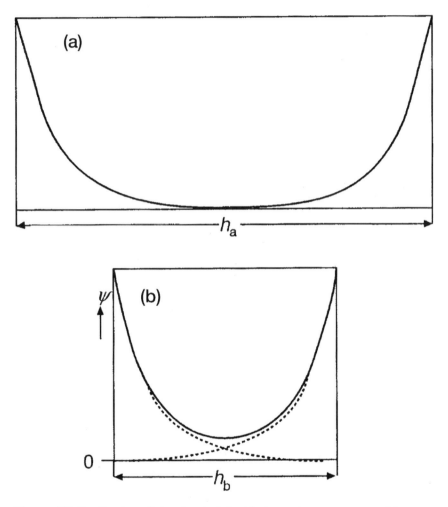

FIGURE 12.3 Overlap of the electrical double layers between two particles at a distance h from each other. In (a) the double layers virtually do not overlap, and the electrostatic potentials ψ equal zero in the middle; in (b) the particles are closer and their electrical double layers do overlap, which causes the potentials to locally increase. Highly schematic.

longer distances than does double layer repulsion. For two spheres of equal size it is given by

$$V_{El,s} = \frac{4\pi \in_0 \in R^2 \psi_0^2}{2R + h} \approx \frac{3.3 \cdot 10^{-10} R^2 \psi_0^2}{2R + h} \tag{12.9}$$

where the part after the $\approx$ sign applies to triglyceride oil as the medium. Assuming, as an example, that $R = 2 \, \mu m$ and $\psi_0 = 25 \, mV$, the repulsive free energy at $h = 0$ amounts to $50 \, k_B T$, which is a strong repulsion. However, at $h = 2R = 4 \, \mu m$, the repulsion is still quite strong at $25 \, k_B T$. Since repulsion acts over such large distances, the droplets sense about the same repulsive force from all sides, unless the emulsion is very dilute (see Fig. 9.6). This comes down to virtually no net repulsion. Equation (12.9) would also apply to repulsion across a layer of (dry) air, with $\varepsilon = 1$.

12.2.3 Total Interaction Energy

The basic idea of the DLVO theory is that the stability of lyophobic colloids in aqueous systems is determined by the combination of van der Waals attraction and electrostatic repulsion and that the two are exactly additive. In other words, the total interaction free energy V_T would at any value of h be given by

$$V_T = V_{vdW} + V_{El} \tag{12.10}$$

This turns out to be very nearly correct, provided that no other interaction forces act.

Some calculated results are plotted in Figure 12.4. It may be noticed that the range over which the interaction energy is of importance seems to be restricted to about 10 nm; closer inspection shows that interparticle distances up to about 20 nm may be relevant. Because the (negative) van der Waals term is proportional to $1/h$, and the (positive) electrostatic term decreases about exponentially with h, attraction will always prevail at very short and at large distances. At large h, however, the attractive energy may be negligibly small. It is generally assumed that aggregation in the primary minimum (at y) is irreversible and that a sufficiently high maximum value of V (at x) can prevent this aggregation. Aggregation in the secondary minimum (at z), if it exists, is supposed to be reversible, since the minimum is generally not very deep.

Important Variables. The results in Figure 12.4 are calculated for a range of conditions as commonly experienced in food systems.

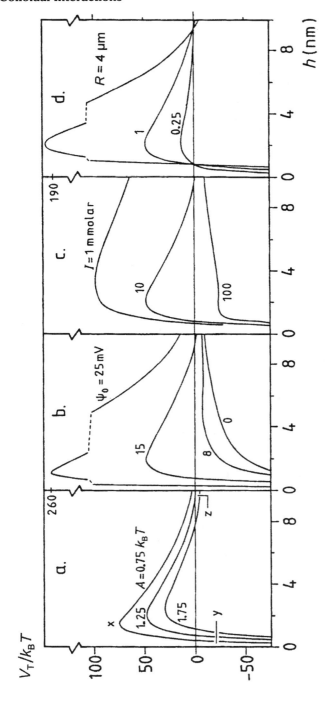

FIGURE 12.4 Examples of the colloidal interaction energy V_T (in units of $k_B T$) as a function of interparticle distance h, calculated according to the DLVO theory for spherical particles in an aqueous phase. Illustrated are the effects of the magnitude of (a) the Hamaker constant A; (b) surface potential $|\psi_0|$; (c) ionic strength I; and (d) particle radius R. Unless indicated otherwise, $A = 1.25\,k_B T$; $|\psi_0| = 15\,\text{mV}$; $I = 10\,\text{mmolar}$; $R = 1\,\mu\text{m}$. x, y, and z denote the maximum, the primary minimum, and the secondary minimum in V_T, respectively.

1. The *Hamaker constant* does not vary greatly in most of the systems considered (see Table 12.2). Nevertheless, the variation may be important, especially because it affects the depth of the secondary minimum if the ionic strength is fairly high.

2. The *surface potential* has a very large effect, since electrostatic repulsion is proportional to ψ_0^2. This means that the pH will have a large effect on aggregation in many food dispersions.

3. Also the *ionic strength* has a large effect. If I is very small, the system is nearly always stable to aggregation (unless $\psi_0 \to 0$), whereas it is unstable at high I values. As discussed before, multivalent ions affect the value of I far more than monovalent ions do, and undissociated salts have no effect. Ionic strength primarily determines the *range* over which repulsion acts, whereas pH (via ψ_0) determines the *maximum* repulsive energy. However, salt may also affect ψ_0, because counterions may associate with ionic groups on the particle surface. This greatly depends on the chemical constitution of the afore-mentioned ionic groups and of the counterions, as well as on the counterion activity. In practice, it is often difficult to differentiate between the two mechanisms.

4. Equations (12.1) and (12.7) show that interaction energy between spheres is proportional to the *radius* at any value of h. As shown in Figure 12.4d, this may imply, for instance, that the larger particles in a dispersion would be stable against aggregation in the primary minimum, whereas the smaller particles are not. In practice, this is often not observed. In other words, the dependence of stability on particle size is smaller than predicted.

Application. To apply the DLVO theory in practice, several pieces of information have to be collected. Particle size (distribution) and shape can generally be experimentally determined. Hamaker constants often are to be found in the literature or can be calculated from Lifshits theory. The surface potential can be approximated by the zeta potential obtained in electrophoretic experiments. The ionic strength is generally known (or can be calculated) from the composition of the salt solution. All the other variables needed are generally tabulated in handbooks. This then allows calculation of $V(h)$. To arrive at an aggregation rate, more information is needed; this is discussed in Section 13.2.

The DLVO theory has been very successful in predicting (in) stability against aggregation for many, especially inorganic, systems. Although developed for lyophobic colloids, the theory can often be usefully applied to lyophilic colloids; these are often found in biogenic systems, including most foods. However, some complications and other interaction forces may come into play.

Complications. Below are some factors of importance:

Particle shape. Calculation of the interaction energy works well for perfect spheres, i.e., for most O–W emulsions. Although equations for a number of simple particle geometries are to be found in the literature, application to real nonspherical particles often involves several difficulties. For instance, the shape may be irregular and variable, and the particles can encounter each other in a number of orientations.

Particle size. The deficiency of the theory was already mentioned. Currently, a number of refinements of the DLVO theory have been proposed to solve this problem, but agreement has not yet been attained.

For real particles, which are often inhomogeneous, the value of the Hamaker constant may be uncertain.

Likewise, real particles may not have a hard and smooth surface, and this may interfere with the equality of surface and zeta potential (the position of the slipping plane not coinciding with the outside of a Stern layer).

Several other interaction forces may act. These are for the most part discussed in the rest of this Chapter.

Question 1

Consider emulsions made of 2% triglyceride oil in 0.1% surfactant solutions that have an ionic strength of 5 millimolar and a pH of 7. The surfactants are (a) sodium dodecyl sulfate; (b) glycerol monolaurate; and (c) β-lactoglobulin. Moreover, to part of each of the emulsions 0.5% NaCl is added, and to another part sufficient HCl to bring the pH down to 5. Nine emulsions are thus obtained; typical droplet sizes are $4\,\mu m$. Which of the emulsions are expected to be stable to aggregation and which are not? For some of them you can predict this without making calculations (which ones?), for others you have to make calculations. You may assume that the Hamaker constant equals $5 \cdot 10^{-21}$ J and that the surface potential equals $-40\,mV$ for (a) and $-15\,mV$ for (c) at pH 7. What calculations would you perform? Why are the concentrations of oil and of surfactant given?

Question 2

In Section 11.2.2, drainage of liquid from a foam is discussed. In a foam with flat lamellae between the bubbles, drainage from a lamella to the adjacent Plateau borders is caused by a difference in Laplace pressure, which equals $\rho g H$, according to Eq. (11.1), where H is the height above the bottom of the foam. For an aqueous foam at $H = 1$ cm, the pressure difference would thus be about 100 Pa. Assuming that the film does not rupture, drainage will go on until the net colloidal disjoining pressure in the film equals the mentioned Laplace pressure difference. Can you calculate at what film thickness this will be the case if repulsion is caused by an adsorbed ionic surfactant? We assume that the Hamaker constant of the film material will be $30 \cdot 10^{-21}$ J (value for water for not very small h), the surface potential -32 mV, and the ionic strength 0.01 molar.

Answer

The DLVO theory would be appropriate for the present case. However, we need pressure, not interaction energy per unit area. The (repulsive) disjoining pressure can be derived from

$$p_{\text{disj}} = -\frac{dV_p}{dh}$$

Using Eq. (12.2) for the van der Waals attraction and Eq. (12.8) for the electrostatic repulsion between flat plates, we arrive after differentiation at

$$p_{\text{disj}} = \frac{-A}{6\pi h^3} + 1.4 \cdot 10^{-9} \psi_0^2 \kappa^2 \exp(-\kappa h)$$

Using Eq. (12.6) for κ and inserting the values given above, the disjoining pressure can be calculated as a function of h. The result is $p_{\text{disj}} \approx 16 \cdot 10^5 \, (e^{-h/3} - h^{-3})$, in Pa for h in nm. The result equals 100 Pa at $h \approx 27$ nm. Actually, the van der Waals attraction will have been overestimated for such a distance due to retardation occurring. Hence, the equilibrium film thickness would be 30 nm or a little more.

Further question: What would be the effect of a higher ionic strength, e.g., 0.2 molar, on the relation between p_{disj} and h?

12.3 ROLE OF POLYMERS

Polymers can strongly affect colloid stability. Many polymers can adsorb onto particles and then cause steric interaction (Section 12.3.1), which is often repulsive and thereby stabilizing, although attractive interaction can also occur. If polymer molecules adsorb on two particles at the same time, they cause bridging (Section 12.3.2), hence aggregation. Polymers in solution can also cause aggregation via depletion interaction (Section 12.3.3), or they can stabilize a dispersion by immobilizing the particles in a gel network.

Chapter 6 gives basic aspects about polymers, and Chapter 7 about proteins, the polymers often used in stabilizing food dispersions. Adsorption of surface active polymers is discussed in Section 10.3.2.

12.3.1   Steric Interaction

The **mode of adsorption** of various polymers is illustrated in Figure 10.12. Polymers can also be "grafted" onto a particle, which means that one end of each molecule is strongly attached to the surface. The conformation and the thickness of a layer of adsorbed or grafted polymer molecules will depend on (a) the number density of the layer, i.e., the number of molecules attached or adsorbed per unit surface area σ (e.g., one per $10 \, nm^2$); (b) the length (degree of polymerization) of the molecules; (c) the number and distribution of segments that have affinity for the particle surface; and (d) the solvent quality. It is difficult to define unambiguously the thickness δ of a polymer layer. It may be taken as the distance from the surface that gives a layer that includes a certain proportion, e.g., 90%, of the polymer segments.

Figure 12.5 gives examples for one polymer length. The "*mushroom*" type is typical for grafted polymers, if solvent quality is good and σ low.

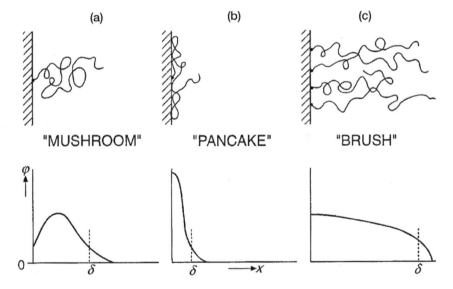

FIGURE 12.5   Various conformations adapted by grafted polymer molecules sticking out in a liquid. Grafting is depicted by a dot. The local polymer segment density φ is indicated as a function of distance from the surface x. The approximate thickness of the polymer layer δ is also indicated.

Conformations similar to "*pancakes*" and "*brushes*" can also occur for adsorbed polymers. To achieve substantial repulsive action, a brush is generally desired; to that end, a good solvent quality and a high number density are needed.

A polymer layer, whether adsorbed or grafted, can cause so-called **steric repulsion**. Two mechanisms can be distinguished, as illustrated in Figure 12.6. If the proximity of a second particle *restricts the volume* in which the protruding polymer chains can be, this means that the number of conformations that a chain can assume is restricted, hence the entropy of these chains is lowered, hence the free energy is increased, hence a repulsive

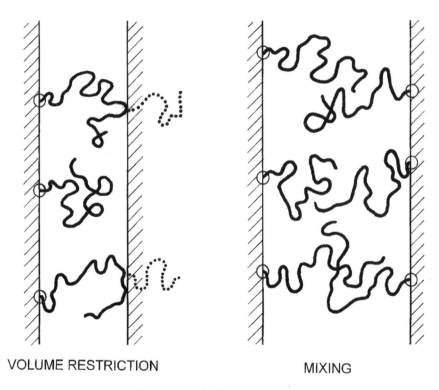

VOLUME RESTRICTION   MIXING

FIGURE 12.6 Two mechanisms involved in steric repulsion due to grafted (or adsorbed) polymer chains (heavy lines). The dotted lines indicate possible conformations of the polymer chain in the absence of volume restriction.

force will act. The repulsive free energy is very large: several times $k_B T$ for each polymer chain involved, and the number of polymer molecules in the gap between two particles will often be of order 10^3. This means that at very close approach—roughly speaking for $h < \delta$—the repulsive energy between the particles will always be positive and large; it increases very steeply with decreasing value of h.

However, when two particles approach each other, *mixing* of both polymer layers will occur for $h < 2\delta$, i.e., before volume restriction comes into play. This means that the mixing entropy will decrease, and this then would also lead to repulsion. It mostly does, but not always, since it depends on the *solvent quality* for the polymer chains. If the quality is fairly low, just high enough to allow the chains to protrude into the liquid, the attractive energy between polymer segments may be large enough to more than compensate for the decrease in mixing entropy, thereby causing attraction.

Another way to explain these mechanisms is by considering the *osmotic pressure* in the liquid between approaching particles. If the pressure increases in the gap between the particles, solvent will be drawn into the gap to lower the osmotic pressure again, hence there will be repulsion. Recalling the equation for the osmotic pressure Π_{pol} of a polymer solution in Section 6.4.1,

$$\Pi_{\text{pol}} = RT \left(\frac{1}{v_p} \varphi + \frac{\beta}{2v_s} \varphi^2 + \frac{1}{3v_s} \varphi^3 + \cdots \right) \qquad (12.11)$$

where φ is the net polymer volume fraction and v the molar volume for polymer (p) and solvent (s); β measures the solvent quality, which is "ideal" for $\beta = 0$ (i.e., a theta solvent), and poor for $\beta < 0$. The values of φ can become very high in the gap between approaching particles; say, 0.02 if $h = 2\delta$ and 0.4 at $h = \delta$. We can in principle use Eq. (12.11) to explain the repulsion between the particles. The first term between brackets does not apply: it stands for the number of species per unit volume, and in the present case we have only two species, i.e., the pair of particles. The third term can be interpreted as being due to volume restriction; it is seen to be always positive and it strongly increases with polymer concentration, i.e., for a closer approach. The second term is governed by the value of β; if the latter is significantly negative, attraction between polymer chains occurs, and this may result in attraction between particles.

The conformations of the molecules in the polymer layer and the resulting steric interaction energy can be calculated by means of a numerical *self-consistent field model*. The free energy of the polymer layers then is minimized by considering all possible conformations (including adsorbed segments) of the chains. We will not discuss the theory because it can rarely be applied to situations in foods: it needs knowledge of the magnitude of a number of variables, which is generally not available. In Chapter 6 (especially Section 6.2.1), variables affecting polymer conformation are discussed: solvent quality and polymer properties like chain length, chain stiffness, and branching. For polyelectrolytes, also density and distribution of electric charges and ionic strength are important variables.

By way of example, some results on the **total interaction free energy** are shown in Figure 12.7, which serves to illustrate trends. Besides the steric

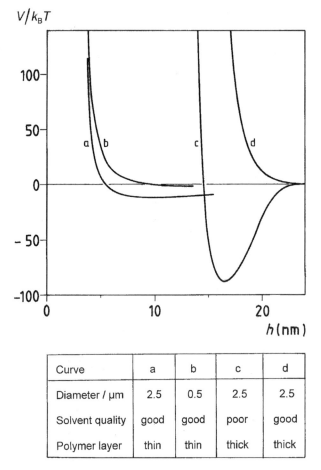

Curve	a	b	c	d
Diameter / μm	2.5	0.5	2.5	2.5
Solvent quality	good	good	poor	good
Polymer layer	thin	thin	thick	thick

FIGURE 12.7 The total interaction free energy V relative to $k_B T$ as a function of interparticle distance h between equal-sized spheres covered by adsorbed layers of a homopolymer. Calculated taking van der Waals attraction and steric interaction into account.

interaction, van der Waals attraction is taken into account. The number density σ of polymer molecules at the surface is kept constant. Three factors have been varied. If the *solvent quality* is poor ($\beta < 0$), a deep energy minimum can occur, causing strong attraction. Curves (c) and (d) refer to the same polymer and only the composition of the solvent is different. Notice that the layer thickness is smaller for a lower solvent quality, which means that the polymer chains are less stretched. In other words, the configuration tends to change from what is depicted in Figure 12.5c in the direction of b, if solvent quality is decreased. Even if the solvent quality is very good, but the *polymer layer thickness* is small due to low molar mass of the polymer, say $\delta < 5$ nm, van der Waals attraction may prevail at some distance, at least if the particles are fairly large (curve a). Van der Waals attraction is proportional to *particle radius*, as discussed, whereas the steric repulsion will hardly depend on particle size (unless the particles are very small: R comparable to δ). The ratio of δ/R thus is an important variable. If the ratio is not very small, σ is high, and the solvent quality is good ($\beta > 0$), repulsion will always be strong and very stable dispersions result.

It may further be noted that polymers that do adsorb onto particles, and then stabilize them against aggregation, are nearly always *copolymers*, where some segments have affinity for the particle surface, while for other segments the solvent quality is good enough for them to prefer being in the continuous phase. It is very difficult to have a homopolymer that adsorbs onto a particle without assuming a conformation as depicted in Figure 12.5b, or an even flatter one, which would generally imply steric repulsion to be weak.

Polyelectrolytes. Several kinds of polymers bear electric charges, and these polyelectrolytes are discussed in Section 6.3. An important characteristic is the charge density $1/b_{ch}$, where b_{ch} is the average linear distance between charged groups. If the charge density is high and the charges are all (or predominantly) of the same sign, an adsorbed or grafted layer of polymer chains of sufficient number density σ readily forms a brush that causes strong electrostatic plus steric repulsion. Another crucial variable is the ionic strength I, and basically three cases can be distinguished.

If I is high, say > 1 molar, the thickness of the electric double layer $1/\kappa$ is very small; electrostatic repulsion then acts over distances far smaller than b_{ch} and the effect of charge is small. The result is a *neutral brush*, i.e., the situation discussed above.

If the magnitude of I is around 0.1 molar, a *salted brush* can be obtained. Due to the *Donnan effect*, discussed in Section 6.3.3, the concentration of ions in the brush is high. An important parameter determining the thickness of the brush then is (σ/b_{ch}^2). If its magnitude is below a critical value, the brush collapses, implying that steric repulsion will be small. Although the magnitude of the critical value cannot be easily predicted—it depends, e.g., on I, chain length, and solvent quality—the effects of charge and number density are clear. For adsorbed polyelectrolytes, the magnitude of σ would decrease with increasing solvent quality; σ may also depend on polymer concentration. In most situations in foods, polyelectrolyte layers will be salted brushes, providing strong "*electrosteric*" repulsion.

If ionic strength is still smaller, say < 0.01 molar, an *osmotic brush* would be obtained. Considering for the moment a polyacid, with carboxyl groups providing the charges, the ions in the brush will be largely restricted to cations, including protons. It turns out that the degree of dissociation of the carboxyl groups now greatly depends on ionic strength. The smaller I, the higher the effective isoelectric pH. This may mean a decreased charge density, hence decreased repulsion.

Proteins. As an illustrative example, we will briefly discuss some aspects of the colloidal stability of **casein micelles**. These are proteinaceous particles to be found in milk, with mean diameter about 120 nm. They consist predominantly of caseins, calcium phosphate, and water. One of the caseins present, called κ-casein, has a very hydrophilic C-terminal part of 64 amino acid residues, containing some acidic sugar groups, net charge -9 to -12. κ-casein is at the outside of the micelles, and the C-terminals stick out into the solvent, forming a "hairy" layer. The value of σ is about 0.03 nm^{-2}; at the ionic strength (75 mmolar) and pH (6.7) of milk, the layer thickness is $\delta \approx 7$ nm. It may be considered as a salted brush of grafted polymer chains. It provides complete stability against aggregation of the casein micelles. The van der Waals attraction between the micelles is not strong, because the particles are fairly small and consist for the most part of water, which implies that the Hamaker constant is small.

These particles can be induced to *aggregate* by various means. One is renneting: the rennet added contains chymosin, a proteolytic enzyme, that specifically cleaves κ-casein, causing its C-terminal end to be split off and going into solution. When σ has been reduced to about 30% of its original value, the particles start to aggregate, and aggregation rate increases as σ is further reduced. (There is no clear evidence of collapse of the brush at a certain stage: it just disappears.) A second cause of casein micelle aggregation is lowering of the pH. Below a pH of about 5, the net charge of the hairs is close to zero, and the brush collapses as predicted. This leads to rapid aggregation. A third cause is the addition of ethanol. This will significantly diminish the solvent quality for the hairs, which would also cause collapse of the brush. However, as long as the chains are charged, their solvent quality cannot be diminished greatly, unless the ethanol concentration is so high as to decrease markedly the dielectric constant and thereby the ionization of ionogenic groups. It is indeed observed that the concentration of ethanol needed to cause aggregation of the casein particles decreases with decreasing pH.

Another example is given by O–W emulsions made with pure **Na-caseinate** in water. The ionic strength is then quite small, and the pH would be, e.g., 6.6, two units above the isoelectric pH of casein. The emulsion droplets then show weak aggregation, which can be undone by adding a little NaCl. In all probability, the caseinate at the surface of the droplets forms an osmotic brush, and at the prevailing low ionic strength many of the carboxyl groups on the casein would be undissociated, implying that the brush is more or less collapsed. Adding salt then increases dissociation and thereby the negative charge. Hence the adsorbed layer will swell and the repulsion will be greatly enhanced. The effect of salt is to some extent comparable to the salting in of proteins discussed in Section 7.3.

There are few proteins like the caseins, which behave more or less like random coils. The prime example for steric repulsion is *β-casein*, which has a strongly amphiphilic character (see Figure 7.1) and which provides a brush upon adsorption (see Figure 10.12d). *Gelatin* can adsorb in a similar manner (at least at temperatures above about 30°C), and possibly some denatured *globular proteins*. All proteins adsorb at apolar surfaces, including O–W and A–W interfaces, but the repulsion that they provide will generally be for the most part electrostatic; which means that a pH near the isoelectric point and a very high ionic strength may cause aggregation. As a rule of thumb, if conditions are such that the protein used is poorly soluble, it will probably not provide sufficient repulsion to, say, oil droplets to prevent their aggregation. Close to the isoelectric pH of the protein there will even be electrostatic attraction at close distance, because the protein will then have both positive and negative groups.

Other Surfactants. Most *polysaccharides* do not adsorb onto apolar surfaces. But those that do adsorb can provide strong steric repulsion. See Section 10.2.2 for adsorbing polysaccharides.

An important class of small-molecule surfactants contains *polyoxyethylene chains*. The prime example used in foods is the Tweens (see Table 10.2). They adsorb strongly at apolar surfaces and the POE chains, if long enough, provide strong steric repulsion, since σ is relatively large. The repulsion hardly depends on pH or ionic strength but becomes greatly reduced at high temperatures, say over 60°C.

Question

In the discussion of Figure 12.7 it is stated that "the steric repulsion will hardly depend on particle size (unless the particles are very small: R comparable to δ)." Can you explain why the particle size has little effect? Can you explain the caveat? Would the small effect of particle size also hold for steric attraction? You may assume that the particles are equal-sized spheres.

Answer

The area over which the polymer layers of two particles overlap, say at h being somewhat below 2δ, will be larger for larger particles. Geometrical consideration shows that the area will be about proportional to R for the same value of h (provided $h \ll R$). This means that for larger R a greater number of polymer chains will be involved, hence stronger repulsion at the same distance. However, the steric repulsion increases so very steeply with decreasing h value that an extremely small decrease of h will increase the repulsion for small particles to the value obtained for large particles. In other words, the effect of particle size on repulsion can hardly be measured. But if the particles are very small, say $R < \delta$, their surface area may be too small to accommodate sufficient polymer to give a reasonably dense layer at a distance δ from the particle surface.

The attractive force occurring when the solvent quality of the adsorbed (or grafted) polymer is poor—i.e., $\beta < 0$—will often be weak, since β may only be very slightly below zero. In that case, the number of chains participating in the interaction would become an important variable. Larger particles would then be less stable to aggregation.

12.3.2 Bridging

Consider two particles with an adsorbed layer of a homopolymer, where part of the chains protrude from the particle surface. When these particles are brought close together, theory predicts that at equilibrium some of the polymer molecules will become adsorbed onto both particles, forming a bridge. Such **bridging by adsorption** is depicted in Figure 12.8a. It implies, of course, that the particles are aggregated. The same situation may occur with several kinds of copolymers (though not for a diblock copolymer with one part of the molecule being hydrophobic and adsorbed and the other part hydrophilic and protruding into the solvent).

In practice, however, the predicted bridging may not happen, because thermodynamic equilibrium is not reached. When two particles meet by Brownian motion, they will be close together for, say, 1 ms, and it will probably take a far longer time for the adsorbed layers to attain an equilibrium conformation (cf. Section 10.4). Hence the dispersion will appear stable. On the other hand, if the particles stay together for a long time, for instance because they are sedimented, bridging may occur in the long run. This has been observed in some systems, but the author is unaware of an unequivocal example involving adsorbed proteins.

Anyway, bridging by adsorbed polymers will often depend on the *history of the system*. Consider a dispersion of small solid particles that tend to aggregate. To stabilize the suspension, an adsorbing polymer is added that can give a maximum surface excess $\Gamma_{plateau}$ of 6 mg $\cdot$ m^{-2}. If the specific particle surface area A is 1 m^2 per ml and 3 mg of a suitable surfactant polymer is added per ml, a Γ value of at most 3 mg per m^2 can result. This value is on the low side, but the particles are nevertheless stabilized; apparently, polymer chains stick out far enough into the solvent to cause steric repulsion. Assume now that to 1 ml of the suspension 6 mg of polymer is added and that subsequently 1 ml of suspension without polymer is added.

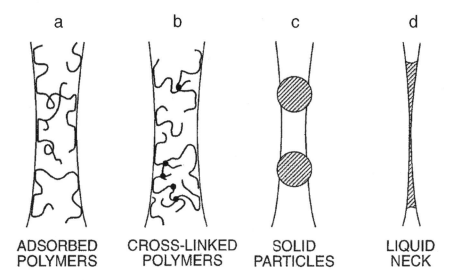

a   b   c   d

ADSORBED   CROSS-LINKED   SOLID   LIQUID
POLYMERS   POLYMERS   PARTICLES   NECK

FIGURE 12.8 Various modes of bridging two colloidal particles. Situation (c) can only occur for emulsion drops (or possibly gas bubbles) and (d) generally for solid particles. Not to scale: the number of polymer molecules involved in situations (a) and (b) would be very much larger than depicted.

Although the average value of Γ would be the same as in the former case, copious aggregation occurs due to bridging. A bare particle that meets a particle with a polymer layer will immediately stick to it, because some segments of a protruding polymer chain can adsorb onto the bare surface.

This example also shows that bridging is more likely if the total amount of polymer is small compared to the amount that can become adsorbed, i.e., $A \cdot \Gamma_{\text{plateau}}$. Since aggregation causes a dispersion to attain a greatly increased viscosity, we can thus have a situation where addition of more polymer (often called a thickening agent) leads to a lower viscosity. Similar situations can be observed if O–W emulsions are made by homogenization with variable amounts of protein, provided that the protein has high molar mass. A small protein concentration in the aqueous phase may then result in a highly aggregated emulsion, whereas a normal emulsion is obtained if more protein is added before homogenization. (Adding it afterwards does not help, since it will take a very long time before equilibrium adsorption layers have been formed.)

One would thus expect that a dispersion of particles having a suitable polymer layer that causes steric repulsion will always be stable against aggregation, as long as no bridging adsorption occurs. This is not always true. Figure 12.8b illustrates another type of bridging, i.e., by **cross-linking**. Steric repulsion does not prevent contact between particles. On the contrary, it only works when the polymer layers overlap, which implies that segments of polymer chains from two particles will frequently touch each other. If these segments can react with each other, a bridge is formed. An example is the formation of —S—S— bridges between protein layers. This may especially happen during heat-treatment of the system, but for some proteins even at room temperature.

Another case concerns particles stabilized by a *polyelectrolyte brush*; suppose that the charges on the chains are negative and that the solvent contains divalent cations, e.g., Ca^{2+}. —Ca— bridges may now be formed, and although such bonds have a short lifetime, the particles will become permanently bridged if the number of bonds per particle pair is sufficiently high. The forces involved can be measured, and an approximate example is illustrated in Figure 12.9. Here the force $F\,(=-dV_T/dh)$ needed to bring the two particle surfaces to a distance h is shown. Starting at point (a), F increases as h is decreased. At (b), —Ca— bridges must have formed, and now the increase of F with decreasing h is much stronger. Moreover, when the distance is subsequently increased, hysteresis occurs: F decreases and then becomes attractive, e.g., at (d). If no external force is acting, the particles will assume a mutual distance corresponding to point (c), and they are thus bridged. When pulling the particles apart with sufficient force, they will, of course, become separated, maybe as depicted by the broken line. The bridging by calcium ions can readily occur for adsorbed protein layers, if the calcium ion activity is sufficiently large, say > 5 millimolar. The casein micelles mentioned in the previous section also aggregate in such a situation.

We finish this section with an important conclusion, which is that considerable **hysteresis** can occur in the force–distance or the free energy–distance curves between approach and moving apart of two particles. Classical DLVO theory proceeds from the assumption that the interaction force is at equilibrium under all conditions, but this is mostly not the case if polymers are involved. The shape of a curve as depicted in Figure 12.9 will also depend on the rate at which the particles are moved.

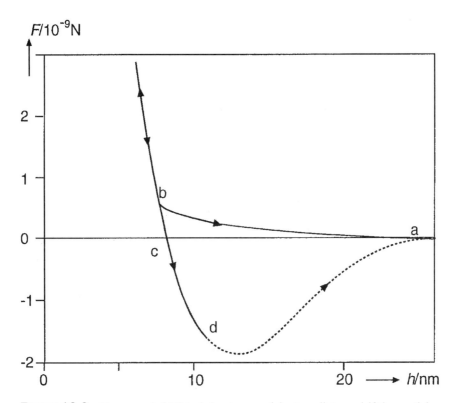

FIGURE 12.9 Force needed (F) to bring two particles to a distance h if the particles are covered by negatively charged polymer layers and the solvent contains divalent cations. The scales are only meant to indicate an order of magnitude. The dotted line is conjectural.

12.3.3 Depletion Interaction

Assume a dispersion of particles of radius r_p that also contains some nonadsorbing polymer; the polymer molecules form random coils, radius of gyration r_g. The situation is depicted in Figure 12.10. The center of mass of the polymer molecules cannot come closer to the particles' surface than at a distance about equal to r_g. This means that a solvent layer around the particle of thickness $\delta \approx r_g$ is not available for (is depleted of) polymer. For the whole dispersion this **depletion layer** would make up a fraction $A \cdot \delta / (1 - \varphi)$ of the solvent, where A and φ are specific surface area and volume fraction of the particles, respectively. Consequently, the presence of the particles causes the polymer concentration, and hence its contribution to the osmotic pressure Π_{pol}, to be increased. If now two particles touch each other, their depletion layers partly overlap, thereby increasing the volume of solvent available for the polymer. Consequently, aggregation of the particles will lead to a decrease in osmotic pressure, hence to a decrease in free energy. This implies that there is a driving force for aggregation. In other words, the presence of the dissolved polymer induces attraction between the particles.

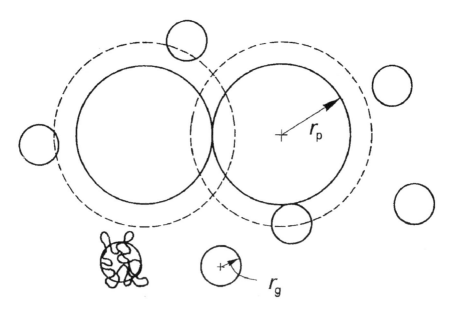

FIGURE 12.10 Illustration of the depletion of polymer molecules, radius of gyration r_g, from a layer around spherical particles, radius r_p, and of the decrease in depletion volume resulting from particle aggregation.

This *depletion free energy* can be calculated for a pair of spherical particles at a distance h from each other by

$$V_{depl} = -2\pi r_p \Pi_{pol}(2\delta - h)^2 \qquad 0 \leqslant h \leqslant 2\delta \qquad r_p \gg \delta$$

$$\Pi_{pol} = RT\left(\frac{c}{M} + Bc^2 + \cdots\right) \tag{12.12}$$

where c is concentration (mass per unit volume) and M molar mass of the polymer, and B is the second virial coefficient. Since it generally concerns fairly dilute solutions, a third virial coefficient is not needed. [Note that the form of the equation differs from Eq. (12.11). The relation between the parameters of the second term is, if water is the solvent, $B = 28 \cdot 10^3 \beta / \rho^2 \approx 0.012\beta$, where ρ is the mass density of the polymer.]

As an example, consider spherical particles of radius $1\,\mu m$ and a polymer concentration c of $1\,kg \cdot m^{-3}$ (i.e., 0.1%). The polymer may be xanthan, radius of gyration $30\,nm$, $M = 10^3\,kg \cdot mol^{-1}$, and $B = 5 \cdot 10^{-4}\,mol \cdot m^3 \cdot kg^{-2}$. It then follows that the depletion interaction at $h = 0$ would be about $-16\,k_B T$. This is a fairly strong interaction, and it will undoubtedly cause particle aggregation, unless a strong repulsive force also exists, e.g., due to an electric charge. It is observed in practice that **depletion flocculation**, as it is commonly called, can often be induced by xanthan concentrations as small as 0.02%. For polysaccharides of smaller r_g, i.e., having a lower molar mass or chains that are less stiff than the xanthan just mentioned, higher concentrations would be needed to induce aggregation.

Complications. In practice, precise calculation of the interaction free energy is not always easy. Eq. (12.12) applies to a pair of identical hard spheres, and even in that case the result may differ from the prediction. The relation $\delta = r_g$ is not exact, even if the polymer is monodisperse, because some chains will protrude beyond r_g. If the particles are somewhat deformable, because they are very soft or because they have a deformable adsorption layer, the depletion interaction forcing them together may cause local flattening, by which $|V_{depl}|$ becomes even larger. Particle shape has a large effect, and for platelets the depletion interaction is far stronger than for spheres of equal volume (can you explain this?).

In practice, depletion flocculation in a dilute dispersion causes the particles to sediment rapidly. Increasing the polymer concentration at first leads to increasing sedimentation (rate), but at still higher concentrations, sedimentation slows down or will be altogether absent. This occurs if the polymer concentration c is larger than the chain overlap concentration c^*; see Section 6.4.2. At such conditions, the particles are *immobilized* in a space-filling, although fluctuating, network of polymer molecules, and flocculation will not occur.

Also other species may cause depletion interaction. A case in point is *surfactant micelles*; for example, if an emulsion has been made with an unnecessary large concentration of surfactant, so that micelles remain after emulsification, this may cause aggregation of the droplets. In foods, however, the surfactant concentrations needed for depletion flocculation to occur are generally unacceptable.

It may finally be noted that depletion flocculation of particles by a soluble polymer is closely related to *segregative phase separation* of two soluble polymers—e.g., a protein and a polysaccharide—as described in Section 6.5.2. It is therefore not surprising that emulsion droplets, covered with a protein layer, can readily show depletion flocculation due to the addition of a nonadsorbing polysaccharide.

Question

Assume that to a dispersion of spherical particles in water a given amount of a linear dextran (see Table 6.1) is added to increase the viscosity. However, the addition of the polymer also tends to cause depletion flocculation, which is undesirable. What would be the best value of the molar mass M of the polymer—large, small, or indifferent—if flocculation is to be prevented, while the viscosity becomes as high as possible?

Answer

Table 6.1 gives for the exponent a in the Mark–Houwink equation (6.6) the value of 0.5. This means that intrinsic viscosity will be proportional to $M^{0.5}$, and viscosity will thus increase with increasing M. The question is now how the depletion interaction V_{depl} will depend on M. The radius of gyration r_g is proportional to M^v, according to Eq. (6.4), and from Eq. (6.6) we derive that $v = (a + 1)/3 = 0.5$. Using Eq. (12.12) for $h = 0$, where the interaction is strongest, we see that the equation will then have a factor δ^2, which will be about equal to $r_g^2 \propto M^{2v} = M$. We further have to know the dependence of Π_{pol} on M. In Section 6.2.1 it is shown that for $v = 0.5$, the solvent quality parameter $\beta = 0$ ("ideal behavior"), which also means that the second virial coefficient $B = 0$. This implies, in turn, that $\Pi \propto M^{-1}$. Thus the value of V_{depl} will be independent of M. In conclusion, the molar mass of the dextran should be as large as possible.

> *Note* If M is so small that Eq. (6.4) does not apply any more, the reasoning given breaks down. Below a certain small value of M no depletion flocculation occurs.

12.4 OTHER INTERACTIONS

Table 3.1 lists some other kinds of intermolecular forces, and most of these can also act between particles.

Hydrogen bonding can certainly occur, but it only acts at a very short distance.

Solvation, i.e., mostly hydration, of groups at a particle surface may cause repulsion. Especially charged groups are hydrated, but these cause electrostatic repulsion anyway. However, dipoles may play a part. Their dehydration is needed for close contact, and that costs free energy. The range over which such a hydration force acts is quite short, however, at most 2 nm. The importance of hydration in stabilizing particles against aggregation is unclear, and probably fairly small.

Apolar particles in water may be subject to attraction caused by the **hydrophobic effect**. Again, the range over which such a force acts is generally small.

If particles are already quite close to each other, hydrogen bonds or hydrophobic interaction may lead to **bond strengthening**. Also the kind of cross-linking discussed in relation to Figure 12.8b may be enhanced after particles become aggregated. The theories of colloidal interaction, as discussed in the previous sections, may well predict whether aggregation will occur, but they tell little about the force needed to break the link between particles. In other words, links that are assumed to be more or less reversible (a few times $k_B T$) may become irreversible after a while.

Figure 12.11 illustrates some other mechanisms of bond strengthening. The *particle rearrangement* shown increases the coordination number. Two particles being attracted by a net free energy of $1 k_B T$ will not stay together for a long time, but a particle attached to 6 other ones by the same energy will not readily come loose. Solid particles often have an uneven surface. If aggregated, such particles may move a little until they have obtained the *closest fit*, implying the strongest van der Waals attraction. Soft or fluid particles may *flatten* upon attraction, whereby the attractive force is generally enhanced. Finally, Ostwald ripening (see Section 10.5.3) may lead to *local sintering* of solid particles, especially crystals. At sharp edges the solubility of the particle material is enhanced; at a gap between particles it will be decreased. This then leads to diffusion of material to the gap and thereby to growth of the junction between particles, as depicted.

Junctions. This brings up another point. It is fairly common to speak of the *bond* between two aggregated particles. However, it rarely concerns one bond on the atomic or molecular scale. Actually, it often

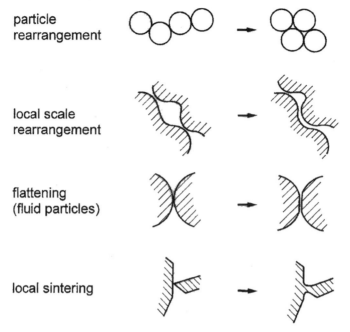

particle
rearrangement

local scale
rearrangement

flattening
(fluid particles)

local sintering

FIGURE 12.11   Illustration of some bond strengthening mechanisms. See text.

involves hundreds of such bonds. These bonds may even be of various kinds. Consequently, it may be advisable to speak of a *junction* between particles as a general term.

Also **bridges**, as depicted in Figure 12.8, are a form of junction. Some additional mechanisms will be mentioned. Figure 12.8c shows bridging of large fluid particles by small solid ones. An example is partially solid emulsion droplets bridging air bubbles, as in whipped cream; or compact protein aggregates bridging oil droplets, if insufficient protein is present to cover fully the O–W interface.

Figure 12.8d shows a liquid neck between solid particles in a fluid, which holds the particles together by capillary forces (see Section 10.6); the attractive interaction energy can be very large. An example is bridging of various particles in cocoa mass ($\approx$ melted chocolate)—where oil is the continuous phase—induced by tiny water droplets; this then gives the cocoa mass a greatly enhanced viscosity.

Finally, it has been observed that air bubbles or oil droplets in water can be bridged by layers of an α-gel of a suitable small-molecule surfactant (see Section 10.3.1), if sufficient surfactant is present. This is used to lend stiffness and stability to some whipped toppings.

Question

A fairly dilute O–W emulsion is made with a protein as the sole surfactant, and a polysaccharide is added to increase viscosity. The emulsions show undesirable aggregation of the droplets. How can you establish the probable cause of the aggregation by simple means?

Answer

The first question may be whether the polysaccharide is involved. This is readily checked by omitting it in the formulation. If that makes no difference, the cause must be that the protein does not give sufficient repulsion. This can in turn have various causes. The pH can be too close to the isoelectric point, or the ionic strength too high to provide sufficient electrostatic repulsion; the checks needed to find this out are obvious. Bridging can result from the protein concentration being too small, and this can be remedied by adding more protein before emulsification; the bridging may be as depicted in Figure 12.8a or c. In the former case, the protein must have an unfolded conformation; in the latter case, it probably implies that the protein has formed fairly large aggregates, and measures that cause better solution would also prevent aggregation. Bridging can be due to cross-linking, e.g., by —S—S— bonds or by —Ca— linkages. This can be established by adding reagents that specifically break such bonds, e.g., dithiothreitol and ethylene diamine tetra-acetate, respectively.

If the polysaccharide is involved in the aggregation, the most likely cause is depletion flocculation, but it may also be caused by adsorption of the polymer onto the protein-covered droplets in such a way that bridging occurs. It may be difficult to find out which mechanism is responsible. An unequivocal criterion is whether the polysaccharide concentration in the serum phase of the emulsion is increased or decreased as compared to the original solution, but determining this cannot be called a simple method. Dilution with the solvent may give a clue, since depletion flocculation tends to depend more strongly on polymer concentration than does bridging. The latter will generally involve electrostatic interaction, say between positive groups on the protein and negative ones on the polysaccharide. This can be established by varying the pH.

12.5 RECAPITULATION

Several colloidal interaction forces can act between particles dispersed in a liquid. If these particles attract each other, they will aggregate, which means that the dispersion is unstable. The interaction forces can also affect the stability of a thin film (e.g., between air bubbles) and the rheological properties of particle gels.

One generally tries to calculate the **interaction free energy** V between two particles as a function of the distance between their surfaces h. *Van der*

Waals forces nearly always cause attraction. The interaction depends on the materials involved, which dependence is given by the Hamaker constant; besides the material of the particles, that of the fluid between them is involved. Generally, the attraction across air is much stronger than that across water. Besides, the geometry of the system affects the interaction. It is strongest for $h \approx 0$, and is inversely proportional to h (spherical particles) or to h^2 (platelets).

Repulsive forces are generally caused by substances adsorbed onto the particles, as illustrated in Figure 12.12. In all cases, the interaction force increases with increasing surface excess (number of adsorbed molecules per unit interfacial area).

Electric charge on the surface induces an *electric potential*. Counterions accumulate near the surface to neutralize the charge, and they thus form an electric double layer. The potential thus decreases when going away from the surface. If the particles become close, their double layers start to overlap, causing a locally increased osmotic pressure, hence a repulsive free energy acts, trying to drive the particles apart. In practice, the repulsion increases with increasing surface charge (as affected, e.g., by pH), and the range over which the repulsion acts decreases with increasing ionic strength. In the *DLVO theory*, van der Waals attraction and electrostatic repulsion are added to calculate the total $V(h)$, from which the stability of the system can be predicted.

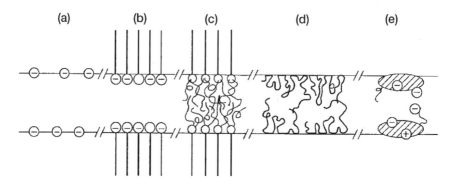

FIGURE 12.12 Illustration of materials being adsorbed onto fluid interfaces (O–W or A–W), thereby causing repulsion between two of such interfaces. (a) Anions, (b) soaps, (c) Tween-like surfactants, (d) neutral polymers, (e) proteins. Highly schematic; the straight lines represent aliphatic chains. The repulsion is electrostatic (a and b), steric (c and d), or mixed (e).

In foods, the DLVO theory is often insufficient to calculate the total interaction. In many cases *steric interaction* is involved, caused by adsorbed or grafted polymers, which provide a layer of protruding polymer chains around a particle. If such layers overlap, their mixing generally causes the osmotic pressure to increase, and repulsion occurs. However, if the solvent quality for the polymer is poor, attraction may prevail, although this is rather exceptional. On still closer approach of the particles, the polymer chains lose conformational freedom by volume restriction, and this leads to very strong repulsion. If the thickness of the polymer layer is not very small compared to the particle size, steric repulsion is generally greater than van der Waals attraction, and stability against aggregation is assured.

If the polymer involved is also charged, the situation is more complicated, and the ionic strength is an important variable. If the latter is neither very small nor very large, the resulting "electrosteric" repulsion tends to be strong.

On the other hand, dissolved polymers may cause **depletion interaction**. This is because the polymer molecules cannot come very close to the particle surface, which amounts to polymer being depleted from part of the solvent. Hence, polymer concentration, and thereby osmotic pressure, is increased. If particles aggregate, the depletion layer decreases in volume due to overlapping, and the polymer concentration decreases, hence the osmotic pressure decreases. This means that an attractive force acts between the particles. It increases with concentration and radius of gyration of the polymer.

Table 12.3 summarizes the interactions mentioned and—in a qualitative sense—the main factors affecting the strength of the interactions. Some other interactions can act, especially **bridging**; this means that a polymer molecule becomes adsorbed onto two particles at the same time. It can especially happen if insufficient polymer is present to cover the particles. Moreover, protruding polymer chains may show cross-linking when the layers of two particles overlap. A case in point is cross-linking of negatively charged groups by divalent cations. Several other interactions can occur. Most of those are fairly rare, or act only at very close distance, like solvation forces.

Finally, after particles are aggregated, slow **strengthening** of bonds often occurs. Actually, most particle pairs are kept aggregated by a "*junction*" that involves a great number of separate bonds, and that number may increase with ageing.

TABLE 12.3 Summary of the Most Important Colloidal Interaction Forces and of Factors Affecting Their Magnitude

Interaction	A/R	d	PC	pH	I	SQ	r_g
van der Waals	A, (R)	+	+	−	−	−	·
Electrostatic	R	+	i	+	+	−	·
Steric	R, A	(+)	i	−	−	+	+
Electrosteric[a]	R	(+)	i	+	+	(+)[b]	+
Depletion[c]	A	+	−	−	−	+	+

A: Attractive
R: Repulsive
d: Particle diameter
PC: Particle composition
I: Ionic strength
SQ: Solvent quality
r_g: Radius of gyration, as affected by stiffness and molar mass
+: Does affect the force
−: Does not affect it
(+): Weak or in some conditions
i: Indirectly, as it can affect adsorption
[a] Polyelectrolyte brush.
[b] SQ primarily affected by pH and ionic strength.
[c] As caused by a neutral polymer.

BIBLIOGRAPHY

All textbooks on colloid science discuss colloidal interaction forces, such as

P. C. Hiemenz, R. J. Rajagopalan. Principles of Colloid and Surface Chemistry, 3rd ed. Marcel Dekker, New York, 1997.
and

R. J. Hunter. Introduction to Modern Colloid Science. Oxford University Press, Oxford, 1993.

A clear and detailed discussion on most of the colloidal interaction forces is also found in

J. N. Israelachvili. Intermolecular and Surface Forces, 2nd ed. Academic Press, London, 1992.

A thorough discussion of some basic aspects is found in

J. Lyklema. Fundamentals of Interface and Colloid Science. Vol. I "Fundamentals" 1991, Vol. II "Solid–Liquid Interfaces", 1995. Academic Press, London.

Especially useful are Vol. I, Chapter 4, for van der Waals forces, and Vol. II, Chapter 3, for electric double layers. A comprehensive monograph on adsorbed and grafted polymer layers, including treatment of steric interaction and depletion interaction, is

G. J. Fleer, M. A. Cohen Stuart, J. M. H. M. Scheutjens, T. Cosgrove, B. Vincent. Polymers at Interfaces. Chapman and Hall, London, 1993.

13

Changes in Dispersity

In Section 9.1 it is stated that all lyophobic dispersions are thermodynamically unstable. They tend to change in such a way that the interfacial free energy will be minimized, causing an increase in particle size. Moreover, some lyophilic dispersions can become unstable if conditions like pH, ionic strength, or solvent quality change. In Chapter 12, the interaction forces involved are discussed, and additional basic information is given in Chapter 10. The instability can lead to a variety of changes, which are the subject of this chapter. The rate at which they will occur and the factors affecting them will be discussed. The particle size distribution, discussed in Section 9.3, is often an important variable.

The changes can also be caused or affected by external forces, especially gravitational and hydrodynamic forces. The latter are briefly discussed in Section 5.1.

13.1 OVERVIEW

Figure 13.1 gives an overview of the various types of change in the state of dispersity. We will briefly discuss them here.

Growth/Dissolution. The driving force is here a difference in the chemical potential between the substance making up the particle and the same substance in solution. If the solution is *under*saturated the particles will dissolve, whereas if it is *super*saturated the particles can grow. For new particles to form, nucleation is needed (Chapter 14). Growth of solid particles generally means crystallization, a complicated phenomenon (Chapter 15). We will not further discuss this type of change in the present chapter.

Ostwald Ripening. The driving force is the difference in chemical potential of the material in particles differing in *surface curvature*, as given by the Kelvin equation (Section 10.5.3).

Type of change		Particles involved	Effect of agitation
a. Growth/dissolution	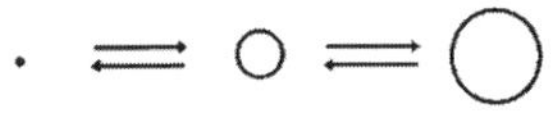 · ⇌ ○ ⇌ ◯	S, L, G	(↑)
b. Ostwald ripening	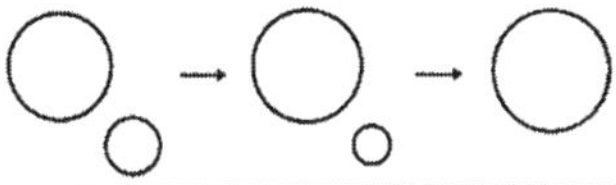 ◯ → ◯ → ◯	S, L, G	(↑)
c. Coalescence	 ◯◯ → ◯◯ → ◯	L, G	↓ or ↑
d. Aggregation	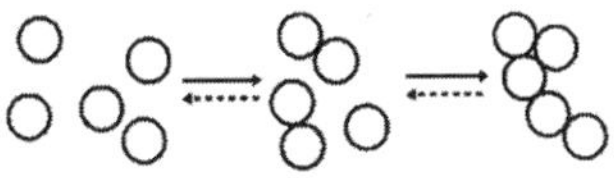 ○ ○ ⇢ 8○ ⇢ 8	S, L	↑ or ↓
e. Partial coalescence	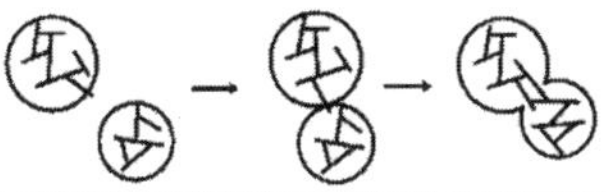 ⊛ → ⊛ → ⊛	L / S	↑
f. Sedimentation	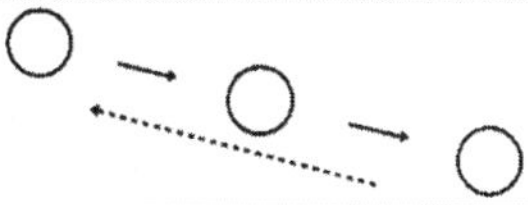 ○ ↘ ○ ↘ ◯	S, L, G	↓

FIGURE 13.1 Illustration of the various changes in dispersity. S = solid, L = liquid, and G = gaseous particle. ← − − − possible in some cases. ↑ enhances, (↑) may be slightly enhancing, and ↓ impedes the change. The solid lines in the particles in (e) denote (fat) crystals.

Coalescence. This is caused by *rupture of the film* between two emulsion drops or two foam bubbles. The driving force is the decrease in free energy resulting when the total surface area is decreased, as occurs after film rupture. The Laplace equation (Section 10.5.1) plays a key role.

Aggregation. The interactions involved are treated in Chapter 12. It follows that the main cause is often *van der Waals attraction* (Section 12.2), as given by the Hamaker equations. Another important cause is *depletion interaction*, where the driving force is increase in mixing entropy of polymer molecules or other small species present in solution (Section 12.3.3).

Partial Coalescence. This is a complicated phenomenon. It can occur in O–W emulsions if part of the oil in the droplets has crystallized. The ultimate driving force is, again, a decrease in interfacial free energy, but the relations given by Hamaker, Laplace, and Young (Section 10.6.1) all are involved.

Sedimentation. This can either be *settling*, i.e., downward sedimentation, or *creaming*, i.e., upward (negative) sedimentation. The driving force is a difference in buoyancy, or in other words the decrease in potential energy of the particles that will occur upon sedimentation.

Figure 13.1 also indicates what *kind of particles* can be involved in the various changes; solid particles are generally not spherical. It is further shown (by backward arrows) whether the change is readily *reversible*, e.g., by changing the physicochemical conditions or by mild stirring. The indicated effects of agitation will be discussed later. It may further be noted that (a) and (b) involve transport of *molecules* through the continuous phase, generally by diffusion. In (d), (e), and (f) the *particles* are transported: by Brownian motion (d), or because of a field force (d–f). In (c), *flow of the particle fluid* occurs, and this is also involved in (e). Many of the changes in dispersity can be arrested by immobilization of the particles.

Consequences. The resulting changes in the dispersion can be of various kinds. In (a–e) *particle size* changes, whereas in (f) the position of the particles becomes less random, leading to *demixing*. Changes (a–c) cause a change in particle size only, whereas (d) and (e) also cause a change in *particle shape*. The latter may lead to the formation of large structures, enclosing part of the continuous phase, and ultimately to a space-filling network (Section 13.2.3). Changes (a–c) can only cause coarsening of the dispersion, and the particles will ultimately become visible by eye.

Mutual Interference. One kind of change can affect the rate or even the occurrence of the other. All changes leading to an increase in particle size, i.e., (a–e), cause an enhanced sedimentation rate. The effect is especially strong for aggregation, since sedimentation in turn tends to enhance the aggregation rate. This means that all of the instabilities would ultimately lead to demixing, unless aggregation results in the formation of a space-filling network. Coalescence can only occur if the emulsion droplets or the gas bubbles are close to each other, and in nearly all cases they have to be close for a fairly long time, say at least a second. This then means that coalescence generally has to be preceded either by aggregation or by sedimentation. Coalescence eventually leads to *phase separation*, e.g., between oil and water.

As mentioned, particle size can increase by various mechanisms (processes), and it depends on several internal and external variables which process is predominant. To find a remedy for an undesirable change, it is necessary to establish which process is occurring. Interference between the various processes may make it difficult to establish the prime cause.

Free Energy Change. It may be enlightening to consider the magnitude of the decrease in free energy (ΔG) involved in some of the changes that can occur. Consider an O–W emulsion, oil volume fraction $\varphi = 0.1$, droplet diameter $d = 1\,\mu m$, O–W interfacial tension $\gamma = 10\,mN \cdot m^{-1}$.

Aggregation. ΔG would amount to 0.5 times the coordination number of the aggregated particles (say, 6), times the droplet number concentration ($= 6\,\varphi/\pi d^3$), times the decrease in G due to the formation of one particle–particle "bond" (for example, 500 $k_B T$). This amounts to about $1\,J \cdot m^{-3}$.

Creaming over a height H (say, 20 cm) would lead to $\Delta G = 0.5 \cdot \varphi \cdot \Delta\rho \cdot g \cdot H$, where ρ is mass density. This amounts to about $10\,J \cdot m^{-3}$.

Coalescence would lead to the release of the interfacial free energy, which will roughly equal $6 \cdot \gamma \cdot \varphi/d$. The result is about $6 \cdot 10^3\,J \cdot m^{-3}$.

Oxidation of the oil is given for comparison. It would occur spontaneously if the emulsion is in contact with air. ΔG would be given by φ times the oil density (about $920\,kg \cdot m^{-3}$) times the heat of combustion (about 37 MJ per kg of triglyceride oil). This result is about $3.4 \cdot 10^9\,J \cdot m^{-3}$.

It may also be illustrative to compare these quantities with the amount of energy needed to heat the emulsion by 1 K, which equals about

$4 \cdot 10^6 \, \mathrm{J} \cdot \mathrm{m}^{-3}$. In other words, full coalescence would lead to an increase of the temperature of the system (provided it is isolated) by $1.5\,\mathrm{mK}$. Hence the free energy decrease involved in changes in dispersity generally is quite small.

The most important conclusion is, however, that the stability is not at all related to the value of ΔG. If the aqueous phase has a viscosity like that of water, creaming will spontaneously occur and be clearly observable within a day. On the other hand, the emulsion may be stable against significant coalescence and oil oxidation for several years. The time needed for visible aggregation to occur may vary from one to 10^6 minutes, or even longer. We thus need to study the kinetics of the various processes. They may be slow because of a high activation free energy $\Delta G^{\ddagger}$, or because the transport process involved is impeded, e.g., owing to a high viscosity. Only in a limited number of cases is the change slow because the driving force is small.

It may finally be noted that in many systems instability—expressed as a rate of change—may vary with time, e.g., owing to some reaction causing a change in pH or viscosity.

13.2 AGGREGATION

Two particles are said to be aggregated if they stay together for a much longer time than they would do in the absence of colloidal interaction forces. In the absence of such forces, the time together (i.e., the time needed for two particles to diffuse away from each other over a distance of about 10 nm) would often be on the order of some milliseconds. It may be noticed that the terms *flocculation* and *coagulation* are also used, where the former is, for instance, used to denote weak (reversible) and the latter strong (irreversible) aggregation.

In this section, the kinetic aspects of aggregation will especially be discussed. Most of the theory derived is valid for the ideal case of a dilute dispersion of monodisperse hard spheres. Most food dispersions do not comply with these restrictions. Where possible, the effects of deviations from the ideal case will at least be mentioned. Some consequences of aggregation are also discussed.

> *Note* Particles can aggregate because of the attractive free energy acting between them. However, aggregation will decrease the translational mixing entropy of the particles, and that would counteract aggregation. It can be derived that the average free

energy involved per pair of particles aggregating is given by

$$\Delta G_{mix} \approx k_B T \varphi (\ln \varphi - 1)$$

This gives for $\varphi = 0.001$ and 0.1, values of 0.008 and 0.33 times $k_B T$, respectively. Since the attractive free energy generally amounts to several times $k_B T$, this means that the loss in mixing entropy is in most cases too small to affect the aggregation significantly. However, for very small particles, say, $d < 0.1 \, \mu m$, the attractive free energy tends to be much smaller, e.g., of the order of $k_B T$, and the change in mixing entropy may then play a part, especially if φ is not very small.

13.2.1 Perikinetic Aggregation

The term *perikinetic* signifies that the particles encounter each other because of their Brownian motion or diffusion.

"Fast" Aggregation. Smoluchowski has worked out a theory for this case, assuming that particles will stick and remain aggregated when encountering each other. He considered particles diffusing to a "central" particle. Because any particle colliding with the central one is, as it were, annihilated, a concentration gradient of particles is formed. By solving Fick's equation (5.17) for spherical coordinates, he obtained for the flux J_{peri} (in s^{-1}) of particles (2) toward a central particle (1)

$$J_{peri} = 4\pi D_{pair} R_{coll} N \tag{13.1}$$

where N is the number of particles per unit volume. D_{pair} is the mutual diffusion coefficient of two particles, and it is assumed to equal $D_1 + D_2$. R_{coll} is the *collision radius*, which equals, for spheres, $a_1 + a_2$.* For equal spheres we can just use a single diffusion coefficient D and a single diameter d. Since, moreover, $D = k_B T / 3\pi\eta d$ [Eq. (5.16)], the particle diameter is eliminated from the result.

Assuming that each pairwise collision reduces the particle number by unity, the change in particle concentration with time is obtained by multiplying J by N and dividing by 2 to prevent counting every collision

---

* In this chapter we will use the symbol a for the radius of a spherical particle.

twice. The result is

$$-\frac{dN}{dt} = \frac{1}{2}J_{\text{peri}}N = 2\pi D_{\text{pair}}R_{\text{coll}}N^2 = 4\pi DdN^2 = \frac{4k_B T}{3\eta}N^2 \tag{13.2}$$

where the last two equalities are valid for equal-sized spheres only. A useful quantity is the *halving time* $t_{0.5}$, i.e., the time needed for halving the number of particles, which is given by

$$t_{0.5} = \frac{2}{J_0} = \frac{3\eta}{4k_B T N_0} = \frac{\pi d_0^3 \eta}{8k_B T \varphi} \tag{13.3}$$

where J_0, N_0, and d_0 are the initial values of J, N, and d, respectively. The second part of the equation is obtained by using the relation $\varphi = \pi d^3 N/6$, φ being the volume fraction of particles.

Of course, the aggregation rate decreases as aggregation proceeds, since N keeps decreasing (and the aggregate size increasing). Nevertheless, the number concentration as a function of time follows a simple relation, viz.,

$$N_t = \frac{N_0}{1 + t/t_{0.5}} \tag{13.4}$$

Incidentally, this relation is valid in every case where $-dN/dt \propto N^2$. Figure 13.2 gives a calculated example of the decrease in N with time, as well as the concentration of doublets, triplets, etc. Notice that doublets do not only form but also disappear again by aggregation.

For spherical particles in water at room temperature, Eq. (13.3) reduces to $t_{0.5} \approx 0.1 \, d^3/\varphi$ seconds, if d is expressed in μm. Assuming $d = 1$ μm and $\varphi = 0.01$, the halving time would be only 10 seconds, implying that aggregation proceeds very fast. Note that the initial aggregation rate strongly increases with decreasing particle diameter for constant volume fraction. The observed results for unhindered perikinetic aggregation generally agree well with the Smoluchowski predictions for the first few aggregation steps.

The relations are different if the collision radius is larger than the hydrodynamic radius determining the diffusion coefficient [although Eqs. (13.1) and (13.4) would still be valid]. Such a difference between radii occurs if the particles are strongly anisometric or irregularly shaped, and the aggregation rate will then be faster than for spherical particles of the same volume. For the same reason, polymer chains that extend from the particles' surface and that can cause interparticle bridging, may markedly increase perikinetic aggregation rate over the Smoluchowski prediction. *Polydisper-*

sity tends to enhance the aggregation rate somewhat, even for spheres, because the effects of particle size on collision radius and mutual diffusion coefficient then do not exactly cancel in Eq. (13.2).

"Slow" Aggregation. There are several reasons why aggregation rate can be slower than that predicted by Eq. (13.1). Smoluchowski introduced a *stability factor* W, by which the calculated result would have to be divided to obtain the experimental result, without giving theory for the magnitude of W. Others speak of the *capture efficiency* α, defined as the probability that two particles stick upon closely encountering each other. It is generally assumed that $W = 1/\alpha$. Equations (13.1) and (13.2) should be multiplied by α, and Eq. (13.3) be divided by it (or be multiplied by W).

An obvious cause for the slowing down of aggregation is the presence of a repulsive free energy V at some distance h between the particles, as discussed in Chapter 12. A particle approaching the "central" particle will

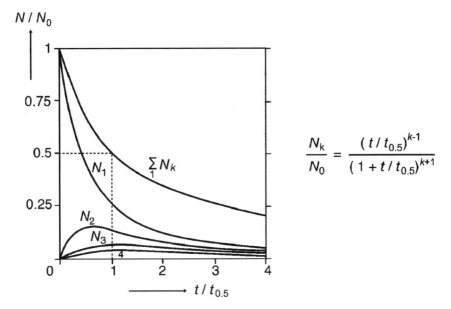

$$\frac{N_k}{N_0} = \frac{(t/t_{0.5})^{k-1}}{(1 + t/t_{0.5})^{k+1}}$$

FIGURE 13.2 Perikinetic aggregation. The relative particle number concentration as a function of the time t after aggregation has started over the halving time $t_{0.5}$. N refers to number, N_0 to initial total number, and the subscripts 1, 2, 3, etc. to the numbers of particle monomers, dimers, trimers, etc. The equation for the number of k-mers is also given.

then often diffuse back again after coming so close that it senses significant repulsion. This implies that the particle concentration gradient near the central particle will be smaller than in the absence of repulsion. Fuchs found a general solution for the resulting retardation for the case of equal-sized spheres. The result is

$$W = \frac{1}{\alpha} = 2 \int\limits_{2}^{\infty} z^{-2} \exp\frac{V(z)}{k_B T}\, dz \tag{13.5}$$

$$z = 2 + \frac{h}{a}$$

where a is particle radius and h interparticle distance.

A calculated example is in Figure 13.3, which gives the magnitude of $V/k_B T$ and of the integrand y of Eq. (13.5) (i.e., the expression between the integral sign and dz) as a function of the relative distance z. Twice the area under the y-curve corresponds to $W \approx 365$, implying that the assumed repulsion would lead to a slowing down of the aggregation by this factor. It may further be noticed that a significant contribution of the repulsion to W is restricted, in the present case, to the range where V is larger than about $10 k_B T$.

It is tempting to take the maximum value of V as an activation free energy for aggregation, which would imply that $W = \exp(V_{max}/k_B T)$, but this is not correct. In the example of Figure 13.3, $V_{max} \approx 12.8\, k_B T$, which would lead to $W \approx 360,000$, almost 1000 times the value according to Fuchs. If V_{max} is only 7 times $k_B T$, Eq. (13.5) predicts $W = 1.2$, i.e., almost negligible, whereas $e^7 = 1100$, which would imply strong retardation.

The reason that W can be small despite V_{max} being large is that the range of z over which the value of y is significant is so small. In other words, the distance between the particles h^* over which repulsion effectively affects the value of W is very small relative to particle radius a. A larger value of h^*/a leads to a higher W-value. Broadly speaking, h^*/a tends to be larger for a smaller value of a, and for very small particles, say one nm, the use of V_{max} as an activation free energy may be reasonable.

The Fuchs theory has been applied to colloids stabilized by **electrostatic repulsion**. It can then be derived that the dependence of the aggregation rate on salt concentration is roughly as given in Figure 13.4. The *critical concentration* m_{cr} is defined as the value above which the aggregation rate does not further increase; presumably, the stability factor then equals about unity. As shown, the magnitude of m_{cr} greatly depends on the valence of the ions, even stronger than expected on the basis of the Debye expression for the ionic strength $(I = (1/2)\Sigma m_i z_i^2)$, i.e., with the square of the valence. For the data in Figure 13.4, the proportionality would be to $z^{3.5}$. The results also depend on the absolute value of the electrokinetic potential, m_{cr}, being greater for a higher potential. Moreover, other variables, especially A and d, affect the result, such as the slope of the curve and the dependence of m_{cr} on valence. Nevertheless, the figure

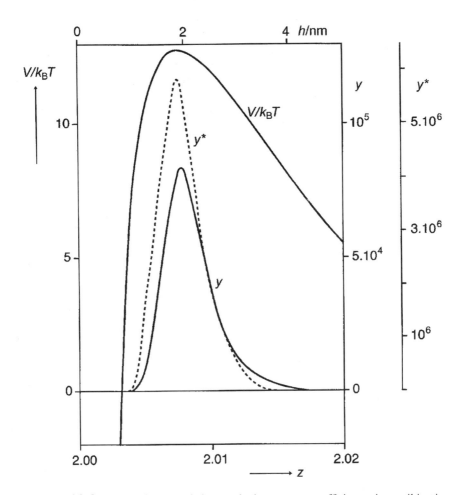

FIGURE 13.3 Basic data needed to calculate capture efficiency in perikinetic aggregation. As an example, part of the colloidal interaction potential V as a function of interparticle distance h is given; it is calculated from DLVO theory [Eqs. (12.1) and (12.7)] for $d = 0.5\,\mu m$, $A = 1.25\,k_BT$, $\psi_0 = 15\,mV$, and $I = 10\,mM$ ($1/\kappa = 3\,nm$). y is the integrand of Eq. (13.5), y^* that of Eq. (13.7); note the different scales.

illustrates the large effect of ionic strength on the stability of colloids stabilized by electric charge.

Hydrodynamic Retardation. Smoluchowski assumed in the derivation of his equations that $D_{pair} = D_1 + D_2$, but this is not true if the diffusing particles are relatively close to each other. When two particles come close, the liquid between them has to flow out of the gap, and this means that (a) the local velocity gradient is increased and (b) the flow type becomes "biaxial elongation" rather than simple shear (see Section 5.1). Both factors cause the effective viscosity to be increased, which means in turn that the mutual diffusion coefficient of the particles is decreased, the more so as the particle separation (h) is smaller. The phenomenon is called the *Spielman–Honig effect*.

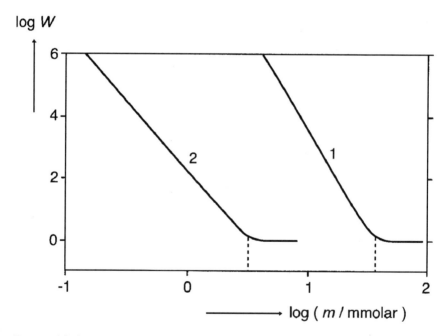

FIGURE 13.4 Idealized curves of the stability factor W in perikinetic aggregation for small particles stabilized by electrostatic repulsion, as a function of the salt concentration m. Numbers near the curves denote the valence of the ions in the solution. The dotted lines indicate the "critical coagulation concentrations" m_{cr}.

Note The theory does not apply for fluid particles in the absence of surfactant, where slip occurs at the particle surface; see Section 10.8.3.

Exact calculation is intricate but has been realized for spheres, and Figure 13.5 gives the result for spheres of equal radius (*a*). An approximate relation for that case is

$$D(u) \approx \frac{6u^2 + 4u}{6u^2 + 13u + 2} D_\infty \tag{13.6}$$

which reduces for small *u* (say, $u < 0.01$) to

$$D(u) \approx 2uD_\infty \tag{13.6a}$$

Here $u = h/a$ and D_∞ is the pair diffusion coefficient at large separation, i.e., twice the particle diffusion coefficient.

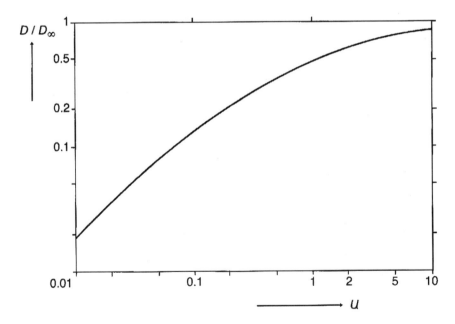

FIGURE 13.5 Value of the particle relative diffusion coefficient D/D_∞ as a function of the relative interparticle distance u ($= h/a$) of two equal-sized spheres. Calculated according to Eq. (13.6).

These relations imply that for $u \to 0$ also $D \to 0$, or in other words the particles would never touch each other. Although the theory breaks down for very small distances, the aggregation rate will be significantly decreased. How then is it possible that it has been observed that particles can aggregate at a rate predicted by Eq. (13.2) for fast aggregation? The answer is that attractive forces acting at small interparticle distances will enhance the mutual diffusion rate, just as repulsive forces decrease the rate. Sample calculations have shown that for particles that attract each other by van der Waals attraction (other interaction forces being negligible), the attraction induced increase and the hydrodynamically induced decrease of the aggregation rate may compensate each other.

If repulsive interaction is significant, Eq. (13.5) should be modified to include the retardation due to hydrodynamic interactions. The result is

$$W \approx 2 \int\limits_{2}^{\infty} \frac{D_\infty}{D(z)} z^{-2} \exp \frac{V(z)}{k_B T} dz \tag{13.7}$$

where $D_\infty/D(z)$ can be obtained from Eq. (13.6) with $u = z - 2$. The result for the case of Figure 13.3, i.e., the integrand y^*, is given in the figure (note the difference in scale). From Eq. (13.7), we now calculate $W = 24{,}000$, i.e., larger than the result according to Eq. (13.4) by a factor 65, but still smaller by a factor 15 than $\exp(V_{\max}/k_B T)$.

Reaction-Limited Aggregation. The validity of Eq. (13.7) can be questioned. It is unlikely to be correct if the particles are not very small (say, $a > 0.1 \, \mu m$), and moreover the Fuchs stability ratio W is of considerable magnitude. The reason is that for a small value of D_{pair}, combined with a significant repulsive barrier, not only the close approach but also the drifting apart of two particles is greatly slowed down. This means that a significant concentration gradient of particles near a "central" particle cannot form. In other words, the system is almost ideally mixed, implying that the aggregation would proceed as in the classical theory for bimolecular reactions, discussed in Section 4.3.3. In principle, a relation comparable to Eq. (4.11) would apply, but it has not yet been worked out.

We now can distinguish three situations (regimes):

1. Diffusion-controlled aggregation according to Smoluchowski, i.e., Eq. (13.1).
2. The Fuchs treatment, if the colloidal interaction curve is known, and as modified by hydrodynamic retardation, i.e., use of Eq. (13.7). This equation can be considered as a solution of the problem discussed in Section 4.3.5.

3. Reaction-limited aggregation. This occurs if W is larger than a critical value (of the order of 100?); presumably, the critical value is smaller for larger particles. The regime has not (yet) been seriously considered by colloid scientists.

Further Complications. The theory as discussed in this section, when applied to *dilute* suspensions (since at not very small φ the diffusion distance of the particles is smaller than presumed) of *small* particles (say, < 1 μm) under *quiescent* conditions (see Section 13.2.3), has been successful in predicting a number of observations, at least for the first few aggregation steps. This concerns (a) the reaction being second order; (b) the value of $t_{0.5}$ being inversely proportional to N_0; (c) the distributions of multiplets obtained; and (d) the effect of viscosity on the aggregation rate. Also the absolute rate is often well predicted (within a factor of two or three) if it is sure that *fast aggregation* occurs; this rate is often called the "Smoluchowski limit." However, the values of the *capture efficiency* often deviate markedly from the prediction, even for smooth spherical particles.

For *electrostatic repulsion*, it has been reasoned that during the approach of two particles, the electric double layer may not be in equilibrium: for the surface potential to remain constant, the surface charge has to change, and this may not happen fast enough, since ions in the double layer have to diffuse out of the gap. The effect is larger for larger particles, and it follows that the influence of particle size on the height of the repulsive barrier is smaller than predicted. Also surface roughness of the particles tends to decrease the effect of particle size on DLVO interaction. If the roughness is of the order of $1/\kappa$, the magnitude of the roughness, rather than the interparticle distance, tends to determine the effective magnitude of h.

Deaggregation of particle doublets may occur, thereby reducing the effective aggregation rate. This can readily occur for DLVO-type interactions if the aggregation is "in the secondary minimum" (see Figure 12.1). It appears that deaggregation can also occur from a primary minimum, if the difference $V_{max} - V_{min}$ is not too large (say, $< 10 k_B T$). This is because, again, the value of h^*/a tends to be small. Deaggregation may also occur for other types of interaction.

For *polymer-stabilized* particles (Section 12.3.1), calculation of W from theory is often not possible (and this may also be the case if some other interactions are involved: see Section 12.4). Moreover, hydrodynamic retardation would not be according to the theory outlined above. In these cases, it is often very difficult to predict capture efficiencies.

The particle surface may have *reactive patches*, often called "hot spots." Emulsion droplets, for instance, may have an adsorbed layer with patches of protein (surface area fraction θ), while the remainder is covered by small-molecule surfactants. If then aggregation will only occur when protein covered patches come very close to each other, the aggregation rate would be θ^2 times the value of fully protein covered particles.

Some other important complications are discussed in Sections 13.2.2–4.

13.2.2 Orthokinetic Aggregation

The term *orthokinetic* signifies that particles encounter each other due to velocity gradients in the liquid. See Section 5.1.1 for types of flow and velocity gradient.

"Fast" Aggregation. Smoluchowski has worked out a theory for the case of simple shear flow, further assuming that particles will stick and remain aggregated when colliding with each other. The situation is illustrated in Figure 13.6 for spheres of equal size, neglecting hydrodynamic interaction forces for the moment. A particle (e.g., A) will meet a reference particle (B) if its center is in a half cylinder of radius a_c (which equals the collision radius $2a$), as depicted. The linear rate of approach of particle A will be proportional to δ $(0 < \delta < a)$. Four subsequent positions of particle A are shown. Near position 3, A and B collide and stick. The doublet formed now rotates in the shear field; in fact, Smoluchowski assumed the particles to fuse into one sphere. From simple geometric considerations it then follows that the decrease in particle number

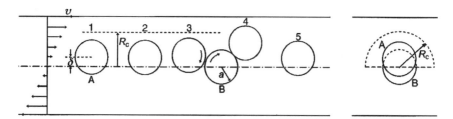

FIGURE 13.6 Orthokinetic aggregation of particles of equal diameter in simple shear flow, as envisaged by Smoluchowski. At left, the flow velocity profile is shown. Particle A moves from left to right (positions 1 to 5), particle B is in a stationary position, but it does rotate. At right a cross section is given that illustrates the geometry of the half-cylinder containing the centers of the particles to the left of B that will "collide" with B.

concentration N will be

$$-\frac{dN}{dt} = \frac{1}{2}J_{\text{ortho}}N = \frac{2}{3}d^3N^2\Psi = \frac{4}{\pi}\varphi N\Psi \qquad (13.8)$$

where Ψ is the velocity gradient, which here equals the shear rate. Integration of (13.8) and putting $N = N_0/2$ yields the halving time

$$t_{0.5} = \frac{\pi\ln 2}{4\varphi\Psi} \approx \frac{0.54}{\varphi\Psi} \qquad (13.9)$$

Comparing (13.8) with perikinetic aggregation [Eq. (13.2)] shows that the particle size dependence is now retained, and that it is to the third power; this means that the equation can be written involving φ, as shown. Since φ remains in principle constant during aggregation, the equation is first-order with respect to time, rather than second-order as Eq. (13.2). Moreover, viscosity is not in the equation; however, in an agitated vessel a higher viscosity often leads to a smaller velocity gradient. The ratio of the initial rates of orthokinetic over perikinetic aggregation is given by

$$\frac{J_{0,\text{ortho}}}{J_{0,\text{peri}}} = \frac{d^3\eta\Psi}{2k_BT} \qquad (13.10)$$

For particles in water at room temperature, this ratio reduces to $0.12\,d^3\,\Psi$ if d is expressed in µm. Taking, for instance, $\Psi = 10\,\text{s}^{-1}$, the ratio is slightly greater than unity for $d = 1\,\mu m$. This would mean that encounters by Brownian motion and by the velocity gradient are about equally frequent. However, because Eq. (13.10) has d^3 in the numerator, the preponderance of orthokinetic over perikinetic aggregation will rapidly increase during ongoing aggregation. It should further be realized that slight temperature fluctuations, as usually occur in a laboratory, readily lead to convection currents with $\Psi = 0.1\,\text{s}^{-1}$ (or higher), which then would lead to significant orthokinetic aggregation for particles over $4\,\mu m$. In other words, orthokinetic aggregation often is more important than perikinetic aggregation in food dispersions.

Capture Efficiency. The value of the capture efficiency α is nearly always smaller than unity, for a variety of reasons. In many cases the magnitude of α can be reasonably predicted, although the theory is complicated. *Hydrodynamic* effects are in principle like those in perikinetic aggregation: they impede close encounter of the particles. Moreover, the velocity gradient exerts a shear stress, order of magnitude $\eta\Psi$, onto the particles. In a situation as in Figure 13.6, position 3, the particles are pressed together, but a moment later, in position 4, the force is pulling at the particles; they then may even separate, leading to position 5.

Even in the absence of colloidal repulsion, where *van der Waals forces* will cause attraction, α will be smaller than unity. Approximately, $<\alpha> = 0.6(A/\eta \, \Psi \, d^3)^{0.18}$, where A is the Hamaker constant; this leads to α-values between 0.1 and 0.5 for most situations. Note that it concerns an average α-value. This is because α will depend on the position at which the particles collide, i.e., on the value of δ/a (see Figure 13.6). Note also that a higher shear stress and larger particles cause a smaller value of $<\alpha>$.

If there is considerable *colloidal repulsion* between the particles, the capture efficiency can be very much smaller. On the other hand, the shear stress may also push the particle pair "over the maximum" if the interaction curve is of the type depicted in Figure 12.1, solid curve.

Other Complications. At *high volume fractions*, the encounter rate will be more than proportional to φ, because the effective volume available for the particles is decreased owing to geometric exclusion. This also applies to perikinetic aggregation. For orthokinetic aggregation a high volume fraction will, moreover, affect the capture efficiency, because the stress sensed by the particles will be greater than $\eta\Psi$; see Section 5.1.2.

Other flow types, especially elongational flow, give other results. Equation (13.8) will not alter very much, but the capture efficiency may be greatly affected. Elongational flow exerts greater stress on a particle pair than simple shear flow (Section 5.1).

Particle shape can also have a large effect. Anisometric particles have an increased collision radius as compared to spheres; see the discussion in relation to Figure 5.3. Moreover, capture efficiency is affected, but prediction of the effect is far from easy.

Sedimentation. If particles are subject to sedimentation, this may also lead to enhanced aggregation rate, since particles of different sizes will move with different velocities through the liquid; see Section 13.3. This implies that a large particle can overtake a smaller one, and a kind of orthokinetic aggregation occurs. This will be the case if (a) there is a substantial spread in particle size and (b) particle motion over a distance equal to its diameter takes less time by sedimentation than by Brownian motion. The condition for the latter is approximately

$$d > 2\left(\frac{40k_B T}{\pi g|\Delta\rho|}\right)^{0.25} \approx 17 \cdot 10^{-6}|\Delta\rho|^{-0.25} \tag{13.11}$$

where $\Delta\rho$ is the density difference between particle and continuous phase. Especially for emulsions with droplets over $2\,\mu m$ in diameter, the effect of sedimentation on aggregation rate can be significant.

Question

What shear rate would be needed in an aqueous dispersion of particles to push a particle pair "over the repulsion barrier" to obtain an aggregate (doublet)? Take as an example the situation depicted in Figure 13.3. Moreover, what shear rate would be needed to pull the particles apart if they are aggregated in the primary minimum?

Answer

The force to be overcome is the derivative of the interaction free energy V with respect to h. The maximum force is where the interaction curve is steepest. For the curve in Figure 13.3, the slope beyond the maximum is steepest where h equals about $3.2\,nm$; the slope then equals $2.8\,k_BT$ per nm, which corresponds to about $10^{-11}\,N$ ($k_BT \approx 4 \cdot 10^{-21}\,N \cdot m$). The shear stress equals $\eta\Psi$. To obtain the force, the stress must be multiplied by the surface area on which it is acting, roughly d^2. Since $d = 0.5\,\mu m$ and $\eta = 1\,mPa \cdot s$, the force will be about $25 \cdot 10^{-17}$ times Ψ, implying that the latter has to be about $40,000\,s^{-1}$ (to obtain $10^{-11}\,N$), which is quite a large value.

The force needed for pulling the particles apart is more difficult to estimate. We need the steepest slope to the left of the maximum, but that value is very uncertain, since very slight surface irregularities and solvation effects can greatly affect the interaction curve at such short distances. Taking the slope at $V = 0$, it is about 10 times the one just calculated, and the force would then also be 10 times larger, as is the shear rate needed. The latter then becomes irrealistic.

> Note   The particles involved are quite small. If the DLVO theory holds, the colloidal interaction force is proportional to d, whereas the force due to the shear stress is proportional to d^2. This then means that the shear rate needed is proportional to $1/d$. This is an example of a fairly general rule, namely that external forces tend to become of greater importance for larger particles.

> Note   The calculations done here may suggest that it can simply be derived whether the capture efficiency equals unity or zero. This is not the case, because of the stochastic variation in α mentioned above.

13.2.3 Fractal Aggregation

When perikinetic aggregation occurs and the particles that become bonded to each other stay in the same relative position as during bond formation, ongoing aggregation leads to the formation of **fractal aggregates**. Figure 13.7a shows an example: it is seen that the aggregate has an open structure, for the most part consisting of fairly long and branched strands of particles. If aggregation occurs under the same conditions, while the number of particles in an aggregate is varied (for instance by varying the duration of aggregation), it turns out that a simple relation is found between the size of an aggregate and the number of particles N_p that it contains:

$$N_p = \left(\frac{R}{a}\right)^D \tag{13.12}*$$

as illustrated in Figure 13.7b. Here a is the particle radius and R any type of radius that characterizes the aggregate. The constant D is called the *fractal*

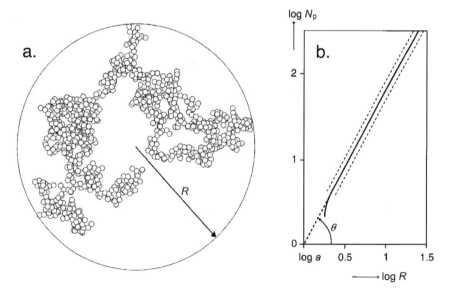

FIGURE 13.7 Fractal aggregates. (a) Side view of a simulated aggregate of 1000 identical spherical particles of radius a (courtesy of Dr. J. H. J. van Opheusden). (b) Example of the average relation between the number of particles in an aggregate N_p and the aggregate radius R as defined in (a). The fractal dimensionality $D = \tan \theta$; its value is 1.8 in this example. The region between the dotted lines indicates the statistical variation to be encountered (about 2 standard deviations).

*dimensionality.** Actually, the equation should also contain a proportionality constant; when R is taken as the radius of the smallest sphere that can contain the aggregate (as in Figure 13.7a), this constant is close to unity. The fractal dimensionality has a noninteger value smaller than three, in the example shown about 1.8. The equation is not precisely obeyed, but that is because of statistical variation. The aggregate is therefore called a *stochastic fractal*.

Figure 13.7 has been produced by computer simulation of the aggregation process. The simulations can mimic the experimental results very well, provided (a) that the particles and clusters† move by Brownian motion, implying that the trajectories have a given randomness; and (b) that clusters form aggregates with other clusters, not merely with single particles ("*cluster-cluster aggregation*").

Consequences. The fractal nature of the aggregates implies two important properties:

1. **Scale invariance.** The structure of the aggregates is scale-invariant or self-similar. The same type of structure is observed (albeit with statistical variation) at any length scale that is clearly larger than a and smaller than R. In other words, observations at various magnifications show roughly the same picture. This applies to different clusters as well as to different regions in one (large) cluster.

2. **Larger clusters are more open** (more tenuous or rarefied). This is shown as follows. Equation (13.12) gives the number of particles in a cluster. We can also calculate the number of sites N_s in the cluster, i.e., the number of particles in a sphere of radius R if closely packed:

$$N_s = \left(\frac{R}{a}\right)^3 \tag{13.13}$$

The volume fraction of particles in a fractal aggregate then is given by

$$\varphi_A \equiv \frac{N_p}{N_s} = \left(\frac{R}{a}\right)^{D-3} \tag{13.14}$$

Since invariably $D < 3$ (in three-dimensional space), this means that the particle density of the cluster φ_A will decrease with increasing values of R or N_p.

---

* Note that the same symbol is used for diffusion coefficient and fractal dimensionality.
† The terms cluster and aggregate are used here as synonyms.

Note This does not imply that the density decreases when going from the center to the periphery of a cluster: that would be incompatible with the scale-invariance.

The decrease in particle density of the clusters as they grow in size has another important consequence, viz:

3. **Gel formation.** The volume fraction of particles in an aggregate φ_A [Eq. (13.14)] decreases until it equals the volume fraction of particles in the whole system φ. At that point, the system can be considered to be fully packed with aggregates, which thus form a space-filling network, spanning the vessel containing the system, implying that a gel results. The average radius of the aggregates at the moment of gelation R_g then is given by

$$R_g = a\varphi^{1/(D-3)} \tag{13.15}$$

The average number of particles in an aggregate of radius R_g is given by

$$N_{p,g} = \varphi^{D/(D-3)} \tag{13.16}$$

$N_{p,g}$ and R_g are smaller for larger φ and smaller D. For example, the values $\varphi = 0.1$ and $D = 2$ yield $N_{p,g} = 100$, and $R_g = 10a$. This implies that a gel would soon be formed after aggregation starts.

The reader should be cautioned that Eqs. (13.12)–(13.16), as well as any derived equations, are in fact *scaling relations*. They should contain numerical constants in order to be precise. These constants generally do not differ greatly from unity.

Another point is that not all aggregation leads to fractal structures. **Conditions** for the formation of perfect fractal aggregates and of a space-filling network are following:

> During aggregation the liquid should be *quiescent* (perikinetic aggregation), since velocity gradients can deform and possibly break up aggregates. In practice, one often adds a reagent to a sol to induce aggregation, e.g., acid, salt, or ethanol, while stirring to disperse it throughout the liquid. In the meantime, aggregation occurs, and fractal clusters formed are maltreated and tend to become compacted. The resulting dense aggregates will generally sediment; in other words, precipitation occurs. To obtain a gel, the transition of the particles from stable to reactive (i.e., prone to aggregation) has to occur in the absence of significant velocity gradients.

The clusters formed should not show significant *sedimentation* before they reach the size R_g, since otherwise demixing would occur. Significant sedimentation will occur if

$$R > R_{sed} = \left(\frac{3k_B T}{2\pi a^{3-D} g |\Delta\rho|} \right)^{1/(D+1)}$$
(13.17)

where R_{sed} is the critical cluster radius for sedimentation and $\Delta\rho$ is the density difference between particle and liquid. R_{sed} will thus be smaller for larger values of a and $|\Delta\rho|$.

If significant *restructuring* of a cluster occurs, its fractal properties may be lost. If this occurs before the clusters have grown out to the size R_g, a gel may still be formed if φ is not very small.

Vessel size. Equation (13.15) implies that a gel will always form, however small φ is, although the gel may be extremely weak at very small φ. (In practice, presence of a yield stress cannot be perceived if φ is below about 0.001.) There is, however, also a geometrical lower limit to the value of φ. If the smallest dimension x of the vessel is smaller than $2R_g$, the vessel does not contain enough particles to make a network spanning the length x. From Eq. (13.15) the condition then becomes

$$x > d\varphi^{1/(D-3)}$$
(13.18)

where d is the particle diameter. Since d is of micrometre scale, this condition is nearly always met for macroscopic vessels. For instance, for $d = 1\,\mu m$, $\varphi = 0.005$ and $D = 2.30$, it follows that $x > 2\,mm$ is needed. However, if the vessel is also of microscopic size, say an emulsion droplet containing aggregating particles, a high value of φ is needed to obtain a network throughout the whole droplet.

The **fractal dimensionality** can vary considerably with the conditions that prevail during aggregation: values between 1.7 and 2.4 have been obtained. This is of importance, since the magnitude of D affects several parameters. For instance, the example given just below Eq. (13.16) yields for $D = 2$ a value of $N_{p,g}$ of 100. For D values of 1.7 and 2.4, the result would be 20 and 10^4 particles per cluster, respectively.

Simulation studies for *diffusion-limited aggregation* (i.e., for $W = 1$) generally yield D values of 1.7–1.75. Careful experiments lead to about the same or a slightly higher value, but only if the particle volume fraction is 0.01 or smaller. For increasing φ, the fractal dimensionality considerably increases, thereby invalidating the basic Eq. (13.12); moreover, the structures formed are not truly fractal any more. However, diffusion-limited aggregation is the exception rather than the rule in fractal aggregation.

If velocity gradients during aggregation have to be avoided, the common situation is that the reactivity of the particles slowly increases, for example owing to lactic acid bacteria producing acid and thereby a lower pH and a decreased electrostatic repulsion between the particles. In such a case, aggregation will begin and be often completed while the W *value* is quite high. In other words, it concerns *reaction-limited aggregation*. Both from simulations (by Brownian dynamics) and from experiment it follows that for large W the value of D is generally 2.35–2.4 and independent of φ (up to about $\varphi = 0.25$). Accurate predictions cannot (yet) be made, and the fractal dimensionality has to be experimentally determined. This can be done by light scattering methods, especially for a dispersion of fractal aggregates, and from gel properties as a function of φ (see Section 17.2.3).

Short-term rearrangement is another factor affecting the magnitude of D. It concerns a change in mutual position of the particles directly after bonding. This is illustrated in Figure 13.8a. The particles roll over each other until they have obtained a *higher coordination number*, which implies a more stable configuration. When aggregation goes on, fractal clusters can still be formed, as illustrated in Figure 13.8b, though the building blocks of the aggregate now are larger. They can be characterized by an effective radius a_{eff} or a number of particles n_p, where $n_p \approx (a_{\text{eff}}/a)^3$. Equation (13.12)

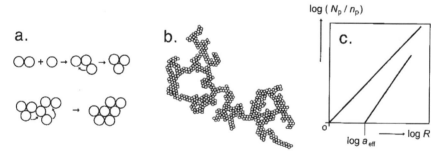

FIGURE 13.8 Short-term rearrangement. (a) Examples of particles rolling over each other so that a higher coordination number is attained. (b) Example of a fractal cluster in two dimensions, where short term rearrangement has occurred (according to Meakin). (c) Schematic example of the relations between particle number N_p and aggregate radius R according to Eqs. (13.12) ($n_p = 1$), upper curve; and (13.19), lower curve.

should now be transformed into

$$\frac{N_p}{n_p} = \left(\frac{R}{a_{eff}}\right)^D \tag{13.19}$$

Moreover, the value of D is generally increased, by an amount up to 0.2, but not above the 2.3–2.4 mentioned above. For small φ this increase becomes manifest. Figure 13.8c illustrates the difference between results according to the two equations.

Short-term rearrangement is a very common phenomenon in perikinetic aggregation, often with n_p values of order 10. Rearrangements on a larger scale can also occur, though generally after a space-filling network has formed. The fractal geometry will then gradually disappear. As mentioned, large-scale rearrangement during aggregation generally leads to a precipitate.

> *Note* It is interesting to compare the discussion in this section with that of Section 6.2.1 on the conformation of random-coil linear polymers. Also in that case a larger molecule, i.e., one consisting of a higher effective number of chain elements n', is more tenuous. Equation (6.4) reads $r_m = b(n')^v$, where r_m may be considered proportional to the parameter R in Eq. (13.12); b then would correspond to a in (13.12), and n' to N_p. For a polymer molecule conformation that follows a "self-avoiding random walk," the exponent v is equal to 0.6. Rewriting of Eq. (6.4) then leads to $n' = (r_m/b)^{1.67}$, which is very similar to Eq. (13.12) with a "fractal dimensionality" of 1.67. Depending on conditions, the exponent can vary between about 1.6 and 2.1.

Question

A food company wants to produce a gelled oil-in-water emulsion. In a laboratory trial an emulsion of 5%(v/v) triglyceride oil in a 1%(w/w) solution of Na-caseinate is made. It is homogenized to obtain an average droplet size $d_{32} = 1\,\mu m$. To cause aggregation, glucono delta-lactone (GDL) is added. This substance slowly hydrolyzes in water to yield gluconic acid, which lowers the pH. Sufficient GDL is added to obtain pH 4.6, the isoelectric pH of casein. It has been established in several studies that caseinate-covered emulsion droplets do aggregate at low pH. Would they form a gel?

Answer

The aggregates formed will be most likely of a fractal nature. The main question is whether sedimentation (creaming) of clusters will occur before gelation. We thus need to know R_{sed} [Eq. (13.17)] in relation to R_g [Eq. (13.15)]. From Sections 13.2.1 and 2 it will be clear that the aggregation will be perikinetic, and because of the slow decrease of pH, the stability factor W will be high. This implies that the fractal dimensionality will be about 2.3. From Table 6.1 we derive that $\Delta \rho = -70 \, \text{kg} \cdot \text{m}^{-3}$. Equation (13.17) would then reduce to $R_{sed} = 72 \cdot 10^{-9}/a^{0.21}$ in SI units; since $a = 5 \cdot 10^{-7}$, we obtain $R_{sed} = 1.5 \, \mu\text{m}$, or $3a$. Equation (13.15) then reads $R_g = a \, \varphi^{-1.42} = 70a$ for $\varphi = 0.05$. In other words, sedimentation will occur well before gelation. (Even if φ is increased to 0.2, still $R_g = 10a$; moreover, the value of a could be decreased to $0.2 \, \mu\text{m}$ by strongly increasing the homogenization pressure, leading to $R_{sed} = 1.8 \, \mu\text{m}$, which then may be just sufficient.)

The answer would thus be no, but in practice a perfect, though weak, gel is formed. The following is the most likely explanation. The emulsion droplets have a specific surface area $A = 6\varphi/d_{32} = 0.3 \, \text{m}^2/\text{ml}$. The surface load would be about 3 mg of casein per m^2 (see Section 11.4.3). This makes up about 1 mg of adsorbed casein per ml. Total casein concentration was 10 mg/ml, implying that 9 mg per ml is left in solution. Na-caseinate tends to associate in solution into particles of about 400 kDa. The number concentration of these particles would then be $N = (9/400) \cdot N_{Av} \approx 10^{22}$ per m^3. These particles will also aggregate at low pH. The number of emulsion droplets will be $6\varphi/\pi d^3$, which equals about 10^{17} per m^3. From Eq. (13.3) we see that the aggregation rate will be proportional to $1/N$. This means that the small caseinate particles will aggregate $10^{22}/10^{17} = 10^5$ times faster than the emulsion droplets. (The casein particles are far too small to show significant sedimentation.) What will thus happen is that a space-filling network (a gel) of casein particles is formed that entraps and binds to the emulsion droplets before the latter can sediment.

Note A 1% Na-caseinate solution does indeed give a gel upon slow acidification.

13.2.4 Aggregation Times

In practice, technologists are often not so much interested in the aggregation rate as in the time it will take before a perceptible change occurs in the dispersion. This may be

1. The emergence of *visible particles*
2. The formation of a *gel*
3. Separation into *layers*, i.e., a large-scale inhomogeneity.

Figure 13.9 illustrates an important aspect. If $t_2 \ll t_1$, particles will immediately coalesce upon aggregation. If so, particles will grow in size and may eventually become visible to the eye (1); the volume fraction of particles φ does not alter. Nearly the same situation arises when the particles upon aggregation immediately rearrange into compact aggregates of dimensionality close to three, although φ will then somewhat increase. If $t_2 \gg t_1$, fractal aggregation occurs, φ markedly increases, and eventually a gel tends to form (2). If large particles or aggregates formed do sediment before (1) or (2) can occur, layer separation (3) is the result. In case (2) a fractal dimensionality applies. Of course, intermediate situations can occur.

Prediction of the aggregation time, i.e., the critical time for occurrence of a perceptible change t_{cr}, is often desirable. The halving time $t_{0.5}$ [e.g., Eq. (13.3)] is often called the coagulation time, assuming that t_{cr} will generally be a small multiple, say by a factor of 10, of $t_{0.5}$. However, that may turn out to be very misleading. The way to calculate t_{cr} is to use the equation for $-dN/dt$, rearrange it into an equation for dR/dt, invert the latter, and calculate

$$t_{cr} = \int_{a}^{R_{cr}} \frac{dt}{dR} \, dR \tag{13.19}$$

Applying this to **perikinetic aggregation** [Eqs. (13.2) and (13.3)] leads to

$$t_{cr} = t_{0.5}(q^D - 1) \approx t_{0.5}q^D \tag{13.20}$$

where $q = R_{cr}/a$. The stability factor W can be included in the expression of $t_{0.5}$. For case (1), $D = 3$ and $R_{cr} = R_{vis} \approx 0.2$ mm. For case (2), D is mostly between 1.8 and 2.4, and $R_{cr} = R_g$, as given by Eq. (13.15). For case (3), see Eqs. (13.11) and (13.17).

A sample calculation may be enlightening. Assume aggregation of spherical particles in water at room temperature; $W = 1$ (rapid perikinetic aggregation), $a = 0.2\,\mu m$, $\varphi = 0.01$, and $\Delta\rho = 0$. The halving time would then be 0.6 s. Aggregation according to case (1), where $D = 3$, would lead to

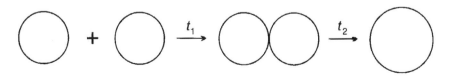

FIGURE 13.9 Aggregation of particles followed by coalescence.

$t_{vis} = 63 \cdot 10^7\,s \approx 20$ years. For case (2), assuming $D = 2$, we obtain $t_g = 6300\,s \approx 1.8\,h$, i.e., smaller than for case (1) by a factor of 10^5. (Make these calculations yourself to check whether you have understood the derivations.)

However, there are several complicating factors, and both results given will be too high:

It is very difficult to avoid *convection currents* with a velocity gradient of the order of $0.1\,s^{-1}$ to occur. As soon as the particles have grown to a size for which Eq. (13.10) equals unity, *orthokinetic aggregation* will take over, which soon proceeds very much faster than perikinetic aggregation.

If fractal aggregation occurs, its *rate constant will increase* during the process, because the value of φ keeps increasing, decreasing the diffusional distances involved. Moreover, the value of W tends to decrease during aggregation.

Sedimentation can readily occur, especially if the particles and $|\Delta\rho|$ are large. In such a case, t_{sed} should be estimated and compared to t_{vis} or t_g.

Altogether, it is far from simple to predict aggregation times. This is because (a) there are so many variables involved (D, a, φ, $\Delta\rho$, Ψ, η, W); (b) it depends on combinations of their magnitudes what equation is valid; and (c) the theory has some uncertainties. Nonetheless, some understanding is useful. Knowledge about trends as to what will happen if certain variables are changed can be important, and that can often be deduced. One trend is that the formation of a gel tends to occur much faster than the formation of visible particles, because it takes far fewer aggregation steps to obtain a large fractal aggregate than a large compact particle, unless the particle size is large and the volume fraction of the particles very small. Another trend is that for large $|\Delta\rho|$, layer separation due to sedimentation is quite likely, unless a is very small and φ is large.

If the capture efficiency is close to unity, aqueous dispersions tend to have short aggregation times, often a few minutes and nearly always below an hour, provided that φ is larger than about 0.02 and the viscosity is not much greater than that of water. It should further be realized that food systems can be subject to slow chemical changes that affect the magnitude of W; for instance, some bacteria can make lactic acid, whereby the pH is decreased, which can decrease the negative charge of particles, which then will significantly decrease electrostatic repulsion.

Question 1

Can you think of methods to decrease convection currents?

Answer

Convection currents in a quiescent liquid arise because temperature fluctuations generally lead to variations in mass density of the liquid and thereby to unbalanced buoyancy forces. To prevent this one can

Apply a very accurate and stable thermostat.

Store at about 4°C, since at that temperature water has $d\rho/dT = 0$, implying that small temperature fluctuations do not lead to density fluctuations.

Increase viscosity; in first approximation, the value of Ψ will be proportional to $1/\eta$.

Perform the experiments in free space, where gravity is virtually zero (a rather expensive method).

Question 2

In a research laboratory, fractal aggregation of emulsion droplets is studied. An O–W emulsion is made with a buffer containing β-lactoglobulin as the continuous phase. To prevent sedimentation of the droplets, part of the oil is brominated to match the density of the oil phase and the aqueous phase. The droplet diameter obtained is $2\,\mu$m, $\varphi = 0.1$. To prevent convection currents, the emulsion is stored at precisely 20°C in a thermostat. The emulsion then forms a gel after 2.5 hours. By use of Eq. (13.20), assuming $R_{cr} = R_g$ and $D = 2$, it is calculated that $W \approx 10$. In order to check the assumptions, it is decided to dilute the emulsion with an equal amount of water; it is expected that the gel time will then be 2^3 times as long, i.e., 20 h. However, this proves incorrect: after a month, the emulsion is still liquid. What can be the explanation?

Answer

The surfactant β-lactoglobulin is a small protein molecule that does not strongly unfold upon adsorption at the O–W interface (Section 10.3.2). Consequently, stability against aggregation will primarily be due to electrostatic repulsion. The dilution with water will lower the ionic strength by about a factor of 2. It can be seen in Figure 13.4 that this will cause a considerable increase in W, e.g., by a factor of 50. Moreover, the larger W value will cause an increase in the value of D, which will also increase the gel time.

13.3 SEDIMENTATION

The Stokes Equation.   By Archimedes' principle, the force F_B due to buoyancy and gravity acting on a submerged sphere of diameter d is given by

$$F_B = \frac{1}{6} \pi d^3 g (\rho_d - \rho_c) \tag{13.21}$$

where g is the acceleration due to gravity ($9.81 \ \text{m} \cdot \text{s}^{-1}$), ρ is mass density, and the subscripts d and c refer to dispersed particles and continuous liquid, respectively. Hence the sphere will move downward (or upward if $\rho_d < \rho_c$) through the liquid and sense a drag force F_S that equals, according to Stokes,

$$F_S = fv = 3\pi d \eta_c v \tag{13.22}$$

where f is the friction factor, v is the linear velocity of the particle with respect to the continuous phase, and η is viscosity. The particle will accelerate until $F_B = F_S$, and from this equality the Stokes velocity is obtained:

$$v_S = \frac{F_B}{f} = \frac{g \, \Delta \rho d^2}{18 \eta_c} \tag{13.23}$$

For instance, an oil droplet in water of $2 \ \mu\text{m}$ and $\Delta \rho = -70 \ \text{kg} \cdot \text{m}^{-3}$ at room temperature would attain a sedimentation rate $v = -0.15 \ \mu\text{m} \cdot \text{s}^{-1}$, i.e., cream by 13 mm per day.

For *centrifugal* sedimentation, the acceleration g should be replaced by $R\omega^2$, where R is centrifuge radius and ω revolution rate ($\text{rad} \cdot \text{s}^{-1}$). The centrifugal acceleration is often expressed in units of g. It should be realized that the magnitude of R, hence the effective acceleration, generally varies throughout the liquid.

Conditions.   For Eq. (13.23) to hold, a number of conditions must be met, including the following:

The particles should be perfect *homogeneous spheres*. If they are inhomogeneous—e.g., small oil droplets with a thick protein coat— a correction can often be calculated. For nonspherical particles of equal volume, the numerical factor in the equation will be smaller than 1/18, the more so for greater anisometry.

The particle *surface should be immobile*. Even for fluid particles this is nearly always the case; see Section 10.7.

The *particle Reynolds number*, given by $dv\rho_c/\eta_c$, must be smaller than about 0.1, since otherwise turbulence will develop in the wake of the sedimenting particle, decreasing its velocity. Putting $v = v_S$, it turns out that for an oil drop in water, the critical particle size is 140 µm.
The particle must not be subject to *other forces* causing their motion. This can involve Brownian motion or weak convection currents. These will disturb sedimentation for particles smaller than about 0.2 µm (assuming $\Psi \leqslant 0.1\,s^{-1}$).
The continuous phase should be a *Newtonian* liquid. This is often not the case.

Non-Newtonian Liquids. If the shear stress σ in a liquid is less than proportional to the velocity gradient Ψ, the liquid is said to be *strain rate–thinning* and the viscosity is an apparent one; see Section 5.1.3. The magnitude of η_a then will depend on the strain rate, or the shear stress, applied; examples are in Figure 13.10. The stress acting on a sedimenting particle is equal to F_B over the particle surface area and is thus given by $\Delta\rho\ g\ d/6$. For an oil drop in water of 5 µm size this stress would then be $6 \cdot 10^{-4}$ Pa. The viscosity to be used in the Stokes equation should thus be determined at this value of σ. Many rheometers cannot give results at a shear stress below 0.1 or even 1 Pa, and the viscosity measured then may be far from relevant, as seen in Figure 13.10. In other words, the sedimentation rate so calculated would be greatly overestimated.

Polydispersity. In practice, we are often interested in the amount of material arriving in the sediment (or cream) layer per unit time. An instrumental relation results if v is divided by the maximum sedimentation distance (height of the liquid) H; by using the Stokes velocity we obtain

$$Q = \frac{v_S}{H} = \frac{g\,\Delta\rho < d^2 >}{18\eta_c H} \tag{13.24}$$

where Q is the volumetric proportion of the particles reaching the sediment per unit time. Since the particles vary in size, an average should be taken. See Section 9.3.1 for an explanation of size distributions. The volume of particles in each size class i is proportional to $n_i d_i^3$, and to obtain its contribution to Q it has to be multiplied by the particle velocity, which is proportional to d_i^2. Summing over the size classes and dividing by the total

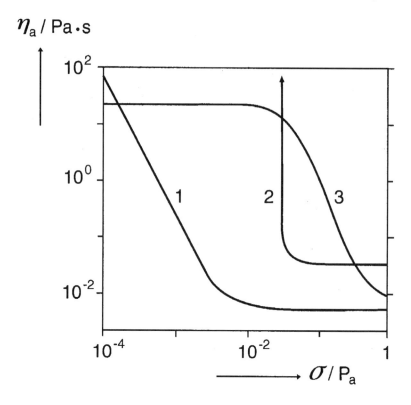

FIGURE 13.10 Schematic examples of the dependence of apparent viscosity η_a on the shear stress σ applied for various non-Newtonian liquids. Curve (1) can represent a slightly aggregating dispersion of very small particles, (2) a solution of polymers that interact with each other to give a weak gel and hence produce a yield stress (of about 0.03 Pa), and (3) a nongelling polymer solution.

volume then yields

$$< d^2 > = \frac{\Sigma n_i d_i^5}{\Sigma n_i d_i^3} = d_{53}^2 \tag{13.25}$$

to be inserted in Eq. (13.24).

Sedimentation Profile. The value of Q generally is not constant during sedimentation. If all particles have exactly the same size, and sedimentation is not disturbed in any way, a concentration profile of particles will develop as depicted in Figure 13.11a for creaming. A sharp boundary between particle concentration = 0 and its original value develops, moving upward at a constant rate. Hence the flux of particles to the cream layer—which also has a sharp boundary—is constant until every particle has arrived. However, nearly all dispersions are polydisperse, and then the largest particles move fastest. At a given moment all of the larger particles can be in the cream layer, whereas most of the smaller ones are still in the subnatant. In other words, the magnitude of d_{53} in the subnatant (i.e., the liquid below the cream layer) will continuously decrease and thereby the particle flux. Figure 13.11b gives an example of how the concentration profile can evolve. The greater the width of the size distribution, the longer it will take before sedimentation is complete (for the same d_{53} value). This is further illustrated in Figure 13.11d.

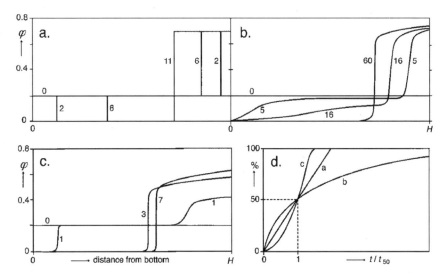

FIGURE 13.11 Examples of the concentration profile developing during creaming of O–W emulsions. Particle concentration is given as volume fraction φ. H is the maximum creaming distance. The numbers near the curves denote time after starting (say, in hours). (a) Calculated for a strictly monodisperse emulsion, assuming Eq. (13.24) to hold. (b) Polydisperse emulsion, no aggregation. (c) Polydisperse emulsion with aggregating droplets. (d) Percentage of the droplets creamed as a function of relative time for the three cases a–c. Highly schematic and only meant to illustrate trends.

Hindered Sedimentation. Strictly speaking, Eq. (13.23) is only valid for one particle in an infinite amount of liquid, and not for finite values of φ. Moreover, it gives the speed relative to the liquid, whereas that relative to the vessel is desired. To correct for this effect, i.e., *displacement of liquid* by the sedimenting particles, v_S has to be multiplied by $(1-\varphi)$. The sedimentation rate decreases much more steeply with φ, however. In Section 5.1.2 we discussed how the *effective viscosity* increases with increasing φ value. An even much larger correction stems from the frictional effect caused by the *counterflow* of displaced liquid. Furthermore, there is a positive contribution to the sedimentation rate by *group sedimentation* (see below). Altogether, theory leads to an equation of the type

$$\frac{v(\varphi)}{v_S} = (1 - \varphi)^n \tag{13.26}$$

where the exponent n would equal 6.55. This would imply that for $\varphi = 0.1$, the velocity is reduced by a factor of 2. The relation has been confirmed for strictly monodisperse suspensions. In practice, higher n values are observed, up to 9. The main cause appears to be polydispersity, which further impedes sedimentation rate.

Group Sedimentation. Sedimenting particles also show Brownian motion. This implies that two particles may for a little while be close together, i.e., form a temporary doublet. If, moreover, the doublet is oriented in the direction of g, it will sediment faster than a single particle: the net gravitational force is doubled, but the frictional force will be less than doubled. If the doublet remains for a period over which significant sedimentation occurs, sedimentation will be enhanced. It is useful to estimate a *Péclet number* giving the ratio of the time scales involved. The numerator is the time it takes for two particles to diffuse away over the distance of a particle diameter, given by $d^2/4D$. The denominator is the time needed for sedimentation over the same distance, to be derived from Eq. (13.23). Using also Eq. (5.16) for the diffusion coefficient, we arrive at

$$\text{Pe} = \frac{\pi d^4 g |\Delta \rho|}{6 k_B T} \tag{13.27}$$

If $\text{Pe} > 1$, the sedimentation rate of the doublet is enhanced, but it will only be quantitatively important if (a) it lasts relatively long, i.e., for $\text{Pe} \gg 1$, and (b) if a sufficient proportion of the droplets is at any time present in doublets. The latter proportion is, in first approximation, proportional to φ.

Figure 13.12 gives results on the creaming of O–W emulsions of various oil contents, both by gravity and by centrifuging. The duration of creaming was chosen so that Eq. (13.24) predicts the same amount of creaming for both series of emulsions. As seen, this was indeed the case for $\varphi \rightarrow 0$. For gravity creaming, Eq. (13.26) was obeyed with $n = 8.6$. The Péclet number is calculated at about 0.4, implying that group sedimentation would have been negligible. Centrifugal creaming occurred at $200 \cdot g$, hence $Pe \approx 80$, implying that considerable group sedimentation must have occurred. This is borne out by the results around $\varphi = 0.03$. For higher φ values, mutual hindrance must have more than offset the effect of group rising. For gravity sedimentation of large and dense particles, hence high Pe values, an increase of settling rate over the Stokes velocity has also been observed (up to a factor of about 2.5) at φ-values around 0.02.

Aggregating Particles. Repulsive colloidal interaction forces between particles hardly affect sedimentation, but the effect of attractive forces can be very strong. Aggregates naturally sediment faster than single particles. Fractal aggregates containing N particles tend to move faster than

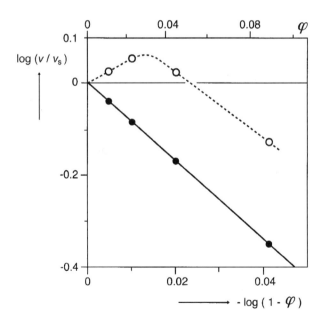

FIGURE 13.12 Creaming rate relative to the Stokes velocity v/v_S of O–W emulsions of various volume fraction φ and of average droplet size $d_{53} = 1.4\,\mu m$. Creaming under gravity (●) or in a centrifuge (○). (Results from the author's laboratory.)

single particles by a factor of about $1.8\,N^{1-1/D}$. The sedimentation also enhances aggregation, since the larger aggregates tend to overtake the smaller ones, whereby larger aggregates are formed, etc. An example is creaming of cold milk. Raw cows' milk contains a cryoglobulin, a large associate of protein molecules, that adsorbs onto and forms bridges between the milk fat globules at low temperature. Upon cooling it takes a while before aggregation starts, but then creaming becomes very fast, to such an extent that formation of a cream layer in a tank of 10 m height occurs almost as fast as in a 10 cm beaker.

Figure 13.11c shows creaming profiles of an emulsion to which a little xanthan gum has been added to induce depletion aggregation (see Section 12.3.3). After aggregates have formed, creaming becomes quite rapid, and a sharp front is formed between an all but empty subnatant and the rest of the liquid, which then "shrinks" to form a sharply defined cream layer. Afterwards, the cream layer becomes somewhat more compacted. Notice that in Figure 13.11b the φ value in the cream is about 0.7, roughly corresponding to dense packing; in frame c φ equals about 0.5, due to the fractal aggregates having an internal φ value well below unity. If the volume fraction of particles in the liquid is much smaller than the value 0.2 in the figure, the φ value in the cream layer may be as small as 0.25. Altogether, aggregation leads to a greatly enhanced sedimentation rate and to a sediment or cream layer that is far less densely packed than for sedimentation of single particles.

The aggregation in the system of Figure 13.11c must have been partly reversible. If more xanthan gum is added, the "bonds" formed between the particles are virtually irreversible, and the whole system tends to gel before significant creaming can occur.

Preventing or Retarding Sedimentation. In many cases, settling or creaming of particles during storage is undesirable, be it in the final product or at some stage during processing. It may not be the sedimentation as such that causes the problem, since it can often be simply undone, for instance by inverting the closed vessel a couple of times. However, the particles in a sediment or cream layer are closely packed and this may lead to gradual formation of strong bonds between them (see Section 12.4, bond strengthening). This then gives rise to a coherent mass that cannot be easily dispersed again. Emulsion droplets in a cream layer may possibly coalesce.

Complete prevention of sedimentation is often impossible, but slowing down to a specified level may then suffice. The following measures may be considered.

1. *Decreasing of particle size.* This is often applied, for instance, to emulsions by use of a high-pressure homogenizer. In many cases, however, it is impossible or insufficient.

2. *Lowering the density difference.* This is rarely practicable.

3. *Increasing the viscosity.* This is often done. A problem may be that the liquid becomes too viscous for handling. This may be overcome by giving the continuous liquid a strongly strain rate–thinning character, as illustrated in Figure 13.10, curves 1 and 3. At very small shear stress, as relevant for sedimenting small particles (e.g., 0.01 Pa), the apparent viscosity then may be high, whereas at higher shear stresses, as prevail during pouring of the liquid (e.g., 100 Pa), the viscosity is small. Addition of a little polysaccharide of high molar mass is a suitable option. However, it should not induce depletion aggregation of the particles; see Section 12.3.3, especially the Question at the end.

4. *Giving the liquid a yield stress* will immobilize the particles. This is discussed below.

5. *Mildly agitating the liquid.* This is generally applied in storage tanks by stirring. A suitable method may be occasional bubbling through of large air bubbles, which is often applied in huge milk tanks.

6. *Preventing aggregation of the particles.* See Chapter 12 for causes of aggregation, hence for measures to prevent it. Also coalescence of emulsion droplets should be prevented; see Section 13.4.

Immobilization of Particles. If the particles involved are immobilized, *sedimentation* cannot occur. Neither can they encounter each other, whereby also *aggregation* is prevented. Since sedimentation or aggregation has generally to occur before fluid particles can coalesce, immobilization will also prevent *coalescence*.

Proceeding with the discussion in point 3, above, one may try to give the liquid not just a strongly strain rate–thinning character, but a yield stress σ_y, as depicted in Figure 13.10, curve 2. Some combinations of polysaccharides give a weak network that imparts a small yield stress at very low concentration. The stress that a particle causes onto such a network is given by the gravitational force [Eq. (13.21)] over the cross-sectional area of the particle, hence

$$\sigma_{sed} = \frac{|F_B|}{(1/4)\pi d^2} \approx g|\Delta\rho|d \tag{13.28}$$

To prevent sedimentation, σ_y should be larger than σ_{sed}. To give an example, for $d = 10\,\mu m$ and $|\Delta\rho| = 100\,kg \cdot m^{-3}$, $\sigma_{sed} = 0.01\,Pa$ results. Such a yield stress would thus be needed to prevent sedimentation. It would be wise to make σ_y somewhat higher, since a yield stress tends to be smaller for longer time scales (see Section 5.1.3).

A yield stress of 1 Pa or less generally goes unnoticed: 1 Pa corresponds to the stress exerted by a column of water of height 0.1 mm. This also implies that slight agitation, e.g., caused by pouring liquid out of the vessel, will overcome the forces holding the network together, whereby the yield stress disappears. For the dispersion to be stable, the network should then be restored before the particles can show substantial sedimentation. It is observed that a resting time of about 15 min often suffices to restore fully the value of σ_y. Taking the same example as just given, and assuming, moreover, that the effective value of η_c equals 0.1 Pa·s, we calculate that $v_S \approx 55$ nm·s^{-1}, which then leads to sedimentation over at most 50 μm in 15 min. This will generally be acceptable.

The particles that would sediment can also themselves form or participate in forming a weak, generally fractal, network. A good example is soya milk, made by wet milling of soaked (whole or dehulled) soya beans. The soya milk so obtained contains small oil droplets and a variety of solid particles, including cell fragments, whole cells and fragments of the hulls. Some of these particles are several μm in size and would certainly sediment in a Newtonian aqueous liquid. However, soya milk tends to be quite stable. Figure 13.13 shows some flow curves, and it is seen that all samples exhibit a yield stress (the intercepts of the curves on the stress axis). It is seen that the magnitude of σ_y depends on pretreatment. Comparing soya milks from whole or dehulled beans, the former contains larger particles, which would demand a higher yield stress, as is indeed the case.

Finally, a gel network containing entrapped particles, or consisting of particles, may still exert a gravitational force onto the liquid. This can result in slow compaction of the network, starting at the bottom or at the top, according to $\Delta\rho$ being negative or positive, respectively. The only way in which this is prevented for very weak gels is by a network that *sticks to the* wall of the vessel, whereby a stress at the wall can counterbalance the gravitational force. It depends on the material and the cleanliness of the vessel whether sticking occurs.

Question

A food company produces a number of different oil-in-water emulsions. It is decided to apply a centrifuge test to predict how fast creaming will occur in the stored emulsions. In the test the percentage of the oil that reaches the cream layer under specified conditions is taken as the criterion. (a) What should the test conditions (like

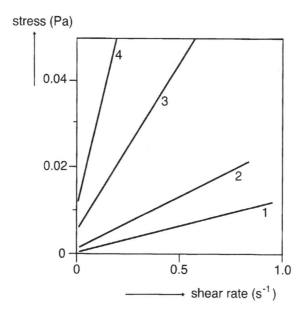

FIGURE 13.13 Flow curves (shear stress versus shear rate) of soya milk. Curves 1 and 2 are for milk made from dehulled beans (about 6% dry matter), 3 and 4 from whole beans (about 7% dry matter). Soaking was overnight at room temperature (curves 1 and 3); or 4 hours at 60°C (2 and 4). (From results by A. O. Oguntunde et al. Proc. 7th World Congr. Food Sci. Technol. Singapore, 1987, p. 307.)

centrifugal acceleration and duration, and tube dimensions) be to make the test meaningful? (b) It is generally observed that the test underestimates the times needed for creaming during storage. In some cases the discrepancy is very large. What can be the possible causes and how can it be established which of these actually holds? (c) In one of the emulsions, with quite a small volume fraction of oil, the creaming during storage is much faster than predicted. What could be the cause and how could it be established?

Answers

(a) According to Stokes, the sedimentation velocity is proportional to the acceleration, i.e., to $R\omega^2$ versus g. However, Eq. (13.24) shows that the creaming rate will be proportional to acceleration over H, the creaming distance. The latter will generally be larger in the vessels in which the emulsions are stored (H_V) than in the centrifuge tubes (H_T). Moreover, the times involved in the gravity creaming and the test should be such that $t_T \cdot R\,\omega^2/H_T = t_V \cdot g/H_V$. In this way the proportion creamed in the test and in the vessel will be equal, provided that Eq. (13.24) holds.

The point is of importance because of polydispersity: d_{53} decreases during creaming, and thereby Q, as is illustrated in Figure 13.11d. Assume, for example, that $R\omega^2/H_T = 450 \cdot g/H_V$, that centrifuging time was 20 min, and that 90% creaming was obtained. Hence 90% creaming would occur in 9000 min in the vessel, but it often is more interesting to know when a little creaming would occur, say 3% of the oil. Then it would be wrong to assume that it would take 3/90 as long, i.e., 5 hours; it may be more like 1 hour if the emulsion is quite polydisperse (see also Figure 13.11d). Finally, it should be checked that the droplet Reynolds number is not over 0.1 in the centrifuge test, but that is rarely the case. (b) Assuming now that the conditions just specified are met, the following may be causes for overestimating the creaming rate:

- Group sedimentation during centrifuging: see Figure 13.12. This will nearly always occur, and it can be checked by calculating the Péclet numbers in both situations. The effect is not very large, rarely over a factor of 2.
- If the continuous liquid is strain rate–thinning, as illustrated in Figure 13.10 (curves 1 and 3), the effective viscosity may be much smaller in the centrifuge than in gravity creaming, since σ_{sed} is by $R\omega^2/g$ times larger. For some systems, the discrepancy may be large, up to a few orders of magnitude. It can be checked by determining η_a as a function of shear stress.
- It may also be that the system has a yield stress (Figure 13.10, curve 2) and that σ_{sed} is larger than σ_y in the centrifuge and smaller under gravity. That means fairly rapid versus no sedimentation, respectively. The magnitude of a possible yield stress can be determined in sensitive rheometers.
- A fairly trivial case is that the liquid in the vessel is subject to convection currents, e.g., due to temperature fluctuations, whereas agitation is negligible in a centrifuge (provided it has a swing-out rotor).

(c) If the emulsion droplets show slow aggregation, this may considerably enhance sedimentation as soon as sizable aggregates have formed. However, this may take some hours, and a centrifuge test can be done soon after the emulsion has been made and can be completed in, say, half an hour. This then means that the test is not suitable to detect aggregation-enhanced sedimentation. Aggregation can readily be detected by microscopy.

13.4 COALESCENCE

Coalescence is induced by rupture of the thin film between close emulsion droplets or gas bubbles; this phenomenon of film rupture will be discussed first. However, the whole coalescence process involves a number of additional variables, and these are rather different for emulsions and foams (cf. Table 11.1), which is the reason that they are discussed separately. Our understanding of coalescence is yet unsatisfactory, because (a) so many variables are involved and (b) some fundamental problems have not been fully resolved. Nonetheless, useful general rules can be obtained.

13.4.1 Film Rupture

To arrive at the rate of aggregation of particles, one has to multiply a frequency factor (f) with a capture efficiency, which is due, in turn, to the existence of a free energy barrier for contact (ΔG); see Section 13.2. A similar situation would exist for film rupture, leading to an equation of the type

$$J_{\text{rup}} = f \exp\left(\frac{-\Delta G}{k_B T}\right) \approx 10^{31} \exp\left(\frac{-\gamma \delta^2}{k_B T}\right) \tag{13.29}$$

giving the number of rupture events to be expected per unit film area per unit time. It can be derived from theory that the frequency factor would approximately equal the sound velocity over the molecular volume of the surfactant at the film interfaces, leading to a value of the order of $10^{31}\,\text{m}^{-2} \cdot \text{s}^{-1}$.

Hole Formation. According to de Vries, the magnitude of ΔG will be as illustrated in Figure 13.14a. Heat motion of molecules in the film will occasionally lead to the formation of a small hole, as depicted. If now the Laplace pressure p_L near position 1 is larger than further away from the hole (position 2), the hole will spontaneously grow in size and the film will rupture. Near position 1, $p_L = 2/\delta - 1/R$ (see Section 10.5.1), and near position 2, $p_L \approx 0$; this implies that rupture can occur for $R > \delta/2$. To obtain this situation, the interfacial area of the film has to be increased by an amount of about δ^2, which implies an increase of interfacial free energy by an amount $\gamma \delta^2$, which then would equal ΔG.

As an example, assume that $\gamma = 5\,\text{mN} \cdot \text{m}^{-1}$ and $\delta = 10\,\text{nm}$. This results in $\Delta G = 5 \cdot 10^{-19}$ J or $125\,k_B T$ and $J_{\text{rup}} = 5 \cdot 10^{-24}\,\text{m}^{-2} \cdot \text{s}^{-1}$; in other words, the film would never rupture. If δ is equal to 3 nm, we arrive at a rupture rate of $10^{26}\,\text{m}^{-2} \cdot \text{s}^{-1}$ or $10^{14}\,\mu\text{m}^{-2} \cdot \text{s}^{-1}$; the film would thus rupture immediately. This does not agree with experience: films can rupture at a far greater thickness than 10 nm, and films of 3 nm can be stable. Moreover, the reasoning given would allow films without surfactant to be stable, which they never are. The de Vries theory is thus insufficient.

Capillary Waves. At a liquid surface, capillary waves will always form, due to heat motion and induced by vibration. On a film, symmetrical (varicose) waves can develop, as depicted in Figure 13.14b. If now the amplitude α becomes almost as large as $\delta/2$, the film will rupture according to the de Vries mechanism. However, the waves may be damped. A theory for film stability has been worked out based on wave formation and damping, for the most part by Vrij and by Sheludko. Two kinds of forces determine whether the wave is damped or not. (a) *Colloidal interaction.* If only van der Waals attraction acts across the film, the amplitude will always tend to grow, because the attractive force is largest where the film is thinnest (see Section 12.2.1). If also repulsive forces act, which is the common situation, growth of the amplitude will be slowed down. (b) The *Laplace pressure* in the film will be greater where the surface is concave than where it is convex. This means that liquid will stream from the thick parts of a film to the thin parts, damping the wave. For the same amplitude, the local curvature of the film surfaces, and hence the difference in Laplace pressure and the damping tendency, will be greater for a shorter wavelength λ. The balance of these forces will determine whether the amplitude will tend to keep increasing or not. Vrij derived that α would grow, and the film be

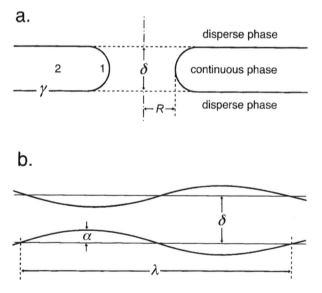

FIGURE 13.14 Cross section through a part of film of (average) thickness δ. (a) Illustration of hole formation. (b) Properties of a varicose wave on the film.

unstable, if the condition

$$\frac{d^2 V(h)}{dh^2} < -\frac{2\pi^2 \gamma}{R_f^2} \qquad (13.30)$$

is fulfilled. Here $V(h)$ is the total colloidal interaction free energy per unit area as a function of distance h (which equals δ in the present case), and R_f is the radius of the film. It should be mentioned that for an interaction profile as depicted in Figure 12.1 (solid curve), the instability criterion may be met at a relatively large value of h (near the secondary minimum), but that upon closer approach the balance shifts again; if so, the film would be stable.

The theory as discussed thus far already leads to some important conclusions. The probability of film rupture will be smaller if

> γ *is larger*. This may look paradoxical: a surfactant is needed to make and stabilize emulsions and foams, and it lowers γ. Moreover, the driving force for coalescence is that it causes the interfacial free energy of the system to decrease, which decrease is greater for a higher value of γ. This is true enough, but the probability of rupture is not determined by the concomitant decrease in interfacial free energy, but by an activation free energy, and the latter is smaller for a lower interfacial tension.
>
> *Colloidal repulsion is stronger* (or δ remains larger). This is as expected.
>
> *The film is smaller*, implying that the maximum possible λ value is smaller. This means that smaller droplets or bubbles will be more stable.

Notice that qualitatively the same effect of the values of γ and δ is also predicted by Eq. (13.29).

Gibbs Elasticity. If no surfactant is present, a film is extremely unstable. This may be due to colloidal repulsion then being very small, but such films are observed to break at far larger δ values than predicted. One shortcoming of the original Vrij theory is that it assumes the interfaces to be immobile in the tangential direction. However, if no surfactant is present, this assumption is not true (see Section 10.7). Immobility requires a certain Gibbs elasticity, or surface dilational modulus E^{SD}. It turns out that for a thin film, the condition is $E^{SD} > A/8\pi\delta^2$, where A is the Hamaker constant. The critical magnitude is virtually never larger than $1\,\mathrm{mN} \cdot \mathrm{m}^{-1}$, a value that is nearly always reached unless the surfactant concentration is very small.

We note in passing that the Vrij theory, as modified by introducing the Gibbs elasticity, is an extension of Gibbs' explanation of film stability (Section 10.7).

Some Other Factors. If the modified Vrij theory predicts stability, the film is indeed nearly always stable. However, some films predicted to be unstable resist rupture for a very long time. Several mechanisms have been suggested to explain this, but some of them do not apply, or are quite uncertain, for the surfactants commonly used in foods.

Surface active polymers, such as proteins, can give very stable films. The main explanation will be a not very small γ value and a strong repulsion acting at a relatively large distance, but there seem to be other factors involved. In some cases, a correlation between film stability and the apparent surface shear viscosity η_a^{SS} of A–W or O–W surfactant layers has been observed, but there are exceptions as well. Molecular size or the thickness of the surface layers may be involved and probably also the layer coherence. Film rupture would also need a kind of disruption of the adsorption layers; presumably, this will readily occur for most surfactants, but a layer of protein molecules that are somehow cross-linked may resist disruption.

Finally, the discussion here has been restricted to static films. In *dynamic* situations as during foam formation, where films are periodically stretched, other factors come into play, as briefly explained in Section 11.2.3.

13.4.2 Emulsions

Figure 13.9 illustrates the two essential steps in coalescence. The droplets have first to encounter each other—by aggregation or in a sediment layer—and come close before the film between them can rupture, leading to their merging into one drop. Nearly always, the first step is much faster than the second. This means that coalescence tends to be a *first-order process*, unlike the aggregation that often precedes it. Only if the droplets immediately coalesce upon encountering each other—which may happen if the surfactant concentration is very small—will the second order aggregation rate be determinant.

Weber Number. When droplets come close, they will either keep their spherical shape or be pressed together, with formation of a flat film between them, as depicted in Figure 13.15a. What will happen is determined by the *ratio of the external stress* that forces the drops together *over the* *internal stress.* The latter is simply given by the Laplace pressure $(2\gamma/a)$. The external stress may be due to colloidal attraction, to hydrodynamic forces, or to gravity as in a sediment layer. Since initially the area over which the external force acts may be much smaller that the cross-sectional area of a droplet (πa^2), the external stress σ_E must be multiplied by a stress concentration factor Λ. Figure 13.15b illustrates that for $a \gg h$ the area of close contact is of order $\pi a h$, which leads to $\Lambda \approx a/h$. Generally, the magnitude of h upon droplet encounter is determined by colloidal repulsion. A dimensionless Weber number can be defined as the quotient of the local external force over the internal one, which would equal

$$\text{We} = \frac{\sigma}{p_L} = \frac{\sigma_E \Lambda}{p_L} = \frac{\sigma_E a^2}{2\gamma h} \tag{13.31}$$

For We < 1, the drops remain almost undeformed, for We > 1, a flat film between them will be formed.

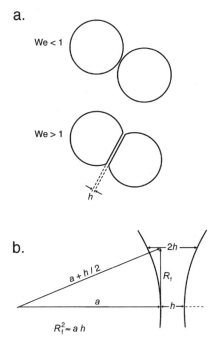

FIGURE 13.15 (a) Formation or not of a flattened film between droplets (thickness h of the film is greatly exaggerated). (b) Derivation of the nominal radius of the film between close spheres.

Note In principle, the "equilibrium" size of the flat film can also be calculated.

Some *sample calculations* are in Table 13.1, where both p_L and σ have been varied. It is seen that for small drops and not very low interfacial tension, We < 1, except in the cream layer in a centrifuge. For large drops and small γ values, We may be larger than unity. It is to be expected that for We < 1, drops are far more stable to coalescence than for We > 1, mainly because the film area is much smaller [cf. Formula (13.30)]. Moreover, small We tends to go along with relatively large γ, which in itself decreases the chance of film rupture.

Few studies have been published in which We has been systematically varied. Nevertheless, it is clear from what is known that for high We coalescence tends to be much faster than for low We. Studies in which interfacial tension was varied have especially shown this. Fundamental studies on coalescence often involve large droplets (some mm in diameter), where clearly We $\gg$ 1. It is very difficult, if possible at all, to translate results so obtained into coalescence in real food emulsions, where generally We < 1. The same holds for centrifugal tests to determine coalescence, aimed at rapid prediction of the stability of the emulsion; as is well known to

TABLE 13.1 Role of Weber Number [Eq. (13.31)] in Coalescence of Emulsion Droplets (Calculated Examples)

Laplace pressure	$p_L = 2\gamma/a$	
$\gamma = 12\,\text{mN} \cdot \text{m}^{-1}$, $\quad a = 0.25\,\mu\text{m}$		$p_L \approx 10^5\,\text{Pa}$
$\gamma = 3\,\text{mN} \cdot \text{m}^{-1}$, $\quad a = 6\,\mu\text{m}$		$p_L \approx 10^3\,\text{Pa}$
Local stress	$\sigma = \text{external stress} \times \Lambda = \sigma_E\, a/h$	
1. Van der Waals attraction[a]	$\sigma = A/12\pi h^3$	
$A = 5 \cdot 10^{-21}$ J, $h = 10\,\text{nm}$		$\sigma \approx 10^2\,\text{Pa}$
$A = 5 \cdot 10^{-21}$ J, $h = 3\,\text{nm}$		$\sigma \approx 5 \cdot 10^3\,\text{Pa}$
2. Hydrodynamic shear stress	$\sigma = \eta \Psi a/h$	
$\eta \Psi = 10^2\,\text{Pa}$, $a/h = 100$		$\sigma \approx 10^4\,\text{Pa}$
3. Stress in a sediment layer	$\sigma = \varphi_{\text{sed}} \Delta\rho\, gHa/h$	
$\Delta\rho = 70\,\text{kg} \cdot \text{m}^{-3}$, $H = 10\,\text{mm}$, $a/h = 100$		$\sigma \approx 5 \cdot 10^2\,\text{Pa}$
Same in centrifuge at $1000\,g$		$\sigma \approx 5 \cdot 10^5\,\text{Pa}$

[a] Here it is assumed that strong repulsion starts at interparticle distance h.

many workers (although hardly ever published), the correlation between the test result and coalescence in practice tends to be quite poor.

Droplet Size. From the above, it may be clear that, other things being equal, an emulsion with larger drops will generally be less stable to coalescence. This is because the film between larger drops will be larger, even if We < 1, and it will be much larger if We > 1. Moreover, larger drops will more readily sediment, which will (a) increase the magnitude of We, (b) lead to a smaller h value, and (c) very much increase the time that a thin layer between drops exists. It has often been observed that an emulsion showing coalescence while standing remained stable if sedimentation was prevented by slowly rotating the vessel end over end. Finally, the larger the drops, the smaller their number, and the smaller the number of coalescence events needed to produce a visible change, e.g., the formation of a continuous layer of the liquid making up the disperse phase.

Effects of Flow. As long as We remains smaller than unity, stirring of an emulsion has little effect on coalescence rate. At high We numbers, however, the drainage of the film between the droplets to a thickness where coalescence can occur is an important variable. In studies on coalescence of emulsions during agitation, it is often observed that the coalescence rate at first increases with increasing velocity gradient, because the encounter rate is proportional to Ψ. For still higher Ψ values coalescence rate often decreases, because the duration of an encounter, and hence the time available for drainage, is inversely proportional to Ψ. The change from increasing to decreasing rate occurs at a lower Ψ value for larger drops, since they make a larger film, and hence need a longer drainage time. Such relations have been observed, for instance, in stirred W–O emulsions to be used for margarine making.

Oil-in-Water Emulsions. An example of the coalescence of *protein-stabilized droplets* with a planar interface is shown in Figure 13.16. It is seen that droplet diameter has a strong effect, presumably because a larger drop gives a larger effective film radius. However, from the protein concentration and the aging time, a value of the surface load Γ of about $0.4 \, \text{mg} \cdot \text{m}^{-2}$ is calculated, i.e., far below the plateau value of about 3, and even considerably below the value of about $1 \, \text{mg} \cdot \text{m}^{-2}$ for extended peptide chains. For conditions where the plateau value was reached, no coalescence was observed. This is a general observation: protein-stabilized O–W emulsions, with droplet size of a few μm or less and a plateau surface load, are very stable against coalescence. If part of the adsorbed protein is wholly or partly displaced by a small-molecule amphiphile

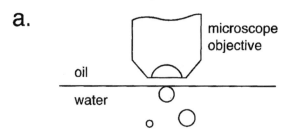

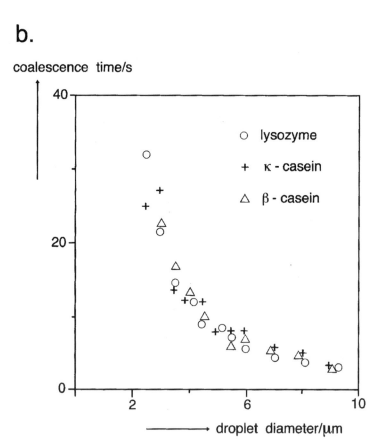

FIGURE 13.16 Coalescence of protein stabilized emulsion drops. (a) Experimental setup. The coalescence time at the planar oil–water interface is observed. (b) Average coalescence time as a function of droplet size for three proteins. Protein concentration 1 g per m^3; aging time of the interface 20 min. (From results by E. Dickinson et al. J. Chem. Soc. Faraday Trans. 1, 84 (1988) 871.)

(Section 10.3.3), the stability may be considerably impaired: the amphiphile gives a lower interfacial tension and may cause weaker repulsion than a protein.

Results on coalescence of emulsions made with *peptides* are shown in Figure 13.17. The amphiphilic peptides were derived from β-casein, and the molar mass was about half that of the casein. (Comparable emulsions made with the unmodified protein showed no coalescence in 10 days.) It is seen that coalescence occurred, especially at the lower peptide concentrations, which corresponded to a low value of Γ. It is also seen that the coalescence rate decreased with time, most likely because coalescence will lead to a higher Γ value, hence greater stability. Finally, the coalescence rate was faster for a higher ionic strength, indicating that the stabilizing action of these peptides is largely due to electrostatic repulsion.

The method employed to obtain the results of Figure 13.17, i.e., *determination of average droplet size* as a function of time, is quite suitable to establish the rate of coalescence in emulsions. It may even be better to

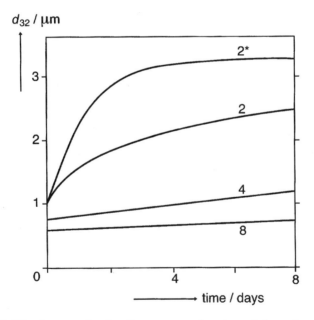

FIGURE 13.17 Average droplet diameter as a function of time after making of O–W emulsions with various concentrations (numbers near the curves in mg per ml) of amphiphilic peptides. Volume fraction of oil 0.2, pH 6.7, ionic strength 75 mmolar or 150 mmolar(*). (Results of P. Smulders et al. Roy. Soc. Chem. Special Public. 227 (1999) 61.)

determine full size distributions. The largest drops in an emulsion are the most susceptible to coalescence, as concluded above. This implies that coalescence tends to increase the width of the size distribution; it may even become bimodal.

> *Note* A change in particle size of an emulsion may also be due to aggregation of the droplets, and this should, of course, be checked, e.g., by microscopic observation.

Finally, a word about *extreme conditions*. Although O–W emulsions made with protein tend to be very stable, they may show marked coalescence at conditions where the droplets are forced close to each other. This can happen in a centrifuge, especially at high acceleration; by agitating an emulsion of high φ value; by freezing an emulsion, where ice crystals press droplets together; or by drying an emulsion.

Water-in-Oil Emulsions. It is difficult to stabilize these emulsions against coalescence, since suitable food grade surfactants are not available. Suitable means sufficiently hydrophobic and providing strong repulsion across an oil film; highly unsaturated monoglycerides provide some stability. In fact, W–O emulsions are very rare in foods. Butter and margarine are not simple emulsions, since the continuous oil phase contains triglyceride crystals. These can provide *Pickering stabilization* as depicted in Figure 13.18. The requirements are (a) that the solid particles have such a contact angle with water and oil that they adhere to the aqueous drops but stick out in the oil (see Section 10.6); (b) that the particles are not very small, say larger than 20 nm [see Eq. (10.14) for the reason]; and (c) that the particles are densely packed at the droplet surface, since otherwise bridging of drops can occur, or even coalescence.

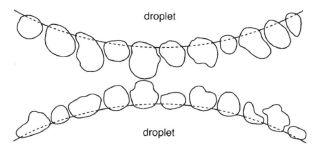

FIGURE 13.18 Pickering stabilization. Small solid particles, preferably wetted by the continuous phase, adsorbed onto emulsion drops.

Actually, the fat crystals, if sufficiently abundant, also form a space-filling network that *immobilizes the drops*. If the volume fraction of water is quite high, as in several low-fat spreads, there may be not enough crystals for Pickering stabilization and formation of a network. If so, stability can often be achieved by "*solidifying*" the aqueous drops, whereby their coalescence becomes impossible. The agent of choice is gelatin, a solution of which is liquid at, say, 30°C, and quickly gels (solidifies) upon cooling.

Question 1

A small food company makes an O–W emulsion with 10% by volume of oil and 1% of a well soluble protein; a powerful stirrer is used. The emulsion shows appreciable creaming, about 20% of the oil in one day. This is considered undesirable, and it is concluded that the oil droplets must be too large. Since more efficient emulsification equipment is not available, 1% of a monoglyceride is added to the oil, on the consideration that this "emulsifier" will produce smaller drops. However, it turns out that the emulsion so made creams faster. What would be a likely explanation?

Answer

Interpreting the results by using Eq. (13.24) suggests that the droplets must have been of the order of a few μm in diameter to explain the creaming. This is a quite reasonable value. The specific surface area ($6\varphi/d$) must thus have been of the order of $0.2\,m^2$ per ml of emulsion. Assuming the plateau surface excess to be $3\,mg \cdot m^{-2}$, there must have been more than sufficient protein to cover the droplets. But 10 mg of monoglyceride per ml of oil is then equivalent to 5 mg per m^2 of oil surface; it will thus adsorb and may to a considerable extent prevent protein adsorption. The monoglycerides may indeed have produced smaller droplets, but their stability to aggregation and subsequent coalescence may have been greatly diminished. This is because monoglycerides hardly cause repulsion across an aqueous film and give a much lower interfacial tension than a protein. The droplets would thus gradually become larger, hence cream faster.

Question 2

A food company makes an O–W emulsion with 10% oil and 0.1% whey protein isolate as the emulsifier. The average droplet size d_{32} is determined at 1.8 μm. After standing for two days, d_{32} has increased to 2.2 μm, and it is concluded that the emulsion is unstable to coalescence. However, during further storage, the droplet size does not increase any more. What would be the explanation?

Answer

It is readily calculated that the specific surface area was $0.33\,m^2$ per mL of oil, and the amount of protein available thus was 3 mg per m^2 of O–W interface. A glance at Figure 11.15b suggests that this would not have been enough to provide the drops with a plateau value surface load. Hence some coalescence may indeed have occurred. This would also lead to a decrease in interfacial area, hence to an increase in surface load and thereby, maybe, to an increase in stability. Stability may also increase owing to slow cross-linking of the adsorbed protein molecules, and it has been shown that β-lactoglobulin, the major component of whey protein, is prone to the formation of intermolecular —S—S— bridges in an adsorbed layer.

13.4.3 Foams

The lifetime of a foam is typically some orders of magnitude smaller than that of an emulsion: say, an hour versus several months. This will primarily be due to the *number of film ruptures* needed to break a foam being smaller by a factor of, say, 10^5 than in an emulsion. Moreover, the *films* between bubbles *are quite large* (We $\gg$ 1) and *nontransient*, greatly promoting film rupture. On the other hand, the *interfacial tension A–W is larger* than the O–W tension by a factor of about 5. Another important aspect is that the various instabilities in a foam may reinforce each other. For instance, Ostwald ripening (Section 13.6) gives larger bubbles and hence larger films and faster drainage, hence a decreased film stability.

It may even be questioned whether there is a close correlation at all between film stability and the lifetime of a foam. Much of the research on film rupture concerns quite large films, several mm in radius, whereas those in food systems are rarely above $50\,\mu m$. Most of the films studied were stabilized by small-molecule surfactants that are never or rarely used in foods. In the author's opinion these studies, however interesting they may be in general, are hardly relevant for foods. Moreover, Ostwald ripening tends to be the dominant instability in most food foams.

When considering **foam film stability**, three rather different cases should be distinguished. Since film thickness is an essential parameter, it is useful to consult Section 11.2.2 on drainage.

1. **Thick films.** These are often young films. They tend to be quite stable according to the Vrij theory, unless the surfactant concentration is very small. Consequently, an option to stabilize a foam against coalescence is to retard film thinning, i.e., drainage. Drainage is counteracted by the development of a surface tension gradient on the film surfaces, and the gradient can be greater for smaller films and for surfactants of high surface

dilational modulus, hence for proteins. Consequently, a foam of small bubble size and made with protein as the surfactant drains relatively slowly.

Drainage is further impeded by the liquid viscosity being high, but this strategy for improving stability has its limits: a very high viscosity will greatly hinder making a foam. What should help is giving the liquid a yield stress. It is, however, not easy to predict what its magnitude should be. In a foam layer of height H, the maximum gravitational stress equals $\rho_{water} g H$ [cf. Eq. (11.1)]. Its magnitude would then be of order 1 kPa, but the stress is to a considerable extent counterbalanced by the surface tension gradients mentioned. The author is not aware of systematic studies. In practice, gelation can be achieved in various ways, leading to more or less "solid foams" (see also Section 11.2.4); for instance, by heat treatment of egg white foams. Another mechanism to counteract drainage is to provide the liquid with small hydrophilic particles, e.g., protein aggregates; the film then cannot become much thinner than the particle size.

2. **Thin films**. It may take a long time before a film drains until the gravitational stress is counterbalanced by the colloidal disjoining pressure, i.e., a thickness of order 10 nm. However, at the top of a foam, film thinning can be due to the evaporation of water, and this can happen very much faster. Thin films may readily rupture, as discussed. Protein layers at the film surfaces appear to provide reasonable stability, especially if the layer is coherent, e.g., due to intermolecular cross-links, as discussed. Also mixtures of proteins that have opposite electric charge, and that thereby give a coherent adsorption layer, appear to be suitable. The presence of some small-molecule surfactant may greatly impair film stability.

3. **Films with hydrophobic particles**. It is often observed, especially in protein-stabilized foams, that the presence of small lipidlike particles is quite detrimental to foam stability. Such extraneous particles are often present, and they can cause rupture of relatively thick films. Possible mechanisms are depicted in Figure 13.19, and it may be useful first to consult Section 10.6.1 on contact angles. In cases (a) and (b) a particle becomes trapped in a thinning film and then makes contact with both air bubbles. Because of the obtuse contact angle θ, the curvature of the film surface where it reaches the particle becomes high, leading to a high Laplace pressure. Hence the water will flow away from the particle, leading to film rupture. In cases (c) and (d), contact with one A–W surface may suffice. An oil droplet reaching the surface will suddenly alter its shape, to a flat lens, which induces flow of the water in the film away from the droplet; if the film is fairly thin this may cause its rupture, as depicted. If the contact angle equals 180°, the situation as depicted in (d) can arise. It may also occur if the particle is partly solid and contains a substance, generally a liquid, that can displace protein from the film surface. For this to occur, the spreading pressure [Eq. (10.12)] has to

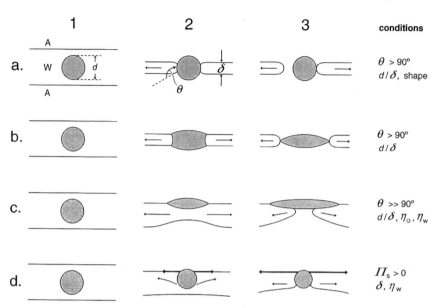

FIGURE 13.19 Possible mechanisms involved in the rupture of aqueous foam films of thickness δ induced by hydrophobic particles; 1, 2 and 3 indicate subsequent stages. (a) Solid particle; (b) and (c) oil droplet; (d) oil droplet or composite particle. Thick arrows indicate spreading of oil or surfactant. A is air, W is water; θ is contact angle as measured in the water phase; η is viscosity of the water phase (subscript W) or the oil phase (subscript O); Π_S is spreading pressure.

be positive. For most combinations of foaming agents and particle material, this condition is not fulfilled, but for proteins at the A–W surface (not a very small γ value) and small-molecule surfactants in the particle, spreading may occur. Mechanism (d) may occur more readily when the film is stretched, as during beating, because then γ_{aw} is higher. Incidentally, extensive stretching of a film will always cause its rupture.

Another condition for all mechanisms in Figure 13.19 to occur is that the thin water film between the particle and the A–W surface has to break for the particle to be "wetted" by the air. Colloidal repulsion will counteract this. However, if the particles are oil droplets containing fat crystals, such wetting may readily occur (see Section 13.5 for the mechanism), making the emulsion a very efficient *foam-breaking agent*.

It is well known that lipids (e.g., from lipstick) can readily break beer foam. It is also known that skimmed milk, which contains very little fat, does foam copiously on beating, but that adding a little whole milk strongly impairs foaming, provided that the fat globules in the milk contain fat crystals. On the other hand, whipping cream contains a very high concentration of fat globules, and it is readily whipped into a quite stable foam. However, the concentration of fat globules in whipping cream is so high that these particles very rapidly cover all of the air bubble surfaces, giving rise to Pickering stabilization (Figure 13.18).

Question

When one splashes beer into a glass, a considerable head of foam is generally formed; beer contains about 0.5% protein. However, beer devoid of alcohol shows far less foam and the bubbles are larger. Addition of about 1% ethanol restores the normal foaming behavior. The stability of the foam is also affected by ethanol concentration, and has a maximum at about 0.25% ethanol. What would be likely explanations?

Answer

A look at Figure 10.4 shows that 1% ethanol added to water (i.e., about 0.22 molar) causes a significant decrease in surface tension of about $5\,mN \cdot m^{-1}$. Half a percent of protein is sufficient for protein-covered foam bubbles to form, but—as discussed in Chapter 11—a small-molecule surfactant is more efficient than protein in the formation of foam and the breaking up of large bubbles into smaller ones. This is because the protein has a low molar concentration and needs to unfold upon adsorption to cause considerable lowering of surface tension, which takes a relatively long time. Hence the ethanol should indeed promote foam formation.

The effect on stability is more difficult to explain, since we do not know what kind of instability is involved. Assuming for the moment that it is coalescence, ethanol may well impair the "coherence" of the layer of adsorbed protein, thereby promoting its rupture: at the prevailing concentration the surface load of ethanol is significant, causing a decreased surface shear viscosity. The optimum concentration may then be explained by the decreased bubble size—which would lower the coalescence rate—which effect is for higher ethanol concentration offset by the smaller coherence.

13.5 PARTIAL COALESCENCE

In most food oil-in-water emulsions, the oil is a triglyceride mixture, and several such oils are partially crystalline at room temperature. This means that the oil droplets can contain crystals; such droplets are better called *fat globules*. The system is not a true emulsion, as it has three phases. Fat globules are subject to partial coalescence or clumping, as depicted in Figure

13.1e. The adjective *partial* signifies that the globules do not coalesce into one drop, although there is oil–oil contact between them.

The mechanism of partial coalescence is illustrated in Figure 13.20a. A fat crystal protruding from the globule surface may pierce the film between close globules (or between a globule and a true droplet). If there is attraction between the globules, or if they are pressed together, the sharpness of the crystal causes considerable stress concentration, facilitating the piercing of the film and, if needed, of the adsorption layer on the globule. The contact angle water–oil–crystal, as measured in the water phase, tends to be obtuse (Figure 13.20a), which means that the crystal is preferentially wetted by the oil. This then leads to a junction between the globules, consisting of an oil

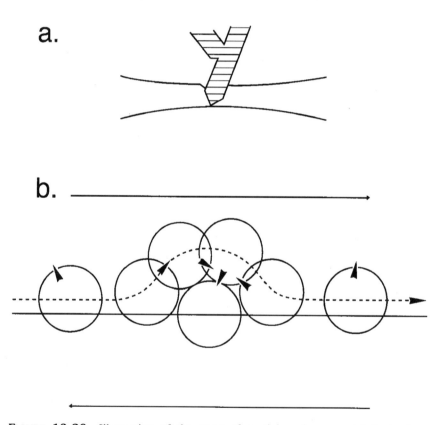

FIGURE 13.20 Illustration of the onset of partial coalescence. (a) Protruding crystal piercing the film between approaching globules. (b) Enhanced probability for piercing of the film when globules roll around each other in a shear field.

neck containing a crystal. Often the junction becomes larger and firmer upon aging.

Consequences. Partial coalescence differs from true coalescence in two important respects.

(a) The process tends to be very much *faster* than true coalescence, especially if the emulsion is agitated. For emulsions with and without crystals, but otherwise similar, coalescence rates often differ by a factor of 10^6 or even more. Moreover, the process proceeds like *orthokinetic aggregation* rather than coalescence. At rest, partial coalescence generally does not occur, although it may happen if the globules are closely packed. An example is mayonnaise, where $\varphi \approx 0.8$; when it is kept in a refrigerator, some crystallization may occur in the oil droplets (depending on oil composition), leading to partial coalescence.

(b) Since *irregular aggregates* or clumps are formed, the effective volume fraction of the disperse phase increases. This causes an increase in viscosity. In some cases, a space filling network of globules may even be formed, although large aggregates are often broken up again during agitation. Increasing the temperature of the aggregated system leads to melting of the crystals and coalescence of the clumps into large droplets.

Partial coalescence is quite common for milk fat globules. It leads to the formation of butter granules during churning of cream. Also in the whipping of cream, and in the beating-annex-freezing of ice cream mix, partial coalescence is essential, as it leads to the formation of a space filling network. In these cases, the process is more complex because air bubbles are generally present.

Aggregation Rate. We will consider primarily the orthokinetic rate, since for the globule sizes involved (generally not below 1 μm) it will be much faster than perikinetic aggregation; see Section 13.2.2. The rate is the product of the encounter frequency and the capture efficiency α. The latter parameter can in principle be determined as the ratio of the observed aggregation rate over the calculated one; the latter would roughly be given by Eq. (13.8) (although a more sophisticated treatment is to be preferred). Values of α between 10^{-6} and about 1 have been observed. The rate of partial coalescence depends on a great number of variables. This is illustrated by Figure 13.21; the presumed underlying mechanisms are discussed below.

1. **Volume fraction of globules**. Other things being equal, one expects the encounter rate, and thereby the aggregation rate, to be proportional to φ^2. This is indeed observed (frame a). For $\varphi > 0.2$ the aggregation rate tends to increase more strongly with φ.

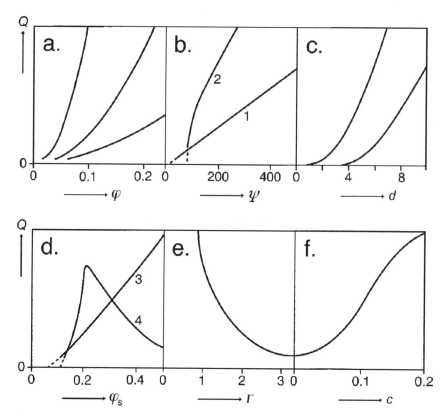

FIGURE 13.21 Effect on the rate of partial coalescence (Q) in protein-stabilized O–W "emulsions" of the variables: (a) volume fraction (φ); (b) shear rate (Ψ/s^{-1}); (c) globule diameter ($d/\mu\text{m}$); (d) fraction of the fat being solid (φ_S); (e) protein surface load ($\Gamma/\text{mg}\cdot\text{m}^{-2}$); and (f) concentration of small-molecule surfactant added ($c/\%$). Only meant to illustrate trends. In some frames results for different systems are shown.

    2. **Agitation**. This has a strong effect, as illustrated in frame (b). The following mechanisms play a part. (a) The *encounter frequency* is proportional to the velocity gradient (shear rate) Ψ. For some systems nearly the same holds for the aggregation rate: see curve 1. (b) The chance that a protruding crystal does indeed reach the film between two globules is greatly enhanced by the globules *rolling* over each other, as they do in a shear field: see Figure 13.20b. The effect is substantial, but it does not depend on the magnitude of Ψ. (c) Generally, the globules repel each other,

and the thickness of the film between approaching globules may then remain greater than the distance over which crystals protrude. A higher Ψ value leads to a greater shear stress, *forcing the globules together*. A threshold value of Ψ for partial coalescence to occur is often observed; above that value, the coalescence rate is about proportional to Ψ: see curve 2. The threshold value and the magnitude of α at higher Ψ vary considerably with composition and structure of the system. For the same shear stress, *larger globules* are pressed together with a greater force and will thus be less stable, which may explain part of the trend in frame (c).

Subjecting the system to *turbulent flow* leads to even faster aggregation.

3. **Protruding crystals**. Actually, it concerns two variables: how many crystals protrude and how large they are. The number is strongly dependent on the fraction solid of the fat (φ_s); cf. frame (d). If φ_S is below a certain threshold value, the crystals cannot form a network spanning the diameter of the globule. This follows from fractal aggregation theory: see Formula (13.18). In practice, the threshold value is of the order of 0.1. Below that value, crystals will generally not protrude very far, and they may even be pushed into the oil on encountering another globule. Presumably, protrusion does especially arise from the shrinking of the globule due to ongoing crystallization after the network has formed. As more fat crystallizes, more crystals will protrude and over a greater distance. The distance may also depend on the size of the crystals in the network. Several other factors will determine the results, including triglyceride composition and crystallization history, which primarily implies temperature history; see Section 15.4. Exceptionally, crystals may grow out of the globules, making very long protrusions. In such a case, the system is extremely unstable: merely pouring it causes immediate formation of very large lumps or even a solid network of globules; presumably $\alpha \approx 1$.

Another factor affecting the effective number of protruding crystals is the *globule size*. This is because the area of the film between approaching spheres is proportional to sphere diameter; see Figure 13.15b. Moreover, large globules may contain larger crystals that protrude farther. The effect of globule size on partial coalescence rate is always large; see frame (c).

4. **Contact angle**. To allow wetting of the crystal by the oil in the second globule, the contact angle θ should be between 90 and 180 degrees. If crystals have the freedom to move to an equilibrium position in the O–W interface, the three interfacial tensions will determine the contact angle: see Eq. (10.10). A smaller θ value will mean farther *protrusion* of the crystal into the aqueous phase, as is illustrated in Figure 10.25. It has indeed been observed that the partial coalescence rate increases when one adds more sodium dodecyl sulfate (as in Figure 10.25) prior to crystallization. For

proteins as the surfactant, generally θ is in the range 120° to 150°. The effect of adding small-molecule surfactants on the aggregation rate, as illustrated in frame (f), may in part be due to a decrease in θ.

5. **Colloidal repulsion**. Strong repulsion between globules will stabilize them against partial coalescence, although a large shear stress may overcome the repulsion barrier, as discussed in point 2. Proteins can provide strong repulsion, although it depends, of course, on the surface load (illustrated in frame e). Moreover, repulsion will depend on such factors as pH, ionic strength, and solvent quality. Small-molecule surfactants may displace proteins from the globules (Section 10.3.3), which can decrease repulsion. This must be an important part of the explanation for the relation depicted in frame (f).

6. **Permanence of junctions**. Junctions just formed can be broken again by the shear stress acting on the doublet, thereby effectively decreasing the capture efficiency. Presumably, the strength of a junction depends on its diameter, hence on the amount of oil contributing to it. If no oil is present, a junction cannot be formed, hence almost fully solidified fat globules will not show partial coalescence. But even at a far smaller fraction solid, it may be difficult to get oil out of the crystal network. This will largely depend on the *size of the pores* in the network, which depends, in turn, on original crystal size. Considering a fat globule for the moment as a rigid sponge filled with oil, local removal of oil from the sponge will cause the oil–water interface in the pores at the outside of the sponge to become curved. This causes a negative Laplace pressure that will resist the removal of oil, the more so for smaller pores.

This mechanism may explain the shape of curve 4 in frame (d), which refers to milk fat, a fat with notoriously small crystals (length often < 1 µm). At higher fractions solid, the pores will then be smaller and the network more rigid, preventing sufficient oil from becoming available to make a strong enough junction. Curve 3 refers to a fat with far larger crystals, and then the aggregation rate keeps increasing up to high values of φ_S.

Kinetics. We have seen now that many variables affect the rate of the process, but the actual situation is even more complex. In some systems, subjecting the liquid to a well-defined shear field leads to a gradual increase of the average particle size (determined after heating to melt the clumps). In this way, a partial coalescence rate can be unequivocally established. Other systems may show a very different pattern. The largest globules are especially prone to partial coalescence, and the resulting aggregates even more. Large clumps are formed that rapidly cream, and heating the liquid leads to the formation of an oil layer, whereas the liquid beneath has a decreased fat content and a decreased average globule size. Moreover, all situations intermediate between the two mentioned can occur. This greatly complicates the determination of an aggregation rate. The reader may have noticed that the graphs in Figure 13.21 do not have a scale for the rate. This is because it would mean something different in different systems. Nevertheless, the trends given have been consistently observed.

Conclusion. Whereas O–W emulsions with small globules (say $< 5\,\mu m$) that are covered with protein are very stable to coalescence, partial coalescence may readily occur if the oil in the globules becomes partly crystalline. The phenomenon is a good example of the complexity of stability problems that can be encountered in food systems. There are so many variables that it is generally not possible to predict quantitatively the rate of partial coalescence. Nevertheless, it is also a good example in that it shows how systematic research, making use of the fundamentals of colloid and surface science, can lead to the unraveling of such a complex problem.

13.6 OSTWALD RIPENING

Most dispersions are polydisperse. According to the **Kelvin equation** (10.9), the solubility of the material making up a particle is greater for a smaller particle. For convenience, the equations are repeated here:

$$\frac{s(a)}{s_\infty} = \exp\left(\frac{x'}{a}\right)$$

$$x' = \frac{2\gamma V_D}{RT} \tag{13.32}$$

where s is solubility, γ is interfacial tension, and V_D is the molar volume of the material in the disperse phase (it equals molar mass over density). If the particle material is at least a little soluble in the continuous phase, part of it will be transported from a smaller to a larger particle, because its concentration in the continuous phase will be larger near a small than near a large particle. The driving force is a *difference in chemical potential* of the particle material caused by the difference in radius. The transport phenomenon can be called *isothermal distillation*, where the molecules move by *diffusion*. The result is a *disproportionation* of the particle size: large particles become larger and small ones become smaller or even disappear. The whole process is called *Ostwald ripening*.

The rate of the process depends on the driving force, hence on the ratio of the characteristic length x' over the particle radius. Some typical examples are given in Table 10.4. The rate is, moreover, proportional to the bulk solubility of the material s_∞ and to its diffusion coefficient D in the continuous phase. Hence a significant solubility is a prerequisite for Ostwald ripening to occur at a perceptible rate. This change in dispersity is especially important in foams (Section 13.6.1), and it can also occur in emulsions (13.6.2). Several kinds of suspensions show it as well, especially small crystals in a saturated solution. Such suspensions are not very common in foods; we will briefly come back to them in Chapter 15.

Ostwald ripening is almost the only change in dispersity that proceeds *faster for smaller particles*: the excess solubility is roughly inversely proportional to particle diameter, and there is generally no free energy barrier. Since the average particle size increases, the process will proceed ever slower. Except in foams, the process is generally not important if the particle radii are over, say, $10 \, \mu m$.

13.6.1 Foams

De Vries Theory.   Most foams have a high volume fraction, and a small bubble will generally be rather close to several large ones: see Figure 11.1. De Vries derived an equation for the disappearance of such a small bubble, assuming it to be separated by an average distance δ from bubbles with negligible curvature. Figure 13.22a illustrates the assumed concentration gradient. Furthermore, de Vries used the Laplace pressure ($p_L = 2\gamma/a$) rather than the Kelvin equation; since the solubility of a gas is proportional to its pressure, the same relation is obtained (for ideal solubility behavior). Then the small particle will shrink with time t according to

$$a^2(t) = a_0^2 - \frac{RTDs_\infty \gamma}{p\delta} t \qquad (13.33)$$

where D is the diffusion coefficient of the gas in water (about $1.5 \cdot 10^{-9} \, m^2 \cdot s^{-1}$ at room temperature), and p is atmospheric pressure (about $10^5 \, Pa$). When putting $a^2(t) = 0$, t equals the lifetime of a bubble of initial radius a_0.

Some sample calculations may be enlightening. Assume that the gas is either N_2 ($s_\infty \approx 7 \, \mu mol \cdot m^{-3} \cdot Pa^{-1}$) or CO_2 ($s_\infty \approx 0.4 \, mmol \cdot m^{-3} \cdot Pa^{-1}$), that the surface tension is $0.05 \, N \cdot m^{-1}$, that the original bubble radius is $50 \, \mu m$, and that δ is $20 \, \mu m$. We then calculate for a nitrogen bubble a lifetime of $3800 \, s$, or about one hour, and for a carbon dioxide bubble $66 \, s$, or about a minute. For a nitrogen bubble of $10 \, \mu m$ and a δ value of $5 \, \mu m$, the lifetime would be $38 \, s$. Small bubbles can thus disappear very fast, especially if the gas is highly soluble, as CO_2 is, which is common in several foods. Experimental results on single bubbles agree reasonably well with the

a.

b.

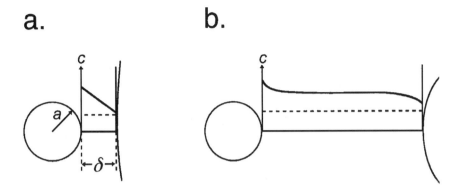

FIGURE 13.22 Concentration c of the particle material in the continuous phase as a function of the distance between two particles of different radii. The broken line gives the concentration corresponding to s_∞. Concentration profile according to de Vries (a) and according to SLW theory (b).

predictions. Equation (13.33) further implies that the shrinkage of a bubble goes ever faster, so that it more or less implodes at the end; see Figure 13.23, curve 2.

Surface Dilational Properties. The calculations just given suggest that all foam will rapidly disappear, but several foams can be fairly persistent, and some can hardly be destroyed. Stabilization to Ostwald ripening (or to disproportionation as foam researchers usually call it) is thus possible.

Consider a shrinking bubble with a surfactant layer. The decrease in radius means a decrease in surface area, hence an increase in surface load, hence a *decrease in surface tension* (cf. Section 10.2.3). This implies a mechanism that counteracts the increase in Laplace pressure. The decrease in γ is expressed in the surface dilational modulus $E^{SD} \equiv d\gamma/d\ln A$ [Eq. (10.20)]. According to Gibbs, the following condition now determines whether the bubble will be prevented from shrinking:

$$\frac{dp_L}{da} = \frac{d(2\gamma/a)}{da} = \frac{2a(d\gamma/da) - 2\gamma}{a^2} = \frac{4E^{SD} - 2\gamma}{a^2} \geqslant 0 \tag{13.34}$$

where use has been made of the relation $d\ln A = (2/a)\, da$. This simply comes down to

$$E^{SD} \geqslant \frac{\gamma}{2} \tag{13.34a}$$

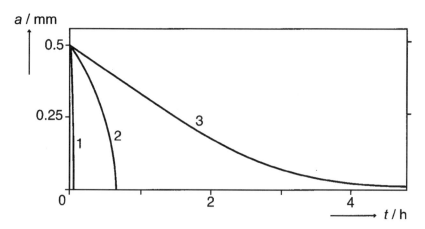

FIGURE 13.23 Shrinkage of a single CO_2 bubble. Radius of bubble (a) as a function of time (t) in three liquids. Curve 1: water, $\eta^{SD} = 0$. Curve 2: beer sample, $\eta^{SD} \approx 0.01$ (d ln A/dt)$^{-0.9}$. Curve 3: beer sample, $\eta^{SD} \approx 0.08$ (d ln A/dt)$^{-0.9}$ (SI units). (After results by A. D. Ronteltap, A. Prins. Colloids Surf. 47 (1990) 285.)

This relation only holds for bubbles differing very little in size; in practice, the condition $E^{SD} > 2\gamma$ is more realistic. As long as this condition is fulfilled, no gas will dissolve from the bubble, so it will not shrink; moreover, no excess gas is available for large bubbles to grow. Even for $E^{SD} = \gamma$, Ostwald ripening tends to be quite slow.

Several surfactants cause a large enough modulus, but the problem is that E^{SD} *tends to decrease with time*; see, e.g., Figure 10.34. The main reason is that the decrease in surface tension implies an increase in surface pressure, and surfactant will thus desorb. For many small-molecule surfactants, E^{SD} will be virtually zero during bubble shrinkage. For proteins, however, the rate of decrease in modulus is often quite slow. It can then be useful to invoke the (apparent) *surface dilational viscosity* $\eta^{SD} \equiv \Delta\gamma/(\mathrm{d} \ln a/\mathrm{d}t)$ [Eq. (10.21)]. Its magnitude can be experimentally determined, and by combining the observed relation with Eq. (13.33), the shrinkage can be calculated (although there is no analytical solution). Some experimental results for single bubbles are given in Figure 13.23, and they agree well with the prediction. It is seen that a high surface viscosity can considerably retard Ostwald ripening. Moreover, "implosion" of the bubble then does not occur: after shrinkage to a certain radius, the shrinkage goes ever more slowly.

Stopping Ostwald Ripening. This is possible by giving the A–W surface a large and permanent E^{SD} value. This can be achieved by cross-linking of protein in the adsorption layer, or by using large protein species, be they molecules or aggregates. This is what happens when one beats egg white, where aggregation and cross-linking occur owing to surface denaturation of ovalbumin. Small *solid particles* of suitable contact angle, that adsorb to give a packed layer, will also prevent shrinkage. The configuration is as depicted in Figure 13.18. A good example is classical whipped cream, where the air bubbles (diameter about 50 µm) are fully covered by largely solid fat globules (about 4 µm). In passing, the fat globules will also partially coalesce during whipping, giving rise to clumps of larger size, which also form a space filling network, lending stiffness to the whipped cream. Some whipped toppings contain a surfactant mixture, such as glycerol lactopalmitates, that tend to give a stiff and persistent α-gel layer around the bubbles.

Another way to make a permanent foam is by giving the continuous phase a *yield stress*. A bubble of 10 µm in diameter has a Laplace pressure of about 20 kPa (0.2 bar), and to prevent the bubble from shrinking, the yield stress should be larger. This value is so large that the material is to be considered a solid. There are solid foods that have permanent small bubbles, such as some types of chocolate. If the material is solid, the system will also be stable if the bubbles are interconnected, i.e., form a sponge rather than a foam, as in bread.

Finally, it should be mentioned that the main instability of *food foams* is generally Ostwald ripening, rather than coalescence. This is because the bubble diameter is relatively small, mostly between 10 and 100 µm. Larger bubbles would give a too weak consistency to be acceptable in a food. For the same reason, one generally tries to prevent strong drainage, which then means that the films between bubbles are not very thin. It is especially large and thin films that are sensitive to rupture, and small bubbles that are prone to Ostwald ripening. All kinds of change in a foam, i.e., film thinning, coalescence, and Ostwald ripening, occur fastest at the top of a foam (can you explain these observations?).

Question

Coming back to the question at the end of Section 13.4.3, can you think of an alternative explanation for the observation that an increased ethanol concentration decreases beer foam stability?

13.6.2 Emulsions

LSW Theory. In most emulsions, the distance between drops is far larger than in foams, which makes the de Vries theory less suitable. Often the LSW theory is used (independently developed by Lifshitz and Slyozov and by Wagner). It is presumed (a) that the concentration of the droplet material in the continuous phase is higher than s_∞ and attains a magnitude corresponding to $s(<a>)$, where $<a>$ is the number average droplet radius; and (b) that concentration gradients only occur near the droplets. These assumptions are illustrated in Figure 13.22b. It can be derived that Ostwald ripening now leads to a size distribution of standard shape, which is thus time invariant. After the distribution has attained this shape, the rate of the process would be given by

$$\frac{d<a>^3}{dt} = \frac{4x'Ds_\infty}{9\rho_D} \tag{13.35}$$

where s is expressed in $kg \cdot m^{-3}$. The theory is well obeyed at very small φ, but for higher φ the rate is faster than predicted (say by a factor of 2 or 3 for $\varphi = 0.2$). This is because the assumption about the concentration gradient does not hold any more and because Brownian motion of the drops is neglected. On the other hand, the rate can be much slower owing to the surfactant layer producing a considerable surface dilational viscosity, as discussed in the previous section.

Most food **oil-in-water emulsions** do not show any Ostwald ripening for the simple reason that the solubility of triglyceride oil in water is negligible. Some plants, especially citrus fruits, contain essential oils, which consist for a considerable part of various terpenes, which are soluble in water. Emulsions of such oils, which are used as flavoring agents, can show distinct Ostwald ripening.

Water is somewhat soluble in triglyceride oils, about $1.3 \, kg \cdot m^{-3}$ at room temperature, and this may cause significant Ostwald ripening in **water-in-oil emulsions**. For example, take pure water droplets of $2 \, \mu m$ diameter in pure triglyceride oil at $20°C$. Then $\gamma_{OW} \approx 0.03 \, N \cdot m^{-1}$ and $D \approx 10^{-11} \, m^2 \cdot s^{-1}$. The time needed for doubling the average droplet volume is given by $<a>^3$ over the right-hand side of Eq. (13.35), which would lead to a value of about 4 days. This is not negligible in practice. (At higher temperature, the rate will be markedly higher, by a factor of about 10 at $50°C$, due to increased water solubility and diffusion coefficient.) In margarine, Ostwald ripening tends to be far smaller for a number of reasons. One is that surfactants are present and that the magnitude of γ_{OW} will thus be lower ($5–10 \, mN \cdot m^{-1}$). A second reason is that solid particles, i.e., fat crystals, may give a kind of Pickering stabilization. A third reason is discussed below.

Compound Droplets. If the droplets contain various components that differ significantly in solubility, Ostwald ripening is slowed down. The reason is that a less soluble compound will leave a small droplet at a slower rate than a more soluble one, which implies that the concentration, and hence the chemical potential, of the less soluble compound in the drop increases. This produces a driving force for the more soluble compound to diffuse back to the small droplet. This would play a part in slowing down Ostwald ripening in the essential oil emulsions mentioned.

Ostwald ripening will even stop if the droplet contains a solute that is fully insoluble in the continuous phase, provided that its concentration exceeds a critical magnitude. This can occur for droplets of an NaCl solution in oil. It is simplest to consider the osmotic pressure of the salt solution, given by

$$\Pi_{osm} = mRT = m_0 \left(\frac{a_0}{a}\right)^3 RT \tag{13.36}$$

assuming ideal behavior. Here m is the osmolarity of the solution in the drop, and the subscript 0 refers to the initial situation. The drop will not shrink if the change in Laplace pressure due to a change in radius is equal to that in osmotic pressure (not opposite and equal, since osmotic pressure is in fact a negative pressure). After differentiation of p_L and Π_{osm} with respect to a, the following condition results;

$$m > \frac{2\gamma}{3aRT} \tag{13.37}$$

Taking as an example a droplet of radius 0.1 μm and $\gamma = 0.01$ N · m^{-1}, the minimum osmolarity should be at least 27 mol · m^{-3}. For the almost fully dissociated NaCl this corresponds to less than 0.1% NaCl. The water droplets in butter and margarine always contain sufficient salt fully to prevent Ostwald ripening.

Altogether, Ostwald ripening is rarely a problem in food emulsions, apart from soft drinks containing essential oils.

13.7 RECAPITULATION

Various types of instability can cause changes in dispersity, i.e., in size or in arrangement of the particles and often in both. These types are illustrated in Figure 13.1, and some particulars are summarized in Table 13.2. Some

TABLE 13.2 Summary of the Changes in Dispersity Resulting from the Various Processes Discussed

Process	Consequences	Reversible?	Remarks
Growth/dissolution	d increases/decreases	By changing concentration	Small particles disappear
Ostwald ripening	d increases	No	Often, especially large particles grow
Coalescence	d increases	No	$\rightarrow$ Particles of irregular shape; η increases
Aggregation	d increases	In principle[a]	Yield stress results
	Continuous network forms	In principle[a]; agitation needed	
Partial coalescence	d increases	Generally not	Irregular lumps form; heating $\rightarrow$ coalescence
Sedimentation	Demixing	By agitation	Rate enhanced by increase of d

[a] Depending on the colloidal interaction forces involved, alter the properties of the continuous phase: pH, ionic strength, solvent quality, dilution.

changes can reinforce other instabilities: an increase in particle size leads to enhanced sedimentation, and coalescence of fluid particles is greatly enhanced if they are for a long time close together—as in an aggregate or a sediment. If a remedy has to be found for a change in dispersity, the type of instability has to be established first; microscopy will often be useful, and particulars like those given in Table 13.2 will also give clues. A way to prevent coalescence, aggregation, and sedimentation is arresting particle motion; this can generally be achieved by giving the continuous phase a small yield stress. Except for Ostwald ripening and growth from solution, a change in dispersity nearly always proceeds faster for larger particles.

Aggregation of particles can only occur after they encounter each other, and this can be due to Brownian motion (*perikinetic* aggregation) or by a velocity gradient (*orthokinetic* aggregation). The rates of these processes can be predicted, especially if the capture efficiency upon encounter is unity; the process is then called fast aggregation. In practice, slow aggregation often prevails because of colloidal repulsion and hydrodynamic interaction. In some cases, the capture efficiency can be predicted from theory. If the particles are large, say over 1 μm, orthokinetic aggregation tends to be faster than perikinetic, even for the small velocity gradients that are induced by (small) temperature fluctuations.

The aggregates have an irregular shape (unless the particles immediately coalesce), and *fractal clusters* mostly emerge when aggregation is not disturbed by stirring or sedimentation. Fractal structures are scale-invariant and are more tenuous if larger. This implies that the joint clusters will take up an ever greater part of the volume as they increase in size, until a network spanning the whole volume is formed, resulting in a *particle gel*. The essential parameter governing the properties of the aggregates is the fractal dimensionality, which is the (average) proportionality factor between log(number of particles) and log(radius) of the clusters. Its magnitude ranges between 1.7 and 2.4, depending on several conditions.

Aggregation time, i.e., the time needed for a visible change to occur, can vary greatly, even for the same aggregation rate. A visible change may be the emergence of large particles, the formation of a sediment, or gelation. It is usually one of the latter two, and gelation then tends to give the shorter time. This is because fractal aggregation needs relatively few aggregation steps before a gel is formed, as compared to the formation of dense aggregates. Calculation of aggregation time is difficult, but some trends can be well predicted.

Sedimentation can be settling or creaming, depending on the sign of the density difference between particles and surrounding liquid. Theory for sedimentation of a *single sphere* is well developed. The simple *Stokes equation* can be used if a number of conditions is fulfilled. The most important one is that the liquid shows Newtonian behavior. Many liquids (e.g., most polymer solutions) are markedly strain rate–thinning, and the shear stress on a sedimenting particle then is (far) smaller than the stress prevailing in most viscometers. The sedimentation rate as predicted from the determined apparent viscosity then is greatly overestimated. Sedimentation in a *whole dispersion* is more complex. At high volume fractions, sedimentation is greatly slowed down. Polydispersity causes the sedimentation profile (particle concentration as a function of height) to evolve in an intricate manner, especially when aggregation of particles occurs; sedimentation rate then is strongly enhanced.

The critical event in **coalescence** is the rupture of the film between close fluid particles. A hole can spontaneously form in a film, but in the presence of some surfactant this can only happen if the film is very thin (a few nm). Local thinning of a film may occur because of the development of capillary waves on the film surfaces. These waves are readily damped if the interfacial tension is large, the repulsion between the particles extends over a large distance, and the film is small. This then implies that protein is a very good stabilizer against film rupture, far more so than many small-molecule amphiphiles.

For coalescence of **emulsion droplets**, an important variable is whether a flattened film between the droplets is formed. This is governed by the ratio of the external stress over the Laplace pressure. The external stress can be due to colloidal attraction (e.g., van der Waals forces), a shear stress, or gravitational forces in a sediment layer. Small protein-stabilized droplets will not deform, except in a sediment layer in a centrifuge, and they are very stable to coalescence. If the drops are large, the interfacial tension is low, and the external stress is high, droplets will deform and coalescence can readily occur. Water-in-oil emulsions cannot be made with protein as the surfactant, and it is often difficult to stabilize them against coalescence, except by a layer of small hydrophobic particles (Pickering stabilization).

Most **foams** are far less stable to coalescence than most emulsions. This is because foam bubbles are large. This means that (a) the number of bubbles is small, implying that relatively few coalescence events are needed to produce a very coarse foam; and (b) the films are large and flat and thereby relatively unstable to rupture. However, in most foods it takes a while for films to drain to a thickness allowing rupture. Fairly thick films can also rupture, but that is due to the presence of (extraneous) small hydrophobic particles; some mechanisms have been proposed to explain the film rupture.

Fat globules, i.e., oil droplets that contain fat crystals, can be subject to **partial coalescence** or clumping. A crystal may protrude somewhat from the globule surface and can pierce the film between two approaching globules. It will then be wetted by oil from the other globule, and oil–oil contact is made, without full coalescence occurring. Irregular clumps are formed that will fully coalesce upon melting of the fat crystals. Partial coalescence is very much enhanced in a shear field and can then readily be a million times faster than the coalescence of the same droplets but without crystals. Several variables affect the rate of the process, and the results mostly cannot be predicted, although the trends are clear. If the globules are quite small and are coated with protein, they tend to be stable. Addition of a small-molecule surfactant, which tends to displace (part of) the adsorbed protein, will then induce instability.

Agitation or, more precisely, the presence of a velocity gradient thus greatly enhances the partial coalescence rate. It can also speed up aggregation. On the other hand, particle aggregates can be disrupted, and the formation of a "fractal" gel can be prevented. Agitation strongly disturbs sedimentation.

According to the **Kelvin equation**, the solubility of the material in a small particle is enhanced over the equilibrium value because of the particle curvature. The smaller the particle, the higher the solubility, leading to diffusion of material from small to large particles, which is called **Ostwald ripening**. The result is that small particles disappear and large ones grow. Prerequisite is that the material of the particles be at least slightly soluble in the surrounding liquid. Ostwald ripening can occur in suspensions, emulsions, and foams.

It is often the most important instability in *foams*, because gases are relatively soluble in water. Small CO_2 bubbles may disappear within a minute, air bubbles within an hour. Ostwald ripening will stop if the surface dilational modulus of the bubble surface is larger than the surface tension. Proteins are suitable to stabilize against Ostwald ripening because they generally give a high modulus. However, the modulus tends to decrease slowly in time, so that Ostwald ripening is only retarded, not prevented. A highly insoluble protein layer, as is formed during whipping of egg white, can virtually stop the process.

Triglyceride oil-in-water *emulsions* do not show Ostwald ripening: the oil is insoluble in water. Water-in-oil emulsions can readily show it, but then it can be stopped by providing the aqueous phase with some salt. Shrinking of a bubble then increases its salt content, and hence its osmotic pressure, which counteracts the pressure increase due to the increase in curvature.

Question

A food company makes a product that is an oil-in-water emulsion. To check on its physical stability, a sample of each day's production is stored overnight in a thermostat and it should then not show visible creaming. One day, significant creaming is observed. Make a list of the possible causes and also give the methods used to establish which of the possibilities was the real cause.

BIBLIOGRAPHY

Most texts on colloid science give ample information on some of the changes discussed, especially on aggregation. The classical treatment of aggregation kinetics still is

J. T. G. Overbeek. Kinetics of flocculation. In H. R. Kruyt, ed. Colloid Science. Vol. I. Irreversible Systems. Elsevier, Amsterdam, 1952, Chapter 7.

A monograph on various aspects of aggregation is

K. J. Ives, ed. The Scientific Basis of Flocculation. Sijthoff and Noordhoff, Alphen aan de Rijn, 1978.

Especially the chapters by Ives on rate theories (pp. 37–61) and by L. A. Spielman on hydrodynamic aspects of flocculation (pp. 63–88) are recommended. An extensive and detailed treatment is in

T. G. M. van de Ven. Colloidal Hydrodynamics. Academic Press, London, 1988.

This especially stresses the hydrodynamic aspects of aggregation and coalescence. An overview of fractal aggregation is

P. Meakin. Fractal aggregates. Adv. Colloid Interface Sci. 28 (1988) 249–331.

The effects on aggregation times of fractal aggregation and of other particulars of the process are treated by

L. G. B. Bremer, P. Walstra, T. van Vliet. Estimations of the aggregation times of various colloidal systems. Colloids Surf. A 99 (1995) 121–127.

A discussion on the various types of instabilities in emulsions is

P. Walstra. Emulsion stability. In: P. Becher, ed. Encyclopedia of Emulsion Technology. Vol. 4. Dekker, New York, 1996, Chapter 1.

Emulsion stability is also treated in

B. P. Binks, ed. Modern Aspects of Emulsion Science. Roy. Soc. Chem., Cambridge, 1998.

Especially Chapters 1 by B. P. Binks (introduction), 7 by A. S. Kabalnov (coalescence), and 9 by J. G. Weers (Ostwald ripening and related phenomena) are recommended.

Coalescence in emulsions, with an emphasis on engineering applications, is in

A. K. Chesters. Coalescence Probability. Trans. Inst. Chem. Eng. 69 (1991) 259–270.

Useful information is also in the textbook

E. Dickinson. An Introduction to Food Colloids. Oxford Univ. Press, Oxford, 1992.

Especially Chapter 5 on foams is recommended. Foam stability is reviewed by

A. Prins. Principles of Foam Stability. In: E. Dickinson, G. Stainsby, eds. Advances of Food Emulsions and Foams. Elsevier Appl. Sci., London, 1988.

An authoritative monograph, going into great detail, but almost leaving out polymeric surfactants, is

D. Exerova, P. M. Kruglyakov. Foams and Foam Films. Elsevier, Amsterdam, 1998.

14

Nucleation

When the temperature, the pressure, or the solute concentration of a homogeneous system is changed, a new phase can be possibly formed. Examples are the formation of a vapor phase when a liquid is heated to above its boiling point and the formation of sucrose crystals when a sucrose solution is cooled to below its saturation temperature. The formation of a new phase is often very slow if the conditions are not far from equilibrium. For the new phase to develop, nucleation has to occur, i.e., the formation of small regions of the new phase that are large enough to grow spontaneously. This is the subject of this chapter. The emphasis is on nucleation of a solid phase in a liquid. Unless mentioned otherwise, the ambient pressure is assumed to be constant at about 1 bar.

Most basic aspects are to be found in Chapters 2 and 10, especially Sections 10.5.3 and 10.6.1.

14.1 PHASE TRANSITIONS

Assume the existence of a phase α which tends upon cooling to change (wholly or partly) into a phase β. At one temperature T_{eq} there is equilibrium between the two phases; this implies that the free energy G of the material in both phases will be equal. At $T < T_{eq}$ the value of G will be

548

smaller for phase β than for phase α. This is illustrated in Figure 14.1a. Remembering that $G = H - T\Delta S$, we derive that

$$\Delta_{tr}G = G(\beta) - G(\alpha) = \Delta_{tr}H - T\Delta_{tr}S \qquad (14.1)$$

Since at $T = T_{eq}$ we have $\Delta_{tr}G = 0$, it follows that

$$\Delta_{tr}H = T_{eq}\Delta_{tr}S \qquad (14.2)$$

Combination of (14.1) and (14.2) now gives

$$\Delta_{tr}G = \Delta_{tr}H\left(1 - \frac{T}{T_{eq}}\right) = \Delta_{tr}S(T_{eq} - T) \qquad (14.3)$$

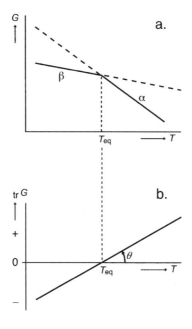

FIGURE 14.1 (a) Free energy G of a material when present in two different phases, α and β, as a function of temperature T; the broken lines are extrapolations beyond the temperature T_{eq} at which both phases are in equilibrium. (b) The change in free energy $\Delta_{tr}G$ upon transition from α to β as a function of temperature. Highly schematic.

This is illustrated in Figure 14.1b, where $\tan \theta = -\Delta_{tr}H/T_{eq}$. It has been implicitly assumed that ΔH and ΔS are independent of temperature; for most systems this is about correct if the temperature range is small (say, a few times 10K). Values of $\Delta_{tr}H$ can generally be obtained by calorimetry. (Note that in the present case (Fig. 14.1, $\alpha \to \beta$) the values of ΔH and ΔS are both negative; for the transition $\beta \to \alpha$, they would both be positive, with $T\Delta S > \Delta H$.)

It thus appears at first that at temperatures even slightly below T_{eq}, phase β will spontaneously form. This however is not the case, since the formation of a region of another phase introduces a phase boundary, which goes along with a specific interfacial free energy $\gamma_{\alpha\beta}$. This introduces a term ΔG^S that is positive. For a very small region of the new phase—which is called an *embryo*—the surface term will generally be dominant. This is because $\Delta_{tr}G$ is proportional to volume ($\pi d^3/6$) and ΔG^S proportional to surface area ($\pi d^2/4$). An example may illustrate this.

In the freezing of water, $\Delta_{tr}H = -330\,\text{MJ} \cdot \text{m}^{-3}$; the interfacial tension γ between water and ice $= 25\,\text{mN} \cdot \text{m}^{-1}$ (Table 10.1). Assuming these values to be correct at $-10°C$, and considering a spherical embryo of ice of diameter 5 nm (which would contain about 2000 molecules), we calculate from Eq. (14.3) for $\Delta_{tr}G$ a value of $-8 \cdot 10^{-19}$ J and for ΔG^S $2 \cdot 10^{-18}$ J. This means that formation of the embryo would cause an increase in free energy by $12 \cdot 10^{-19}$ J, or about 300 times $k_B T$, which is extremely unlikely to occur. Even if such an embryo were present, it would immediately dissolve, since that will decrease the total free energy. An embryo that can and does grow to form a new phase is called a *nucleus*. To obtain a nucleus, a much deeper undercooling (supercooling) would be needed. This is further discussed in Section 14.2.1.

The reasoning given above applies to most kinds of phase transition, for instance also for $\beta \to \alpha$ in the same system. An overview (not exhaustive) is given in Table 14.1; it also specifies the equilibrium temperature and the transition enthalpy. It should be added that it here concerns so-called *first-order (phase) transitions*. In section 16.1 the difference between first- and second-order transitions will be discussed.

The formation of a new phase *inside a solid phase* is very difficult, because the transition generally implies a change in density, hence in volume. This leads to a change in pressure, and thereby to an additional, generally large and positive, term in free energy for nucleus formation. Except for some solid → solid transitions where the change in density is small, this tends to prevent nucleation; any formation of a new phase will occur at the boundary of the system. Sublimation then does not need nucleation: the new phase (gas) is already present. The same holds for a solid phase (crystals) in a solution. When crystals in air tend to melt, a liquid phase is not yet present; nevertheless, a phase transition occurs readily, as is explained in Section 14.2.2.

Transitions *inside a liquid phase* can be of various kinds. The whole liquid can freeze, or—in the case of a solution—the liquid can separate into a less concentrated solution and a phase made up by the solute. The latter phase may be either solid (crystals), a liquid (droplets), or a gas (bubbles). A liquid → liquid transition is rare in foods, except when it concerns aqueous polymer solutions.

Although deposition of material—especially water—from a *gas phase*, be it as a liquid or a solid, frequently occurs in foods, it seems never to involve a phase transition, since the other phase is already present.

> *Note* A stable dispersion of small solid or liquid particles may also show a kind of phase separation when conditions in the liquid are changed in such a way that attractive forces between the particles become dominant. A separation into a condensed phase (high volume fraction of particles) and a very dilute dispersion would then result. The interfacial tension between these phases is very small, e.g., a few $\mu N \cdot m^{-1}$. Conditions for this to occur are (a) that the particles are about monodiperse and of identical shape and (b) that the attractive forces do not become large (because that would lead to fractal aggregation; see Section 13.2.3). Since these conditions are rarely met in food systems, we will not further discuss the phenomenon. Nevertheless, phenomena like depletion flocculation (Section 12.3.3) show some resemblance to a phase separation.

Question

Would you consider the precipitation of protein from a solution, e.g., due to a change in pH, a (first-order) phase transition?

Answer

No. Precipitation generally involves the formation of irregularly shaped aggregates, and a layer of these aggregates forms due to sedimentation. A phase transition will only occur if protein crystals are formed.

TABLE 14.1 Various Kinds of Phase Transitions and the Role of Nucleation

Phase change	Nucleation?	Theory[a] useful?	Occurs in foods?	$\Delta_{tr}H =$	$T_{eg} =$
Solid→solid	Yes	No	?	Transition heat[b]	Transition point
Solid→liquid	No or yes	—	Yes	—	Melting point
Solid→gas	No	—	Yes	—	Sublimation point
Liquid→solid "freezing"	Yes	Yes	Yes	Heat of fusion	Melting point
Liquid→solid "saturation"	Yes	Yes	Yes	Heat of dissolution	Saturation point
Liquid→liquid "classical"[c]	Yes	Yes	Rare	Heat of dissolution	Saturation point
Liquid→liquid "polymers"[d]	Yes or no	No	Yes	—	—
Liquid→gas	Yes	No	Yes	—	Boiling point
Gas→liquid	Yes	Yes	No	Heat of evaporation	Boiling point
Gas→solid	Yes	Yes	No	Heat of sublimation	Sublimation point

[a]Theory as outlined in Section 14.2.
[b]The difference between the heats of fusion of the two phases.
[c]Two different liquids that are partly miscible (e.g., water and 1-butanol).
[d]The separation of an aqueous polymer solution into two aqueous phases.

14.2   NUCLEATION THEORY

Nucleation theory is presently in a state of confusion, and quantitative predictions of nucleation rate cannot generally be given. Nevertheless, the basic principles are enlightening, and trends can be well predicted. Experimental determination of the nucleation rate is discussed in Section 14.3.

14.2.1   Homogeneous Nucleation

In a homogeneous system, i.e., in the absence of phase boundaries, nucleation is said to be homogeneous. A new phase β can form if conditions (temperature, pressure, or solute concentration) are such that its chemical potential is smaller than that of phase α. It will then frequently happen that a cluster of molecules in phase α orient themselves by chance as they would be in phase β. If the new phase is a liquid, the solute (or vapor) molecules have merely to associate to form such an embryo; if the new phase is a solid, the molecules also have to attain a mutual orientation as in the crystals to be formed.

Embryos.   The excess free energy of a spherical embryo of radius r over that of the same volume of phase α is given by

$$\Delta G_{emb} = \frac{4}{3}\pi r^3 \Delta G^V + 4\pi r^2 \Delta G^S \tag{14.4}$$

where ΔG^V is the free energy change of the material as given by Eq. (14.3) expressed per unit volume. ΔG^S is the interfacial free energy per unit surface area between the two phases; it equals the interfacial tension γ. Assuming γ to be independent of temperature, which is not too far from reality for a small temperature difference, we obtain

$$\Delta G_{emb} = \frac{4}{3}\pi r^3 \Delta H^V \left(1 - \frac{T}{T_{eq}}\right) + 4\pi r^2 \gamma \tag{14.5}$$

Examples of the relation between ΔG_{emb} and the embryo radius r are given in Figure 14.2. As long as $r < r_{cr}$, the embryo tends to dissolve, since that gives a decrease in free energy. When $r > r_{cr}$, the embryo will spontaneously grow and hence become a nucleus. By differentiating Eq. (14.5) and putting the result equal to zero, which corresponds to $\Delta G_{emb} = \Delta G_{max}$, the value of the *critical radius for nucleation* r_{cr} is obtained:

$$r_{cr} = -\frac{2\gamma}{\Delta H^V(1 - T/T_{eq})} \tag{14.6}$$

Inserting this into Eq. (14.5) we obtain for the *maximum of ΔG*

$$\Delta G_{max} = \frac{16\pi\gamma^3}{3[\Delta H^V(1 - T/T_{eq})]^2} = \frac{4}{3}\pi r_{cr}^2\gamma \tag{14.7}$$

It may be noted that the value of ΔG_{max} equals 1/3 of the interfacial term in Eq. (14.5) for $r = r_{cr}$.

It follows that the lower the temperature, the smaller will be r_{cr}, as is illustrated in Figure 14.2. This can also be reasoned by use of the *Kelvin*

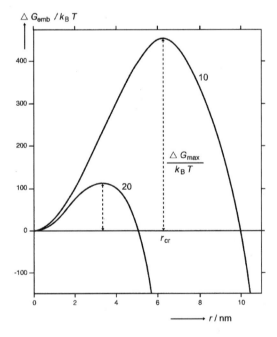

FIGURE 14.2   The excess free energy ΔG_{emb} of a spherical embryo of a new phase as a function of embryo radius r, calculated by use of Eq. (14.5) for two cases: undercooling by 10 or 20 K.

equation (10.9), which can be written as

$$RT \ln \left(\frac{s_r}{s_\infty} \right) = \frac{2\gamma V}{r} \tag{14.8}$$

where s is solubility and V molar volume. It gives the equilibrium radius for a very small particle of the new phase (an embryo), where s_r/s_∞ is the supersaturation ratio. The lower the temperature, the greater the supersaturation and the smaller the equilibrium radius. Putting the latter equal to r_{cr}, we can write the Kelvin equation as

$$\Delta\mu = RT \ln \beta = \frac{2\gamma V}{r_{cr}} \tag{14.9}$$

where $\Delta\mu$ is the difference in chemical potential, β is the *supersaturation ratio*, and $\ln \beta$ is called the *supersaturation*. Equation (14.9) can be shown to be identical to Eq. (14.6) in the present case. The larger the supersaturation is, the smaller the equilibrium value of r, i.e., r_{cr}. It may be noted that Eq. (14.9) has general validity, whatever the cause of the difference in chemical potential.

Nucleation Rate. The value of ΔG_{max} is often considered to be an activation free energy for nucleus formation. Moreover, it is generally assumed that there is also a certain barrier to incorporation of a molecule into an embryo, owing to diffusional resistance; this is usually expressed in a molar activation free energy for transport $\Delta G*$. Classical nucleation theory then stipulates that the nucleation rate (number of nuclei formed per unit time per unit volume) will be given by

$$\begin{aligned} J_{hom} &= N \frac{k_B T}{h_P} \exp\left(-\frac{\Delta G*}{RT} \right) \exp\left(-\frac{\Delta G_{max}}{k_B T} \right) \\ &= N \frac{k_B T}{h_P} \exp\left(-\frac{\Delta G*}{RT} \right) \exp\left[-\frac{16\pi\gamma^3 T_{eq}^2}{3(\Delta H^V)^2 (\Delta T)^2 k_B T} \right] \end{aligned} \tag{14.10}$$

where N is the number per unit volume of molecules that can form the new phase. As mentioned in Section 4.3.3, the factor $k_B T/h$ (of order $6 \cdot 10^{12}$ s^{-1}), where h_P is Planck's constant, is the maximum reaction rate possible, i.e., in the absence of free energy barriers.

Examples. Some examples of results, experimentally observed and interpreted by means of Eq. (14.10), are in Table 14.2. An important conclusion that can directly be drawn from the equation is that the second

TABLE 14.2 Examples of Homogeneous Nucleation Rates Observed for Different Systems

---

Freezing of water

$T_{eq} = 273.15 \text{K}$ $\Delta H^V = -330 \text{ MJ} \cdot \text{m}^{-3}$ $\gamma = 23.3 \text{ mN} \cdot \text{m}^{-1}$

This results in $J_{hom} = 10^{41} \exp\left[-45{,}000/(273.15 - T)^2\right]$

A better fit to the experimental results is

    $J_{hom} = 10^{36} \exp\left[-82{,}000/(273.15 - T)^2\right] \text{m}^{-3} \cdot \text{s}^{-1}$

Examples:

    At $-32°C$: $J_{hom} = 10 \text{ m}^{-3} \cdot \text{s}^{-1}$, i.e., 1 per liter per min

    At $-40°C$: $J_{hom} = 10^{14} \text{ m}^{-3} \cdot \text{s}^{-1}$, i.e., 1 per s in 10^4 μm^3

At $-40°C$, $r_{cr} = 1.85$, i.e., a nucleus of ~ 200 molecules.

Crystallization of sucrose from a saturated solution

$T_{eq} = 353 \text{ K}$ $\Delta H^V = -25 \text{ MJ} \cdot \text{m}^{-3}$ $\gamma = 5 \text{ mN} \cdot \text{m}^{-1}$

This results in $J_{hom} = 10^{39} \exp\left[-93{,}000/(353 - T)^2\right]$

A better fit to the experimental results is

    $J_{hom} = 10^{32} \exp\left[-33{,}000/(353 - T)^2\right] \text{m}^{-3} \cdot \text{s}^{-1}$

Examples:

    At $58°C$: $J_{hom} = 250 \text{ m}^{-3} \cdot \text{s}^{-1}$, i.e., 15 per liter per min

    At $52°C$: $J_{hom} = 10^{14} \text{ m}^{-3} \cdot \text{s}^{-1}$

At $52°C$, $r_{cr} = 2.4 \text{ nm}$, i.e., a nucleus of ~ 160 molecules.

Crystallization of tristearate from paraffin oil

$T_{eq} = 317 \text{ K}$ $\Delta H^V = -150 \text{ MJ} \cdot \text{m}^{-3}$ $\gamma = 10 \text{ mN} \cdot \text{m}^{-1}$

This results in $J_{hom} = 10^{38} \exp\left[-19{,}000/(317 - T)^2\right]$

A better fit to the experimental results is

    $J_{hom} = 4 \cdot 10^{25} \exp\left[-18{,}000/(317 - T)^2\right] \text{m}^{-3} \cdot \text{s}^{-1}$

Examples:

    At $25°C$: $J_{hom} = 10^4 \text{ m}^{-3} \cdot \text{s}^{-1}$, i.e., 10 per liter per s

    At $18°C$: $J_{hom} = 10^{14} \text{ m}^{-3} \cdot \text{s}^{-1}$

At $18°C$, $r_{cr} = 1.7 \text{ nm}$, i.e., a nucleus of ~ 14 molecules.

The results are quite approximate and meant to illustrate trends.

exponential is much more dependent on temperature than the other factors. This is because a small change in T gives a relatively large change in $(\Delta T)^2$. Thus experimental results are often given as a constant "preexponential factor" times an exponential factor involving $(\Delta T)^2$.

It will be clear from the results that the nucleation rate depends very strongly on temperature. This is further illustrated in Figure 14.3 for ice nucleation. Near $-40°C$, the magnitude of J_{hom} increases by a factor of 10 for a temperature decrease of about 0.5 K. Consequently, one often speaks of the *homogeneous nucleation temperature*, meaning the temperature at which nucleation becomes noticeable in practice. Rather arbitrarily, a value of $J_{\text{hom}} = 10^{14} \, \text{m}^{-3} \cdot \text{s}^{-1}$ (e.g., corresponding to about 1 in $10 \, \mu\text{m}^3$ during 15 min) is often taken. Then T_{hom} would equal about $-40°\text{C}$ for water. At $T = T_{\text{hom}}$, we may put $r_{\text{cr}} = r_{\text{hom}}$, the radius of nuclei characteristic for homogeneous nucleation.

It is also seen that there are great differences between systems, both in the preexponential factor—which primarily affects the overall level of the nucleation rate—and in the exponential factor—which primarily determines the steepness of the curve with respect to temperature. The difference $T_{\text{eq}} - T_{\text{hom}}$ roughly equals 40 K for ice, 28 K for sucrose, and 26 K for tristearate in paraffin oil. In natural fats, where triglycerides crystallize from a triglyceride oil, the temperature difference is only about 20 K, presumably because the interfacial tension between oil and crystal is far smaller (about 4 rather than $10 \, \text{mN} \cdot \text{m}^{-1}$).

Complications. It is also seen in Table 14.2 that large differences are observed between theoretical and experimental results, in both the preexponential and the exponential factor. For highly polar materials, such

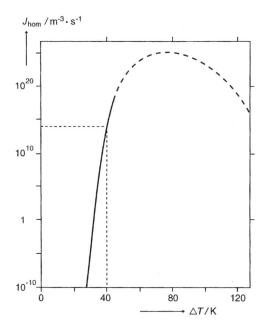

FIGURE 14.3 Rate of homogeneous nucleation J_{hom} for the formation of ice in pure water at ambient pressure as a function of the undercooling ΔT. The broken line is a rough estimate. The dotted lines indicate the "homogeneous nucleation point."

as most salts, the discrepancies tend to be even greater. In fact, many workers consider the theory to be essentially wrong, and try to develop other theories of nucleation rate.

Considering the exponential factor, the basic problem is that the magnitude of γ generally cannot be determined, and it is therefore often used as a fitting factor. Moreover, the value of ΔH^V may be quite uncertain near T_{hom}, because of nonideality; this is especially important for the sucrose/water system in Table 14.2, since tabulated values of the enthalpy of dissolution refer to the addition of a small amount of solute to pure solvent, whereas a saturated sucrose solution contains about 65% w/w solute. More fundamental is the criticism that an embryo, i.e., a very small region, cannot be treated as constituting a phase, with a clear phase boundary, especially when it concerns formation of a crystalline phase. The results on r_{hom} for ice and sucrose lead to nuclei of about 200 molecules, and such a region is possibly large enough to be considered a phase. However, a nucleus of tristearate would only contain 14 molecules, which implies that the assumption is questionable. On the other hand, in the case of tristearate the agreement between theory and experiment for the exponential factor is quite good, presumably because the solution properties are almost ideal in this case.

There are also great differences between predicted and observed values of the preexponential factor. Assuming that the treatment is fundamentally sound (which can be questioned on the same grounds as discussed in Section 4.3.5), it must be the magnitude of ΔG^* that is much larger than predicted from diffusional resistance. Nevertheless, the latter is of importance, since nucleation rate tends to decrease again at very low temperature. This is illustrated in Figure 14.3 (broken line). A strong decrease occurs near the glass transition temperature, as is discussed in Section 16.1. Presumably, the nucleation rate for ice is negligibly small at about $-140°C$. In passing, it is extremely difficult to cool water at such a rate that no nucleation occurs before reaching this temperature.

Another factor that presumably contributes to ΔG^* is the loss of *orientational and conformational entropy* of the molecules that become incorporated in an embryo. It is seen that the difference between theory and experiment of the preexponential factor increases when going from ice (five orders of magnitude) to sucrose (seven) to tristearate (twelve), and this is at least in qualitative agreement with the loss in entropy, which will be larger for more anisometric and more flexible molecules. Especially tristearate molecules will lose very much conformational freedom upon crystallization.

In conclusion, classical nucleation theory is insufficient, and the results on nucleation rate may be off by as much as ten orders of magnitude.

Moreover, the errors vary with the kind of phase transition. Nevertheless, substantial homogeneous nucleation does only occur at very strong supersaturation, and it depends very steeply on temperature or concentration, as predicted. As a rule of thumb, the values for r_{hom} are generally about 1 or 2 nm, and the nucleus then generally contains between 50 and 250 molecules.

Question

Consider air saturated with water vapor at 25°C. To what temperature should the air be cooled before water droplets would spontaneously form? The enthalpy of evaporation of water equals about $2.45 \, GJ \cdot m^{-3}$ and the surface tension about $75 \, mN \cdot m^{-1}$ at low temperature.

Answer

Assuming homogeneous nucleation of water droplets, we need to find T_{hom}. Further assuming r_{hom} to be equal to the value observed for ice nucleation, i.e., 1.85 nm (Table 14.2), Eq. (14.6) directly yields $T_{hom} = 15°C$. This would be a small undercooling.

In practice, however, homogeneous nucleation is observed to occur at about 5°C, at a supersaturation $\ln \beta \approx 1.43$. Inserting the latter value in Eq. (14.9), we derive $r_{hom} = 0.82$ nm, which corresponds to a nucleus containing 75 water molecules, and which is far smaller than an ice nucleus. Equation (14.6) in fact shows that for a smaller value of $\gamma/\Delta H^V$, r_{cr} will be smaller at the same temperature ratio. For water nucleation to occur, this ratio is therefore much smaller than for ice nucleation (cf. the values in Table 14.2). It is likely that also the value of r_{hom} is smaller. Hence it is not allowed to take a fixed value for r_{hom} and to derive results from it.

14.2.2 Heterogeneous Nucleation

In the previous section is was stated that the (homogeneous) nucleation temperature of ice in pure water is about −40°C. It is common experience, however, that ice starts to form at far higher temperatures, sometimes even at −1°C. How is this to be explained? The following experiment may be enlightening.

A suitable substance, e.g., a triacylglycerol, is melted, and above the melting temperature it is emulsified into an aqueous solution with a suitable surfactant (say, a protein). The emulsion then is brought under a microscope and slowly cooled. As soon as a droplet crystallizes, which can be seen when using polarized light microscopy, the temperature T and the droplet diameter d are noted. After several repeats, a graph like Figure 14.4 can be constructed. There is considerable scatter in the crystallization temperature of droplets of a given size, but the average T is lower for a smaller d. Moreover, below a certain value of d, the lowest temperature observed does not decrease any further. It is concluded that this must be the homogeneous nucleation temperature, in the case of trimyristin 21 K below T_{eq} (this concerns the α-modification: see Section 15.4.2).

It is further concluded that at temperatures above T_{hom}, nucleation is induced by small particles called **catalytic impurities**. Their number

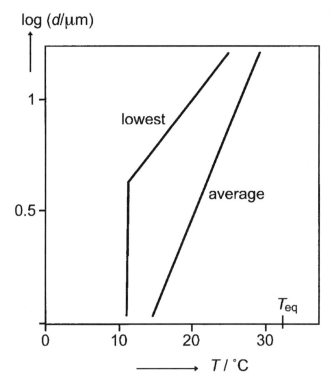

FIGURE 14.4 Crystallization temperature T observed in droplets of various diameter d in an O–W emulsion of trimyristin (tri-tetradecylglycerol). The average temperature and the lowest temperature are given. T_{eq} is the equilibrium temperature (clear point) of trimyristin. See text. (After results by L. W. Phipps. Trans. Faraday Soc. 60 (1966) 1126.)

concentration would be limited, and the larger a droplet is, the larger the probability that it contains at least one catalytic impurity, which would be sufficient to induce crystallization. Droplets devoid of catalytic impurities would then freeze at the homogeneous nucleation temperature.

To understand how catalytic impurities may work, it will be useful to consult Section 10.6.1 first. In Figure 14.5a an embryo is shown that may lead to homogeneous nucleation, if it is small enough. If now a surface of a material κ is present, an embryo may be formed on that surface, as depicted in Figure 14.5b, provided that $\cos \theta$ is finite. Its value depends on the three interfacial tensions (specific interfacial free energies) according to the *Young equation* (10.10), which can be written as

$$\cos \theta = \frac{\gamma_{\alpha\kappa} - \gamma_{\beta\kappa}}{\gamma_{\alpha\beta}} \tag{14.11}$$

At the same supersaturation (which generally means at the same temperature), the curvature will be the same according to Eq. (14.9), and this implies that the volume of the embryo is smaller. The volume of the embryo divided by that of a sphere of the same radius is given by

$$f_{\text{cat}} = \frac{1}{4}(2 + \cos \theta)(1 - \cos \theta)^2 \tag{14.12}$$

which value is smaller for smaller θ. The value of $\Delta G_{\max}$ as given in Eq. (14.7) should now be multiplied by f_{cat}.

Heterogeneous Nucleation Rate. Inserting f_{cat} into Eq. (14.10) then would give the rate of heterogeneous nucleation J_{het}. This would apply for a high concentration of identical catalytic impurities. In practice, this is rarely the case. The concentration of impurities will generally determine the magnitude of the preexponential factor. The particles present may be of various materials, leading to a range of θ values, as illustrated in Figure 14.5b, frames 1 and 2. The particles may also have different shapes, and frame 3 shows how that can affect the size of the embryo. Moreover, as depicted in frame 4, the boundary α–β may even be flat (depending on the shape of the indentation), which would mean that nucleation can occur at zero undercooling. The effectiveness of particles in catalyzing nucleation will thus show considerable statistical variation, which implies that the lower the temperature (or the higher the supersaturation), the greater the number of impurities that are catalytic.

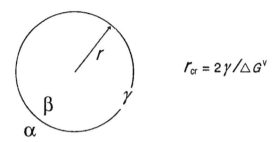

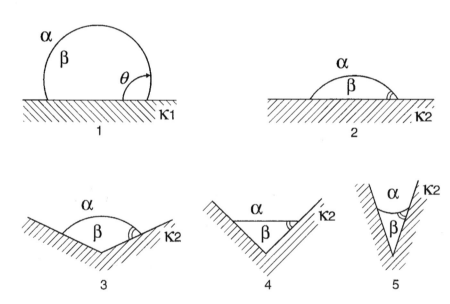

FIGURE 14.5 Homogeneous and heterogeneous nucleation. Embryos (radius r) of phase β formed from phase α (a) in the absence of foreign particles, and (b) at the surface of foreign particles of material $\kappa 1$ or $\kappa 2$; in frames 2–5, the contact angle $\theta = 45$ degrees.

It is often observed that the number of catalytic impurities depends on temperature according to

$$\log N_{cat} \approx A - BT \tag{14.13}$$

within a limited temperature range; A and B are constants. This relation can be interpreted as N_{cat} being about proportional to the supersaturation $\ln \beta$. Some experimental results are in Figure 14.6, which also shows results on the initial heterogeneous nucleation rate J_0 in the same system. (J_{het} tends to decrease with time in an emulsified material; see Section 14.3.) The nature of the catalytic impurities can vary widely and depends on the composition of the system and on its purity. No general rules can be given. It should further

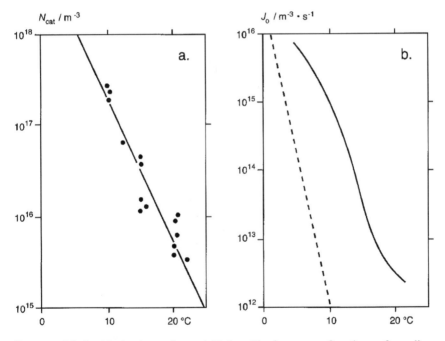

FIGURE 14.6 Nucleation of emulsified milk fat as a function of cooling temperature. (a) Concentration of catalytic impurities N_{cat}. (b) Initial heterogeneous nucleation rate J_0; the broken line gives a rough estimate of the homogeneous nucleation rate. (After results by P. Walstra, E. C. H. van Beresteyn. Neth. Milk Dairy J. 29 (1975) 35.)

be mentioned that macroscopic surfaces (vessel walls, etc.) can catalyze nucleation by the same mechanism as shown in Figure 14.5b.

Epitaxy.   Equation (14.12) shows that θ has to be small for φ_{cat} to become really small. For $\theta = 90$ degrees, f_{cat} will equal 0.5, which will not greatly affect nucleation temperature, but for $\theta = 15$ degrees, $f_{cat} \approx 0.0009$, and this would make an enormous difference, e.g., reducing the undercooling needed for fast nucleation from 30 to 1 K. (Try to calculate this for one of the systems given in Table 14.2.) It is, however, very unlikely that such a thin spherical segment of crystalline material would form, since it would only be one or two molecules thick. However, something like it can occur if the crystal lattice of the material to crystallize is very similar to that of one of the crystal faces of the impurity. In such a case, a monolayer of the new phase can be deposited onto the impurity, and grow out. This is called epitaxy. A well-known example is the nucleation of ice on silver iodide crystals in cold air supersaturated with water vapor.

In foods something similar occurs during the crystallization of fats. Natural fats have a wide compositional range, implying that several different kinds of crystals have to be formed. This needs undercooling, but as soon as the first crystals have formed, other crystals nucleate on the existing ones. This must be due to epitaxy, since the various crystals are very similar in lattice structure. It is indeed observed that the undercooling needed is only by about 2 K.

A very small contact angle tends to be formed also if the solid is a crystal, phase α is air, and β is the liquid phase of the crystal material. This implies that the crystal acts as an effective catalytic impurity for the formation of its own melt. This explains why very little overheating generally suffices to induce the phase transition crystal → melt, as was mentioned in Section 14.1.

Memory.   Figure 14.5b, frame 5, shows that in a sharp crevice in a particle or the vessel wall, a concave interface between crystalline material (β) and solution or melt (α) can exist, as seen from phase α. This can occur if the contact angle θ is small. It implies that phase β will spontaneously grow. Moreover, if the system is heated to a temperature above T_{eq}, some crystalline material may remain in the crevice. If now the system is cooled again, crystallization may occur without significant undercooling being needed, since the crystalline phase is already present. This is called a memory effect. The memory of the system can be "destroyed" by increasing the temperature to several degrees above T_{eq}. Often about 5K will suffice, but there are examples of a more persistent memory.

It should be mentioned that frames 3–5 in Figure 14.5b are two-dimensional and the real (three-dimensional) situations will generally be different, with two principal radii of curvature; see Section 10.5.1. However, trends will often be as mentioned.

Question

To what temperature above 0°C should ice be heated before water starts to form at its surface? The surface tension of water at low temperature equals about $75 \, \text{mN} \cdot \text{m}^{-1}$ and γ(water-ice) $\approx 25 \, \text{mN} \cdot \text{m}^{-1}$ (Table 10.1). The surface tension of ice is unknown, but you may assume a value of $95 \, \text{mN} \cdot \text{m}^{-1}$.

Answer

Assume the nucleus to be a spherical water cap on the ice surface. From Eq. (41.11) we calculate $\cos \theta = (95 - 25)/75 = 0.93$ (corresponding to $\theta = 21°$). Inserting this value into Eq. (14.12) yields $f_{cat} = 0.0036$. In the situation envisaged, all nuclei would be the same, i.e., have the same radius and contact angle. This would mean that we may insert f_{cat} in ΔG_{max}, i.e., in the second exponent of Eq. (14.10). Table 14.2 gives J_{hom} for ice nucleation, and the same results would apply for water nucleation since the absolute values of γ and ΔH will be the same. We thus obtain for the present case $J_{het} = 10^{36} \exp(-82{,}000 f_{cat}/(\Delta T)^2)$. Putting this equal to 10^{14}, we calculate for ΔT a value of 2.3 K. Actually, the preexponential term may be higher than 10^{36}, say 10^{41}, because the transition ice $\rightarrow$ water does not involve a decrease but an increase in entropy, but this would only cause a very small difference in the value of ΔT. Altogether, water would form at about 2°C.

Subsequent Question

In nature it is commonly observed, however, that ice becomes wet at, say, 0.3°C. On the other hand, it can stay dry at, say, 5°C. How is this possible?

Answer

In the first place, the result obtained may be wrong, because of uncertainties in the theory and in the γ values. Nevertheless, overheating by a few degrees would be needed on an ideal flat ice surface. However, "natural" ice will always contain little crevices, and water can readily form in these, as illustrated in Figure 14.5b.

Furthermore, it was implicitly assumed that the air above the ice is exactly saturated with water vapor at 0°C. At significant supersaturation of the air, water can condense on the ice (beginning in crevices). The air may also be much drier, i.e., its relative humidity can be well below 100%. In that case, water will be removed from the ice by desublimation (especially when a wind is blowing).

14.2.3 Other Aspects

The importance of the magnitude of the rate of nucleation is (a) that it determines the supersaturation needed to realize the desired phase transition and (b) that it is the main factor determining the size of the regions of the new phase obtained. The latter is specially important in crystallization.

Crystal Size. The size of crystals may be an important quality variable in foods. In liquid foods it will often determine whether settling of crystals occurs, which generally is undesirable. In several liquid and soft solid foods it is undesirable that crystals be felt in the mouth, a sensation often described as *sandiness*. To prevent this, crystals generally have to be smaller than about 10 μm (the critical size can also depend on crystal shape). Examples are ice crystals in ice cream, crystals of amino acid salts in some cheeses, and lactose crystals in some concentrated milk products. The rheological properties of a plastic fat depend, among other factors, markedly on fat crystal size. Moreover, if crystallization is aimed at harvesting the solid, as in sugar boiling or fat fractionation, it is usually desirable to obtain large crystals, to ensure their efficient separation from the mother liquid.

Figure 14.7 gives very roughly the effect of the value of the supersaturation on the rates of nucleation and crystal growth. The scales on the graph and the shapes of the curves will greatly vary with the system considered, but the main point nearly always holds: nucleation rate depends far more strongly on supersaturation (hence, on temperature) than does growth rate. Consequently, many crystals are formed when supersaturation is high (especially if catalytic impurities are absent), and many crystals implies small crystals. At a low concentration of impurities that are catalytic at small supersaturation, only a few, hence large, crystals are formed. If the increase in supersaturation is very fast, which generally means fast cooling, small crystals will nevertheless result.

In some situations *seed crystals*, i.e., crystals of the material to be formed, are added after a given supersaturation is reached. The seed crystals then grow, without nucleation being needed. When adding a few crystals at low supersaturation, large crystals will be formed, albeit slowly. On the other hand, a large number of tiny seed crystals can be added at high supersaturation, to ensure rapid formation of small crystals. This method is used, for instance, if it is difficult to attain conditions at which homogeneous nucleation occurs. Very small seed crystals are generally obtained by dry grinding.

Secondary Nucleation. It is often observed that agitation during crystallization enhances the number of crystals formed, which presumably implies a higher nucleation rate. This may be due to a number of mechanisms. When heterogeneous nucleation occurs at the surface of (parts of) the equipment, agitation will greatly increase the volume of liquid that comes into contact with such surfaces. Intensive stirring may lead to breaking of newly formed crystals upon impact, or in "scraping off" small pieces from a crystal. *True secondary nucleation* means that nuclei are formed in the *vicinity* of a crystal of the same phase (not at the *surface*). It can occur in the absence of agitation.

A clear example of such quiescent secondary nucleation is provided by some natural fats, i.e., mixtures of several triglycerides. Consider Figure 14.6a. At 15°C, the number concentration of catalytic impurities observed was about $3 \cdot 10^{16} \, \text{m}^{-3}$, which means one impurity per $30 \, \mu\text{m}^3$. A droplet of

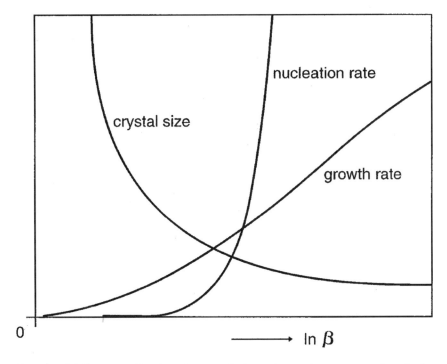

FIGURE 14.7 Crystallization from solution. Approximate example of the nucleation rate, the average linear crystal growth rate, and the average crystal size resulting, as a function of supersaturation $\ln \beta$. Arbitrary linear scales.

diameter 4 µm has about that volume. Upon cooling an emulsion of the same milk fat, one would thus predict that droplets of about 4 µm contain one crystal. Actually it would, because of statistical variation, be zero, one, or a few crystals. The proportion of droplets containing no crystals is about as predicted. However, the other droplets do not contain one or two crystals, as predicted, but many, at least 100. This must be due to secondary nucleation. In other fats something similar is observed, though the number of crystals often is smaller. In emulsions of paraffin mixtures, or of a solution of hexadecane in oil, one does observe droplets with one or two (large) crystals and no droplets with many crystals, in accordance with the presence of a limited number of catalytic impurities, hence no secondary nucleation would have occurred.

The phenomenon is of considerable practical importance for the size of the crystals obtained. There is no generally accepted theory for secondary nucleation. It appears to occur at high supersaturation in systems where (nevertheless) crystal growth is slow.

Spinodal Decomposition. Liquid–liquid phase separation has been briefly discussed for polymer solutions in Section 6.5; see especially Figure 6.17. The theory applied also yields an alternative treatment for nucleation of the new phase. This will be briefly discussed for the simplest case, i.e., a binary mixture (solution).

The change in free energy upon phase separation only involves a mixing term, since a transition in the physical state (crystallization or evaporation) does not occur. In other words, $\Delta_{tr}G = -\Delta G_{mix}$. For an ideal solution (see Section 2.2), phase separation cannot occur, since it implies that ΔH_{mix} equals zero, and ΔS_{mix} is always positive. For nonideal solutions, it depends on the shape of the free energy curve whether phase separation can occur.

Figure 14.8 gives a hypothetical example. In (a) a phase diagram is shown (as in Figure 6.17, but the variable is now temperature rather than solvent quality). The coexistence line or binodal, where $\Delta G_{mix} = 0$, bounds the region where phase separation will occur. Also the spinodal is given. For the condition $T = T_1$, the free energy curve is given in (b). Part of this curve is concave towards the m-axis. If now the composition of the mixture is between A and F, phase separation into two liquids of composition A and F can occur, since this will lead to a lower G value. The total free energy then is given by the broken line, which is tangent to the curve at two points. These points, A and F, correspond to points at T_1 on the binodal.

The question is now whether the phase separation will occur spontaneously. Consider the second derivative of the free energy $G'' \equiv (\partial^2 G/\partial m^2)_{T,p}$. The points where $G'' = 0$ denote inflection points, in

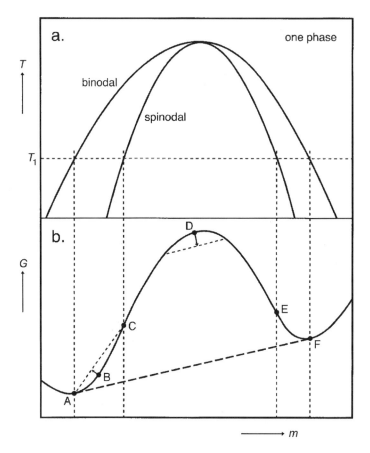

FIGURE 14.8 Liquid–liquid phase separation. Hypothetical example of (a) the effect of temperature T on the binodal and the spinodal and (b) the mixing free energy G at temperature T_1, of a binary liquid system as a function of the mole fraction m of one of the components.

this case C and E. Between these points $G'' < 0$ and, as illustrated for composition D, any demixing, however small, will lead to a decrease in G. Phase separation will thus occur spontaneously; it is called *spinodal decomposition*. C and E correspond to the points on the spinodal curve at $T = T_1$. Between A and C and between E and F, $G'' > 0$ and, as shown for composition B, demixing initially leads to an increase in G. Hence *nucleation* is needed to initiate phase separation in the composition range between the binodal and the spinodal.

It should be noted that Figure 14.8 gives just one example and that the relations can be very different. The dependence of solubility on temperature may be very weak (in many polymer solutions) or even opposite. Other variables may be important, notably solvent quality. For many systems, the curve of G versus m is convex towards the m-axis over the whole range, which implies that phase separation cannot occur. For demixing of solutions of two polymers, the same considerations hold, but the treatment is more complicated. The treatment is not restricted to polymer solutions. For instance, mixtures of a higher alcohol and water may give relations very similar to those in Figure 14.8.

In a sense, the spinodal gives the composition and hence the supersaturation needed for homogeneous nucleation. People have tried to develop more general nucleation theories from the thermodynamics of mixtures along these lines. It is, however, difficult, if not impossible, to predict free energy curves with sufficient accuracy. When working with polymer solutions or aqueous polymer mixtures, one often has experimentally to estimate the spinodal, to find the conditions needed for spontaneous demixing.

Question

Sweetened condensed milk is supersaturated with lactose, and the lactose crystals formed tend to settle and also give the product a sandy mouth feel. To prevent this, crystal size should be at most 8 μm, and one generally tries to achieve this by adding crystalline seed lactose, about 0.3 g per kg product. Calculate what the maximum size of the seed particles may be. One kg of sweetened condensed milk contains about 110 g lactose, 440 g sucrose, and 260 g water. Consult also Table 2.2.

Answer

The product contains $440/260 = 1.69$ kg sucrose per kg water and, according to Table 2.2, the solubility of lactose then is close to 200 g per kg water. This implies that about 58 g of lactose will eventually crystallize. Assuming that the number of crystals formed equals the number of seed crystals added, the volume of a seed crystal must be at most $0.3/58$ times the volume of a crystal eventually formed. Assuming the crystals to be of the same shape, the maximum diameter of a seed crystal should then be $(0.3/58)^{1/3} \times 8\,\mu m = 1.38\,\mu m$.

14.3 NUCLEATION IN A FINELY DISPERSED MATERIAL

When a bulk material contains about one catalytic impurity per mm^3 at a given supersaturation, nucleation will generally be fast enough to ensure a rapid phase transition, say crystallization. If the material is divided into emulsion droplets of 100 μm^3, only about 1 in 10^7 droplets will contain an impurity that can induce crystallization, which is a negligible amount. The emulsified material thus needs a far greater supersaturation, possibly corresponding to homogeneous nucleation, before most of it can crystallize (cf. Figure 14.4). This is because a phase transition in one drop cannot induce a transition in another drop (see further on for an exception). A material in a finely dispersed state thus can be undercooled far deeper than a bulk material, and its nucleation rate will generally be much slower. Moreover, the amount crystallized will depend on the temperature history. This is of practical significance in several oil-in-water emulsions. It is especially important in natural foods, where most of the water is in cells: most cells do not contain effective catalytic impurities. Consequences will be discussed in Section 16.3.

Figure 14.9a gives a few examples. A highly purified emulsified tristearate can be cooled to almost the *homogeneous* nucleation temperature before crystallization occurs (curve 1); it thus contains virtually no catalytic impurities. Upon heating the cold emulsion, it shows no sign of melting until it almost reaches the equilibrium temperature (curve 3). There is thus a large hysteresis between cooling and melting, about 28 K in this case. A nonpurified sample shows less undercooling, and the cooling curve (2) is far less steep. Here, nucleation must have been almost completely *heterogeneous*. The catalytic impurities for triglycerides must be for the most part composed of monoglycerides (inverse micelles?), since N_{cat} correlates well with the concentration of these substances.

Kinetics. Emulsions have long been used to determine crystal nucleation kinetics, i.e., to find values of N_{cat} and of J_{het} or J_{hom} as a function of supersaturation or other conditions. The material is emulsified in a suitable medium, e.g., water in silicone oil or oil in water, with a suitable surfactant, and the droplet size distribution is determined. To obtain useful results, a series of emulsions differing in average droplet size should be made. The emulsions are cooled to various temperatures T_c below T_{eq}, and after a given time the amount of crystalline material is determined, e.g., by a change in density, or from the heat of fusion, or by means of some spectroscopic method. The same method is applied to the bulk material starting at T_c, and from the ratio of the results the value y is calculated, which is the volume fraction of the material that is in droplets containing crystals. This is not necessarily the fraction frozen: if the material is a mixture of components—as is nearly always the case in practice—it will generally have a melting range, implying that part of each drop can remain liquid. If y does not increase any more with time, it is assumed that the maximum value y_{max} has been reached. In a temperature region where y_{max} increases with decreasing temperature, nucleation will be predominantly heterogeneous. Assuming that the distribution of catalytic impurities over the droplets is random, i.e., obeys Poisson statistics, N_{cat} can now be derived from

$$y_{max} = 1 - \exp(-vN_{cat})$$

$$v = \frac{\pi d_{63}^3}{6}$$

(14.14)

where v is volume-weighted average droplet volume (see Section 9.3 for size distributions). For very small values of vN_{cat}, we have $y_{max} = vN_{cat}$; for

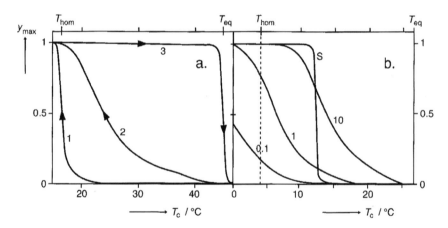

FIGURE 14.9 Nucleation in oil-in-water emulsions. y_{max} is the volume fraction of droplets containing crystals after 24 hours keeping at temperature T_c; T_{hom} is the approximate homogeneous nucleation temperature and T_{eq} the equilibrium temperature (clear point). (a) Tristearate in paraffin oil. 1 and 2, cooling curves of highly purified and not purified tristearate, respectively; 3, heating curve. (b) Mixture of triglycerides. Calculated cooling curves for heterogeneous nucleation of emulsions with various average droplet volume v, indicated in μm³. S denotes an example of surface nucleation.

$\nu N_{cat} > 5$, $y_{max} \approx 1$. The results shown in Figure 14.6a have been obtained in this way; by using emulsions of various ν values, the validity of Eq. (14.14) can be checked.

From results on y as a function of time t at T_c, the nucleation rate J can be derived (assuming the time needed for crystallization in a droplet once a nucleus has been formed to be relatively short). However, it has been observed that J decreases with time, approximately according to

$$J = J_0 (1 - \frac{y}{y_{max}}) \qquad (14.15)$$

A decrease in J is to be expected, since the catalytic impurities will show variation in efficiency and the most efficient ones tend to act first. Actually, this makes the concept of a fixed N_{cat} value at a given temperature questionable, since the value will depend on the duration of the experiment.

Equation (14.15) is an empirical one; it is well obeyed in one kind of system (triglyceride mixtures), but whether it applies in the same form to other systems is uncertain. Assuming it to be correct and combining it with Eq. (14.14), integration leads to

$$y = \frac{1 - \exp(-\nu N_{cat})}{1 + 1/J_0 \nu t} = \frac{y_{max}}{1 + 1/J_0 \nu t} \qquad (14.16)$$

Note that the number of catalytic impurities per droplet νN_{cat} will determine what proportion of the droplets will eventually contain crystals; the initial rate of formation of nuclei per droplet $J_0 \nu$ will determine the value of y/y_{max} as a function of time. Naturally, the values of J_0 and of N_{cat} are correlated in a given system; see e.g., Figure 14.6. The time needed to obtain a certain ratio of y over y_{max} greatly varies, e.g., between a minute and some hours for a ratio of 0.8. According to Eq. (14.16), the time will increase with decreasing value of ν for a given value of T; it also tends to increase with decreasing supersaturation (increasing temperature) for a given value of ν.

Figure 14.9b. gives some examples of $y_{max}(T)$ for various values of ν, calculated for heterogeneous nucleation. Note that near 5°C, homogeneous nucleation will take over. It is seen that droplet size has a very large effect on the extent of crystallization at a given temperature.

Surface nucleation. Thus far, we have only considered volume nucleation, implying that the number of catalytic impurities in a droplet is proportional to its volume. Another possibility is nucleation at the inside of the droplet boundary, if that surface is catalytic for nucleus formation. This is called heterogeneous surface nucleation. An example of the resulting relation between y_{max} and T_c is given in Figure 14.9b, curve S. Over a very small temperature range all droplets still devoid of crystals then (partially) crystallize. For triglyceride oils, this especially occurs if the surfactant used has a long and saturated paraffin chain and these chains are closely packed in the interface. For tristearate, surface nucleation temperatures of 1–20K above T_{hom} have been observed, according to surfactant type.

A curve as steep as shown in the figure is, however, not very common. It needs a pure surfactant giving a fully packed interfacial layer, and it works best at temperatures below the chain crystallization temperature (see Section 10.3.1). Most surfactants are mixtures (or are at least impure), or the concentration is too small to obtain a plateau value of the surface load. In such cases, the interfacial layer may contain patches that are catalytic for nucleation. This then means that the droplets contain catalytic impurities, but their number would be proportional to droplet surface area rather than volume. In the simplest case, Eq. (14.14) is to be modified by replacing vN_{cat} by $N_{cat}^S \pi d_{53}^2/4$, where the superscript S indicates that the number is per unit area. The curves obtained then are rather similar to those for volume nucleation, and it needs painstaking experiments to determine which of the two mechanisms prevails.

Seed Crystals. Crystals inside droplets may occasionally stick out of the surface over several nanometers. If such a drop encounters another one by Brownian motion (see Section 13.2.1), the protruding crystal may occasionally pierce the surface of that droplet. If the latter is still fully liquid, the crystal may act as a seed and induce crystallization. This has been observed to occur in emulsions of hexadecane in water, where part of the drops were solid and part liquid (undercooled). It is a slow process, for instance taking two weeks for completion. It has been calculated that about one in 10^7 or 10^8 encounters was effective in such a case.

It may well be that the same mechanism can act in crystallizing triglyceride emulsion droplets, but the author is unaware of clear evidence.

Question 1

An emulsion of milk fat in a Na–caseinate solution is made and the average droplet size is determined at $d_{32} = 1.8\,\mu m$. The emulsion is cooled to 15°C and kept at that temperature. What will roughly be the value of y_{max} reached, and how much time will it take to obtain values of $y = 0.1$ and 0.9? If the milk fat contains no catalytic impurities, to what temperature should the emulsion then be cooled to obtain the same cooling times?

Answer

What is first needed is the magnitude of $v = \pi d_{63}^3/6$. Assuming the emulsion to have been made in a high-pressure homogenizer, its size distribution will be rather wide (see, e.g., Figure 11.9a); presumably, d_{63} will then be 3 to 4 times d_{32}, so we may estimate v at about $120\,\mu m^3$. Figure 14.6a indicates $N_{cat} \approx 3 \cdot 10^{16}\,m^{-3}$. Inserting these values in Eq. (14.14) yields $y_{max} = 0.925$. Figure 14.6b indicates that $J_0 \approx 3 \cdot 10^{13}\,m^{-3} \cdot s^{-1}$. Equation (14.16) then yields times of about half a minute and one hour for $y = 0.1$ and 0.9, respectively.

In the absence of catalytic impurities, homogeneous nucleation is needed, which would presumably lead to $y \approx y_{max}$. Equation (14.16) shows that we need to know the values of y/y_{max}; these are pretty close to the values just applied since we took $y_{max} = 0.925$. We can now insert the value of J_{hom} to obtain $y(t)$. This means that we only have to find the temperature at which $J_{hom} = J_0(15°C)$, and Figure 14.6b shows this to be about $7°C$.

It makes no sense to do more precise calculations, since the results will anyway be approximate because of natural variation in the values of J and N_{cat}.

Question 2

Suppose that you want to keep a solution of an enzyme in an aqueous buffer at $-30°C$ without freezing (because the freeze concentration would lead to irreversible enzyme inactivation). How can this be achieved? Assume the number of catalytic impurities at $-30°C$ to be $10^{11}\,m^{-3}$.

Answer

The best way is to make a W–O emulsion. It is advisable to have an oil phase that freezes at low temperature, since that prevents settling and possible coalescence of the drops. The oil should not affect the protein. Tetradecane seems suitable (freezing point about $6°C$). An emulsifier has to be added and it should (a) have a low HLB value (why?), (b) not give rise to surface nucleation at $-30°C$, and (c) not affect the protein. It may take some trial and error finding a suitable surfactant. Assume that "without freezing" implies that y_{max} should become at most 0.01. By use of Eq. (14.14) it follows that the average droplet diameter d_{63} then must be below $57\,\mu m$ to prevent heterogeneous nucleation, which can easily be realized. Figure 14.3 shows that at $-30°C$ J_{hom} is about $1\,m^{-3} \cdot s^{-1}$, which gives negligible nucleation in $57\,\mu m$ drops over several centuries.

14.4 FORMATION OF A GAS PHASE

Liquid–gas transitions are common during food processing or storage, but they often occur at the boundary of a condensed phase, as mentioned in Section 14.1. In some situations, a gas phase forms inside a liquid or solidlike material. Examples are the formation of CO_2 bubbles in beer after the release of the pressure in the bottle or can; slow formation of CO_2 bubbles in a bread dough due to sugar fermentation by yeast cells; the formation of "eyes" in some types of cheese by a similar mechanism; and the change of a whipping cream that is under high pressure in an aerosol can into a foam when the cream leaves the can through a nozzle.

It may be useful to recall a few relations about gases. The first is *Henry's law* about the equilibrium distribution of a substance over gas and liquid phases. It reads

$$p_A = k_H m_A \tag{14.17}$$

where p_A is the partial pressure of component A in the gas phase and m_A its mole fraction in the liquid phase. The magnitude of the Henry constant k_H depends for every A on the solvent (liquid phase) and on the temperature. It is assumed that the solute behavior is ideal, i.e., the activity of A in the liquid phase equals m_A. We further recall the *ideal gas law*, $pV = nRT$. In the following, we assume ideal behavior; Eq. (14.17) then implies that the solubility of a gas, say air, in a liquid is proportional to its pressure. The solubility of air and most other gases decreases with increasing temperature. For N_2O and CO_2, the gas law does not hold precisely at high pressures. Moreover, CO_2 can form H_2CO_3 upon dissolution in water, and this may dissociate and form salts, like $CaCO_3$; in other words, its solute behaviour may be far from ideal.

Nucleation of gas bubbles is notoriously difficult, and the following calculation may explain it. Assume that a gas embryo of 2 nm radius is formed in a liquid at atmospheric pressure. The interfacial tension will generally be about $0.07 \, N \cdot m^{-1}$. The Laplace pressure in the embryo will then be about $2 \times 0.07/2 \cdot 10^{-9} = 7 \cdot 10^7$ Pa, which equals 700 bar. The supersaturation ratio of the gas should then be about 700 for such a small bubble to survive, and that is very unlikely to be the case. In a beer bottle the pressure is a few bar, in an aerosol can with N_2O about 7 bar. Moreover, the number of gas molecules inside the embryo would be about 560 (try to make the calculation), more than could possibly associate by chance. Homogeneous nucleation can therefore not occur. By similar reasoning, it can be derived that heterogeneous nucleation on a surface, as depicted in Figure 14.5b, frames 1 and 2, is not possible either, even if θ is quite small.

Nucleation in a crevice (as in frames 4 and 5) cannot be ruled out; see, however, the discussion below under "Seeding."

A practical example is given by the existence of trees. A tree contains vascular channels for transport of water from the roots to the leaves. At the surface of the leaves water evaporates, providing the driving force for water transport. Some trees are as high as 120 m. Taking into account that a 10 m water column will exert a pressure of about one bar under gravity, this means that the pressure in the channels at the top of the tree would be $1 - 120/10 = -11$ bar. Despite this negative pressure, the water column does not "break" more frequently than once in 10 or 100 years (breaking of a water channel is irreparable; if it occurs frequently, the tree will die). Hence nucleation of water vapor bubbles must be extremely difficult, at least below about $40°C$.

Seeding. It must be concluded that gas nucleation generally does not occur. What can happen is that bubbles form by growth of tiny gas pockets (mostly air) already present, i.e., by a seeding mechanism. Possible situations are

Entrapment of air by agitation. This occurs, for instance, during kneading of bread dough, where the entrapped cells cannot escape because of the very high viscoelasticity of the dough. The number of cells in a dough is about 10^{14} m^{-3}, or one cell in about 20 μm cubed. Also when beer is splashed into a glass, air bubbles are entrapped that then grow, because CO_2 is supersaturated and diffuses to the existing bubbles.

Persistent remnants of air bubbles that have almost disappeared by Ostwald ripening (Section 13.6.1). Some bubbles contain sufficient solid particles on their surface to form on shrinkage a closely packed layer, as depicted in Figure 14.10a. These tiny bubbles have been called "aphrons". They are quite stable, and they occur in most natural waters, albeit in very small numbers.

Gas pockets that have been left, for instance in tiny crevices. When a soft drink supersaturated with CO_2 is poured into a plastic beaker, bubbles are generally formed at many sites on the wall. A crevice contains some air, and CO_2 diffuses to the air pocket, which thereby grows. A gas bubble then forms on the wall, and it will be dislodged by the buoyancy force acting on it, leaving a little gas pocket in the crevice; this is illustrated in Figure 14.10b. The process repeats itself numerous times, leading to a train of bubbles. A prerequisite is that the contact angle wall–water–air be large. When beer is gingerly poured into a glass, gas bubbles hardly form, at most at one or a few places on the wall. This is because the contact angle at a glass surface is so small (in fact it is zero) that the Laplace pressure in the air is positive, and an air pocket will thus disappear; see Figure 14.10c. However, a few crevices may be contaminated with lipid material, which would greatly increase the contact angle and hence allow the retention of an air pocket.

Besides bread and cake, in which yeast or baking soda is used to obtain gas cells, other baked products have gas inclusions without CO_2

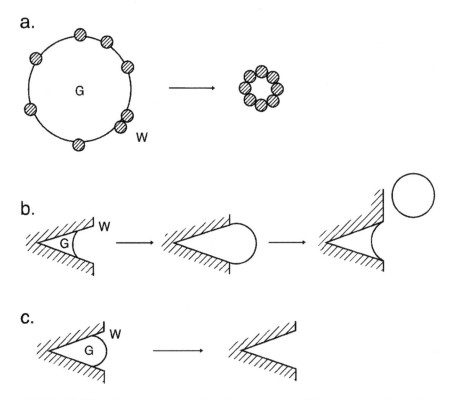

FIGURE 14.10 Gas pockets in a liquid. G is a gas, W is water, and hatching indicates a solid. (a) Formation of a persistent air bubble due to adsorbed hydrophobic particles and dissolution of most of the air. (b) Growth of a gas pocket if the contact angle, as measured in the water phase, is large and the water is supersaturated with the gas, and subsequent formation and dislodgement of a bubble. (c) Disappearance of a gas pocket in a crevice if the contact angle is small.

production. This concerns especially puff pastry, and also several types of biscuits (cookies). The gas cells are for the most part due to the expansion of air and the evaporation of water. They can be seeded by entrapping air bubbles during kneading, as in bread dough, but fat often plays an essential role. Use is made of butter or a vegetable shortening (a partly hydrogenated oil). These fats contain triglyceride crystals, especially at low temperature, and it appears that tiny air cells adhere to these crystals and are thereby carried into the dough or paste. The contact angle fat–water–air is quite large, and, especially if the crystals, or lumps of crystals present, have some concave surface regions, adhering air cells can be very persistent. Using oil, rather than a plastic fat in the recipe, does indeed result in a dense product.

14.5 RECAPITULATION

Table 14.1 gives an overview of **phase transitions**. The formation of a new phase generally demands undercooling, overheating, or supersaturation beyond the equilibrium temperature or the concentration at saturation. A very small region of the new phase, called an embryo, can always be formed by chance, but it will generally dissolve. Although the free energy of the material will be smaller in the new phase, there is also an interface between the phases, which goes along with an increase in interfacial free energy. Only if the decrease mentioned exceeds the increase mentioned can an embryo grow out and form a new phase; this means that it has become a nucleus. This often needs considerable undercooling, etc. A phase transition is thus subject to nucleation. Nucleation theory is quite uncertain in a quantitative sense, but the trends can be well predicted. The factors involved are discussed in general, but nucleation of crystals has the main emphasis.

The various **nucleation mechanisms** are indicated in Figure 14.11. Homogeneous nucleation occurs in the absence of any surface, and it needs considerable undercooling, mostly by 20–40 K. The smaller the transition enthalpy (e.g., the heat of fusion), and the higher the interfacial tension between the phases, the deeper the undercooling needed. Nucleation rate is very strongly dependent on temperature (or supersaturation). In practice, nucleation can occur at much smaller supersaturation. Then nucleation occurs at a surface, be it of foreign particles (called catalytic impurities when they cause nucleation) or the vessel wall. The efficiency of the impurities depends (a) on the contact angle of the new phase on the foreign surface (which depends on the nature of the materials involved) and (b) on the shape of the surface (a crevice being far more effective than a flat surface). The number of catalytic impurities as well as the rate of nucleation increase with increasing supersaturation. Heterogenous nucleation rate is generally much less dependent on the degree of supersaturation than is homogeneous nucleation. In some systems, copious secondary nucleation can occur in the vicinity of a crystal of the new phase (this is not seeding: see below).

A finely **dispersed material**, like water in cells or oil in emulsion droplets, generally needs far more undercooling for crystallization to occur than for the same material in bulk. This is because of the limited number of catalytic impurities: every cell or droplet has to contain at least one of these for crystallization to occur (unless the temperature is low enough for homogeneous nucleation), and the smaller the droplets, the smaller the chance that such is the case. This implies that in O–W emulsions, for

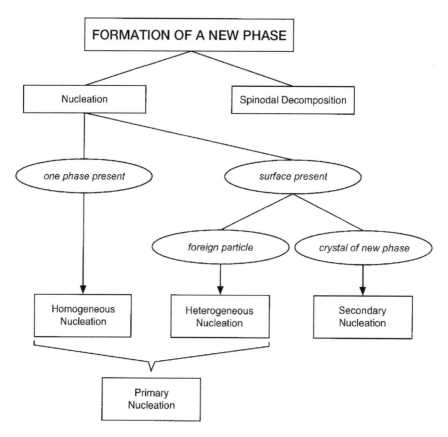

FIGURE 14.11 Scheme of the various mechanisms that can be involved in the initiation of a new phase.

instance, considerable hysteresis is observed between cooling and heating curves of the fraction of the fat being solid. Emulsions of known droplet size distribution are often used for determining the concentration of catalytic impurities and the nucleation rate as a function of temperature or supersaturation.

Nucleation rate greatly affects crystal size, since it depends far more strongly on supersaturation than does crystal growth. Hence at low supersaturation a low number of large crystals tends to be formed, at high supersaturation many small ones. A phase transition can also be achieved by seeding, i.e., by adding small particles of the phase to be formed. Nucleation then is not necessary, since the particles added can just grow out. This is often used in crystallization to regulate crystal size.

Nucleation of a **gas phase** inside a liquid or soft solid is virtually impossible. The Laplace pressure in a very small bubble is so high that the supersaturation needed to realize formation of such a nucleus cannot be reached. This means that initiation of the new phase occurs by seeding, i.e., by the outgrowth of little pockets of air. These can be present in small crevices at the vessel wall, or at the surfaces of fat crystals. Air can also be entrapped by agitation, e.g., when one kneads a dough.

Figure 14.11 also mentions **spinodal decomposition**. This occurs especially in polymer solutions. Beyond a certain supersaturation, any small fluctuation in composition leads to a local decrease in free energy and will spontaneously grow to form a large region. This means that nucleation is not needed. For a smaller supersaturation, nucleation is needed for phase separation to occur.

BIBLIOGRAPHY

There is a profusion of literature on nucleation phenomena, but it may cause more confusion than understanding, except for specialists in the field. Most textbooks on physical chemistry hardly discuss the subject, but books on surface chemistry give basic information, for instance

A. W. Adamson, A. P. Gast. Physical Chemistry of Surfaces, 6th ed. John Wiley, New York, 1997.

A classical monograph is

A. C. Zettlemoyer, ed. Nucleation. Marcel Dekker, New York, 1969.

although this book is to some extent outdated. Especially Chapters 1 (introduction) and 5 (nucleation in liquids and solutions) are useful.
     More up to date is

R. H. Doremus. Rates of Phase Transformations. Academic Press, Orlando, FL, 1985.

especially Chapters 5 (nucleation from condensed phases) and 6 (phase separation of liquids). The latter includes a brief discussion of spinodal decomposition.

An authoritative, though far from easy, discussion on classical and newer theories and their limitations is

D. W. Oxtoby. Nucleation. In: D. Henderson, ed. Fundamentals of Inhomogeneous Fluids. Marcel Dekker, New York, 1992, Chapter 10.

Nucleation phenomena in food systems are reviewed in Chapter 5 of

R. W. Hartel. Crystallization in Foods. Aspen, Gaithersburg, MD, 2001.

15

Crystallization

Crystals are often formed in foods and during food processing. It concerns ice, sugars, salts, triacylglycerols, and other, generally minor, components. Crystallization can have large effects on food properties, especially on consistency, mouth feel, and physical stability. Crystallization is mostly preceded by nucleation, which is discussed in Chapter 14. Effects on consistency are given in Chapter 17. Freezing of foods, i.e., freezing most of the water in a food, has several ramifications; it is the main subject of Chapter 16.

We will only consider "true," i.e., solid and three-dimensional, crystals. Liquid crystalline phases can be formed in aqueous solutions of some amphiphilic molecules (briefly mentioned in Section 10.3.1), and some materials in a densely packed adsorption layer can attain a two-dimensional crystalline order. Unless mentioned otherwise, crystallization from solution is implied in this chapter.

15.1 THE CRYSTALLINE STATE

Order. A crystal consists of a material in a solid state in which the building entities—molecules, atoms or ions—are closely packed so that the free energy of the material is at minimum. To achieve this, all of the entities have to be subject to the same interaction forces (magnitude and direction), i.e., be in the same environment. As a result the entities are arranged in a regularly repeating pattern or lattice. This is illustrated in Figure 15.1a.

In principle, the building entities have a *fixed, perfectly ordered, and permanent position*. In practice, this is not fully true, and the following points can be mentioned:

   The molecules (or atoms, or ions) are subject to *heat motion*. Hence only the average positions will be fixed.

   Occasionally, two molecules may interchange positions. In other words, *diffusion* can occur in a crystalline material, but the time scales involved are centuries rather than seconds.

   Crystals are virtually never perfect, but contain defects, often due to incorporation of a foreign molecule. A defect generally leads to a *dislocation* in the crystal lattice, as is illustrated in Figure 15.1b.

   Some solid materials are "*polycrystalline*", i.e., they are composites of many small crystalline domains of various orientations; see Figure 15.1c.

Noncrystalline solid states are discussed in Section 16.1.1.

Bonds Involved.   Crystals can be classified according to the nature of the main bonds keeping the building entities together. The following list is very roughly in the order of decreasing bond strength, heat of fusion, and melting temperature.

1.   *Covalent*. The prime example is carbon in diamond or graphite, where each carbon atom is covalently bonded to four others. Another example is quartz, $(SiO_2)_n$; in a large crystal—in fact one giant molecule—n can be as large as 10^{24}.

2.   *Metallic*. Metal crystals consist of an array of cations, through which electrons can freely move.

3.   *Ionic*. Salts typically form ionic crystals, i.e., a regular packing of cations and anions. In several of these, e.g., NaCl, the ions simply alternate in a cubic array.

4.   *Molecular*. This concerns, for instance, triglycerides and sugars. The bonds involved are due to van der Waals attraction and often also hydrogen bonds.

The crystal bonds in foods are almost exclusively of Types 3 and 4. Often, a combination of ionic and hydrogen bonds and van der Waals forces is involved. Consequently, we will consider the building entities of a crystal to be molecules, unless stated otherwise. Some crystals contain a fixed molar proportion of solvent molecules, e.g., potassium oxalate $(K_2C_2O_4 \cdot 2H_2O)$;

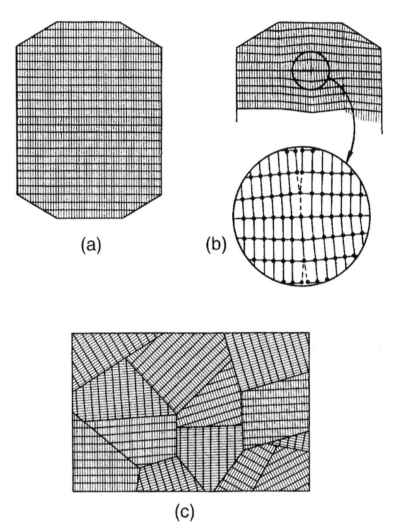

FIGURE 15.1 Two-dimensional illustration of crystalline order. (a) Example of a crystal lattice with perfect order. (b) Example of a defect in the crystal leading to a dislocation in the lattice. (c) A polycrystalline material. (From L. S. Dent Glasser. Crystallography and Its Applications. Van Nostrand Reinhold, Wokingham (Berks), 1977.)

the water molecules present allow a closer packing and/or a greater number of hydrogen bonds.

Crystal Systems. To understand several aspects of crystallization, it is useful to know about a few of the rudiments of *crystallography*.

Crystallographers study the geometrical structure of crystals. They consider an idealized picture of a crystal, ignoring the nonidealities mentioned above. A crystal then is defined as an *infinite real homogeneous discontinuum*. The infinity is introduced to keep the mathematics simple. The crystal is discontinuous at the scale of the building entities or smaller, but is homogeneous in the sense that at a larger scale all regions are identical.

To characterize a crystal lattice, straight lines are drawn through fixed points, such as the centers of the molecules or of a particular atom in each molecule. Plane faces then are constructed through parallel lines. The perfect order of an ideal crystal lattice guarantees that such straight lines and plane faces can be constructed. The planes will have a variety of directions and will thus enclose volume elements. The smallest volume element that contains all geometrical information about the lattice is called a *unit cell*. The whole crystal can be considered as a perfect three-dimensional stacking of unit cells. Figure 15.2a gives an example where the unit cell is a cube. There are molecules at each of the eight corners of the cube. Nevertheless, the unit cell contains the equivalent of one molecule, since each of the molecules is part of eight adjacent cells. This is not the only manner in which molecules can form a cubic unit cell. The two other possibilities are also shown. In (b) the unit cell contains two, in (c) four molecules.

It can be shown that 14 different unit cells or *Bravais lattices* can be distinguished (three of which are in Figure 15.2). By convention these are grouped into seven crystal systems (or by some authors into six). These are

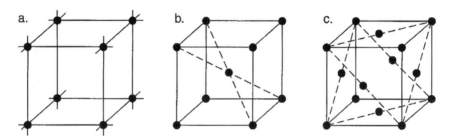

FIGURE 15.2   Cubic unit cells. The dots represent building entities (often single and identical molecules). The dotted lines are drawn to clarify the position of dots not at the corners. (a) Simple cubic. (b) Body-centered cubic. (c) Face-centered cubic.

illustrated in Figure 15.3, which gives the *primitive cells*, i.e., only the shape; the number of Bravais lattices in each system is indicated. The hexagonal system differs from the others in that the unit cell can be considered trigonal (if one allows $a = b \neq c$ in the trigonal system); however, three of these cells can be stacked to form a hexagonal prism, as illustrated.

As shown, the unit cells are characterized by three axes of length a, b, and c; and by three angles between the axes, α, β, and γ. The dimensions (lengths and angles) of the unit cell can generally be determined by x-ray diffraction. In principle, the orientation of the molecules in the cell can also be established.

A unit cell may contain one or more molecules, for instance, a sugar and a water molecule. In such a case, the pair of molecules is considered to be the building entity. If the cell contains two identical molecules, these do not have identical orientation: otherwise, the unit cell could also be defined as containing one molecule, and be half as large.

Anisotropy.   Since the distances between atoms or molecules, as well as the bond strengths between them, varies with direction (with respect to the axes of the unit cell), physical properties of a crystal may vary with direction. This is called anisotropy (cf. Section 9.1). For instance, the elastic modulus or the breaking strength of a crystal may depend on the direction of the force. Anisotropy tends to be more pronounced for crystal lattices in which the unit cell is more asymmetric.

A well known example is optical anisotropy, especially *birefringence*. The latter implies that the refractive index of the material depends on direction. When a narrow light beam passes through the crystal, it will be split into two beams, or even into three. It also means that a linearly polarized beam of light will attain elliptical polarization; such a material can be seen in a polarized light microscope if the specimen is between crossed polarizers. These phenomena do not occur if the light beam is parallel to the optical axis. As given in Figure 15.3, some crystal systems have one optical axis, others are biaxial; cubic crystals are isotropic and hence do not cause birefringence.

Crystal Morphology.   Since a crystal can be seen as a stack of unit cells, the simplest shape that a crystal can have directly follows from the unit cell: the *angles* between the faces are the same. Since the stacking need not amount to the same number of units cells in each direction, the relative sizes of the various faces van vary. Moreover, other (plane) faces can form, as illustrated in Figure 15.4 for the two-dimensional case. The lines denoted (10) and (01) are parallel to the axes a and b, respectively, but other lines can be drawn through corner points of unit cells. The same applies to the three-dimensional case, where plane surfaces rather than straight lines are

UNIT CELL	N	NAME EXAMPLES	AXES ANGLES	OPTICAL ANISOTROPY
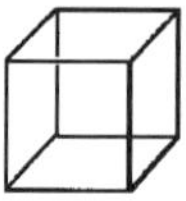	3	Cubic diamond, NaCl	$a = b = c$ $\alpha = \beta = \gamma = 90$	isotropic
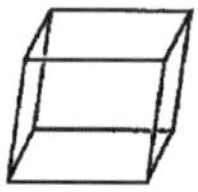	1	Rhombohedral (trigonal) calcite ($CaCO_3$)	$a = b = c$ $\alpha = \beta = \gamma \neq 90$	one axis
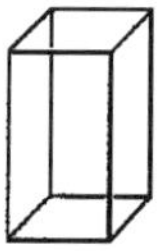	2	Tetragonal $NaClO_3$	$a = b \neq c$ $\alpha = \beta = \gamma = 90$	one axis
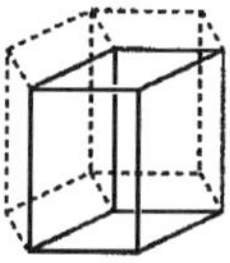	1	Hexagonal ice I, α-tripalmitate	$a = b \neq c$ $\alpha = \beta = 90, \ \gamma = 120$	one axis
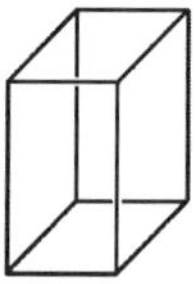	4	Orthorhombic β'-tripalmitate	$a \neq b \neq c$ $\alpha = \beta = \gamma = 90$	two axes
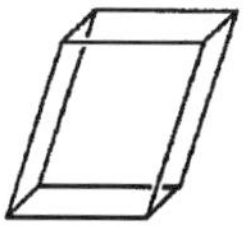	2	Monoclinic most sugars	$a \neq b \neq c$ $\alpha = \gamma = 90, \beta \neq 90$	two axes
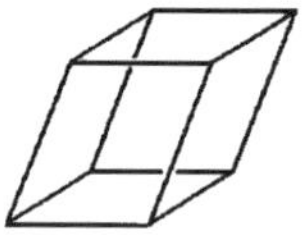	1	Triclinic β-tripalmitate	$a \neq b \neq c$ $\alpha \neq \beta \neq \gamma \neq 90$	two axes

FIGURE 15.3 Crystal systems. Shown is the shape of the unit cell for each system, the number N of different Bravais lattices in each system, geometrical characteristics of the unit cell, and optical anisotropy (birefringence). Some examples of crystals of various materials are given.

envisaged. Hence, cubic crystals can have angles between faces that do not equal 90° but also, for instance, 135°. The angles are measured inside the crystal; they are rarely larger than 180°.

To characterize the various faces, several systems have been devised. The most common one is the **Miller index** (*hkl*); the values of *h*, *k*, and *l* are established as follows. The coordinate system used is directly derived from the unit cell; this implies that the angles between axes can be different from 90°. The plane considered cuts off from the axes lengths of *A*, *B*, and *C* times the corresponding lengths of the unit cell (*a*, *b*, and *c*); these numbers can be $1, 2, 3, \ldots, \infty$. The Miller index of the face is now given by (N/A N/B N/C), where *N* is chosen such that *h*, *k*, and *l* are the smallest possible natural numbers, including zero (N/∞). This is illustrated in Figure 15.4 for two dimensions. Miller indices of crystal faces are, for instance, (100), (110) or (021); numbers larger than 3 rarely occur. It should further be noted that different faces can have the same index, because the axes have a sign and a face may cut through an axis on the positive side ("above" the origin) or on the negative side; cf. lines (21) and ($\bar{2}$1) in the figure (if need be, the negative

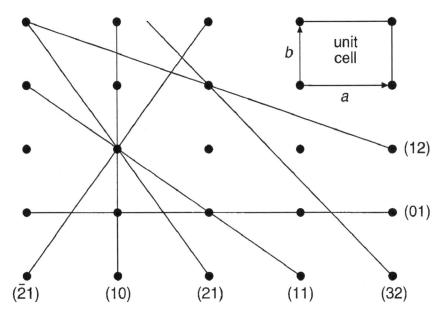

FIGURE 15.4 Illustration of various faces that can occur for a simple crystal lattice, and of the determination of the Miller indices (the numbers between brackets). Two-dimensional case.

value is indicated by a dash above the number). If a crystal has a face (110), it likely also has the faces ($\bar{1}$10), (1$\bar{1}$0), and ($\bar{1}\bar{1}$0), for symmetry reasons. Finally, the sense (and in some cases also the name) of the axes can be taken arbitrarily, but various workers use the same convention for a given crystal.

Crystals show enormous variation in **external shape** or **habit**, although all these shapes arise from ordered stacking of unit cells. This is illustrated in Figure 15.5, which speaks for itself. However, for a given unit cell, the angles between faces that can exist are fixed. This thus allows determination of the crystal system and of the dimensions of the unit cell from a crystal (if it is "perfect" and not too small). Figure 15.5 shows that considerable variation in shape can be encountered. Even far more intricate shapes are observed in ice crystals present in snow (which are formed by desublimation of water from the air). The variation in shape is caused by variation in the growth rate of the various faces of a crystal, which rates often depend on the composition of the solution; see Section 15.2.

In a perfect crystal, the edges (where two faces meet), and especially the corners, have a very *high curvature*. According to the Kelvin equation (10.19), this means that at the edge the solubility of the crystalline material is greatly enhanced. It is often observed by microscopy that corners and edges are not sharp but rounded. Small ice crystals (some micrometers in diameter) in foods such as ice cream are often seen to be almost spherical. In large crystals, curved faces are rare but not impossible. Some needlelike crystals have a slight twist, for instance.

Polymorphism. Polymorphic literally means multiform, but the term does not refer to variation in external shape. It indicates that crystals of the same molecules have different unit cells, be it of the same or of a different crystal system. The phenomenon is quite common. There are two types of polymorphism. *Enantiotropic polymorphs* each are stable within a certain range of temperature and pressure. Consequently, a phase diagram of the various polymorphs can be made. The prime example is ice (Section 15.3.1). If *monotropic polymorphs* exist, all but one of these are unstable. There is no phase diagram and, given time, only the most stable form will remain. The prime examples are compounds with long paraffinic chains, including most lipids (especially acylglycerols), where three main polymorphs exist (α, β', and β).

Strictly speaking, a material that can crystallize with and without solvent molecules, such as $CaCO_3$ and $CaCO_3 \cdot 6H_2O$, does not show polymorphism: the composition of both types of crystals is not the same. Neither is it polymorphism when the α- and the β-anomer of a reducing sugar, such as glucose, can crystallize, since these are different molecules.

Compound Crystals. Several molecular crystals contain more than one component. Substances crystallizing with solvent (mostly water) molecules have already been mentioned. There are several other examples, such as crystals containing α- and β-lactose in the stoichiometry $\alpha_5\beta_3$. If the composition of the crystals can vary continuously, be it in all proportions or within a limited range, one speaks of *solid solutions*. These can occur if the molecules involved are very similar and their packing in the crystal is not very tight. The prime example is, again, acylglycerols.

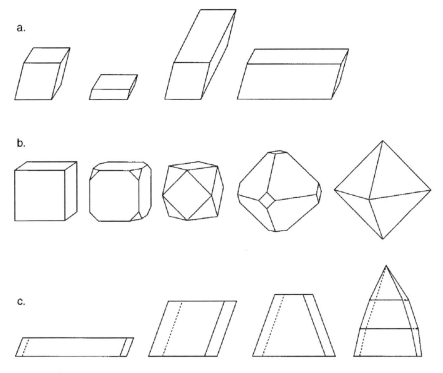

FIGURE 15.5 Variation in crystal morphology for identical unit cells. (a) Crystals in the rhombohedral system that only have faces present in the unit cell: (001), (010), and (100). (b) Cubic system; the leftmost picture is a cube, the rightmost one a regular octahedron; in between are intermediate shapes. (c) Examples of the various shapes that an α-lactose monohydrate crystal (monoclinic) can assume in practice. In (a) and (b) the shapes are shown in perspective. In (c) we have projections (all in the same direction with respect to the axes of the unit cell).

Question

Does the absence of certain crystal faces necessarily imply that such faces do not grow at all?

Answer

No. It can also be due to some faces growing faster than others. Referring to Figure 15.5b, the cubic shape at left results when the faces involved grow faster than the slanting faces depicted, whereby the latter disappear. If the relative growth rates are reversed, the octahedron shape at right results. For about equal growth rates of all of these faces, the shape depicted in the middle is formed.

> *Note* A face can also exist if it has not grown at all. An example is discussed at the end of Section 15.2.2.

15.2 CRYSTAL GROWTH

The rate of crystal growth can vary by several orders of magnitude. It is therefore important to understand the factors involved that control *crystallization rate*, which is often desirable during the processing or storage of foods. Moreover, the *shape* that a crystal will attain depends on the relative growth rates of the various crystal faces, which may vary markedly. Growth rate can be expressed in various ways. The growth of a given crystal face can be given in $kg \cdot m^{-2} \cdot s^{-1}$ (mass basis), or as the linear growth rate in $m \cdot s^{-1}$ (volume basis). The overall growth rate of a dispersion of crystals can be expressed in $kg \cdot m^{-3} \cdot s^{-1}$, or as $d\varphi/dt$, where φ is the volume fraction of crystals in the system.

Growth occurs by attachment (binding) of molecules (or ions, etc.) to a crystal surface. On the other hand, molecules will also become detached. There is thus a continuous transport of molecules across the crystal surface in both directions. The resultant of these two processes (reactions) determines growth rate.

Supersaturation. The rate of growth of a crystal face will always increase with the difference in chemical potential $\Delta\mu$ between crystal and solution or melt. For crystallization from solution we have

$$\Delta\mu = RT \ln a - RT \ln a_{eq} = RT \ln \beta \qquad (15.1)^*$$

where a is the activity of the solute and a_{eq} the activity at saturation, assumed to equal that at the crystal surface; $\ln \beta$ is called the supersaturation. According to Eq. (15.1) the *supersaturation ratio* $\beta = a/a_{eq}$, but it is usually taken as c/c_{eq}, where c is concentration. The latter ratio may significantly differ from the real value of β, since the solution may be far from ideal. Common causes of *nonideality* can be (a) that the properties of solute and solvent molecules are very different, and (b) that the solution is far from dilute. It is in most cases difficult, if at all possible, to determine the activities. Anyway, the supersaturation, and thereby the crystallization, rate will be zero if the concentration ratio c/c_{eq} equals unity. In principle, one would expect the linear growth rate of a crystal face to be proportional to the supersaturation, but experiments show this to be the exception rather than the rule.

It may further be noticed that for β close to unity, $\ln(c/c_{eq}) \approx (c - c_{eq})/c_{eq}$. The latter ratio is also called supersaturation (often expressed as a percentage) by some authors.

15.2.1   Growth Regimes

During growth, molecules diffuse to a crystal surface, and some of these are incorporated into the crystal lattice. Assuming (a) thermodynamic equilibrium at the crystal surface and (b) that each molecule arriving at the surface becomes incorporated, so-called normal growth will occur. Then the linear growth rate L_C(in m·s^{-1}) is proportional to $\beta - 1$, and the maximum (or Wilson–Frenkel) rate can be calculated; an example is in Figure 15.6.

However, there is generally no thermodynamic equilibrium at the crystal surface. Moreover, the chance that a molecule becomes and stays attached to a plane surface often is very slight, since the free energy gain per molecule is small. Substantial growth thus involves other processes. In principle, three regimes can be distinguished:

1. Thermodynamic roughening
2. Smooth surfaces with a number of subregimes
3. Kinetic roughening

In each of these the relations governing the incorporation of molecules (or ions, etc.) in the crystal are different. The regimes will be briefly discussed, without giving the full rate theories. (Actually, theoretical results generally are obtained from Monte Carlo simulations.) It is assumed that there is only one solute present.

Thermodynamic Roughening. A crystal in a solution that is precisely saturated may have faces that are not flat and smooth on a microscopic scale. This means that there is a transition zone of some molecules thick between the solid and the liquid phase. This roughness persists at increasing supersaturation. Whether roughening occurs depends on the surface entropy parameter

$$\alpha = \frac{4\varepsilon}{k_B T} \tag{15.2}$$

where ε is the net binding energy of a growth unit (often a molecule) to a flat crystal surface. If α is smaller than about 3—which will generally be the case for small molecules that are bound by van der Waals forces only—the surface tends to be rough. At such a surface, the growth rate is distinctly

growth rate

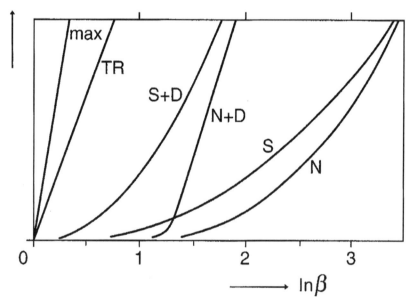

FIGURE 15.6 Normalized linear growth rate of crystal faces as a function of supersaturation $\ln \beta$. Calculated curves for thermodynamic roughening (TR), surface nucleation and growth (N), and spiral growth (S); for the last two cases additional calculations including the effect of surface diffusion (D) are also shown. The curve denoted "max" gives the hypothetical maximum for growth rate. Only meant to illustrate trends. (From results of Bennema; see in Bibliography.)

faster than at a smooth surface for a given supersaturation. The sticking probability of a molecule at a smooth surface is small because it involves only one intermolecular contact. At a rough surface, however, attachment of a molecule can involve contacts, i.e., bonds, with more than one molecule, considerably enhancing the chance that the molecule remains attached.

Growth rate in the regime of thermodynamic roughening is proportional to the value of $\ln \beta$ (see Figure 15.6, curve TR) and higher for a lower α value, because that means stronger roughening. Eq. (15.2) shows that the extent of roughening will also increase with increasing temperature, as would intuitively be expected. In practice, thermodynamic roughening is observed for a limited number of systems.

Smooth Surfaces. If the value of α is larger than 3.5, the crystal faces tend to be smooth and hard. This is the most common situation. Growth now has to occur via **surface nucleation**, followed by two-dimensional growth. This is illustrated in Figure 15.7. The activation free energy for attachment (adsorption) of a growth unit (often a single molecule) is given by

$$\Delta G^* = \frac{\gamma_x^* \gamma_y^* N_{AV}}{\Delta \mu} = \frac{\gamma_x^* \gamma_y^* N_{AV}}{k_B T \ln \beta} \tag{15.3}$$

where γ^* is the increase in edge free energy in the x- or y-direction upon adsorption of a growth unit. These quantities can be calculated (by use of

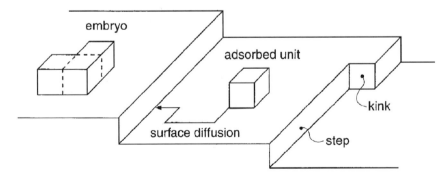

FIGURE 15.7 Illustration of attachment of growth units onto nonroughened crystal faces.

connected-network theory) from the various crystal bond energies, if they are known. It follows that ΔG^* will generally be different for each crystal face.

Note The parameter γ_x^*/x (in newtons), where x is the length of the growth unit in the x-direction, can be seen as a line tension, i.e., the one-dimensional analogue of surface tension.

At small supersaturation ln β, ΔG^* will thus be large, generally much larger than $k_B T$. This implies that adsorption of a single unit is unlikely and is soon followed by its desorption. If a greater number of growth units is adsorbed to form a two-dimensional *embryo*, the total edge free energy per unit will be smaller, because there is no edge between two adjacent units. On the other hand, the negative adsorption free energy will be proportional to the number of growth units in the embryo. If the latter has reached a certain size, further growth will lead to a decrease in total free energy, and two-dimensional growth will occur. This is analogous to the situation described in Section 14.2.1. The larger ln β, the smaller the radius of the nucleus, and the greater the rate of surface nucleation.

All this means that growth will primarily occur at the "step" formed by the nucleus (see Figure 15.7), since it is less likely that an adsorbed molecule will desorb if it is adjacent to two molecules on the surface rather than one. The step will thus move over the surface (from left to right in the figure). As depicted, a "kink" may also form, where a molecule can form bonds on three sides, being even more strongly bound. Hence the kink will also move. Taking these phenomena into account, modeling of the growth can be done, and an example (for a case where $\alpha = 8$) is shown in Figure 15.6, curve N. It is seen that the growth rate would be very much smaller than in the regime of thermodynamic roughening.

However, some modification will be needed. It has been shown that *surface diffusion* of an adsorbed molecule can occur and that it can meet a step (as illustrated in Figure 15.7) before it desorbs. Overall growth will then be much faster, as is illustrated in Figure 15.6, curve N+D.

Spiral Growth. When growth is two-dimensional, it will stop after a full monomolecular layer has formed. A flat surface then results and nucleation is needed again to induce growth. Especially at low values of ln β, where the nucleation rate is low, this will greatly slow down crystal growth. However, steps on a crystal face may also be caused by *dislocations*, most of which are due to inclusion of a foreign molecule or tiny particle in the crystal lattice. Some of these defects, called *screw dislocations*, tend to remain while molecules are attached in the step, to form a *growth spiral*, illustrated in Figure 15.8; such spirals can often be observed by microscopy, e.g., on growing sucrose crystals. The spirals greatly enhance crystal growth rate. A calculated curve (S) is in Figure 15.6 (the calculation is for the combined effect of spontaneous nucleation and spiral growth). The number of screw dislocations greatly depends on the purity of the system, and it is well known that very pure solutions tend to give very slow crystal growth. The number also depends on the crystal face.

Also for spiral growth, the rate can be enhanced by surface diffusion. Figure 15.6 gives a calculated example (curve S+D), but it should be noted that there is no agreement among researchers on the importance of this effect. Spiral growth is quite common in practice, and the growth rate then would be probably somewhere between the curves N+D and S+D. It should be added that Figure 15.6 gives examples and that the actual growth curves within each regime vary widely among systems, as discussed in the next section.

Kinetic Roughening. If the value of ΔG^* becomes of order $k_B T$, which means $\ln \beta \approx \gamma_x^* \gamma_y^* / (k_B T)^2$, the nucleus size becomes about one growth unit. This implies that many nuclei form on the crystal face, which thereby becomes rough. The growth rate of such a face is thereby enhanced. Another

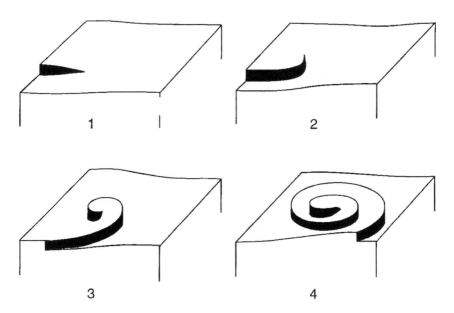

FIGURE 15.8 Illustration of the development of a growth spiral on a crystal face; stages 1–4.

consequence is that the face does not remain flat on a μm scale, but becomes *curved* (convex). For most crystal faces a very high supersaturation is needed to achieve kinetic roughening. It has been observed for some of the faces of paraffin crystals at high $\ln \beta$. The shape of these crystals then becomes almost like a very flat cylinder.

15.2.2 Factors Affecting Growth Rate

The linear growth rates of crystal faces vary enormously. Some approximate examples of the average rate for crystallization from solution are

Sucrose	$\ln \beta = 0.1$	Rate $\approx 100\,\text{nm} \cdot \text{s}^{-1}$
α-Lactose hydrate	$\ln \beta = 0.4$	Rate $\approx 1\,\text{nm} \cdot \text{s}^{-1}$
Compound fat	$\ln \beta = 0.4$	Rate $\approx 0.05\,\text{nm} \cdot \text{s}^{-1}$

Moreover, the rates generally vary between the faces of one crystal, and it is not uncommon that some faces do not grow at all under some conditions.

Below, the main variables affecting growth rate are discussed.

Supersaturation. This has been discussed above. Note that not only the rate at a certain value of $\ln \beta$ varies but also the kind of relation between rate and $\ln \beta$. Approximating the relation to $L_C \propto (\ln \beta)^n$, the exponent varies considerably, say from 1 to 4, with $n \approx 2$ being a fairly common value.

Transport to the Crystal Surface. It is the supersaturation at the crystal surface that is determinant and that may be below the average value. If the rate of attachment to the crystal is high, the diffusional transport may be too slow to keep up with the local depletion of solute, and the rate of crystallization will be *diffusion limited*. See Section 5.2.2 for theory on diffusion. If (mild) stirring enhances the crystallization rate, there is diffusion limitation. A clear example is crystallization of sucrose during sugar boiling. Even when stirring, diffusion has to occur through a laminar boundary layer, thickness about 10–100 μm.

If the rate of attachment is small, the system tends to be ideally mixed, and the crystallization rate then is *reaction limited* or interface controlled. Of course, intermediate situations can occur. In food processing and storage, most crystallization is reaction limited. However, in foods of high viscosity or with narrow pores through which the solute has to move, diffusion may be the limiting step.

Crystallization Regime. This has been discussed in the previous section. It should be realized that properties of the crystal primarily determine which regime applies. This concerns (a) the magnitude of the various bond energies between the molecules in the crystal and (b) the concentration and the nature of dislocations present, screw dislocations in particular. The latter greatly depends on the impurities present. The density of screw dislocations tends to vary between crystal faces.

Fitting Difficulty. When argon crystallizes (say, at $-200°C$), there is no difficulty of incorporating the perfectly spherical atoms in the crystal lattice. If an atom arrives at a vacant site it will always fit. For most molecular crystals the situation is quite different. Consider a sugar molecule, say a disaccharide. If it arrives at a vacant site on a crystal, it has to be in a specific *orientation* to become incorporated; otherwise it will probably diffuse away. This will markedly slow down the growth rate. Another property of many, especially large, organic molecules is that they can assume various *conformations*, and only one of these will fit in a vacant site. This phenomenon also decreases growth rate.

A clear example of fitting difficulty is given by *triglyceride* molecules, which can assume many conformations. The loss in entropy upon crystallization is thus very large. Figure 15.9 gives data on the growth and the dissolution rates of (a face of) a trilaurin crystal. Dissolution does not involve any fitting difficulty. If the same were true for growth, the growth curve would just be an extrapolation of the dissolution curve, but it clearly is not. For small values of $|\ln \beta|$, the difference in slope is by about three orders of magnitude.

Note At $\ln \beta = 0$, the rates at which molecules become incorporated into and dissolve from a crystal face is equal and opposite. This means that the curve must exhibit a gradual change in slope near the origin. The accuracy of the experiment is not nearly good enough to check this.

Most likely, triglyceride molecules do not have to be in an exactly correct conformation to become attached to the crystal: that would take too long. Presumably, they become at first partly attached and then adsorb further segment by segment. Comparison of theory and experiment leads to the conclusion that about 80% of a molecule needs be in the correct conformation before it is permanently attached (at least if $\ln \beta > 0.4$).

This also means that the molecule does not have to arrive in a perfect orientation, although some orientation is certainly needed. Long-chain molecules in solution near a crystal surface tend to be oriented more or less parallel to this surface. If they have to become incorporated in a direction about perpendicular to the surface, this misorientation causes an additional decrease in growth rate. This may be the main reason why triglyceride crystals tend to be rather flat platelets, since the molecules in the crystal are about perpendicular to the largest, i.e., the slowest growing, faces.

Desolvation. As discussed in Section 3.2, solute molecules or ions can be solvated, which means that one or more solvent molecules are associated with the solute. This is especially true for ions or ionized groups in aqueous solutions; the solvation (in this case hydration) then is due to ion–dipole interactions, which are fairly strong, particularly for positive groups (cations). Before an ion can be incorporated into a crystal lattice, desolvation has to occur. This implies a temporary increase in free energy, causing an activation barrier for crystallization. This will slow down the crystallization rate.

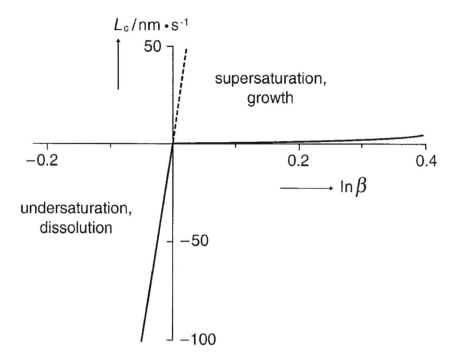

FIGURE 15.9 Rate of growth and dissolution L_C of a certain crystal face of a β-crystal of trilaurin in triglyceride oil as a function of supersaturation (ln β). (After results by W. Skoda, M. van den Tempel. J. Crystal Growth 1 (1967) 207.

Competition. It often happens that some kinds of "foreign" molecules fit more or less in a vacant site in the crystal. These molecules then compete with those of the crystallizing species for attachment sites. Generally, an incorporated foreign molecule locally prevents further crystallization, and it has to move out before a truly fitting molecule can fill the site. The overall result is slowing down the crystallization.

We will illustrate the phenomenon for the crystallization of *lactose*. Lactose is a reducing sugar, and in solution the α and β anomers are in equilibrium with each other; the ratio β over α is about 1.6. The crystallization of α-lactose monohydrate, the most common crystalline form, has been studied in detail. Figure 15.10 depicts a crystal as formed at

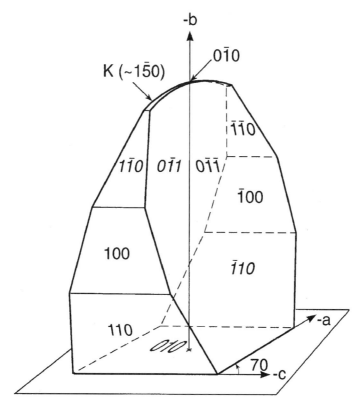

FIGURE 15.10 Common shape of an α-lactose monohydrate crystal. The crystallographic axes (a, b, and c) and the Miller indices of the faces are given.

low supersaturation. It has a very strange shape. The crystal has started to grow at the apex i.e., at the top in the picture, and the $(0\bar{1}0)$ face does not form. Also the $(0\bar{1}1)$ and $(0\bar{1}\bar{1})$ faces do not grow, but they emerge due to the growth of other faces. The main explanation is competition by β-lactose; presumably, its adsorption on the faces mentioned almost completely prevents incorporation of α-lactose. The concentration of β-lactose can be varied and then kept roughly constant for a time that is long enough to allow determination of growth rates. Some of such results are given in Figure 15.11, and it is seen that β-lactose can strongly reduce growth. This is part of the explanation of the very slow growth of lactose crystals as compared to sucrose crystals: sucrose has no anomers.

The extent of growth reduction varies widely with the crystal face, thereby strongly affecting **crystal shape** (habit). In the example given, especially both (010) and both $(0\bar{1}1)$ faces grow fast in the absence of β-

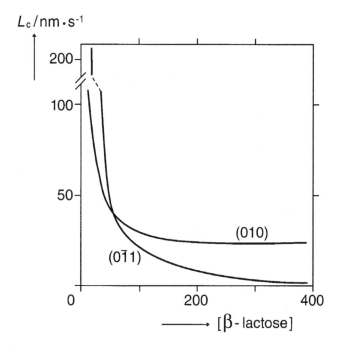

FIGURE 15.11 Growth rate L_C of two faces (Miller indices indicated) of a crystal of α-lactose monohydrate as a function of the concentration of β-lactose (in g per kg water). Supersaturation of α-lactose: $\ln \beta \approx 1.0$. (After results by A. van Kreveld. Neth. Milk Dairy J. 23 (1969) 258.)

lactose. In Figure 15.5c, the shapes indicated from left to right in fact represent a series of crystals grown at increasing concentrations of β-lactose.

Competition will generally be stronger when more competing molecules are present. A prime example is crystallization of a multi-component fat, which contains several molecules that are only slightly different from each other, often in relatively high concentrations.

Inhibition. Some substances, however, can inhibit growth of a crystal face when present in quite low concentrations. The simplest explanation is that such components are surface active and adsorb onto crystal faces. The adsorbed molecules then have to desorb before molecules fitting the crystall lattice can be incorporated. If the inhibition is so strong as to prevent growth of one or more crystal faces, the phenomenon is called *poisoning*. Examples are given in Table 15.1 for two components strongly inhibiting growth of some faces of α-lactose. Many other cases have been observed, but a specific molecular explanation is usually lacking. A "poisoning" material generally does *not* inhibit nucleation.

Finally, Table 15.1 confirms that an increase in supersaturation has a very large effect on growth rate, and that the relative increase in rate significantly varies between different faces. The value of ln β can thus affect crystal shape.

Question

When studying crystal growth in a dispersion of crystals in a supersaturated solution, it can be observed that the linear growth rate is greater for small than for large crystals. In other cases, the opposite is observed. Can you propose explanations?

Answer

(a) If the growth is diffusion limited, lowering of the concentration of solute will be stronger near a large than near a small crystal, because of the difference in surface area, hence in the amount of solute that can be incorporated in the crystal. The local supersaturation then is greater for a small crystal, which will thus grow faster. A prerequisite is that the average distance between crystals be substantially larger than crystal size. (b) If the growth rate is reaction limited, another effect can become dominant. The material in a small crystal has a higher solubility than that in a large crystal, according to the Kelvin equation (10.9), because of its stronger average curvature. Hence, local c_{eq} will be higher and $c - c_{eq}$ is smaller, whereby supersaturation is smaller near a smaller crystal and growth will be slower. Prerequisites are that the small crystals be quite small (say, radius $< 0.1\ \mu m$), and that ln β be low. Why?

TABLE 15.1 Rate of Growth of Some Faces of α-Lactose Monohydrate Crystals for Various Solvent Compositions

Supersaturation		Growth (nm·s⁻¹) of face[a]				
$\ln \beta$	Remarks on composition	010	110	100	$1\bar{1}0$	$0\bar{1}1$
0.44	–	1.05	0.92	0.36	0.08	0.00
0.44	+ 10 ppm gelatin	0.33	0.28	0.28	0.11	
0.44	+ 100 ppm riboflavin	0.75	0.00	0.00	0.00	0.00
0.79	–	12	9.4	5.8	3.3	1.9

[a]Indicated by Miller indices.
Source: After results by A. S. Michaels and A. van Kreveld. Neth. Milk Dairy J. 20 (1966) 163.

15.2.3 Overall Crystallization Rate

In practice, one is generally not so much interested in the linear growth rate L_C as in the amount (mass or volume) of crystals formed per unit volume and unit time. In principle, the latter can be given as $L_C \times A_C$, where A_C is the specific surface area of the crystals. However, both factors tend to vary with time. Generally, L_C decreases because crystallization implies depletion of solute and hence a decrease of supersaturation. Moreover, release of the heat of fusion may cause the temperature to increase significantly, hence the solubility to increase, hence $\ln \beta$ to decrease. A_C increases because (a) each crystal increases in size, and (b) more crystals are formed if nucleation goes on. Several, often complicated, growth rate theories have been worked out for various conditions. We will only touch on a few aspects.

Nucleation and Growth. A common situation is when a homogeneous solution (or melt) is rapidly cooled to a constant temperature where $\ln \beta$ is large enough to induce nucleation. Then the semiempirical **Avrami equation** often applies:

$$\varphi(t) = 1 - \exp(-Kt^n) \approx Kt^n \qquad (15.4)$$

where φ = volume fraction of crystals, t = time, and K and n are about constant for a given system. The part after the $\approx$ sign applies for small φ, say < 0.1. The exponent n ranges between 0.5 and 4. It is larger for reaction-limited than for diffusion-limited crystallization; larger for constant-rate than for depletion-dependent nucleation; and smaller for a more anisometric crystal shape.

For the case of (hypothetical) spherical crystals and reaction-limited crystallization, the Avrami equation takes the form

$$\varphi(t) = \frac{1}{3}\pi L_C^3 J t^4 \tag{15.5}$$

where the growth rate L_C and the nucleation rate J are assumed to be constant; the equation will thus only apply in the very beginning of the process. When the dependence of L_C and J on $\ln \beta$ is known, a more realistic numerical relation between φ and t can be derived.

An example of experimentally observed growth curves, for various values of the *initial* supersaturation $\ln \beta_0$, is given in Figure 15.12a; under the conditions of the experiment, the hardened palm oil will behave almost like a single solute (approximately like tripalmitate). It is seen that there is a kind of *induction time* t_{ind} for crystallization to take off. In principle, $t_{ind} \propto 1/J$, whether homogeneous, heterogeneous, or secondary nucleation occurs. The growth rate then rapidly increases with time, for the most part owing to an increase of A_C. At higher values of φ the rate slows down because the decrease of $\ln \beta$ becomes overriding. A recalculation of the results to the dependence of L_C on $\ln \beta$ is given in Figure 15.12b, assuming

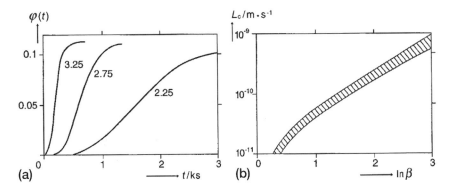

FIGURE 15.12 Rate of crystallization (in the β'-polymorph) of 12% hardened palm oil in sunflower oil. (a) Dependence of the volume fraction of crystals $\varphi(t)$ on time t for various values of the initial supersaturation $\ln \beta_0$ (indicated near the curves), varied by varying temperature. (b) Recalculated linear growth rates L_C during the process as a function of supersaturation $\ln \beta$; the hatched region contains all the curves for $\ln \beta_0 = 2.25$–3.50. (After results by W. Kloek. Ph.D. thesis, Wageningen University, 1998.)

spherical crystals and taking $J(\ln \beta)$ from independent measurements. It is seen that the results almost fit a single curve, as is to be expected.

Crystal Size Distribution. Crystal size was briefly mentioned in Section 14.2.3; Figure 14.7 gives a rough indication of the effects of supersaturation on growth rate, nucleation rate, and crystal size.

If crystallization occurs in a steady state so that supersaturation is kept constant—which can, e.g., be realized in a continuously operating crystallizer—the following equation for the number frequency of crystals of radius r will hold:

$$f(r) = \frac{J}{L_C} \exp\left(\frac{-r}{L_C \tau}\right) \tag{15.6}$$

See Section 9.3.1 for the mathematics of size distributions. τ is a characteristic time scale; it would be the mean residence time in the crystallizer mentioned. A plot of $\log f(r)$ versus r is a straight line of slope $-1/L_C \tau$. This generally implies a wide size distribution. The number average radius r_{10} is simply given by $L_C \tau$. In a steady-state crystallizer, the average size is thus independent of nucleation rate.

However, in many situations τ would be a kind of induction time for nucleation and hence be proportional to $1/J$. In such a case, the average r then is proportional to L_C/J, as may be intuitively expected. In most situations, the relations will be more complicated, although the trends are as given. If copious secondary nucleation occurs, the crystals formed will be quite small, and it is not simple to achieve formation of large crystals.

Agitation. Often, crystallization from solution occurs while the mixture is stirred or otherwise agitated. This generally increases the overall crystallization rate, and it may also affect crystal size. The following effects can be involved:

Sedimentation and aggregation of crystals is generally prevented.
For diffusion-controlled crystal growth, crystallization rate is enhanced.
More effective removal of the heat of fusion results in a higher supersaturation near the crystal surface.
Nucleation rate is generally enhanced because (a) the effective value of $\ln \beta$ is larger; (b) any heterogeneous nucleation at macroscopic surfaces will be enhanced; and (c) secondary nucleation can be enhanced, be it due to contact nucleation, breaking off protrusions from crystals, or "sweeping off" clusters of oriented molecules from growing crystal faces.

Agitation also tends to decrease the width of the crystal size distribution, because local variation in supersaturation is reduced.

Question

Assume a solution supersaturated with one crystallizable solute. To part of the solution, a crystal growth inhibitor is added. It is established that in both cases the relation $L_C = a(\ln \beta)^2$ holds, only the constant a being different. Assume now that both solutions have the same initial, rather weak, supersaturation and that crystallization is started by adding tiny seed crystals. Would the ratio of the overall crystallization rates $d\varphi/dt$ now be the same as the ratio of both a values?

Answer

Assuming for the moment L_C to remain constant in time, and the crystals to have spherical shape, we may assume that

$$\frac{d\varphi}{dt} = N\pi r^2 L_C$$

where N is the number of crystals per unit volume. r is crystal radius, and it will of course increase during crystallization, its magnitude being given by $L_C t$ (neglecting the radius of the seed crystal). Insertion and integration then yields

$$\varphi = \frac{1}{3}\pi N L_C^3\, t^3 = \frac{1}{3}\pi N a^3 [\ln \beta(t)]^2\, t^3$$

The growth rate would thus be proportional to a^3. Since β decreases with time in a manner depending on the value of a, the relation will become more complicated, but the growth rate will be always more than proportional to a.

15.3 CRYSTALLIZATION FROM AQUEOUS SOLUTIONS

Crystallization in aqueous systems is by far the most common case in foods and food processing. Either water or the solute(s) can crystallize, and in some cases both. We will in this section primarily consider crystallization of a single solute. Moreover, the formation and properties of ice are briefly discussed.

15.3.1 Ice

Solidification. When pure water is cooled to below 0°C and heat is continuously removed, all of the water will crystallize (assuming for the moment that ice nucleation occurs readily). This is a case of *crystallization from the melt*, often called solidification. Crystallization is a cooperative transition and it completely occurs at the equilibrium temperature, i.e., 0°C. That means that we cannot speak of a supersaturation. Proceeding from Eq. (14.3), we derive the driving force for crystallization, i.e., the difference in chemical potential between ice and water, as

$$\Delta\mu = \Delta_{L \to S}G = \Delta_{L \to S}H(1 - \frac{T}{T_{eq}}) \approx 22(T_{eq} - T) \qquad (15.7)$$

where the part after the $\approx$ sign is valid for ice formation, if expressed in SI units. It is assumed that ΔH and ΔS are independent of temperature, which is not precisely true. See Appendix I for further data.

In all foods, the water contains solutes, but the ice formed will have the same crystal structure as that formed from pure water.

Crystal Structure. Water can solidify in eight different stable modifications or enantiotropic polymorphs. Seven of these form at very high pressures—generally above 200 MPa—and need not be considered here. The ice formed under common conditions is called ice I; the crystal system is hexagonal. The unit cell contains four water molecules; these are for the most part kept together by hydrogen bonds, leading to a fairly open structure, as is reflected in the low density. The other crystal modifications have far higher densities.

Hydrogen bonds are fairly strong; according to Table 3.1 they are of the order of 10 RT or 22 kJ per mole. In ice, every molecule is H-bonded to four neighbors, in water to about three; this would mean a decrease of 0.5 bond per molecule, corresponding to about 11 kJ · mol^{-1}. This is of the same order as the enthalpy of fusion: 6 kJ · mol^{-1}. The latter value is high, and also the melting temperature of ice is high, for such a small molecule. The relatively open structure (low density) of ice is also due to hydrogen bonding. The energy of these bonds is strongly dependent on bond angle, and to minimize the free energy the molecules in ice are so oriented that the bonds (O—H···O) are almost linear.

"Anomalies." Figure 15.13 gives some properties of water at low temperature. It is seen that water has its maximum density at 4°C, and that also below 0°C density keeps decreasing with decreasing temperature. The density of ice is lower, as explained above. Also the strong increase of the viscosity of water at decreasing temperature, despite the decrease in density, is uncommon. So is the strong increase in c_p at low temperatures. All these phenomena have to do with the particularities of hydrogen bonding, but we will not go further into possible explanations.

The anomalies have some important consequences. The expansion of the system due to ice formation can cause high pressures. Nearly everybody knows about the bursting of water pipes that become frozen. Some

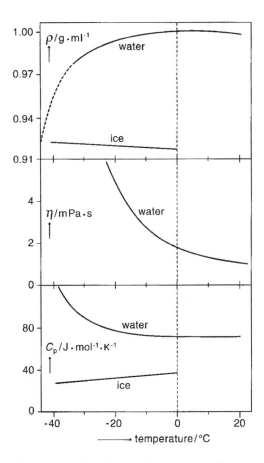

FIGURE 15.13 Some properties of ice and (undercooled) water as a function of temperature. Given are mass density ρ, viscosity η, and specific heat at constant pressure c_p, all at atmospheric pressure.

consequences for foods will be discussed in Section 16.3. Another consequence of the density difference is that dispersed ice crystals tend to cream.

The expansion tends to be smaller if solutes are present. The main reasons are that generally (a) not all of the water will freeze and (b) that (most of the) freezing occurs at lower temperatures, where $\Delta\rho$ will be smaller (see Figure 15.13). For highly concentrated solutions, e.g., of sucrose, the expansion can even be negligible.

The high values of the thermal conductivity and the thermal diffusivity of ice (see Appendix I) imply that ice can conduct heat better than (immobilized) water.

Impurities. Water is never absolutely pure. For one thing, it contains H^+ (or H_3O^+, rather) and OH^-. Upon crystallization, most of these ions form H_2O, but some ions remain; the concentration of each in ice is about 10^{-10} molar. This gives ice a finite electrical conductivity. Air is another impurity that is nearly always present. Upon freezing, the remaining water becomes gradually more supersaturated, often resulting in the formation of air bubbles in the ice.

Particulate impurities are needed for heterogeneous nucleation, and these are always present (natural waters tend to start freezing at, say, $-1°C$). The formation of air bubbles is also dependent on impurities, some of which may contain a tiny air pocket (see Section 14.4); some natural ice is opaque, due to the numerous small bubbles it contains. Other impurities, be they dissolved substances or particulate matter, can greatly affect the size and morphology of the ice crystals.

Growth Rate. The growth of ice from "pure" water is not diffusion limited: the crystal is bathed in water molecules. The growth rate then is expected to be proportional to $-\Delta\mu$, hence to $(T_{eq} - T)$. Moreover, there is no competition, and the fitting difficulty of water molecules is only slight. Growth will thus be very fast, e.g., some $cm \cdot s^{-1}$ at 10 K undercooling.

In the presence of solutes, growth rate can be much reduced. Most substances that are well soluble in water do this, and at high concentration (as will occur when most of the water has become frozen) the growth rate may be decreased by, say, two orders of magnitude. Causes can be

The solute causes a *freezing point depression*, and at the same temperature the undercooling will be smaller for a higher molar solute concentration.

The absolute value of $\Delta\mu$ will be smaller, since now the *water activity* is decreased, the more so for a higher molar solute concentration.

At high solute concentration and low temperature, the viscosity of the
system may be greatly enhanced, whereby the diffusion coefficient of
water is reduced. *Diffusion limitation* may now occur.

Gums present in small concentrations can reduce growth rate by one
to two orders of magnitude, but only if they form a firm gel. Gums
that merely increase viscosity have very little effect.

Some solutes may cause *inhibition*, i.e., reduce growth at quite low
concentrations. Several, generally polymeric, substances can adsorb
onto ice crystals and thereby reduce the freezing rate. This includes
many proteins. Especially the *antifreeze peptides*, found in many
plants and cold-blooded animals that have been subjected to cold
stress, strongly decrease growth rate at fairly low concentrations.
The presence of solutes and particulate material can also affect the
shape of the ice crystals.

In practice, the rate of ice formation during freezing of foods is nearly
always limited by the rate of *heat removal*. Even at a few degrees of
undercooling, a considerable portion of the water can freeze, and this
produces as much as 334 J of heat per g water frozen. Moreover, the system
will soon become more or less solid, preventing convection and agitation.
Hence the undercooling at the ice crystal surface is reduced to small values
(often much below one kelvin). Linear crystallization rate then is rarely
more than some micrometers per second; this is, however, still fast
compared to the rate obtained for crystallization of most substances from
solution.

Dendritic Growth.   Occasionally, ice crystals exhibit the formation
of slender protrusions or *dendrites*. A crystal surface will often be slightly
irregular, and a local bulge may occur. In most cases, heat of crystallization
will be removed predominantly through the ice, because of its relatively high
thermal diffusivity. This means that the bulge will be at a somewhat higher
temperature than the adjacent flat surface, causing its growth to become
retarded; thus the interface will become flat again. However, if the ice crystal
face is in contact with undercooled water (or solution), an opposite
temperature gradient is formed. At the crystal surface the temperature
equals T_{eq}; further into the liquid it is lower by a few kelvins. Now a bulge
will grow faster, because its heat of crystallization can be more readily
removed than that at the flat surface. This means that the bulge develops
into a longer protrusion, thereby further increasing the difference in
temperature and hence in rate of growth. In this way several dendrites are

formed at the crystal surface. This is a kind of roughening, though at a much larger scale than the roughening phenomena discussed in Section 15.2.1; the diameter of a dendrite may be several μm, the length, e.g., 0.1 mm.

Dendritic growth will mostly stop after a short while, since the additional crystallization causes the release of more heat of melting, leading to evening out of the temperature gradient. In aqueous solutions the situation is more complex, but if the freezing point depression is smaller than the local temperature difference, dendritic growth can occur. Another condition is that the linear growth rate has to be very high, a situation that is not very common during the freezing of foods.

Crystal Size and Shape. As mentioned, ice I crystallizes in the hexagonal system, and ice crystals can occur in a wide variety of shapes, with a typical threefold symmetry. Such shapes are primarily found in snowflakes; the explanation is somewhat similar to that for the formation of dendrites. During the freezing of liquid foods, simpler forms generally result. Hexagonal plates or prisms can be observed, but small crystals tend to have rounded shapes. The crystals often are fairly small; for example, a typical ice crystal in ice cream may be 40 μm. This may be due to substances inhibiting growth rate: since these generally do not inhibit nucleation, as mentioned, many nuclei can form before much ice has formed, and small crystals result.

Sintering. The Hamaker constant for ice in water $(A_{11(3)})$ is small, about 0.04 times $k_B T$. Nevertheless, for crystals of several μm in radius, the van der Waals attraction would be large enough to cause their aggregation. Moreover, ice crystals will cream, forming a layer. Touching ice crystals will readily sinter when crystallization is not yet complete. They can then form a polycrystalline mass, such as depicted in Figure 15.1c.

In many dispersions of ice crystals, however, the latter are clearly separate from each other; see for example Figure 9.1, depicting ice cream. This may be due to colloidal repulsion between crystals. Several substances can presumably adsorb onto ice, like proteins and ions, providing steric and electrostatic repulsion. However, this has not been systematically studied to the author's knowledge.

15.3.2 Phase Diagrams

Eutectics. Figure 15.14 gives a partial phase diagram of the binary mixture D-fructose and water. The solid curves given are called *coexistence lines*. Such a line gives the boundary between two phases; on any point on the line these phases are in equilibrium with each other. The diagram shows four different phases and five regions. The region below the broken line denotes a mixture of two phases: ice crystals and fructose $\cdot 2H_2O$ crystals. The diagram also has two coexistence points (C and D) where three phases are in equilibrium.

Suppose we have a solution of composition and temperature given by point A. If we now decrease the temperature, it will at some time reach that of point A′. At that moment ice can begin to form. Ice formation does not begin at 0°C, since the fructose causes depression of the freezing point: see Section 2.2.4, Eq. (2.16). As more ice is formed, the solution becomes more concentrated and the freezing point will be further depressed. The composition of the solution and the temperature of ice and solution will change along the coexistence line until point C is reached. Then the whole system will freeze, forming ice and sugar crystals. C is called the *eutectic point*, characterized by the eutectic temperature T_e and the eutectic composition ψ_e.

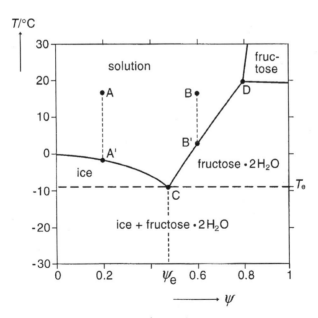

FIGURE 15.14 Partial phase diagram of the mixture water and D-fructose. T is temperature; ψ is mass fraction of fructose; the subscript e refers to the eutectic point.

Suppose now that we evaporate water from the initial system to reach point B. Cooling then causes the formation of fructose $\cdot$ 2H$_2$O crystals when point B$'$ is reached, whereby the solution becomes more dilute, and the coexistence line is followed until C is reached, when also ice starts to form. In other words, cooling a solution of composition $\psi < \psi_e$ leads to formation of ice (as is desired in freeze concentration), whereas sugar crystallizes for $\psi > \psi_e$ (as is desired in sugar manufacture). At temperatures below T_e only crystals are present; upon heating to above T_e melting occurs.

Most water-soluble solids, like salts and sugars, show eutectic phase behavior, which means that they have phase diagrams of the type depicted for fructose. Some eutectic points are as in the table.

Fructose	$T_e = -9°C$	$\psi_e = 0.48$
Sucrose	$T_e = -14°C$	$\psi_e = 0.63$
NaCl	$T_e = -23°C$	$\psi_e = 0.245$

Both T_e and ψ_e vary widely among solutes. Further details of the phase diagram can also vary. For instance, it is seen in the figure that fructose has two crystalline phases (with and without water), but other systems are different, being either simpler or having more phases. Some mixtures do not show eutectic behavior; see Section 15.4.3.

It should be stressed that a phase diagram depicts an *equilibrium* situation, and it will take time, sometimes a long time, before equilibrium is reached after conditions are changed. These aspects are discussed below.

Cooling Curves. Figure 15.15 gives examples of the temperature evolution resulting when heat is removed at a constant rate. The graph would apply to a small amount of material and slow heat transfer rate, since otherwise significant temperature gradients in the sample will result, leading to complex relations.

We will first consider **pure water** (curve W). Starting at A, the temperature will fall almost linearly with time, since the specific heat of water hardly changes over the temperature range involved (see Figure 15.13). Cooling goes on to a point B where T is clearly below the freezing point (0°C). This is because undercooling is needed to cause ice nucleation. It will greatly depend on the purity of the water and on the rate of cooling at what temperature nucleation, hence freezing, will start. Because of the release of heat of crystallization, T will increase to 0°C (point C), and it will remain at that value until all of the water is frozen (F). Subsequently, the ice will cool at a rate about twice that of water, because of its smaller specific heat.

Consider now curve S for a **solution** of one compound, e.g., a sugar, in water. The heat capacity will generally differ somewhat from that of pure water, leading to a different slope. Cooling will proceed to a point B′, generally at a lower temperature than B, since the initial freezing temperature T_f is lower than that of water. Again, nucleation has to occur, and after that ice is formed (at B′) and heat is given up until the equilibrium freezing curve is reached at point C′. Notice that this is at a temperature below T_f, because the solution is already somewhat concentrated. Ice formation, and hence the concentrating of the remaining solution, and temperature decrease go on until a point D′. The system is then undercooled with respect to solute crystallization, again because nucleation has to occur first. After that, heat of crystallization is given up, the temperature rises to E′, after which equilibrium crystallization of both solute and water occur, until everything is frozen (point F′). Finally, the mass is further reduced in temperature.

Upon *slow heating* of the frozen system, about the same curve results in the opposite direction, but now the equilibrium lines will be followed. Hence, the "dips" near D′ and B′ will not occur. Neither will nucleation cause delay of melting, since initiation of solid–liquid transitions generally needs negligible overheating (Sections 14.1 and 14.2.2).

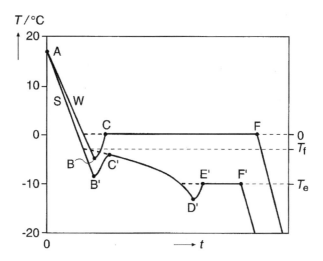

FIGURE 15.15 Examples of cooling curves for water (W) and an aqueous solution (S). The rate of heat removal ($J \cdot kg^{-1} \cdot s^{-1}$) is constant. T is temperature, t is time. T_e is eutectic and T_f initial freezing temperature of the solution. Broken lines denote hypothetical relations if no undercooling were to occur.

The situation becomes more complicated when more solutes and especially when particulate material is present. It is not uncommon that crystallization of solutes does not occur within a reasonable time. We will come back to this in Chapter 16.

Question

It is often useful to know the freezing point depression of a solution. An example is given by the detection of adulteration of milk due to dilution with water. This is possible because milk as it comes from the cow has a very constant freezing point depression. Can you think of a method for accurately determining the freezing point of a solution?

Answer

From Figure 15.15, curve S, we see that noting the temperature at the moment that the first ice is formed (point B′) gives a result that greatly differs from the equilibrium value T_f, because of the delay involved for ice nucleation to occur. It is better to observe the maximum temperature reached after the first ice has formed (C′), but this temperature is also too low; moreover, the deviation will depend on such uncontrollable factors as the concentration of catalytic impurities in the sample. Still better is it to watch the further temperature evolution, and then extrapolate the curve C′–D′ back to the line A–B′. In practice, even that turns out to be insufficiently accurate. However, this method can be used to obtain a first estimate of T_f. In a second experiment, the solution is cooled to slightly below this value and then seeded with a tiny ice crystal. The maximum temperature observed after seeding will then be very close to T_f.

15.3.3 Crystallization of Sugars

When crystallization from solution is desired, the first thing one wants to know is the solubility of the solute as a function of temperature. For ideal solutions, this relation can be calculated by the Hildebrand equation (2.9) from the heat of fusion and the fusion temperature of the solute. However, sugar solutions are strongly nonideal, and solubilities have to be determined experimentally.

Figure 15.16 gives some **solubility–temperature curves**. The relation can be simple, as for sucrose, but for most sugars it is more complicated. Consider *glucose*: either the α- or the β-anomer can crystallize, the former as a monohydrate or in anhydrous form. When starting with an equilibrium mixture of 80% glucose at 40°C, part (about 30%) of the glucose will be crystalline, and that as α-monohydrate. When increasing the temperature rapidly to 60°C, the solubility of the monohydrate would probably still be surpassed, as indicated by the extrapolated broken line. However, the solubility of the anhydrous form is lower. What will happen is that the hydrate dissolves and that some glucose crystallizes in the anhydrous form (provided that nucleation occurs within a reasonable time), until all of the hydrate has disappeared. Cooling again will lead to the opposite: the anhydrous form dissolves and monohydrate crystallizes. Similar changes can occur near 90°C, where changes from the α- to the β-anomer, and vice versa, are possible.

Figure 15.17a gives further details for *sucrose*. It is useful to express the concentration in kg (or moles of) solute per kg water, especially if more solutes are present. By plotting log concentration against temperature, approximately straight lines are often obtained. The curves divide the

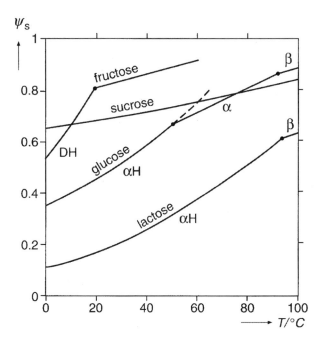

FIGURE 15.16 Solubility, expressed as mass fraction ψ_S, of some sugars in water as a function of temperature T. α and β refer to different anomers, H to monohydrate, and DH to dihydrate.

diagram into various domains. For a supersaturation ratio $\beta < 1$, the solution is undersaturated and no crystallization will occur. For $\beta > 1.3$, in the domain designated "labile," crystallization occurs spontaneously. In the metastable domain, crystallization does not occur within a reasonable time, and in the intermediate domain crystallization depends on the time available and on the purity of the solution.

The simplest interpretation is that for $\beta > 1.3$ homogeneous nucleation occurs, that in the intermediate domain heterogeneous nucleation occurs, and that in the metastable domain virtually no catalytic impurities are present. The results for sucrose in Table 14.2 agree well with homogeneous nucleation occurring at a supersaturation ratio of about 1.3. It is seen that an undercooling by about 23 K will be needed to obtain crystals rapidly and reliably. This is often not suitable, and to ensure fast crystallization at lower supersaturation, especially in the metastable domain, seeding with small crystals is generally applied. Remember that sucrose crystals tend to grow quite fast even at low supersaturation, which means that nucleation is often the limiting step for fast crystallization.

We now turn to *lactose*, where the situation is more complicated. As mentioned, several sugars, including lactose, occur in two forms, the anomers denoted α and β. In solution, the two are in equilibrium with each other, i.e., $\alpha \rightleftarrows \beta$. The equilibrium ratio varies with temperature: $R = [\beta]/[\alpha] = 1.64 - 0.0027 \, (T - 273.15)$. The reaction leading to equilibrium—generally called *mutarotation*—is rather slow, the first-order rate constant at 30°C being about $1 \, h^{-1}$. Both α-lactose, as a monohydrate, and anhydrous

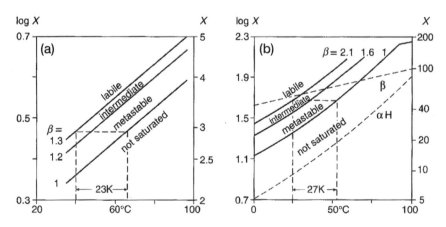

FIGURE 15.17 Solubility and supersaturation of (a) sucrose and (b) lactose. X is concentration in kg per kg water, β is supersaturation ratio.

β-lactose can crystallize, and their (estimated) solubility curves are given in Figure 15.17b (broken lines). The solid lines give results for total lactose concentration, after equilibrium between α and β has been reached. Notice that the solubility curve (labeled 1) has a sharp bend at 93.5°C. At that point on the line both α- and β-lactose are saturated. Since here $R = 1.39$, this means that β-lactose is more soluble than α, as seen in the diagram. At $T > 93.5$, the solubility of α increases more strongly with temperature than that of β, implying that now β-lactose will crystallize. Below 93.5°C, α-lactose monohydrate crystallizes, which is the common form.

Notice also the differences in scale between the diagrams for sucrose and lactose, sucrose having a much higher solubility. Lactose also needs a higher supersaturation for spontaneous crystallization to occur than sucrose, ln $\beta = 0.74$ and 0.26, respectively. On the other hand, owing to the marked difference in slopes of the curves, the undercooling needed is not greatly different.

Question 1

You have a bottle containing finely crystallized lactose, but you don't know whether it is α-lactose hydrate or β-lactose. How can you find out? You have no other equipment than laboratory glassware and a balance.

Answer

Add at room temperature 40 g of the lactose to 100 g of water and stir. If it is α-lactose, it will not fully dissolve; if it is β-lactose it will, but after some time crystals will form in the solution, since β will be transformed into α, and the latter becomes supersaturated. As a further check, you can add 15 g to 100 g of water. β-lactose will then readily dissolve, but α-lactose only after long stirring.

Question 2

In Section 15.2.2 it was argued that the linear growth rate of α-lactose hydrate crystals is very much smaller than that of sucrose crystals, because in the former case fierce competition with β-lactose occurs, and nothing like that can occur for sucrose. Can you now give additional causes for the crystallization rate of a lactose solution being far smaller than that of a sucrose solution at the same supersaturation, especially at low temperature? Tip: also consult Section 2.2.

Answer

1. As shown in Figure 15.17, lactose needs a much higher supersaturation for reasonably fast nucleation to occur than sucrose.

2. When lactose crystallizes, α-lactose is depleted from the solution, decreasing the supersaturation. Then β-lactose is converted into α, to make up for part of the depletion, but this is a slow process, especially at low temperatures. In other words, the supersaturation of α-lactose is on average smaller than the diagram leads us to expect.

3. The supersaturation ratio β as given in the figures is based on concentrations, whereas it should be based on activities. Figure 2.3 shows that near the saturation concentration of sucrose (about 65% by mass), the slope of the relation between activity and concentration is strongly increasing with increasing concentration. This means that the true supersaturation ratio for sucrose will be much higher than the concentration ratio, leading to much faster crystallization. Such a strong effect is not expected for lactose, since the saturation concentration is far smaller (about 20%), so that the difference between mole fraction ratio and activity ratio will be quite small.

Incidentally, the greatly enhanced activity ratio of sucrose must also be the explanation of cause 1.

15.4 FAT CRYSTALLIZATION

Fat and oils are mixtures of triacylglycerols—or triglycerides, for short—where oils are liquid at room temperature and fats are apparently solid. In fact, a fat is always a mixture of triglyceride crystals and oil. Many foods contain fat, and fat crystallization is an important phenomenon. It largely determines the following **important properties**.

> *Mechanical properties* of "plastic" fats and of high-fat products (butter, margarine, chocolate) during storage and handling. This concerns especially yield stress—in relation to stand-up during storage and firmness or spreadability during handling—and fracture properties.
>
> *Eating properties.* This may concern, again, yield or fracture properties, and also stickiness (observed when small fat crystals do not melt in the mouth), and coolness (due to the heat of fusion consumed during melting in the mouth).
>
> *Physical stability* of some foods. This concerns, for instance, formation and sedimentation of crystals in oil, oiling off in plastic fats and related products (see the Question in Section 5.3.1), (prevention of)

coalescence of aqueous droplets in butter and margarine, and partial coalescence in some O–W emulsions (see Section 13.5).

Visual appearance, such as the appearance of bloom on chocolate (tiny white fat crystals appearing on the surface); gloss of chocolate and margarine; and turbidity in oils.

All of these properties can change during storage.

Fat crystallization is a highly **complex phenomenon**. Before a more detailed discussion, we mention

Triglycerides are large and highly anisometric molecules that can assume many conformations. This gives rise to *polymorphism*; the main polymorphs are called α, β', and β, in order of increasing melting point and stability.

Natural fats are always mixtures of a (great) number of different triglycerides and therefore have a *melting range*, generally spanning several times 10 K.

Several of the different triglycerides are nevertheless quite similar in molecular structure. This leads to the extensive formation of *compound crystals* (solid solutions).

The first two causes mentioned are also responsible for the *slowness* of fat crystallization. All of them contribute to the *nonequilibrium* state that is common in nearly all fats.

Nucleation, including that of fat crystals, has been discussed in Chapter 14, nucleation of emulsified fat in Section 14.3. Growth rate is discussed in the present chapter, especially in relation to Figures 15.8 and 15.11. Mechanical properties will be discussed in Section 17.3, proceeding from the discussion in the present section.

15.4.1 Melting Range

Natural fats always contain several different fatty acid residues. If the number of these is given by n, the possible number of different triglyceride molecules (including stereoisomers) equals n^3. Since several natural fats contain well over a hundred different residues (especially milk fat and fish oils), the number of different triglycerides can be enormous, but most of these are present in trace quantities. More realistically, the number of triglycerides in significant quantities will mostly be 10–100. In most fats, the distribution of fatty acid residues over the molecules is neither fully random nor fully ordered.

Pure Triglycerides. To understand something of natural fat crystallization, we need to know some properties of pure triglycerides. Some data are given in Table 15.2.

We will start with *uniform triglycerides*, in which the molecules contain only one species of fatty acid residue. The table gives melting points of the constituent fatty acids as well as of the triglycerides. Notice that there is a fair correlation between the two, although the range is markedly wider for the triglycerides (about 150 K). There is also a fair correlation between melting point and heat of fusion. The latter values vary more widely if the molar heat of fusion is taken. The variation in melting properties then follows from the variation in the fatty acid molecules. The most important variables are

TABLE 15.2 Some Properties of Fatty Acids and Triglycerides[a]

Fatty acid	Code	Notation	$T_m °C$	Triglyceride	$T_m(\alpha)$, °C	T_m, °C	ΔH_f, $J \cdot g^{-1}$
Butyric	B	C 4:0	−8	BBB		−75	
Caproic	Co	C 6:0	−4	CoCoCo		−25	
Caprylic	Cy	C 8:0	16	CyCyCy	−54	8	148
Capric	C	C10:0	32	CCC	−10	32	170
Lauric	L	C12:0	44	LLL	14	46	186
Myristic	M	C14:0	54	MMM	31	58	197
Palmitic	P	C16:0	63	PPP	46	66	205
Stearic	S	C18:0	70	SSS	55	73	212
Elaidic	E	C18:1, *t*	44	EEE	16	42	150
Oleic	O	C18:1	16	OOO	−32	5	113
Linoleic	Li	C18:2	−6	LiLiLi		−13	85
Linolenic	Ln	C18:3	−13	LnLnLn		−24	
				PPB	21	44	155
				PSP	47	67	195
				POP	18	37	190
				PPO	18	34	
				SOS	23	43	194
				SOO	1	23	

[a]In the "notation" the first number is that of carbon atoms in the molecule; after the colon is the number of double bonds, followed by *t* if it concerns a double bond in the trans configuration (cis is far more common). Melting points T_m and heats of fusion ΔH_f are given in the most stable polymorph (generally β). $T_m(\alpha)$ refers to the α polymorph.

1. The *number of carbon atoms*. The higher it is, the higher are T_m and ΔH_f. The larger the molecule is, the greater the gain in conformational entropy upon melting, and the greater the melting enthalpy.

2. The presence and the *number of double bonds*. Crystallization of paraffinic compounds is difficult but occurs easiest if the chains can assume a linear zigzag conformation. A chain with a cis double bond cannot do this: it tends to form a bend (see Figure 15.19b, SOS). This makes crystallization more difficult. The effect will be greater for more double bonds. Cf. the series SSS–OOO–LiLiLi–LnLnLn.

3. *Double bond configuration*. A trans double bond allows the formation of an almost straight chain and hence causes a smaller difference in melting properties than a single cis bond. Cf. SSS, EEE, and OOO.

4. *Position of the double bonds* in the chain. The most important effect is that two conjugated cis double bonds (—CH=CH—CH=CH—) can give a more nearly straight chain than two nonconjugated ones (—CH=CH—CH$_2$—CH=CH—), the common form in natural fats. Hence the former one crystallizes more readily (higher melting point).

5. *Branching* of the chain tends to cause an appreciably lower melting point.

Moreover, modified fatty acid residues occasionally occur, e.g., containing an —OH group.

Table 15.2 also gives data for some *mixed triglycerides* (not to be confused with a mixture of triglycerides). The melting point and the heat of fusion are generally smaller than the "average" of the corresponding uniform triglycerides. (Only if it concerns saturated chains that do not differ by more than two C-atoms in length, the results agree well with the average.) It is also seen that symmetric molecules, like POP, crystallize more readily than asymmetric ones, like PPO. Large differences in chain length (as in PPB) have a strong effect.

It should finally be mentioned that the data on the enthalpy of fusion are subject to uncertainty; various authors give values that may differ by several percent.

Simple Mixtures. In natural fats, triglycerides do not crystallize from the melt but from solution. The higher melting species are dissolved in the lower melting ones. The solubility is given by the Hildebrand equation (2.9), which is repeated here:

$$\ln x_s = \frac{\Delta H_f}{R}\left(\frac{1}{T_m} - \frac{1}{T}\right) \tag{15.8}$$

where x_s is the mole fraction soluble, ΔH_f is the heat of fusion of the pure solute, and T_m its melting temperature. Figure 15.18 gives a calculated example for a mixture of equal parts of tristearin (SSS) and triolein (OOO). The final melting point is not the 73°C of the SSS, but 69°C. At 45°C the solubility of SSS has decreased to virtually zero. (Cooling to 5°C would lead to crystallization from the melt of OOO, assuming equilibrium to be reached.) Experimental results agree well with the calculations, indicating that triglyceride mixtures can show virtually ideal solution behavior.

Assume that we have a natural fat that has a final melting point— often called *clear point*—of 40°C. Further assuming that PPP is the highest melting component present, we may calculate its molar proportion in the mixture. Table 15.2 yields, after some recalculation, T_m = 339 K and ΔH_f = 165 kJ · mol^{-1}. From these data we calculate that $x_s = 0.008$ at 313 K; in other words, the mixture would contain less than 1% PPP. In practice, the situation is more complex, since we do not have a binary mixture. For

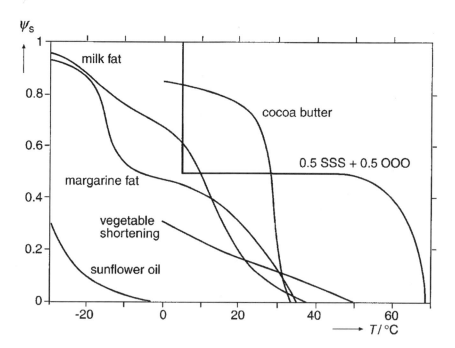

FIGURE 15.18 Melting curves. Mass fraction solid ψ_s as a function of temperature for a number of fats and for an equimolar mixture of tristearin and triolein. The curves are observed when warming up slowly after the fats have been cooled to a low temperature.

ternary mixtures, fairly simple equations are available, but the relations are intricate for more components.

Natural Fats. Figure 15.18 gives some examples of melting curves. The natural fats are cacao bean fat, generally called cocoa butter, and milk fat. Although they have about the same clear point, they vary greatly in melting curve. Cocoa butter consists for a very large part of high-melting triglycerides (fatty acid pattern: >60% P+S, >30% O), leading to a narrow melting range for most of the fat: about 70% over 10 K. This gives cocoa butter a special property: it behaves like a solid at room temperature (e.g., 25°C) and it is fully liquid at body temperature. This is very desirable when handling and eating chocolate (consisting of cocoa particles and sugar crystals embedded in cocoa butter); it also makes the fat quite suitable as a base for lipstick. Milk fat has a far wider compositional range (fatty acid pattern from C4:0 to C18:0 + about one third unsaturated). Consequently, its melting range spans about 80 K, and its properties vary gradually with temperature.

These fats represent two extremes. Many animal body fats have a melting pattern more or less comparable to milk fat. Some vegetable fats, like palm oil and coconut oil, behave like cocoa butter (here the word oil is a misnomer since the clear points are about 40 and 30°C, respectively). A third type includes most vegetable oils, like sunflower, soybean, and peanut oil, that contain a high proportion of polyunsaturated fatty acid residues; marine oils also are in this category. They generally have clear points in the range −5 to 5°C.

Melting curves can vary greatly for one and the same type of fat; shifts in the curve over 10 K are often observed. Following are the main causes.

Natural variation in triglyceride composition, due to variation in cultivar and in growth conditions. Also the type and concentration of impurities can vary.

The *temperature history* of the fat, including the rate of heating during the experiment.

Inaccuracy of the analytical results, which varies with the method applied.

The latter two points will be discussed later on.

Modification. Natural fats can be modified to give them desirable crystallization properties. The most applied methods are as follows.

Hydrogenation (or hardening), i.e., saturation of (part of) the double bonds. This is frequently applied to vegetable and marine oils. An example is the vegetable shortening in Figure 15.18. It is seen that ψ_S changes weakly with temperature, which points to a very wide compositional range. This is indeed the case, also because simple hydrogenation not only saturates double bonds but also induces shifts in double bond position and cis–trans isomerization. More directed hydrogenation is also possible, and an example is the margarine fat in the figure. Here the aim is to have the proportion solid at a constant level of about 0.4 in the range of 5–25°C, this to keep the firmness constant. It is seen that this is indeed better the case than for the shortening or the milk fat.

Interesterification leads to a (more) random distribution of fatty acid residues over and in the triglyceride molecules. Depending on the natural configuration of triglycerides, it can lead to a higher or to a lower fraction solid at room temperature.

Fractionation can be achieved by letting a fat partly crystallize at a given temperature and then separating the crystals from the remaining oil. Of course, the opposite is also applied, i.e., the blending (mixing) of various fats and/or oils.

Question

What would be the molar entropy of fusion of tristearate in its most stable form? Try to derive this value in two completely different ways.

Answer

1. At the temperature of fusion (melting), the difference in free energy between the molten and the crystalline state is zero. Hence we can readily derive that $\Delta S_f = \Delta H_f / T_m$. From Table 15.2 we obtain $T_m = 73°C$ and $\Delta H_f = 212$ J·g^{-1}. The chemical formula of SSS is $C_{57}O_6H_{110}$, leading to a molar mass of 891 Da. Hence $\Delta H_f = 189$ kJ·mol^{-1}; $T_m = 346$ K. We thus obtain $\Delta S_f = 546$ J·mol^{-1}·K^{-1}.

2. Equation (2.1) states that $S = k_B \ln \Omega$, where Ω is the number of ways in which a system can be arranged. For a triglyceride this includes the number of conformations that the molecule can assume, which causes virtually the only difference in entropy between the crystalline and the molten state. From the formula of SSS we derive that there are 59 C—C and C—O bonds. Assuming that for every bond three different conformations can occur, we arrive at $\Omega = 3^{59}$ possible conformations. For the entropy per mole we have to take $R (= 8.314$ J·mol$^{-1})$ rather than k_B, and we calculate that $\Delta S_f = 521$ J·mol^{-1}. This comes quite close to the precise value derived above from measurable parameters, although the value derived for Ω is a crude approximation.

15.4.2 Polymorphism

Imagine the following experiment: a fat is melted, e.g., at 40°C, and a little of the molten material is sucked into a thin capillary, which is then put in ice water. It is observed that the contents turn turbid within a minute; this is due to crystallization of part of the fat. The capillary is now brought to a higher temperature, e.g., 20 or 25°C, and it is observed that the contents become clear (implying that the crystals have melted); but they soon become turbid again. The crystals now formed melt at a higher temperature, e.g., 35°C. This seems to indicate that the material shows double crystallization and melting. It should be realized, however, that it concerns a natural fat, containing several triglycerides, and some of these may have a low clear point while others have a high one.

Therefore the result has to be checked by experiments on pure triglycerides, which indeed behave in about the same manner. They may even show three times crystallizing and melting. It has long ago been realized that this means that a triglyceride can crystallize in three *monotropic modifications*. These are currently designated the α-, β'- (pronounced beta prime) and β-form, in order of increasing melting point and stability.

Polymorphism is of great practical significance, and it has been extensively studied. Although the old division into three (main) forms is still useful, it has been shown that the crystallization behaviour is more complex.

Crystal structure. All substances with long paraffinic chains (e.g., fatty acids, alcohols, and mono, di, and triglycerides) show about the same geometries of chain packing in crystals. These are characterized by a **subcell**, i.e., a part of the unit cell containing two C-atoms of each chain in it. The most common subcells are illustrated in Figure 15.19a. In the *hexagonal* one (typical for α crystals), each chain is surrounded by six others at equal distances. Here the chains have some freedom to move, both rotationally and by "wiggling"; there is thus partial disorder. The *orthorhombic* subcell (common in β' crystals) shows denser and more perfect packing. Notice that the (vertical) planes through the zigzag of the chains are perpendicular to each other. In the *triclinic* subcell (always found in β crystals) the packing is densest. The zigzag planes are now in the same direction, but the direction of the C—C bonds at the same horizontal level alternates. Actually, other subcells can occasionally be observed; the total seems to be nine. The subcells are characterized by the repeat distances of the chains, and these can be derived from x-ray diffraction (also called x-ray scattering). It concerns the so-called *short spacings*, determined by wide-angle x-ray diffraction (WAX).

Triglyceride molecules have three chains and a glycerol, residue, and—for the same subcell—**packing of the molecules** can occur in several ways. Some of these are shown in Figure 15.19b. In all configurations, the molecules form regularly stacked double layers. The variation concerns:

The tilt of the chains with respect to the plane of the layers. In α crystals the angle is 90°.

The double layers can span two or three of the paraffin chains, denoted L2 and L3. Both configurations can occur in β′ and β crystals.

If one of the three fatty acid residues is (much) shorter than the other two, an L3 conformation is obtained, as illustrated for SCS.

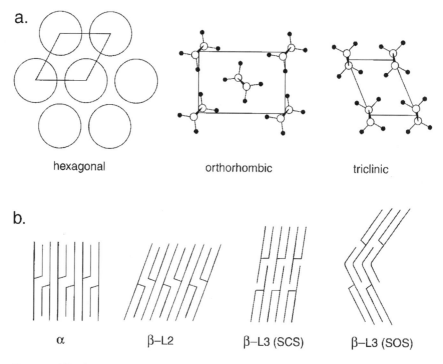

FIGURE 15.19 Crystal structure in various polymorphs. (a) Packing modes of paraffinic chains. The straight lines indicate the cross section of the subcell. (b) Packing modes of triglyceride molecules.

If one of the chains is an oleic acid residue, a bended (L3) conformation is often assumed for these chains, as illustrated for SOS.

For several mixed triglycerides, especially those containing unsaturated chains, or chains that all differ in length, other forms of molecular packing can occur.

The distances between the bilayers or *long spacings* can be determined by small-angle x-ray diffraction (SAX).

A wide variety of polymorphs has been observed. Often two forms of β' and of β can be observed, generally differing in tilt angle (apart from the differences between the L2 and L3 forms). These are, e.g., designated β_1 and β_2, the former being more stable than the latter. Another variety is designated γ or sub-α or pre-β'. It tends to have an orthorhombic subcell, whereas the molecular packing is like that in the α-form (no tilt). Some data of the three main forms of tristearate are given in Table 15.3. See Table 15.2 for melting points of various α and β crystals.

Transitions. Table 15.3 gives the molar entropy of fusion of the three polymorphs of SSS. It is seen that the values increase when going from α to β' to β, which presumably means that the conformational freedom of the chains decreases in that order (probably being about zero for the β-form). For triglycerides containing unsaturated chains and for mixed triglycerides, the variation in ΔS_f among polymorphs is even greater. All this means that α crystals tend to form easiest (at least if the temperature is

TABLE 15.3 Some Properties of the Three Main Polymorphic Modifications of Tristearin

Property	Modification		
	α	β'	β
Crystal unit cell	hexagonal	orthorhombic	triclinic
Main short spacings, nm	0.415	0.38; 0.42	0.46
Melting point, °C	55	63	73
Enthalpy of fusion, kJ $\cdot$ mol^{-1}	110	150	189
Entropy of fusion, J $\cdot$ mol$^{-1} \cdot$ K^{-1}	335	446	546
Melting dilatation, cm$^3 \cdot$ kg^{-1}	119		167

below the α clear point); this is because the constraints imposed on the molecules upon incorporation in the crystal lattice are less severe than for the other polymorphs.

However, the molar free energy of the other crystals is lower than that of α: see Figure 15.20a. Only the β-form will be thermodynamically stable. This results in the α-form generally being transformed into one of the more stable polymorphs, even at constant temperature, while giving up heat. Only *exothermic transitions* can thus occur: α → β′, α → β, and β′ → β, as well as β₂ → β₁, etc. *Endothermic transitions* can only be from the crystal— whatever the polymorph—to the liquid state, never to a "lower" polymorphic form.

In some cases, the β polymorph is not observed, e.g., for PPO. In other cases, some polymorphs can only form under special conditions. On the other hand, several triglycerides show more than three forms. An example is given in Figure 15.20b for SOS. Here also a γ-form is seen and two different β-forms. It depends on conditions, especially the cooling regime, what will happen in practice.

Note In other systems, a γ-form is observed having a higher G value than α, at least at low temperature. A reversible transition γ ⇄ α may then occur. These phenomena need further study.

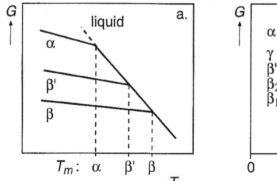 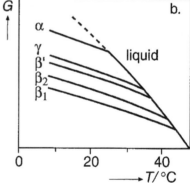

FIGURE 15.20 Gibbs free energy G versus temperature T for liquid and various crystal forms of triglycerides. (a) General trend for the main polymorphs. (b) Observed trends for SOS. (Results taken from K. Sato. Progr. Colloid Polym. Sci. 108 (1988) 58.)

A typical example of the changes occurring upon cooling and heating is given in Figure 15.21 for PPP. Upon cooling (curve 1), only the α-form results (but if the cooling is done slowly, the main form obtained will be β'). On directly heating again (curve 2), the α crystals melt, subsequently β crystals form (+ a little β'), and finally β melts. If the heating is halted at 46°C for 15 min, the α-form disappears, presumably changed into β' and some β. Heating then (curve 3) causes β' to melt (note the double peak, hence presumably two β'-forms), more β is formed, and finally β melts. It will thus depend on the triglyceride(s) present and on the temperature–time regime applied what precisely will happen, but the same principles apply.

Multicomponent fats tend to form compound crystals, and this significantly affects polymorphic forms and transitions. In some cases, however, a group of closely related triglycerides can almost behave like a single triglyceride, especially in the α- and β'-forms.

An example is given by *fully hydrogenated palm oil*. 97% of its fatty acid residues are either stearic or palmitic acid. When mixing it with a fully liquid oil, its solubility nicely follows the Hildebrand equation (15.8), provided that the parameters ΔH_f and T_m are experimentally determined

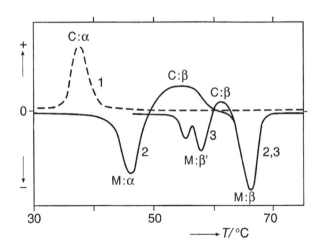

FIGURE 15.21 Differential scanning calorimetry of tripalmitate. The vertical axis gives the heat flow (e.g., in $(J \cdot kg^{-1} \cdot s^{-1})$, + indicating exothermic and – endothermic flow. Curve 1: cooling from 80 to 30°C. Curve 2: heating from 30 to 80°C. Curve 3: after heating from 30 to 46°C, the sample is kept for 15 min at 46, then heated to 80°C. C: crystallization peak. M: melting peak. (After results by K. Sato et al. Progr. Lipid Res. 38 (1999) 91.)

(generally by differential scanning calorimetry). Results are shown in Figure 15.22 (curves marked $\ln \beta = 0$) for the α and β' polymorphs; curves for various supersaturations are also given. It is seen that conditions can be chosen such that nucleation and crystallization occur either in the α- or in the β'-form. The β' crystals of the present system are quite stable, and it takes a very long time before the transition to the β-form will occur. The α crystals, however, are quite unstable and soon transform into β. The latter tend to be large crystals that form a very weak network, whereas the much smaller β' crystals form a firm fat, even at low concentrations. This can be of great practical importance, for instance in margarine.

Another example is **cocoa butter**. It consists for 80–85% of the triglycerides POP, POS, and SOS, which are very similar molecules that form compound crystals, even in the β-form; the molecules will then be packed as depicted in Figure 15.19b, SOS. Figure 15.18 gives an example of the melting curve, and it is seen that the curve is very steep above 25°C, as is to be expected if most of the crystals have the same composition. The shape of the curve can, however, markedly depend on the temperature history. Moreover, the composition of the fat can vary.

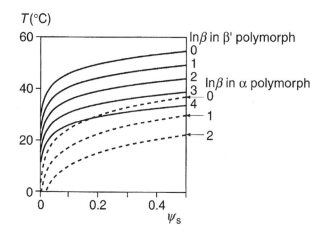

FIGURE 15.22 Solutions of fully hydrogenated palm oil in sunflower oil, mass fraction ψ_S. Calculated curves for various supersaturations ($\ln \beta$) as a function of temperature T for the α and the β' polymorph. Parameters used for α, $\Delta H_f = 98$ kJ · mol^{-1} and $T_m = 315$ K; for β', $\Delta H_f = 160$ kJ · mol^{-1} and $T_m = 330.5$ K. (After W. Kloek. Ph.D. thesis, Wageningen University, 1998.)

TABLE 15.4 Polymorphic Forms of Cocoa Butter

Code	Unit cell	Melting point[a], °C
I	γ	5–18
II	α	16–24
III	β_2'-L2	20–27
IV	β_1'-L2	25–28
V	β_2-L3	26–33
VI	β_1-L3	32–36

[a]Peak temperature on DSC curve.

Six polymorphic forms are generally observed, and some properties are given in Table 15.4. The common notation is from I to VI, in order of increasing melting point. In the present case, melting *point* is a reasonable term, since the crystals in one polymorph give a pretty sharp melting peak (on a DSC scan), almost like that of a single triglyceride; the γ-form tends to give a melting range. I–V can form from the melt and by transition from a less stable polymorph; the latter is the only route to obtain form VI. It should be added that cocoa butter also contains 2–3% trisaturated triglycerides, which crystallize separately from the main group. Most other triglycerides remain liquid, even below 20°C.

Cocoa butter is an essential component of chocolate, and its crystallization is of substantial importance for the quality of the product, which should be firm and smooth (both visually and in the mouth) and melt rapidly to provide a sensation of coolness. A problem can be the formation of bloom: this consists of whitish patches on the surface, which are due to the slow formation of largish crystals (consumers may think that it is mold growth). It arises from slow recrystallization, especially due to the transition from form IV to V. The way to circumvent this is *tempering*. In the present context this implies (a) rapidly cooling the liquid (the "chocolate mass") to 15°C or lower and keeping it there for some time, which ensures that nucleation is fast and that many small crystals form; (b) warming to a temperature just below the melting point of form V and keeping it there for at least an hour; and (c) bringing the chocolate to room temperature. Such a process leads to small crystals, nearly all of which are in form V. Some bloom can still slowly form via the transition V → VI, particularly when the storage temperature is above 25°C or when temperature fluctuations occur.

Question

A company producing olive oil has occasionally a lot that is unfit for normal use owing to off-flavors. Purification processes remove the off-flavors but also the highly desirable natural flavor of the oil. The company wants to find another outlet for these lots and considers that the very special composition of olive oil may lead to a useful solid fat by complete hydrogenation. The company's laboratory is asked to do some trials and they make blend of 20% fully hydrogenated and 80% unmodified oil, to study crystallization behavior. Suppose that you have to do experiments in which the aim is to obtain (a) very small crystals, (b) rapid crystallization, or (c) fully stable and largish crystals. What temperature regime would you apply in each of these cases? The fatty acid composition of the oil is 70% O, 15% Li, 10% P, and 5% others, of which about half is C18.

Answer

The triglycerides obtained after hydrogenation will be for the most part SSS, the rest being largely made up of S and P. This means that the composition of the fat is quite similar to that of the solid fraction in the system discussed with reference to Figure 15.22. Consequently, we can directly read off the clear points for the α- and β'-forms from that figure at $c = 0.2$ (mass and molar fractions will be virtually identical for the system studied); the values are 30 and 50°C, respectively. For the β-form, we may assume that the clear point is given by the highest melting component, i.e., SSS. Assuming that the distribution of fatty acid residues over the triglycerides is random, SSS will make up about $0.87^3 = 0.66$ of the hydrogenated fat; hence, its molar fraction in the total mixture is about 0.13. Applying the Hildebrand equation (15.8) and the data for SSS in Table 15.3, we obtain for the β clear point 63°C. The questions asked can now roughly be answered. (a) To obtain small crystals, nucleation has to be fast, which needs undercooling to well below the α clear point. For instance, mixing one part of the melted hydrogenated fat, temperature, e.g., 75°C, with 4 parts oil of 10°C yields a temperature of 23°C, i.e., 7 degrees below the α clear point. When the mixture then is continuously stirred, many small crystals will result. (b) To assure fast crystallization, the specific surface area of the crystals has to be large (hence many small crystals), and the supersaturation has to be high. Initial cooling thus can be as mentioned under (a). However, owing to crystallization, the supersaturation will decrease; this is aggravated by the temperature rise caused by the release of the heat of fusion. Assuming the latter to be about $170\,kg \cdot kJ^{-1}$, and taking into account that the specific heat of oil equals about $2.1\,kJ \cdot kg^{-1} \cdot K^{-1}$ (Table 9.2), the heat release will lead to a temperature increase by $0.2 \times 170/2.1 = 16$ K. It will thus be desirable to cool the crystallizing mixture. (c) To ensure stable crystals, the crystallizing temperature should be above the β' and below the β clear point, i.e., between 50 and 63°C. At such a temperature, nucleation will be extremely slow, and it would thus be necessary briefly to precool the system to a lower temperature (to be experimentally estimated).

Note What the investigators should do first is to determine by DSC the needed parameters and check by x-ray diffraction which of the polymorphs is involved.

15.4.3 Compound Crystals

As mentioned, triglycerides that are closely similar in molecular structure may form compound (mixed) crystals, i.e., more than one type of molecule occurs in the same crystal lattice.

Binary Mixtures. Compound crystals have been best studied for binary mixtures in the most stable polymorph, generally β, occasionally β'. Examples of *phase diagrams* are given in Figure 15.23. The most common situation is a *eutectic* mixture, where no compound crystals are formed, as depicted in (a). The situation is the same as was shown in and discussed in relation to Figure 15.14. If the two kinds of molecules are very similar, as in Figure 15.23b, a *solid solution* can be formed, i.e., compound crystals in a range of compositional ratios, which range may be from 0 to 1. This situation will be further described below. A third situation is illustrated in Figure 15.23c. Here the two components behave as a eutectic mixture, except for molar ratios very close to unity, where compound crystals in the β-form can form. For completeness, in cases like (a) and (c), compound

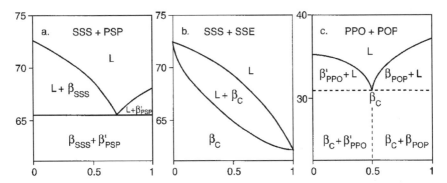

FIGURE 15.23 Approximate examples of phase diagrams of binary mixtures of triglycerides, indicated in each frame. The vertical axes give temperature (°C), the horizontal axes the mole fraction of the second triglyceride indicated. L is liquid, crystals are indicated by β or β', and subscript C signifies compound crystals.

crystals can form during crystallization in a less stable polymorph, but they will soon transform into the most stable polymorph and then give the phase diagrams shown.

In Various Polymorphs. In practice, the examples just discussed are not very interesting. Compound crystals form more readily and are much more common in the β′-form and tend to be abundant in the α-form, provided that the temperature is below the respective clear points. Table 15.3 gives the melting dilatations of α and β crystals of SSS, and it follows that the density of the former crystals is significantly smaller than that of the latter; β′ crystals will have an intermediate density. This means that α crystals can more readily accommodate molecules of somewhat different form than β′ crystals, and that this is generally impossible in the β polymorph. Most likely, the ends of the three chains of the triglyceride molecules in α crystals are not rigid but "liquid," as it were, allowing various chain lengths in one crystal.

Multicomponent Fats. Compound crystals will be particularly abundant in fats of a wide compositional range. The following argument may explain this. Assume that a fat contains 10 different, but closely similar, species, that each has about the same mole fraction (say, 0.02), enthalpy of fusion (say, $100 \, kJ \cdot mol^{-1}$) and melting point (say, 320 K). Putting these data into Eq. (15.8) yields a clear point of 290 K. For each of these components, the supersaturation $\ln \beta = 0$ at that temperature, but for the joint components in a compound crystal (mole fraction 0.2) the supersaturation will be $\ln \beta = \ln (0.2/0.02) = 2.3$, which is a considerable value. In other words, the driving force for crystallization is far greater for the compound. The reasoning given is not fully correct, since the melting point and the molar heat of fusion of the compound will be somewhat lower than presumed, but its supersaturation will nevertheless be much larger than the average values of its components.

In principle, it is even possible that a given mixture of triglycerides is at a given temperature supersaturated as a compound in the β′-form, but not in the β-form, since that does not allow compound crystal formation. In other words, for this triglyceride mixture, the β′ polymorph will appear to be the most stable form. At a much lower temperature, however, the saturation concentration of the single components will also be surpassed, and the β-form will then become the most stable one.

Polymorphic Transitions. In many cases, α crystals form first upon cooling of a liquid fat, because nucleation is easiest in the α-form. In fats of relatively homogeneous composition, the α-form is short-lived; it may even within a minute be transformed into β'. The β'-form generally has a longer *lifetime*. In multicomponent fats, α crystals tend to live much longer, and it is no exception that (part of the) β' crystals remain almost indefinitely. Again, the greater supersaturation for compound crystals causes the driving force for transition to be small or even negative. Moreover, the transitions $\alpha \rightarrow \beta' \rightarrow \beta$ must go along with a change in crystal composition, since compound β' crystals can host fewer different molecules than α, and compound β crystals hardly exist. This also implies that the polymorphic transitions are not true solid-state transitions: they have to proceed via the liquid state. This will also hinder changes in polymorphic form, especially if very little liquid phase is left (i.e., at low temperature).

Solid Solutions. It is now well established that in a multicomponent fat true solid solutions involving many different triglycerides can be present. The phase diagram would be comparable to that depicted in Figure 15.23b. The number of components n will, however, be (much) larger than two, and the phase diagram would thus be n-dimensional. This does not necessarily imply that it concerns one continuous series of compound crystals over the whole temperature range. Crystals containing different groups of triglyceride may be present at the same time or at various temperature ranges. For instance, there may be a group of saturated triglycerides of relatively constant chain length forming β'-L2 compound crystals; another group of triglycerides incorporating one short chain per molecule may form β'-L3 crystals (like SCS in Figure 15.19); a third group may incorporate one oleyl chain per molecule and form other β'-L3 crystals (like SOS in the figure); finally, some other molecules may form compound α-crystals. Moreover a solid solution will not always be formed over the full compositional range possible.

The presence of solid solutions has some important implications, as will be illustrated for a highly simplified example of two components, depicted in Figure 15.24. Assume that we have a liquid fat of composition a_3 at temperature T_1. Cooling it to T_2, crystals will form (provided that nucleation occurs). Now a liquid phase of composition a_2 will result and a solid phase (crystals) a_5. The molar ratio of solid to liquid will be given by the ratio $(a_3 - a_2)/(a_5 - a_3)$. Assume now that the mixture is cooled further to T_3. The crystals of composition a_5 remain, and the liquid a_2 will separate into a liquid and crystals of composition a_1 and a_5, respectively. Note that the composition of the system differs from that resulting from cooling directly to T_3. In other words, there is no equilibrium (i.e., within the polymorphic form, probably β'). There is, however, a "surface equilibrium": the crystals a_4 will probably be formed at and around the existing crystals

a_5. Hence the surface of the crystals will be in local equilibrium with the liquid phase. There is no equilibrium at the interface between a_4 and a_5, but it would need solid state diffusion (which is quite slow) to achieve it.

Other temperature regimes can be analyzed in a similar manner, and predictions can be made about the influence on the solids content. Although the values obtained cannot be precise—the form of the phase diagram can only be guessed—the predicted trends are well observed, at least for milk fat, the only fat that has been extensively studied.

Consequences. An example of experimental results is given in Figure 15.25. Here the fat had been cooled in two steps, i.e., about the regime just discussed. It is seen that the differential melting curve has two maxima, a few kelvins above the crystallizing temperatures applied. It is also seen that these two maxima disappeared upon storage of the fat. This cannot—or more precisely, cannot only—have been due to polymorphic

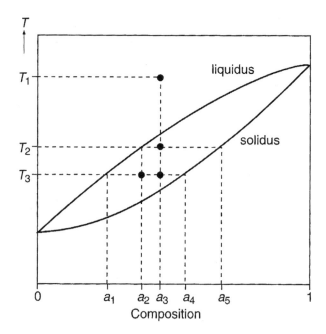

FIGURE 15.24 Hypothetical phase diagram of a binary mixture forming a solid solution over the whole compositional range. The X-axis gives the mass fraction of the higher melting component. For situations between the liquidus and the solidus curve, a mixture of liquid and solid (compound crystal) material is present.

transitions, since these invariably cause the melting temperature to increase. In the figure we see that the new peak is at a lower temperature than the original second peak. It is also seen that the area under the curve, i.e., the amount of fat melting, increased during storage.

The following phenomena can be seen as consequences of the occurrence of solid solutions in the fat.

1.   The melting range is narrowed. If two multicomponent fats of different melting range are mixed in the liquid state and then cooled, the differential melting curve of the mixture is observed to be narrower than the average of the melting curves of the original fats. The initial melting point is higher and the clear point is lower, and the shape of the curve tends to be significantly altered.

2.   The temperature at which most of the fat melts depends on the temperature at which crystallization took place, as discussed.

3.   Crystallization in steps of decreasing temperature gives less solid fat than direct crystallization at the lowest temperature. Taking the example of Figure 15.24, crystallization at T_3 gives a fraction solid of

$$\frac{(a_3 - a_1)}{(a_4 - a_1)} = 0.72$$

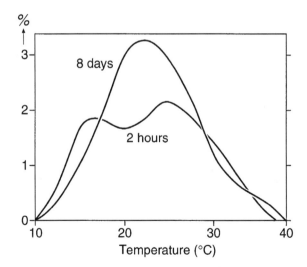

FIGURE 15.25   Differential melting curve (expressed as % of the fat melting per K temperature increase) for a milk fat sample. The fat was held for 1 day at 19°C and then at 10°C for 2 h or 8 days. (From data in H. Mulder, P. Walstra. The Milk Fat Globule. Pudoc, Wageningen, 1974.)

whereas crystallizing in steps at T_2 and T_3 would give

$$\frac{a_3 - a_2}{a_5 - a_2} + \frac{a_5 - a_3}{a_5 - a_2} \times \frac{a_2 - a_1}{a_4 - a_1} = 0.23 + 0.77 \times 0.44 = 0.57$$

By the same reasoning, slowly cooling to a low temperature yields less solid fat upon melting than fast cooling to the same temperature.

   4.   Precooling to a lower temperature before bringing to the final temperature gives more solid fat than direct cooling to the latter. This can be reasoned by making a similar calculation as in point 3, and it is discussed below with respect to Figure 15.26b.

   5.   Unstable polymorphs have a (much) longer lifetime than those of pure triglycerides. Polymorphic transitions go along with changes in crystal composition.

   6.   Within one polymorph (usually β'), changes in crystal composition can occur during storage at constant temperature (Figure 15.25). Such changes proceed (much) slower at a lower temperature.

Question
With reference to Figure 15.24, what would be the fraction solid when going from T_1 to T_2, and what when going to T_2 via T_3?

Answer
0.24 and 0.37, respectively.

15.4.4   Nucleation, Growth, and Recrystallization
Now that we have learned about the various phenomena occurring during fat crystallization, as well as about several factors affecting it, it may be useful to discuss some aspects in the light of the knowledge gained.

   Nucleation.   This is the subject of Chapter 14, and nucleation of triglyceride crystals is particularly discussed in Section 14.2.2. We may conclude that nucleation will generally be *heterogenous* and occur in the α-*form*. However, when an oil is cooled to, and kept for a while at, a temperature between the α and the β' clear point, nucleation in the β' form will generally occur. The nucleation rate is greatly dependent on monoglyceride content, and these substances presumably form catalytic impurities. Monoglycerides may readily form inversed micelles in the presence of a trace of water, and the micelles are possibly transferred into a crystals. Whether micelles or crystals form the catalytic impurities is unknown. Monoglycerides in fats arise for the most part from lipolysis, i.e., hydrolytic splitting off fatty acids from tri and diglycerides. The fatty acids are often removed from industrial fats by washing with an alkaline solution, but this leaves the monoglycerides.

An example involving nucleation is in Figure 15.26. In frame (a) considerable hysteresis is observed between cooling and heating curves, and the explanation must be the limited probability of a catalytic impurity being present in a small droplet. Below 10°C, i.e., about 10 K below the α clear point, significant nucleation occurs, and it strongly increases with decreasing temperature. Results for bulk fat are in Figure 15.26b, and it is seen that the melting curve is almost like that in frame (a) but the cooling curve greatly differs. Apparently, 24 h keeping suffices to induce sufficient nucleation for almost all of the crystallizable fat to crystallize. The hysteresis is very small between 20 and 35°C, by about 1 K. This indicates an important fact, namely that triglyceride crystals are very effective catalytic impurities for

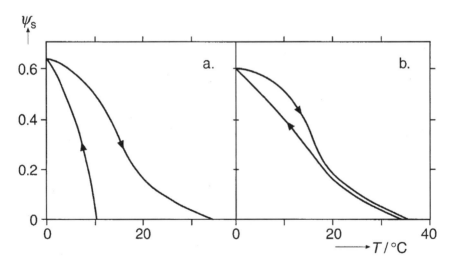

FIGURE 15.26 Cooling and heating curves of milk fat: mass fraction solid (ψ_s) versus temperature (T). Cooling at each temperature (5 K intervals) for 24 h; after keeping for 24 h at 0°C, heating to the higher temperatures for 0.5 h. The α clear point of the fat was about 20°C. (a) Fat emulsified (droplet volume about 1 μm³); (b) the same fat in bulk. (After results by P. Walstra, E. C. H. van Beresteyn. Neth. Milk Dairy J. 29 (1975) 35.)

other triglyceride crystals, and presumably also for crystals in another polymorph.

The hysteresis at lower temperatures has quite a different explanation: as discussed in Section 15.4.3, precooling leads to an increased solid fat content, provided that compound crystals form.

In Section 14.2.3 *secondary nucleation* is discussed, with particular reference to triglycerides. It is obvious that copious secondary nucleation can occur in multicomponent fats, at least at high supersaturation. The condition seems to be very slow crystal growth, while nevertheless the growth regime is "kinetic roughening"; the latter implies a high supersaturation. This situation can arise in multicomponent fats owing to the fierce competition between different molecules for incorporation into the crystal. Secondary nucleation appears to be abundant in milk fat, occurs to a lesser degree in a typical margarine fat, and is presumably absent in cocoa butter. The phenomenon is quite important in practice, since it greatly reduces average *crystal size* formed.

Crystal Growth. Triglyceride crystal growth tends to be *very slow*: see, e.g., Figures 15.9 and 15.12b. To obtain reasonably fast crystallization, the supersaturation needs to be high. Fortunately, this can often be realized (see, e.g., Figure 15.22), because of the high values of the molar heat of fusion and of the molar mass of the solvent. (Can you explain the relation?) It may be realized that $\ln \beta = 4$ corresponds to a supersaturation ratio of 55. At high $\ln \beta$ values, the crystals formed tend to be small and of platelet shape (e.g., length : width : thickness $= 10 : 3 : 1$, or even more slender). Although crystal growth tends to be slower for a wider range of crystallizing triglycerides, the extent of secondary nucleation tends to be higher as well, still causing the crystals formed to be small. At very high $\ln \beta$ values, crystal growth can be in the regime of kinetic roughening for some of the crystal faces, which then will become curved.

At small values of $\ln \beta$, larger and more isometric crystals tend to form, especially for crystallization in the β-form. At very low supersaturation, a special kind of (large) spherulites often form; they consist of narrow platelets growing out from one point, and they bifurcate when growing longer. A kind of spherical, prickly sponge is formed, partly consisting of liquid fat.

Recrystallization. Equilibrium is rarely reached in partially solid fats. It is quite common that changes occur after the crystallization apparently is finished (i.e., no measurable change in fraction solid within, say, an hour). Such recrystallization tends to be slower for a smaller fraction of liquid fat, which often means a lower temperature. Recrystallization can involve change in polymorph, change in the composition of compound crystals, or both. Two extremes on a scale of situations will be discussed.

In a fat of very *simple composition*, like cocoa butter (only a few crystallizing triglycerides of very similar molecular structure), crystallization phenomena closely resemble those of a pure triglyceride in oil. Recrystallization primarily involves a change in polymorph(s), as indicated in Figures 15.20 and 21. Generally, the α-form is very short lived, but further transitions tend to be slower. The β-form is often reached. Presumably, the changes are solid-state transitions, which implies that crystal size remains as it was. In some fats, especially if the triglyceride has chains of different lengths, the β'-form is persistent, except—of course—when the temperature comes close to the β' clear point or above.

Multicomponent fats, like milk fat, have a wide melting range, and secondary nucleation is common, as discussed. Moreover, the α polymorph may at low storage temperature persist for a long time. The β polymorph forms hardly, if at all. All this means that most of the crystalline fat is generally in compound β' crystals. As discussed, equilibrium (within this polymorph) will rarely exist, and even if it is reached (which may take weeks or even longer), it is disturbed when the temperature is changed. It is commonly observed that temperature fluctuations lead to significant changes in the melting curve.

Many fats, e.g., shortening and margarine fats, behave in an *intermediate* way. Both polymorphic and compound crystal changes occur, often concomitantly. The recrystallization has important practical consequences. The transition from β' to β, especially when it occurs slowly, causes the formation of large crystals, like the spherulites discussed above. This gives the fat an undesirable texture: too soft, grainy, and subject to oil separation. Some fats are strongly "β-tending," especially those with little spread in fatty acid length (like partly hydrogenated soybean oil). The presence of diglycerides then counteracts the transition β' → β; presumably, the diglyceride acts as a strong growth inhibitor for β-crystals. Changes in compound crystal composition tend to increase the firmness of the fat. This is discussed in Section 17.3.

Estimation of the Fraction Solid. We have seen that the proportion of the fat that is solid, ψ_S, can greatly depend on the temperature history of the fat. It can also markedly depend on the method of estimation applied. A classical method is *calorimetry*, where the heat of melting per unit mass (i.e., $-\Delta H_f$) is measured. But this value varies significantly among triglycerides, as shown in Table 15.2; it roughly parallels the melting point. Since a constant (average) value is taken, this means that the resulting ψ_S is significantly biased: at the low temperature end of the melting range ψ_S is overestimated, at the high end underestimated. The value of ΔH_f also depends on the polymorph: see Table 15.3. It may thus be that a transition $\beta' \rightarrow \beta$ occurs without change in ψ_S, whereas calorimetry shows an increased heat of melting. Moreover, compound crystals have a lower value of ΔH_f than the average value of the components.

Another classical method is *dilatometry*, based upon measurement of the melting dilatation. The magnitude of the latter is subject to a similar variation as the melting heat, and a similar bias as in calorimetry results. Moreover, the method is time-consuming. The most convenient and rapid method is calorimetry in a differential scanning mode (DSC), but this gives the additional problem that the base line (hence the values of the specific heat to be subtracted) is often quite uncertain, especially for a wide melting range.

A newer method is pulsed wide-line *proton NMR*. It can be executed in various modes, but the best one usually is direct estimation of the fraction liquid, which appears to be virtually unbiased. Incidentally, the trends observed for changes in ψ_S due to changes in compound crystal composition have been confirmed by using this method.

Question 1

As mentioned, cocoa butter contains a very small fraction of trisaturated triglyceride. Can this substance nevertheless play a role in chocolate manufacture? You may, for the sake of simplicity, assume that the fat contains 75% SOS and 3% SSS as the only crystallizable substances.

Answer

Yes. Assume that nucleation occurs in the α-form. From Table 15.3 we obtain for α tristearate $T_m = 328$ K and $\Delta H_f = 110$ kJ·mol^{-1}. By use of Eq. (15.8) this gives, for $x = 0.03$, an α clear point of 302 K, or 29°C. According to Table 15.2 the α melting point of SOS is 23°C and, since it makes up only 0.75 of the triglyceride mixture, its α clear point will be somewhat lower, say by 3 K. This then means that SSS nucleation will occur at a clearly higher temperature than that of SOS. The presence of SSS thus allows cooling to be less deep to obtain a sufficient number of crystals.

Note A similar situation occurs in quite a number of (partly) hydrogenated oils.

Question 2

Milk fat as it occurs in milk, i.e., in globules of a few μm in diameter, yields upon cooling always very small crystals. Can you (a) explain this and (b) think of a method to obtain larger crystals in the globules?

Answer

(a) Because of the dispersed state of the fat, considerable undercooling is needed to ensure nucleation. This implies that crystal growth occurs at high supersaturation, which means in turn extensive secondary nucleation and hence small crystals. (b) Heat the cooled milk to slightly below the β' clear point, say 31°C. Most of the fat will melt, and only a few crystals will remain. Subsequent slow cooling will then cause these crystals to grow and become large. (Fast cooling would, again, induce secondary nucleation.)

> *Note* The temperature treatment just mentioned makes the globules quite unstable to partial coalescence (Section 13.5).

15.5 RECAPITULATION

It is mostly water, sugars, salts, and triglycerides that crystallize in foods during processing or storage. It nearly always concerns crystallization from a supersaturated solution, not from the melt. The building blocks of these crystals are molecules or ions. The forces holding them together are van der Waals interactions, hydrogen bonds, or ionic bonds.

The Crystalline State. In principle, the packing of molecules in a crystal shows perfect order. Seven different packing types or crystal systems can be distinguished, based on the geometry of the unit cell. This is the smallest volume element that comprises all geometric information; it usually contains one or two molecules. A unit cell is characterized by the length of its three axes and by the angles between them. The shape (habit) of a crystal results from the stacking of units cells. In this way the crystal obtains various faces, which are identified by their Miller indices. Which faces are formed and how large they become is determined by the relative rate of growth of each. The shape of a crystal can thus vary widely for one and the same unit cell, but the angles between faces are constant. Some properties of a crystal, particularly its birefringence, are determined by the shape of the unit cell.

Several substances show *polymorphism*, which means that they can crystallize in various systems (or with various unit cells in the same system).

Monotropic polymorphism involves one or more unstable forms and a stable one; this occurs in fats. Compound crystals contain two or more different substances. They may be solute and solvent, usually water as in $NaCl \cdot 2H_2O$, or a range of very similar molecules.

Crystal Growth. The linear growth rate of a crystal face depends on several factors. The first one is the growth regime. Incorporation of molecules on a smooth face is difficult and needs two-dimensional nucleation. However, the presence of certain types of imperfections or dislocations in a crystal gives rise to the formation of growth spirals, whereby the growth rate is greatly enhanced; this is a rather common situation. In other regimes roughening of the crystal surface occurs, which also greatly enhances growth. Growth rate naturally increases with increasing supersaturation ln β, where β is the supersaturation ratio, but the relation varies with growth regime. Moreover, nonideality of the solution may strongly affect the effective supersaturation ratio, which is an activity ratio. Some variables that retard growth rate are (a) the difficulty of fitting in the crystal lattice, especially occurring for large anisometric and flexible molecules; (b) competition with very similar molecules that almost fit the crystall lattice; and (c) the presence of molecules that strongly inhibit growth by adsorbing onto the crystal face. All of these factors cause the growth rate to vary by some orders of magnitude for the same value of the supersaturation. Moreover, they greatly affect crystal shape, since the effect on growth rate may greatly vary among crystal faces.

The *overall crystallization* rate also depends on (a) the total crystal surface area, which will be larger if more crystals are formed and hence if nucleation is faster; (b) stirring, which increases the rate if diffusion of molecules to the crystal surface is growth limiting; and (c) removal of the released heat of fusion, since an increase in temperature causes a decrease in supersaturation. Naturally, the growth rate will decrease when most of the crystallizable material has crystallized.

Crystallization from Aqueous Solutions. Pure water can freeze very fast after nucleation has occurred, because there is virtually no retardation mechanism acting. The freezing of water is exceptional in that it goes along with an increase in volume rather than a decrease. The physical properties of water show a number of anomalies at low temperature.

Generally, water freezes from an aqueous solution. A *phase diagram* (temperature versus composition) then gives the phase transitions that will occur, assuming equilibrium. Most solutes give a eutectic diagram characterized by a eutectic temperature (T_e) and solute concentration (ψ_e). Below ψ_e, water will freeze upon cooling, and the solution becomes more concentrated until the eutectic point (ψ_e, T_e) is reached, where also the solute will freeze. At high initial concentration, the solute will crystallize first, etc. The relation between temperature and time during heat removal is complicated and variable, since nucleation requires (a variable extent of) undercooling. Some solutes can markedly decrease ice crystal growth rate, such as gums that form a firm gel, and some proteins adsorbing onto ice (especially antifreeze peptides).

Crystallization of most *sugars* (but not sucrose) gives some complications because (a) crystals with or without water can form, and (b) because of the presence of two crystallizable anomers (α and β) showing mutarotation. The sugar will crystallize in the form that shows the highest supersaturation, which depends on temperature and composition. The presence of another, not crystallizing, anomer can greatly retard crystal growth by competition, and also strongly affect the crystal shape obtained. Lactose is a typical example.

Fat Crystallization. The crystallization of fats, i.e., mixtures of triglycerides, shows a number of complications. In the first place, fats have a *range of similar components* of various melting properties, where the high melting species can dissolve in the low melting ones. The solutions show near ideal behavior, and the Hildebrand equation well predicts solubility. Melting temperature and heat of fusion are closely related. In pure triglyceride they depend mainly on (a) the chain length of the fatty acid residues, (b) the number and configuration (cis or trans) of double bonds in the chain, and (c) distribution of fatty acid residues in the triglyceride molecules, a more uneven and asymmetric distribution giving a lower melting point. Consequently, fats have a melting range. Natural fats vary greatly (a) in their clear point (final melting point), below $0°C$ for most oils and up to $40°C$ in plastic fats; and (b) in the wideness of the melting range, where milk fat (melting range from -40 to $40°C$) and cocoa butter (most of the fat melting between 22 and $32°C$) are extreme examples. Moreover, fats can be modified, the most common treatment being hydrogenation (saturation of double bonds) of oils.

In the second place, triglycerides can crystallize in a number of *monotropic polymorphs*. These can vary in (a) the manner of packing of the paraffinic chains, and (b) the conformation and packing of the whole molecules. This gives rise to a bewildering variety of polymorphs, but the main types—primarily varying in chain packing—are α, β', and β. This is also the order of increasing melting point, heat of fusion, density, and stability, and transitions can occur to a more stable form. The α-form is mostly short-lived, but the β'-form can persist for longer times; in some fats, the β-form generally does not form. All this means that a natural fat may have a number of increasing "final" melting points (2 to 6), for each polymorph formed.

In the third place, triglycerides of similar molecular structure can form *compound crystals*. In multicomponent fats these can take the form of solid solutions, where one type of crystal can contain a continuously varying range of components. This occurs readily in the α-form, where the packing of the molecules is not very dense, but this form is generally short-lived. It also occurs, though with less variation in composition, in the β'-form, and hardly or not at all in the β-form. Polymorphic transitions thus imply a change in crystal composition and must therefore proceed via the liquid state. Owing to the formation of solid solutions, the temperature history of the fat has a marked effect on the proportion solid and on the shape of the melting curve.

Altogether, *partially solid fats* are almost never in thermodynamic equilibrium and show slow recrystallization on storage. Two extremes can be distinguished. A fat of relatively simple composition, like cocoa butter, can almost behave as a system of one or two pure triglycerides crystallizing in oil. Recrystallization primarily involves polymorphic transitions. In a multicomponent fat, like milk fat and some margarine fats, compound crystals of a wide compositional range are formed, which composition will change during storage and temperature fluctuation; transitions to another polymorph tend to be slow.

Fat crystallization tends to be very slow, unless the supersaturation is high; the latter can generally be realized. Nucleation is for the most part heterogeneous, where existing fat crystals are very effective catalytic impurities for other crystals. In a multicomponent fat, considerable secondary nucleation occurs, which results in quite small crystals. Slow recrystallization can lead to the formation of large crystals, and possibly to oil separation from the mass of crystals.

BIBLIOGRAPHY

The basics of crystallography and crystal properties are treated in most textbooks on physical chemistry. A useful introduction into many aspects of crystallization is in

R. H. Doremus. Rates of Phase Transformations. Academic Press, Orlando, FL, 1985.

See especially Chapter 9, Crystal growth from solution. An authoritative and detailed description of several aspects of crystal growth is

P. Bennema. Growth and morphology of crystals. In: D. T. J. Hurle, ed. Handbook of Crystal Growth. Vol. 1. Elsevier, Amsterdam, 1993.

Introductory discussions related to foods are found in

J. M. V. Blanshard, P. Lillford. Food Behaviour and Structure. Academic Press, London, 1987.

This especially concerns Chapters 3, General principles of crystallization by J. Garside; 4, Ice crystallization by J. M. V. Blanshard, F. Franks; and 5, Fat crystallization by P. Walstra.

A monograph covering many aspects of crystallization in foods, including engineering aspects, is

R. W. Hartel. Crystallization of Foods. Aspen Gaithersburg, MD, 2001.

A monograph covering almost all aspects of triglyceride crystallization is

N. Garti, K. Sato, eds. Crystallization Processes in Fats and Lipid Systems. Marcel Dekker, New York, 2001.

16

Glass Transitions and Freezing

Many foods of low water content are wholly or partly in a glassy (vitreous) state. This is of great importance for the mechanical properties and the physical and chemical stability of the food. A glassy state can also form in foods of high water content when the food is frozen, causing removal of liquid water by freeze concentration. Moreover, freezing can cause other changes that affect properties and stability. These phenomena are the subject of this chapter.

Some basic aspects are found in Chapters 5 (diffusion and rheology), 8 (water relations), 14 (nucleation of crystals), and 15 (crystallization of water and phase diagrams).

16.1 THE GLASSY STATE
16.1.1 Fundamentals

Definition.   A *glass* is an *amorphous solid* showing a *glass transition*. Here a solid is defined as a material having an apparent *viscosity* (at the time scale involved) larger than a specified value, often 10^{12} Pa $\cdot$ s, i.e., 10^{15} times that of water. A solid can be crystalline or amorphous. An amorphous material does not show a *regular periodicity* in atom or molecule density, as illustrated in Figure 16.1. This can be established by x-ray diffraction: the diffractogram of crystalline sugar shows many sharp peaks, that of amorphous sugar none. The *glass transition* phenomenon will be explained for a pure substance (a monomolecular liquid) with reference to Figure 16.2.

Assume that a liquid is slowly cooled, so that the system remains in equilibrium. In Figure 16.2a we see that the *specific volume* $(1/\rho)$ gradually decreases until the melting point T_m is reached, where $1/\rho$ sharply decreases because the material crystallizes. In the crystalline state the value further decreases with decreasing temperature, though at a slower rate. However, if cooling proceeds extremely fast, crystallization may not occur—since it takes time to get started—and then $1/\rho$ keeps decreasing as depicted, until its value approaches that of the crystalline solid. At this temperature, T_g, a *glass* is said to be formed. The specific volume curve shows a sharp bend, and its value now remains close to, though somewhat larger than, that of a crystal. The excess of specific volume over that in the crystalline state can be seen as a measure of the extent to which molecules can show free translational and rotational motion. In the glassy state such freedom is nearly zero, as in a crystal.

The change from a liquid with great freedom of motion to the state at T_g where molecular motion is arrested is not a sharp one. Since the amount of molecular motion varies inversely with the *viscosity*, this can be illustrated by the change in viscosity, as in Figure 16.2b. Its value increases ever

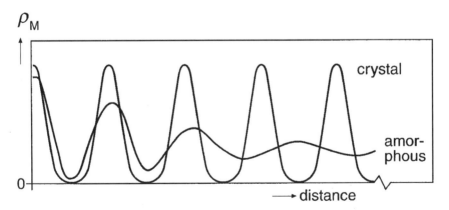

FIGURE 16.1 Molecule density ρ_M, or the probability of finding the center of gravity of a molecule as a function of the distance from a central molecule *in a given direction*, for a crystalline and an amorphous solid consisting of the same (small) molecules. Highly simplified.

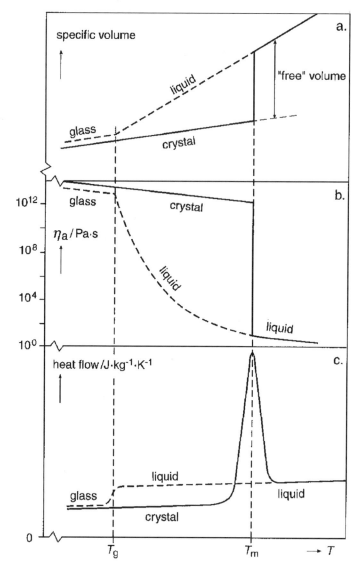

FIGURE 16.2 Illustration of the glass transition, as compared to a melting/crystallizing transition, of a pure nonpolymeric substance. The solid curves indicate equilibrium, the broken ones nonequilibrium states. (a) Specific volume (e.g., in $m^3 \cdot kg^{-1}$) as a function of temperature T. (b) Apparent viscosity η_a. (c) Heat (enthalpy) flow or effective specific heat; the positive direction signifies an endothermic change during temperature increase. T_g indicates glass transition, T_m melting/crystallization. Only meant to illustrate trends.

stronger upon approaching T_g, where a sharp bend in the curve occurs. Notice that η_a is given on a log scale. The decrease when going from T_g to $T_g + 10$ K is, for example, by a factor of 300, whereas in most pure liquids (above the melting point) a 10 K temperature change causes a change in viscosity by a factor between 1.2 and 2.

Changes in enthalpy of the system follow the same pattern as the changes in specific volume given in Figure 16.2a. Frame (c) gives the *first derivatives of the enthalpy curves*. The sharp discontinuity in the enthalpy curve for crystallization or melting at T_m is seen in frame (c) as an (endothermic) peak. (Actually, the melting peak will be much narrower and far higher than drawn.) This is characteristic for a first-order transition (Section 14.1). At the glass transition, T_g, the enthalpy curve is continuous, but the first derivative shows a sudden step. The second derivative will show a peak. We then speak of a *second-order transition*, which is thus characteristic for a glass.

It may be added that formation or disappearance of a glass is not a thermodynamic phase transition, since a glass does not represent an equilibrium state; nor is the value of T_g precisely constant. At equilibrium, crystals will form upon cooling. A glass is only obtained at very fast cooling rates, over 10^5 K $\cdot$ s^{-1} in highly fluid liquids like water. By the same token, heating a glass above T_g will lead to the formation of crystals (showing up in an exothermic peak), unless heating is extremely fast.

Like melting points, *glass transition temperatures* vary greatly among pure substances. Examples are given in Table 16.1, and it is seen that the difference $T_m - T_g$ does not vary greatly. (The values for T_g' are discussed in Section 16.2.)

Polymers. As mentioned, the formation of a glass from a "simple" liquid, which means a liquid consisting of small molecules, is very difficult. However, polymer glasses can form readily. As mentioned in Section 6.6.1 for starch, several polymer melts, as well as highly concentrated polymer solutions, can form crystallites—i.e., microcrystalline regions—below a certain temperature. Further cooling then does not lead to a greater proportion of crystalline material: as the crystalline fraction increases, it becomes ever more difficult for a (part of a) polymer chain to become incorporated in a crystallite, because its conformational freedom becomes ever more constrained. See also Figure 6.22. In practice, the crystalline portion will often be of the order of one-third. Still further cooling will then lead to a glass transition. Below T_g, part of the material will thus be crystalline, the remainder being glassy.

Starch is a common polymer in foods. In its native state it contains crystallites, and it shows a melting transition during gelatinization. After

TABLE 16.1   Estimated Values of the Melting Point T_m and the Glass Transition Temperature T_g of Pure Substances of Various Molar Masses M, As Well As the Special Glass Transition Parameters (T_g' and ψ_w') for Maximally Freeze-Concentrated Solutions

Compound		Dry systems			Freeze concentration	
Name	M, Da	T_m,°C	T_g, °C	$(T_g/T_m)^a$	T_g', °C	ψ_w'
Water	18	0	−137	0.50		
Glycerol	92	18	−93	0.62	−65	
Fructose	180	125	7	0.70	−42	0.14
Glucose	180	158	31	0.71	−43	0.16
Sucrose	342	192	70	0.74	−33	0.17
Maltose	342		87		−32	0.17
Lactose	342	214	101	0.77	−28	0.15
Maltohexose	991		134		−14	
Gelatin	10^6	25^b			−12	
Starch^c	$>10^7$	$(255)^b$	(122)	0.8	−6	0.26

^{a}Absolute temperatures.
^{b}Of crystallites.
^{c}Gelatinized.

cooling, crystallites form again (retrogradation). In many foods, starch is in its gelatinized form. Figure 16.3a gives an example of T_g and T_m values. Native starch tends to have a higher T_g value at the same water content; compare Figures 16.3a and 16.5.

In the glassy state, the polymer backbone is considered to be almost completely *immobile*, causing the material to be brittle. Between T_g and T_m the system is not to be considered as a highly viscous liquid but as an elastic material; the transition near T_g is called a *glass–rubber* transition. The crystallites act as cross-links between flexible stretches of the polymer chains, causing the material to have a rubbery consistency, with an appreciable elastic modulus. Above T_m a viscous liquid is formed. Figure 16.3b illustrates the rheological relations.

For glassy polymer systems without crystallites or other cross-links, a glass–liquid transition occurs. For systems with permanent, i.e., covalent, cross-links, the elastic modulus keeps decreasing with increasing temperature until a plateau value is reached.

Mixtures.   Figure 16.3a concerns mixtures of starch and water. In such a case, the solvent (water) acts as a *plasticizer*. It is seen that T_g (and T_m as well) strongly decreases with increasing water content, which also means that at a given temperature the viscosity will markedly decrease with increasing water content. Such behavior is also observed for glucose, the monomer of the starch polymer. This is primarily due to water molecules being smaller, hence more mobile, than glucose molecules. It is further seen that the values of T_g for the monomer are by about 75 K lower than for the polymer. Difference in mobility will, again, be the main cause. Generally, T_g of polymers increases with increasing degree of polymerization, n, but a plateau value tends to be reached for $n \approx 20$.

In these examples, starch or sugar is the component responsible for forming the glass and water is the diluent. Most foods contain several substances that can be in the glassy state. A glass then is formed much more readily—which means, in practice, at much slower cooling rates—than for a pure system, because crystallization is hindered. A glass can be a true, albeit noncrystalline, solid solution of many components, whereas crystallization must go along with phase separation. Moreover, different components tend to crystallize at different temperatures. Generally, some components will hinder the crystallization of others; see Section 15.2.2. Finally, the viscosity

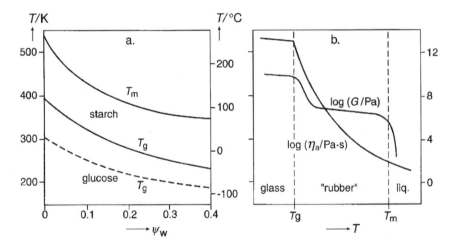

FIGURE 16.3 Glass transitions in polymeric systems. (a) Melting temperature T_m and glass transition temperature T_g of gelatinized potato starch as a function of mass fraction of water ψ_w. The values at very small ψ_w are extrapolated. A T_g curve of glucose is also given. (b) Approximate relations between rheological properties of polymeric systems and temperature; elastic shear modulus G for a system where crystallites melt at T_m, and apparent viscosity η_a for a system without a melting transition.

of the system will become so high as to hinder greatly the diffusion of crystallizable components.

Mixtures of a polymer, say starch, with a small molecule, say glucose, can also be made. To be sure, such a mixture will also contain water, since (a) starch, whether native or gelatinized, contains water; (b) glucose has to be dissolved before a homogeneous mixture can be made; and (c) it is virtually impossible to remove all water from the mixture. Having made a mixture with a low water content, the *glucose* then acts as a *plasticizer*, meaning that it lowers the apparent viscosity as compared to a system without glucose at the same starch–water ratio. This is, again, because glucose molecules are much smaller, hence more mobile, than those of starch.

In many mixtures, the value of T_g is uncertain: T_g is not an equilibrium parameter and its value will depend to some extent on the temperature history of the system. Moreover, it is often difficult to determine the T_g of mixed systems (more about this in Section 16.2). Several equations have been proposed for calculating T_g of mixtures from the data for pure systems. A simple equation directly relates it to the mass fractions ψ_i of each component i, according to

$$\frac{1}{T_g} \approx \sum \frac{\psi_i}{T_{g,i}} \tag{16.1}$$

This equation well describes the relation between water content of potato starch and T_g as given in Figure 16.3a; the relation is less perfect for several other binary mixtures. Many low-moisture foods are mixtures of polymers, water, and several other small-molecule components. Prediction of T_g from composition then is generally uncertain; in practice a (narrow) transition range rather than a sharp transition point is often observed.

For pure substances and their mixtures with water, the ratio T_g/T_m varies between about 0.85 and 0.50. See further Table 16.1. For polymer–water systems that form crystallites, the ratio tends to be smaller than for dry systems, e.g., 0.7 for gelatinized starch with 30% water as compared to 0.8 for dry starch.

Viscosity Relations. Several equations have been proposed to describe the dependence of the viscosity of the system on temperature. For polymer systems the *Williams–Landel–Ferry* (WLF) equation is often used. It reads

$$\log\left(\frac{\eta}{\eta_g}\right) = \frac{C_1(T - T_g)}{C_2 + (T - T_g)} \tag{16.2}$$

where η_g is the viscosity at T_g (T is absolute temperature). C_1 and C_2 would be constants for each system, and often fixed values are used, viz. -17.4, and 51.6 K, respectively. The curve for η_a in Figure 16.3b is calculated according to Eq. (16.2) with these values for the constants and $T_m - T_g = 90$ K. It is seen that the decrease in viscosity is very strong, by a factor of about 670 for the first 10 K increase over T_g. The equation often fits results for polymer systems well, although the two constants may vary significantly. Moreover, the viscosity measured is an apparent one and will depend on measuring conditions.

For other systems, the relations tend to be different and variable, and there is currently no consensus about the theory. The most important point is that the decrease in viscosity with increasing temperature tends to be weaker, and in some cases much weaker, than predicted by the WLF equation, especially near the glass transition. To be sure, even then the temperature dependence is strong as compared to that for "simple" liquids, which tend to follow an Arrhenius type relation $[\eta \propto \exp(C/T)]$. For complex mixtures, prediction of the temperature–viscosity relation from theory is currently impossible.

Molecular Mobility. In a glass the mobility of molecules presumably is very small. In a glass of one component, this appears indeed to be the case. Recalling Eq. (5.16), i.e., $D = k_B T/6\pi\eta r$, we calculate that a molecule of radius $r = 0.5$ nm, in a glass of viscosity 10^{12} Pa · s, has a diffusion coefficient $D \approx 10^{-24}$ m^2 · s^{-1}, which is extremely small. It would mean that it takes a molecule 300 centuries to diffuse over 1 µm distance. It originally had been assumed that these relations were generally valid. This would imply that the translational motion of molecules in a glassy food cannot occur, making the food completely stable to all changes involving diffusion, which includes nearly all chemical reactions. Moreover, it was sometimes assumed that above T_g, WLF viscosity could be used to calculate D via Eq. (5.16). These assumptions have not been confirmed: diffusivities tend to be far greater.

A rather trivial reason may be that the system can be *inhomogeneous*, possibly containing tiny cracks, that allow much faster diffusion. Moreover, Eq. (5.16) is poorly obeyed for highly concentrated systems, as discussed in Section 5.3.2. A more fundamental reason seems to be that the glasses always contain *more components*. Consider, for example, a sugar glass that also contains water. At or below T_g, the sugar molecules are presumably fully immobilized. However, smaller molecules, like water, can still move in the spaces between the sugar molecules. An example is given in Figure 16.4a. It is seen that well above T_g, the calculation according to the WLF equation is reasonably well obeyed, but that the discrepancy becomes very large close

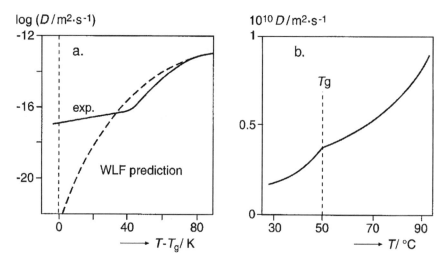

FIGURE 16.4 Effect of temperature on diffusion coefficients (D) in systems near the glass transition (T_g). (a) Experimental values and calculation according to the WLF theory for diffusion of fluorescein in a sucrose–water system. (From results by D. Champion et al. J. Phys. Chem. B 10 (1997) 10674.) (b) Diffusion of water in 81% pullulan (a polysaccharide) in water. (From results by S. Ablett et al. In: J. M. V. Blanshard, P. Lillford, eds. The Glassy States in Foods. Nottingham Press, 1993, p. 189.)

to T_g, up to about 5 orders of magnitude. Nevertheless, the diffusivity is very much smaller than at high temperature. Figure 16.4b gives an extreme example. Here the glass is formed by a polysaccharide that apparently leaves relatively large pores for water to diffuse through. At T_g the value of D for water is only by a factor of about 100 smaller than in pure water. At the glass transition, there is only a weak bend in the curve. On the other hand, the diffusivity of the polymer is immeasurably small.

Altogether, the diffusivity of small molecules like water is greatly reduced in a glass, but it is generally not negligible. Prediction of the value of D from theory is currently not possible.

Question

Is it possible that, in a mixture of two biopolymers containing, say, 30% water, one of the polymers is in a glassy and the other in a rubbery (or liquid) state?

Answer

Provided that some conditions are fulfilled. The first is that the polymers are not intermingled on a scale much smaller than the polymer length, since it is difficult to envisage how small parts of a molecule could be glassy and adjacent small parts not. This means that they should be phase-separated (see Section 6.5.2), which is the most likely situation at such high concentrations. On the other hand, at low water content, phase separation will not occur spontaneously at a reasonable rate, hence the water content must have been much higher than 30%, or the temperature very much higher than T_g, at some stage. The water can move faster, even close to T_g, which implies that the water activity will soon be the same everywhere. If now the two phases, if taken apart, have clearly different values of T_g, one of them will be glassy and the other rubbery in the range between the two T_g values. Such relations have been observed, e.g., for casein and gluten. The situation becomes even more complicated if one or both of the polymers forms crystallites.

16.1.2 Applications

Glassy Foods.   Several dry foods are wholly or partly in a glassy state. A simple example is a *high-boiled sweet*. It looks and feels like a piece of glass. It consists of a mixture of sugars (sucrose and oligomers of glucose) and 2–3% water. Some other sugar-based confectionary is also in this category.

Somewhat similar is *dried skim milk* (skim milk powder). Its main component is lactose (> 50%), and it has some 4% water. Figure 16.5 gives the T_g of lactose (a mixture of α- and β-lactose) as a function of water content; the curve for dried skim milk is almost identical (at low ψ_w). The powder particles consist of a glass of lactose and some other low-molar-mass substances, like salts (almost fully in associated form), in which casein micelles and globular proteins are embedded. Figure 16.5 gives curves for lysozyme, a typical globular protein, and gluten, a mixture of various, mostly nonglobular, proteins. It is seen that these can readily form a glassy state.

In most dry foods, starch is the main component responsible for a glassy state. This concerns several dry *cereal products*, like hard biscuits and various breakfast cereals. Figure 16.5 gives T_g for wheat starch, but if low-molar-mass components, generally sugars, are also present, the curve is shifted to lower temperatures, somewhere between those for starch and those for lactose. Another example is *pasta*, in the dry form in which it is commonly sold, which consists for the most part of wheat starch. Pasta has an impressive shelf life in the glassy state, and it readily takes up water upon cooking, to attain a soft rubbery texture.

Processes. To obtain a food in a glassy state, water generally has to be removed without crystallization of solutes occurring. In some products this is easy, in others difficult. The following methods can be distinguished.

Sugar boiling, as in the manufacture of boiled sweets. Evaporating water from a solution of sucrose will lead to its crystallization. By mixing sucrose with an equal amount (on a dry mass basis) of glucose syrup, crystallization can be prevented.

Air drying at ambient temperature, or at a higher temperature followed by cooling. This works well for many starchy foods, like pasta. Toasting a slice of bread, if done slow enough to allow evaporation of much of the water, also results in a glassy material (mainly consisting of partly gelatinized starch and gluten). When

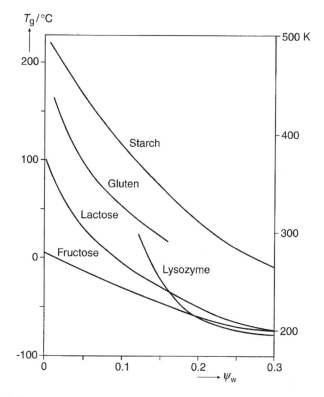

FIGURE 16.5 Glass transition temperature T_g as a function of mass fraction of water ψ_w, for some substances (indicated near the curves; "starch" means native wheat starch).

pieces of vegetables are dried, the cell walls can acquire a glassy state, leading to a hard and possibly brittle product.

Baking of a dough is one of the most common processes. At high temperature water can readily be removed and cooling then can lead to a glassy state, as in a hard biscuit.

High-pressure *extrusion* involves comparable changes. The product becomes hot and vapor bubbles are formed in it. The bubbles expand when the material leaves the extruder, and water vapor can readily escape. Upon cooling, a solid foam is formed, i.e., a glass containing many air cells. This is applied in manufacturing dry snacks.

Spray drying is applied to liquid foods like skim milk. Drying must proceed very fast, since otherwise lactose will crystallize. By dispersing the milk into fine droplets that are brought into hot air, the drying can be almost completed within a few seconds, whereby crystallization is prevented.

Also *freeze drying* can be applied to solutions. Water evaporates (desublimates) at low temperature and low pressure from the ice crystals present, leaving a highly porous structure of dry solutes, which often is in a glassy state.

Freeze concentration is a very common process; it is the main subject of Section 16.2.

Texture.   A hard biscuit has a crisp or brittle texture. This implies that it deforms in a fully elastic manner upon application of a force, until it breaks (snaps) at a relatively small deformation. Breakage goes along with a "snapping" sound. It appears from empirical observations that a crisp material has an apparent viscosity of at least 10^{13} or 10^{14} Pa $\cdot$ s. The water content or temperature above which *crispness is lost* closely corresponds to T_g. Sensory evaluation shows that an increase in water content by 2 or 3 percentage units, or in temperature by 10 or 20 K, can be sufficient to change a crisp food into a soft (rubbery) material.

A comparable phenomenon is that a solid material obtained by freeze drying can change into a highly viscous liquid upon increasing water content or temperature; this is generally called *collapse*, since the desirable porous structure of the freeze-dried product is lost. The collapse temperature tends to be a few kelvins above T_g.

If the glass consists in substantial part of low-molar-mass components, like sugars, the material tends to become quite *sticky* somewhat above T_g. This can occur with several spray-dried powders that take up water from the air; the stickiness then causes the powder particles to form a coherent mass, a phenomenon called *caking*. On the other hand, powder particles can be made to agglomerate—a process applied to enhance the dispersibility of powders in water—by shaking the powder for a short while at a temperature where it is sticky and then cooling or drying it.

Stability. In the glassy state, *molecular mobility* is greatly reduced, hence the food will have greater stability (shelf life). The dependence of various reaction rates on water content is discussed in Section 8.4.

An example is *crystallization* of sucrose in a system with low water content. Figure 8.6 shows that this starts, albeit very sluggishly (at room temperature), above a water content of 3%. At that value, T_g equals about 25°C. Skim milk powder with too high a water content will allow crystallization of α-lactose hydrate; according to the curve for lactose in Figure 16.5, this will occur above 30°C for 5% water, which fits the experimental results. Another example is *staling* of systems containing gelatinized starch, which is due to the formation of crystallites; see Figure 6.27a. It has been shown that the slow rate of staling at a temperature of, say, −20°C is due to the system being near the glassy state, where the molecular mobility of amylopectin molecules is already greatly reduced. In glassy systems with a high water content, which generally means that the glass is for the most part made up of polymers, freezing of water can occur a few kelvins above T_g.

Reaction rates in foods as a function of water content have been discussed in Section 8.4.2. We may add that even below the glass transition, small molecules usually have finite molecular mobility, allowing some chemical reactions, such as lipid oxidation, to proceed, albeit it slowly. Microbial growth (Section 8.4.3) generally stops at water contents far above T_g.

Question 1

It was mentioned above that the crystallization of lactose can occur at a critical water content, just above the glass transition. It was further (implicitly) assumed that this would happen at the same mass fraction of water ψ_W in skim milk powder. Experiments show that this is not precisely correct but that the critical conditions for crystallization are at the same water activity. Does this imply that the glass transition is determined by a_W rather than ψ_W?

Answer

When making lactose water mixtures, ψ_W can be precisely known. This is not the case for the lactose in skim milk powder. However, if a_W is the same in both systems (and an equilibrium water distribution has been reached, which may take several days), we also know that the water content of the lactose in the powder is the same as it is in the lactose–water system. In other words, it may still be ψ_W that is determinant (as is assumed by theorists).

Question 2

Assume that you want to develop a crispy breakfast cereal in the form of flakes in which the main component is wheat starch. To start with, you use the simplest model system of native starch and water. How would you proceed, and in particular what should be the course of water content and temperature to obtain a product with the desired physical properties? Tip: Also consult Section 6.6.2.

Answer

A glance at Figure 16.5 shows that to obtain a crispy product, i.e., one for which T_g is well above room temperature, the ψ_W value should be below about 0.17. In order to obtain a coherent dough (and also to obtain a product that is well digestible), the starch should be gelatinized. To that end, you should add water to a ψ_W value of about 0.5 and heat to above the gelatinization temperature (about 70°C). Then the mass should be cooled down to a lower temperature, but above T_g, to obtain a dough consistency that allows making flakelike shapes. The flakes are then heated to evaporate water, taking care that the temperature remains between T_g and T_m (why?), until the desired water content is obtained. Then cool to room temperature.

16.2 THE SPECIAL GLASS TRANSITION

State Diagrams. As mentioned, a glass transition point can be reached in a solution by *freeze concentration*. An example is given in Figure 16.6, a state diagram of the sucrose–water system. What will happen when a solution is slowly cooled has been discussed in relation to Figures 15.14 and 15. Assuming equilibrium, ice starts to crystallize upon reaching the phase boundary curve T_f. The solution becomes concentrated due to further freezing, and its composition follows T_f. When the eutectic point E is reached, sucrose should start to crystallize. However, considerable supersaturation can occur if cooling proceeds quickly. The system then further follows T_f, which now is not an equilibrium curve. At some stage, the line for homogeneous nucleation of sucrose will be crossed, but the viscosity of the system may by then be high enough virtually to prevent nucleation. Nevertheless, freezing of water goes on. (As mentioned, water can have a much greater mobility than sucrose near the glass transition point.) The diagram also gives the glass transition curve T_g for sucrose–water mixtures. Below and to the right of the curve the system is in a glassy state. When freeze concentration has proceeded to the point $(T_g{'}; \psi_s{'})$, a glassy state is reached and no further freezing occurs.

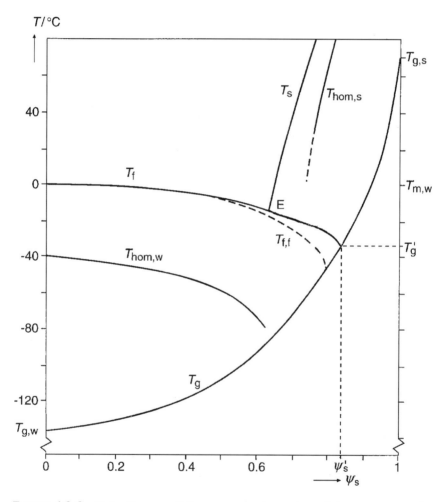

FIGURE 16.6 State diagram of the sucrose–water system. T is temperature, ψ_S mass fraction of sucrose. T_f gives the freezing temperature of water and T_S the solubility of sucrose. T_g is the glass transition temperature and T_{hom} the homogeneous nucleation temperature.

T_g' (generally read as "tee gee prime") may be called the *special glass transition* temperature; it characterizes the glass transition in a maximally freeze concentrated system. The corresponding value $\psi_W' = 1 - \psi_s'$ is the *residual water content* (often denoted W_g'), i.e., the proportion of water that will not freeze, however low the temperature. Some time ago, it was generally assumed that the residual water does not freeze because it is "bound" to the solute. As discussed in Section 8.3, this assumption is no longer tenable. The explanation is that the freezing rate of the water becomes infinitesimally small after the glassy state has been reached.

Figure 16.6 also gives a curve for the homogeneous nucleation of water. It is seen that T_{hom} decreases ever more below the value of pure water ($-40°C$) with increasing solute concentration. As a rule of thumb, $T_{hom} \approx -40 - 2\Delta T$ in $°C$, where ΔT is the freezing point depression caused by the solute. It may be clear that it is generally impossible to bring the solution without freezing to a temperature below T_g, except at a high sucrose concentration (above the eutectic point), with very rapid cooling or with very rapid drying.

Stability. As long as the temperature remains below T_g', the composition of the system is virtually fixed. This implies physical stability: crystallization, for instance, will not occur. As mentioned, some chemical reactions may still proceed, albeit very slowly because of the high viscosity and the low temperature. The parameters T_g' and ψ_s are, however, not invariable: they are not thermodynamic quantities. Their values will depend to some extent on the history of the system, such as the initial solute concentration and the cooling rate. The curve in Figure 16.6 denoted $T_{f,f}$ (for fast freezing) shows what the relation may become if the system is cooled very fast. The T_g curve is now reached at a lower ice content, so the apparent T_g' and ψ_s' values are smaller. However, the system now is physically not fully "stable": water can freeze very slowly until the "true" ψ_s' is reached.

Various Systems. Table 16.1 gives values of T_g' and ψ_W' for some substances. It is seen that the values of T_g' follow the same pattern as the T_g of the pure substance, e.g., increasing with increasing molar mass. However, the temperature range is not nearly as wide, although the values of ψ_W' do not vary greatly. The main variation seems to be in the steepness of the T_g curve between $T_{g,s}$ and T_g', at least for sugars.

In *mixtures* of solutes there is not just one eutectic, and so in principle more than one solute can crystallize. On the other hand, a wider compositional range tends to make it easier to reach a glassy state, just as in unfrozen systems. The values of T_g' and ψ_s' are, however, more dependent on temperature history.

Table 16.1 gives two example of *biopolymers*. It is seen that T_g' is not far below zero, especially for starch. Again, the glassy state is reached easily for polymers, although it may be difficult to establish a clear glass transition point. This is illustrated in Figure 16.7, which shows freezing curves of some complex systems. It is seen that at a considerable freeze concentration the curves go steeply down, and near a ψ_s value of about 0.8 the curves tend to become vertical. This means that no more ice crystallizes; presumably, the system then is in a glassy state. Newer studies indicate that the "proper" values of T_g' can be substantially higher than those suggested by the figure.

Uncertainties. The value of T_g' is generally determined by differential scanning calorimetry (DSC), and Figure 16.8a gives an example of a curve near the glass transition. It is commonly seen that not one but two second-order transitions occur, the first one being small compared to the second. The existence of two transitions is generally ascribed to two different physical relaxation mechanisms occurring in the glass; currently, there is no agreement about the explanation. The results for T_g' given in Table 16.1 represent the second, major, transition. Other workers assume the first transition to represent properly the glass transition. Also the "overshoot" often seen in the curve, which can be more prominent in some systems, is still posing questions, since it seems to indicate a first-

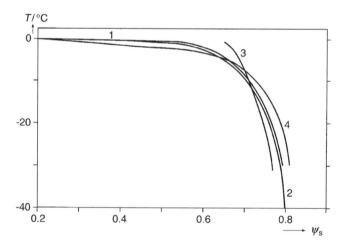

FIGURE 16.7 Freezing curves of various biological systems. T is temperature, ψ_s mass fraction solid. Curve 1, human blood; 2, washed yeast cells in water; 3, collagen; and 4, muscle tissue. (From results by A. P. MacKenzie. In: R. B. Duckworth, ed. Water Relations of Foods. Academic Press, London, 1975, p. 477.)

order transition. Altogether, some basic problems have not yet been resolved, and it may well be that there is not just a single classical, second-order, transition. The values obtained and published for T_g' thus are uncertain by a few degrees.

DSC scans of mixtures of solutes, and particularly of mixtures including biopolymers, do not show a clear second-order transition. An easier method of determining T_g' then is by dynamic rheology in small samples as a function of temperature. Figure 16.8b gives an example. Again, two points can be chosen, at the bend in the curve of the storage modulus, or the optimum temperature for the loss modulus. Usually, the latter point is taken, if only because it can be established with greater accuracy. The values obtained also depend on the rate of temperature increase and on the deformation frequency during the dynamic measurement (usually 1 Hz).

The greatest uncertainty is in the value of the *residual water content* ψ_W' as determined by DSC. It is obtained by determining the area under the DSC curve between T_g' and T_m and above the (uncertain) base line (cf. Figure 16.2). It has been shown that this method tends to involve a number of errors, which altogether lead to considerable overestimation of the value of ψ_W'. For instance, most values published for sucrose are around 0.36, whereas 0.17 is now considered to be the best estimate. It must be assumed that most of the published values for pure substances are too high by a substantial amount. The best way to obtain the value of ψ_W' is to find T_g', e.g., by DSC, and then separately determine T_g versus ψ_s for mixtures of low

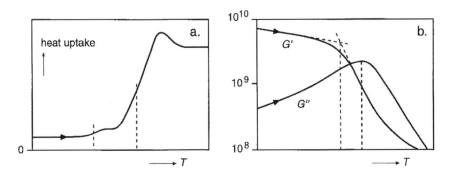

FIGURE 16.8 Methods of determining the special glass transition temperature T_g'. (a) Scan obtained by DSC (differential scanning calorimetry), giving the heat uptake (e.g., in $J \cdot kg^{-1} \cdot K^{-1}$) versus temperature T. (b) Scan of the storage (G') and loss modulus (G'') in Pa versus temperature. (After examples given by Champion et al. [see Bibliography].)

water content. This can be done accurately for simple mixtures, since the water content can be unequivocally determined. For mixtures of solutes, the problem remains. As a rule of thumb, the value of ψ'_W is generally not far removed from 0.2.

Conclusion. Most foods that are aqueous solutions and/or dispersions can be freeze-concentrated without crystallization of solutes and so become mixtures of ice crystals and a glassy material. The glass is of mixed composition, and roughly 20% of its mass is unfrozen water. The frozen food is stable to all physical changes and to many chemical reactions, provided that the temperature is kept below T_g'.

Question

Suppose that a company wants to produce "aromatic sugar," i.e., sucrose containing aroma (volatile flavor) compounds. The product should be in a fine granular or powdered form, readily dissolvable in cold water, and be stable, in particular to loss of aroma. What manufacturing methods would be suitable? Give particulars, especially on the temperature–water content sequence.

Answer

Grinding crystalline sucrose and adding the aroma compounds will not work: most of these compounds are somewhat hydrophobic and will not at all or only slightly adsorb onto sucrose. Even if sufficient "aroma" can be adsorbed, it soon will be lost by evaporation.

The aroma compounds should therefore be incorporated into a sucrose glass. (The chance that compound crystals of sucrose and aroma components can be formed is very small.) The sugar can be melted, mixed with the aroma compounds, and then cooled to below T_g. However, owing to its high melting temperature (192°C: Table 16.1), the sucrose will show caramelization and browning, which is presumably undesirable; moreover, aroma compounds may be lost or broken down.

Manufacture must then start by making an aqueous sucrose solution and adding the aroma. The sucrose concentration can be high, say $\psi_s = 0.5$, but it must be checked that sufficient water is present to dissolve the aroma compounds. A convenient process then would be spray-drying, which makes a powdered, glassy material. A considerable proportion of the aroma compounds can often be retained in spray-drying, since the outside of a drying drop very soon becomes glassy, greatly reducing diffusion of the compounds, whereas water diffusivity is still considerable. However, as seen in Figure 16.6, the drying temperature should be at most 40°C, to leave a mere 2% water and still be beyond T_g. It will be quite difficult to dry fast enough to such a low water content at such a low temperature.

Another option is freeze-drying. Starting with a concentrated solution, freezing will soon lead to a mixture of ice crystals and a sucrose-aroma glass: see Figure 16.6. Drying should then be done at high vacuum, taking care that the temperature remains below T_g. The porous glassy product obtained can readily be ground to a fine powder.

Note Probably, the manufacturer will not use pure sucrose but admix other sugars, such as a dextrose syrup, which will greatly reduce the tendency of sucrose to crystallize. That would allow the application of spray-drying at higher temperatures.

16.3 FREEZING OF FOODS

In the food industry, freezing is applied to make ice cream and comparable products (edible ices or frozen desserts) or as a process step in freeze-concentration or freeze-drying. The main purpose is to prevent or delay deterioration of the food. This concerns undesirable changes due to microbial growth, enzyme action, chemical reactions, or physical processes, usually involving mass diffusion. Such freezing is applied to natural products (fish, fruits, vegetables, etc.) as well as to a variety of fabricated foods. Essential primary changes occurring owing to freezing are

Lowering of the temperature
Freeze-concentration of the aqueous solution
Greatly increasing viscosity and possibly attaining a glassy state

An important question is, of course, how stable the product is and what factors determine stability. A problem can be that freezing greatly alters the consistency of the product and—often in combination with the thawing of the frozen material—leads to damage of its structure; another question is thus how to minimize such damage.

Low Temperature. Most aspects were discussed before, especially in Chapters 4 and 8. Briefly summarizing, the following effects of lowering the temperature can be mentioned.

The rates of nearly all chemical reactions decrease.
Chemical equilibria can change, since the rates of the reactions leading to the equilibrium may decrease to a different extent. Some

examples concerning ion association are in Figure 16.9. Most association constants increase upon cooling, but to a different extent. The pH of *pure* water is about 8.0 at $-20°C$.

Uncoupling of consecutive reactions can occur, leading to a changed product mix. This is discussed in Section 4.4.

Hydrophobic interactions decrease in magnitude and may even become repulsive below $0°C$. See Figure 3.4. This may, e.g., cause dissociation of proteins.

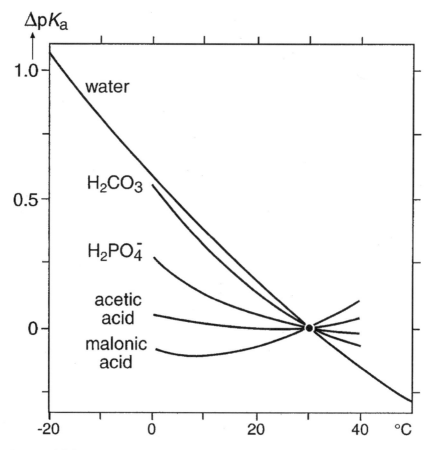

FIGURE 16.9 The effect of temperature on the change in association constant, relative to its pK_a value at $30°C$, of various acids and water. (Mainly after Franks [see Bibliography].)

Globular proteins can denature at low temperature, for the most part because of decreased hydrophobic interaction; see Figure 7.5. The denaturation is generally reversible, but enzymes are inactive in the denatured state.

Most microorganisms cannot grow at temperatures below zero, although their fermentation activity may slowly proceed at some degrees below zero.

Freeze-Concentration. Freeze-concentration has been discussed. Some examples of the proportion of water frozen are in Figure 16.10. By and large, the smaller the initial freezing point depression of the product, the higher the proportion of water that can freeze, although it also depends on the water content at which a glass transition is reached, which depends in

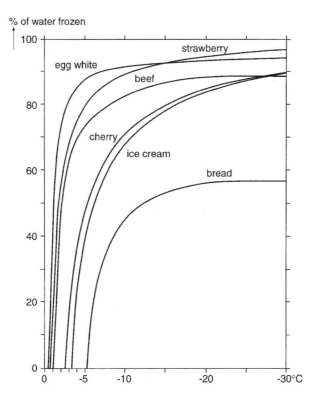

FIGURE 16.10 Examples of the proportion of water frozen as a function of temperature in some foods.

turn on the type and concentration of biopolymers present. For beef and bread, the curves become horizontal above $-30°C$, which implies that T_g' is reached.

The *concentration of solutes increases* upon freezing. As long as only water crystallizes, the molality of a solute after a mass fraction ψ_i of the water has frozen is given by the original molality divided by $(1 - \psi_i)$. Moreover, the activity coefficients (γ) may change (Chapter 2): for most neutral solutes, γ increases with concentrating, for ionic solutes, it will initially decrease.

The concentrating effect implies that *reactions can increase in rate*. For a simple second-order reaction of the type $A + B \rightarrow$ reactant (s), the rate of change of A will be given by

$$\frac{d[A]}{dt} = -k(T)\frac{m(T)}{m_0}[A]_0[B]_0 \tag{16.3}$$

(if the activity coefficients equal unity). Here k is the second-order rate constant, m is the molar concentration, and subscript 0 refers to conditions before freezing. Provided that k decreases relatively less with decreasing T than m increases, the reaction rate will increase. This is often observed, at least for moderate freeze-concentration. The relation is comparable to that shown in Figure 8.10a.

A first-order reaction may increase in rate if it is catalyzed by a substance that is concentrated. An example is the mutarotation of reducing sugars. The reaction is virtually first order (as long as a_W does not alter greatly) and is catalyzed by protons, for example. Figure 16.11a gives an example for glucose (the reactant) and HCl (providing the catalyst). It is seen that the rate constant decreases with decreasing temperature until the freezing point of the solution is reached. The freeze concentration upon further cooling then causes an increase in rate.

Concentrating the solutes implies that all *colligative properties* will increase in magnitude, including freezing point depression, osmotic pressure, and $-\ln a_W$. We will discuss this for *water activity*. Equation (8.6) states that $a_W = p_{v,solution}/p_{v,water}$, where p_v is water vapor pressure. At equilibrium, moreover, $p_{v,ice} = p_{v,solution}$, which then leads to

$$a_W = \frac{p_{v,ice}}{p_{v,water}} \tag{16.4}$$

Since the vapor pressure of pure water and of ice are only dependent on temperature (at constant pressure), water activity of a partly frozen system at equilibrium will only depend on temperature. The relations are illustrated in Figure 16.12a; Figure 16.12b gives the dependence of a_W on T. Notice

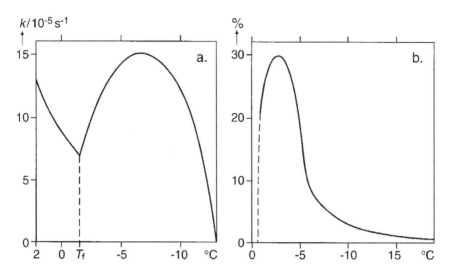

FIGURE 16.11 Effects of freezing and temperature on reaction rates in foods. (a) The reaction rate constant k for mutarotation of glucose in a solution of 100 g glucose per liter in 0.1 molar HCl. $T_f =$ freezing point. (Results of T. E. Kiovsky, R. E. Pincock. J. Am. Chem. Soc. 88 (1966) 4704.) (b) The percentage of the protein in the moisture expressed from red meat that has become insoluble, after 3 months storage of the beef at various temperatures. (Results of D. B. Finn. Proc. Roy. Soc., Ser. B, 111 (1963) 715.)

that very deep freezing is needed to decrease the water activity greatly. According to Eq. (8.3), the osmotic pressure is given by $\Pi = -46 \cdot 10^4 T \ln a_W$.

Generally, freeze-concentration leads to an *increase in ionic strength*. The effect is quite strong if salts are the main solutes (cf. Figure 16.14). An important consequence is that it causes salting out of proteins (Section 7.3), which may subsequently cause them to become insoluble (denatured?). This has been observed, e.g., for egg white and muscle protein, and it is illustrated for the latter in Figure 16.11b. Upon further freeze concentration, the ion activity coefficients will decrease and the ions will associate. Eventually, the ionic strength will decrease again.

Freeze-concentration may lead to *crystallization of some solutes*. Solutes that are poorly soluble can reach their solubility before the viscosity of the liquid becomes so high as to impede crystallization strongly. If the crystals make up only a small fraction of the material, they become

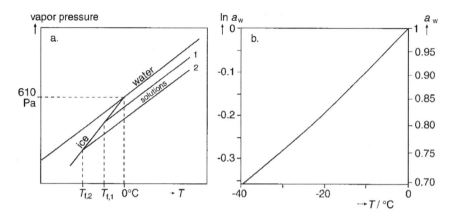

FIGURE 16.12 (a) The equilibrium vapor pressure of water, ice, and two aqueous solutions differing in molar concentration, as a function of temperature. Highly schematic and not to scale. (b) The water activity of aqueous solutions in the presence of ice as a function of temperature, assuming equilibrium.

entrapped in the glass formed at further freeze-concentration. The consequence of specific substances crystallizing is that the reaction mixture is altered, which can affect reaction rates and pathways.

An example is the effect of crystallization of buffer components on the *pH*. Consider a buffer of NaH_2PO_4 and Na_2HPO_4, giving a pH of 6.8. Upon cooling to $-0.5°C$, the eutectic point of NaH_2PO_4 is reached; the pH then is 6.6. At $-10°C$, where the other phosphate starts precipitating, the pH then drops to 3.6. The changes depend, of course, on the type of buffer. For instance, the pH of potassium phosphate buffers is less sensitive to freeze-concentration. The pH change will also depend on the ionic strength, the presence of proteins, and other compositional variables. Altogether, it is very difficult to predict pH values after freeze-concentration.

Finally, ongoing freeze-concentration leads to a *strong increase of viscosity* and possibly the formation of a glass. Besides a direct effect on the consistency of the product, it goes along with slowing down, and possibly stopping, of reactions, as discussed. The decrease of reaction rates at low temperatures shown in Figure 16.11 will at least be partly due to this effect. Moreover, the association of ionized components may play a part.

To *ensure stability*, it is often desirable to keep the frozen product at a temperature below T_g'. Approximate values are

Fruits and fruit juices	$-40°C$
Starchy vegetables (maize, potato)	$-12°C$
Green vegetables	-15 to $-25°C$
Muscle (fish, red meat)	$-12°C$
Ice cream	$-32°C$

However, these values vary with product source and also with pretreatment and freezing rate. In heterogeneous foods, T_g' may vary with location.

Freezing Damage. As discussed in Section 15.3.1, the freezing of water causes an *increase in volume*. Consequently, ice crystals formed in a disperse system can cause locally increased pressures, which can in turn cause mechanical damage. This occurs, for instance, when freezing an oil-in-water emulsion, especially if freezing is slow, the volume fraction of oil is large, and the oil droplets are not very small. At the prevailing low temperature, part of the oil will generally crystallize, which then means that partial coalescence can occur when fat globules are pressed together (Section 13.5). This becomes manifest upon thawing of the emulsion, when clumps of fat globules appear that will melt to form large oil droplets at higher temperature.

In *natural foods*, i.e., in animal or vegetable tissues, the situation is more complicated. If undercooling is at most by a few degrees, which generally means slow freezing, ice crystals are nearly always formed outside the cells. This means that the extracellular liquid will be freeze-concentrated, whereby its osmotic pressure increases. That will cause *plasmolysis*, i.e., the osmotic dehydration of the cell. The outer cell membrane or plasmalemma is permeable to water, but not, or poorly, to most solutes. Hence water will leave the cells, which will shrink considerably: see Figure 16.13. The intracellular liquid becomes highly concentrated, although the ice formation occurs outside the cell. This causes all the changes due to freeze-concentration, and it can possibly lead to damage of the cell membrane.

It appears that most cells do not contain internal surfaces that act as catalytic impurities for ice nucleation, until the temperature is as low as a few degrees above the homogeneous nucleation temperature (about $-40°C$). In tissues, however, ice crystals are observed inside cells at temperatures below -10 to $-15°C$. This is presumably due to the penetration of extracellular crystals through the plasmalemma into the cell. This will

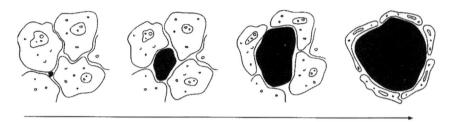

FIGURE 16.13 Development of an extracellular ice crystal (black) and the resulting plasmolysis occurring in a piece of tissue upon slow cooling. Highly schematic. (Adapted from H. T. Meriman. Federation Proc. 22 (1963) 81.)

decrease plasmolysis, but generally damage the plasmalemma and possibly intracellular organelles. The latter may upon thawing cause the release of enzymes into solution, whereby they become active. Mechanical damage to cell walls, i.e., in vegetable tissues, is also possible; this may result in a product of poor, e.g., sloppy, texture upon thawing. Fast freezing, which leads to the formation of more and smaller ice crystals, often tends to reduce freezing damage. During frozen storage, temperature fluctuations can cause partial thawing and refreezing, generally resulting in an increase of crystal size and increased freezing damage.

There is a vast literature on preventing or minimizing *freezing injury* to living organisms. Most organisms can protect themselves by a number of metabolic adaptations to freezing conditions, a subject outside the scope of this book. Killing of vegetative bacteria by freezing can occur to some extent, but it is never complete. Bacterial spores are very resistant to freezing.

Cryoprotectants. These are substances that reduce undesirable effects caused by freezing. Addition of nonionic solutes *reduces ionic strength* of freeze-concentrated solutions containing ionic substances. Since a high ionic strength can be very damaging to proteins, addition of solutes may be quite a useful measure. At a given temperature, the magnitude of the water activity (as well as that of the osmotic pressure) is fixed in a partly frozen system (Figure 16.12). The higher the initial molar solute concentration, the smaller the proportion of water freezing at a given temperature, hence the lower the concentration of the ionic species. This is illustrated in Figure 16.14, and it is seen that the salt concentration can be considerably reduced at a given temperature. The lower the molar mass of the cryoprotectant, the greater its efficiency for a given mass fraction.

NaCl concentration factor

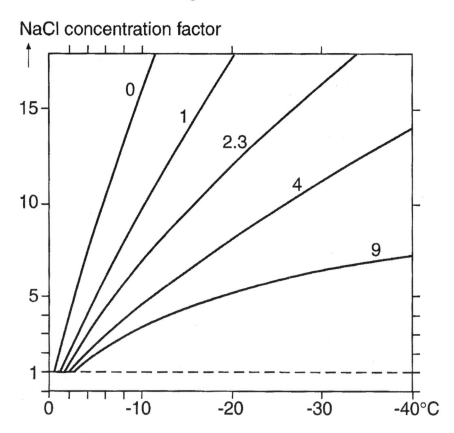

FIGURE 16.14 Concentration factor due to freeze concentration of a 0.85% NaCl solution containing various concentrations of glycerol. The numbers near the curves denote the mass ratio of glycerol to NaCl. (After results by J. Farrant. Lab. Practice 15 (1966) 402.)

Moreover, it should be readily soluble. Glycerol is thus quite effective ($M = 92$ Da, eutectic temperature $-46°C$), but it is not acceptable in substantial concentrations in most foods. Consequently, sugars and polyols (e.g., sorbitol) are often used.

Some added solutes can also reduce plasmolysis, because they can readily pass through the cell membrane. However, there is considerable variation in permeability for a given substance among species, and even among specific cells within a species. Most microorganisms let glycerol pass into the cell; see Section 8.4.3.

Mechanical damage depends on the extent of *volume expansion* caused by ice formation. To reduce mechanical damage, it is often desirable that ice crystals remain small; small crystals may also be needed to obtain a desirable texture and mouth feel. To that end, the *antifreeze peptides* briefly discussed in Section 15.3.1 can be quite effective in a product like ice cream. They strongly reduce ice crystal growth at low or moderate undercooling, but they do not impede nucleation. Consequently, more, and thereby smaller, crystals will form.

In the manufacture of ice cream, *biopolymers* (polysaccharides, gelatin) are often added as "stabilizers." These can affect the size and shape of the ice crystals formed, presumably via their effect on agitation during the freezing process, but also during quiescent crystallization. The stabilizers do not significantly affect nucleation. Neither do they substantially retard crystal growth when present in solution, since they do not markedly affect the diffusion rate of small molecules (see Figure 5.14). However, if they can form a gel, i.e., an elastic polymer network, they may be able to reduce ice crystal growth rate. Locally, the polymer concentration can be quite high, because of (a) freeze-concentration; (b) phase separation between the "stabilizer" and the milk proteins present; and (c) local concentration near a growing crystal face. That will cause the gel to become quite strong and have quite small pores. This will eventually prevent ice growing into the network, by mechanical resistance, causing growth to stop and the crystals to remain small. Moreover, the gel may hinder mutual contact, hence sintering, of ice crystals. The effect varies among biopolymers and increases with increasing biopolymer concentration. Although the mechanisms involved are not yet fully resolved, the general effect is clear. (Some workers assume that some polysaccharides can adsorb onto ice crystals and act like the antifreeze peptides just mentioned.)

In ice cream, as well as in other frozen products, considerable *recrystallization* can occur, particularly when the storage temperature is relatively high and fluctuates. It involves the disappearance of small ice crystals and the growth of large ones. This suggests that the process is due to Ostwald ripening (Section 13.6), but it seems to be more complex. Gel-forming biopolymers, as well as antifreeze peptides, can also retard recrystallization.

In conclusion, the factors involved in cryoprotection are not yet fully understood. It is often far from clear why one substance works better than a similar one. Also the effect of the freezing rate on damage is variable. In some instances fast freezing, and in other cases slow freezing, gives the best results, and often the freezing rate has little effect. For frozen vegetables, very fast freezing is generally considered desirable.

Question

Consider a solution of 8.5 g NaCl and 1.7 g glucose per liter. What would be (a) the freezing point of the solution and (b) the ionic strength of the freeze-concentrated solution at $-12°C$? You may assume that the glucose does not crystallize. Consult also Section 2.3.

Answer

(a) According to Eq. (2.16); the freezing point depression of a solution is given by $-103 \ln x_W$ (in K), where x_W is the mole fraction of water. A kg of solution contains $8.5/58.5 = 0.145$ mole NaCl, $17/180 = 0.0944$ mole glucose, and $974.5/18.015 = 54.1$ mole water. Assuming that NaCl is fully dissociated, we obtain $x_W = 0.993$, which yields a freezing point of $-0.72°C$. (b) Assuming that glycerol will act like glucose, and taking into account that the molar mass of glucose is twice that of glycerol, the curve noted 1 in Figure 16.14 would apply to the present solution. This then implies that at $-12°C$ the concentration factor equals 11.4. A kg of concentrated solution then contains 709 g of water and $69.9 \text{ g} = 1.66$ mole of NaCl. The salt molality then is $1.66/0.709 = 2.34$. This is not the ionic strength of the solution, since substantial association of Na^+ and Cl^- occurs at high concentration. Rewriting Eq. (2.21) for the association constant $K_A \ (= 1/K_D)$ yields

$$K_A = \frac{[NaCl]}{\gamma_{\pm}^2 [Na^+][Cl^-]}$$

assuming the activity coefficient for undissociated NaCl to equal unity. According to Table 2.4, the association constant for NaCl equals about $1 \, L \cdot mol^{-1}$, hence also about $1 \, kg \cdot mol^{-1}$, and according to Figure 2.11, the ion activity coefficient $\gamma_{\pm}$ for NaCl will equal about 0.7. Solution of the equation then yields $[Na^+] = [Cl^-] = 1.4$ mole per kg water. This is also the ionic strength.

    The reader will by now be aware that calculation of the data asked is not simple and could only be done because essential basic quantities are given in the book. It is therefore advisable to find an independent check of the result. This can be done via the water activity as a function of temperature given in Figure 16.12b. It can readily be calculated that the mole fraction of water at $-12°C$ is 0.89 (you may check this). For a not too concentrated and simple solution as considered, we have $x_W \approx a_W$, and the figure gives indeed $a_W = 0.89$ at $-12°C$.

16.4 RECAPITULATION

The Glassy State. A glass is an amorphous solid that shows a glass transition upon heating. A crystalline material melts upon heating, taking up heat at the melting temperature. A glass changes into a highly viscous liquid or into a rubbery material upon heating; this also occurs at a pretty sharp glass transition temperature T_g, where the specific heat of the material suddenly increases. A glass is not in an equilibrium state, and the value of T_g depends somewhat on temperature history. T_g is always below the melting temperature, often by 100 K or more. It is very difficult to transform a pure liquid into a glass. Some mixtures, however, particularly those containing biopolymers (starch, protein), can readily form a glass; polymers may be partly crystalline, partly glassy. Examples of glassy foods are high-boiled sweets, dried pasta, hard biscuits, some breakfast cereals, and dried skim milk.

The lower the water content of a food, the higher its glass transition temperature. This is because water acts as a plasticizer. As long as the temperature is below T_g, a dried food tends to be brittle and crisp. One function of the glassy state then is to provide crispness. Crispness can suddenly disappear on a temperature increase of a few degrees or a mere 1 to 2% increase in water content.

The other important function of the glassy state is to provide stability to physical and chemical changes. The viscosity of a glassy material is extremely high, e.g., 10^{14} Pa $\cdot$ s. This effectively stops all molecular motion and hence all change. However, in a glass of mixed composition, molecules that are smaller than those of the component(s) responsible for the glassy state can still diffuse, although greatly hindered. Hence complete chemical stability is usually not reached, but physical changes, like crystallization, are generally impossible below the glass transition temperature.

A glass can be made by drying a liquid or semisolid food, including such methods as baking, extrusion, and freeze-drying.

The Special Glass Transition. Drying can also be achieved by freeze concentration. When a solution is cooled, water can crystallize, thereby increasing the solute concentration in the remaining liquid. If freezing is fast, crystallization of solutes often fails to occur, and the dry matter content of the solution becomes so high as to reach the glassy state. This may be called the special glass transition, which occurs in a maximally freeze-concentrated system. It is characterized by a temperature denoted T_g' and a mass fraction unfrozen water ψ'_W. This water does not freeze because its diffusivity is virtually zero, not necessarily because it is "bound." The values of T_g', and to a lesser extent of ψ'_W, depend to some extent on the freezing rate and the initial water content of the system.

Many foods show the special glass transition upon freezing. For most of these, T_g' ranges between -10 and $-40°C$, whereas ψ'_W is generally about 0.2. Frozen storage at a temperature below T_g' then greatly enhances stability. Physical changes do not occur anymore, and most chemical changes become negligible; some reactions, e.g., lipid oxidation, may proceed very slowly. It is often difficult to determine the special glass transition point with accuracy.

Freezing of Foods. Freezing of foods has many ramifications. The decrease in temperature causes virtually all chemical reactions to decrease in rate. Globular proteins tend to unfold at very low temperatures, e.g., causing enzymes to become inactive. Moderate freeze-concentrating may increase reaction rates, because the concentration of reactants, or that of a catalyst, increases. Also the ionic strength increases, if salts are present, which can cause salting out of proteins and possibly irreversible changes (e.g., in muscle tissue). At deeper freezing, the viscosity of the remaining solution greatly increases, slowing down all reactions, and most of them stop below T_g'.

Freezing of tissues (vegetable or animal) generally causes some damage. At temperatures above $-10°C$, ice crystals only form outside the cells. This causes freeze concentration of the extracellular liquid, hence to an osmotic pressure difference between intra- and extracellular liquid and hence to osmotic dehydration of the cells (plasmolysis). At lower temperatures, ice crystals tend to penetrate the cells, whereby intracellular crystallization occurs. This reduces plasmolysis but tends to increase mechanical damage, possibly leading to a soft texture of the tissue after thawing.

Cryoprotectants are additives that can reduce adverse effects of freezing. One of these can be a high ionic strength. Water activity and osmotic pressure depend on temperature only, as soon as ice has formed. For instance, at $-23°C$, $a_W = 0.80$, independent of food composition. By adding nonionic solutes, such as sugars, the extent of freeze concentration at a given temperature is thus reduced, and thereby the ionic strength. This reduces damage to proteins. The smaller the molar mass of the solute, the more effective a given weight of solute is.

Solutes can also decrease mechanical damage, since less ice is formed at any given subfreezing temperature. Freezing causes a volume increase, which can cause local pressure differences and hence mechanical damage. Less ice and smaller crystals tend to give less damage. Substances that reduce crystal size, like some biopolymers (that give a strong gel in the freeze-concentrated solution) and antifreeze peptides, also reduce damage. Moreover, large crystals may give an undesirable texture.

BIBLIOGRAPHY

Much of what is discussed in this chapter is also treated, with an emphasis on practical usefulness, by

O. R. Fennema. Water and ice. In: O. R. Fennema, ed. Food Chemistry, 3rd ed. Marcel Dekker, New York, 1996, Chapter 2.

An extensive review on the glass transition is by

J. M. V. Blanshard. In: S. T. Beckett, ed. Physico-Chemical Aspects of Food Processing. Blackie, London, 1995. Ch. 2

The book also contains some other chapters related to the topic. A more recent review, stressing some fundamental aspects, is

D. Champion, M. le Meste, D. Simatos. Towards an improved understanding of glass transition and relaxations in foods. Food Sci. Technol. 11 (2000) 41.

An interesting monograph, giving much information on fundamental aspects, is

F. Franks. Biophysics and Biochemistry at Low Temperatures. Cambridge Univ. Press, Cambridge, 1985.

A very useful monograph, giving a wealth of information, although it is somewhat outdated as to fundamental aspects, still is

O. R. Fennema, W. D. Powrie, E. H. Marth. Low-Temperature Preservation of Foods and Living Matter. Marcel Dekker, New York, 1973.

A more recent book is

L. E. Jeremiah, ed. Freezing Effects on Food Quality. Marcel Dekker, New York, 1996.

Especially Chapter 1, by M. E. Sahagian and H. D. Goff, on fundamental aspects of the freezing process, is recommended.
       An enlightening discussion on the role of "stabilizers" in controlling ice formation is

A. H. Muir, J. M. V. Blanshard. Effect of polysaccharide stabilizers on the rate of growth of ice. J. Food Technol. 21 (1986) 683.

but some authors have more recently come to somewhat different conclusions.

17

Soft Solids

Many foods can be considered soft solids: bread, cheese, margarine, peanut butter, meat, several fruits, jam, puddings, mousse, aspic, boiled potatoes, etc. The term soft solid is ill-defined, as is the word semisolid, which is also used. The (implicit) meaning of the word solid does not comply with the definition given in Section 16.1 (a material with a viscosity over a given value, e.g., 10^{12} or 10^{14} Pa·s). A "solid" as meant in this chapter is a material that primarily exhibits elastic deformation upon applying a stress. The word soft then signifies that a relatively small stress is needed to obtain a substantial deformation; this may be due to the elastic modulus of the material being low (bread, soft fruits), or the yield stress being small (margarine, jam). The examples mentioned represent a wide variety of physical properties.

Virtually all soft solids are *composite materials*, which implies that they are inhomogeneous on a mesoscopic or even macroscopic scale. Their properties depend on this physical structure, and a structural classification can be useful. Main types are

> *Gels*. These are systems that consist mainly of solvent (mostly water), with the solid character being provided by a space-filling network. Some idealized types of network are provided by (a) long and

flexible polymer chains that are cross-linked and (b) aggregated particles.

Closely packed systems, in which deformable particles make up by far the largest volume fraction, whereby they deform each other. The interstitial material can be a liquid or a weak gel.

Cellular materials. Most vegetable tissues are in this category. They are characterized by connected, fairly rigid, cell walls, enclosing a liquidlike material. Several man-made cellular materials contain gas-filled cells.

Not all soft solids fit in this classification, e.g., meat and some types of cheese, and intermediate types occur.

Mechanical and other properties of some fairly simple soft solid materials will be discussed.

17.1 RHEOLOGY AND FRACTURE

Mechanical properties are essential attributes of soft solids, and this concerns primarily consistency during handling or eating. Moreover, physical stability often depends on these properties. Basic aspects of rheology are discussed in Section 5.1.3. This section is primarily about large deformation, including the phenomena of yielding and fracture.

17.1.1 Rheology of Solids

Elastic Moduli. A modulus is defined as the ratio of stress over strain (relative deformation); see Figure 5.8. There are various modes of deformation, as illustrated in Figure 17.1, corresponding to various types of modulus. Important types are

G = *shear* modulus; γ = shear strain.
E_u = Young's modulus or *uniaxial elongational* modulus; ε_u = uniaxial elongational strain.
E_b = *biaxial elongational* modulus; ε_b = biaxial elongational strain.

Figure 17.1c illustrates that uniaxial compression (vertical in this case) leads to biaxial extension in the directions perpendicular to the compression.

The various moduli are related to each other. For instance,

$$E_u = 2G(1 + \mu)$$
$$E_b = 4G(1 + \mu) \tag{17.1}$$

where the *Poisson ratio* μ is given by

$$\mu = \frac{1}{2}(1 - \frac{\mathrm{d}\ln v}{\mathrm{d}\varepsilon})$$ (17.2)

The second term between parentheses is a measure of the change in volume v upon application of a tensile stress on the material. The relative volume change is quite small for most solid foods, implying that the Poisson ratio is close to 0.5, leading to $E_u = 3G$, etc. Foods of a spongy nature, like bread crumb, may have μ close to zero.

It should further be understood that the deformation will often be of an intermediate type, although elongational components tend to be dominant. Pure biaxial elongation occurs around a gas bubble that expands in a semisolid food. Also the flow of material between two approaching particles is largely biaxial elongation. For undefined flow types, we will use E for modulus and ε for strain.

Prediction of the magnitude of the modulus from the properties of the structural elements, and the geometry of the network that they form, is desirable but quite difficult to achieve. Consider a simple system, consisting of a network of identical structural elements. An external force F_{ex} is applied in the x direction, leading to deformation. The cross section of the specimen perpendicular to the direction of force equals A. The deformation causes a reaction force, and the condition of force balance leads to

$$F_{ex} = -AN\frac{\mathrm{d}F_{in}}{\mathrm{d}h}\Delta x$$ (17.3)

where N is the number of connections between structural elements per unit

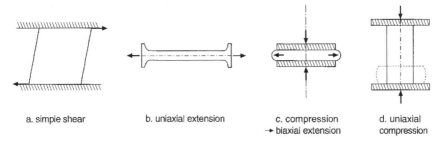

a. simple shear b. uniaxial extension c. compression → biaxial extension d. uniaxial compression

FIGURE 17.1 Various modes of deformation. Cross sections through test pieces; —·—·—indicates an axis of revolution. Arrows indicate forces. In (d) the broken line depicts a compressed test piece.

cross section perpendicular to x, and Δx is the deformation of the bond between two structural elements; $\mathrm{d}\,F_{in}$ is the change in the interaction force between two structural elements when moved with respect to each other over a distance $\mathrm{d}h$. Further, the external stress is $\sigma = F_{ex}/A$, and we postulate that $\Delta x = C\varepsilon$; C is a constant of dimension length, whose value will depend on the geometry of the network. Finally, we have $F_{in} = -\mathrm{d}V/\mathrm{d}h$, where V is the interaction free energy between two structural elements (discussed in Section 12.1). Assuming that $\mathrm{d}F_{in}/\mathrm{d}h$ is constant, which generally implies a very small $\mathrm{d}h$ and hence a very small deformation, we arrive at

$$E = -CN\frac{\mathrm{d}^2V}{\mathrm{d}h^2} \qquad (17.4)$$

Notice that V is a (Gibbs) free energy. It can thus be due to enthalpic factors (as in a fat crystal network) or to entropic factors (as in a rubberlike material). Often, both enthalpy and entropy are involved.

Although Eq. (17.4) is rigorous if the prerequisites are met, even then application is generally not easy. Analysis of micrographs may yield an estimate of the variable N. It is more difficult to estimate the value of C, since precise knowledge of the structure is needed. The magnitude of V as a function of h needs to be precisely known, and this is rarely the case (see Chapter 12). Nevertheless, in a few simple cases, reasonable predictions can be made or, in reverse, experimentally established relations between E and some variable—say, volume fraction of the network material—can be used to derive information about the network structure.

However, most real systems do not comply with the presumptions made in the derivation of Eq. (17.4). Generally, more than one type of interaction force will act, and the structural elements often vary in type or size, which implies a spectrum of interaction forces; we will see examples of this in the following sections. The contributions to the modulus of the various forces involved are not additive, primarily because the bonds vary in direction. Moreover, such materials are generally not fully elastic, which implies that the modulus will be complex [see Eq. (5.12)] and depend on deformation rate; virtually all soft-solid foods show viscoelastic behavior of some type. Finally, some systems are quite inhomogeneous, which further complicates the relations.

Large Deformations. For most soft solids, the direct proportionality between stress and strain only holds up to a very small strain, rarely over 0.01. One may, of course, calculate an apparent modulus $E_a = \sigma(\varepsilon) / \varepsilon$, which is for most foods smaller than the true modulus (cf.

Figure 17.2), but it makes more sense to determine the full deformation curve. Most solid foods eventually fracture upon increasing the stress. A simple example is given in Figure 17.2, which also defines the relevant parameters. These are

Modulus E or G, expressed in $N \cdot m^{-2} = Pa$. This has already been discussed. A suitable common term for the modulus is "stiffness."

Fracture stress σ_{fr} (in Pa) is often an important parameter. It is related to what is generally called the "strength" of a material. (It should be noted, however, that some authors speak of "gel strength" when they mean the modulus of the gel, which can readily cause confusion.)

Fracture strain ε_{fr} (dimensionless) can be called "longness," but this term is rarely used. "Shortness," also called "brittleness," is a common term, and it may be defined as $1/\varepsilon_{fr}$. If ε_{fr} is large, it may be called "extensibility."

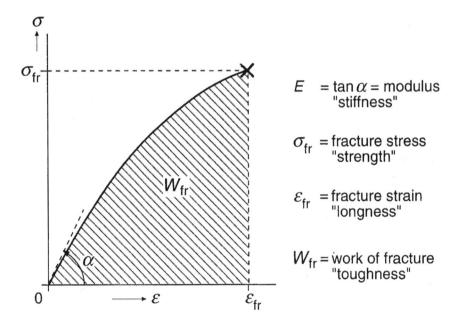

FIGURE 17.2 Illustration of various textural properties that can be determined when a solid test piece is deformed until it fractures (indicated by a ✕). W_{fr} is given by the area under the curve.

The *specific work of fracture* of the material W_{fr} (in $J \cdot m^{-3}$) can be called "toughness." It is given by

$$W_{fr} = \int_0^{\varepsilon_{fr}} \sigma \, d\varepsilon \qquad (17.5)$$

A curve like that in Figure 17.2 can be obtained in various ways, generally not leading to the same result. To name the most important points: (a) either the stress is increased (in a controlled manner) or the strain; (b) the rate of doing so, generally to be expressed as a strain rate Ψ, can be varied; and (c) the deformation mode can be like any depicted in Figure 17.1 or yet otherwise (bending, penetration, etc.). One should always choose these variables in accordance with the situation of interest. For instance, if the quantity to be determined is the resistance of a bread dough to the growth of a CO_2 bubble in it, it makes little sense to study deformation in shear, since it is biaxial extension that occurs around the bubble; moreover, it makes little sense to do experiments at a strain rate of, say, $10 \, min^{-1}$, since increase of bubble radius by a factor of two takes a far longer time than $0.1 \, min$ during a typical baking process.

Many rheological tests yield results that depend on the size and shape of the test piece. To obtain true *material properties*, some conditions must be fulfilled. First, the deformation should be homogeneous, i.e., the same everywhere in the test piece; this generally implies that the material has to be homogeneous above the scale of the smallest structural elements. Second, true stress and strain should be calculated. Consider the compression of a test piece, as depicted in Figure 17.1d. The cross-sectional area A will increase with increasing compression, which means that the stress ($\sigma = F/A$) will not be proportional to the force F, and a correction is needed. Moreover, the strain is often given as the linear or *Cauchy* strain $\varepsilon_C = (L - L_0)/L_0$, where L is length and L_0 original length. As discussed in Section 5.1.1, this is not a good representation of the strain, and the natural or *Hencky* strain should be used. It is defined as

$$\varepsilon_H = \int_{L_0}^{L} \frac{1}{L} \, dL = \ln\left(\frac{L}{L_0}\right) \qquad (17.6)^*$$

The strains often are given as the absolute values, and then for compression $\varepsilon_H > \varepsilon_C$, and for extension the other way around.

An example is given in Figure 17.3, and it is seen that the effect is very large for large deformation. Even the shape of the curve can be strongly affected. Just plotting force against linear deformation can readily lead to wrong conclusions. For very small strains, as applied when measuring moduli, the discrepancy is usually negligible.

Figure 17.4a shows a wide variety in deformation behavior; such, or even larger, differences are readily observed when comparing different systems. Curves 1 and 2 show the same modulus, and even the fracture stress is almost the same, but the fracture strains differ widely; also the toughness differs greatly (by a factor of about 15). Material 1 is stiff and short, like chocolate, whereas 2 is much "longer," like a semihard cheese. Curve 5 shows a much smaller modulus, while fracture stress and strain are like those of curve 2; however, the material is less tough, and the shape of the curve is as observed for some polymer gels. Curves 3–6 have the same modulus, but they are otherwise quite different. Curve 3 is of a weak and brittle material, e.g., a rusk. Curve 6 shows no fracture; instead, it shows yielding; see below.

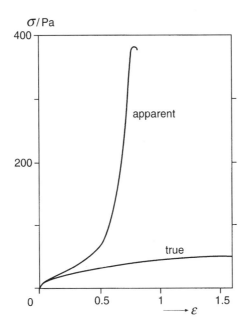

FIGURE 17.3 Apparent and true deformation parameters of a test piece (of semihard cheese) under compression (as in Figure 17.1d). σ is stress, ε is strain. Apparent stress is force over initial cross-sectional area; apparent strain is change in height over initial height (i.e., Cauchy strain).

It is often (implicitly) assumed that the modulus of a material is a measure of its strength. Figure 17.4 shows that this can be very wrong. Even within one structural class, i.e., polymer gels, the relation can be quite poor, as illustrated in Figure 17.4b.

Structure Breakdown. When a linearly elastic material is deformed and then allowed to relax, the stress–strain curve is fully *reversible*, as illustrated in Figure 17.5, frame (a). For a larger deformation, the curve is generally not linear, and perceptible *hysteresis* tends to occur, as depicted in frame (b). Nevertheless, the deformation is fully reversible. This means that deformation/relaxation has left the structure unaltered; repeating the test on the same specimen leads to an identical result. The hysteresis is due to energy dissipation, caused by flow of solvent through the gel network if it concerns a gel, as mentioned in Section 5.1.3.

Most soft solids undergo an *irreversible* change in structure upon large deformation; examples are in Figure 17.5, frames (c–e). In frame (c), relating to a starch gel, for instance, the hysteresis is considerable. Part of the

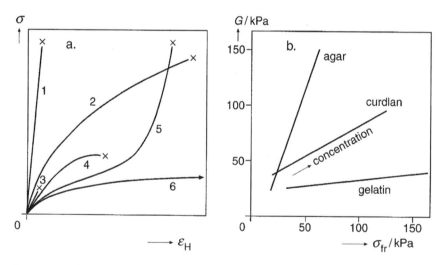

FIGURE 17.4 Variation in textural properties. (a) Stress (σ) versus strain (ε_H) curves and fracture points ($\times$). (b) Relation between elastic shear modulus G and fracture stress (σ_{fr}) for some types of gels of various concentrations of the network-forming material. (From approximate results by H. Kimura et al. J. Food Sci. 38 (1973) 668.)

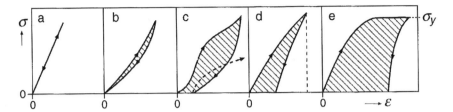

FIGURE 17.5 Examples of stress σ versus strain ε for various materials that are first compressed, then decompressed (both at constant strain rate). Examples are meant to illustrate various kinds of behavior. The hatched areas indicate the deformation energy that is dissipated (not recoverable). The scales are generally different for the various frames.

deformation is permanent, and a second deformation (broken line) leads to significantly smaller stresses. Frame (d) is a rather common type for several soft solids; its shape implies that the elastic deformability is limited. The hatched area between the compression and the decompression curve is a measure of the energy that is dissipated into heat, largely due to structure breakdown. The area under the decompression curve (bounded by the vertical dotted line) represents recoverable energy.

Frame (e) illustrates a system that shows *yielding*. This means that above a certain stress, called the *yield stress* σ_y, flow occurs; see also Figure 5.11. In such cases the structure is strongly altered by the deformation. Some materials exhibit partial or full recovery of the original structure after some time, a phenomenon called *thixotropy*; see also Section 5.1.3. This occurs typically in moderately concentrated dispersions in which the particles attract each other, though with forces that are not overly strong (can you explain this?). Systems for which the yield stress is rather high, say between 1 and 10^3 kPa, are often called *plastic solids*. The yield stress then appears to be a good measure for what is generally called *"firmness"*.

Time Effects. When one deforms a viscoelastic material, the stress depends not only on the strain but also on the *strain rate*. This is illustrated in Figure 17.6a; it is seen that at a given strain, σ increases with increasing Ψ value. This is because bonds that have come under stress owing to the deformation can break after some time; generally, new bonds are also formed, though not as many as had been broken. These phenomena cause *stress relaxation*. A material has a characteristic stress *relaxation time* which becomes manifest when it is kept at a given strain: see Figure 5.10 and Eq. (5.13). If Ψ is small, deformation time is relatively long, and much of the stress in the material can relax before a considerable strain is attained; this means that σ can not attain a high value. If deformation is fast, there is little time for relaxation and a relatively high stress results. At small Ψ values, the material thus behaves predominantly liquidlike (high tan δ value), at high Ψ values, more solidlike (lower tan δ value).

Another consequence of stress relaxation is that, for a higher stress applied to a material, it takes a shorter time before a high strain is attained and a shorter time before failure occurs (i.e., yielding or fracture, depending on the material). This is illustrated in Figure 17.6b.

Figure 17.6c relates to a material that exhibits *stress overshoot* (indicated in the figure for the upper curve) before yielding is complete. The magnitude of the overshoot increases with increasing Ψ value. Again, a higher stress can then be reached before much structure breakdown has occurred. The relaxation will occur over roughly the same time span, i.e., the relaxation time, for various strain rates. At quite low Ψ values, stress overshoot tends to be negligible.

It should be emphasized that various soft solids behave much differently in a quantitative sense, although the time dependence is qualitatively the same for most materials. It primarily depends on the Deborah number [Eq. (5.14)] how strong the time effects will be. Furthermore, stiffness, strength, and shortness of the material affect the result.

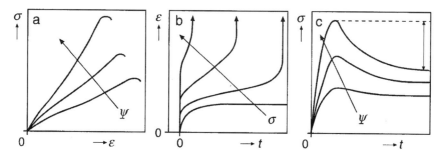

FIGURE 17.6 Time effects in deformation. The arrows through the curves indicate increasing values, generally by two or three orders of magnitude, of the parameter indicated. σ is stress, ε is strain, Ψ is velocity gradient (strain rate), and t is time after starting deformation. In (a) and (c), the strain rate is kept constant during the test, implying that ε is proportional to t for each value of Ψ; σ is measured. In (b), the stress is kept constant and ε is measured as a function of t.

Inhomogeneity. For large deformation, most rheological relations, including time scale dependence, can be quite different from what has been discussed so far if the material is inhomogeneous. What is meant is inhomogeneity at length scales larger than those of the smallest building blocks that affect the rheological properties. The deformation of a test piece will then always result in an inhomogeneous stress distribution throughout the sample. In other words, local *stress concentration* will occur, which will result in *inhomogeneous deformation*. It is generally impossible to predict the distribution of stress or strain throughout the test piece. Interpretation of the measured values in rheological or structural terms then is quite difficult.

The problem can be aggravated by the testing itself. Because the material is inhomogeneous, structure breakdown can occur in specific local regions, thereby increasing the inhomogeneity. In several cases, a kind of yielding zones or planes are formed throughout the test piece. A well-known phenomenon is *slip*, i.e., formation of a yielding plane close to the surface of the measuring body holding the test piece. The measured stresses then are much smaller than would have been obtained in the absence of slip. In other words, the test results may be useless (although the observation that slip occurs may in itself be significant). A comparable problem is that handling of a soft solid prior to testing, e.g., by forcing a specimen into the measuring body of the rheometer, may cause considerable structure breakdown, by which the rheological properties of the material are greatly altered.

Note In an experiment as illustrated in Figure 17.1c, the material must slip over the platens if we are to measure its elongational viscosity. In such a case, lubrication by a thin liquid that does not diffuse into the sample is often applied.

Question

An important property of a product like margarine is its consistency or, more precisely, its "spreadability", that is, can it readily and evenly be spread onto a slice of bread? Presumably, the yield stress of the material σ_y is essential in this respect: if it is too low, the product may flow under its own weight during storage, if too high, spreading takes too much effort and the bread will crumble. An empirical test that is often done to assess the consistency of spreads is cone penetration. A sharp cone of given mass and cone angle is clamped in a vertical position above the test piece, the cone tip just touching the surface, and released. The penetration depth L_p of the cone in the test piece then is measured.

How would the relation be between σ_y and L_p? Do you think that it is a good test for the yield stress? What criticisms can be made? How would the magnitude of the half cone angle θ and of the cone mass m affect the value of L_p for a given σ_y value? What are the advantages of the test?

Answer

As the cone penetrates the material, it will at first exert a stress of considerable magnitude; but this stress becomes ever smaller as the cone goes deeper, and when the stress is decreased to the value of σ_y the cone will stop. The force exerted by the cone would approximately equal $m \cdot g$ and it acts, in first approximation, over an area πR_p^2, where R_p is the radius of the cone at a distance L_p from its tip. Hence $\sigma_y = mg/\pi R_p^2$. Since $R_p = L_p \tan \theta$, we would obtain $\sigma_y \propto L^{-2}_p$. Moreover, for a given value of σ_y then $L_p \propto m^{0.5}$ and $L_p \propto \cot \theta$.

These quantitative relations are very approximate. The force exerted by the cone involves momentum as well as weight: the cone falls into the sample. The reaction force of the material involves friction exerted by the material and depends on the volume of material that does yield, which may not depend in a simple manner on L_p; also the buoyancy force of the material on the "submersed" part of the cone may not be negligible; finally, the phenomenon of stress overshoot (see Figure 17.6c) may upset the relations. Careful comparison for a number of margarine samples with a well-defined test for the yield stress led to the relation $\sigma_y \propto L_p^{-1.6}$.

The advantages of cone penetration include the simplicity of the test; that the test piece can be used without any previous deformation (which would affect the value of σ_y); and that the deformation time (of the order of a second) is about the same as during the spreading of margarine. Altogether, the test results correlate well with the subjectively evaluated spreadability.

17.1.2 Fracture Mechanics

Failure.   When a stress, however small, is applied to a Newtonian liquid, it flows, implying that all bonds between the constituting molecules frequently break, while new ones are formed. When an increasing stress is applied to a piece of an elastic solid, it becomes deformed, and when a certain stress is reached, the test piece starts to fracture. *Macroscopic fracture* is characterized by

> *simultaneous breaking in one or more macroscopic planes throughout the specimen of all bonds between structural elements of the solid*

which results in the specimen falling into pieces. This implies that (most of) the broken bonds do not reform. The structural elements can be atoms, molecules, or particles, and the bonds that break are generally those between the largest elements. The word macroscopic means that the size of the fracture plane is much larger than that of these structural elements:

fracture is generally visible. In soft solids the situation is more complicated, since fracture is generally preceded by local structure breakdown.

In several cases the material does not fracture but shows *yielding*. Also during yielding a great number of bonds in some planes are broken, but at the same time new bonds form (these are not necessarily of the same nature as the original bonds). The specimen then is strongly deformed but remains a continuous mass. This is often because liquid that is present between the solid structural elements immediately flows to the tiny cracks formed.

Fracture and yielding both cause a sudden and significant change in the mechanical properties of the specimen put under stress. Such a change is often termed *failure*. Other types of failure can occur. Consider, for instance, the breaking of the stem of a flower without the stem being severed. Buckling of two-dimensional structures is another example. It can occur with the cell walls in cellular material when it is put under compressive stress (see Section 17.5).

In the present section, we consider fracture mechanics. The theory distinguishes three *regimes*: linear-elastic, plastic-elastic, and time-dependent fracture.

Linear-Elastic Fracture. This is also called brittle fracture. When a homogeneous isotropic elastic test piece is put under stress, and the magnitude of the stress applied is larger than the fracture stress of the material, the test piece can break. For crystalline materials, the fracture stress can be predicted from the known bond strengths and the geometry of the crystal structure. However, it is generally observed that the experimentally established fracture stress is much smaller than the theoretical one, say, by two orders of magnitude. The discrepancy is primarily due to the material being inhomogeneous, i.e., containing imperfections or even tiny cracks at various sites. Virtually all materials contain such *defects*. They give rise to **stress concentration**.

Consider the example illustrated in Figure 17.7a, i.e., a flat strip of material put under elongational stress. To simulate a defect a small notch has been applied. At the tip of the notch—and in principle the same applies near a small imperfection—the stress will be larger than the overall stress σ_{ov} in the specimen at some distance away from the notch. The local stress then is given by

$$\sigma_{loc} \approx \sigma_{ov}\left(1 + 2\left[\frac{L}{R}\right]^{0.5}\right) \qquad (17.7)$$

where L is notch length and R is the radius of the tip of the notch. The stress concentration can be considerable, the more so for a larger size and a sharper tip of the notch. Supposing that the notch is 100 μm deep and that R = 10 nm, σ_{loc}/σ_{ov} will equal about 200. A well-known example is the linear notch (scratch) applied on a glass pane by a diamond needle, which enormously decreases the force needed to break the pane. A defect in a crystal is generally of molecular dimension, i.e., $R \approx 1$ nm.

Figure 17.7a relates in fact to a two-dimensional system, but also in three dimensions the stress concentration around a small hole can in principle be calculated. The figure is drawn for an experiment in which the test piece is subject to uniaxial extension. Other loading modes can be used. Uniaxial compression is frequently applied to solid foods; as mentioned, it may lead to biaxial extension. Other tests invoke bending (as applied when breaking a glass pane) or shearing. Although the mathematical relations are somewhat different, the principles of fracture mechanics remain the same.

Crack Propagation. That σ_{loc} is higher than σ_{fr} is a necessary condition for fracture to occur, but not a sufficient condition. It suffices for *fracture initiation*, i.e., the formation of a small crack, but not for spontaneous *propagation* of the crack. To explain this, *energy relations*

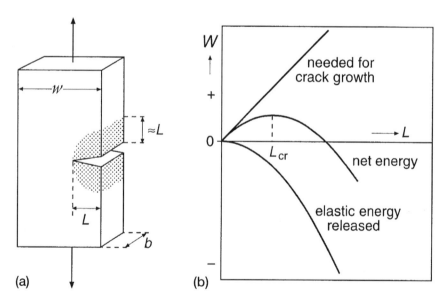

(a) (b)

FIGURE 17.7 Fracture in tension of a test piece containing a notch or crack. (a) Geometry of the test piece. (b) Energies involved (W) as a function of crack length (L).

must be taken into account. Fracture means an increase in surface area and hence an increase in surface free energy, and this energy must somehow be provided. When an elastic material is put under stress, the energy applied to do so (W_{el}) is stored in the test piece and can be available for realizing fracture. In the situation as given in Figure 17.7a (assuming now that the crack depicted was formed by putting the test piece under tension), the energy that had been stored in the shaded region was available to cause the fracture by being transported to the tip of the crack. Its amount is roughly proportional to L^2. The total fracture energy or work of fracture W_{fr} would be proportional to L. The crack will now be propagated, i.e., fracture will spontaneously proceed, if the differential energy released surpasses the differential energy required. The condition thus is

$$\frac{1}{2}\pi b[(L + dL)^2 - L^2]W_{el}^V > bW_{fr}^A dL \tag{17.8}$$

The symbol W is used for the total amount of work done or energy released, and the specific values are denoted by the superscripts A or V for amount per unit area and volume, respectively.

The energies involved are illustrated in Figure 17.7b, and it is seen that a maximum in free energy occurs at a crack length L_{cr}, beyond which the crack will grow spontaneously if the test piece remains under stress. From Formula (17.8) it follows that the *critical crack length* is given by

$$L_{cr} = \frac{W_{fr}^A}{\pi W_{el}^V} = \frac{2W_{fr}^A E}{\pi \sigma_{fr}^2} \tag{17.9}$$

where the part after the second equals sign only holds for an ideally elastic solid, where $W_{el}^V = \frac{1}{2}\sigma\varepsilon$ and $E = \sigma/\varepsilon$. Since the modulus (E) and the specific work of fracture are material constants, this means that in this case σ_{fr} is proportional to $L^{-0.5}$. It should be noted that σ_{fr} is defined as the *overall stress*—i.e., the force divided by $w \times b$—acting on the specimen at fracture.

In linear-elastic fracture, crack propagation (after L_{cr} has been reached) proceeds roughly at the speed of sound in the material (of order $1\,km \cdot s^{-1}$). This gives rise to a snapping sound, typical for fracture of brittle materials.

Elastic-Plastic Fracture. Many elastic solids can show yielding at high stress. During fracturing, the local stress near the crack tip can be very high, as mentioned. This then may lead to local yielding, which has several consequences. (a) *Local plastic deformation* occurs, which means that the pieces remaining after fracture do not precisely fit to each other, as they would do after linear-elastic fracture. (b) *The tip of the crack becomes* *blunted* [i.e., a larger value of R in Eq. (17.7)], which means that the stress concentration becomes smaller and a higher overall stress is needed to achieve fracture. (c) *The work of fracture increases*, since it now includes the energy dissipated owing to the local yielding. It may be clear that "soft solids" by their nature cannot exhibit linear-elastic fracture, but some do show plastic-elastic fracture.

Grinding. An important aspect is often whether such a material can be ground, and how small then are the particles obtained. It can be derived from the theory that the thickness of the zone near a crack in which plastic deformation (yielding) occurs in a homogeneous isotropic material is given by

$$z = \frac{32}{3} \frac{W_{fr}^A E}{\sigma_y^2} \tag{17.10}$$

where σ_y is the yield stress. If the diameter of a particle of the material is smaller than z, it cannot be broken up and hence it cannot be ground any finer. Expressed in another way, not enough elastic energy can be stored in the particle to provide the total work needed for fracture propagation. Values for z mostly range between 1 and 100 μm. For materials that exhibit plastic-elastic fracture, decreasing the temperature leads to a smaller z value, since the yield stress tends to markedly decrease while the modulus is not strongly affected (the work of fracture will also decrease). If the temperature decrease causes a glass transition, the regime even changes to linear-elastic fracture, greatly facilitating the grinding process.

The **specific work of fracture** is in the simplest case given by 2γ, where γ is the surface tension of the material; the factor 2 is because fracture leads to the formation of two new surfaces. For foods, γ will rarely be larger than the value for water ($70 \, \text{mN} \cdot \text{m}^{-1}$), which would imply that $W_{fr}^A \approx 0.1 \, \text{J} \cdot \text{m}^{-2}$. However, values observed are generally between 1 and $10 \, \text{J} \cdot \text{m}^{-2}$, one to two orders of magnitude higher. Possible reasons are (a) that the fracture surface is often uneven, causing the surface area of the crack to be larger than it appears on a macroscopic scale; (b) that a crack is often accompanied by small side cracks or a few small cracks about parallel to the main one, which also increases the effective surface area; and (c) that local yielding causes energy to be dissipated during fracture, and the increase of the work of fracture will be about proportional to the value of z in Eq. (17.10). Prediction of the magnitude of the excess is very difficult, because the rheological behavior of the yielding material is highly nonlinear.

Values of W_{fr}^A thus have to be determined experimentally. This can be done by means of various cutting tests, but it is often difficult to obtain reliable results.

Time-Dependent Fracture. (This is not an established term.) In this regime, which is frequently observed in soft-solid foods, the whole test piece shows lasting deformation and hence structural breakdown before fracture occurs. Because of this, the fracture phenomena will depend on time scale. The resulting fragments do not fit each other at all.

We will first consider the **energy relations** for fracture. The total work done upon deformation of the test piece would be given by

$$W_{tot} = W_{el} + W_{ve} + W_{str} \ (+W_{fr}) \tag{17.11}$$

Here W_{el} is the *elastic energy stored*; its amount is proportional to the size of the test piece for a given stress. The last term, i.e., the net work of fracture, is put between parentheses, because it is derived from W_{el}, which thereby will decrease. In linear-elastic fracture, the sum of W_{el} and W_{fr} is even constant as soon as the fracture propagates spontaneously. In time-dependent fracture, where fracture propagation can be slow, the sum is not constant (a) because the material is viscoelastic and part of the energy is dissipated during its transport to the crack tip, and (b) some elastic energy will be added to the system after fracture has started, since deformation is still going on.

The other two terms in the equation refer to the *energy dissipated into heat*; its amount is larger for a larger test piece. Somewhat arbitrarily, it has been split into two terms. W_{ve} refers to what happens in a *homogeneous viscoelastic material*. As discussed in Section 17.1.1 under "Time effects," the stress *relaxation time* τ is an essential parameter if its value is in the range of the values of $1/\Psi$ applied. At low values of Ψ, there is enough time for part of the stress to relax, implying that σ remains low and also that σ_{fr} is small. See Figure 17.6a, where the maximum in each curve roughly coincides with fracture. At low Ψ values, the proportion of the energy that is dissipated (i.e., W_{ve}/W_{tot}) is larger; hence the amount of energy stored W_{el} is smaller for a given value of ε; so fracture will occur at a larger ε value. This is seen in Figure 17.6a, and also in frame (b), since a higher stress causes a larger Ψ value. All these effects are stronger for materials having a larger value of tan δ (at a given Ψ value).

The term W_{str} becomes important for *inhomogeneous materials* at large deformation. Presumably, the energy dissipation is mainly caused by friction between structural elements due to inhomogeneous deformation. Moreover, immediate irreversible bond breaking can contribute. It is difficult to predict how total energy dissipation will depend on strain rate. This is because the deformation will change the structure and thus the rheological properties of the material. Altogether, increasing values of Ψ nearly always cause an increase in σ_{fr}, albeit to a widely variable extent.

Several soft solids show **yielding** at low and fracture at high strain rates, although the critical strain rate for fracture may vary widely. Most plastic fats, including butter and margarine, show yielding at high stress under most conditions, but at quite high deformation rates, say $\Psi > 10\,s^{-1}$, fracture can occur. Another example is given in Figure 17.8 for two types of cheese. The "short" cheese (1) shows virtually no dependence of ε_{fr} on the strain rate, whereas the much "longer" cheese (2) exhibits flow at strain rates of about 10^{-4} s^{-1} or smaller. This flow induces irreversible structural changes, which means that yielding occurs. On the other hand, some materials can show very slow fracture when a relatively small constant stress is applied, as is illustrated in Figure 17.6b; this may represent the behavior of some starch gels. Even the lowest curve, which suggests that an equilibrium deformation has been reached, may start to increase in height and finally become vertical when fracture occurs, which may be after quite a long time (say, one hour).

This brings us to variation in **fracture mode**. In the case just mentioned, small cracks presumably form in many places in the test piece and slowly grow in size until they become connected over larger distances, leading to visible fracture. (Differentiation between yielding and fracture then is almost arbitrary.) In most cases, test pieces fracture either in shear or

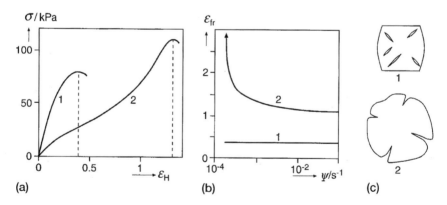

FIGURE 17.8 Fracture behavior of two cylindrical cheese samples subjected to uniaxial compression. Sample (1) concerns a hard and fairly brittle cheese, (2) a semihard green cheese that is quite extensible. (a) Stress σ versus Hencky strain ε_H, resulting upon deformation at a strain rate Ψ of 10^{-2} s^{-1}; the dotted lines indicate the strain at fracture ε_{fr}. (b) Values of ε_{fr} obtained at various Ψ. (c) Fracture mode as seen in cross sections through the test pieces at the moment of maximum stress; for (1) it is a vertical cross section, for (2) a horizontal one.

in tension. The two samples of cheese discussed give good examples, illustrated in Figure 17.8c. The "short" sample (1) fractures in shear. Before fracture is seen at the outside, cracks have formed inside, in planes that are about 45° from the direction of the compression. The "long" sample (2) shows cracks at the bulging surface after the test piece has been considerably compressed. Fracture occurs obviously in tension, and it starts at the outside, where the elongational strain is largest.

It will be clear now that the relations governing fracture of viscoelastic materials are far more complex than those for elastic solids. The discussion above gives some qualitative relations that generally hold. Quantitative prediction of the behavior from first principles is mostly not possible. It all becomes even more complex for inhomogeneous materials.

Notch Sensitivity. Consider testing a material in the geometry as depicted in Figure 17.7a. Test pieces of the same shape, but containing horizontal notches of various lengths L, are used. The notches have been made before applying the stress. Assuming for the moment that there is no stress concentration at the tip of the notch, the stress in the plane of the notch would be given by $\sigma_0 w/(w - L)$, where σ_0 gives the overall stress. This implies that fracture will occur at an overall stress proportional to $(1 - L/w)$, as shown by curve 1 in Figure 17.9. Most soft solids do not show such a relation, since stress concentration will occur, leading to notch sensitivity.

To explain this we consider the overall stress needed to *propagate* a crack σ_{pr} (not the fracture stress σ_{fr}) in a homogeneous isotropic elastic material. Equation (17.9) gives the value of L_{cr}, further illustrated in Figure 17.7b. The equation can also be "inverted": assuming that a notch of sufficient length has been applied, we then obtain for the crack propagation stress

$$\sigma_{pr} = \left(\frac{2W_{fr}^A E}{\pi L}\right)^{0.5} \qquad L > L_{cr}^0 \qquad (17.12a)$$

where L_{cr}^0 is the critical crack length for fracture propagation in the absence of a notch. A relation like that of curve 2 in Figure 17.9 is obtained: the curve starts at L_{cr}^0 and then steeply decreases. The equation will hold as long as L is significantly smaller than w. Since at $L = L_{cr}^0$, we have $\sigma_{pr} = \sigma_{pr}^0$ (the stress needed for fracture propagation in the absence of a notch), Eq. (17.12a) can be rewritten as

$$\sigma_{pr} = \sigma_{pr}^0 \left(\frac{L_{cr}^0}{L}\right)^{0.5} \qquad (17.12b)$$

It follows that the curve will decrease more steeply with increasing L for a smaller value of L_{cr}^0.

The fracture propagation stress may thus very strongly depend on notch length. Figure 17.9 also contains a curve for the *modulus* as a function of notch length for the same material as for curve 2. It hardly decreases with increasing L. Since moduli are measured at very small strains, the ratio between stress and strain—i.e., the modulus—will virtually be constant throughout the test piece. This illustrates that *inhomogeneities in a sample have a far greater effect on strength than on modulus*.

The latter phenomenon is also involved in the shape of curve 3, which is of a quite common type. It is frequently observed that σ_{pr} does not decrease with increasing L until it reaches a certain value that is greater than L_{cr}^0. The length scale at which the curve starts to decrease is commonly

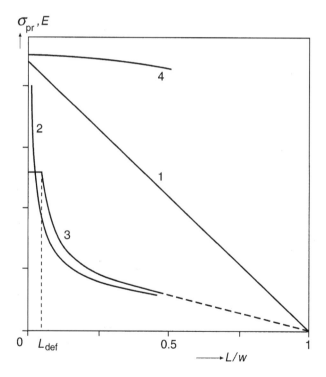

FIGURE 17.9   The effect of the length of a notch L in a test piece on the stress σ_{pr} needed for fracture propagation in the piece. Configuration as depicted in Figure 17.7a; w is test piece width. The vertical scale is linear and arbitrary. Curve 4 shows the modulus E as a function of L. Meant to illustrate trends.

interpreted as representing the inherent effective *defect length* L_{def} of the material, i.e., the size of inhomogeneities that are responsible for stress concentration. In several cases, good agreement between L_{def} and microscopically observed inhomogeneities has been obtained. For instance, the defect length in concentrated starch gels agrees well with the size of the partly swollen starch granules in the gel (e.g., about 0.1 mm).

Equation (17.12) is valid for an ideally elastic material. Some sugar glasses do fit the prerequisites and show an even steeper curve than curve 2 in Figure 17.9. But most soft solids behave differently. Some gels come close to the ideal relation: for instance, curves 2 and 4 fit reasonably well for a 10% potato starch gel. Many materials give relations of the general shape of curve 3. That means that these materials are notch sensitive and show defects of a length far greater than molecular. However, Eq. (17.12) cannot be used to predict relations.

The **extent of notch sensitivity**, i.e., the steepness of decrease of σ_{pr} with notch length, varies considerably. The factors affecting it are not fully understood. Some factors that decrease notch sensitivity are following:

> *A larger inherent defect length.* This is because the curve starts at a higher L value. Compare curves 2 and 3 in Figure 17.9.
>
> *A lower yield stress.* As mentioned, a low yield stress leads to a relatively thick zone around the crack where yielding occurs, hence to blunting of the crack tip, hence to decreased stress concentration.
>
> *More pronounced anisotropy.* The anisotropy meant implies that the forces keeping the material together vary with direction. Consider a fibrous material like muscle tissue. If it is put under tension in the direction of the fibers, and a small notch is made perpendicular to the fibers, hardly any stress concentration occurs. This is because the breaking stress in the direction of the fibers is quite high, whereas the fibers can easily be torn apart if the stress is perpendicular to their direction. Thus only a small fraction of the elastic energy applied to one fiber can be transmitted to other ones, and each fiber will have to break separately. The relation between σ_{fr} and L then will be close to that of curve 1 in Figure 17.9.

Question

Consider an isotropic material that exhibits elastic-plastic fracture. Will its yield stress be higher or lower than its fracture stress, or is the difference unimportant?

Answer

This depends on what is meant by the fracture stress. The yield stress must be larger than the overall fracture propagation stress, since otherwise the specimen put under stress will undergo yielding rather than fracture. But the yield stress must be smaller than the local (or "true") fracture stress, since otherwise blunting of the crack tip will not occur and pure elastic fracture results.

17.1.3 Fracture in Elongational Flow

Consider a long and thin cylinder of a viscoelastic material upon which an extensional stress is acting, which stress is larger than the yield stress. Elongational flow will then occur. This happens during spinning of, say, a highly concentrated protein solution into a thread. In such a configuration, somewhere in the thread a slightly thinner spot or "neck" will frequently form by chance, and this may lead to breaking of the thread. The force F acting on and in the direction of the thread will be everywhere the same. Since $F = \sigma A$, the stress will be higher for a smaller value of the cross-sectional area A, if the value of the Young modulus E_u is constant. This means that the thread will be extended faster at the neck than at other places, hence the thin part will become ever thinner, and the thread will break. Nevertheless, several materials can be spun. How is this possible?

It will only succeed if the reaction force (stress times area) in the thin part can be larger than that in the thick part of the cylinder. Let the cross-sectional area at the neck be $A + dA$, where dA is small and negative. The condition then becomes

$$(A + dA) \cdot \sigma(A + dA) > A \cdot \sigma(A)$$

which can be reduced to

$$\sigma \cdot dA + A \cdot d\sigma > 0$$

Dividing by $A \cdot d\sigma$ results in

$$d \ln A + d \ln \sigma > 0$$

Since $\varepsilon_u = \ln(L/L_0)$, where L is specimen length and the subscript 0 denotes the original value, and $L \cdot A$ is constant, we have $d\varepsilon_u = -d \ln A$. Inserting this we obtain the relation

$$-d\varepsilon_u + d \ln \sigma > 0$$

which can be rewritten as

$$\frac{d \ln \sigma}{d\varepsilon_u} > 1 \tag{17.13}$$

The condition for stability is thus that the stress in the material increases with the strain, a phenomenon called *strain hardening*, and that the relative increase is more than proportional to the increase in the uniaxial Hencky strain. Of course, when elongation goes on, the thread will finally break, but at a far larger strain value than would occur in the absence of strain hardening.

A comparable situation occurs for **biaxial extension** of a thin film. This is, for instance, important in a bread dough. During fermentation and in the beginning of baking, many large gas cells are formed, which soon deform each other. The thin film between them is being extended and may readily break, leading to coalescence. Again, the formation of thin spots can be prevented by sufficient strain hardening. The condition for stability now becomes

$$\frac{d \ln \sigma}{d\varepsilon_b} > 2 \tag{17.14}$$

The magnitude of the derivative is called the *strain hardening index*. A good correlation is observed between its value (between 0.5 and 2.5 in most doughs) and the strain at rupture. There is also a good correlation between the index and the gas holding capacity of the bread, since early coalescence of the cells leads to loss of gas to the atmosphere.

Strain hardening is generally ascribed to materials containing strongly entangled polymers, which become stretched upon extension because they resist disentanglement. This can especially happen with polymers of which the backbone contains large (groups of) side chains. This is the case, for instance, for wheat gluten and xanthan gum, materials that exhibit strong strain hardening. There are some additional factors of importance. First, the strain hardening also depends on strain rate, and bread dough tends to be *strain rate thinning*; consequently, the strain hardening index will be smaller, to an extent that depends on the rate of increase of the strain rate during deformation. Second, the apparent viscosity of the material must be between certain limits at the conditions applied: at low η_a values, the extension will proceed too fast, while at high values the stress needed for extension is too high. Third, for a very thin film, say $< 10\,\mu m$, the Gibbs mechanism of film stability (see Figure 10.29c) can become significant, provided that the material is homogeneous at the scale of the film thickness (which is not the case in bread dough).

Question

In Emmentaler ("Swiss") cheese, gas formation causes the formation of holes in the cheese mass. The gas is CO_2, produced from the lactate in the cheese by propionic acid bacteria. If al goes well, spherical openings with a smooth surface ("eyes") are formed, diameter about 2 cm; the time over which the fastest growth of the holes occurs is about 10 days. Sometimes, slits rather than eyes are formed; the slits or cracks are shaped like a flat disk, with a rough surface and a sharp edge. Two experiments are done with respect to hole formation. (1) Two cheeses are made from the same milk in the same manner, but one has a higher concentration (number per unit volume) of the bacteria mentioned. For the lower concentration, eyes are formed, for the higher one slits. (2) Two cheeses are made, of which one has a shorter consistency than the other. The short one gives slits, the "long" one eyes. Explain in a qualitative sense the differences in (1) and in (2). (3) Why does a slit not grow throughout the whole cheese? See also Question 2 in Section 5.1.3.

Answer

(1) Gas is formed and will diffuse to any small hole present in the cheese. This means that gas pressure in the hole may become larger than atmospheric, by an amount σ_s. If now $\sigma_s > \sigma_y$, the hole will expand. Since σ_y is quite small in cheese, this will readily occur; the cheese mass around the hole will flow, and an eye is formed. However, if the gas production is very fast (more bacteria), the stress in the hole will become higher and $\sigma_s > \sigma_{fr}$ may be reached; consequently, fracture occurs, leading to a slit. This is also likely because a higher gas production generally implies a faster growth rate. The elongational strain rate around the hole is given by $\Psi = d \ln R / dt$. For the situation given, this means $\Psi \approx 10^{-6} \, s^{-1}$, and for faster growth it may perhaps be by 10 times higher. A glance at Figure 17.8b shows that a higher strain rate may cause a much smaller fracture strain and hence—for the same material—a smaller fracture stress. (2) Now the gas production rate will be the same, and the σ_s value in the holes will differ little between the cheeses. As illustrated in Figure 17.8, however, a short cheese will have a lower σ_{fr} value than a long one. (3) If $\sigma_s > \sigma_{fr}$, a roughly circular fracture plane is formed and the hole becomes larger, since the two faces can bulge farther. This implies that σ_s sharply decreases, and fracture will stop; in other words, insufficient energy is available for slit growth. Naturally, the CO_2 pressure can build up again, and the process may be repeated a few times. After a while, the lactate will be fully converted and no further CO_2 will be formed, generally before the slits can grow quite large.

17.1.4   Texture Perception

The texture of a food, i.e., its consistency and physical inhomogeneity as perceived by the consumer, often is an important quality mark. When handling the food, this may relate to spreading, slicing, grinding, mixing, etc. The most important is, of course, texture as perceived during eating, briefly called *mouth feel*. Any sensory perception, be it (mainly) by eye, by ear, by touch, or by the chemical senses, is a highly complex phenomenon. The result depends on the individual and his or her experience, and on a number of conditions during the perception. Thus perception of the same attribute in the same food is subjective and quite variable.

Texture perception is poorly understood. The sensation that a person experiences when eating is the resultant of an interaction between the nervous impulses evoked in the mouth and the (almost fully unknown) processing of the resulting signals in the brain. Here the dynamic aspect is essential: without motion of jaws or tongue, texture perception is virtually impossible. Even "spatial information," i.e., food inhomogeneity, is largely derived from "temporal information." Moreover, it is not only the consistency and the physical inhomogeneity of the food that is responsible for the perceived mouth feel; other variables can also contribute, for instance flavor, flavor inhomogeneity, sharpness, coolness, and any sound emitted due to fracture of a piece of food. All of this is in the realm of "psychorheology," a discipline that is clearly outside the scope of this book.

However useful it would be to know what makes a food "crispy" or "creamy" or "sticky," most of it is still uncertain. Food manufacturers have to rely on correlations between evaluations of trained panels and experimental test results. A prerequisite then is that these tests should be done under conditions that mimic what happens in the mouth. This concerns mainly the following aspects:

> *Deformation mode.* Deformation processes in the mouth vary from swallowing, to pressing (e.g., between tongue and palate), to chewing and biting. The phenomena occurring then range from flow, to yielding, to fracture. This should be mimicked in the test. Where it concerns flow or yielding, the deformation in the mouth is predominantly (biaxial) elongation. Especially for viscoelastic materials, the difference in apparent viscosity as measured in elongational or in simple shear flow can be large, owing to variation in the value of the Trouton number (Section 5.1.3). For hard solid foods, biting may have to be imitated.
>
> *Strain rate.* For most foods, rheological and fracture parameters are time scale dependent. Generally, the effective strain rate in the mouth during biting or chewing is of the order of $2\,s^{-1}$, and the same rate should be applied in the test.
>
> *Inhomogeneity* of the food poses problems. If it contains "soft" and "firm" regions, this may affect perception; it will also affect

mechanical test results, but probably in a different manner. If the food is strongly anisotropic, as is the case with most kinds of meat, additional uncertainties arise.

The conditions during the test should also in other respects be similar to those in the mouth. *Temperature* can be readily adjusted, but the dilution with *saliva*, or its lubricating effect, may be more difficult to mimic.

17.2 GELS

Apart from a discussion of functional properties, this section will be divided according to the network structure of gels. Flory* gave the following structural classification:

1. Well-ordered lamellar structures, including gel mesophases
2. Covalent polymeric networks; completely disordered
3. Polymer networks formed through physical aggregation; predominantly disordered, but with regions of local order
4. Particulate disordered structures

Types 2–4 are illustrated in Figure 17.10, frames (a), (b), and (c), respectively.

We will here only consider types 3 and 4—called *polymer gels* and *particle gels*, respectively—which are both common in foods. Type 1 can occasionally occur in foods, but the concentrations of small-molecule surfactants needed to obtain such gels are generally too high to be acceptable. Type 2 does not occur, although some covalent bonds between structural elements occasionally contribute to gel structure, such as —S—S— bridges between protein molecules. Several intermediate gel types occur in foods, and some of these will be discussed.

17.2.1 Functional Properties of Gels

Food gels are made for a specific purpose, which generally means that the gel should have one or more specific physical properties. When making a pudding, for instance, it is desirable that it should not sag under its own weight during keeping, a property often called "stand-up." Neither should the pudding fracture, e.g., fall into pieces during transport. Another stability aspect concerns leakage of liquid from the gel, which tends to be very

---

*P. J. Flory. In: *Faraday Discussions* **57** (1974) 8.

undesirable. In some puddings, flavoring and coloring substances are not evenly distributed—when pieces of fruit are present, for instance—and diffusion of these substances throughout the pudding during storage may be detrimental for appearance and flavor. A good eating quality involves that pieces of the pudding can readily be scooped with a spoon, and be subsequently deformed or minced in the mouth to allow swallowing. Mouth feel further includes physical inhomogeneity and (lack of) stickiness. Finally, perception of flavor components depends on gel structure, although the mechanisms involved are poorly understood. Of course, the properties desired may vary among types of pudding.

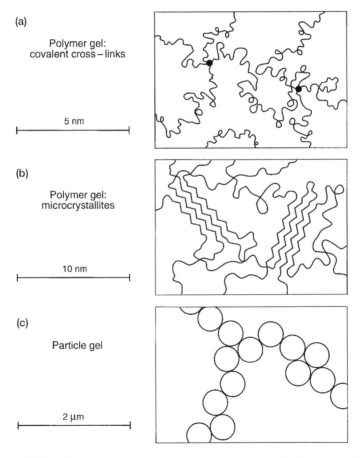

(a)

Polymer gel: covalent cross – links

5 nm

(b)

Polymer gel: microcrystallites

10 nm

(c)

Particle gel

2 μm

FIGURE 17.10 Illustration of three types of gel structure, highly schematic. The dots in frame (a) denote the cross-links. Note that the scale may greatly differ.

Some important functionalities are mentioned in Tables 17.1 and 2, and the physical properties needed to realize these qualities are specified. It is seen that widely varying qualities may be desired, often in combination, as illustrated above. It also follows that a gel can generally not be characterized by its modulus: for hardly any of the functionalities mentioned is it an essential parameter. The message of these tables is that one should always first find the properties needed and then find methods to determine those properties; then it should be possible to develop formulations and processes to realize the qualities desired. The tables are not exhaustive: special products often require special properties. Moreover, it is often desirable that a property should depend in a given way on external conditions. For instance, a food, or parts of it, may have to be "solid" at ambient temperature but melt in the mouth.

Table 17.1 gives mechanical properties, and these have already been discussed for the most part. *Shaping* may need some elaboration. In the table, shaping of a gel is what is meant, i.e., a gel is present and it has to be forced into another shape. To that end it must exhibit yielding; if yielding then leads to a low apparent viscosity, the time needed before the solid character is restored is of importance. Shaping more commonly involves pouring of a (thick) liquid into a mold and then letting it gel, for instance by lowering temperature (gelation of gelatin), or by changing pH (some spinning operations).

Table 17.2 indicates measures to be taken against physical instability of various kinds. To prevent particle motion, a very weak gel generally suffices. This is discussed in Section 13.3 under "Immobilization of Particles." In several systems, the particles themselves can form the gel;

TABLE 17.1 Desired Mechanical Characteristics of Gels Made for a Given Purpose

Property desired	Essential parameters	Relevant conditions
"Stand-up"	Yield stress (or modulus, if quite low)	Time scale
Firmness	Modulus, or yield stress, or fracture stress	Time scale, strain
Shaping	Yield stress, restoration time	Several
Handling, slicing	Fracture stress and work of fracture	Strain rate
Eating characteristics	Yield and/or fracture properties	Strain rate
Strength, e.g., of a film	Fracture properties or strain hardening	Stress; + time scale or strain rate

TABLE **17.2** Gel Properties Needed to Provide Physical Stability

Prevent or impede	Gel property needed
Motion of particles	
Sedimentation	Significant yield stress + short restoration time
Aggregation	Significant yield stress
Local volume changes	
Ostwald ripening	Very high yield stress
Motion of liquid	
Leakage	Small permeability + significant yield stress
Convection	Significant yield stress
Motion of solute	
Diffusion	Very small pores, high viscosity of continuous liquid

see Section 17.2.3. To prevent leakage of liquid from the gel, its permeability as appearing in the Darcy equation (5.24) should be small. This is discussed in Section 5.3.1. A significant value of the yield stress may also be needed, since otherwise liquid may be pressed from the gel under its own weight (cf. the Question at the end of Section 5.3.1). Hindered diffusion is discussed in Section 5.3.2.

Finally, the gel network itself should be stable, implying that it should not show some undesirable change in structure. Since the kind of changes that can occur and the variables involved greatly depend on gel type, these aspects will be discussed further on.

17.2.2 Polymer Gels

Basic aspects of polymers are discussed in Chapter 6; see especially Section 6.2.1.

The Rubber Theory of Elasticity. The simplest structure that a polymer gel can have is depicted in Figure 17.10a: long linear polymer molecules that are cross-linked at various places along the chains. We will denote the part of a molecular chain from one cross-link to the next one a *cord*. A similar structure occurs in vulcanized rubbers. The rheological properties of such a system are supposed to be determined by the conformational entropy of the cords. It is assumed that the distribution of the monomers—or, more precisely, of statistical chain elements—in each cord is Gaussian ("normal"), which will be true if their number n' is large enough. Any change in the end-to-end distance of the cords due to deformation of the gel will lead to a *decrease in entropy*, hence to an increase in free energy. Flory showed that the shear modulus of an entropic gel is simply given by

$$G = \nu k_B T \tag{17.15}$$

where ν is the number of cords (or twice the number of cross-links) per unit volume. Further assumptions about the structure are that (a) the number of cords per original polymer molecule is large (cf. Figure 6.2); (b) all cords have about the same contour length; and (c) the network contains negligible numbers of entanglements (see Figure 6.16), closed loops, and loose ends.

Moreover, the equation can only be accurate for small strains, since considerable change in the end-to-end distance of the cords would distort the Gaussian distribution of statistical chain elements. This happens more readily for a smaller value of n'. It also implies that at increasing strain, the chemical bonds in the primary chain become increasingly distorted. Consequently, the increase in elastic free energy is due not merely to a decrease in conformational entropy but also to an increase in bond enthalpy. If the value of n' is quite small, even a small strain will cause an increase in enthalpy. (In a crystalline solid, only the increase in bond enthalpy contributes to the elastic modulus.)

Theory has been developed that takes these aspects into account. Figure 17.11 gives calculated relations for the elongation of a gel specimen. For the calculation some assumptions about the network structure have to be made, but the trends given are generally observed. It is seen that for small n', the curve readily becomes vertical, implying that the cords are fully stretched; further stress increase would then lead to breaking of chains or cross-links.

Although food gels are never of the rubber type discussed, the relations given provide some qualitative insight in the factors governing the rheological properties of polymer gels.

Junctions. Most food polymer gels have cross-links in the form of junctions as schematically depicted in Figure 17.10b. Here no chemical (covalent) bonds are formed between molecules, but microcrystalline regions that involve a great number of (mostly weak) "physical bonds". Van der Waals attraction, hydrophobic interaction, and hydrogen bonding may contribute; for charged polymers, ionic bonds can be involved. The simple type shown in the figure, i.e., a microcrystallite of stretched polymer chains, is not very common. It appears to occur in gels of galactomannans and possibly of xanthan. It may be that in amylose gels crystallites of straightened single helices can provide junctions. Since nearly all of these junctions are made of quite weak bonds, a significant number of bonds have to be formed at the same time to become more or less permanent; in other words, junction formation involves a cooperative transition.

A common type of junction is shown in Figure 17.12a. Many polysaccharides form *double helices* below a given temperature and at given physicochemical conditions. Apparently, the helices often involve two molecules. This may be difficult to achieve because of geometrical constraints: the parts of a molecule not incorporated in the helix then would also become twisted to the same extent (but in the opposite sense), which is largely prevented by the entanglements in the system. However, in many polysaccharides, complete rotation (i.e., by 360°) about the bonds between monomers appears to be possible, at least at some positions along

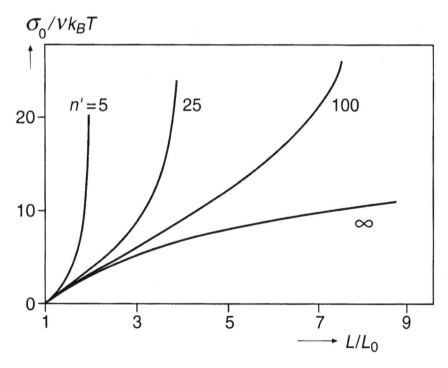

FIGURE 17.11 The effect of n' (the number of statistical chain elements in a cord between cross-links) on the relation between stress and strain of a polymer gel in elongation. σ_0 is the force divided by the original cross-sectional area of a cylindrical test piece, v is twice the cross-link density, L is the length, and L_0 the original length of the test piece. (After calculations by L. R. G. Treloar. The Physics of Rubber Elasticity. Clarendon, Oxford, 1975.)

the chain (probably stretches devoid of bulky side groups). This then prevents the parts of the chains near a double helix to become twisted. Nevertheless, each junction tends to contain several helices that align to form a microcrystallite, as depicted in Figure 17.12a. It is often observed that the helices form in a very short time upon cooling (as inferred by circular dichroism spectroscopy), whereas the modulus develops at a much slower rate after helix formation. It may be that the helices are unstable unless they are part of a microcrystallite.

Figure 17.12 shows two other types of junctions, those involving *triple helices* (b) as in gelatin, and the so-called *egg-box junctions* (c) that appear in some ionic polysaccharides. Gels with junctions of the latter type do not melt above a given temperature, as do gels with junctions involving helices. Presumably, the egg boxes can also form stacks, further enhancing the stiffness of the gel.

In most polymer gels, the junctions contain a substantial proportion of the polymer material, say 30%. This implies that the length of the cords between junctions is limited, and hence these cords do not have a truly Gaussian conformation. Consequently, the *enthalpic contribution* to the modulus tends to be considerable. Moreover, the gels contain numerous *entanglements*. These structural aspects also cause the "linear region"—i.e., the critical strain above which the (apparent) modulus starts to depend significantly on the strain—to be (much) smaller than for a rubberlike gel. Table 17.3 gives some examples of critical strain values.

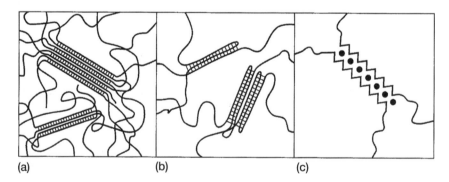

(a)   (b)   (c)

FIGURE 17.12 Various types of junctions in polymer gels. (a) Stacked double helices, e.g., in carrageenans. (b) Partly stacked triple helices in gelatin. (c) "Egg-box" junction, e.g., in alginate; the dots denote calcium ions. Highly schematic; helices are indicated by hatching.

TABLE 17.3 Approximate Values
of the Critical Hencky Strain ε_{cr}[a] for
Some Materials

Material	ε_{cr}
Rubber	1.6
Gelatin gel	0.7
Agarose gel	0.25
Alginate gel	0.20
Xanthan gel	0.05
Acid casein gel	0.03
Plastic fat, e.g.	0.0005

[a]Above which the stress is no longer
proportional to the strain

Gelatin. Gelatin is a breakdown product of collagen, the main constituent of tendon and several other connective tissues. According to source and method of preparation, gelatin properties can vary significantly. In its native state, collagen exists as a triple helix, strongly resembling a triple polyproline helix. Gelatin is readily soluble in water at temperatures above 40°C, forming a viscous solution of random-coiled linear polypeptide chains. On cooling a gelatin solution to, say, 20°C, collagenlike helices are formed, albeit not very long ones and including only part of the material. This results in a gel.

A peptide chain, as in gelatin, does not allow twisting about the peptide bond. Hence helix formation as described above can hardly occur, if at all. The formation of junctions presumably proceeds in the following manner. At a beta turn in the peptide chain at first a "hairpin" is formed. Both ends wind around each other and so a double helix of limited length results, i.e., a kind of stump. This stump can wind itself into another chain, forming a triple helix. If the third chain in the helix belongs to another molecule, a junction has formed. If a triple helix is intramolecular, which appears to occur frequently, it can nevertheless contribute to the gel properties. This is because also in gelatin stacks of helices, i.e., microcrystallites, can be formed, enhancing gel stiffness. Figure 17.12b illustrates the structure. Junction formation is a slow process, and the modulus can keep increasing for several days. All this depends on the cooling regime and on the gelatin concentration, since these variables affect the junction formation process.

A gelatin gel resembles an ideal rubber-type gel to some extent. As given in Table 17.3, the linear region in a rubber may extend to a strain of about 1.6, versus, say, 0.7 for gelatin (i.e., extension by a factor of 2). This is possible because the gelatin chain is highly flexible, the length of a statistical chain element probably being about 2 nm. On the other hand, Eq. (17.15) is not obeyed at all, but this is because the amount of material participating in junctions markedly increases with decreasing temperature.

Figure 17.13a gives an example of the modulus as a function of gelatin concentration. The minimum concentration for gel formation is about 1%. As mentioned, several factors affect gel stiffness. Figure 17.14a gives an example of the temperature dependence, and it is seen that considerable *hysteresis* occurs between cooling and heating curves, although gelation is completely thermoreversible. The extent of hysteresis greatly depends on cooling rate. The fact that a gelatin solution gives a gel at low, e.g., room, temperature, but that the gel melts at body temperature, can of course be utilized, especially in food systems. The slowness of gelation on cooling provides the possibility to make a system quite homogeneous by stirring after its has obtained a high viscosity but not yet a significant modulus; gel formation can then occur under quiescent conditions.

Polysaccharides. There is considerable variation among gelling polysaccharides in molecular structure, in type of junctions, and in the dependence of gel properties on external conditions. Treatment of all of these systems and their particulars is outside the scope of this book, so the discussion will be restricted to some general aspects. An important

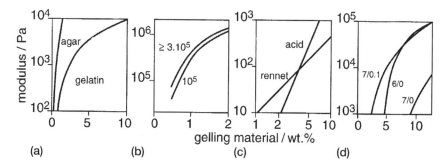

FIGURE 17.13 Effect of concentration of the material making the gel matrix on the shear modulus of various gels. (a) Agar and gelatin gels. (b) Gels of κ-carrageenan of two molar masses (indicated) in 0.1 molar KCl. (c) Casein gels made by slow acidification or by renneting. (d) Heat-set gels of bovine serum albumin; figures near the curves denote pH/added NaCl (molar).

characteristic of most food polysaccharides is the presence of various, in some cases quite bulky, side groups along the polymer chain, unlike the simple polymers primarily discussed in Chapter 6. The distribution of the side groups over the chain is quite variable, and in many polysaccharides *"hairy" regions*—with side groups being near to each other—and *"nonhairy" regions*—devoid of side groups—are distinguished. The distribution of such regions can vary, even among molecules of the same polysaccharide.

One thing that nearly all polysaccharide gels have in common is that they are relatively *stiff* gels, the elastic modulus being for a large part due to change in enthalpy. This is because the length of a statistical chain element b is relatively large: see Table 6.1. For most gelling polysaccharides $b = 15$–60 nm, as compared to 2–3 nm for most polypeptide chains. The "stiffness" of the primary chain is often enhanced by the presence of bulky side groups. This means that the cords between junctions contain but a limited number of statistical chain elements. Consequently, the linear region tends to be markedly smaller than that of a gelatin gel: see Table 17.3.

Polysaccharide gels contain various types of *junctions*; generally, these can only form in the nonhairy stretches of the molecules. The junctions often involve helices: agar (and its main component agarose), the carrageenans, gellan, and possibly xanthan. Some anionic polysaccharides form egg-box junctions (Figure 17.12c) if divalent cations are present: alginate, and pectin under some conditions. Junctions formed by micro-

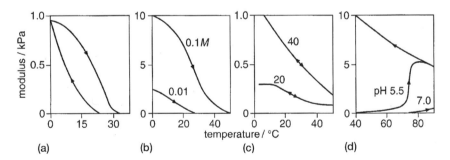

FIGURE 17.14 Effect of the measurement temperature on the shear modulus of various gels. The arrows indicate the temperature sequence. (a) Gelatin (2.5%). (b) κ-Carrageenan (1%) for two concentrations of $CaCl_2$ (indicated). (c) Acid casein gels (2.5%), made and aged at two temperatures (indicated). (d) β-Lactoglobulin (10%) at two pH values (indicated). The results may greatly depend on heating or cooling rate and on other conditions.

crystallites of stretched chain sections are present in gels of some galactomannans and possibly xanthan.

Another categorization distinguishes *"strong"* and *"weak"* gels. A strong gel, e.g., agar or alginate, is characterized by a relatively high modulus and a low value of tan δ (generally < 0.1). The deformation is linear up to a strain of about 0.25 and at a somewhat larger strain the gel will fracture. This means that the gel type may perhaps better be named *"brittle."* A weak gel, e.g., xanthan, has a lower modulus and a higher tan δ value; its deformation is linear up to a far smaller strain, but it will yield rather than fracture when the strain is increased.

Pectin is somewhat special, since it can make two types of gel. A high-methoxyl, high-molar-mass pectin, present in a highly concentrated sugar solution of low pH (to neutralize most carboxyl groups), can form a typical weak gel, where the junctions contain stacked helices. This is what is obtained in traditional jam. A low-methoxyl pectin, i.e., with relatively many carboxyl groups, can form egg-box junctions in the presence of calcium ions. A typical brittle gel results. In both cases, only nonhairy regions on the pectin chain participate in the junctions.

The **magnitude of the modulus** can be affected by several variables:

Gelation time. As mentioned for carrageenans, it may take a while (e.g., an hour) before the gel is fully developed after the gelation temperature has been reached. An increase of the modulus with time is observed for many polysaccharide systems (and also for gelatin), but the time needed varies widely among systems and with process conditions (temperature, concentration, pH, etc.). All of this means that "gelation time" is a poorly defined property. "Weak gels," as formed, e.g., by xanthan, may show yielding upon mild agitation. It will then take some time, e.g., 15 min, before the yield stress is restored.

Concentration of gelling material. Examples are given in Figure 17.13a and b, and it is seen that the relations can vary widely. The minimum concentration at which a gel is formed is in principle given by the chain overlap concentration φ_{ov} [Eq. (6.12)] or c^*: see Sections 6.4.2 and 6.4.3. The value can vary significantly among polymers, as illustrated in Table 6.5. In practice, the lowest concentration giving a gel tends to be somewhat higher than c^*.

Molar mass, or degree of polymerization, of the polysaccharide. This is because for a higher molar mass, the value of c^* will be smaller. Consider a series of samples of the same material but varying in molar mass. If the concentration range is well above c^* for all samples, the effect of molar mass is in principle zero; if the concentration is in the range of the c^* values, the modulus will be higher for a larger molar mass. An example is in Figure 17.13b.

Temperature. For all polysaccharides where gelation depends on helix formation, there is a critical temperature T_m at which the gel will melt. The modulus of such gels increases in magnitude with decreasing temperatures (below T_m). An example is given in Figure 17.14b. The value of T_m may greatly depend on other variables, especially those mentioned below. Gels of polysaccharides forming egg-box junctions generally do not melt below 100°C. A few chemically modified polysaccharides can form reversible gels *above* a certain temperature. This mainly concerns some cellulose ethers, especially methylcellulose, which contains —OCH_3 groups. It forms a gel at temperatures above 50–90°C, depending on concentration, degree of methylation, and further structural details. Presumably, the gel formation is mainly due to hydrophobic bonding.

Ionic composition. For anionic polysaccharides, ion composition and strength, including pH, can have a large effect on the modulus. Egg-box junctions need divalent cations, Ca^{2+} generally being the most effective. The gelation of the various carrageenans is also enhanced by cations (see Figure 17.14b). Presumably, the ions screen the negative charge on the polymer, and the effect greatly depends on the ion involved. For κ-carrageenan, for example, K^+ ions are more effective in inducing gelation than either Na^+ or Ca^{2+} ions at the same molar concentration.

Solvent quality. A good example is provided by the gelation of pectin in jam. The concentrated sugar solution is a poor solvent for pectin, markedly increasing its activity coefficient and thereby lowering its solubility.

Mixed polysaccharide gels are briefly discussed in Section 17.2.5.

Fracture. Actually, we have not much to add to what has been discussed in Section 17.1.2, also because little systematic study has been made of the fracture of various food polymer gels. These gels show either yielding ("weak gels") or time-dependent fracture ("brittle gels") at large deformation. Breakage of covalent bonds almost never occurs, but *"unzipping" of junctions* does. Brittle gels are generally notch-sensitive. If the gel is clear (gelatin, agar, some alginates), the presence of large defects is unlikely. The inherent defect length then will be about equal to the distance between junctions and hence quite small. A relation like that given by curve 2 in Figure 17.9 tends to be obeyed. In actual food systems, however, pure polymer gels almost never occur, and far larger defects are generally present.

Swelling and Syneresis. Again, little systematic study has been made of these phenomena. According to Flory, the equilibrium volume of a covalently cross-linked polymer gel would be proportional to $(m - 1)^{-0.6}$, where m is the average number of cross-links in which each polymer molecule is involved. Increasing the number of cross-links after a gel has been made would thus lead to shrinking. However, when a solution of gelatin or some polysaccharide is made to gel, the whole volume turns into a gel, and this generally does not change upon aging, even though the modulus of the gel may markedly increase, and probably the number of junctions as well. This means that soon after a gel is formed, its shape is more or less fixed.

On the other hand, by altering the solvent quality, i.e., the parameter χ or β discussed in Section 6.2.1, swelling—i.e., taking up of solvent by the gel—or shrinkage—i.e., syneresis or the expulsion of liquid from the gel—can occur. For polyelectrolytes, the same phenomena can happen when altering the pH or the ionic strength. However, these processes tend to be very slow, because of the very small permeability of the gels: generally $B = 10^{-17}-10^{-16}\,\mathrm{m}^2$. This causes a very strong resistence to the flow of liquid through the gel; see Section 5.3.1.

Question

Classical jam manufacture is achieved by mixing fruit pulp with a large proportion of sucrose and then evaporating water by boiling until the mass ratio of sucrose to water equals about 2:1. As mentioned, the pectin of the fruit can form a typical "weak" gel at these conditions, leading to a jam of desirable consistency, i.e., with a significant, but not overly large, yield stress. Currently, jam is often made with far less sugar, to reduce "calories." The gel obtained then is too weak, for the most part because the solvent quality for pectin is too good. What measures can you think of to improve the consistency of such a jam?

Answer

Several possibilities can be considered.

1. Add more pectin (or use fruit varieties with a higher pectin content) and lower the pH to neutralize carboxyl groups on the pectin. This helps to some extent, but many consumers consider the jam too acid.

2. Lower the solvent quality by addition of poorly digestible saccharides, like mannitol. This gives a good consistency, but the mannitol reaching the large intestine attracts large amounts of water to decrease osmotic pressure, often leading to diarrhea.

3. Add another substance that gives a weak gel, such as xanthan. This seems to be too expensive.

4. Add a suitable low-methoxyl pectin, ensure that the pH is not too low, and add some $CaCl_2$. This gives a sufficiently stiff, but somewhat brittle, consistency.

17.2.3 Fractal Particle Gels

As discussed in Chapter 13, aggregating small particles form fractal aggregates or clusters. These become ever more tenuous, and at some stage the clusters fill the whole system, forming a gel, provided that no disturbance (agitation, sedimentation) occurs. There is no lower limit to the concentration, but gels have not been observed for a particle volume fraction $\varphi < 0.001$; in practice $\varphi \approx = 0.01$ is a minimum, since the gel would otherwise be very weak. The gel network consists of strands of particles that are linked to other strands, leaving pores of various sizes; often, the strands have an average thickness corresponding to a few particle diameters. Here we will further discuss the properties of particle gels, proceeding on the treatment of fractal aggregation in Section 13.2.3.

The gels obtained are often called "fractal," but this is to some extent misleading. Figure 17.15 gives an example of the number of particles and the *fractal dimensionality* D as a function of length scale, i.e., the distance from a central particle. Inside the building block of the gel, i.e., at a scale smaller than a_{eff}, $D = 3$ or nearly so. Also at macroscopic scales, which means clearly larger than R_g, D must equal 3: if we take two cubes of gel with edges of 1 and 2 cm, the latter cube will, of course, contain $2^3 = 8$ times as much gelling material (particles) as the former. Between a_{eff} and $R_g = a_{eff}\,\varphi^{1/(D-3)}$, the dimensionality is < 3 and hence fractal. This means that the fractal range will decrease with increasing value of φ. For instance, for $D = 2.34$, $\varphi = 0.03$ gives $R_g = 200\,a_{eff}$, and for $\varphi = 0.3$ this is reduced to $6\,a_{eff}$. Because of the transition zones between the regimes for D, this means that above $\varphi \approx 0.3$, the fractal nature of the gel is lost.

Fractal aggregation can occur in many food dispersions; some examples are discussed in Sections 17.2.4 and 17.3.1. However, the formation of truly fractal gels is not common. Most casein gels are fractal, and they have been well studied. Consequently, we will concentrate on these systems in this section. We will first briefly describe the particles.

Milk contains proteinaceous particles, called **casein micelles**. These are roughly spherical, volume-surface average diameter about 90 nm; they contain besides the various caseins (α_{S1}, α_{S2}, β, and κ) some undissolved calcium phosphate (about 8% of the dry matter), and are swollen (voluminosity about 2.2 ml per g dry casein). They are protected against aggregation by a layer of protruding parts of the κ-casein chains. Upon addition of rennet, which contains the proteolytic enzyme chymosin, these chains are slowly cut off, causing the micelles to aggregate, and a "rennet gel" is formed. The physiological pH of milk is 6.7. Acidification can also cause aggregation: at low pH the calcium phosphate goes into solution, and

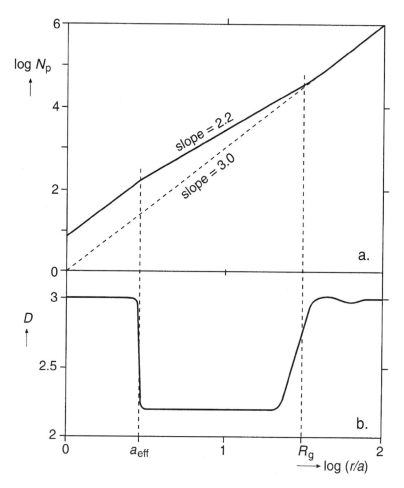

FIGURE 17.15 Effects of length scale (distance r from a "central" particle of radius a) on the fractal properties of a particle gel. (a) Log of the particle number N_p in a sphere of radius r/a (log scale). (b) Fractal dimensionality D as a function of $\log(r/a)$. Hypothetical and somewhat simplified example.

many of the properties of natural micelles are significantly changed, but the particles roughly keep their size and shape. Near the isoelectric pH (4.6), the protruding chains collapse, and aggregation occurs. If the acidification proceeds slowly, an "acid gel" is formed. A gel can also be obtained by dissolving sodium caseinate in a buffer (neutral pH, ionic strength about 0.07 molar), followed by slow acidification. Neither acid nor rennet gels are formed at temperatures below 10°C.

The **permeability** B, as defined in the Darcy equation (5.24), is a useful parameter to characterize the fractal nature of particle gels (apart from being important in practice). It can be readily measured under conditions where the flow through the gel does not alter gel structure. The largest holes in the gel have a radius close to R_g. According to Eq. (5.23), the flux through a pore of that radius would be proportional to R_g^4. The number of these large pores per unit cross section of the gel will be proportional to R_g^{-2}, leading to B being proportional to R_g^{-2}. Furthermore, R_g is proportional to a_{eff}, and by use of Eq. (13.15) we obtain

$$B = \frac{a_{eff}^2}{K} \varphi^{2/(D-3)} \tag{17.16}$$

The value of the proportionality parameter K will depend on structural details. It must be larger than unity, since we have taken the largest pores as a reference; in practice, $K = 50-100$ is observed.

Linearity between $\log B$ and $\log \varphi$ is well obeyed for casein gels. From the slope of the line, D can be calculated, and for acid gels, values between 2.35 and 2.4 are observed, in good agreement with simulation results. For rennet gels, the slope yields $D = 2.2-2.25$.

Short-term rearrangements, i.e., changes in particle arrangement occurring before gel formation, cause an increase in a_{eff} according to Eq. (13.19) [$N_p/n_p = (R/a_{eff})^D$, where n_p is the number of primary particles in a building block] and hence in B. For acid casein gels (about 3% casein), the extent of rearrangement greatly depends on the temperature of formation. For instance, at 20°C, $B = 0.13\,\mu m^2$, and $n_p = 1$ was observed, i.e., no rearrangement; at 30 and 40°C, n_p was 25 and 45, respectively, hence considerable rearrangement, with a corresponding increase in B (about 1 and 3.5 μm^2, respectively). For rennet gels made at 30°C, B was smaller (0.25 μm^2) with $n_p \approx 8$. However, if renneting takes place at a lower pH, the permeability tends to be higher: see Figure 17.16a.

Long-Term Rearrangement. This concerns changes after the gel has formed. Figure 17.16a shows that such changes occur in rennet gels, since the B value increases with time after gel formation. This can go along with the occurrence of **syneresis**, i.e., the expulsion of liquid from the gel, which is undesirable in most gels. On the other hand, it is an essential process in cheese making, where most of the liquid (whey) in the rennet gel has to be removed. The process can be explained as follows.

Since the particles forming a fractal gel are "reactive" for aggregation over their total surface, the free energy of the system would decrease when the coordination number of the particles increases, or in other words when more junctions between particles would be formed. However, since the particles are more or less immobilized in the network, this is hardly possible, especially if the gel is constrained in a vessel and sticks to the vessel wall. However, if φ is not very high, some strands in the gel have a certain freedom to move (Brownian motion), so that a new junction between particles may occasionally form. This causes tension in the strand(s)

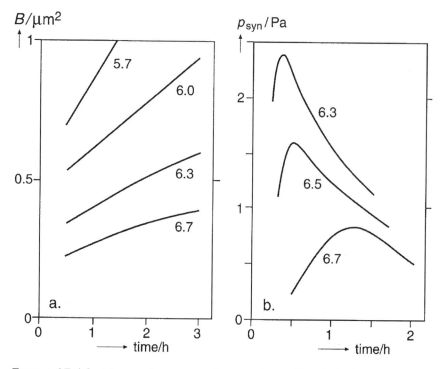

FIGURE 17.16 Changes in rennet gels at various pH values (indicated) as a function of time after rennet addition, at 30°C. (a) Permeability B. (b) Endogenous syneresis pressure p_{syn}. (After results by H. J. C. M. van den Bijgaart. Ph.D. thesis, Wageningen University, 1988.)

involved. If the junction strength is not too high, it may lead to breaking of a strand; so loose ends are formed, which can form new junctions. In this way the gel tends to become more compact and "tries" to expel liquid. These events are illustrated in Figure 17.17.

Consequently, a particle network in which junctions can occasionally be broken will build up a pressure on the liquid, called *endogenous syneresis pressure*, p_{syn}. If the liquid can indeed flow out, the Darcy equation can be used to obtain the rate of the process. We write it in the form

$$Q = A \frac{B}{\eta} \frac{p_{syn}}{L} \tag{17.17}$$

where Q is the volume flow rate, A the surface area of the (piece of) gel, and L the distance over which the liquid has to flow through the gel. Application of the equation is not simple, because A, B, p_{syn}, and L change with ongoing syneresis. Moreover, even if no liquid can flow out, as in the middle of a piece of gel or in a constrained gel, both B and p_{syn} change with time: see Figure 17.16. This always leads to the formation of larger holes (pores) in the gel, a process called *microsyneresis*. Because syneresis can only occur if a gel has formed, p_{syn} increases with the buildup of the gel. The decrease afterwards is explained below. Syneresis rate greatly increases with increasing temperature and hardly occurs below 20°C.

In an acid casein gel (such as set yoghurt), syneresis hardly occurs, and the value of B remains virtually constant during storage. Presumably, breaking of junctions is hardly possible in these gels. Only for gels made at high temperature, say 40°C, some syneresis is often observed.

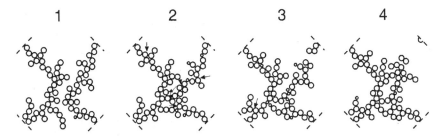

FIGURE 17.17 Schematic representation of strands of particles in a gel forming new junctions, leading to shrinkage and to breaking of a strand.

Rheological Properties. We have seen that the permeability of a particle gel is related to the fractal structure in a simple way. This is not the case for the rheological properties, because these are affected by several variables, of which the quantitative effect often cannot be readily established.

There is, however, one basic principle, which derives from the fact that fractal clusters are *scale invariant*. When a gel is formed, clusters of size R_g make bonds with each other via strands at the periphery of the cluster, and the average number of the strands involved per cluster will *not* depend on cluster size. Since the apparent surface area of a cluster scales with R_g^2, the number of junctions between clusters per unit area of cross section of the gel will scale with R_g^{-2}. By using the equation for R_g we arrive at the following equation for the shear *modulus* of a fractal particle gel:

$$G = K'\left(\frac{a_{\text{eff}}}{a}\right)^{-\alpha}\varphi^{\alpha/(3-D)} \tag{17.18}$$

Here K' is an unknown constant, which contains among other things the rheological properties of the primary particles. Other variables being equal (which they usually are not), the modulus will be larger if a_{eff} and α are smaller and D is higher.

The magnitude of α depends on the structure of the gel-forming clusters, as is illustrated in Figure 17.18. The variables are the length of the stress-carrying strands L and their elastic constant C (in $N \cdot m^{-1}$). According to the values of L and C, four regimes can be distinguished.

1. *Fractal strands.* This is the original, and in principle the most accurate, model, which proceeds on the assumption that the stress-carrying strands are fractal. The value of x can vary between almost 0 and 1. The model implies that the clusters are highly rarefied and that the strands are relatively long and slender. This means quite a low φ value, and also D should be relatively small. These conditions are rarely met, and if they are, the very weak gel obtained is prone to restructuring, invalidating the model.

2. *Hinged strands.* This model is more realistic. Now the bending modulus of the strands, which is relatively small, is a factor in the parameter K'. The value of α would equal 3, and this is often observed. Casein gels formed by slow acidification fit this regime.

3. *Stretched strands.* When microsyneresis occurs in the gel, this leads to local shrinkage and hence to stretching of the originally hinged strands. This occurs in rennet casein gels. Stretched strands can be observed in an acid gel if it has been made by acidification in the cold (2°C), followed by slowly heating up: a gel is then formed at, say, 12°C, and upon further heating the primary particles shrink to a significant degree, causing the

strands to stretch. We now have $\alpha = 2$, and the parameter K' includes the Young modulus of the primary particles. Its value tends to be much higher than that of the bending modulus.

4. *Weak link*. Here the clusters are considered to be rigid. Upon deformation of the gel, the clusters remain undeformed, and only the relatively weak links between the clusters are strained. This generally means that both φ and D will be high, and that the primary particles are quite rigid. Such systems have rarely been observed in foods.

It may be concluded that the value of α will generally be 2 or 3, although intermediate values, or values slightly larger than 3, may also occur. The dependence of the modulus on φ is strong; for $D = 2.35$, and α values of 2 and 3, the exponent of φ equals 3.1 and 4.6, respectively. Such relations are shown in Figure 17.13c (page 716). Because of the uncertainties in the relation between modulus and volume fraction, it is better to derive D values from permeability results.

Regime	Structure	$L \propto$	$C \propto$	$\alpha =$
Fractal strands	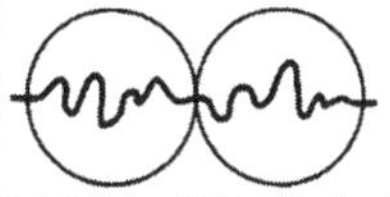	R_g^x	$R_g^{-(2+x)}$	$4 + x$
Hinged strands	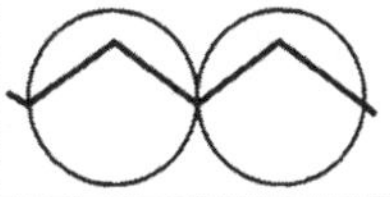	R_g^0	R_g^{-1}	3
Stretched strands	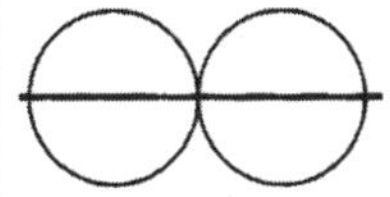	R_g^0	R_g^0	2
Weak link	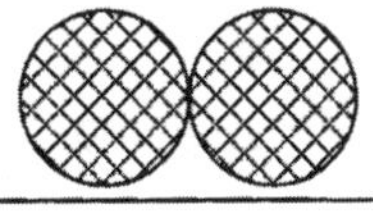	-	R_g	1

FIGURE 17.18 Various regimes, based on structural models, for the scaling relations between rheological parameters (especially G') and the volume fraction of particles φ making up the gel network. The circles schematically denote the clusters forming the gel (of radius R_g); inside only the stress-carrying strands (of length L) are shown. It is assumed that the stress on the system is in the horizontal direction. C is the elastic constant of the strands; the parameter α is given in Eq. (17.18). Highly schematic.

It is also seen that the **properties of the primary particles** play a large role. This is borne out by results of dynamic measurements obtained for casein gels. The value of tan δ ranges between 0.1 and 0.6, depending on several conditions, and the value of G' increases with increasing frequency ω. For a fractal network of rigid particles tan δ would almost equal zero and G' would be independent of ω, since at the very small deformations applied, bonds between particles would not break (and if they break they would not reform). Hence these observations on casein gels must be explained by deformation of the casein particles.

This also plays a part in large deformation and fracture. Before fracture occurs, the stress-carrying strands have become stretched. That means that Eq. (17.18) can also be used for the fracture stress with $\alpha = 2$, albeit with a very different value for K'. The strain at fracture ε_{fr} will strongly depend on the deformability of the primary particles; it will also be larger for a smaller value of φ.

The most important effect of the particle material on the mechanical properties of the gel may be that it determines the *strength of the junctions*. The junctions inside a cluster and between clusters are in principle identical. Each junction can involve a great number of bonds, and the number can vary with time and with physicochemical conditions. It is observed that casein gels increase in modulus during aging after the network has fully developed. The rate of increase depends on several conditions, but it is always faster for rennet gels (e.g., by a factor 5 in 20 min) than for acid gels (e.g., by a factor 5 in 20 h). The main cause of the increase appears to be that the junction zone between two particles increases in area with aging. This must also be the explanation for the value of the syneresis pressure p_{syn} decreasing after the gel has fully formed, as illustrated in Figure 17.16b. After all, junctions must break for substantial syneresis to occur.

In Table 17.4 some properties of three types of casein gels are compared. Two have stretched strands ($\alpha = 2$) and one hinged strands ($\alpha = 3$), but otherwise the spatial structure is much the same (values of B and D). The two acid gels are built of virtually the same primary particles. It is seen that the stretching of the strands (type 2) leads to a much higher modulus, but not to a higher fracture stress; the fracture strain is much higher for the hinged strands. All these results are in qualitative agreement with the structure models (can you explain them?). The rennet gel has a relatively low modulus, despite having stretched strands. This is probably due to the primary particles being much more deformable (having a smaller Young modulus) than those in the acid gels. In agreement with this, the fracture strain is very high. The fracture stress is much smaller than for the acid gels. This would agree with the rennet gels being far more prone to syneresis, since syneresis involves the breaking of strands.

TABLE 17.4 Some Properties of Various Casein Gels Made of Skim Milk

Property	Acid gel, type 1[a]	Acid gel, type 2[b]	Rennet gel[c]
Strands are	Hinged	Stretched	Stretched
Permeability $B/\mu m^2$	0.15	0.15	0.25
Fractal dimensionality D	2.35	2.4	2.25
Syneresis	no	no	yes
Modulus G'/Pa	20	180	30
Fracture stress σ_{fr}/Pa	100	100	10
Fracture strain ε_{fr}	1.1	0.5	2–3[d]

[a]Made by slow acidification at 30°C.
[b]Made by acidification at 2°C and slowly heating to 30°C.
[c]Made by renneting at 30°C.
[d]Depends substantially on strain rate.
Approximate results for well-developed gels tested at 30°C.

It may be added that not all variables have been discussed, such as pH, ionic strength, and solvent quality, or the time scale of deformation. See Figure 17.14c for the effects of formation temperature and measurement temperature, which are opposite in a sense.

A general **conclusion** may be that fractal particle gels follow some simple scaling laws that provide much insight. However, it needs much more than a knowledge of the values of φ, D, and a_{eff} to predict quantitatively the mechanical properties. For every gel type, careful study at a range of conditions is needed to obtain a full picture.

Question

Consider the making of curd (green cheese) from milk. The milk is made to gel by the addition of rennet. Then whey has to be removed by syneresis. What measures can you think of to enhance the rate of syneresis?

Answer

As is mentioned above, the syneresis rate increases with increasing temperature and decreasing pH. A look at Eq. (17.17) shows that the total outflow of whey is proportional to the surface area of the gel and inversely proportional to the distance over which the whey has to flow. Cutting the gel into small pieces thus very much enhances the syneresis rate (in fact, syneresis is almost imperceptible before the gel is cut). The equation contains the endogenous syneresis pressure p_{syn}. However, for a system that can show structural rearrangement leading to syneresis, any externally applied pressure will, of course, speed up the process. Since p_{syn} is quite small—of the order of 1 Pa—external pressures can have a very strong effect. Finally, a look at Figure 17.16b will show that one should not wait too long before starting syneresis (i.e., before cutting the gel into pieces), since then p_{syn} rapidly decreases.

17.2.4 Heat-Set Protein Gels

When a protein solution is heated, a gel may be formed. This occurs with well-soluble *globular proteins*, if the protein concentration is above a critical value c_0. The gel is only formed after at least part of the protein has been *heat denatured*—e.g., as inferred from a change in the DSC curve—and the gel formation is irreversible upon cooling; see Figure 17.14d. Gel formation is a relatively slow process, taking at least several minutes and possibly hours.

These observations are almost generally valid, but further details about formation and properties of the heat-set gels are widely variable among proteins. This is because globular proteins vary greatly in molecular structure and conformational stability, as was discussed in Chapter 7; especially Section 7.2 gives useful information in relation to denaturation. The situation is even more complex because heat-set gels are generally made of protein mixtures, such as isolates of whey proteins or soya proteins, where the various species react in a different way during heat treatment and often mutually react as well. The bonds keeping the denatured protein molecules together in the gel network are of various natures. Presumably, —S—S— bridges and hydrophobic interactions often play major roles. Moreover, H-bonds can be involved in intermolecular junctions between β-strands.

Gel Structure. The gels come in two main types, although the difference is rather gradual.

1. *Fine-stranded.* These are clear gels, consisting of relatively fine strands, that are branched to form a network. Strand thickness generally is between 20 and 50 nm, i.e., clearly larger than molecular size (4–8 nm). These gels typically form at a pH far away from the isoelectric point (pI) at low ionic strength.

2. *Particle.* These are turbid gels, with a coarse microstructure. The building blocks are roughly spherical particles, diameter generally between 0.1 and 4 μm; the larger ones would contain well over a million protein molecules. These gels typically form at a pH close to the isoelectric point at high ionic strength.

Many proteins form a coarse precipitate rather than a gel if the pH is very close to the isoelectric point, combined with a high ionic strength. Under these conditions the net charge on the protein is quite small and the thickness of the electric double layer is also small, allowing rapid aggregation. Presumably, some electrostatic repulsion between the (denatured) protein molecules is needed for a gel, rather than a precipitate, to be formed. This may in part explain the slowness of gel formation.

The following factors affect the coarseness of the gel:

Type of protein. For example, fine-stranded gels are far more readily formed from myosin (an elongated molecule) or α-lactalbumin (a small molecule that does not form —S—S— bridges) than from β-lactoglobulin or soya protein.

The effects of *pH and ionic strength* have already been discussed.

Divalent cations, especially Ca^{2+}, are very effective in causing coarse gels to be formed (at a pH above pI), because they strongly promote protein aggregation. This may be due to the strong binding of Ca ions to several proteins, thereby lowering electric charge, or to the formation of intermolecular Ca bridges.

For a higher *protein concentration*, the particles making up the gel often are larger.

The *rate of heating* of the protein solution can have a large effect. For example, heating a 10% solution of β-lactoglobulin at pH 5.3 at a rate of $5 \, K \cdot min^{-1}$ or faster gave particles of about $3 \, \mu m$, whereas a rate of $0.1 \, K \cdot min^{-1}$ resulted in particles of about $15 \, \mu m$ diameter.

The *mechanism* by which the relatively large particles are formed is still a matter of debate. It has been proposed that they are the result of phase separation, presumably between unfolded and native molecules, which leads to small volume elements of a high concentration of denatured molecules, which subsequently form roughly spherical gel particles. Anyway, it has become clear that the particles themselves are not quite rigid and do not (always) consist of closely packed protein molecules. However, other mechanisms cannot be ruled out and the question is far from settled. Nor is it clear why the strands in clear gels can be relatively long and of almost constant diameter.

Gel formation involves a number of consecutive reactions: (1) protein molecules become denatured; (2) denatured molecules aggregate to form (roughly spherical, or elongated) particles; and (3) these particles then aggregate further to form a space-filling network. After a little while, all of these reactions proceed at the same time, unless the temperature is much higher than the denaturation temperature of the protein, when denaturation can be a very fast reaction.

An example is given in Figure 17.19. It is seen that a gel was formed after about 30% of the protein had been denatured, i.e., at a denatured protein concentration of about 1.4%. Then a permeability and a shear modulus could be measured. However, the permeability soon reached a kind of plateau value, although it kept slowly decreasing, until almost all protein had been denatured. The modulus, however, strongly increased with time. This can most likely be interpreted as the formation of a space-filling network that is initially more or less fractal, after which denatured protein molecules or protein particles are deposited onto the existing network. The network then will keep about the same overall geometry, although the pores become somewhat smaller, whereas the junctions between the particles will be greatly strengthened, and the network strands will become much stiffer.

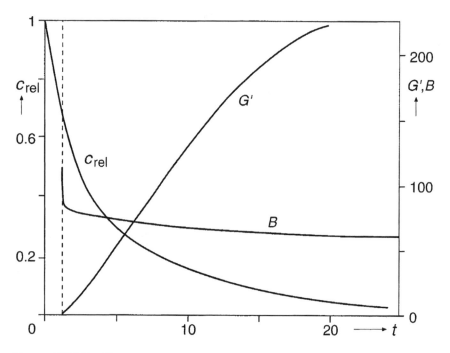

FIGURE 17.19 Changes occurring when keeping a 4.5% solution of a whey protein isolate (in 0.4 M NaCl) at 68.5°C for various times t (h). Shown are fraction of the protein undenatured (c_{rel}), gel permeability ($B/10^{-15}\,m^2$), and dynamic shear modulus (G'/Pa). (From results by M. Verheul, S. P. F. M. Roefs. Food Hydrocolloids 12 (1998) 17.)

Presumably, about the same happens during the formation of most heat-set protein gels, although often at a faster rate. (In the example given, the temperature was such that denaturation was quite slow, allowing us to distinguish the various stages.) However, the rates of each of the processes will greatly vary, according to the protein(s) involved and depending on such conditions as protein concentration, pH, ionic strength, and especially heating temperature or rate of temperature increase.

Fractality. The *permeability* of heat-set protein gels ranges for the most part between 10^{-15} and 10^{-12} m²; the main variables are protein concentration and size of the particles. This allows determination of the relation between $\log B$ and $\log c$, while keeping the physicochemical conditions constant. If, moreover, the concentration is taken as that on the onset of gelation, as in Figure 17.19, and a linear relation results, the initial fractal dimensionality can be established. For systems as given in the figure, a value of $D \approx 2.4$ resulted, well in agreement with the values obtained for the truly fractal acid casein gels (Section 17.2.3). However, the ongoing aggregation will upset the fractal properties of the final gel.

For soya protein gels made at a low pH (3.8), relations between $\log B$ and $\log c$, as well as between $\log G'$ and $\log c$ (for $\alpha = 2$, i.e., stretched strands), D values of 2.3 resulted. In many cases, however, the relations do not agree well with those for fractal particle gels. In conclusion, fractal aggregation is generally essential in obtaining a heat-set protein gel, but the final structure is mostly not fractal any more.

Rheological Properties. Some relations between the *dynamic shear modulus G'* (at a frequency of about 1 Hz) and protein concentration c are given in Figure 17.13d. At higher c values, G' is often proportional to c^2; assuming the gel to be fractal and that $\alpha = 2$ [see Eq. (17.18)], this would imply $D = 2.0$. This is not realistic, since D should be higher, and α probably as well; moreover, these are incompatible. The value of G' also depends on the time elapsed after incipient gelation (Figure 17.19), pH, ionic strength, etc. G' values are rarely over 100 kPa.

Generally, curves of G' (or its logarithm) versus c are vertical at some low c value. In other words, there is a *critical concentration c_0*, below which no modulus can be measured. For example, β-lactoglobulin solutions gave c_0 values ranging from 0.4% (high salt, pH close to pI) to 8% (high salt, pH far from pI). However, for other heating rates or waiting times, other c_0 values can be observed. The value that the modulus assumes depends on many conditions, and for each protein system studied, relations between parameters like B, G', c_0, and process variables have to be established and if possible explained.

As illustrated in Figure 17.14d the value of the modulus increases with decreasing *temperature*, once a gel has formed. This is as observed for casein gels, but the explanation for heat-set gels is uncertain.

Large deformation properties also vary widely. The strain at fracture ε_{fr} is mostly between 0.3 and 1.0; the fracture stress can be up to 25 kPa. β-Lactoglobulin gels formed at a pH close to the pI are brittle, whereas rubberlike gels are formed at high pH; this qualitatively agrees with the difference in gel structure discussed. Again, relations have to be established separately for each protein system.

Question

Consider the situation given in Figure 17.19 at the moment that a gel is formed. (a) What would be the size of the particles making the fractal gel, i.e., what is the value of a_{eff}? (b) Assuming that unhindered perikinetic aggregation occurs (i.e., $W = 1$), what would be the aggregation time needed to form a gel? (c) Is the latter value a reasonable one?

Answer

(a) According to Eq. (17.16), $B = (a_{eff}^2/K)\varphi^{2/(D-3)}$. We assume $K \approx 80$ and $D = 2.4$, as were given above. The figure gives $B \approx 10^{-13}\,\text{m}^2$. The concentration of aggregating material was about 1.4%; taking into account that the molecules are denatured and hence have an increased voluminosity, and remembering that the molecules in the particles are probably not closely packed, a mass fraction of 0.014 may correspond to a volume fraction of about 0.05. This then would yield $a_{eff} \approx 20$ nm. (Electron micrographs show that the particles have a diameter of at least 50 nm, i.e., about the same.) (b) Equation (13.20) gives $t_{cr} = t_{0.5}q^D$. Equation (13.3) gives an expression for $t_{0.5}$, and assuming that the viscosity was 1 Pa·s, we arrive at $t_{0.5} = 0.13$ ms. $q = R_{gel}/a_{eff} = \varphi^{1/(D-3)}$, according to Eq. (13.15). Hence $q \approx 150$ and we arrive at $t_{cr} \approx 0.02$ s. (c) The value for t_{cr} is not reasonable, simply because we assumed $W = 1$, whereas it was concluded above that gel formation will not occur unless the particles show some mutual repulsion, hence $W \gg 1$. Even for $W = 10^4$, for instance, t_{cr} would be 200 s, i.e., small compared to the time needed to form a sufficient number of particles.

17.2.5 Mixed Systems

In practice, various mixtures of gel forming materials are applied, giving rise to gels of widely varying properties. We will briefly mention some types; the discussion is for the most part qualitative.

Polymer–Polymer. The following alternative situations may occur.

1. The two polymers form a *homogeneous solution.* If now the conditions are changed so that polymer A would form a gel, e.g., by lowering temperature, it will also do so in the mixture, and the rheological properties will not be greatly different from those of a gel of polymer A at the same concentration. The permeability will be decreased, owing to polymer B forming a viscous solution in the pores of the gel network. Such systems are quite rare, however.

2. Phase separation occurs because of *incompatibility*; see Section 6.5.2. This is a very common situation, unless the polymer concentrations are quite low. Consider, for instance, a mixture of gelatin and a nongelling polysaccharide, e.g., dextran. The mixture is made at a temperature above the gel point of the gelatin, and phase separation sets in. Eventually two layers will be formed, but that takes a long time, and meanwhile one of the polymers is (predominately) in a continuous phase, the other in drops. If gelatin is in the continuous phase (which depends on the relative concentrations of the two polymers), and the mixture is cooled, gelation occurs, and the nonequilibrium situation is "frozen." The modulus of the gel will be markedly higher than for a gelatin solution of the same overall concentration, because its concentration c in the continuous phase is significantly increased, and the modulus is, e.g., proportional to c^2. The fracture stress would be higher owing to the higher concentration, but lower because the presence of the drops decreases the actual fracture surface; consequently, a general rule cannot be given. If gelatin forms the disperse phase, cooling leads to a dispersion of gel particles in a dextran solution.

3. *Weak attraction* between the polymers may cause gel formation, even if both are nongelling. Such mixtures are often used to form weak gels, e.g., to prevent sedimentation of particles. An example is given by xanthan and some galactomannans, such as locust bean gum. The mixture is first heated and then cooled. Mixed junctions are formed, presumably involving a xanthan helix and a stretched part devoid of bulky side groups (i.e., a straight zigzag) of the locust bean gum chain. In this way 0.1% of each polymer may jointly form a gel with a yield stress of order 10 Pa. Another example is given by the addition of, say, 0.03% κ-carrageenan to milk. It then interacts with the κ-casein in the casein micelles, forming a weak gel; the interaction is due to electrostatic attraction.

4. Stronger mutual attraction of the polymers or higher concentrations may lead to phase separation by formation of a *complex coacervate.* This often results in lumps of coacervate in a dilute solution, a system that is generally useless.

Polymers with Heat-Setting Proteins. Consider a solution of a globular protein at a pH above its isoelectric point that will give a gel upon heating. Now make solutions also containing a polysaccharide, where the total concentration of protein + polysaccharide is the same. If the polysaccharide is anionic (e.g., carrageenan), a homogeneous solution is often formed; upon heating, a gel results whose modulus (G') is higher than that of the protein alone. If the polysaccharide is neutral (e.g., dextran), phase separation due to incompatibility occurs. After heating the modulus is lower than that in the absence of polysaccharide, which is presumably due to the inhomogeneity of the system. These are rules of thumb and exceptions may be possible.

Polysaccharides with Surfactant Micelles. Consider a solution of a fairly hydrophobic polysaccharide, such as a cellulose ether. The hydrophobic groups cause a weak attractive interaction, leading to a somewhat increased viscosity at low shear rates. If an anionic small-molecule surfactant is added, say SDS (sodium dodecyl sulfate), at a concentration above the CMC (critical micellization concentration), micelles are formed that interact with the polymer; more specifically, one or a few polymer chains can pass through a micelle. In this way, polymer chains can be cross-linked. If now the polymer concentration c is below c^* (the chain overlap concentration), mainly intramolecular junctions are formed. If $c > c^*$, however, a gel results. In this manner, viscoelastic gels can be made with a modulus of the order of 10 Pa.

Filler Particles. We will consider gels containing particles that are much larger than the pores in the gel network. The primary gel (i.e., without particles) may be a polymer gel, a fractal particle gel as obtained with casein, or a heat-set protein gel. We will consider the effect of *"filler particles"* on gel properties, especially the elastic modulus, which has been studied best. The following factors are known to affect the properties.

    1. Particle *concentration*, generally expressed in the particle volume fraction φ. The modulus will either increase or decrease with increasing φ; see below.

    2. Particle *bonding*. The particles can be bonded or not to the gel matrix, as schematically depicted in Figure 17.20, frames 2 and 1. As shown in Figure 17.20a, bonded particles increase the value of the modulus, which is logical since they effectively increase the amount of gel material; nonbonded particles tend to decrease the modulus, since they effectively behave like voids at very small deformation.

Bonding can occur if the gel material will adsorb onto the particle surface. Filler particles can be emulsion droplets, and a gel is often made by

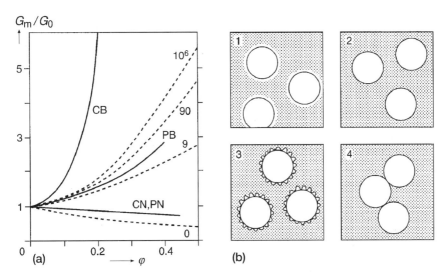

FIGURE 17.20 The effect of filler particles on gel properties. (a) Relative modulus (G_m/G_0) as a function of particle volume fraction (φ). The broken lines are calculated for various values of the ratio G_p/G_0, indicated near the curves. The drawn lines are average experimental values for acid casein gels (C) and polymer gels (polyvinyl alcohol, P), with emulsion droplets that are either bonded (B) or nonbonded (N) to the gel matrix. (b) Highly schematic pictures of the gel structure. Shaded area denotes primary gel. Particles are nonbonded (1); bonded (2); bonded but with intermediate layer (3); bonded and aggregated (4). (Adapted from T. van Vliet. Colloid Polymer Sci. 266 (1988) 518.)

emulsifying oil in a surface active polymer solution, say gelatin, after which the system is allowed to set, e.g., by cooling. Bonding then is ensured, since the particles will via their gelatin-coated surfaces directly participate in the gelation. If the gel material does not adsorb, but the gel is quite stiff, a little surface roughness of the particles causes effective bonding by friction, at least at the very small strains as applied during measurement of a modulus. If the gel is made of a dilute nonadsorbing polymer, bonding does not occur; there will even be a thin layer depleted of polymer around the particles (frame 1).

3. The *particle modulus.* For bonded particles, the van der Poel theory can be used to predict the relative modulus of the mixed system, i.e., G_m/G_0, as a function of φ and the relative particle modulus G_p/G_0; here G_0 is the modulus of the primary gel. In Figure 17.20a calculated curves are given for four values of G_p/G_0. For $G_p = 0$, the result applies also to nonbonded particles, whatever their modulus. If the particle is an emulsion droplet or

gas bubble, its shear modulus is simply given by the Laplace pressure

$$G_p = p_L = \frac{4\gamma}{d} \tag{17.19}$$

For a protein-covered oil droplet of diameter $d = 1\,\mu m$ and with interfacial tension $\gamma = 10\,mN \cdot m^{-1}$, the modulus then is $40\,kPa$, i.e., quite high. Effects of droplet size on G_p and thereby on G_m will thus be considerable and have been experimentally confirmed.

The predicted results often do not correspond to the experimental results, as shown in Figure 17.20a. The results for nonbonded particles are too high; as mentioned, this may have to do with the particles being effectively bonded to a limited extent. The two curves for bonded particles should coincide with the theoretical curve for $G_p/G_0 = 90$, but they differ widely. This is discussed below.

4. *Intermediate gel layer.* Even for adsorbing flexible polymers, a kind of depletion layer tends to form around a particle. In this intermediate layer the polymer concentration, and thereby the modulus, is decreased, as schematically depicted in frame 3 of Figure 17.20. According to a modified van der Poel theory, this can explain the curve for PB in Figure 17.20a.

5. *Particle shape.* Anisometric particles tend to give a higher G_m value than spherical ones at the same value of φ, although the effect may be small. Large deformation properties are greatly affected by particle anisometry, but published results appear to be conflicting.

6. *Aggregation.* Gels made of casein or a heat-setting protein generally show a very strong increase in modulus with filler particle concentration, as illustrated by curve CB in Figure 17.20. Microscopical observation then shows the particles to be aggregated; this would materially increase the effective particle volume fraction and also increase anisometry; see frame 4 of Figure 17.20. Whether these effects can fully explain the large discrepancy between theory and results, remains to be established.

Large deformation properties of gels with various filler particles have also been studied, but the results vary widely and explanations are largely lacking.

Question

A food technologist wants to make heat-set protein gels filled with emulsion droplets, using as little protein as possible and a high oil concentration. Solutions are made of 1.5 and 3% protein, which give gels on heating with shear moduli G of about 0.1 and 1 kPa, respectively. In each solution oil is emulsified at a volume fraction φ of 0.4, using a homogenizer at 25 MPa. Earlier experiments gave at this homogenizing pressure droplets of $d_{32} \approx 0.6\,\mu m$, which would make droplets of a quite high shear modulus. From a graph like that in Figure 17.20a, it is expected that the gels formed from these solutions will have a modulus of at least four times that of the solutions without oil, and we hope much higher, since some aggregation of the droplets would occur. However, the emulsion made with the 1.5% protein solution did not give a gel at all; the 3% one gave a gel, but the modulus was only 1 kPa. What can have been wrong in the reasoning?

Answer

(a) Because protein is needed to cover the emulsion droplets made, there will be less left in solution. The amount needed to cover the droplets would be $6\Gamma\varphi/d_{32}$, and assuming $\Gamma = 3\,mg \cdot m^{-2}$, this would amount to 12 mg per ml emulsion, whereas the "1.5% protein" emulsion would only contain $15(1 - \varphi) = 9$ mg per ml. In other words, there is not enough protein present. This means that the drops obtained will be larger and that the aqueous phase will contain very little protein, insufficient to make a gel. On heating, the protein-covered droplets will aggregate, and the aggregates will cream, thereby disturbing any very weak gel that may have formed. For 3% protein, about 30−12=18 mg of protein per ml is left in solution, and such a solution would make a gel, presumably of a modulus of about 0.2 kPa. The emulsion gel had $G \approx 1$ kPa, i.e., about 5 times as large. (b) The ratio of G_p/G_0 will depend on the droplet size and on the value of G_0, both of which were not constant. Since the ratio will be quite high in all cases, variation would not greatly affect the results, as suggested by Figure 17.20a. (c) The size and the number of any aggregates of droplets in the gel may well depend on variables like protein content and droplet size.

17.3 PLASTIC FATS

A plastic fat consists of a space-filling network of triglyceride crystals, where the continuous phase is triglyceride oil. At room temperature, the volume fraction of the crystals is generally below 0.5, and a plastic fat thus is a gel according to the definition given at the beginning of this chapter. However, the properties of a plastic fat are rather different from those of most gels, and the formation is a complex process; this is the reason why it is given a separate treatment in this chapter. The treatment proceeds from Chapter 15, and knowledge of the matter in Section 15.4 is especially needed to understand what follows.

Properties of plastic fats are very important in some foods:

The *consistency* of the fat determines its shape retention during storage, plastic deformability during handling (e.g., when spreading on a slice of bread, or mixing it into a dough) and during eating.

This concerns especially butter, margarine, shortening, and cocoa butter. The key rheological parameter is mostly the yield stress.

Oiling off is generally undesirable; it can occur if the permeability of the fat is high and its yield stress is low (see the question at the end of Section 5.3.1).

The crystal network can *immobilize* other structural elements in the system. A good example is given by margarine, where the network prevents coalescence of the aqueous droplets.

17.3.1 Network Formation

When making a plastic fat, the triglyceride mixture is brought to a temperature where it is fully liquid, and it is then cooled to a temperature where at least part of the triglycerides are greatly supersaturated. Then nuclei form, crystal growth occurs, and crystals start to aggregate, in that order. After a little while, however, all three processes (and some others, see below) happen simultaneously. This will be illustrated by an example, shown in Figure 17.21.

The example concerns a fairly simple system, 12% fully hydrogenated palm oil in sunflower oil. As seen in Figure 15.22, this mixture can be cooled to such a temperature that it is greatly supersaturated in the β'-form, but not in the α-form. In this manner β'-crystals are formed that do not (or do only very sluggishly) transform into β-crystals. Results are given for two values of the initial supersaturation. By and large, the various events occur very soon after cooling. Crystal growth starts almost immediately. A soon as the nominal crystal radius is about 25 nm, the van der Waals attraction is sufficiently large to induce aggregation, since there is no repulsion (except hard core repulsion): see Question 2 in Section 12.2.1. Fractal aggregates are formed, the dimensionality being about 1.7. This low value implies that virtually no rearrangement occurs in the aggregates (see Section 13.2.3). From Eqs. (13.15) and (13.20), the time needed for gel formation is calculated at 60 and 250 s for $\ln \beta_0$ values of 3.5 and 2.75, respectively, in good agreement with the observed times needed for a measurable modulus to develop.

Conditions change during the process, because $\ln \beta$ decreases with increasing fraction solid. The initial supersaturation of 3.5 is after 200 s already decreased to a value of 2.1 (or in terms of the supersaturation ratio from 33 to 8.2). This has a large effect on growth rate and especially on nucleation rate.

The initial value of the supersaturation has a large effect; see especially the development of the modulus with time. It is also seen that for the higher

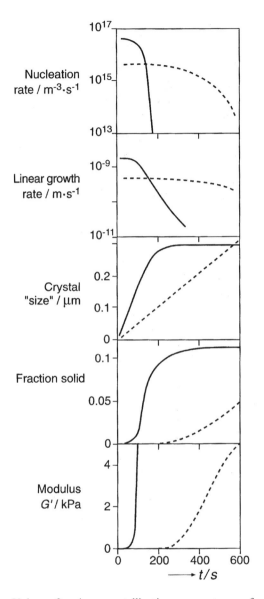

FIGURE 17.21 Values of various crystallization parameters as functions of time t after cooling of a mixture of 12% hydrogenated palm oil in sunflower oil at two values of the initial supersaturation: ———, $\ln \beta_0 = 3.5$; – – –, $\ln \beta_0 = 2.75$. The crystal size is given as the diameter of a sphere of equal volume. (Partly calculated from experimental results on nucleation and growth rate by W. Kloek. Ph.D. thesis, Wageningen University, 1998.)

In β_0 value, the initial crystal size is larger, but not any more at a longer time scale. However, some reservations have to be made: (a) the crystals are not spheres but platelets, ratio of length to width to thickness, e.g., 50 : 10 : 1, though this ratio will depend on fat composition and supersaturation, which change during the process. (b) The crystals have soon formed a network, and the development of crystal size after that is poorly defined (see below). (c) Figure 17.21 concerns crystals of a "narrow" composition, only a few closely similar triglycerides. This fat did not show significant secondary nucleation. However, many natural fats have a far wider compositional range, and these may give copious secondary nucleation at high supersaturation. This means that smaller crystals will result at such conditions.

A Fractal Network? It may appear logical to assume that the fat crystal network eventually formed will also be fractal. However, the results shown in Figure 17.22 counteract this assumption. Plots of $\log B$ or of $\log G'$ versus $\log \varphi$ should be linear, which they are over the short range of φ values available, but the slopes are too high: 5.2 and 6.0 in frame (a) and 4.5 and 7.1 in frame (b). These values would correspond to fractal dimensionalities of 2.5–2.7, i.e., far higher than the value of 1.7 observed for the fractal aggregates. Moreover, extrapolation to $\varphi = 0.01$, at which concentration a space-filling network is already formed, leads to quite unrealistic values. For a B value of $10^{-6}\,\mathrm{m}^2$, the largest pores will be at least 2 mm in diameter ($B = R_g^2/K$, where K is always > 1) and hence be visible to the naked eye, which they are certainly not. A permeability of $10^{-8}\,\mathrm{m}^2$ is about the highest one could expect, and the value may be significantly smaller. The thin lines in the figure give the range of what would be reasonable values; here, it is taken into account that ongoing deposition of solid fat onto the existing fat crystal network would somewhat decrease the B value, roughly as for heat-set protein gels (see Figure 17.19). The experimental results are very different. Something similar applies to the values of the modulus. Near the vertical axis, estimated values at φ values near 0.01 are shown, and these differ widely from the extrapolated values.

We must thus conclude that *the network is not fractal*. It appears more likely that the network forms as is schematically depicted in Figure 17.23. Frame 1 shows that nuclei form and grow into small crystals. In frame 2 the crystals have become larger and started to form small (fractal) aggregates. In the mean time, nucleation and growth go on. In frame 3, the aggregation has led to a space-filling fractal network, but not nearly all of the crystallizable material is incorporated into the network. Nucleation and subsequent aggregation of newly formed crystals go on, and in frame 4, we see again small fractal aggregates. In frame 5 these "*secondary aggregates*" have grown out to a size comparable to that of the larger pores in the original network, thereby blocking these pores, greatly diminishing the permeability. Because the crystals aggregate unhindered ($W = 1$) and crystal growth is relatively slow, despite the high supersaturation, such events can occur. The second generation aggregates will also become bonded to the original network, thereby making it much stiffer.

Moreover, the crystals in the network will grow by deposition of crystallizing triglycerides, which can lead to *sintering*; this is illustrated by the two details given of frames 2 and 4. The occurrence of sintered bonds is borne out in that the moduli observed are significantly larger than is calculated on the basis of van der Waals attraction between crystals.

Some Variables. Figure 17.22 shows that for a higher initial supersaturation, the values for B are smaller and those for G' are larger at the same φ. This is in qualitative agreement with the sequence of events outlined above: for a higher value of $\ln \beta_0$, the supersaturation at the moment of gel formation is higher, allowing more extensive nucleation, whereby more second generation aggregates can be formed; this will result

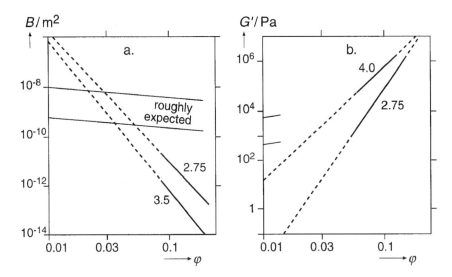

FIGURE 17.22 Properties of fats consisting of hydrogenated palm oil in sunflower oil, crystallized at various initial supersaturation $\ln \beta_0$ (indicated near the curves) as functions of the volume fraction solid φ. (a) Permeability B. (b) Storage shear modulus G'. Heavy solid lines give experimental results. (From results by W. Kloek. Ph.D. thesis, Wageningen University, 1998.)

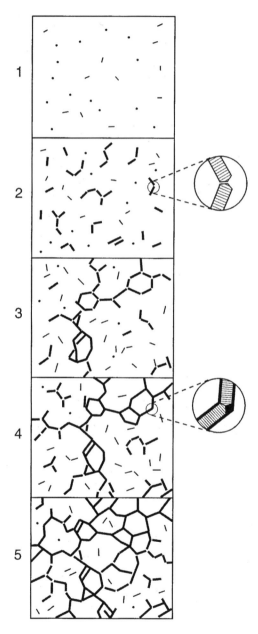

FIGURE 17.23 Various stages in the formation of a fat crystal network. Approximate scale: the edge of a frame would be several μm.

in a more even and compact network. The same mechanism may explain the steepness of the relations with $\log \varphi$, since for a higher value of φ, the supersaturation will be less reduced at the gel point.

So far, we have only considered network formation in one system. It appears likely that for other fat compositions or other process conditions the same phenomena occur, although additional changes may occur and quantitative relations will be different. The exponents of φ in the relations between B or G' and φ are generally about the same, although the absolute values of B or G' may vary significantly. Some important considerations are

1.   Fats of a *wider compositional range* tend to give more secondary nucleation and hence smaller crystals; the crystals also tend to be more anisometric (more slender). Moreover, slow changes can occur into another polymorph, which generally goes along with a change in compound crystal composition. Such changes can greatly enhance sintering between crystals. In systems where only one triglyceride, e.g., tripalmitate, can crystallize, sintering does not occur, and the modulus of the fat tends to be relatively small.

2.   Often the *supersaturation is smaller*, leading to larger crystals and a weaker network. Isothermal recrystallization, e.g., into another polymorph, generally occurs at very low supersaturation, and it may cause the formation of larger and more isometric crystals (e.g., spherulites) that make very weak networks.

3.   It has been implicitly assumed so far that crystallization occurs *isothermally* in a quiescent system. In practice, this is generally impossible. Either the *temperature rises*, due to release of the enthalpy of fusion, which will decrease the supersaturation; or the system is *agitated*, which breaks up networks formed. The latter point will be briefly discussed later on.

17.3.2   Rheological Properties

The effects of some variables on the modulus have already been mentioned, and to some extent these variables also affect large deformation properties, although other factors will also play an important role. It must be assumed that the bonds in the network are due to van der Waals attraction (relatively weak) and to sintering of touching crystals (often quite strong). Moreover, crystals can be bent under stress.

Linearity.   The strain at which the modulus is no longer proportional to the applied stress is always quite small for plastic fats. Defining the linear region as that over which the change in modulus is at most 1%, the linearity extends to, for instance, $\varepsilon = 10^{-4}$, $3 \cdot 10^{-4}$ and $14 \cdot 10^{-4}$, for hydrogenated palm oil crystals in oil, margarine fat, and butter, respectively. Presumably, the differences are largely due to differences in the anisometry of the crystals, which tends to increase in the same order; in other words, the crystals in butterfat can be bent more readily than those of the other fats. For strains larger than about 10^{-3}, the deformation tends to become partly irreversible, and the extent of irreversibility increases in the reverse order. For increasingly larger deformations, the changes become increasingly irreversible.

Deformation Regimes. Figure 17.24 illustrates what will happen when a typical margarine fat is subjected to increasing strain. The following regimes can be distinguished:

1. *Linear.* The strain is proportional to the stress and the loss tangent is small (< 0.1). The deformation is fully *elastic* and hence reversible, which implies that no bonds between crystals are broken.

2. *Nonlinear.* Note that this region extends over a strain range about 200 times that of the linear region. With increasing strain, the deformation becomes more and more *viscoelastic*, and the apparent loss tangent increases up to a value of about unity. This also means that the shape of the curve will depend on strain rate (cf. Figure 17.6). Bonds are broken, the more so for greater strains. Presumably, van der Waals bonds can reform on removal of

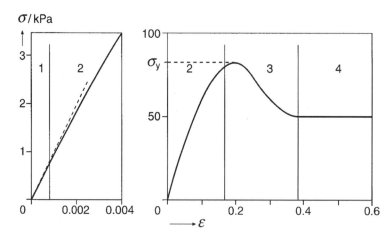

FIGURE 17.24 Schematic diagram of an example of the relation between stress (σ) and strain (ε) during deformation of a typical margarine fat. σ_y is the yield stress.

the stress, whereas the sintered bonds that are broken do not reform, at least not within a short time.

3. *Stress overshoot.* Here, much of the network structure is broken down, and the system shows *yielding*. It should be realized that the inherent inhomogeneity of the structure, with "weak" spots or planes, causes the structure breakdown to be very uneven; in several small regions the original structure may almost remain. The magnitude of the stress overshoot tends to be larger for a higher strain rate and for a more brittle fat (smaller linear range).

4. *Plastic flow.* After yielding and the accompanying stress overshoot, the material flows, having a very high Bingham *viscosity*, often between 5 and 50 kPa·s; it is somewhat strain rate thinning and exhibits some elasticity. These phenomena are probably explained by the irregular, spiky shape of the network fragments present, causing strong frictional forces, and possibly to colloidal interaction forces between structural elements (cf. Section 17.4).

Other plastic fats show similar behavior, but the scales of the curve may vary significantly.

Yield Stress. Because the spreading, cutting, and shaping of a plastic fat all involve permanent deformation, the yield stress, as illustrated in Figure 17.24, is an essential parameter. It correlates well with the values obtained by some practical tests and with the sensory perceived firmness.

Figure 17.25a gives a few examples of σ_y as a function of the fraction solid. The latter is an essential variable, but whereas the modulus scales with φ to a power of 4 to 7 for nearly all plastic fats, the scaling exponent is generally about 2 for the yield stress. This agrees with the network structure having undergone extensive change before yielding occurs. Like the modulus, the yield stress tends to be higher if the fat has crystallized at a higher value of $\ln \beta_0$. When the temperature is lowered, σ_y tends to increase strongly (often by a factor of about 10 for a 10 K decrease), not only because φ increases but also because of increased sintering. The latter effect is especially strong for a multicomponent fat, and if the cooling rate is slow (can you explain this?). Also temperature fluctuations tend to increase the firmness of the fat, as well as its brittleness: the fat that melts on temperature increase tends to cause extensive sintering upon cooling, increasing stiffness as well as firmness of the fat.

Figure 17.25b illustrates the phenomenon of *work softening*. When a fat is strongly worked (kneaded) at constant temperature, its yield stress will markedly decrease (generally by 30–70%). The firmness starts to immediately increase again upon storage: at first crystals and network remnants will aggregate, being kept together by van der Waals forces, and then sintering can occur, which will be a slow process. The original value of σ_y is not reached.

Especially a multicomponent fat exhibits *setting*, i.e., it increases in firmness with time, be it after working or without. This is because it is not in equilibrium: upon storage slow transitions take place in polymorphic form and in compound crystal composition. Since such recrystallization goes via the liquid state, it can cause further sintering. However, if slow recrystallization causes the formation of β-crystals—which especially occurs if nearly all fatty acid residues have 18 carbon atoms—quite large crystals may form, and the firmness then markedly decreases. This phenomenon occurs upon storage of some margarines. The transition to β can be counteracted by the presence of diacylglycerols.

Crystallization Under Agitation. To obtain a firm fat and to speed up operation, crystallization at a high supersaturation is desirable. This implies that a space-filling network is rapidly formed, which greatly hinders the removal by diffusion of the heat of fusion released, unless the distance over which it has to occur is quite small. In practice, e.g., in margarine making, a scraped-surface heat exchanger is used. The fat is rapidly cooled at the cold surface, which soon results in the formation of a

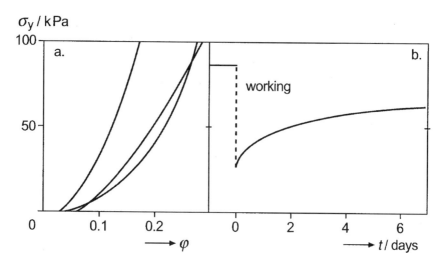

FIGURE 17.25 The yield stress σ_y of plastic fats. (a) Examples of the dependence of σ_y on the fraction solid φ. (b) Example of work softening of a plastic fat and the subsequent increase in σ_y with time t.

solid crystal network. This is regularly scraped off the surface, whereby it is strongly worked and hence made soft. If now the fat leaves the heat exchanger after crystallization is about complete, it will have a low yield stress; upon storage it will set to some extent, about as shown in Figure 17.25b. If the fat is not fully crystallized when leaving the heat exchanger, further crystallization will cause the network to become more homogeneous, comparable to what is shown in Figure 17.23, and also sintering will generally occur. Within an hour or so, the yield stress can reach a value more than twice that of a fat that was fully crystallized when leaving the heat exchanger, but it will not exhibit significant setting upon further storage.

In conclusion, many different processes occur during fat crystallization, at widely varying rates, all depending on temperature (history) and on fat composition. This makes it very difficult to predict the mechanical properties of plastic fats. However, the various processes involved have been identified, and their dependencies on composition and several external variables have been established. This means that trends can often be predicted.

Question

Consider a fat that is fully liquid above 40°C and that contains numerous different triglycerides. Upon cooling from 50 to 20°C a plastic fat is formed. The yield stress is determined at 20°C under various conditions: (a) 1 hour or 1 day after cooling; (b) after fast or slow cooling to 20°C; (c) cooling first to 4°C before bringing to 20, or direct cooling to 20°C; (d) keeping a few hours at 30°C before bringing to 20, or direct cooling to 20°C; (e) after one day at 20°C, keeping it a day at 10, then a day at 20°C; or keeping three days at 20°C; (f) after one day at 20°C, keeping it a day at 25, then a day at 20°C; or keeping three days at 20°C; (g) measure the yield stress while slowly or rapidly increasing the strain. For which of each pair is the measured yield stress higher than the other and why? Tip: Also consult Section 15.4.

Answer

(a) After a day: then sintering will be strong and the fraction solid may be higher; (b) fast cooling: smaller crystals generally give a firmer and more homogeneous network; (c) precooling: for about the same reasons as given in b; moreover, the fraction solid will be higher in a system forming compound crystals; (d) direct cooling: the same arguments as given in c; (e) probably little difference; the increased firmness at 10°C will be lost after warming again; (f) probably the temporary stay at 25°C: this gives much sintering as the melted fat crystallizes again; (g) fast deformation: see Figure 17.6 and the explanation given for the effect.

17.4   CLOSELY PACKED SYSTEMS

Several types of food systems may be considered as highly concentrated dispersions of soft solid particles that touch each other; the volume fraction of the particles generally exceeds 0.8. To a first approximation, the modulus of such a system would equal the modulus of the particles; in practice, it will often be substantially smaller.

Attractive forces may act between the particles, or the continuous phase may be a weak gel, but this is not needed to obtain a solid consistency. Consider a vacuum-sealed pack of ground coffee. Although the packing material is flexible, the pack behaves as a hard solid. This is because the particles are solid and are pressed together by atmospheric pressure and cannot move relative to each other. Punching a hole in the pack leads to influx of air and the contents change into a free flowing powder. It thus is the *volume restriction* that makes a closely packed system a (soft) solid.

In this section we will describe some closely packed systems and give some relations for rheological properties in ideal cases. Some of the structures occurring are illustrated in Figure 17.26.

Foams and Emulsions.   Most polyhedral foams are soft solids, as are some emulsions, like mayonnaise. The structure is illustrated in Figure 17.26a. The value of the volume fraction must be above the critical value for close packing, φ_{cr}. For dispersions of perfect spheres that are not excessively polydisperse, φ_{cr} tends to be about 0.71.

In Section 11.2.1 it is discussed how a foam is formed; see especially Figure 11.1. Due to gravity the bubbles start to deform each other, and

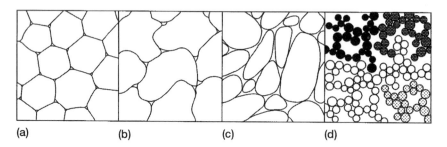

(a)   (b)   (c)   (d)

FIGURE 17.26   Structure of various closely packed systems. (a) Polyhedral foam or emulsion. (b) Concentrated potato starch gel. (c) Concentrated dispersion of anisometric gel particles ("fluid gel"). (d) Dispersion of fragments of a particle gel (different fragments differ in grayness). Highly schematic. The scales can vary considerably.

drainage of the foam layer then leads to a high volume fraction of gas. A highly concentrated O–W emulsion is more difficult to make: a high ratio of oil to water is needed and a high concentration of a suitable surfactant. Agitation then has to provide the forces needed to obtain permanently deformed droplets. An alternative is to subject a less concentrated emulsion to centrifugation. Repulsive colloidal interaction forces between the bubbles or droplets prevent breaking of the thin films between them.

The particles thus do not necessarily attract each other. If a lump of a foam or an emulsion is put in water or, preferably, in a solution similar to the continuous phase, the lump often will slowly disperse. The slowness is largely due to the channels through which the water has to penetrate being quite narrow.

On deformation of the system, the bubbles are deformed, which increases their Laplace pressure p_L. Moreover, some films between particles are stretched and others are compressed, causing surface tension gradients to form, which also needs energy. Above a certain stress, yielding may occur, which means that bubbles (or drops) start to slip past each other, which generally occurs in planes about parallel to the direction of flow. Calculation of the *shear modulus* and the *yield stress* from first principles is virtually impossible because of the intricacy of the problem for a three-dimensional and polydisperse system, but trends can be predicted. One relation is that these parameters are proportional to the *average apparent Laplace pressure*

$$p_{L,a} = \frac{4\gamma}{d_{32}} \tag{17.20}$$

where the values of γ and d refer to undeformed particles. Figure 17.27 gives experimentally established relations.

For a typical food foam with $d_{32} = 20\,\mu m$ and $\gamma_{AW} = 50\,mN \cdot m^{-1}$ we have $p_{L,a} = 10^4\,Pa$. The same result is obtained for a typical O–W emulsion made with protein and having $d_{32} = 4\,\mu m$ and $\gamma_{AW} = 10\,mN \cdot m^{-1}$. Assuming $\varphi = 0.9$, we then obtain from Figure 17.27 $G = 1600\,Pa$ and $\sigma_y = 160\,Pa$. Putting a heap of this foam of height h on a table, the gravitational stress at the bottom will be $gh\,\Delta\rho$, and if this value is greater than the yield stress, the lump will start to subside under its own weight. Assuming $\Delta\rho = 100\,kg \cdot m^{-3}$, the critical height for yielding to occur is 16 cm. For the emulsion, where $\Delta\rho$ is about nine times as high, the critical height will be less than 2 cm.

In practice, most systems are more complicated. *Whipped toppings*, for instance, come in various types. Fairly simple are toppings with a gas volume fraction $> \varphi_{cr}$. Since they also contain fat globules, the effective φ

value is higher and the topping has a sufficiently high yield stress. Other toppings contain gelling agents, so that φ can be smaller. Classical whipped cream has $\varphi(\text{air}) \approx 0.5$ and $\varphi(\text{fat globules})$ is at most 0.2. Firmness is here obtained in another way. Fat globules have become attached to the air bubbles and are, moreover, clumped (partially coalesced), so that they make a space-filling network; this is comparable to (part of) the structure of ice cream, shown in Figure 9.1b.

Concentrated Starch Gels. Starch is discussed in Section 6.6. Especially gelatinization (in Section 6.6.2) and retrogradation (in 6.6.3) are aspects of importance.

When native starch grains are heated in an excess of water, gelatinization occurs, which implies that amylose leaches from the granules and that the latter greatly swell and eventually fall apart. If the system then is cooled and the amylose concentration is above the chain overlap concentration c^*, corresponding to 2–4% starch (depending on starch type), a gel is formed. This is mainly due to the formation of microcrystallites of

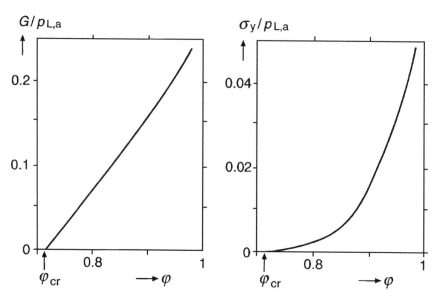

FIGURE 17.27 Concentrated foams or emulsions. Values of (a) the modulus G and (b) the yield stress σ_y, each divided by the apparent Laplace pressure of the particles $p_{L,a}$, as a function of the particle volume fraction φ. (After results given by H. M. Princen. In: R. D. Bee et al., eds. Food Colloids. Roy. Soc. Chem., Cambridge, 1989, p. 14.)

amylose and the gel resembles a polymer gel of the general type depicted in Figure 17.10b.

When the starch concentration is markedly higher, full swelling of starch granules cannot occur, and the system becomes closely packed; for potato starch this occurs above 5% starch, for most other starches at distinctly higher concentrations (up to 15%). Upon cooling, a gel is formed, consisting of closely packed granules, with a thin layer of amylose gel as a kind of glue between the granules. For most types of starch, including wheat starch, the swollen granules are more or less isometric, but in potato starch they lock into each other, almost as in a jigsaw puzzle: see Figure 17.26b. In either case, the modulus of the gel is given by the modulus of the swollen, gellike granules, which is higher for a higher starch concentration (i.e., less swelling) and for longer keeping at low temperature (i.e., stronger retrogradation). For wheat starch, the fracture stress is primarily given by the fracture stress of the amylose "glue," since the swollen granules can slip past each other without being greatly deformed. For potato starch, it is primarily the stiffness of the grains that determines the fracture stress, since it needs considerable deformation (or even fracture) of the grains before they can slip past each other.

To give some idea of the magnitude of the parameters, we will consider 30% starch gels, aged about 2 weeks. We then have roughly

Starch type	Wheat	Potato
Young's modulus	300	900 kPa
Fracture stress	100	400 kPa
Fracture strain	0.3	0.5

For a younger gel, the modulus and the fracture stress are much smaller, the fracture strain being larger. Especially for a very young potato starch gel, elastic deformability is high, ε_{fr} being of the order of 3.

Doughs.   A dough is made by kneading a mixture of grain flour and water (and possibly a litle salt). The water content is, e.g., 45%. The dough contains partly swollen starch grains, making up, e.g., 70% of the volume. The structure is thus like that of a concentrated starch gel, but there are differences as well. The continuous "phase" is a more or less homogeneous viscoelastic liquid, consisting of water, protein, other nonstarch solids, and part of the amylose. The whole dough then is a *viscoelastic* system, and it generally has a negligible yield stress. Some rheological properties at a time scale of about 1 s would be very roughly

Dynamic shear modulus	$G' = 10\text{–}20\,\text{kPa}$
Loss tangent	$\tan \delta = 1\text{–}0.5$
Apparent viscosity	$\eta_a = 10\,\text{kPa}\cdot\text{s}$

The systems are markedly strain rate thinning, meaning that η_a is much higher at lower strain rates.

If the dough is made of wheat flour, it shows marked *strain hardening*, i.e., an increase in modulus with increasing strain (despite its strain *rate* thinning character). The same applies, though to a much lesser extent, to a dough made of rye flour. The importance of these properties was discussed in Section 17.1.3.

Fluid Gels. The name seems like a contradiction in terms. What is meant is a concentrated dispersion of gel particles in a liquid. These systems can be made in various ways, for instance by making a hot solution of a suitable polymer, which is then cooled in a scraped-surface heat exchanger. The cooling causes gel formation and the vigorous stirring breaks it into fragments. In this way a polydisperse and quite concentrated suspension of anisometric gel particles is formed; see Figure 17.26c. It turns out that anisometry of the particles is an essential property.

The effect of the volume fraction φ on the modulus of such systems is illustrated in Figure 17.28. The value of φ can be varied by diluting with water or by centrifuging the dispersion and removing the supernatant. Curve (a) refers to almost *perfect monodiperse spheres*. If φ is below the critical value for close packing φ_{cr}, the system will be liquid; the viscosity tends to go to infinity as φ approaches φ_{cr}. At that concentration, the system obtains a stiffness: a modulus can be measured. A very slight increase in φ gives a large increase in modulus, the increase leveling off at slightly higher φ, where the stress becomes large enough to deform the particles (the particle modulus was about 40 kPa in this case).

The other curves relate to *anisometric polydisperse systems*. At $c_{rel} = 1$, the systems do contain some free liquid, since an agar gel tends to show some syneresis upon fracturing. The main difference with curve (a) concerns the significant shear modulus at lower concentrations. This is because the anisometric particles get in each other's way as soon as the system is being sheared. (It may be noted that the relation between Young's modulus and concentration will probably be different.) It is also seen (curve b) that concentrating the system further increases the modulus, presumably until it would reach the value of a homogeneous gel of the same agar concentration.

It may be concluded that the most important variables affecting the modulus are

The *volume fraction* of the particles.

The *anisometry* of the particles. For a greater anisometry, the modulus will be higher at the same φ, and a measurable modulus will be obtained at a lower value of φ.

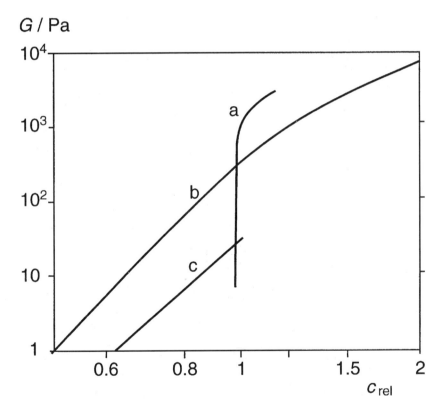

FIGURE 17.28 Shear modulus of concentrated dispersions of agar gel particles in water as functions of relative concentration of agar, c_{rel}. Curve a, spherical monodisperse particles of 2% agar; $c_{rel} = 1$ for close packing. Curve b, anisometric particles of 1.75% agar gel; $c_{rel} = 1$ for an unaltered sheared agar gel (G unsheared = 27,000 Pa). Curve c, same, 0.75% agar gel (G unsheared = 3700 Pa). (After results by W. J. Frith et al. In: Gums and Stabilisers for the Food Industry 11. Spec. Publ. 278, Royal Soc. Chem., 2002, p. 95.)

The *modulus* of the particles. This is clearly shown by the difference between curves (b) and (c), where the particles have about the same anisometry but differ in agar concentration and hence in modulus.

The systems discussed here also have a yield stress; unfortunately, results on large deformation rheology of the systems seem to be lacking.

There are also *liquid systems* of a structure similar to "fluid gels." A case in point is stirred yogurt. The original set yogurt is effectively an acid casein particle gel. By controlled stirring, a thick viscoelastic liquid is obtained, i.e., without a significant yield stress. The rheological properties are for the most part due to the presence of network fragments that are highly anisometric and where protrusions on these fragments may hook into each other: see Figure 17.26d.

Pastes. Fluid gels may to some extent be suitable models for various foods that can be described as pastes, i.e., systems that have a yield stress but are readily deformable and that contain finely dispersed material. Anyway, the qualitative relations listed above for fluid gels also apply to most pastes. We will briefly mention some of the numerous foods in this category.

Quark and comparable fresh cheese varieties. This product is similar to the stirred yoghurt just mentioned, but at a lower water content. This gives it a significant yield stress.

Liver pâté and comparable meat products. It consists of small tissue fragments of various kinds, some separate cells, fat particles, etc. Presumably, interaction forces between the particles, e.g., due to the presence of some gelatin, contribute to the firmness.

Fruit purées, such as apple sauce and tomato purée. These products consist for the most part of tissue fragments and whole cells in an aqueous liquid. Most vegetable tissues are pretty stiff, because the cell walls are stiff and because of cell turgor. However, the material is heated to high temperatures and this process leads to loss of turgor and cell wall rigidity; see Section 17.5, "Plant Tissues." Especially in tomato products, the cells become highly deformable, which is the reason that tomato purée is made of concentrated "juice," to realize a high packing density. Apple sauce tends to be firmer, especially if not cooked too intensively.

Peanut butter is a concentrated suspension of particles obtained by grinding roasted peanuts in the oil that leaks out of the fragments upon grinding. The fragments are relatively large, say a millimeter, anisometric, and rigid, which gives the product some firmness despite the relatively high fraction of free oil. To enhance firmness, a little highly saturated fat is sometimes added, to form a fat crystal network in the oil.

Cheese. Several fully ripened cheese varieties can be considered closely packed systems. When freshly made, the structure resembles that of a paracasein micelle gel as discussed in Section 17.2.3, although with quite a high particle volume fraction; moreover, the "gel" contains unbound filler particles, i.e., the milk fat globules. In a few days, the micelles disappear, and now a more or less continuous mass of much smaller proteinaceous particles, size of the order of 10 nm, can be observed. Presumably, these particles attract each other at least somewhat, since the cheese mass can be considerably deformed without yielding or fracturing: see Figure 17.8, cheese 2. Upon further aging, considerable proteolysis occurs, resulting in a mass of peptides and even smaller molecules. The system now may be considered as a paste, albeit with very small building blocks (and, of course, filler particles). If the moisture content of the cheese is relatively high, a viscous liquid results, as can be observed in several well-ripened soft cheeses with a surface flora. For a lower moisture content (semihard and hard cheeses), the consistency becomes more like that of a brittle solid, as illustrated in Figure 17.8 for cheese 1.

17.5 CELLULAR SYSTEMS

Several foods have cellular structures: nearly all fruits and vegetables, bread and cake, products made by high temperature extrusion, some types of candy bars, etc. A cellular system can be defined as a collection of closely fitting cells. The cells are enveloped by a soft-solid matrix and are closely packed with gas or liquid. Generally, *the matrix provides the stiffness*, this in contrast to the closely packed systems of the previous section, where the solid character is primarily due to the particles.

The cells of the structure can be either closed or open. The former type is like a foam, in that the matrix can geometrically be compared with the continuous phase of the foam, which consists of thin lamellae and Plateau borders. Here we have thicker lamellae or *walls*, and *beams* (struts, ribs) where two lamellae meet; if the cells are filled with gas we can call the system a *solid foam*. Open cells occur when the lamellae contain holes; now we speak of a *sponge*. Some types of sponge structures merely consist of beams. Most *plant tissues* (see, e.g., Figure 9.4) consist of closed cells that are mainly filled with an aqueous liquid.

The volume fraction of cells φ is defined as one minus the volume fraction of the matrix material. Generally, φ varies between 0.5 and 0.85. In most cases, the cells are not very anisometric and not very polydisperse. This makes it easier to develop theory for the rheological properties.

Some Theory. We will only consider air-filled systems. Development of theory generally begins with a simple two-dimensional array of hexagonal cells, as depicted in Figure 17.29a, and then is extended to more complicated systems. The most important structural parameter is the *relative density* of the system, defined as ρ/ρ_m, where ρ is the density of the system and ρ_m that of the matrix material. The simplest case then is prediction of the *modulus*, often Young's modulus E.

For a *sponge* in which nearly all matrix material is in beams of thickness b and length L, $\rho/\rho_m \approx (b/L)^2$. Applying a force F to a beam, it will be deflected by an amount δ, and it is derived from the theory of the bending modulus that $\delta \propto FL^3/E_m b^4$. Taking into account that $E = \sigma/\varepsilon$ (where ε is the Hencky strain), $\sigma \approx F/L^2$ and $\varepsilon = \delta/L$, we arrive for open cells at $E \approx F/L\delta$, which results in

$$E(\text{sponge}) \propto E_m \left(\frac{b^4}{L^4}\right) \propto E_m \left(\frac{\rho}{\rho_m}\right)^2 \tag{17.21}$$

In an ideal *solid foam* all of the matrix material is in walls of thickness b and edge L, leading to $\rho/\rho_m = b/L$. Furthermore, the deflection of a wall is then given by $\delta \propto FL^2/b^3$, and we arrive at

$$E(\text{foam}) \propto E_m \left(\frac{b^3}{L^3}\right) \propto E_m \left(\frac{\rho}{\rho_m}\right)^3 \tag{17.22}$$

Generally, Eq. (17.21) is well obeyed, but for most solid foams the exponent is smaller than three, sometimes even as small as two, because even for a closed-cell structure most of the matrix material may be in the beams rather than in the walls. The theory can be extended for other rheological parameters, such as the yield stress, but these relations are mostly not well obeyed. As a general rule, however, one may state that for any rheological parameter Z

$$Z = K Z_m \left(\frac{\rho}{\rho_m}\right)^n \tag{17.23}^*$$

a scaling law that is mostly well obeyed within the range $\rho/\rho_m = 0.5$–0.15 ($\varphi = 0.5$–0.85); the values of the proportionality constant K and the scaling exponent n cannot be readily predicted and vary widely among systems. In systems of φ values below 0.5 the exponent is generally smaller, approximating unity for $\varphi < 0.2$.

What will happen at **large deformation** will depend on the properties of the matrix material. If it is purely *elastic* and not very stiff (i.e., rubberlike), buckling of beams and cell walls will occur, as illustrated in Figure 17.29b.

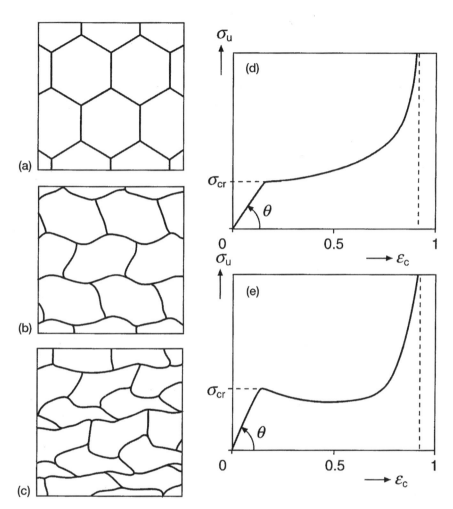

FIGURE 17.29 Uniaxial compression of regular cellular structures; the cells are presumed to be open. (a) Structure of the undeformed system. (b) Structure after buckling under a compressive stress in the vertical direction for a purely elastic matrix. (c) Same, after yielding, for a matrix that is a plastic solid. (d) Stress–strain relation for elastic deformation, as in situation (b). (e) Same, for plastic deformation, as in situation (c). σ_u is the uniaxial compression stress; ε_C is the Cauchy strain; σ_{cr} is the critical stress for buckling (d) or yielding (e); tan θ equals the Young modulus of the system. The vertical broken line denotes the maximum strain possible (then $\varphi \approx 0$). (Adapted from Gibson and Ashby [see Bibliography].)

The relation between stress and strain is in principle as given in frame (d). It shows a sharp buckling point, beyond which the apparent modulus is greatly decreased. For most practical, i.e., far less regular, systems the transition will be more gradual. It is also seen that the stress starts to increase strongly at high strain values. Compression then causes *densification*, i.e., an increase of the relative density of the system, and eventually the rheological properties become equal to those of the matrix material.

If the matrix consists of a *plastic* material, implying that it has a yield stress, yielding will occur, e.g., as illustrated in Figure 17.29c and e. Some stress overshoot then is common. After yielding, the stress does not greatly increase until appreciable densification occurs. Also for a yielding material, the overall yield stress will not be so sharply defined for a less even cellular structure. What often happens is that a layer of one or a few cells in thickness suddenly yields, to be followed by other layers at a higher overall strain: the layer mostly is about perpendicular to the direction of the stress. Such phenomena also occur when a shear stress is applied, but then the orientation of the layer tends to be at 45 degrees to the direction of shear.

If the matrix material is *brittle*, large deformation will lead to fracture, but then we are outside the realm of soft solids.

The **variables** affecting the results discussed so far are φ, open or closed cells, and rheological properties of the matrix material. Other aspects are

> The *properties of the material* are often unknown, because the material may strongly change during formation or growth of the cellular structure. The walls and beams may well be markedly anisotropic.
>
> The *geometry of the matrix* can vary considerably. This concerns cell size and especially shape; the distribution of matrix material over beams and walls; and the overall structure. The latter often is uneven: it may vary from place to place (e.g., from the outside to the inside of a specimen), it may be anisotropic, and it may have weak spots or regions.
>
> At large strains, the *mode of deformation*, such as uniaxial compression, uniaxial extension, shear, or bending, may considerably affect the shape of the stress–strain curve.

All of these variables affect the mechanical properties of the system, but few general rules can be given.

Bread and Comparable Baked Products. These are made by baking a leavened dough. The high temperature causes considerable expansion of the gas cells and (after some time) rupture of the lamellae between cells. Upon cooling, a soft solid sponge is obtained.

Figure 17.30a shows results on the modulus of wafers, made at various gas contents. Equation (17.21) is precisely obeyed. Wafers have a very low water content, and the material is in a glassy state, which explains the very high values of the modulus (upper curve). Upon storing the wafers in air of a high relative humidity, the wafers become soft: the matrix material takes up water and obtains a rubbery state (cf. Figure 16.3). It is seen that the modulus then is markedly decreased. Upon large deformation of the original wafers, fracture occurred at a stress of about 400 kPa. Deformation of the "soft" wafers led to buckling at a stress of about 15 kPa.

Figures 17.30b and c show results on a model wheat bread. (This was made of wheat starch, xanthan gum, water, and yeast, but normal white bread made of wheat flour gives much the same results.) In frame (b), fresh

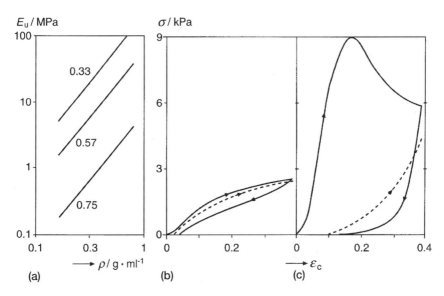

FIGURE 17.30 Mechanical properties of baked cellular products. (a) The Young (compression) modulus E_u as a function of product density φ of wafers kept in air of various water activities (indicated near the curves). (After results by G. E. Attenburrow et al. J. Cereal Sci. 9 (1989) 61.) (b) Compression and decompression (indicated by arrows) of the crumb of a model wheat bread ($\rho/\rho_m \approx 0.35$), giving the stress σ versus the Cauchy strain ε_C. Complete densification would presumably be reached at $\varepsilon_C \approx 0.7$. The solid line shows the first compression and decompression, the broken line the second compression. The bread was 4 hours old. (c) Same experiment, same bread, but now 7 days old. (After results by C. J. A. M. Keetels et al. J. Cereal Sci. 24 (1996) 15.)

bread is deformed. The matrix material is rubbery, and the stress–strain curve is comparable to that in Figure 17.29d, although a clear buckling point cannot be identified (presumably because of the unevenness of the cellular structure). Decompression shows some, though not much, hysteresis, and a second compression curve (after the first decompression) is quite similar to the first one. Frame (c) relates to the same bread after keeping it at room temperature for 7 days, when it was considered very stale. Staling is due to retrogradation, i.e., forming of starch microcrystallites, which makes the matrix much stiffer (Section 6.6.3). The compression curve is now typical for yielding ($\sigma_y \approx 9\,\text{kPa}$), although the stress overshoot is so large that considerable local fracture must have occurred as well. The hysteresis between compression and decompression is very large, and the second compression curve confirms that the first compression has caused considerable irreversible structure breakdown.

Extrusion cooking is used to make several foods of a cellular structure, such as crispbreads, various snacks, and some breakfast cereals. These are primarily starch based products. In the extruder the ingredients are highly compressed and made to flow—which results in a continuous mass—and heated to a high temperature. This implies that the water present is strongly superheated; upon leaving the extruder through a die, whereby the pressure is suddenly released, steam forms and the material strongly expands. A cellular structure is generally formed. The matrix often has such a low water content that it becomes brittle upon cooling, which then means that the product is not a *soft* solid.

The relative density is generally between 0.6 and 0.25. Equation (17.23) tends to be approximately obeyed. Although micrographs suggest that the cells are closed, the relation between modulus and density is like that of a sponge, the scaling exponent being about equal to 2. Deviations from theory seem to be largely due to unevenness of the structure: the cells often show a wide size distribution, and the density may vary, especially between outside and inside of the specimen.

Plant Tissues. Most tissues obviously are cellular soft solids. Plant cells have cell walls, many of which are fairly rigid. However, three quite different factors contribute to the stiffness of plant tissues:

> The rigidity of the *cell wall*
> The strength of the *middle lamella*, which mainly consists of pectins and acts as a "glue" between the cells
> The *turgor* in the cell, i.e., the osmotic pressure inside the cell being higher than outside

The turgor is due to the presence of a cell membrane containing various transport enzymes, which can keep the molality of the cellular liquid at a high level. This mechanism needs energy, and when the cell metabolism stops, energy cannot be supplied any more and the turgor disappears. This is what we observe when a leaf of lettuce wilts, which shows that the turgor can strongly contribute to the stiffness of a tissue. But the other two factors also contribute, to a variable extent. All three factors are generally affected when the tissue is heat treated: the turgor disappears, the middle lamella can become greatly weakened, and the cell walls will soften owing to partial degradation of their structural polysaccharides. Altogether, the relative contribution of the three factors varies widely among tissues, during storage and upon processing.

The theory as briefly touched on in this section cannot be applied to plant tissues. Their mechanical properties are highly specific and need to be studied separately in each case, as is often the case with natural products.

17.6 RECAPITULATION

Soft solids are always composite materials. Their softness means that a small stress leads to a relatively large deformation, either because the elastic modulus is quite small, or because the material readily yields (shows lasting deformation).

Rheology. Determination of mechanical properties can be done in various deformation modes (shearing, compression, extension), leading to different results. Of the parameters to be determined, the modulus or stiffness of materials has been studied and explained best; results can give information about structure.

In practice, large deformation properties are far more useful, and determination of the full relation between stress and strain gives the best information. Materials vary widely in their "linear region," i.e., the strain range over which stress and strain remain proportional. Deforming it much farther, the material may eventually break. Relevant parameters then are fracture stress or "strength," fracture strain or "shortness," and work of fracture or "toughness." The correlation between fracture parameters and the modulus is often poor. Since many soft solids exhibit viscoelastic behavior, the values of these parameters can depend, often markedly, on the strain rate.

Some materials break at high strain rates but yield at small rates. Several soft-solid foods yield under the conditions normally applied, as during spreading on bread. The magnitude of the yield stress then correlates well with "firmness." Another variable is the inhomogeneity of a material, which gives rise to local stress concentration. This hardly affects the modulus, but may strongly affect the large deformation behavior.

Fracture Mechanics.   In a purely elastic material, the overall stress needed to cause fracture is greatly lowered (by up to a factor 100) by local stress concentration: near a tiny crack or other defect, the local stress can readily surpass the fracture stress. However, surpassing the fracture stress is insufficient, since energy is needed for fracture to proceed, i.e., for crack propagation. The energy is derived from the elastic energy stored upon deformation. Each system has a critical crack length above which fracture is propagated.

Soft solids show more complex behavior, and fracture is often "elastic–plastic," implying that near the tip of the crack, where the stress is highest, yielding occurs; the pieces remaining after fracture do not precisely fit to each other. The local yielding increases the energy needed, i.e., the specific work of fracture.

For viscoelastic soft solids, the whole test piece shows lasting deformation before and during the event of fracture, making the relations even more complicated. The fracture parameters now are markedly time-scale dependent. The specific work of fracture is increased, since much energy is dissipated during deformation.

Materials may break in shear or in tension; the former occurs in "short" (e.g., brittle), the latter in "long" (e.g., rubbery) materials. Another material property is notch sensitivity. In a test piece that is put under tension, notches can be applied, and in several materials the ensuing stress concentration greatly lowers the overall stress needed for fracture propagation. The notch sensitivity is smaller if the material contains defects of the size of a small notch, or if the bonds between structural elements are much stronger in the direction of the applied stress than in a perpendicular direction, as in many fibrous systems.

Elongational flow, due to uniaxial or biaxial tensional stress (e.g., spinning of a thread, or blowing of a film, respectively), tends to lead to fracture: the stress will generally be largest where the test piece is thinnest, so that the thinnest part will experience an ever increasing stress. This need not occur if the material is strongly strain hardening. More precisely, the stress should increase with increasing strain to a certain extent. Some concentrated protein solutions show this behavior, so that they can be spun, and wheat flour doughs do the same, preventing rupture of thin dough films between gas bubbles upon rising of the dough in the oven.

Texture perception is briefly discussed, and pitfalls in relating sensory perceived texture to instrumentally measured properties are pointed out.

Gels. Gels consist for the greater part of liquid; the solid character is provided by a space-filing structure (the gelling material). Nearly all food gels are man-made; the main purpose is to provide one or more specific physical properties. Desirable mechanical properties, be they for storage, handling, or eating, mostly involve the yield stress. In some cases fracture properties are more important. Gels are also made to provide physical stability to a product; again, the yield stress often comes into play, but also the permeability can be an important characteristic.

Gels can be of various types. Properties of some of the main types in foods are summarized in Table 17.5.

Polymer Gels. These are very common. They are made of long polymer molecules that are cross-linked at some places. In food gels, these are rarely chemical bonds; rather they are junctions, each of which involves a number of weak physical bonds. They are often microcrystallites, i.e., stacks of helices or other straightened polymer sections. Most of these gels are thermoreversible: they melt upon heating and (re)form upon cooling. Some gels of anionic polymers can form egg-box junctions in the presence of calcium ions; these gels generally do not melt at high temperatures.

A gelatin solution gels on cooling by forming triple helices and stacks of these; it may take a long time before it is fully set. Because of the high flexibility of the gelatin chains, the gel is more or less rubberlike, with a large linear region. The modulus is largely entropy determined.

TABLE 17.5 Comparison of Some Properties of a Typical Polymer Gel, a Typical Fractal Particle Gel, and a Heat-Set Protein Gel.

Gel type	Polymer	Particle	Heat-set protein
Appearance	Clear	Turbid	Clear–turbid
Permeability B/m^2	10^{-16}	10^{-12}	10^{-15}–10^{-12}
Modulus at 3%[a] G'/Pa	10^3–10^6	10^2	0–10^2
Fracture strain ε_{fr}	> 1[b]	< 1[c]	< 1
Stability of the gel[d]	Good	Variable	Good

[a]Percent (w/w) of gelling material.
[b]Widely variable.
[c]For rigid particles < 0.1.
[d]At constant conditions.

Polysaccharide molecules tend to have much stiffer chains; conse-quently, the linear region of a polysaccharide gel tends to be much smaller, and its modulus is for a considerable part enthalpy determined (as in an elastic solid). The gelling polysaccharides vary widely in properties, and gelation often involves quite specific aspects. Factors determining gel stiffness, deformability, and strength generally are concentration of gelling material, gel time, temperature of gelation and of measurement, and solvent quality. For ionic polysaccharides, pH and ionic strength are important variables, and for some ionic composition is important. Often weak and brittle (or strong) gels are distinguished. Ideally, the former type has a low modulus and shows yielding at a relatively small strain; the latter type has a higher modulus and does not yield but fractures, at a relatively large strain.

Particle Gels.   These gels generally form by fractal aggregation. Two types of casein gels, made by acidification and by renneting, make about ideal fractal gels. The fractal dimensionality mostly is 2.3–2.4. Considerable short-term rearrangement can occur, which means that the small fractal aggregates initially formed change into more compact particles (containing, say, 4–40 primary particles), which then form the building blocks of the larger fractal aggregates, which ultimately form the gel.

The gel structure is determined by the volume fraction of particle material, the size of the building blocks, and the fractal dimensionality. Simple scaling laws are derived for the permeability and for rheological properties as functions of particle concentration. The rheological para-meters also depend on those of the particles, especially the extent of the linear range.

Long term rearrangement, i.e., after gel formation, can occur under some conditions. In first instance, it leads to straightening of strands of particles in the gel, which causes an increase in modulus, a weaker dependence of the modulus on particle concentration, and a decrease in fracture strain; the fracture stress and the permeability are hardly affected. Stronger rearrangement does lead to an increase in permeability, and syneresis can readily occur. All these changes depend on gel type, formation temperature, storage temperature, pH, etc.

Heat-Set Protein Gels.   When a solution of a globular protein is heated, the protein molecules denature and subsequently aggregate to form a gel. The gelation is irreversible: upon cooling the modulus increases. The aggregation appears to proceed in two steps. At first roughly spherical aggregates form, size 20 nm to 4 μm, and these then aggregate to form a gel. At a pH close to the isoelectric point and a high ionic strength, large aggregates are formed, resulting in a coarse gel. Small aggregates lead to a fine stranded gel, which forms at low ionic strength and further away from the isoelectric pH.

Upon heating, a weak gel tends to form already when only part of the protein is denatured, and this gel is of a fractal nature, having relatively large pores; the fractal dimensionality is about 2.4. Subsequently, aggregating protein is deposited onto the network, which does not greatly alter its overall structure; the permeability slightly decreases, but the modulus strongly increases. Many variables, for instance temperature or heating rate, affect all of these processes, hence the gel properties. However, the results greatly vary among proteins, and few general rules can be given.

Mixed Gels. A wide variety of mixed gels is produced, often with specific (combinations of) properties. A mixture of a gelling and a nongelling polymer may show phase separation due to incompatibility, and the mixture can then be made to gel. If the gelling polymer forms the continuous phase, the presence of the other polymer increases gel stiffness. Mixtures of polymers that show weak mutual attraction, often of an electrostatic nature, may give a gel at concentrations at which neither of them gels by itself.

Adding polysaccharides to a protein solution that is subject to heat setting may either increase or decrease the modulus at constant concentration (protein + polysaccharide).

A solution of a moderately hydrophobic polysaccharide can often be made to gel by adding a small-molecule surfactant at a concentration above the CMC. The surfactant then makes micelles through many of which one or more polysaccharide chains pass, thereby forming cross-links. A fairly weak gel results.

The presence of filler particles can markedly affect the properties of various gels. If the particles are not bonded to the gelling material, their effect is to decrease the modulus. If they are bonded, generally because the gelling material also adsorbs onto the particles, the modulus may strongly increase, the more so as the particles themselves have a higher modulus; the modulus of a fluid particle equals its Laplace pressure. In protein–emulsion gels, the gelling material (e.g., gelatin, casein, or heat-setting proteins) is also used as the emulsifying agent for the oil droplets. Quite high moduli can so be obtained for low protein concentrations.

Plastic Fats. A plastic fat is a space-filling oil-containing network of triglyceride crystals. It can thus be considered a gel, although the mechanical properties are very different from those of most gels. When oil is cooled, so that part of the material becomes supersaturated, crystal nucleation and growth occur, and the crystals rapidly aggregate because of van der Waals attraction, forming a fractal network of low dimensionality (about 1.7). If the initial supersaturation is high, nucleation, growth, and aggregation go on, and second-generation fractal aggregates form that fill the pores of the initial network. In this way the network becomes much more even, its permeability greatly decreases, and its stiffness greatly increases. The fractal structure is lost, and the scaling exponents with crystal concentration for permeability and modulus are quite high, generally 4–7.

Fats of a wider compositional range show more secondary nucleation and hence smaller crystals and a higher modulus. These fats are also further removed from equilibrium, more recrystallization occurs and hence more sintering; the latter is enhanced by temperature fluctuations. However, if β-crystals can be formed, slow (re)crystallization can lead to the formation of quite large crystals and a weak network.

In practice, large deformations are generally applied. The linear region tends to be very small, up to a strain of 10^{-4}–10^{-3}. With increasing strain, the system becomes more and more viscoelastic, and its structure gradually breaks down; then yielding occurs, accompanied by a stress overshoot, and the system finally shows plastic flow. The yield stress, or firmness, is the most important quantity in practice, and it approximately scales with solid fat content squared. Other variables affecting the modulus have a comparable effect on the yield stress.

Working of a fat causes work softening, i.e., it considerably decreases the yield stress. Upon standing, the yield stress increases again, since at first aggregation and then (much more slowly) sintering occurs, but the original value is not reached. During the making of a plastic fat, cooling generally goes along with vigorous agitation, to allow removal of the heat of fusion. If the product now leaves the agitator after fat crystallization is more or less complete, it will have undergone strong work softening, and it keeps increasing in firmness for some time. On the other hand, if the fat leaves the agitator at a stage where crystallization is not yet complete, extensive sintering occurs within a relatively short time, causing the fat to obtain a yield stress that is significantly higher than is obtained by the other procedure.

It should be stressed that quantitative relations are quite complex and depend strongly on fat composition and on the temperature history.

Closely Packed Systems. These are very concentrated dispersions of soft solid particles in a liquid. The particles are closely packed and deform each other. The modulus of the particles primarily determines the modulus of the system. The coherence of the system is due to the restricted volume. Several systems can be distinguished.

Polyhedral foams and emulsions: if they are not very polydisperse, theory for the modulus and the yield stress is available; the only variables are particle volume fraction and Laplace pressure.

Concentrated starch gels consist of partly swollen starch granules that fill virtually the whole volume. The modulus roughly equals that of the starch granules, and thereby greatly increases with starch retrogradation. Relations for other mechanical properties are more complex. Wheat flour doughs are somewhat similar, but the particle volume fraction is much lower, and the continuous liquid, which contains gluten, is highly viscoelastic.

"Fluid gels" consist of anisometric gel particles in a continuous liquid, made, e.g., by cooling a suitable polysaccharide solution while agitating it. The modulus increases with increasing volume fraction, anisometry, and modulus of the particles. The systems generally have a yield stress. Several kinds of pastes are similar to fluid gels.

Cellular Systems. Superficially, these closely packed systems are much like some of those just discussed, but the stiffness now derives from the rigidity of the continuous mass rather than the particles. Three main types are distinguished: solid foams; gas-filled sponges (where the cells are in open contact with each other); and plant tissues (where the cells are largely filled with liquid).

For gas-filled systems, theory has been developed for various rheological properties. The main variable is the relative density, about equal to one minus the volume fraction of gas. In most systems, the modulus roughly scales with the density squared. What occurs at large deformation depends on the properties of the cell walls: buckling if they are rubberlike, yielding if they are plastic solids, and fracturing if they are rigid solids.

Bread and related products, or expanded products formed by extrusion, reasonably fit the theory. This is not so for plant tissues. Here the modulus depends on three main factors: the rigidity of the cell walls, the turgor of the cells, and the strength of the material cementing the cells. All of these parameters greatly decrease upon cooking.

BIBLIOGRAPHY

A recent book covering many aspects discussed in this chapter is

A. J. Rosenthal, ed. Food Texture: Measurement and Perception. Aspen, Gaithersburg, 1999.

Chapters 4 and 5 discuss rheological and other mechanical properties; Chapters 6–10 various types of products.

Another interesting though somewhat outdated book covering much relevant matter, with emphasis on physical structure, is

J. M. V. Blanshard, J. R. Mitchell, eds. Food Structure: Its Creation and Evaluational. Butterworths, London, 1988.

Fracture mechanics of soft solids is discussed by

T. van Vliet, P. Walstra. Large deformation and fracture behaviour of gels. Faraday Discussion 101 (1995) 359.

Although it concerns biological materials rather than foods, very interesting aspects are treated in

J. F. V. Vincent. Structural Biomaterials. Revised Edition. Princeton Univ. Press, Princeton, 1990.

The following gives much basic information on the rheology of gels:

S. E. Hill, D. A. Ledward, J. R. Mitchell, eds. Functional Properties of Food Macromolecules, 2nd ed. Aspen, Gaithersburg, 1998.

Especially Chapters 3, Gelation of globular proteins, by A. H. Clark, and 4, Gelation of polysaccharides, by V. J. Morris, who also discusses mixed polysaccharide gels, are recommended.

The gelation of gelatin is discussed by D. A. Ledward in the first edition of the same book (Chapter 4).

J. R. Mitchell, D. A. Ledward, eds. Elsevier Applied Science, London, 1986.

It also contains a chapter (8) by P. J. Lillford, on the texturization of proteins.

An authoritative and detailed discussion on the gelation of gelatin is given in Section 10, pages 160–193, in a review by

K. te Nijenhuis. Thermoreversible networks. Adv. Polymer Sci. 139 (1992) 1–265.

A wealth of information on a wide range of polysaccharides, albeit it little on physical properties, is in

A. M. Stephen, ed. Food Polysaccharides and Their Applications. Marcel Dekker, New York, 1995.

An introduction to fractal casein gels is given by

P. Walstra, T. van Vliet, L. G. B. Bremer. On the fractal nature of particle gels. In: E. Dickinson, ed. Food Polymers, Gels, and Colloids. Royal Soc. Chem., Cambridge, 1991, p. 369.

Gels made with the aid of surfactants are discussed by

B. Lindman et al. Polysaccharide-Surfactant Systems: Interactions, Phase Diagrams, and Novel Gels. In: E. Dickinson, P. Walstra, eds. Food Colloids and Polymers. Royal Soc. Chem., Cambridge, 1993, p. 113.

Structure and mechanical properties of plastic fats are discussed by

P. Walstra, W. Kloek, T. van Vliet. Fat Crystal Networks. In: N. Garti, K. Sato, eds. Crystallization Processes in Fats and Lipid Systems. Marcel Dekker, New York, 2001, Chapter 8.

Rheological properties of polyhedral foams and emulsions are discussed by

H. M. Princen. The mechanical and flow properties of foams and highly concentrated emulsions. In: R. D. Bee et al., eds. Food Colloids. Royal Soc. Chem., Cambridge, 1989, p. 14.

The same volume also contains articles on cellular solids by P. J. Lillford (p. 1) and A. C. Smith (p. 56). A classical monograph on cellular systems is

L. J. Gibson, M. F. Ashby. Cellular Solids: Structure and Properties. Pergamon, Oxford, 1988.

More about the dairy systems discussed in this chapter can be found in

P. Walstra et al. Dairy Technology: Principles of Milk Properties and Processes. Marcel Dekker, New York, 1999.

Appendix A

Frequently Used Symbols for Physical Quantities

Between parentheses the units—generally SI units—for the quantity are given, unless the dimensions can vary; (–) means dimensionless.

Unfortunately, the number of physical quantities used is far greater than the number of symbols available. Many symbols are used with subscripts and/or superscripts, to identify the quantity further.

Latin

A	area	(m^2)
	Hamaker constant	(J)
	specific surface area	(m^{-1})
a	particle radius	(m)
	thermodynamic activity	
a_{w}	water activity	$(-)$
B	permeability	(m^2)
	second virial coefficient	$(\mathrm{mol}\cdot\mathrm{m}^3\cdot\mathrm{kg}^{-2})$
b	length of statistical chain element	(m)
b_{ch}	distance between charged groups along chain	(m)
C	constant	
c	concentration	
c^*	chain overlap concentration	
c_{sat}	solubility	

c_p	specific heat at constant pressure	$(\mathrm{J \cdot kg^{-1} \cdot K^{-1}})$
c_n	relative standard deviation of order n	$(-)$
D	diffusion coefficient	$(\mathrm{m^2 \cdot s^{-1}})$
	fractal dimensionality	$(-)$
	relative deformation	$(-)$
d	diameter	(m)
E	modulus (in elongation or unspecified)	(Pa)
E^{S}	surface modulus	$(\mathrm{N \cdot m^{-1}})$
E_{a}	activation energy for a chemical reaction	$(\mathrm{J \cdot mol^{-1}})$
e	electronic charge	$(1.602 \cdot 10^{-19}\,\mathrm{C})$
F	force	(N)
f	frequency factor	
G	Gibbs (free) energy	
	shear modulus	(Pa)
G'	storage (shear) modulus	(Pa)
G''	loss (shear) modulus	(Pa)
g	acceleration due to gravity	$(9.807\,\mathrm{m^2 \cdot s^{-1}})$
H	enthalpy	
	height	(m)
h	distance from a surface; interparticle distance	(m)
h_{P}	Planck's constant	$(6.626 \cdot 10^{-34}\,\mathrm{J \cdot s})$
I	ionic strength	(molar)
i	as subscript: indicates class number	$(-)$
J	aggregation rate; particle flux	
	nucleation rate	$(\mathrm{m^{-3} \cdot s^{-1}})$
K	constant	
	stability constant; equilibrium constant	
k	reaction rate constant	
k_{B}	Boltzmann constant	$(1.381 \cdot 10^{-23}\,\mathrm{J \cdot K^{-1}})$
L	length	(m)
l	eddy size	(m)
M	molar mass	$(\mathrm{kg \cdot kmol^{-1}})$
m	mass	(kg)
	molar concentration	$(\mathrm{kmol \cdot m^{-3}})$
N	number concentration	$(\mathrm{m^{-3}})$
	number per unit cross-sectional area	$(\mathrm{m^{-2}})$
N_{AV}	Avogadro's number	$(6.022 \cdot 10^{23}\,\mathrm{mol^{-1}})$
n	degree of polymerization	$(-)$
	number of moles	
	reaction order	$(-)$
	refractive index	$(-)$
n'	number of statistical chain elements in a polymer chain	
p	pressure	(Pa)
p_{L}	Laplace pressure	(Pa)

p_v	vapor pressure	(Pa)
Q	volume flow rate	$(m^3 \cdot s^{-1})$
q	amount of heat	(J)
R	gas constant	$(8.315 J \cdot K^{-1} \cdot mol^{-1})$
	radius; radius of curvature; radial coordinate	(m)
r	radius; radial coordinate	(m)
r_g	radius of gyration	(m)
r_m	root-mean-square end-to-end distance of polymer chain	(m)
S	entropy	
S_n	n-th moment of a (size) distribution	
s	solubility	$(mol \cdot m^{-3})$
T	absolute temperature	(K)
T_g	glass transition temperature	(K)
t	time	(s)
$t_{0.5}$	time needed to halve the value of a parameter	(s)
U	internal energy	
u	eddy velocity	$(m \cdot s^{-1})$
V	volume	(m^3)
	colloidal interaction free energy	$(J; J \cdot m^{-2})$
v	droplet volume	(m^3)
	linear velocity	$(m \cdot s^{-1})$
	molar volume	$(m^3 \cdot mol^{-1})$
W	retardation factor in aggregation, etc.	(–)
	(specific) work	
w	mass fraction of water	(–)
w'	kg water per kg dry matter	(–)
x	mole fraction	(–)
x,y,z	linear coordinates	(m)
z	valence; net number of charges	

Greek

α	angle	(rad)
	capture efficiency (in aggregation)	(–)
	degree of ionization	(–)
β	excluded volume parameter	(–)
	supersaturation ratio	(–)
Γ	surface excess (surface load)	$(mol \cdot m^{-2}; kg \cdot m^{-2})$
γ	activity coefficient	(–)
	interfacial or surface tension	$(N \cdot m^{-1})$
	shear strain	(–)
$\gamma_{\pm}$	free ion activity coefficient	(–)
Δ	root-mean-square displacement in diffusion	(m)
δ	film or layer thickness	(m)
	phase angle ($\tan \delta = G'' / G'$)	(rad)

ε	power density (energy dissipation rate)	$(W \cdot m^{-3})$
	relative deformation or strain	$(-)$
	relative dielectric constant	$(-)$
ε_0	dielectric permittivity of vacuum	$(8.854 \cdot 10^{-12}\, C \cdot V^{-1} \cdot m^{-1})$
ε_H	natural or Hencky strain	$(-)$
η	viscosity	$(Pa \cdot s)$
η_a	apparent viscosity	$(Pa \cdot s)$
$[\eta]$	intrinsic viscosity	
θ	(contact) angle	(rad)
	surface fraction (covered)	$(-)$
κ	reciprocal Debye length	(m^{-1})
Λ	stress concentration factor	$(-)$
λ	thermal conductivity	$(W \cdot K^{-1} \cdot m^{-1})$
	wavelength	(m)
μ	chemical potential	$(J \cdot mol^{-1})$
	Poisson ratio	$(-)$
ν	cross-link density	(m^{-3})
	frequency	(s^{-1})
ξ	relative surface expansion rate	(s^{-1})
Π	osmotic pressure	(Pa)
	surface pressure	$(N \cdot m^{-1})$
Π_S	interfacial spreading pressure	$(N \cdot m^{-1})$
ρ	mass density	$(kg \cdot m^{-3})$
σ	number density of polymer chains	(m^{-2})
	stress	(Pa)
σ_{fr}	fracture stress	(Pa)
σ_y	yield stress	(Pa)
τ	characteristic time (scale)	(s)
	relaxation time	(s)
Φ	hydrophobicity	
φ	volume fraction	$(-)$
χ	solvent-segment interaction parameter	$(-)$
Ψ	velocity gradient, strain rate	(s^{-1})
ψ	mass fraction	$(-)$
	electrostatic potential	(V)
ψ_0	electrostatic surface potential	(V)
Ω	number of degrees of freedom	$(-)$
ω	revolution rate; (angular) frequency	$(rad \cdot s^{-1};\ s^{-1})$

Other

$[A]$	molar concentration of substance A
pI	isoelectric pH
pK	log(stability constant)
pK_a	log(association constant)
	$=$ pH of 50% dissociation

De	Deborah number	(–)
Pe	Péclet number	(–)
Re	Reynolds number	(–)
Tr	Trouton ratio	(–)
We	Weber number	(–)

Appendix B

TABLE Some Frequently Used Abbreviations

CMC	critical micellization concentration
DLVO	
	Deryagin–Landau–Verwey–Overbeek (colloidal interaction)
DSC	differential scanning calorimetry
HLB	hydrophile–lipophile balance
PIT	phase inversion temperature
r.p.m.	number of revolutions per minute
SDS	sodium dodecyl sulfate
WLF	Williams–Landel–Ferry (viscosity equation)

Appendix C

Some Mathematical Symbols

$\equiv$	is by definition equal to
$\approx$	is approximately equal to
$\propto$	is proportional to
∞	infinity
$\sim$	about
$\rightarrow$	goes to; approaches
Δa	increment of a
da	differential of a
∂a	partial differential of a
$|a|$	absolute value of a
$f(a)$	a function of a
$\langle f(a) \rangle$	numerical average of $f(a)$
$\langle a^2 \rangle^{0.5}$	root-mean-square value of a
$\ln a$	natural logarithm of a
e	base of natural logarithms $= 2.7183$
π	$= 3.1416$
i	$= \sqrt{(-1)}$
$a - bi$	complex number
$|a - bi|$	$= (a^2 + b^2)^{0.5}$

Appendix D

SI Rules for Notation

Symbols for (physical) quantities, be they variables or constants, are given by a single character (generally Latin or Greek letters) and are printed in italics, e.g., F (force), p (pressure), μ (chemical potential), k (Boltzmann constant). Further differentiation is achieved by the use of subscripts and/or superscripts; these are printed in italics if it concerns the symbol of a quantity, otherwise in roman type, e.g., c_p (specific heat at constant pressure), $h_{\hat{p}}$ (Planck's constant), E^{SD} (surface dilational modulus). For clarity, symbols are generally separated by a (thin) space, e.g., $F = m\,a$, not ma. Some generally accepted exceptions occur, such as pH, as well as symbols (or two letter abbreviations, rather) for the dimensionless ratios frequently used in process engineering, like Re for Reynolds number and Tr for Trouton ratio (in roman type).

Symbols for operators are given in roman (upright) characters, e.g., log, ln, $\triangle$, d (differential), sin. Preferably, they are separated by a thin space from the quantity symbol, e.g., $\ln \varphi$.

Symbols for units are also given in roman type, e.g., m, Pa. Separation of units is (preferably) by an elevated dot, e.g., $Pa \cdot s$ and $kg \cdot m^{-3}$—or by a space: Pa s and kg m^{-3}—(but not Pas, etc.). Division is indicated by a negative sign in the exponent, such as $m \cdot s^{-1}$; m/s is also allowed, but not m/kg/s, or m/kg s, since these notations are equivocal. Numerical prefixes are put directly before the main symbol, e.g., μm, kPa, GV. For compound quantities, only the first symbol can contain a prefix—e.g., $kPa \cdot m^{-1}$, not $Pa \cdot mm^{-1}$—except kg, e.g., $m^3 \cdot kg^{-1}$.

Numbers. The decimal marker is a comma—e.g., $\pi = 3{,}1416$—but in English language texts a dot is also allowed (3.1416). Do not use a comma or a dot for separation between factors of a thousand, but—if desired—a thin space can be used, e.g., 30 000. The latter can also be written as $3 \cdot 10^4$.

This text has been modified from SI rules and has been edited to the publisher's style.

Appendix E

The SI Units System

SI stands for "Système International," and the SI System of Units concerns the internationally accepted and standardized rules for the values and the notation of units for physical quantities. One basic rule is that

Quantity = number × unit

Furthermore, it is a *metric* system, all units for the same quantity differing by one or more factors of 10.

Base Units

These concerns the units for some measurable, dimensionally independent quantities from which all other units can be derived. The magnitude of each base unit has been unequivocally and precisely defined.

Quantity	Name	Symbol
Length	meter	m
Mass	kilogram	kg
Time	second	s
Electric current	ampere	A

Quantity	Name	Symbol
Temperature	kelvin	K
Amount of substance	mole	mol
Luminous intensity	candela	cd

Supplementary Units

These concern the units for a plane angle, the radian, symbol rad; and the solid angle, the steradian, symbol sr.

Derived Units

Derived units can be made as desired, but some have been given names and symbols, and only these will be given here (in sofar as relevant).

Quantity	Name	Symbol	SI base units	
Amount of electricity	coulomb	C	$A \cdot s$	
Electric capacity	farad	F	$A^2 \cdot s^4 \cdot kg^{-1} \cdot m^{-2}$	(C/V)
Electric potential	volt	V	$kg \cdot m^2 \cdot s^{-3} \cdot A^{-1}$	(W/A)
Electric resistance	ohm	Ω	$kg \cdot m^2 \cdot A^{-2} \cdot s^{-3}$	(V/A)
Energy (work, amount of heat)	joule	J	$kg \cdot m^2 \cdot s^{-2}$	$(N \cdot m; C \cdot V)$
Force	newton	N	$kg \cdot m \cdot s^{-2}$	
Frequency	hertz	Hz	s^{-1}	
Molar mass	dalton	Da	$kg \cdot kmol^{-1}$	(g/mol)
Power	watt	W	$kg \cdot m^2 \cdot s^{-3}$	(J/s)
Pressure, stress	pascal	Pa	$kg \cdot m^{-1} \cdot s^{-2}$	(N/m^{-2})

Numerical Prefixes

The most important numerical prefixes are

Multiplication factor	Name	Symbol
10^9	giga	G
10^6	mega	M
10^3	kilo	k
10^{-1}	deci[a]	d
10^{-2}	centi[a]	c
10^{-3}	milli	m

Multiplication factor	Name	Symbol
10^{-6}	micro	μ
10^{-9}	nano	n
10^{-12}	pico	p

[a]Only to be used for some volume and area units.

Some Other Units

The % symbol is often used and, unless stated otherwise, it is often meant to signify kg/100 kg.

For the temperature *scale*, SI rules allow the use of degrees Celsius, symbol °C, but not for temperature intervals; nor can °C it be used with a prefix. For example: cool the liquid to $-15°C$ at a rate of $10 \, mK \cdot s^{-1}$.

For angles, often the "degree" is used, symbol °.

For volumes, the unit liter can be used ($10^{-3} \, m^3$), symbol l. However, the symbol L often is used for liter without a prefix, to avoid confusion with the numeral 1; however, use ml rather than mL, etc.

Confusion can readily arise for concentrations of chemical substances. The unit "molar" (often used symbol M) means mol/L or $kmol \cdot m^{-3}$, not $mol/m,^3$ which may be considered a violation of SI rules.

Appendix F

Some Conversion Factors

1 ångström (Å)	$= 10^{-10}$ m (0.1 nm)
1 atmosphere (atm)	$= 101\ 325$ Pa
1 bar	$= 10^5$ Pa ($= 0.987$ atm)
1 calory	$= 4.184$ J
1 centipoise (cP)	$= 1$ mPa $\cdot$ s
1 Debye unit (D)	$= 3.336 \cdot 10^{-30}$ C $\cdot$ m
1 dyne	$= 10^{-5}$ N
1 dyne/cm	$= 1$ mN $\cdot$ m^{-1}
1 dyne/cm^2	$= 0.1$ Pa
1 erg	$= 10^{-7}$ J
1 erg/cm^2	$= 1$ mJ $\cdot$ m^{-2}
1 foot	$= 0.3048$ m
1 gallon (USA)	$= 0.003785$ m^3
1 gallon (UK)	$= 0.005461$ m^3
1 inch	$= 0.02540$ m
1 molar (M)	$= 10^3$ mol $\cdot$ m^{-3}
1 ounce (av.)	$= 0.02835$ kg
1 pound (av.)	$= 0.4536$ kg
1 pound/inch^2 (psi)	$= 6895$ Pa
1 radian	$= 57.296$ degrees ($57°\ 18'$)

Appendix G

Recalculation of Concentrations

Assume a *binary mixture* of components 1 (solvent) and 2 (solute). The mass fraction of the solute is given by ψ. Molar mass M is given in daltons, ρ is the density of the solution. We now have

Molar concentration (mol/L) $\quad m = \rho \dfrac{\psi}{M_2}$

Molality (mol per kg water) $\quad m^* = \dfrac{\psi}{1-\psi} \cdot \dfrac{10^3}{M_2} = \dfrac{10^3 m}{\rho - mM_2}$

Mole fraction $\quad\quad\quad\quad\quad x = \dfrac{\psi/M_2}{(1-\psi)/M_1 + \psi/M_2}$

If the system contains more solutes, the first equation will still hold for Component 2, but not the other two equations.

For dispersed systems, the relations become far more complicated. Moreover, in such systems part of the solvent may not be "available" for a solute; see Section 8.3, item 2, "Nonsolvent water."

Appendix H

Physical Properties of Water at 0–100°C

Mass density (ρ), viscosity (η), refractive index at a wavelength of 589 nm (n_D), relative dielectric constant (ε), vapor pressure (p_v), and surface tension (γ) of pure water at various temperatures (T).

T (°C)	ρ kg.m^{-3}	η mPa·s	n_D (–)	ε (–)	p_v kPa	γ mN·m^{-1}
0	999.9	1.787	1.3346	87.9	0.610	75.6
5	1000.0	1.519	1.3346	85.9	0.872	74.8
10	999.7	1.307	1.3343	84.0	1.228	74.1
15	999.1	1.139	1.3338	82.1	1.705	73.3
20	998.2	1.002	1.3333	80.2	2.34	72.6
25	997.1	0.890	1.3329	78.4	3.17	71.8
30	995.7	0.798	1.3323	76.6	4.24	71.0
40	992.2	0.653	1.3309	73.2	7.38	69.4
50	988.1	0.547	1.3293	69.9	12.33	67.7
60	983.2	0.466	1.3275	66.8	19.92	66.0
70	977.8	0.404	1.3255	63.8	31.16	64.3
80	971.8	0.355	1.3231	60.9	47.34	62.5
90	965.3	0.315	1.3209	58.2	70.1	60.7
100	958.4	0.282	1.3182	55.6	101.3	58.9

Appendix I

Thermodynamic and Physical Properties of Water and Ice

Properties at 273.15 K and 100 kPa

Enthalpy of fusion	$-\Delta_{L \to S}H$	$= 6012$	$J \cdot mol^{-1}$
		$= 334$	$kJ \cdot kg^{-1}$
Entropy of fusion	$-\Delta_{L \to S}S$	$= 22.01$	$J \cdot mol^{-1} \cdot K^{-1}$
Expansion upon solidification	$\Delta_{L \to S}V$	$= 90.6$	$ml \cdot kg^{-1}$
(Enthalpy of sublimation		$= 50.9$	$kJ \cdot mol^{-1}$)
Specific heat, water	c_p	$= 4218$	$J \cdot kg^{-1} \cdot K^{-1}$
ice	c_p	$= 2101$	$J \cdot kg^{-1} \cdot K^{-1}$
Thermal conductivity, water	λ	$= 0.56$	$W \cdot m^{-1} \cdot K^{-1}$
ice	λ	$= 2.24$	$W \cdot m^{-1} \cdot K^{-1}$
Thermal diffusivity, water	D_H	$= 1.3 \times 10^{-7}$	$m^2 \cdot s^{-1}$
ice	D_H	$= 11.7 \times 10^{-7}$	$m^2 \cdot s^{-1}$
Volume compressibility,[a] water		$= 5.0 \times 10^{-10}$	Pa^{-1}
ice		$= 1.2 \times 10^{-10}$	Pa^{-1}
Dielectric constant, water	ε	$= 87,9$	
ice	ε	≈ 90	
Refractive index,[b] water	n	$= 1.3346$	
ice	n	$= 1.3104$	
Hamaker constant[c] ice/water	$A_{11(3)}$	$= 0.043$	$k_B T$

[a]Isothermal.
[b]At wavelength $= 589$ nm.
[c]Calculated from Lifshits theory.

Appendix J

Some Values of the Error Function

$$\text{erf } y = \frac{2}{\sqrt{\pi}} \int_0^y \exp(-z^2)\, dz$$

where z is an integration variable.

y	erf y	y	erf y
0.1	0.112	0.8	0.742
0.2	0.223	1.0	0.843
0.3	0.329	1.2	0.911
0.4	0.428	1.4	0.952
0.5	0.520	1.6	0.976
0.6	0.604	2.0	0.995
0.7	0.678	2.5	0.9997

Index

Note: A page number printed in boldface refers to the place where a concept is defined.